# GENETIC INFLUENCES on NEURAL and BEHAVIORAL FUNCTIONS

## Edited by

**Donald W. Pfaff, Ph.D.**
*Laboratory of Neurobiology and Behavior,*
*The Rockefeller University,*
*New York*

**Wade H. Berrettini, M.D., Ph.D.**
*Department of Psychiatry,*
*Center for Neurobiology and Behavior,*
*University of Pennsylvania*
*Philadelphia, Pennsylvania*

**Tong H. Joh, M.D., Ph.D.**
*The W.M. Burke Medical Research Institute,*
*Cornell University Medical College*
*White Plains, New York*

**Stephen C. Maxson, Ph.D.**
*Department of Psychology,*
*University of Connecticut,*
*Storrs, Connecticut*

**CRC Press**
Taylor & Francis Group
Boca Raton  London  New York

CRC Press is an imprint of the
Taylor & Francis Group, an **informa** business

CRC Press
Taylor & Francis Group
6000 Broken Sound Parkway NW, Suite 300
Boca Raton, FL 33487-2742

First issued in paperback 2019

© 2000 by Taylor & Francis Group, LLC
CRC Press is an imprint of Taylor & Francis Group, an Informa business

No claim to original U.S. Government works

ISBN-13: 978-0-8493-2688-2 (hbk)
ISBN-13: 978-0-367-39955-9 (pbk)

Library of Congress Card Number 99-21820

**Library of Congress Cataloging-in-Publication Data**

Genetic Influences on neural and behavioral functions / edited by
    Donald W. Pfaff .... [et al.].
        p.  cm.
    Includes bibliographical references and index.
    ISBN 0-8493-2688-5  (alk. paper)
    1. Mental illness--Genetic aspects.  2. Central nervous system-
Diseases--Genetic aspects. 3. Behavior genetics.  I. Pfaff,
Donald W., 1939-
    [DNLM:   1. Mental Disorders--genetics. 2. Nervous System Diseases-
-gentics.      WM 140 G327 1999]
    RC455.4.G4G435     1999
    616.8'04442—dc21
    DNLM/DLC
    for Library of Congress                                                                            99-21820
                                                                                                                  CIP

**Visit the Taylor & Francis Web site at
http://www.taylorandfrancis.com**

**and the CRC Press Web site at
http://www.crcpress.com**

# Preface

The human genome project and the mouse genome project are providing a flood of information about genes of potential importance to brain research. The coding sequences of these genes will eventually yield a new wave of efforts in protein chemistry, and the regulatory elements of these genes will give neurobiologists clues as to how they are controlled in nerve cells. But among the tremendous number of genes being explored, which, in fact, are the most significant for the central nervous system? Since the brain is the most complicated organ in the body, the chance to use new findings about a relatively simple biochemical, DNA, to help us in our scientific quest represents a real opportunity. The genes in question might be crucial for the normal control of behavior, endocrine function, or autonomic physiology, and these will be studied as novel elements in well-reasoned programs of physiological research. In contrast, whenever the gene in question is damaged as part of a neurological or psychiatric disease, the gene abnormality stimulates a range of clinical and molecular investigations. To help neurobiologists (both experimental and clinical) keep track of at least a small portion of the new genomic information, we have begun to cover some of the most interesting areas in this field. As this area of work rapidly burgeons, the lacunae are almost as impressive as the topics are well covered. Therefore, subsequent editions, published electronically, will be necessary to keep the treatment up-to-date.

# Editors

**Dr. Donald Pfaff** was educated at Harvard and MIT and has spent the bulk of his scientific career at The Rockefeller University in New York, where he is Professor and Head of the Laboratory of Neurobiology and Behavior. His lab unraveled the circuitry for a simple mammalian reproductive behavior, the first discovered among vertebrates. During the last few years, he and his associates have worked on genes regulated by steroid hormones in nerve cells using mice with gene "knockouts" to demonstrate essential behavioral roles in hormone-regulated responses. Summarizing a large number of primary research publications are two books: *Estrogens and Brain Function* (Springer Verlag, 1980), and *Drive* (The MIT Press, 1999). In 1992 and 1994, Dr. Pfaff was elected a Fellow in the American Academy of Arts and Sciences and a member of the U.S. National Academy of Sciences, respectively.

**Dr. Wade Berrettini** attended Dickinson College, where he received a B.S. in Chemistry in 1973, and Jefferson Medical College, where he received his M.D. in 1977 and his Ph.D. in Pharmacology in 1979. For nearly a decade, he studied the genetics of behavioral disorders at the National Institute of Mental Health (NIMH) in Bethesda, MD. In 1997 he accepted his current position as Director of the Center for Neurobiology and Behavior at the University of Pennsylvania. His major scientific achievements include mapping a gene for susceptibility to manic-depressive illness to the short arm of chromosome 18. In 1996, Dr. Berrettini received the Selo Prize from NARSAD for outstanding achievement in the genetics of depression. He has continued to explore this area of research along with other projects, including a genetic study of anorexia nervosa and animal model studies of alcoholism and epilepsy. The author of over 100 scientific publications, Dr. Berrettini is a member of the editorial boards of *Neuropsychopharmacology*, *Psychiatric Genetics*, *Addiction Biology*, *Molecular Psychiatry*, and *Biological Psychiatry*. He resides in Haverford, PA, with his wife, Chris, and their three children, Sarah (age 10), Jack (age 8), and Nicholas (age 6).

**Dr. Stephen C. Maxson** is Professor of Psychology in The Graduate Degree Program in Biobehavioral Sciences at the University of Connecticut. He received a B.S. (1960) and a Ph.D. (1966) in biopsychology from the University of Chicago. His major research interest is the genetics of aggressive behavior in mice with an emphasis on the Y chromosome. He is the 1998 recipient of the Excellence Research Award from the AAUP of the University of Connecticut and of the Dobzhansky Memorial Award for outstanding lifetime scholarship from the Behavior Genetic Association.

**Dr. Tong H. Joh** received his Ph.D. from New York University. He is Professor of Neuroscience at Joan and Sanford Weil Medical College and Graduate School of Medical Science of Cornell University, a position he has held since 1972. He is also Director of the Laboratory of Molecular Neurobiology at the W.M. Burke Medical Research Institute. Dr. Joh is one of the pioneers in establishing the field of biochemistry, pharmacology, and molecular biology of catecholamine, serotonin, and other small molecular weight neurotransmitters. His initial work, focused on the purification and immunochemical and immunocytochemical analysis of all enzymes in catecholamine and serotonin biosynthesis, led the field in establishing monoamine pathways in the central nervous system and in identifying characteristic properties of these enzymes. His recent publications, concerning gene expression and regulation of these neurotransmitter enzymes, are certainly leading publications in the field.

# Contributors

**Armen N. Akopian, Ph.D.**
Department of Biology
University College
London, England

**John K. Belknap, Ph.D.**
Portland Alcohol Research Center
Veterans Administration Medical Center
Portland, Oregon

**Wade H. Berrettini, M.D., Ph.D.**
Center for Neurobiology and Behavior
Department of Psychiatry
University of Pennsylvania
Philadelphia, Pennsylvania

**Stephen D. M. Brown, MA, Ph.D.**
MRC Mammalian Genetics Unit and
UK Mouse Genome Centre
Harwell, Oxon, England

**Kari J. Buck, Ph.D.**
Portland Alcohol Research Center
Veterans Administration Medical Center
Portland, Oregon

**Russell J. Buono, Ph.D.**
Center for Neurobiology and Behavior
Department of Psychiatry
University of Pennsylvania
Philadelphia, Pennsylvania

**Webster K. Cavenee, Ph.D.**
Ludwig Institute for Cancer Research and
Department of Medicine and
Center for Molecular Genetics
University of California-San Diego
La Jolla, California

**Frank J. Coufal, M.D.**
Division of Neurosurgery and
Ludwig Institute for Cancer Research
University of California-San Diego
La Jolla, California

**John C. Crabbe, Ph.D.**
Portland Alcohol Research Center
Veterans Administration Medical Center
Portland, Oregon

**Josep Dalmau, M.D., Ph.D.**
Department of Neurology and
Cotzias Laboratory of Neuro-Oncology
Memorial Sloan-Kettering Cancer Center
New York, New York

**James Eberwine, Ph.D.**
Department of Pharmacology
University of Pennsylvania Medical Center
Philadelphia, Pennsylvania

**Mary-Anne Enoch, M.D.**
Laboratory of Neurogenetics
National Institute on Alcohol Abuse and
   Alcoholism/National Institutes of Health
Bethesda, Maryland

**Thomas N. Ferraro, Ph.D.**
Center for Neurobiology and Behavior
Department of Psychiatry
University of Pennsylvania
Philadelphia, Pennsylvania

**Jonathan Flint, Ph.D.**
Institute of Molecular Medicine
John Radcliffe Hospital
Oxford, England

**David Goldman, M.D.**
Laboratory of Neurogenetics
National Institute on Alcohol Abuse and
   Alcoholism/National Institutes of Health
Bethesda, Maryland

**Dorothy E. Grice, M.D.**
Center for Neurobiology and Behavior
Department of Psychiatry
University of Pennsylvania
Philadelphia, Pennsylvania

**H.-J. Su Huang, Ph.D.**
Ludwig Institute for Cancer Research and
Department of Medicine
University of California-San Diego
La Jolla, California

**Tong H. Joh, M.D., Ph.D.**
The W. M. Burke Medical Research Institute
Cornell University Medical College
White Plains, New York

**Maria Karayiorgou, M.D.**
Laboratory of Human Neurogenetics
The Rockefeller University
New York, New York

**James A. Knowles, M.D., Ph.D.**
Department of Psychiatry
Columbia University College of Physicians and
   Surgeons and
New York State Psychiatric Institute
New York, New York

**Frank H. Koegler, Ph.D.**
Department of Obesity and Metabolism
Pennington Biomedical Research Center
Louisiana State University
Baton Rouge, Louisiana

**Hisatake Kondo, M.D., Ph.D.**
Department of Cell Biology
Graduate School of Medical Sciences
Tohoku University
Sendai, Japan

**C. J. Krebs, Ph.D.**
Laboratory of Neurobiology and Behavior
The Rockefeller University
New York, New York

**Douglas F. Levinson, M.D.**
Department of Psychiatry
MCP-Hahnemann University
Philadelphia, Pennsylvania

**Sheree F. Logue, Ph.D.**
Institute for Behavioral Genetics
University of Colorado
Boulder, Colorado

**Sharon S. Low-Zeddies, B.Sc.**
Department of Neurobiology and Physiology
Northwestern University
Evanston, Illinois

**Andrea I. McClatchey, Ph.D.**
MGH Cancer Center
Charlestown, Massachusetts

**Stephen C. Maxson, Ph.D.**
Department of Psychology
University of Connecticut
Storrs, Connecticut

**Emmanuel Mignot, M.D., Ph.D.**
Department of Psychiatry and Behavioral
   Sciences
Stanford Center for Narcolepsy
Stanford University
Stanford, California

**Bryan J. Mowry, M.B., F.R.A.N.Z.C.P.**
Department of Psychiatry
University of Queensland
Brisbane, Australia

**Dana J. Niehaus, M.B.**
Psychiatric Genetics Program
MRC Unit for Research on Anxiety and Stress
   Disorders
University of Stellenbosch
Cape Town, South Africa

**Sonoko Ogawa, Ph.D.**
Laboratory of Neurobiology and Behavior
The Rockefeller University
New York, New York

**Bruce F. O'Hara, Ph.D.**
Department of Biological Sciences
Stanford University
Stanford, California

**Elizabeth H. Owen, Ph.D.**
Institute for Behavioral Genetics and
Department of Psychology
University of Colorado
Boulder, Colorado

**Adam J. W. Paige, ARCS, B.Sc., DIC, Ph.D.**
ICRF Medical Oncology Unit
Western General Hospital
Edinburgh, England

**Gavril W. Pasternak, M.D., Ph.D.**
Department of Neurology
Memorial Sloan-Kettering Cancer Center
New York, New York

**Donald W. Pfaff, Ph.D.**
Laboratory of Neurobiology and
  Behavior
The Rockefeller University
New York, New York

**Myrna R. Rosenfeld, M.D., Ph.D.**
Department of Neurosurgery
Weill Medical College of Cornell University
New York, New York

**Lori K. Singer, Ph.D.**
Pennington Biomedical Research Center
Louisiana State University
Baton Rouge, Louisiana

**Christina Sobin, Ph.D.**
Laboratory of Human Neurogenetics
The Rockefeller University
New York, New York

**Jin H. Son, Ph.D.**
Department of Neurology and Neuroscience
  and The W. M. Burke Medical Research
  Institute
Cornell University Medical College
White Plains, New York

**Dan J. Stein, M.B.**
MRC Research Unit on Anxiety and Stress
  Disorders
University of Stellenbosch
Cape Town, South Africa

**Joseph S. Takahashi, Ph.D.**
Howard Hughes Medical Institute
Northwestern University
Evanston, Illinois

**Jeanne M. Wehner, Ph.D.**
Institute for Behavioral Genetics
University of Colorado
Boulder, Colorado

**Y. S. Zhu, Ph.D.**
Laboratory of Neurobiology and Behavior
The Rockefeller University
New York, New York and
Department of Medicine and Endocrinology
Cornell University Medical College
New York, New York

# Table of Contents

# 1 Genes and Brain Function: An Introduction

*Donald W. Pfaff*

## CONTENTS

## INTRODUCTION TO THE BOOK

A book like this is both necessary and necessarily incomplete. One reason it cannot be absolutely up to date is obvious — the rapid and accelerating flood of new information from the human and mouse genome projects is yielding important insights for neurobiologists almost weekly. In fact, a main purpose of this book is to allow basic experimental neurobiologists and clinicians to see some of the main topics benefiting from this new information.

At the same time, a book cataloging new information is required because the output from the human and mouse genome projects has no intrinsic organization. This is partly because of the pleiotropy, etc. of genetic effects and because of the subtle dependencies of gene effects on brain function. Given those facts of life about how gene products register their effects on brain function, all we can do is to touch on some of the best subjects extant at the time the book was organized. Electronic updating will allow the treatments to expand as the field expands.

Even with the rapid and accelerating rate of development of genetics, genomics, and molecular biology during the second half of this century, it is clear that discerning genetic contributions to molar behaviors will be a difficult task. That is, with gene sequences in hand and increasingly sophisticated biophysics and biochemistry of proteins now available to the neurobiologist, we must ask how molecular biological knowledge and techniques will further the studies of brain mechanisms for behavior. What will be the content and form of a successfully completed molecular genetics of behavior (normal and abnormal), and how will they be achieved? All too many discussions about the relationships of genes to behavior have already occurred, laden with false dichotomies: Genes are said to both "control behaviors" and to not. Extreme claims, sometimes politically motivated, have led to arguments more philosophical in their content than scientific. Even learned discussions, in some cases, are restricted to genetic diseases of the nervous system. As an introduction to this book, we are inclined to postulate that all behaviors, even complex human behaviors, are influenced by genes to some extent; the important questions boil down to the mechanisms, direct or indirect, through which a given gene acts — and the share of the variance of an individual behavior accounted for by that gene. For most behaviors, multigenically influenced interactions among gene products will be a major subject of study; most important, charting the interactions between gene products and immediate environmental influences will be required to explain any individual behavior.

## PLEIOTROPY, REDUNDANCY, AND PENETRANCE

Despite a long history of brilliant work on genetically amenable organisms such as *Drosophila*, we still cannot easily trace the causal routes and mechanisms by which genes could influence behavior. There are at least four reasons for this state of affairs. First, the pleiotropy of genetic actions (Hall, 1994) dictates that any one gene may have many effects, relationships among which may be difficult to discern (see Figure 1.1). Second, overlap among functions of different genes precludes a simple demonstration that a given gene contributes to a given behavior. Third, incomplete and variable penetrance of a dominant allele render the statistical analysis of behavioral results harder to explain. Fourth and most important, any map of possible mechanistic routes, direct and indirect, from genes to behavior must be as complex as the physiology of the organs contributing to the behavior. Since these will always include the central nervous system, the complexity of the mammalian brain virtually guarantees that the task of discerning gene/neuron/behavior relations will not be finished quickly.

The difficulties of functional genomics have been recognized by leading geneticists (Lander and Schork, 1994; Smithies and Maeda, 1995). With this said, the central nervous system possesses degrees of complexity and adaptability that are likely to add twists that even these experts might not have anticipated. In addition, few scientists would expect mammalian behaviors to depend only on a small number of genes — they are prime examples of multigenic traits. Thus, we should expect to see, for any given knockout mouse, prominent influences of genetic background on the magnitude of the effect of any given gene (Gerlai, 1996).

## LESSONS FROM NEUROENDOCRINE MECHANISMS

Where can we find, given our current state of knowledge (or ignorance), enough data on gene/neuron/behavior relationships to give us a clue as to the opportunities and difficulties laying therein? Clearly, mice, among all mammals, have shown the greatest facility for genetic manipulation, especially when, for biochemical and physiological analysis, subsequent molecular techniques had to be applied to brain tissue. Among neurobiological subjects pursued so far with mice, those functions that are hormone-dependent have offered many advantages. The chemical simplicity of steroid hormones, their analogues and their antagonists, the molecular status of steroid receptors as transcription factors whose genes are studied intensively, and, generally, the ability to bring the information and experimental tools of molecular endocrinology to bear on neurobiological topics all have rendered neuroendocrine results particularly telling for the questions asked above. Thus, it is useful here to recite some of the achievements delineating gene/behavior relations in the neuroendocrine domain, and *pari passu*, some of the complexities revealed during these experiments.

### DIRECT AND INDIRECT ROUTES OF GENETIC INFLUENCE ON BRAIN FUNCTION

A straightforward chart lays out various causal routes for genetic influences on brain function and behavior in a logically complete fashion (see Figure 1.2; Ogawa et al., 1996). The simplest and easiest examples to prove derive from direct actions of steroid hormones on mammalian behaviors. These follow from the classic demonstrations of steroid hormone effects by Frank Beach, William Young, Robert Goy, and their collaborators. At the beginning of modern work in this field, it was

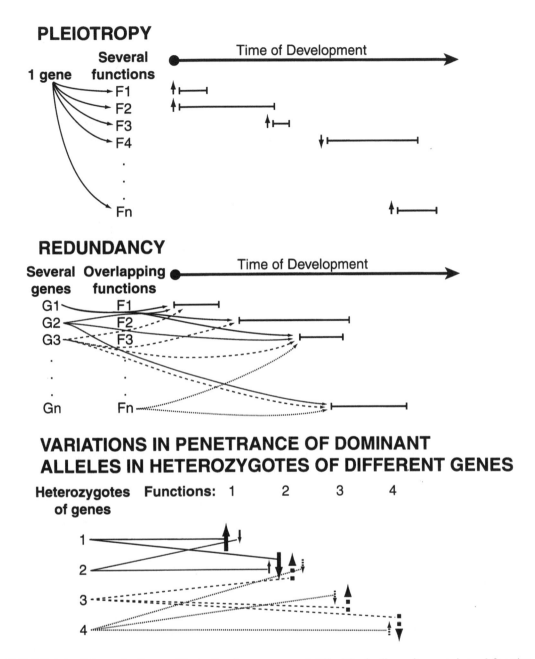

**FIGURE 1.1** Schematic diagram illustrating three sources of difficulties in discerning gene/neural function causal relationships during attempts to reason "from the gene on out."

quite easy to predict that estrogens and progestins acting in the brain would promote a variety of reproductive behaviors. Indeed, localization of the target cells for radioactive estrogens in the rat brain (Pfaff and Keiner, 1973) strongly indicated that estrogen receptors in limbic and hypothalamic neurons would be involved in the facilitation of estrogen-dependent female-typical behavioral responses. As a consequence, deleting the gene for the estrogen receptor should provide a clear example of a direct effect of a gene on mammalian behavior. Results from the estrogen receptor-deficient (ERKO) female mouse prove just that (Ogawa et al., 1996a; Ogawa et al., 1996b). Interestingly, at least three physiological changes seem to underlie the total lack of female-typical

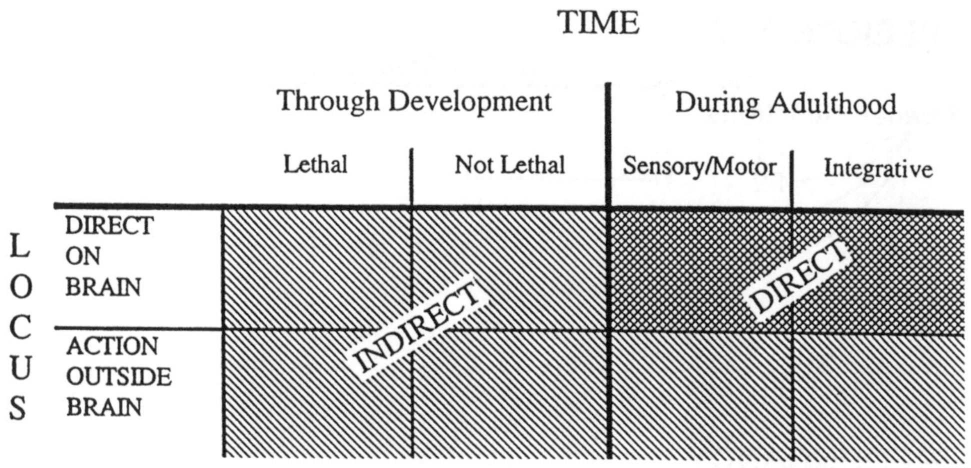

## TIME

FIGURE 1.2   A chart identifying all the various types of causal routes for genetic influences on brain function and behavior. *Note:* Considering adult behavior, all actions of genes on development, including mutations of hormone receptors and alterations of enzymes for steroid hormone synthesis and metabolism would be considered *indirect*. *Direct* effects would include genes expressed in the brain in adulthood necessary for a particular behavior, including genes for hormone receptors and hormone metabolism. (Modified from Ogawa et al., 1996a).

reproductive behavior: (1) the female behaves like and is treated as a male; (2) the ERKO females tend not to be mounted properly by stud males because of the females' behavior; and (3) even given the application of strong cutaneous stimuli as might produce lordosis behavior, ERKO females do not exhibit immobilization and vertebral dorsiflexion as frequently (Ogawa et al., 1996b). Analysis of the circuitry underlying this behavior (Pfaff et al., 1994) shows that gene expression for the estrogen receptor in medial hypothalamic neurons is necessary for the normal estrogenic facilitation of this behavior and, once stimulated, these neurons activate the neural circuitry for lordosis.

The gene for the progesterone receptor provides an example of a direct effect of genes on reproductive behavior (Ogawa et al., 1996a). In the natural induction of the progesterone receptor, the correlation of this gene's messenger RNA with lordosis behavior is clear (Romano et al., 1989; Lauber et al., 1991). Hormonal treatments of progesterone receptor knockout female mice (PRKO) (Lydon et al., 1995; Ogawa et al., unpublished data) can be set up to show the clear necessity of the progesterone receptor for the effect of progesterone on lordosis behavior. This result agrees with previous findings using the antisense DNA technique (Pollio et al., 1993; Ogawa et al., 1994) in which a targeted disruption of progesterone receptor messenger RNA function was associated with reduced lordosis behavior and greatly reduced courtship behaviors. In turn, all of these molecular approaches agree with the effects of the anti-progestin RU-486 (Brown and Blaustein, 1984; Vathy et al., 1989) in which the use of this progesterone receptor blocker significantly reduced female rat reproductive behavior.

**Indirect Effects.** Sex differences in simple mammalian behaviors offer examples of indirect effects of a gene on a wide variety of instinctive responses. Expression of the SRY gene (Lovell-Badge and Hacker, 1995) results eventually in the production of a Mullerian-inhibiting substance (MIS) (Haqq et al., 1994) that, with other factors, allows the development of testes. Secretions from the testes, particularly androgens but not necessarily limited to androgens, reach the brain where, aromatized to estrogens and active during brain development, they defeminize (see Figure 1.3) and masculinize certain behavioral responses.

Products of the estrogen receptor gene clearly contribute to the process of the sexual differentiation of behavior, as documented in the examples given in "Examples of Complex Dependencies."

We note that the phenomena of behavioral sexual differentiation are not limited to laboratory rodents, since even the social play behavior of non-human primates is strongly sexually differentiated, with genetic males showing a much higher frequency of rough-and-tumble play than females (Goy, 1968). The relative shares of basic biological mechanisms and social/cultural influences in the determination of sexually differentiated behaviors offer topics for continuing debate.

**Additional examples.** In addition to the clear examples given above, other cases drawn from the recent neuroendocrine literature can be adduced. Estrogen induces transcription from the gene for the oxytocin receptor and, in turn, the protein encoded is necessary for a full display of female reproductive behavior (Kow et al., 1991; Quinones-Jenab et al., 1997; Bale and Dorsa, 1995a, 1995b, 1997; Richard and Zingg, 1990; Breton and Zingg, 1997; Mohr and Schmitz, 1991). The importance of this gene for lordosis behavior was revealed by antisense DNA microinjections in the ventromedial nucleus of the hypothalamus (McCarthy et al., 1994).

Deletion of part of the gene for gonadotropin-releasing hormone (GnRH) in the mouse (Cattanach et al., 1977; Charlton, 1986) blocks reproductive behavior in both sexes but by different mechanisms. The adult female lacking GnRH simply will not show an estrus cycle, so the ovarian hormone levels required for sex behavior are absent. Upon replacement of GnRH neurons to the brain of the female, reproductive behavior ensues (Gibson et al., 1987). In contrast, the lack of reproductive behavior in the male can be laid to the absence of the normal masculinization of the brain during development.

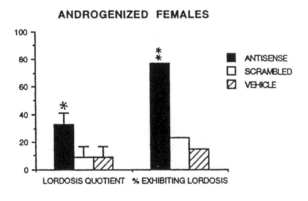

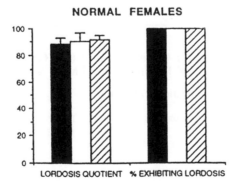

**FIGURE 1.3** Administration of antisense DNA directed against the messenger RNA for the estrogen receptor, to the medial hypothalamus, partially prevented the masculinizing influence of neonatal testosterone ("androgenized females"). The scrambled sequence control DNA administration to the medial hypothalamus had no effect, nor were there any effects in normal control females (McCarthy et al., 1993).

**TABLE 1.1**
**Alterations in Sexual and Aggressive Behaviors of Estrogen Receptor Knockout (ERKO) Female Mice**

|  | Incidence of Female Reproductive Behavior | Incidence of Female–Female Aggression |
|---|---|---|
| Wild type | Normal | 2/21 mice |
| Estrogen Receptor Knockout | None | 10/25 mice* |

* Aggression exhibited by ERKO females; mainly offensive attacks typical of intermale aggression.

Based on data from Ogawa et al., 1996b.

In patients, damage to various genes involved in the synthesis of cortisol leads to congenital adrenal hyperplasia (New, 1985). As a result, there is hypersecretion of adrenal androgenic hormones which, in turn, fosters a male gender identity even among genetically human females (Money and Lewis, 1982).

In men, abnormalities of the gene for the androgen receptor (similar to mouse mutations described by Ohno et al., 1973a, 1973b) could lead to behavioral changes due to reduced androgen binding in the brain (Attardi et al., 1976, 1978). Likewise, mutations in a gene for the enzyme responsible for the conversion of testosterone to 5-alpha dihydrotestosterone (Wilson et al., 1995) result in an androgen insensitivity syndrome. In many such cases, human males are raised as females and show correspondingly appropriate social behaviors (Money, et al., 1984; Gooren and Money, 1995).

## EXAMPLES OF COMPLEX DEPENDENCIES

As noted above, the behavior of the estrogen receptor knockout (ERKO) mouse has proven that classical ER gene expression is indeed required for estrogenic effects on lordosis (Ogawa et al., 1996b). In fact, ERKO yields a female mouse that is more *masculinized* (see Table 1.1), behaving less like a genetic female and actually treated as a male. In dramatic contrast, the behavior of ERKO males is certainly not more masculinized; it is more *feminized* (see Table 1.2). ERKO males show virtually no intromissions or ejaculations even though several indices of their sexual motivation appear normal (Ogawa et al., 1997). Overall, such estrogen-deficient males show marked decreases in a subset of their masculine-typical behaviors and a trend toward a more feminine type

**TABLE 1.2**
**Estrogen Receptor Gene Knockout (ERKO) Causes Behavioral Feminization of Male Mice**

Alterations in ERKO ♂s

| Male-specific intromissions and ejaculations | LESS RARE |
|---|---|
| Aggression　Offensive attacks | LESS RARE |
| Open-field (emotional) responses | FEMINIZED |

Adapted from data in Ogawa et al., 1997.

---

**TABLE 1.3**
**Estrogen Receptor Gene Knockout (ERKO) Causes**
**Behavioral Masculinization of Female Mice**

**Alterations in ERKO ♀s**

| | |
|---|---|
| Maternal behavior (including infanticide) | REDUCED |
| | |
| Aggressive behavior toward females | INCREASED |
| toward males | INCREASED |
| | |
| Female reproductive behaviour | ABOLISHED |

From Ogawa et al., 1996b.

---

of behavioral profile. Their aggressive behaviors were dramatically reduced and, in particular, they showed absolutely no male-typical offensive attacks. Their emotional responses to the open field test were demasculinized. Thus, we are left in a situation in which the ERKO gene alteration renders the behavior of genetic female mice more masculinized yet renders genetic male mice more feminized.

These behavioral results are mirrored in some recent data (Simerly et al., 1997) in which a sexually dimorphic population of dopaminergic neurons in the anterior hypothalamus and preoptic area was measured. Ordinarily, wild-type females have many more of these neurons in a restricted zone of the anterior hypothalamus than do wild-type males. The effect of the disruption of the estrogen receptor gene in the ERKO animals was to reduce the female values toward those expected for normal males, but to increase the male values toward those expected for normal females (Simerly et al., 1997). Thus, as for the behavioral results, the ERKO genetic disruption moved the values for genetic males and genetic females in polar opposite directions along the continuum between normal masculine and feminine values. *The lesson from both of these reports is that as far as a gene like the estrogen receptor is concerned, its effect on neural development depends upon the gender of the animal in which it is expressed.*

Analyses of hormone-dependent behaviors also demonstrate how the effect of a gene on a specific behavior depends upon exactly where and when that gene is expressed. For example, even though the ERKO female is more *masculinized* in its behavior (Ogawa et al., 1996b), a temporary antisense DNA-caused interruption, specifically of hypothalamic ER messenger RNA during neonatal testosterone administration to female rats actually *prevented full masculinization* of the rat brain (McCarthy et al., 1993). That is, Figure 1.3 demonstrated that ER gene product disruption limited to the hypothalamus and applied only on one neonatal day actually reduced the masculinization of forebrain and behavior due to experimentally administered testosterone. In dramatic contrast, the ER gene disruption in ERKO females is limited neither in time nor space, and this genetic maneuver led to more masculinized behavior (see Table 1.3). *The message is that the effect of a genetic alteration on the nervous system and behavior can depend upon exactly where in the CNS and for exactly how long this genetic alteration is applied.*

Furthermore, detection of the effects of a genetic manipulation depends exclusively on the precise conditions of assay. Even slight increases in the complexity of the behavior analyzed can lead to a corresponding complexity of interpretation, as illustrated by a social behavior such as maternal behaviors in female rats. Female rats were not sensitive to the effects of oxytocin when they were unstressed or when they were severely stressed, but at intermediate levels of mild stress, oxytocin facilitated their maternal behavior (Fahrbach et al., 1986).

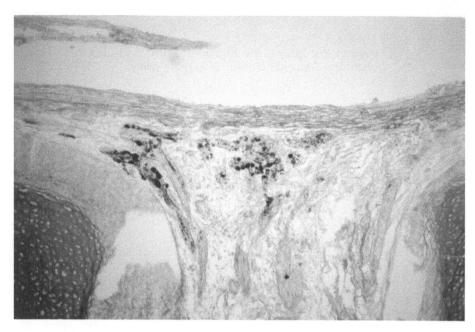

**FIGURE 1.4**   GnRH neurons dammed up in the olfactory apparatus of a human fetus with genetic damage at Xp-22.3, as would lead to Kallmann's Syndrome. In a normal individual, these GnRH neurons migrate into the basal forebrain during development, but in this case, no GnRH neurons reached the brain (Schwanzel-Fukuda et al., 1989).

## Example from a Human Syndrome — Kallmann's Syndrome

A human syndrome provides an example of how subtle and indirect the relationships between genes and social behavior can be. Kallmann's Syndrome, hypogonadotropic hypogonadism coupled with anosmia (Crowley and Jameson, 1992), afflicts men with a striking behavioral change: absence of libido as part of a disinterest in the opposite sex. Causation of X-linked Kallmann's is now understood in light of the surprising findings that GnRH-producing neurons are not born, as expected, near the brain's ventricular surfaces. Instead, they are born in the olfactory epithelium (Schwanzel-Fukuda and Pfaff, 1989). That is, they must migrate from the nose to the brain. Indeed, X-linked Kallmann's Syndrome, caused by genetic damage at Xp-22.3, has been correlated with a failure of migration of GnRH neurons (Schwanzel-Fukuda et al., 1989). In this case, the GnRH neurons were dammed up in the olfactory apparatus (see Figure 1.4) and never reached the brain. In turn, this neuronal migration disorder is rationalized by the fact that damage at Xp-22.3 disrupts a specific gene (Legouis et al.,1991; Franco et al., 1991) whose damage causes X-linked Kallmann's, and this gene codes for a cell surface protein present during migration of GnRH neurons into the brain (Dellovade et al., 1998).

Putting these and related facts in logical order leads to a clear example of complex participation by an individual gene in human behavior through its actions during neural development. That is, in males suffering from X-linked Kallmann's Syndrome, behavioral libido is reduced *because* of low testosterone levels, in turn *because* of reduced gonadotropins LH and FSH, which in turn are low *because* there is no GnRH coming from the brain to enter the pituitary, *because* there are no GnRH-producing neurons in the brain, *because* GnRH neuronal migration has failed, *because* of the absence of the protein produced by a gene at Xp-22.3.

In summary, these data demonstrate a genetic influence on an important human social behavior, but they also illustrate the complicated and indirect nature of that effect.

## INFERENCE

Because of the pleiotropy and redundancy of causal relationships between genes and behavior and because of the complex dependencies illustrated previously, reasoning in a simple-minded fashion "from the gene on out" will not work. Instead, for a systematic analysis, using a geometric style of logic and "reasoning backward" from the behavior toward the logical requirement for genetic contributions will be more successful. That is, we have to envision each physiological mechanism that serves an essential biological function. We must then "reason backward" to deduce how genetic effects feed in at many levels, from neural development, through "housekeeping" maintenance functions, through specific behaviorally relevant regulation.

## SYSTEMATIC LOGICAL TREATMENTS OF EXPERIMENTAL RESULTS: PHYSICS VS. BIOLOGY

Given the difficulties of analysis as illustrated in the previous sections, how do we summarize and conceptualize new findings about genes and brain function? Do we simply make lists? No. We are obliged to do better than that.

Suppose we were trying to reason in a systematic fashion "from the gene on out to brain functions." As argued above, that strategy is not likely to work well. We can see why it will not work very easily by reflecting on the complicated structure of Figure 1.1. Therefore, comparing our field to classical physics, we in biology need a shift of paradigm in order to organize and systematically understand new biological knowledge with respect to genes and brain function. That is, physics yields equations. It is timely to explore the notion that in addition to those equations, neurobiologists will need another, overarching mathematical approach that includes biophysical equations but is not limited to them. In this approach, in line with the reasoning in the previous two sections, we should be able, in a geometric fashion, to "reason backward" from the axiomatic requirements for survival of individuals and preservation of the species to deduce specific physiological functions and their brain mechanisms. Incidentally, beyond that, we will have the capacity to use linear programming to test for optimization of the physiologic integration and their mechanisms involved. In summary, one can predict that we will be using a form of deductive logic to reason from axiomatic biological need to physiological operation to neural operation, and only then, to a systematic appreciation of each direct or indirect genetic contribution.

**Criteria for proof.** What will be the patterns of evidence sufficient to convince a skeptical reader of a genetic effect on any particular behavior? Essentially, the requirements for proof will parallel those used in pharmacology, and the data yielded will form a molecular pharmacology of the central nervous system. First, the gene under investigation must be expressed in the tissue postulated at the proper time for the proposed mechanisms to occur. Second, supplementing for the normal gene by direct gene transfer or by inserting its expected product should produce the behavioral effect in question. The magnitude of the behavioral effect should be correlated with the concentration of the candidate gene's product. Experimental manipulations that alter or synergize with that gene's product should, to the same degree, alter the behavior in question. Third, other genes with unrelated products should not yield the same behavioral result. However, other genes that are redundant with the first should produce the same behavioral effect. Fourth, natural mutation or experimental gene knockout, blocking synthesis of the gene product, or blocking its receptor, all should interfere with the proposed behavioral effect.

How are these investigations likely to proceed? Positional cloning to characterize natural mutations is a proven strategy. Beyond that, the QTL approach (quantitative trait loci) is increasingly popular as an open-ended form of search for multiple unknown genes (Lander and Schork, 1994). Perhaps more efficient, for behaviors whose neurobiology is relatively simple, would be targeted approaches, including knockouts (Smithies and Maeda, 1995) that can be made inducible (Kuhn et al., 1995). Also, in experimental animals, selected gene products can be manipulated with either of two relatively new techniques: neurotropic viral vectors for gene transfer to selected neuronal groups in the brain (Kaplitt et al., 1993, 1991) and antisense DNA microinjections in the brain (Ogawa and Pfaff, 1996c; Nicot and Pfaff, 1997). Compared to knockouts, the antisense technique has the advantage of high temporal and spatial resolution and the disadvantage of requiring multiple experimental controls against non-specific effects. In all cases, testing for behavioral specificity of the effect is also required.

**Summary.** Rigorous and systematic organization of genetic information in our field requires (a) a clear delineation of the axiomatic biological need served by any given brain function; (b) use of a "geometric" type of logic — a comprehensive list of neuronal operations essential for that brain function; and (c) a deduction of how a given gene product serves either the development or the adult performance of one or more of those neuronal operations.

## CONTENTS OF THE BOOK

Some chapters of this book (i.e., Chapter 15 by Son et al.) distinguish neural operations that are not easily linked to specific behaviors from frank behavioral manifestations. Other chapters (i.e., Chapter 5 by Sobin and Karayiorgou) distinguish normal neural operations from abnormal (disease) conditions. Many chapters represent areas of work that are "mushrooming" in the best sense. The precision of behavioral analysis born of psychophysics and the experimental analysis of behavior finds its full employment in showing the contributions of specific genes at this highest level of neuronal integration. This book is intended to start updating clinicians and basic brain scientists with respect to some of the new developments that depend on the human and mouse genome projects and are important for brain function. Electronic publication of this and subsequent volumes will help us keep track of important developments in this dynamic field.

## REFERENCES

Attardi, B., Geller, L.N., and Ohno, S., Androgen and estrogen receptors in brain cytosol from male, female and testicular feminized (Tfm/y hermaphrodite) mice, *Endocrinology*, 98, 864, 1976.

Attardi, B. and Ohno, S., Physical properties of androgen receptors in brain cystosol from normal and testicular feminized (Tfm/Y hermaphrodite) mice, *Endocrinology*, 103, 760, 1978.

Bale, T. and Dorsa, D., Cloning, novel promoter sequence, and estrogen regulation of a rat oxytocin receptor gene, *Endocrinology*, 138, 1151, 1997.

Bale, T.L. and Dorsa, D.M., Sex differences in and effects of estrogen on oxytocin receptor mRNA expression in the ventromedial hypothalamus, *Endocrinology*, 136, 27, 1995a.

Bale, T.L., Dorsa, D.M., and Johnston, C.A., Oxytocin receptor mRNA in the ventromedial hypothalamus during estrous cycle, *J. Neurosci.*, 15, 5058, 1995b.

Breton, C. and Zingg, H.H., Expression and region-specific regulation of the oxytocin receptor gene in rat brain, *Endocrinology*, 138, 1857, 1997.

Brown, T.J. and Blaustein, J.D., Inhibition of sexual behavior in female guinea pigs by a progestin receptor antagonist, *Brain Res.*, 301, 343, 1984.

Cattanach, B.M., Iddon, C.A., Charlton, H. M., Chiappa, S.A., and Fink, G., Gonadotropin-releasing hormone deficiency in a mutant mouse with hypogonadism, *Nature*, 269, 338, 1977.

Charlton, H.M., The physiological actions of LHRH: evidence from the hypogonadal (hpg) mouse, in *Neuroendocrine Molecular Biology*, Fink, G., Harmar, A.J., and McKerns, K.W., Eds., Plenum, New York, 1986, 47.

Crowley, W.F. and Jameson, J.L., Clinical counterpoint: Gonadotropin-releasing hormone deficiency: perspectives from clinical investigation, *Endocrine Rev.*, 13(4), 635, 1992.

Dellovade, T., Hardelin, J.-P., Soussi-Yanicostas, N., Pfaff, D., Schwanzel-Fukuda, M., and Petit, C., Anosmin-I immunoreactivity during embryogenesis in a primitive eutherian mammal, *PNAS*, submitted.

Fahrbach, S.E., Morrell, J. I., and Pfaff, D.W., Effect of varying the duration of pre-test cage habituation on oxytocin induction of short-latency maternal behavior, *Physiol. Behav.,* 37, 135, 1986.

Franco, B., Balabio, A. et al., A gene deleted in Kallmann's Syndrome shares homology with neural cell adhesion and axonal path-finding molecules, *Nature,* 353, 529, 1991.

Gerlai, R., Gene-targeting studies of mammalian behavior: is it the mutation or the background genotype? *Trends Neurosci.,* 19, 177, 1996.

Gibson, M.J., Moscovitz, H.C., Kokoris, G.J., and Silverman, A.J., Female sexual behavior in hypogonadal mice with GnRH containing brain grafts, *Horm. Behav.,* 21, 211, 1987.

Gooren, L. and Money, J., Normal and abnormal sexual behavior, in *Endocrinology* Vol. 2, 3rd ed., DeGroot, L.J., Ed., W.B. Saunders, Philadelphia, 1995, p. 1978.

Goy, R. W., Organizing effects of androgen on the behavior of rhesus monkeys, in *Endocrinology and Human Behavior*, Michael, R.P., Ed., Oxford University Press, London, 1968, 12.

Greenspan, R.J. and Tully, T., Group Report: How do genes set up behavior, in *Flexibility and Constraint in Behavioral Systems,* Greenspan, R.J. and Kyriacou, C.P., Eds., John Wiley & Sons Inc., New York, 1994, 65.

Hall, J.C., Pleiotropy of behavioral genes, in *Flexibility and Constraint in Behavioral Systems,* Greenspan, R.J. and Kyriacou, C.P., Eds., John Wiley & Sons Inc., New York, 1994, 15.

Haqq, C.M., King, C-Y., Ukiyama, E., Falfasi, S., Haqq, T.N., Donohoe, P.K., and Weiss, M.A., Molecular basis of mammalian sexual determination: activation of Mullerian-inhibiting substance gene expression by SRY, *Science,* 266, 1494, 1994.

Kaplitt, M.G., Rabkin, S., and Pfaff, D.W., Molecular alterations in nerve cells: direct manipulation and physiological mediation, in *Current Topics in Neuroendocrinology,* Vol. II, Imura, M., Ed., Springer-Verlag, Berlin, 1993, 169.

Kaplitt, M.G., Pfaus, J.G., Kleopoulos, S.P., Hanlon, B.A., Rabkin, S.D., and Pfaff, D.W., Expression of a functional foreign gene in adult mammalian brain following in vivo transfer via a herpes simplex virus type 1 defective viral vector, *Mol. Cell. Neurosci.,* 2, 320, 1991.

Kow, L.-M., Johnson, A.E., Ogawa, S., and Pfaff, D.W., Electrophysiological actions of oxytocin on hypothalamic neurons, in vitro: neuropharmacological characterization and effects of ovarian steroids, *Neuroendocrinology,* 54, 526, 1991.

Kuhn, R., Schwenk, F., Aguet, M., and Rajewsky, K., Inducible gene targeting in mice, *Science,* 269, 1427, 1995.

Lander, E. and Schork, N., Genetic dissection of complex traits, *Science,* 265, 2037, 1994.

Lauber, A.H., Romano, G.J., and Pfaff, D.W., Sex difference in estradiol regulation of progestin receptor mRNA in rat mediobasal hypothalamus as demonstrated by in situ hybridization, *Neuroendocrinology,* 53, 608, 1991.

Legouis, R., Hardelin, J.P., Petit, C. et al., The candidate gene for the X-linked Kallmann Syndrome encodes a protein related to adhesion molecules, *Cell,* 67, 423, 1991.

Lovell-Badge, R. and Hacker, A., The molecular genetics of *Sry* and its role in mammalian sex determination, *Philos. Trans. R. Soc. London*, 350, 205, 1995.

Lydon, J.P., DeMayo, F.J., Funk, C.R., Mani, S.K., Hughes, A.R., Montgomery, C.A., Shyamala, G., Conneely, O.M., and O'Malley, B.W., Mice lacking progesterone receptor exhibit pleiotropic reproductive abnormalities, *Genes Dev.,* 9, 2266, 1995.

McCarthy, M.M., Kleopoulos, S.P., Mobbs, C.V., and Pfaff, D.W., Infusion of antisense oligodeoxynucleotides to the oxytocin receptor in the ventromedial hypothalamus reduces estrogen-induced sexual receptivity and oxytocin receptor binding in the female rat, *Neuroendocrinology,* 59, 432, 1994.

McCarthy, M.M., Schlenker, E., and Pfaff, D.W., Enduring consequences of neonatal treatment with antisense oligonucleotides to estrogen receptor mRNA on sexual differentiation of rat brain, *Endocrinology,* 133, 433, 1993.

Mohr, E. and Schmitz, E., Functional characterization of estrogen and glucocorticoid responsive elements in the rat oxytocin gene, *Mol. Brain Res.,* 9, 293, 1991.

Money, J. and Lewis V.G., Homosexual/heterosexual status in boys at puberty: idiopathic adolescent gynecomastia and congenital virilizing adrenocorticism compared, *Psychoneuroendocrinology,* 7, 339,1982.

Money, J., Schwartz, M., and Lewis, V.G., Adult erotosexual status and fetal hormonal masculinization and demasculinization, *Psychoneuroendocrinology,* 9, 405, 1984.

New, M., Congenital adrenal hyperplasia, in *Endocrinology*, Vol. 2, 3rd ed., DeGroot, L.J. et al., Ed., W.B. Saunders, Philadelphia, 1985, 1813.

Nicot, A. and Pfaff, D.W., Antisense oligodeoxynucleotides as specific tools for studying neuroendocrine and behavioral functions: some prospects and problems, *J. Neurosci. Methods*, 71, 45, 1997.

Ogawa, S., Gordan, J., Taylor, J., Lubahn, D., Korach, K., and Pfaff, D.W., Reproductive functions illustrating direct and indirect effects of genes on behavior, *Horm. Behav.*, 30, 487, 1996a.

Ogawa, S., Lubahn, D.B., Korach, K.S., and Pfaff, D.W., Behavioral effects of estrogen receptor gene disruption in male mice, *Proc., Natl. Acad. Sci.*, 94, 1476, 1997.

Ogawa, S., Olazabal, U.E., Parhar, I.S., and Pfaff, D.W., Effects of intrahypothalamic administration of antisense DNA for progesterone receptor mRNA on reproductive behavior and progesterone receptor immunoreactivity in female rat, *J. Neurosci.*, 14, 1766, 1994.

Ogawa, S., Taylor, J., Lubahn, D.B., Korach, K.S., and Pfaff, D.W., Reversal of sex roles in genetic female mice by disruption of estrogen receptor gene, *Neuroendocrinology*, 64, 467, 1996b.

Ogawa, S. and Pfaff, D.W., Application of antisense DNA method for the study of molecular bases of brain function and behavior, *Behav. Genet.*, 26(3), 279, 1996c.

Ohno, S., Christian, L., and Attardi, B., Role of testosterone in normal female function. *Nat. (London) New Biol.*, 243, 119, 1973a.

Ohno, S., Christian, L., and Attardi, B., and Kan, J., The modification of expression of the testicular eminization (Tfm) gene of the mouse by a "controlling element" gene, *Nat. (London) New Biol.*, 245, 92, 1973b.

Pfaff, D.W. and Keiner, M., Atlas of estradiol-concentrating cells in the central nervous system of the female rat, *J. Comp. Neurol.*, 151, 121, 1973.

Pfaff, D.W., Schwartz-Giblin, S., McCarthy, M.M., and Kow, L.-M., Cellular and molecular mechanisms of female reproductive behaviors, in *The Physiology of Reproduction*, 2nd ed., Knobil, E. and Neill, J., Eds., Raven Press, New York, 1994, 107.

Pollio, G., Xue, P., Zanisi, M., Nicolin, A., and Maggi A., Antisense oligonucleotide blocks progesterone-induced lordosis behavior in ovariectomized rats, *Mol. Brain Res.*, 19, 135, 1993.

Quiñones-Jenab, V., Jenab, S., Ogawa, S., Adan, R.A.M., Burbach, P.H., and Pfaff, D.W., Effects of estrogen on oxytocin receptor messenger ribonucleic acid expression in the uterus, pituitary and forebrain of the female rat, *Neuroendocrinology*, 65, 9, 1997.

Richard, S. and Zingg, H.H., The human oxytocin gene promoter is regulated by estrogens, *J. Biol. Chem.*, 265, 1, 1990.

Romano, G.J., Krust, A., and Pfaff, D.W., Expression and estrogen regulation of progesterone receptor mRNA in neurons of the mediobasal hypothalamus: an *in situ* hybridization study, *Mol. Endocrinol.*, 3, 1295, 1989.

Schwanzel-Fukuda, M. and Pfaff, D.W., Origin of luteinizing hormone-releasing hormone neurons, *Nature*, 338, 161, 1989.

Schwanzel-Fukuda, M., Bick, D., and Pfaff, D.W., Luteinizing hormone-releasing hormone (LHRH)- expressing cells do not migrate normally in an inherited hypogonadal (Kallmann) syndrome, *Mol. Brain Res.*, 6, 311, 1989.

Simerly, R.B, Zee, M.C., Pendleton, J.W., Lubahn, D.B., and Korach, K.S., Estrogen receptor-dependent sexual differentiation of dopaminergic neurons in the preoptic region of the mouse, *Proc. Natl. Acad. Sci.*, 94, 14077, 1997.

Smithies, O. and Maeda, N., Gene targeting approaches to complex genetic diseases: atherosclerosis and essential hypertension, *Proc. Natl. Acad. Sci.*, 92, 5266, 1995.

Vathy, I., Etgen, A., and Barfield, R., Actions of RU 38486 on progesterone facilitation and sequential inhibition of rat estrous behavior: correlation with neural progestin receptor levels, *Horm. Behav.*, 23, 43, 1989.

Ward, B. and Charlton, H.M., Female sexual behavior in the GnRH-deficient hypogonadal (hpg) mouse, *Physiol. Behav.*, 27, 1107, 1981.

Wilson, J.D., George, F.W., and Renfree, M.B., The endocrine role in mammalian sexual differentiation. *Recent Progr. Horm. Res.*, 50, 349, 1995.

# 2 Genetics and Opioid Pharmacology

*Gavril W. Pasternak*

## CONTENTS

## INTRODUCTION

Opioids are best known for their ability to relieve pain, an action that is quite novel and unique.[1,2] Local anesthetics block the transmission of nerve impulses, leading to the complete loss of sensation. In contrast, opiates interfere with the suffering component of pain, leaving objective sensations intact. Thus, opioids act on the integrated sensation of pain. Patients receiving opioids often report that the pain is still present but that it no longer hurts. Although the ability to relieve pain is often considered their most important characteristic, opiates have many other actions, including the modulation of respiratory function, gastrointestinal transit, and a wide variety of effects on the endocrine system. Indeed, the highest concentrations of opioid peptides in the body are in the adrenal medulla, where they are co-released with epinephrine in periods of stress. Similarly, β-endorphin is generated from β-lipotropin, the same peptide precursor which generates ACTH, and both are secreted together from the pituitary. The past 30 years have yielded many important new insights into the opioid systems, which consist of a large family of endogenous peptides and their receptors.

## OPIOID RECEPTORS AND THEIR PHARMACOLOGY

Opioid receptors were originally proposed based upon the strict structure-activity relationships of the opioid analgesics. The concept of receptor multiplicity was first raised based upon interactions

**TABLE 2.1**
**Opioid Receptor Subtypes and Their Actions**

| Receptor | Ligand | Analgesia | Other |
|---|---|---|---|
| **Mu** | | | |
| $Mu_1$ | Morphine | Supraspinal | Prolactin release |
| | DAMGO | | Acetylcholine release in the hippocampus |
| | Naloxonazine** | | Feeding |
| $Mu_2$ | Morphine | Spinal | Respiratory depression |
| | DAMGO | | Gastrointestinal transit |
| | β-Funaltrexamine** | | Dopamine release by nigrostriatal neurons |
| | | | Guinea pig ileum bioassay |
| | | | Feeding |
| M6G | M6G | | Spinal and supraspinal analgesia |
| | Heroin* | | |
| | 6-acetylmorphine* | | |
| | Fentanyl* | | |
| | 3-Methoxynaltrexone** | | |
| **Kappa** | | | |
| | Ketocyclazocine | | Psychotomimesis |
| | | | Sedation |
| $Kappa_1$ | Dynorphin A | Predominantly | Diuresis |
| | U50,488H | Spinal | Feeding |
| | nor-Binaltorphimine** | | |
| $Kappa_2$ | | Unknown | Unknown |
| $Kappa_3$ | NalBzoH | Predominantly | |
| | Nalorphine | Supraspinal | |
| **Delta** | | | |
| | Enkephalins | | Mouse vas deferens bioassay |
| | Naltrindole** | | Dopamine turnover in the striatum |
| | | | Feeding |
| $Delta_1$ | | Supraspinal | |
| $Delta_2$ | | Spinal and supraspinal | |

* These drugs also have activity at traditional mu ($mu_1$ and $mu_2$) receptors.
**Antagonists.
*Note:* Some actions assigned to general families have not yet been correlated with a specific subtype. NalBzoH refers to naloxone benzoylhydrazone.

between morphine and nalorphine and has greatly expanded with the identification of the opioid peptides and the development of selective agents able to identify the receptors in binding assays.[3,4] Pharmacologically, the actions of these different receptors can be differentiated through the use of highly selective antagonists. The general classifications are given above (see Table 2.1).

## Mu Receptors

The mu receptors are the traditional opioid receptors activated by morphine.[1-3] Pharmacologically, these receptors produce analgesia as well as constipation and, at higher doses, respiratory depression. In receptor binding studies, mu receptors have high affinity for morphine and morphine-related compounds, which include a wide variety of alkaloids and peptides, including most clinical drugs such as methadone, oxycodone, codeine, and even heroin. Mu drugs can produce analgesia when administered either spinally or supraspinally. Supraspinally, a number of brain stem regions are capable of eliciting mu analgesia, including the periaqueductal gray, the nucleus raphe magnus, the locus ceruleus, and the nucleus reticularis giganto cellularis. Spinally, opioids act superficially in the posterior horn. In addition, the importance of peripheral mu mechanisms have become increasingly appreciated. Understanding these various systems in the production of analgesia is important. Although each can function alone, in vivo, they do not act independently. Rather, they interact in a complex synergistic system resulting in profound analgesia. The first indication of interacting mu receptor systems came from work by Young and Rudy[5] who observed that co-administration of morphine both spinally and supraspinally were markedly synergistic compared with doses given to either region alone. Even among brain stem regions, synergy exists between different mu receptor systems.[6,7] The situation is further complicated with the potent modulation of central actions by peripheral mechanisms.[8] Thus, the mu mechanisms of analgesia are extraordinarily complex.

A variety of antagonists can reverse mu actions.[1,2] The first pure opioid antagonist was naloxone, which blocks all mu receptors and even other opioid receptor subtypes at higher doses. Naltrexone has a similar antagonist profile, although it is more potent. Selective mu antagonists also have been developed, including $\beta$-funatrexone ($\beta$-FNA) and CTAP, which block all the mu actions. However, another mu antagonist, naloxonazine, has provided further insights into mu receptor actions. Unlike $\beta$-FNA and CTAP, naloxonazine acts against only a subpopulation of mu receptors, termed $mu_1$. $Mu_1$ receptors mediate supraspinal morphine analgesia, while $mu_2$ receptors are responsible for respiratory depression and constipation. $Mu_2$ receptors also can elicit analgesia, mediating the spinal actions of morphine.

Morphine-6$\beta$-glucuronide (M6G) is an interesting morphine metabolite. Unlike the 3-glucuronide, M6G is active.[9] Indeed, when given centrally to avoid the blood-brain barrier, it is 100-fold more potent than morphine.[10,11] Evidence from binding studies and pharmacological approaches now indicates that M6G acts through yet another mu receptor subtype.[12-16] This is most readily demonstrated in studies comparing the analgesic actions of opioids in CXBK mice. Unlike typical mouse strains in which morphine and M6G are both analgesics, CXBK mice are insensitive to morphine. Despite their lack of response to supraspinal morphine, these animals respond normally to M6G. Thus, the actions of M6G are clearly mediated through mechanisms distinct from those of morphine. It became even more interesting when heroin and its active metabolite 6-acetylmorphine were examined. Both agents retained their analgesic activity in the CXBK mice, confirming that heroin acts differently than morphine.

## DELTA RECEPTORS

The delta receptors[17] were first demonstrated soon after the identification of the enkephalins.[18] The enkephalins were the first members of the endogenous opioids identified and contain the prototypic sequence found in all the opioid peptides, **Tyr-Gly-Gly-Phe,** followed by either methionine or leucine.[1,2] Like many peptides, they are extremely labile and are rapidly degraded enzymatically. A wide variety of stable derivatives have been synthesized with vastly different selectivity profiles. A number are highly selective agonists, while others are potent antagonists of the delta systems. The most widely used delta-selective antagonist, however, is naltrindole.

Delta analgesia is most readily demonstrated at the spinal level. Here, enkephalins such as DPDPE enkephalin and deltorphan produce robust analgesic responses that are readily reversed by naltrindole, a highly selective antagonist. Supraspinally, delta analgesia often requires higher doses. Furthermore, a variety of studies now suggests that supraspinal delta analgesia involves more than one subtype of delta receptor. Spinally, $delta_2$ receptors mediate the response while supraspinally evidence suggests a role for both $delta_1$ and $delta_2$ receptor systems.[19-21] These delta systems are readily distinguished pharmacologically from mu receptor systems. Although both can produce analgesia, each can be activated independently of the other and selective antagonists can block the actions of one system without interfering with the other. However, the delta and mu systems can interact synergistically when both are activated simultaneously.[7] Thus, the complexity of the delta system is becoming far greater as our ability to explore potential subtypes improves.

## KAPPA RECEPTORS

The kappa receptor family is quite broad. Within the kappa family, two general classes can be defined: those sensitive (kappa$_1$) and those insensitive (kappa$_2$ and kappa$_3$) to U50,488.[22] All three subtypes have been defined in binding studies, but only the kappa$_1$ and kappa$_3$ receptors have been characterized pharmacologically.[1,2] Without a selective agonist or antagonist, it is very difficult to define kappa$_2$ pharmacology *in vivo*.

Dynorphin A is the endogenous ligand for the kappa$_1$ system,[23] although it has been characterized pharmacologically by U50,488 and a selective antagonist, norbinaltorphimine (norBNI).[24,25] Kappa drugs are potent analgesics that are typically more active spinally than supraspinally. When given systemically, their actions are predominantly at the spinal level as well. As with the other subtypes, highly selective kappa$_1$ drugs such as U50,488 produce analgesia through a distinct receptor system unrelated to the other subtypes. Highly selective kappa$_1$ drugs show no cross-tolerance with the other receptors, and they have their own antagonist selectivity profile. Thus, kappa$_1$ receptors define a unique analgesic system. Kappa$_1$ agonists also produce an intense diuresis in both animal models and people.

Kappa$_3$ receptors were initially defined on the basis of binding studies with naloxone benzoyl-hydrazone (NalBzoH),[22,26] which has subsequently proven valuable in their pharmacological characterization *in vivo* as well.[27,28] Kappa$_3$ analgesia is most prominent supraspinally, and it shows its own unique antagonist sensitivity profile. As with the others, kappa$_3$ analgesia is mediated independently of the other systems.

## SIGMA RECEPTORS

Sigma receptors originally were considered as opioids, but it is now clear that they represent a novel receptor class unrelated to the opioid receptors.[29,30] The sigma$_1$ receptor has now been cloned,[31-33] and it shows no homology towards the opioid receptor clones. Indeed, the sigma$_1$ receptor does not even conform to a traditional G-protein coupled receptor. Activation of sigma$_1$ receptors leads to a reduction in the analgesic activity of a wide variety of opioid ligands.[34-37] Conversely, sigma$_1$ receptor blockade enhances opioid analgesic responses, particularly those of kappa agents. This ability of sigma antagonist to potentiate opioid analgesia implies a tonic activity of the sigma system. The level of this tonic activity varies among strains of mice. For example, BALB/c mice are far less sensitive toward kappa analgesics than CD-1 mice. Sigma$_1$ receptor blockade potentiates the kappa responses in both strains, making them both equally sensitive to the drugs. Thus, some of the genetically induced differences in sensitivity among mouse strains can be explained by differences in the tonic activity of an anti-opioid sigma system.

It is also interesting that the modulation of opioid analgesia by the sigma system does not extend to other opioid actions. The inhibition of gastrointestinal transit induced by morphine, for example, is not sensitive to sigma receptor blockade. Thus, blocking sigma receptors increases the therapeutic index of drugs, increasing their analgesic potency without increasing their ability to induce side effects.

## MOLECULAR BIOLOGY OF OPIOID RECEPTORS

The first opioid receptor was cloned from a mouse neuroblastoma cell line expressing delta receptors.[38,39] This clone, termed DOR-1, displays binding characteristics when expressed in cell lines that are very similar to those of delta receptors defined with binding studies in brain. This discovery was quickly followed by the identification of clones encoding a mu (MOR-1)[40-47] and a kappa$_1$ (KOR-1) receptor.[40,48-53] The opioid receptors are members of the G-protein coupled receptor family. They have the anticipated seven transmembrane regions and strong homology to other cloned G-protein receptors. The receptors typically act through pertussis toxin sensitive G-proteins (G$_i\alpha$ and G$_o\alpha$), and when expressed, they inhibit adenylyl cyclase activity. They also can modulate potassium channels and calcium fluxes.

The MOR-1 and KOR-1 clones are highly homologous to the DOR-1, particularly within the transmembrane regions. Like DOR-1, MOR-1 and KOR-1 encode receptors with the anticipated binding characteristics of mu and kappa$_1$ receptors, respectively. The gene structure of the MOR-1 differs from the others in that it has four coding exons, while the others have only three.[54] The first exon of all the clones encodes the extracellular NH$_2$-terminus and the first transmembrane region. The second exon continues until the second extracellular loop, which contains the splice site with the third exon. The fourth exon in MOR-1 encodes only a few amino acids at the intracellular COOH-terminus and two alternatively spliced variants have been identified.[41,55]

## CORRELATING CLONES WITH BEHAVIOR

Pharmacological studies have identified a number of subtypes within each opiate receptor class, as discussed previously. Yet, cloning studies have identified only a single gene for each of the major classes of opioid receptors, raising questions about the correlation of the receptors defined pharmacologically with those defined on the basis of their molecular biology. It remains crucial to correlate the cloned receptors with specific pharmacological functions. Correlating behavior with a specific protein at the molecular level can be difficult.[56] One approach is to target the gene encoding the protein of interest and knocking it out. This approach has an advantage in that the loss of the gene in question is unequivocal and the resulting phenotype accurately reflects the loss of this gene. A problem with this approach is the possibility that the loss of the gene product during development may result in compensation through alternative systems. In addition, it is very time- and resource-dependent. Finally, there always is the possibility that the gene of interest is essential and its absence may prove lethal.

Another approach is to down-regulate the protein of interest by targeting the mRNA with antisense oligonucleotides.[57-64] The advantage of this technique is that it can be performed in adult animals where developmental issues are no longer a problem, and it is readily performed. However, the down-regulation of the protein is typically far from complete.[65,66] Stability of the antisense probe also can be an issue,[60,64,67-69] although this is less of a problem within the central nervous system. Diffusion of the probes is limited, so studies have to be carefully designed to assure access of the probe to the region of interest. Oftentimes the antisense has to be administered directly into the region of interest. Although this can prove to be an advantage in certain situations, it more often limits the utility of the approach. Finally, extensive controls are needed to ensure the specificity of the response. Typically, mismatch probes are used. Changing the sequence of as few as four bases in a 20 mer can abolish the antisense response. In addition, it is desirable to document the down-regulation of the mRNA in question and target protein.

### ANTISENSE APPROACHES

Antisense approaches were the first to correlate the cloned opioid receptors with opioid behavior. Antisense approaches involve the use of complementary sequences of bases that selectively target a specific mRNA.[57,70] Typically, most antisense probes range from 18 to 25 bases long. Shorter probes lose some of the selectivity and effectiveness, while longer ones may have problems entering cells and are more expensive to synthesize. By forming a tight complex with the mRNA, the antisense oligodeoxynucleotide downregulates the mRNA in question, leading to a corresponding decrease in protein levels. The mechanisms through which antisense probes act remain unclear at this time and may involve the blockade of translation and/or the increased degradation of the duplex through RNAse H. The specificity of the antisense probe is extraordinarily high. Maintaining the same base composition and simply switching the sequence of 4 of the 20 bases can typically eliminate the activity of the antisense probe. When dealing with normal body tissues, antisense probes composed of normal DNA are quite labile due to enzymatic cleavage. However, within the central nervous system their stability seems to be somewhat better.[66,68,71]

Antisense probes are taken up into neuroblastoma cells and the brain and, once within the cell, they are relatively stable.[66] The accumulation of oligodeoxynucleotides by various cells can differ widely, making it necessary to test each system separately.[67,72,73] Once inside the cell, these antisense oligodeoxynucleotides are stable for at least three days.[66] This has important implications in terms of their dosing schedule. The duration of treatment is very dependent upon the turnover rate of the protein being targeted. Although antisense presumably will down-regulate mRNA levels quickly, this may not be immediately reflected in decreased levels of the protein. Sufficient time must be given for the pre-existing protein to turn over, resulting in an increase in their levels.

## ANTISENSE TARGETING OPIOID RECEPTORS

The initial opioid antisense approaches clearly associated DOR-1 with the delta receptor mediating spinal analgesia.[65,66] Using antisense probes 18 to 21 bases long, the receptor could be downregulated in tissue culture and *in vivo* following intrathecal injections in mice. The *in vivo* paradigm involved the administration of the antisense on days 1, 3, and 5, followed by the behavioral testing on day 6. This provides sufficient time for the pre-existing receptor to be cycled away. Using this approach, delta receptor binding was lowered in treated spinal cords. This drop in binding corresponded to a significant loss of analgesic activity for delta ligands. The response gradually returned over the next few days, indicating that the treatment was reversible. In contrast to the decrease in delta analgesia, mu and kappa analgesia was unchanged in the treated animals, confirming the selectivity of the response. Similar approaches have been employed by a number of laboratories in a wide variety of paradigms.[74-89]

The subsequent cloning of MOR-1[40-47] was quickly followed by antisense approaches correlating it with morphine analgesia. Although the interspecies homology at the amino acid level of the different opioid receptors is extremely high, nucleotide sequences can vary significantly. The need for an accurate match, therefore, usually requires that the antisense studies be performed in the same species as the cloned cDNA. In these first studies, an antisense probe targeting the first exon of MOR-1 was microinjected directly into the periaqueductal gray using the same treatment schedule described previously.[90] Following this treatment, the response to morphine administered into the same region was dramatically decreased. Similar results were observed when the antisense was administered directly into the ventricular system or lumbar space[12,14,65,75,88,91-94] or using murine antisense sequences in mice.[12] Finally, an antisense oligo probe targeting KOR-1 blocks spinal U50,488 analgesia as well.[95,96]

Although the ability of spinal administration of antisense to block spinal analgesia was anticipated, additional studies have confirmed that spinal antisense treatments also block peripheral opioid actions.[8,88,97,98] Presumably receptor synthesis in the dorsal horn neurons is knocked down, implying that peripheral analgesia is mediated on these dorsal horn neurons. Thus, the three major clones correlate well with their predicted analgesic actions. Equally important, they displayed a stringent selectivity that exceeded most of the antagonists currently available.

## ANTISENSE MAPPING

Early studies assumed that antisense probes had to target the initiation site of translation to be effective. However, this appears not to be the case. In a tissue culture system, five different antisense probes targeting each of the three exons of DOR-1 all down-regulated delta receptor binding to the same extent.[65] These observations imply that any site along the mRNA can be targeted by antisense, after taking into consideration the $T_m$ of the probe and the potential of secondary mRNA structure. All these probes also blocked spinal delta analgesia.[65,99] This ability to target individual exons suggested that the approach could be used to explore the functional significance of alternate splicing.

## Neuronal Nitric Oxide Synthase

The utility of antisense mapping in the functional evaluation of splice variants is well illustrated by a recent study exploring two splice variants of the neuronal nitric oxide synthase (nNOS) gene.[100] nNOS is important in opioid actions, since its blockade by enzymatic inhibitors prevents the development of morphine tolerance.[100-108] In addition to the major form,[109] a variant of low abundance lacking exons 9 and 10 has also been reported.[110,111] By carefully designing antisense oligodeoxynucleotides, it was possible to separately down-regulate either or both forms, as documented by RT-PCR.[100] Functionally, the two isoforms were quite different. Down-regulating the major component blocked the development of morphine tolerance, much like the traditional inhibitors such as $N^G$-nitro-L-arginine, implying that this form has an inhibitory effect on morphine analgesia. In contrast, the minor isoform facilitates morphine analgesia. Down-regulation of this second isoform diminishes the morphine response. Thus, antisense mapping is capable of selectively downregulating individual splice variants and defining their functions.

## DOR-1 and Delta Receptor Actions

Antisense mapping approaches have been applied to the opioid receptor clones. As indicated earlier, all five antisense probes targeting the three exons of DOR-1 blocked spinal delta analgesia,[99] indicating that the DOR-1 corresponds to the pharmacologically defined delta$_2$ receptor, which mediates spinal delta analgesia. When antisense mapping was explored at the spinal level, all three exons appeared to be important in the production of delta analgesia. Antisense probes targeting each of the three coding exons all blocked delta analgesia. Similarly, supraspinal deltorphin analgesia was also blocked by the same probes. In contrast, the supraspinal actions of DPDPE showed a markedly different profile. The two antisense probes targeting exon 3 still blocked supraspinal DPDPE analgesia, which pharmacologically had been classified as a delta$_1$ action. However, none of the probes targeting exons 1 or 2 was effective despite their activity in the same model against the delta$_2$ ligand deltorphin (see Figure 2.1). This result confirms prior pharmacological studies indicating that supraspinal DPDPE and deltorphin act through different delta receptor subypes. Yet, the identity of the receptor actually responsible for the supraspinal DPDPE actions remains an open question. The two leading possibilities include a closely related gene or a splice variant of DOR-1 with alternative exons 1 and 2. Future work is needed to distinguish between them.

## MOR-1 and Mu Receptors: A Novel Heroin/Morphine-6β-Glucuronide Receptor

The MOR-1 clone has also been explored in great detail with antisense mapping.[8,12,13,70,112,113] As noted earlier, an antisense probe targeting the 5′ (ft) untranslated region of MOR-1 blocked supraspinal morphine analgesia in both rats and mice. However, more detailed antisense mapping studies in the mouse revealed a far more complex picture. Like the initial probe, two additional antisense oligodeoxynucleotides targeting exon 1 also blocked morphine analgesia, as did a fourth probe targeting the coding region of exon 4. In contrast, five of six additional probes based upon exons 2 and 3 were inactive (see Figure 2.2).[12] This distinction between exons 2 and 3 and exon 1 became more prominent when the actions of a very potent morphine metabolite, morphine-6β-glucuronide (M6G) were examined.

# Antisense Mapping DOR-1 in the mouse

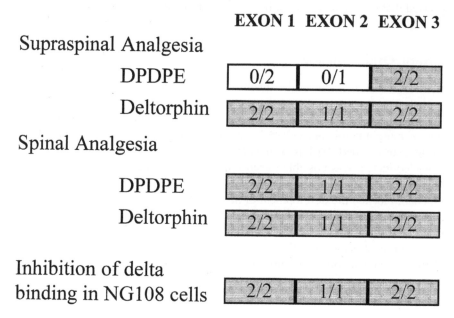

**FIGURE 2.1 Antisense mapping DOR-1.** Schematic of the effects of antisense oligodeoxynucleotides targeting DOR-1 on the indicated activities. This is a summary of results from the text.[65,66,69] Shaded boxes indicate that the antisense probes were active; nonshaded boxes indicate that the probes were not. The numbers represent the number of active probes/the number of probes tested.

Evidence from pharmacological approaches had suggested that M6G might have its own receptor. CXBK mice are insensitive to supraspinal morphine but still respond to M6G, indicating different receptor mechanisms of action.[14] In the antisense mapping paradigm, M6G analgesia was unaffected by the three antisense probes targeting exon 1. However, all six probes based upon exons 2 and 3 potently blocked M6G effects. The fact that every antisense probe examined was active against one of the drugs rules out the possibility that technical issues were responsible for the differential effects of the probes against these two opioids. Thus, M6G and morphine analgesia are mediated through distinct mechanisms. Yet, the activity of a number of antisense probes against morphine and M6G analgesia demonstrate that both receptor mechanisms are intimately related to MOR-1. This relationship, however, remains unclear. It is interesting to consider the possibility that they may represent splice variants of a common gene with alternative exons. Indeed, two splice variants of MOR-1 already have been reported,[41,43] but they do not explain the different pharmacologies noted above. This area needs further study to resolve these questions.

These studies also have opened another interesting aspect of opioid pharmacology. Heroin is the most widely abused opioid, leading some to question whether its actions may be different from morphine. For years many had maintained that pharmacokinetic differences between the two compounds were responsible for the greater abuse liability of heroin compared to morphine.[114-117] Heroin, 3,6-diacetylmorphine, is rapidly deacetylated *in vivo* to yield 6-acetylmorphine, which is thought to be the active component.[117-120] Both heroin and 6-acetylmorphine traverse the blood-brain barrier faster and to a greater extent than morphine. However, heroin and 6-acetylmorphine both are active analgesics in the CXBK mice, in marked contrast to morphine.[14] Furthermore, an exon 2 antisense probe inactive against morphine blocks heroin analgesia.[121] Thus, heroin appears to act, to a large degree, through the M6G receptor.

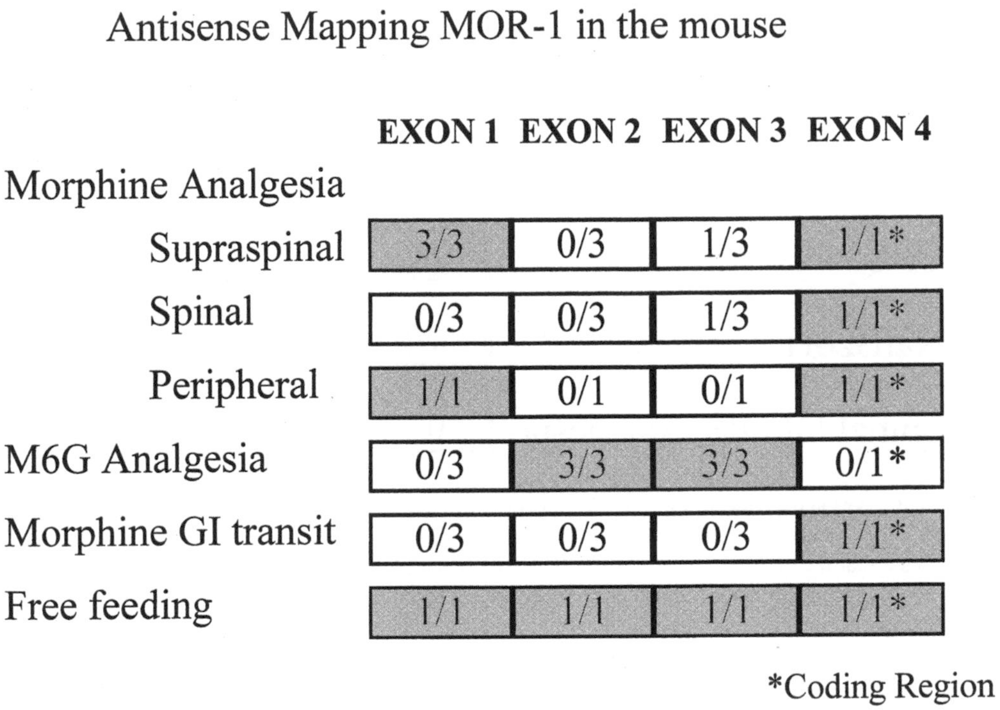

# Antisense Mapping MOR-1 in the mouse

| | EXON 1 | EXON 2 | EXON 3 | EXON 4 |
|---|---|---|---|---|
| **Morphine Analgesia** | | | | |
| Supraspinal | 3/3 | 0/3 | 1/3 | 1/1* |
| Spinal | 0/3 | 0/3 | 1/3 | 1/1* |
| Peripheral | 1/1 | 0/1 | 0/1 | 1/1* |
| M6G Analgesia | 0/3 | 3/3 | 3/3 | 0/1* |
| Morphine GI transit | 0/3 | 0/3 | 0/3 | 1/1* |
| Free feeding | 1/1 | 1/1 | 1/1 | 1/1* |

*Coding Region

**FIGURE 2.2  Antisense mapping MOR-1.** Schematic of the effects of antisense oligodeoxynucleotides targeting MOR-1 on the indicated activities. This is a summary of results from the text.[12,14,112,113] Shaded boxes indicate that the antisense probes were active; nonshaded boxes indicate that the probes were not. The numbers represent the number of active probes/the number of probes tested.

## ORPHANIN FQ/NOCICEPTIN (OFQ/N) AND ITS RECEPTORS

Kappa$_3$ receptors have been well defined pharmacologically.[1,2,22,26,27,70,122-128] Although the kappa$_3$ receptor does not correspond to the fourth cloned member of the opioid receptor family, it appears to be closely related.[94,129-133] This orphan receptor, termed ORL-1 or KOR-3, is highly homologous to the other opioid receptors and encodes a traditional G-protein-coupled receptor, however traditional opioids have a very poor affinity for it.[134-136] Two groups have identified a novel heptadecapepide, termed orphanin FQ[134] or nociceptin[135], which has very high affinity for this site and which appears to be its endogenous ligand. OFQ/N has an unusual and highly complex pharmacology. Some reports suggest that it is hyperalgesic,[134,135,137] while others have noted its ability to block supraspinal opioid analgesia.[138-140] Finally, OFQ/N also has analgesic actions.[137,141-143]

Several lines of evidence have associated this receptor with the kappa$_3$ receptor. A monoclonal antibody generated against the kappa$_3$ receptor in a neuroblastoma cell line[132] recognized the KOR-3 product.[130] Yet, the kappa$_3$ and the OFQ/N receptor differ. This is well-illustrated by antisense mapping KOR-3 (see Figure 2.3). Kappa$_3$ analgesia is easily elicited by naloxone benzoylhydrazone (NalBzoH), which has a relatively poor affinity for the orphanin receptor expressed in cell lines. However, six different antisense probes targeting the second and third exons of KOR-3, the mouse homologue of ORL-1, all blocked kappa$_3$ analgesia.[129-131] Conversely, five of the six additional probes targeting exon 1 were inactive. This clear distinction between the role of exon 1 and exons 2 and 3 is similar to the distinctions seen with other opioid classes, including MOR-1 and DOR-1. Antisense mapping of OFQ/N actions also reveals a complex series of actions. OFQ/N analgesia is blocked by antisense probes targeting exons 2 and 3, but not by a probe based upon exon 1. However, the hyperalgesic actions of OFQ/N are reversed by the exon 1 probe, but not those

# Antisense Mapping KOR-3 in the mouse

### EXON 1  EXON 2  EXON 3

OFQ/N Hyperalgesia | 1/1 | 0/1 | 0/1

NalBzoH Analgesia | 1/6 | 1/1 | 5/5

Spinal OFQ/N Analgesia | 0/1 | 1/1 | 1/1

**FIGURE 2.3  Antisense mapping KOR-3.** Schematic of the effects of antisense oligodeoxynucleotides targeting KOR-3 on the indicated activities. This is a summary of results from the text.[130,137,141,143] Shaded boxes indicate that the antisense probes were active; nonshaded boxes indicate that the probes were not. The numbers represent the number of active probes/the number of probes tested.

targeting exons 2 or 3. Again, probes targeting each exon are effective in at least one behavioral paradigm. Thus, the complex sensitivity profiles cannot be explained by technical issues. As with the other members of the opioid receptor family, the underlying reasons for these observations are not clear.

## CONCLUSION

Opioid pharmacology has a long history. The extensive structure-activity relationships raised the possibility of receptors long ago. The opioid receptors were among the first neurotransmitters identified in binding studies. Cloning them has proven more difficult. At the present time, only four members of the opioid receptor family have been identified. It will be interesting to follow the field over the next few years to determine how these clones will be correlated with their pharmacologically defined counterparts.

## REFERENCES

1. Reisine, T. and Pasternak, G.W., Opioid analgesics and antagonists, in *Goodman & Gilman's: The Pharmacological Basis of Therapeutics,* Hardman J.G. and Limbird L.E., Eds., 9th ed., McGraw-Hill, New York, 1996, 521.
2. Pasternak, G.W., Pharmacological mechanisms of opioid analgesics, *Clin. Neuropharmacol.,* 16, 1, 1993.
3. Martin, W.R., Eades, C.G., Thompson, J.A., Huppler, R.E., and Gilbert, P.E., The effects of morphine and nalorphine-like drugs in the nondependent and morphine-dependent chronic spinal dog, *J. Pharmacol. Exp. Ther.,* 197, 517, 1976.
4. Martin, W.R., Opioid antagonists, *Pharmacol. Rev.,* 19, 463, 1967.
5. Yeung, J.C. and Rudy, T.A., Multiplicative interaction between narcotic agonisms expressed at spinal and supraspinal sites of antinociceptive action as revealed by concurrent intrathecal and intracerebroventricular injections of morphine, *J. Pharmacol. Exp. Ther.,* 215, 633, 1980.
6. Rossi, G.C., Pasternak, G.W., and Bodnar, R.J., Synergistic brainstem interactions for morphine analgesia, *Brain Res.,* 624, 171, 1993.

7. Rossi, G.C., Pasternak, G.W., and Bodnar, R.J., μ and δ opioid synergy between the periaqueductal gray and the rostro-ventral medulla, *Brain Res.,* 665, 85, 1994.

8. Kolesnikov, Y.A., Jain, S., Wilson, R., and Pasternak, G.W., Peripheral morphine analgesia: synergy with central sites and a target of morphine tolerance, *J. Pharmacol. Exp. Ther.,* 279, 502, 1996.

9. Shimomura, K., Kamata, O., Ueki, S., Ida, S., Oguri, K., Yoshimura, H., and Tsukamoto, H., Analgesic effect of morphine glucuronides, *Tohoku J. Exp. Med.,* 105, 45, 1971.

10. Paul, D., Standifer, K.M., Inturrisi, C.E., and Pasternak, G.W., Pharmacological characterization of morphine-6β-glucuronide, a very potent morphine metabolite, *J. Pharmacol. Exp. Ther.,* 251, 477, 1989.

11. Pasternak, G.W., Bodnar, R.J., Clark, J.A., and Inturrisi, C.E., Morphine-6-glucuronide, a potent mu agonist, *Life Sci.,* 41, 2845, 1987.

12. Rossi, G.C., Pan, Y.-X., Brown, G.P., and Pasternak, G.W., Antisense mapping the MOR-1 opioid receptor: evidence for alternative splicing and a novel morphine-6β-glucuronide receptor, *FEBS Lett.,* 369, 192, 1995.

13. Rossi, G.C., Leventhal, L., Pan, Y.-X., Cole, J., Su, W., Bodnar, R.J., and Pasternak, G.W., Antisense mapping MOR-1 in the rat: distinguishing between morphine and morphine-6β-glucuronide antinociception, *J. Pharmacol. Exp. Ther.,* 281, 109, 1997.

14. Rossi, G.C., Brown, G.P., Leventhal, L., Yang, K., and Pasternak, G.W., Novel receptor mechanisms for heroin and morphine-6β-glucuronide analgesia, *Neurosci. Lett.,* 216, 1, 1996.

15. Brown, G.P., Yang, K., Ouerfelli, O., Byrd, D., and Pasternak, G.W., ³H-Morphine-6β-glucuronide binding in brain: evidence for a novel binding site, *J. Pharmacol. Exp. Ther.,* 282, 1291, 1997.

16. Brown, G.P., Yang, K., King, M.A., Rossi, G.C., Leventhal, L., Chang, A., Standifer, K.M., and Pasternak, G.W., 3-Methoxynaltrexone, a selective heroin antagonist: evidence for a unique heroin/morphine-6β-glucuronide opioid receptor, *FEBS Lett.,* 412, 35, 1997.

17. Lord, J.A.H., Waterfield, A.A., Hughes, J., and Kosterlitz, H.W., Endogenous opioid peptides: multiple agonists and receptors, *Nature,* 267, 495, 1977.

18. Hughes, J., Smith, T.W., Kosterlitz, H.W., Fothergill, L.A., Morgan, B.A., and Morris, H.R., Identification of two related pentapeptides from the brain with potent opiate agonist activity, *Nature,* 258, 577, 1975.

19. Jiang, Q., Takemori, A.E., Sultana, M., Portoghese, P.S., Bowen, W.D., Mosberg, H.I., and Porreca, F., Differential antagonism of opiate delta antinociception by [D-Ala²,Cys⁶]enkephalin and naltrindole-5′-isothiocyanate: evidence for subtypes, *J. Pharmacol. Exp. Ther.,* 257, 1069, 1991.

20. Mattia, A., Vanderah, T., Mosberg, H.I., and Porreca, F., Lack of antinociceptive cross tolerance between [D-Pen²,D-Pen⁵]enkephalin and [D-Ala²]deltorphin II in mice: evidence for delta receptor subtypes, *J. Pharmacol. Exp. Ther.,* 258, 583, 1991.

21. Ossipov, M.H., Kovelowski, C.J., Nichols, M.L., Hruby, V.J., and Porreca, F., Characterization of supraspinal antinociceptive actions of opioid delta agonists in the rat, *Pain,* 62, 287, 1995.

22. Clark, J.A., Liu, L., Price, M., Hersh, B., Edelson, M., and Pasternak, G.W., Kappa opiate receptor multiplicity: evidence for two U50,488-sensitive kappa₁ subtypes and a novel kappa₃ subtype, *J. Pharmacol. Exp. Ther.,* 251, 461, 1989.

23. Chavkin, C., James, I.F., and Goldstein, A., Dynorphin is a specific endogenous ligand of the k-opioid receptor, *Science,* 215, 413, 1982.

24. VonVoightlander, P.F., Lahti, R.A., and Ludens, J.H., U50,488: a selective and structurally novel non-mu (kappa) opioid agonist, *J. Pharmacol. Exp. Ther.,* 224, 7, 1983.

25. Chang, A.-C., Takemori, A.E., Ojala, W.H., Gleason, W.B., and Portoghese, P.S., Kappa opioid receptor selective affinity labels: electrophilic benzeneacetamides as kappa-selective opioid antagonists, *J. Med. Chem.,* 37, 4490, 1994.

26. Price, M., Gistrak, M.A., Itzhak, Y., Hahn, E.F., and Pasternak, G.W., Receptor binding of ³H-naloxone benzoylhydrazone: a reversible kappa and slowly dissociable μ opiate, *Mol. Pharmacol.,* 35, 67, 1989.

27. Gistrak, M.A., Paul, D., Hahn, E.F., and Pasternak, G.W., Pharmacological actions of a novel mixed opiate agonist/antagonist, naloxone benzoylhydrazone, *J. Pharmacol. Exp. Ther.,* 251, 469, 1990.

28. Paul, D., Bodnar, R.J., Gistrak, M.A., Pasternak, G.W., Different μ receptor subtypes mediate spinal and supraspinal analgesia in mice, *Eur. J. Pharmacol.,* 129, 307, 1989.

29. Quirion, R., Bowen, W.D., Itzhak, Y., Junien, J.L., Musacchio, J.M., Rothman, R.B., Su, T.-P., Tam, S.W., and Taylor, D.P., A proposal for the classification of sigma binding sites, *Trends Pharmacol. Sci.,* 13, 85, 1992.

30. Walker, J.M., Bowen, W.D., Walker, F.O., Matsumoto, R.R., De Costa, B., and Rice, K.C., Sigma receptors: biology and function, *Pharmacol. Rev.,* 42, 355, 1990.

31. Hanner, M., Moebius, F.F., Flandorfer, A., Knaus, H.G., Striessnig, J., Kempner, E., and Glossmann, H., Purification, molecular cloning, and expression of the mammalian sigma$_1$-binding site, *Proc. Natl. Acad. Sci. USA,* 93, 8072, 1996.

32. Kekuda, R., Prasad, P.D., Fei, Y.-J., Leibach, F.H., Ganapathy, V., and Fei, Y.J., Cloning and functional expression of the human type 1 sigma receptor (hSigmaR1), *Biochem. Biophys. Res. Commun.,* 229, 553, 1996.

33. Pan, Y.-X., Mei, J., King, M.A., Chang, A., and Wan, B.-L., Cloning and pharmacological characterization of a sigma$_1$ receptor, *J. Neurochem.,* 70, 2279, 1998.

34. Chien, C.-C. and Pasternak, G.W., Functional antagonism of morphine analgesia by (+)-pentazocine: evidence for an anti-opioid $\sigma_1$ system, *Eur. J. Pharmacol.,* 250, R7, 1993.

35. Chien, C.-C. and Pasternak, G.W., Selective antagonism of opioid analgesia by a *sigma* system, *J. Pharmacol. Exp. Ther.,* 271, 1583, 1994.

36. Chien, C.-C. and Pasternak, G.W., Sigma antagonists potentiate opioid analgesia in rats, *Neurosci. Lett.,* 190, 137, 1995.

37. Chien, C.-C., Pasternak, G.W., (–)-Pentazocine analgesia in mice: interactions with a $\sigma$ receptor system, *Eur. J. Pharmacol.,* 294, 303, 1995.

38. Kieffer, B.L., Befort, K., Gaveriaux-Ruff, C., and Hirth, C.G., The δ-opioid receptor: isolation of a cDNA by expression cloning and pharmacological characterization, *Proc. Natl. Acad. Sci. U.S.A.,* 89, 12048, 1992.

39. Evans, C.J., Keith, D.F., Morrison, H., Magendzo, K., and Edwards, R.H., Cloning of the delta opioid receptor by functional expression, *Science,* 258, p. 1952, 1992.

40. Uhl, G.R., Childers, S., and Pasternak, G.W., An opiate-receptor gene family reunion, *Trends Neurosci.,* 17, 89, 1994.

41. Bare, L.A., Mansson, E., and Yang, D., Expression of two variants of the human µ opioid receptor mRNA in SK-N-SH cells and human brain, *FEBS Lett.,* 354, 213, 1994.

42. Bunzow, J.R., Zhang, G., Bouvier, C., Saez, C., Ronnekleiv, O.K., Kelly, M.J., and Grandy, D.K., Characterization and distribution of a cloned rat µ-opioid receptor, *J. Neurochem.,* 64, 14, 1995.

43. Zimprich, A., Simon, T., and Hollt, V., Cloning and expression of an isoform of the rat µ opioid receptor (rMOR 1 B) which differs in agonist induced desensitization from rMOR1, *FEBS Lett.,* 359, 142, 1995.

44. Thompson, R.C., Mansour, A., Akil, H., and Watson, S.J., Cloning and pharmacological characterization of a rat µ opioid receptor, *Neuron,* 11, 903, 1993.

45. Wang, J.B., Johnson, P.S., Persico, A.M., Hawkins, A.L., Griffin, C.A., and Uhl, G.R., Human µ opiate receptor: cDNA and genomic clones, pharmacologic characterization and chromosomal assignment, *FEBS Lett.,* 338, 217, 1994.

46. Zastawny, R.L., George, S.R., Nguyen, T., Cheng, R., Tsatsos, J., Briones-Urbina, R., and O'Dowd, B.F., Cloning, characterization, and distribution of a µ-opioid receptor in rat brain, *J. Neurochem.,* 62, 2099, 1994.

47. Chen, Y., Mestek, A., Liu, J., Hurley, J.A., and Yu, L., Molecular cloning and functional expression of a µ-opioid receptor from rat brain, *Mol. Pharmacol.,* 44, 8, 1993.

48. Zhu, J., Chen, C., Xue, J.-C., Kunapuli, S., Deriel, J.K., and Liu-Chen, L.-Y., Cloning of a human kappa opioid receptor from the brain, *Life Sci.,* 56, PL201, 1995.

49. Yakovlev, A.G., Krueger, K.E., and Faden, A.I., Structure and expression of a rat kappa opioid receptor gene, *J. Biol. Chem.,* 270, 6421, 1995.

50. Minami, M., Toya, T., Katao, Y., Maekawa, K., Nakamura, S., Onogi, T., Kaneko, S., and Satoh, M., Cloning and expression of a cDNA for the rat kappa-opioid receptor, *FEBS Lett.,* 329, 291, 1993.

51. Nishi, M., Takeshima, H., Fukuda, K., Kato, S., and Mori, K., cDNA Cloning and pharmacological characterization of an opioid receptor with high affinities for kappa-subtype-selective ligands, *FEBS Lett.,* 330, 77, 1993.

52. Yasuda, K., Raynor, K., Kong, H., Breder, C., Takeda, J., Reisine, T., and Bell, G., Cloning and functional comparison of kappa and δ opioid receptors from mouse brain, *Proc. Natl. Acad. Sci. USA,* 90, 6736, 1993.

53. Meng, F., Xie, G.-X., Thompson, R.C., Mansour, A., Goldstein, A., Watson, S.J., and Akil, H., Cloning and pharmacological characterization of a rat kappa opioid receptor, *Proc. Natl. Acad. Sci. USA,* 90, 9954, 1993.

54. Min, B.H., Augustin, L.B., Felsheim, R.F., Fuchs, J.A., and Loh, H.H., Genomic structure and analysis of promoter sequence of a mouse μ opioid receptor gene, *Proc. Natl. Acad. Sci. USA,* 91, 9081, 1994.

55. Mayer, P., Schulzeck, S., Kraus, J., Zimprich, A., and Höllt, V., Promoter region and alternatively spliced exons of the rat μ-opioid receptor gene, *J. Neurochem.,* 66, 2272, 1996.

56. Gold, L.H., Integration of molecular biological techniques and behavioural pharmacology, *Behav. Pharmacol.,* 7, 589, 1996.

57. Wahlestedt, C., Antisense oligonucleotide strategies in neuropharmacology, *Trends Pharmacol. Sci.,* 15, 42, 1994.

58. Akhtar, S. and Agrawal, S., *In vivo* studies with antisense oligonucleotides, *Trends Pharmacol. Sci.,* 18, 12, 1997.

59. Branch, A.D., A Hitchhiker's guide to antisense and nonantisense biochemical pathways, *Hepatology,* 24, 1517, 1996.

60. Crooke, S.T. and Bennett, C.F., Progress in antisense oligonucleotide therapeutics, *Annu. Rev. Pharmacol. Toxicol.,* 36, 107, 1996.

61. Robinson-Benion, C. and Holt, J.T., Antisense techniques, *Methods Enzymol.,* 254, 363, 1995.

62. Scanlon, K.J., Ohta, Y., Ishida, H., Kijima, H., Ohkawa, T., Kaminski, A., Tsai, J., Horng, G., and Kashani-Sabet, M., Oligonucleotide-mediated modulation of mammalian gene expression, *FASEB J.,* 9, 1288, 1995.

63. Stein, C.A., Does antisense exist? *Nat. Med..* 1, 1119, 1995.

64. Wagner, R.W., Gene inhibition using antisense oligodeoxynucleotides, *Nature,* 372, 333, 1994.

65. Standifer, K.M., Chien, C.-C., Wahlestedt, C., Brown, G.P., and Pasternak, G.W., Selective loss of δ opioid analgesia and binding by antisense oligodeoxynucleotides to a δ opioid receptor, *Neuron,* 12, 805, 1994.

66. Standifer, K.M., Jenab, S., Su, W., Chien, C.-C., Pan, Y.-X., Inturrisi, C.E., and Pasternak, G.W., Antisense oligodeoxynucleotides to the cloned δ receptor, DOR-1: uptake, stability and regulation of gene expression, *J. Neurochem.,* 65, p. 1981, 1995.

67. Beltinger, C., Saragovi, H.U., Smith, R.M., LeSauteur, L., Shah, N., DeDionisio, L., Christensen, L., Raible, A., Jarett, L., and Gewirtz, A.M., Binding, uptake, and intracellular trafficking of phosphorothioate-modified oligodeoxynucleotides, *J. Clin. Invest.,* 95, 1814, 1995.

68. Harrison, P., Antisense: into the brain, *Lancet,* 342, 254, 1993.

69. Mirabelli, C.K., Bennett, C.F., Anderson, K., and Crooke, S.T., *In vitro* and *in vivo* pharmacologic activities of antisense oligoneucleotides, *Anti-Cancer Drug Design,* 6, 647, 1991.

70. Pasternak, G.W. and Standifer, K.M., Mapping of opioid receptors using antisense oligodeoxynucleotides: correlating their molecular biology and pharmacology, *Trends Pharmacol. Sci.,* 16, 344, 1995.

71. Whitesell, L., Gelelowitz, D., Chavany, C., Fahmy, B., Walbridge, S., Alger, J.R., and Neckers, L.M., Stability, clearance and disposition of intraventricularly administered oligodeoxynucleotides: implications for therapeutic application with the central nervous system, *Proc. Natl. Acad. Sci. USA,* 90, 4665, 1993.

72. Ogawa, S., Brown, H.E., Okano, H.J., and Pfaff, D.W., Cellular uptake of intracerebrally administered oligodeoxynucleotides in mouse brain, *Regul. Pept.,* 59, 143, 1995.

73. Loke, S.L., Stein, C.A., Zhang, X.H., Mori, K., Nakanishi, M., Subasinghe, C., Cohen, J.S., and Neckers, L.M., Characterization of oligonucleotide transport into living cells, *Proc. Natl. Acad. Sci. USA,* 86, 3474, 1989.

74. Tseng, L.F., Collins, K.A., Antisense oligodeoxynucleotide to a δ-opioid receptor given intrathecally blocks i.c.v. administered b-endorphin-induced antinociception in the mouse, *Life Sci.,* 55, PL127, 1994.

75. Tseng, L.F., Collins, K.A., and Kampine, J.P., Antisense oligodeoxynucleotide to a δ-opioid receptor selectively blocks the spinal antinociception induced by δ-, but not μ- or kappa-opioid receptor agonists in the mouse, *Eur. J. Pharmacol.,* 258, R1, 1994.

76. Lai, J., Bilsky, E.J., Rothman, R.B., and Porreca, F., Treatment with antisense oligodeoxynucleotide to the opioid δ receptor selectively inhibits δ$_2$-agonist antinociception, *Neuroreport*, 5, 1049, 1994.

77. Lai, J., Bilsky, E.J., Bernstein, R.N., Rothman, R.B., Pasternak, G.W., and Porreca, F., Antisense oligodeoxynucleotide to the cloned delta opioid receptor selectively inhibits supraspinal, but not spinal, antinociceptive effects of [D-Ala$^2$, Glu$^4$]deltorphin, *Regul. Pept.*, 54, 159, 1994.

78. Bilsky, E.J., Bernstein, R.N., Pasternak, G.W., Hruby, V.J., Patel, D., Porreca, F., and Lai, J., Selective inhibition of [D-Ala$^2$,Glu$^4$]deltorphin antinociception by supraspinal, but not spinal, administration of an antisense oligodeoxynucleotide to an opioid delta receptor, *Life Sci.*, 55, 37, 1994.

79. Cha, X.Y., Xu, H., Rice, K.C., Porreca, F., Lai, J., Ananthan, S., and Rothman, R.B., Opioid peptide receptor studies. 1. Identification of a novel δ-opioid receptor binding site in rat brain membranes, *Peptides*, 16, 191, 1995.

80. Bilsky, E.J., Calderon, S.N., Wang, T., Bernstein, R.N., Davis, P., Hruby, V.J., McNutt, R.W., Rothman, R.B., Rice, K.C., and Porreca, F., SNC 80, a selective, nonpeptidic and systemically active opioid *delta* agonist, *J. Pharmacol. Exp. Ther.*, 273, 359, 1995.

81. Lai, J., Bilsky, E.J., and Porreca, F., Treatment with antisense oligodeoxynucleotide to a conserved sequence of opioid receptors inhibits antinociceptive effects of delta subtype selective ligands, *J. Recept. Res.*, 15, 643, 1995.

82. Cha, X.Y., Xu, H., Ni, Q., Partilla, J.S., Rice, K.C., Matecka, D., Calderon, S.N., Porreca, F., Lai, J., and Rothman, R.B., Opioid peptide receptor studies .4. Antisense oligodeoxynucleotide to the delta opioid receptor delineates opioid receptor subtypes, *Regul. Pept.*, 59, 247, 1995.

83. Mizoguchi, H., Narita, M., Nagase, H., and Tseng, L.F., Antisense oligodeoxynucleotide to a δ-opioid receptor blocks the antinociception induced by cold water swimming, *Regul. Pept.*, 59, 255, 1995.

84. Lai, J., Riedl, M., Stone, L.S., Arvidsson, U., Bilsky, E.J., Wilcox, G.L., Elde, R., and Porreca, F., Immunofluorescence analysis of antisense oligodeoxynucleotide-mediated 'knock-down' of the mouse δ opioid receptor in vitro and in vivo, *Neurosci. Lett.*, 213, 205, 1996.

85. Mizoguchi, H., Narita, M., Nagase, H., Suzuki, T., Quock, R.M., and Tseng, L.F., Use of antisense oligodeoxynucleotide to determine δ-opioid receptor involvement in [Δ−Aλα$^2$]deltorphin II-induced locomotor hyperactivity, *Life Sci.*, 59, PL69, 1996.

86. Wang, H.Q., Kampine, J.P., and Tseng, L.F., Antisense oligodeoxynucleotide to a δ-opioid receptor messenger RNA selectively blocks the antinociception induced by intracerebroventricularly administered δ-, but not μ-, ε- or kappa-opioid receptor agonists in the mouse, *Neuroscience*, 75, 445, 1996.

87. Kest, B., Lee, C.E., Mogil, J.S., and Inturrisi, C.E., Blockade of morphine supersensitivity by an antisense oligodeoxynucleotide targeting the delta opioid receptor (DOR-1) blockade of morphine supersensitivity by an antisense oligodeoxynucleotide targeting the delta opioid receptor (DOR-1), *Life Sci.*, 60, PL155, 1997.

88. Bilsky, E.J., Wang, T., Lai, J., and Porreca, F., Selective blockade of peripheral delta opioid agonist induced antinociception by intrathecal administration of delta receptor antisense oligodeoxynucleotide, *Neurosci. Lett.*, 220, 155, 1996.

89. Sánchez-Blázquez, P., García-España, A., and Garzón, J., Antisense oligodeoxynucleotides to opioid *mu* and *delta* receptors reduced morphine dependence in mice: role of delta-2 opioid receptors, *J. Pharmacol. Exp. Ther.*, 280, 1423, 1997.

90. Rossi, G.C., Pan, Y.-X., Cheng, J., and Pasternak, G.W., Blockade of morphine analgesia by an antisense oligodeoxynucleotide against the mu receptor, *Life Sci.*, 54, PL375, 1994.

91. Chen, X.H., Adams, J.U., Geller, E.B., Deriel, J.K., Adler, M.W., and Liu-Chen, L.-Y., An antisense oligodeoxynucleotide to μ-opioid receptors inhibits μ-opioid receptor agonist-induced analgesia in rats, *Eur. J. Pharmacol.*, 275, 105, 1995.

92. Chen, X.H., Liu-Chen, L.Y., Tallarida, R.J., Geller, E.B., De Riel, J.K., and Adler, M.W., Use of a μ-antisense oligodeoxynucleotide as a μ opioid receptor noncompetitive antagonist in vivo, *Neurochem. Res.*, 21, 1363, 1996.

93. Chen, X.H., Geller, E.B., Deriel, J.K., Liu-Chen, L.Y., and Adler, M.W., Antisense confirmation of μ- and kappa-opioid receptor mediation of morphine's effects on body temperature in rats, *Drug Alcohol Depend.*, 43, 119, 1996.

94. Pasternak, G.W., Pan, Y.-X., and Cheng, J., Correlating the pharmacology and molecular biology of opioid receptors: cloning and antisense mapping a Kappa$_3$-related opiate receptor, *Ann. NY Acad. Sci.*, 757, 332, 1995.

95. Chien, C.-C., Brown, G., Pan, Y.-X., and Pasternak, G.W., Blockade of U50,488H analgesia by antisense oligodeoxynucleotides to a kappa-opioid receptor, *Eur. J. Pharmacol.*, 253, R7, 1994.

96. Tseng, L.F., Collins, K.A., Narita, M., and Kampine, J.P., The use of antisense oligodeoxynucleotides to block the spinal effects of kappa$_1$ agonist-induced antinociception in the mouse, *Analgesia*, 1, 121, 1995.

97. Kolesnikov, Y., Jain, S., Wilson, R., and Pasternak, G.W., Peripheral kappa$_1$-opioid receptor-mediated analgesia in mice, *Eur. J. Pharmacol.*, 310, 141, 1996.

98. Khasar, S.G., Gold, M.S., Dastmalchi, S., and Levine, J.D., Selective attenuation of μ-opioid receptor-mediated effects in rat sensory neurons by intrathecal administration of antisense oligodeoxynucleotides, *Neurosci. Lett.*, 218, 17, 1996.

99. Rossi, G.C., Leventhal, L., and Pasternak, G.W., Antisense mapping DOR-1: evidence for delta receptor subtypes, *Brain Res.*, 719, 78, 1997.

100. Kolesnikov, Y., Pan, Y.-X., Babey, A.M., Jain, S., Wilson, R., and Pasternak, G.W., Functionally differentiating two nNOS isoforms through antisense mapping: Evidence for opposing NO actions on morphine analgesia and tolerance, *Proc. Natl. Acad. Sci. U.S.A.*, 94, 8220, 1997.

101. Kolesnikov, Y.A., Pick, C.G., Ciszewska, G., and Pasternak, G.W., Blockade of tolerance to morphine but not to kappa opioids by a nitric oxide synthase inhibitor, *Proc. Natl. Acad. Sci. USA*, 90, 5162, 1993.

102. Babey, A.M., Kolesnikov, Y., Cheng, J., Inturrisi, C.E., Trifilletti, R.R., and Pasternak, G.W., Nitric oxide and opioid tolerance, *Neuropharmacology*, 33, 1463, 1994.

103. Kolesnikov, Y.A., Pick, C.G., and Pasternak, G.W., N$^G$-Nitro-L-arginine prevents morphine tolerance, *Eur. J. Pharmacol.*, 221, 339, 1992.

104. Pasternak, G.W., Kolesnikov, Y.A., and Babey, A.M., Perspectives on the N-methyl-D-aspartate nitric oxide cascade and opioid tolerance, *Neuropsychopharmacology*, 13, 309, 1995.

105. Vaupel, D.B., Kimes, A.S., and London, E.D., Nitric oxide synthase inhibitors — Preclinical studies of potential use for treatment of opioid withdrawal, *Neuropsychopharmacology*, 13, 315, 1995.

106. Dunbar, S. and Yaksh, T.L., Effect of spinal infusion of L-NAME, a nitric oxide synthase inhibitor, on spinal tolerance and dependence induced by chronic intrathecal morphine in the rat, *Neurosci. Lett.*, 207, 33, 1996.

107. Zhao, G.M. and Bhargava, H.N., Nitric oxide synthase inhibition attenuates tolerance to morphine but not to [D-Ala$^2$,Glu$^4$] deltorphin II, a δ$_2$-opioid receptor agonist in mice, *Peptides*, 17, 619, 1996.

108. Hall, S., Milne, B., and Jhamandas, K., Nitric oxide synthase inhibitors attenuate acute and chronic morphine withdrawal response in the rat locus coeruleus: an in vivo voltammetric study, *Brain Res.*, 739, 182, 1996.

109. Bredt, D.S., Hwang, P.M., Glatt, C.E., Lowenstein, C., Reed, R.R., and Snyder, S.H., Cloned and expressed nitric oxide synthase structurally resembles cytochrome P-450 reductase, *Nature*, 714, 1991.

110. Ogura, T., Yokoyama, T., Fujisawa, H., Kurashima, Y., and Esumi, H., Structural diversity of neuronal nitric oxide synthase mRNA in the nervous system, *Biochem. Biophys. Res. Commun.*, 193, 1014, 1993.

111. Hall, A.V., Antoniou, H., Wang, Y., Cheung, A.H., Arbus, A.M., Olson, S.L., Lu, W.C., Kau, C.-L., and Marsden, P.A., Structural organization of the human neuronal nitric oxide synthase gene (*NOS1*), *J. Biol. Chem.*, 269, 33082, 1994.

112. Rossi, G.C., Standifer, K.M., and Pasternak, G.W., Differential blockade of morphine and morphine-6β-glucuronide analgesia by antisense oligodeoxynucleotides directed against MOR-1 and G-protein α subunits in rats, *Neurosci. Lett.*, 198, 99, 1995.

113. Leventhal, L., Bodnar, R.J., Cole, J.L., Rossi, G.C., Pan, Y.-X., and Pasternak, G.W., Antisense oligodeoxynucleotides against the MOR-1 clone alter weight and ingestive responses in rats, *Brain Res.*, 719, 78, 1996.

114. Oldendorf, W.H., Hyman, S., Braun, L., and Oldendorf, S.Z., Blood-brain barrier: penetration of morphine, codeine, heroin, and methadone after carotid injection, *Science*, 178, 984, 1972.

115. Inturrisi, C.E., Max, M.B., Foley, M., Schultz, M., Shin, S., and Haud, R.W., The pharmacokinetics of heroin in patients with chronic pain, *N. Engl. J. Med.*, 310, 1213, 1984.

116. Martin, W.J. and Fraser, H.F., A comparative study of physiological and subjective effects heroin and morphine administered intravenously in post-addicts, *J. Pharmacol. Exp. Ther.*, 133, 388, 1961.

117. Inturrisi, C.E., Schultz, M., Shin, S., Umans, J.G., Angel, L., and Simon, E.J., Evidence from opiate binding sites that heroin acts through its metabolites, *Life Sci.*, 3, 773, 1983.

118. Rady, J.J., Aksu, F., and Fujimoto, J.M., The heroin metabolite, 6-monoacetylmorphine, activates delta opioid receptors to produce antinociception in swiss-webster mice, *J. Pharmacol. Exp. Ther.,* 268, 1222, 1994.

119. Way, E.L., Young, J., and Kemp, J., Metabolism of heroin and its pharmacological implications, *Bull. of Narcotics,* 17, 25, 1965.

120. Rady, J.J., Baemmert, D., Takemori, A.E., Portoghese, P.S., and Fujimoto, J.M., Spinal delta opioid receptor subtype activity of 6-monoacetylmorphine in Swiss Webster mice, *Pharmacol. Biochem. Behav.,* 56, 243, 1997.

121. Standifer, K.M., Rossi, G.C., and Pasternak, G.W., Differential blockade of opioid analgesia by antisense oligodeoxynucleotides directed against various G-protein $\alpha$ subunits, *Mol. Pharmacol.,* 50, 293, 1996.

122. Standifer, K.M., Cheng, J., Brooks, A.I., Honrado, C.P., Su, W., Visconti, L.M., Biedler, J.L., and Pasternak, G.W., Biochemical and pharmacological characterization of *mu, delta* and *kappa*$_3$ opioid receptors expressed in BE(2)-C neuroblastoma cells, *J. Pharmacol. Exp. Ther.,* 270, 1246, 1994.

123. Paul, D., Levison, J.A., Howard, D.H., Pick, C.G., Hahn, E.F., and Pasternak, G.W., Naloxone benzoylhydrazone (NalBzoH) analgesia, *J. Pharmacol. Exp. Ther.,* 255, 769, 1990.

124. Pick, C.G., Paul, D., and Pasternak, G.W., Nalbuphine, a mixed kappa$_1$ and kappa$_3$ analgesic in mice, *J. Pharmacol. Exp. Ther.,* 262, 1044, 1992.

125. Cheng, J., Standifer, K.M., Tublin, P.R., Su, W., and Pasternak, G.W., Demonstration of kappa$_3$-opioid receptors in the SH-SY5Y human neuroblastoma cell line, *J. Neurochem.,* 65, 170, 1995.

126. Paul, D., Pick, C.G., Tive, L.A., and Pasternak, G.W., Pharmacological characterization of nalorphine, a kappa$_3$ analgesic, *J. Pharmacol. Exp. Ther.,* 257, 1, 1991.

127. Standifer, K.M., Cheng, J., Biedler, J.L., and Pasternak, G.W., Expression of $\mu$, $\delta$, and kappa$_3$ binding sites in a human neuroblastoma cell line, *Society for Neuroscience,* 17, 362, 1991.

128. Tive, L.A., Ginsberg, K., Pick, C.G., and Pasternak, G.W., Kappa$_3$ receptors and levorphanol analgesia, *Neuropharmacology,* 31, 851, 1992.

129. Pan, Y.-X., Cheng, J., Xu, J., and Pasternak, G.W., Cloning, expression and classification of a kappa$_3$-related opioid receptor using antisense oligodeoxynucleotides, *Regul. Pept.,* 54, 217, 1994.

130. Pan, Y.-X., Cheng, J., Xu, J., Rossi, G.C., Jacobson, E., Ryan-Moro, J., Brooks, A.I., Dean, G.E., Standifer, K.M., and Pasternak, G.W., Cloning and functional characterization through antisense mapping of a kappa$_3$-related opioid receptor, *Mol. Pharmacol.,* 47, 1180, 1995.

131. Pan, Y.-X., Xu, J., and Pasternak, G.W., Structure and characterization of the gene encoding a mouse kappa$_3$-related opioid receptor, *Gene,* 171, 255, 1996.

132. Brooks, A.I., Standifer, K.M., Rossi, G.C., Mathis, J.P., and Pasternak, G.W., Characterizing kappa$_3$ opioid receptors with a selective monoclonal antibody, *Synapse,* 22, 247, 1996.

133. Pan, Y.-X., Xu, J., Ryan-Moro, J., Mathis, J., Hom, J.S.H., Mei, J.F., and Pasternak, G.W., Dissociation of affinity and efficacy in KOR-3 chimeras, *FEBS Lett.,* 395, 207, 1996.

134. Reinscheid, R.K., Nothacker, H.P., Bourson, A., Ardati, A., Henningsen, R.A., Bunzow, J.R., Grandy, D.K., Langen, H., Monsma, F.J., and Civelli, O., Orphanin FQ: a neuropeptide that activates an opioidlike G protein-coupled receptor, *Science,* 270, 792, 1995.

135. Meunier, J.C., Mollereau, C., Toll, L., Suaudeau, C., Moisand, C., Alvinerie, P., Butour, J.L., Guillemot, J.C., Ferrara, P., Monsarrat, B., Mazargull, H., Vassart, G., Parmentier, M., and Costentin, J., Isolation and structure of the endogenous agonist of the opioid receptor like ORL$_1$ receptor, *Nature,* 377, 532, 1995.

136. Mathis, J.P., Ryan-Moro, J., Chang, A., Hom, J.S.H., Scheinberg, D., and Pasternak, G.W., $^{125}$I[Tyr$^{14}$]Orphanin FQ/nociceptin binding in mouse brain: evidence for heterogeneity, *Biochem. Biophys. Res. Commun.,* 230, 462, 1997.

137. Rossi, G., Leventhal, L., Boland, E., and Pasternak, G.W., Pharmacological characterization of orphanin FQ/nociceptin and its fragments, *J. Pharmacol. Exp. Ther.,* 753, 176, 1997.

138. Mogil, J.S., Grisel, J.E., Zhangs, G., Belknap, J.K., and Grandy, D.K., Functional antagonism of $\mu$-, $\delta$- and kappa-opioid antinociception by orphanin FQ, *Neurosci. Lett.,* 214, 131, 1996.

139. Mogil, J.S., Grisel, J.E., Reinscheid, K.K., Civelli, O., Belknap, J.K., and Grandy, D.K., Orphanin FQ is a functional anti-opioid peptide, *Neuroscience,* 75, 333, 1996.

140. Grisel, J.E., Mogil, J.S., Belknap, J.K., and Grandy, D.K., Orphanin FQ acts as a supraspinal, but not a spinal, anti-opioid peptide, *Neuroreport,* 7, 2125, 1996.

141. Rossi, G.C., Leventhal, L., and Pasternak, G.W., Naloxone-sensitive orphanin FQ-induced analgesia in mice, *Eur. J. Pharmacol.,* 311, R7, 1996.
142. Xu, X.J., Hao, J.X., and Wiesenfeld-Hallin, Z., Nociceptin or antinociceptin: potent spinal antinociceptive effect of orphanin FQ/nociceptin in the rat, *Neuroreport,* 7, 2092, 1996.
143. King, M.A., Rossi, G.C., Chang, A.H., Williams, L., and Pasternak, G.W., Spinal analgesic activity of orphanin FQ/nociceptin and its fragments, *Neurosci. Lett.,* 223, 113, 1997.

# 3 The Search for Susceptibility Genes in Bipolar Disorder

*Wade H. Berrettini*

## CONTENTS

## INTRODUCTION

Bipolar (BP) disorders are common, chronic, recurrent, and episodic mood disturbances associated with variable dysfunctions in sleep, appetite, libido, activity, and cognition. These disorders are typically so severe that they impair occupational functioning. Bipolar disorders are characterized by recurrent episodes of mania and depression, both of which are defined subsequently.

Mania represents a state of persistently elevated (predominantly euphoric) mood with increased activity, intrusive social behavior, irritability (unpredictable angry outbursts are common), decreased need for sleep, grandiosity, excessive energy, increased libido, spending sprees, racing thoughts, and poor judgement (inability to perceive possible adverse consequences of dangerous behavior). Mania represents a more severe syndrome than hypomania, and it is often accompanied by psychotic symptoms, including hallucinations and delusions. Hypomania is a less severe form of mania. Mania causes impairment in functioning, whereas hypomania (by definition) does not. Untreated episodes of mania or hypomania are typically one to three months in length, although this duration is quite variable.

Depression represents persistent and pervasive sadness, accompanied by crying spells, decreased energy, suicidal ideation, decreased libido, anhedonia (inability to experience pleasure), decreased cognitive ability, sleep dysfunction (insomnia or hypersomnia), and appetite disturbance (with or without weight change). Duration of an untreated episode of depression is typically six to nine months.

Bipolar (BP) disorder is characterized by repeated manic or hypomanic episodes and recurrent depressive episodes. Two subtypes of BP disorder are recognized: the BPII category is reserved for persons who have never had a episode of frank mania but have experienced hypomania with recurrent episodes of depression; the BPI category describes individuals with the full syndrome of manic and depressive episodes. Individuals with BP disorder have a median ten episodes of illness during a lifetime, even with treatment. The diagnosis of unipolar disorder (UP) describes individuals who have recurrent episodes of depression but no (hypo) manic episodes.

Persons with UP illness have a median four episodes during a lifetime. The mean age at onset for BP disorders is about 25 and for UP disorders it is about 35, although onset in adolescence is becoming increasingly common among generations born after World War II (Klerman et al., 1985; Klerman and Weissman, 1989; Gershon et al., 1987; Weissman, 1987; Joyce et al., 1990). UP illness affects females twice as often as males, but BP illness affects both sexes equally. BP illness affects about 1% of the general population, while UP illness occurs in about 10% of people (Weissman and Myers, 1978). Suicide is the sole reason for shortened life expectancy among BP and UP individuals, and suicide occurs in about 10% of cases (Klerman, 1987).

## GENETIC EPIDEMIOLOGY OF BIPOLAR DISORDERS

Twin, family, and adoption studies have indicated the existence of genetic predisposition for BP disorder. Monozygotic twins are concordant for BP illness (including UP diagnoses) about 65% of the time, but dizygotic twins show a concordance rate of about 14% (see Table 3.1). The heritability of BP illness may be as high as 80%.

Modern twin studies (Kendler et al., 1992 and 1993; McGuffin et al., 1996), conducted with operationalized diagnostic criteria, validated semi-structured interviews, and blinded assessments, also describe significantly greater MZ twin concordance. The MZ twin concordance rate (about 65%) indicates decreased penetrance of inherited susceptibility or the presence of phenocopies (non-genetic cases). Among MZ twin pairs concordant for mood disorder, when one twin has a BP diagnosis, UP illness is present among 20% of the ill co-twins (Bertelsen et al., 1977; Allen et al., 1974). This suggests that BP and UP syndromes share some common genetic susceptibility factors. This may have clinical relevance in that it provides a heuristic model to support the use of lithium for prophylaxis of recurrent UP illness (Souza and Goodwin, 1991).

Family studies of BP illness show that a spectrum of mood disorders is found among the first-degree relatives of BP probands: BPI, BPII with major depression (hypomania and recurrent UP illness in the same person), schizoaffective disorders, and recurrent unipolar depression (Gershon et al., 1982; Weissman et al., 1984; Baron et al., 1983; Winokur et al., 1982, 1995; Heltzer and Winokur, 1974; James and Chapman, 1975; Johnson and Leeman, 1977; Angst et al., 1980; Maier et al., 1993).

Mendlewicz and Rainer (1977) reported a controlled adoption study of BP probands, including a control group of probands with poliomyelitis. The biological relatives of the BP probands had a

---

### TABLE 3.1
### Concordance Rates for Affective Illness in Monozygotic and Dizygotic Twins*

| Study | Monozygotic twins Concordant Pairs | | Dizygotic twins Concordant Pairs | |
|---|---|---|---|---|
| | Total Pairs | Percent | Total Pairs | Percent |
| Luxemberger, 1930 | 3/4 | 75.0 | 0/13 | 0.0 |
| Rosanoff et al., 1935 | 16/23 | 69.6 | 11/67 | 16.4 |
| Slater, 1953 | 4/7 | 57.1 | 4/17 | 23.5 |
| Kallman, 1954 | 25/27 | 92.6 | 13/55 | 23.6 |
| Harvald and Hauge, 1965 | 10/15 | 66.7 | 2/40 | 5.0 |
| Allen et al., 1974 | 5/15 | 33.3 | 0/34 | 0.0 |
| Bertelsen, et al., 1977 | 32/55 | 58.3 | 9/52 | 17.3 |
| Totals | 95/146 | 65.0 | 39/278 | 14.0 |

* Data not corrected for age. Diagnoses include both bipolar and unipolar illness.

---

31% risk for BP or UP disorders, as opposed to 2% in the relatives of the control probands. The risk for affective disorder in biological relatives of adopted BP patients was similar to the risk in relatives of BP patients who were not adopted (26%). Adoptive relatives do not show increased risk compared to relatives of control probands.

Wender et al. (1986) and Cadoret et al. (1986) studied UP and BP probands. Although evidence for genetic susceptibility was found, adoptive relatives of affective probands had a tendency to excess affective illness, compared with the adoptive relatives of controls. Von Knorring et al. (1983) did not find concordance in psychopathology between adoptees and biological relatives when examining the records of 56 adoptees with UP disorders. Heritable factors may be more evident in BP syndromes than in UP disorders.

The twin, adoption, and family studies have provided impetus to systematic searches of the human genome for BP susceptibility loci, using multiplex BP kindreds and microsatellite genotypes (Weber and May, 1989) in linkage analyses. These reports are reviewed next.

## BIPOLAR MOLECULAR LINKAGE STUDIES — GENERAL CONSIDERATIONS

The human genome consists of about 3.3 billion base pairs of DNA. A strand of DNA consists of a sugar (deoxyribose) phosphate backbone, each sugar bonded to one of four nucleotides in a linear manner. The linear sequence of the nucleotides (guanine, cytosine, thymine, and adenine) is the genetic code. DNA is naturally found as a double helix, in which two complementary (in terms of nucleotide sequence) strands are intertwined. The DNA is organized into 22 pairs of autosomal chromosomes, numbered according to physical size, and a pair of sex chromosomes, X and Y. Each chromosome is constituted by a two complementary strands of DNA, in a double-helix conformation. Physical distance along the chromosomes can be expressed in terms of base pairs. Alternatively, distance can be expressed in terms of centiMorgans (cM), reflecting the frequency of recombination. One cM is about one million base pairs of DNA.

Molecular linkage studies of BP disorder have been conducted using highly polymorphic DNA markers: microsatellites (Weber and May, 1989). These DNA sequences differ in length among individuals because they contain a variable number of a simple repetitive sequence (usually consisting of two, three, or four nucleotides). The most common repetitive sequence in microsatellites is (CA), although (GATA) and others are frequently encountered. Many of these microsatellites have ten or more sizes, each different size constituting an allele that can be traced through a family to determine if the allele segregates with illness. Consider the following kindred, in which the father has BP disorder and the mother is unaffected. At some anonymous DNA marker, the father has alleles 1,2 and the mother has alleles 3,4. It can be seen that allele 1 is transmitted with illness and allele 2 is transmitted to the unaffected children. The probability that the father will transmit allele 1 to each child is 50%. A LOD score statistic assesses the probability that, within a family, co-segregation of illness and a marker allele has occurred randomly, vs. the probability that the co-segregation of illness and a marker allele has occurred because the marker allele is located near a disease gene on the same chromosome, such that the two are transmitted together more often than expected by chance (= 50%) (see Figure 3.1).

LOD score calculations require specification of the disease allele frequency in the population, the mode of inheritance (dominant or recessive or some intermediate model), and the penetrance. If the mode of inheritance is misspecified, then the LOD score may not detect linkage when it is present (Clerget-Darpoux et al., 1986). For BP disorders, of course, none of these parameters is known. In practice, investigators usually calculate LOD scores under dominant and recessive models of inheritance with reduced penetrance. A LOD score numerical value of three occurs one to two times randomly whenever the entire genome is searched for linkage (Lander and Kruglyak, 1995).

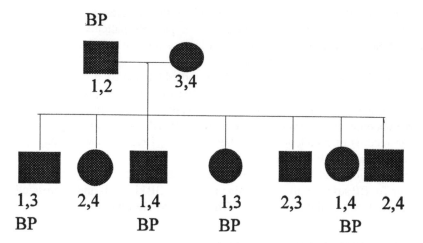

**FIGURE 3.1**   A hypothetical, bipolar pedigree is shown where an affected father transmits allele 1 at a marker locus to affected children and allele 2 to unaffected children, so that the kindred is consistent with linkage between a bipolar susceptibility gene and the marker locus.

Another useful statistic in complex trait analysis is the affected sibling pair (ASP) calculation. This statistic relies on the fact that pairs of siblings will share 50% of their alleles randomly. The distribution of this allele sharing randomly assumes the following pattern:

| Number of alleles shared: | 0 | 1 | 2 |
|---|---|---|---|
| Percent of all sibling pairs: | 25% | 50% | 25% |

Pairs of affected siblings will tend to share alleles to a greater extent when the DNA marker alleles are located near a disease gene that contributes to the illness in the affected siblings pairs. Consider the affected siblings in the previous pedigree diagram. Four affected sibling pairs share one allele and two pairs share two alleles, but none shares zero alleles. This skewing of the expected random distribution of allele sharing toward greater sharing is consistent with the hypothesis that the DNA marker is located near a BP susceptibility gene (i.e., linkage is present). This method can be extended to all pairs of affected relatives (Weeks and Lange, 1988, 1992; Kruglyak et al., 1996). These statistics do not require specification of the mode of inheritance, penetrance, or disease allele frequency, as is necessary for the LOD score method. Because these affected relative statistics do not require specification of these parameters, they are often described as non-parametric methods.

In genetic linkage studies of complex traits, validity is conferred only by demonstrating the underlying DNA sequence variants that explain the linkage statistics or through independent confirmation of the original linkage report in a second group of pedigrees. Statistical guidelines for judging validity of linkage reports in complex disorders have been suggested (Lander and Schork, 1994; Lander and Kruglyak, 1995). These guidelines suggest thresholds for an initial report of "significant" linkage (LOD score = ~ 3.6 or nominal p = ~ 0.00002) and for confirmation (LOD score = 1.2 or p = ~ 0.01). These guidelines should limit false positives to less than 5%. It should be remembered that these guidelines refer to analysis of a single phenotypic definition (e.g., BPI and BPII disorders). If multiple (overlapping) phenotypes are analyzed, some statistical adjustments for multiple hypothesis testing may be necessary.

An associated critical issue is the power of a confirmation study to detect the effect size initially described. Effect sizes are often expressed as the increased relative risk (Risch, 1990) due to a specific genetic locus (Risch, 1990). This increased relative risk refers to the ratio of the risk to a BP proband's relative (e.g., sibling) to develop the disorder divided by the risk for the general population. For BP disorder, family studies suggest that the relative risk for siblings is increased by a factor of about 8–9; for example, see Gershon et al., 1982). Because BP disorder is almost certainly an oligogenic syndrome, in which at least several loci contribute to the increased relative risk, locus-specific relative risk (the increased risk due to a single locus) is expected to be much less than nine. For complex traits, such as hypertension, diabetes, and BP disorder, loci that increase risk by factors greater than two are unusual. One such locus is near the HLA locus for insulin-dependent diabetes mellitus (relative risk is about three (Davies et al., 1994), another is the Apolipoprotein E locus in late onset Alzheimer's Disease (for review see Sandbrink et al., 1996)).

If three loci of equal effect size are used in an interactive, multiplicative model to explain the increased relative risk in BP disorder (each locus increases relative risk by about 2), then these three hypothetical interactive loci explain most of the relative risk ($2 \times 2 \times 2 = 8$). Thus, loci that increase risk for BP disorder will have minor to moderate effects. Substantial sample sizes are required to detect such loci of minor effect. As Hauser and Boehnke (1997) have shown, about 400 affected sibling pairs are needed to have >95% power to detect initially (LOD >3) loci that increase risk by a factor of 2, while 200 pairs are needed to have >95% power to provide confirmation (p < 0.01) of a previously detected locus.

## REVIEW OF BIPOLAR MOLECULAR LINKAGE STUDIES

Molecular methods have been used in BP linkage studies to localize susceptibility genes. A linkage study of Old Order Amish pedigrees described evidence (LOD score > 4.0) for a BP locus on 11p15 (Egeland et al., 1987), but this evidence has been weakened by failure to confirm the finding in numerous other pedigrees (Detera-Wadleigh et al., 1994; Mitchell et al., 1991; Hodgkinson et al., 1987; Mendlewicz et al., 1991; Curtis et al., 1993; Gill et al., 1988; Coon et al., 1993; Stine et al., 1997) and by evaluation of newly ascertained individuals in the original pedigree (Kelsoe et al., 1989). However, this hypothesis (that a BP susceptibility gene exists on the tip of the short arm of chromosome 11) remains viable and interesting. The LOD score in the original Old Order Amish pedigree 110 is ~2.0, and similar weakly positive LOD scores are reported for this region by other investigators (Gurling et al., 1995; Smyth et al., 1996). Furthermore, several reports have described evidence for association of tyrosine hydroxylase (located in 11p15) with BP disorder (Leboyer et al., 1990; Kennedy et al., 1993; Meloni et al., 1993 and 1995; Verga et al., 1993; De Castro et al., 1995; Serretti et al., 1998), although other groups have not confirmed this observation (Nothen et al., 1990; Korner et al., 1990 and 1994; Gill et al., 1991; Inayama et al., 1993; Rietschel et al., 1995; Oruc et al., 1997; Kunugi et al., 1996; Todd et al., 1996; Cavazzoni et al., 1996). The existence of an 11p15 locus of small effect on risk for BP illness remains a tenable hypothesis.

Xq28 was reportedly linked to BP illness (Winokur et al., 1969; Mendlewicz and Fleiss, 1974; Baron, 1977; Baron et al., 1987; Mendlewicz et al., 1980 and 1979; Del Zompo et al., 1984) in studies employing clinically assessed color blindness and G6PD deficiency. Molecular studies have not confirmed these "pre-molecular reports" (Berrettini et al., 1990; Del Zompo et al., 1991; Smyth et al., 1997; Stine et al., 1997). The link to color blindness and G6PD deficiency in the most recent positive report (Baron et al., 1987) was not confirmed in those same pedigrees by molecular methods employing relevant Xq28 DNA markers (Baron et al., 1993). There is no published molecular linkage study consistent with an Xq28 BP susceptibility locus.

The complex inheritance of BP illness and the failure of multiple genome-wide scans to detect major gene effects indicate that BP susceptibility loci represent small to moderate effects. Novel statistical methods to detect loci of small effect (Weeks and Lange, 1992; Goldgar et al., 1993; Kruglyak et al., 1996) and development of dense highly polymorphic marker maps (Gyapay et al., 1994; Dib et al., 1996) have provided the necessary tools to conduct the large-scale, definitive studies.

Suarez et al. (1994) simulated initial detection of linkage, and subsequent independent confirmation of the originally detected locus, in a complex disease caused in part by six equally frequent independent (unlinked) disease loci. A larger sample size and an extended waiting period is likely for confirmation of a previously detected locus. This is intuitively reasonable because of sampling variation. Independent pedigree samples might detect one of the other five loci, as opposed to the one locus initially detected. This simulation study (Suarez et al., 1994) suggests that universal agreement regarding BP linkage studies will not occur. If two or more independent investigators find significant evidence for linkage in independent series of pedigrees, it is reasonable to assume validity (Lander and Kruglyak, 1995; Thompson, 1994). It is reassuring to note that several groups have reported putative BP susceptibility loci that have been confirmed independently. This suggests that genetic dissection of BP disorders will proceed from established linkages, as has been the case with Alzheimer's disease (for review see Sandbrink, 1996).

Berrettini et al. (1994 and 1997) reported significant evidence for a BP susceptibility locus on chromosome 18 using affected sibling pair (ASP) and affected pedigree member (APM) methods ($p = 10^{-4}$ to $10^{-5}$), obtained in 22 Caucasian kindreds of European ancestry. Independent confirmation of this finding was reported by Stine et al. (1995) and others, as noted in Table 3.2. Evidence for linkage appears to be more prominent in those families with paternally transmitted illness (Stine et al., 1995; Gershon et al., 1996; Nothen et al., 1999; Knowles et al., 1998).

Schwab et al. (1998) studied 59 multiplex German and Israeli schizophrenia pedigrees, in which there were only two BP cases. Their analyses involved a broad affection status model in which 23 recurrent UP cases were included. When these data were analyzed by a multipoint identical-by-descent (IBD) statistic, the maximum LOD score was 3.2 at D18S53 (Schwab et al., 1998) also describe linkage disequilibrium ($p = \sim 0.0001$) with the 124 BP allele of a microsatellite in the Golf gene, a candidate gene in 18p11.2. This region of 18p may contain a gene that increases risk for psychotic disorders of varying syndromal form. The possibility that BP and schizophrenic disorders might share some of the same susceptibility factors is consistent with family studies of schizophrenia, which report an increased risk for schizoaffective and unipolar disorders among the first degree relatives of schizophrenic probands (Gershon et al., 1988; Maier et al., 1993). Similarly, increased risks for schizoaffective and UP disorders are found among the first-degree relatives of BP probands, compared to first-degree relatives of controls (Gershon et al., 1982; Maier et al., 1993; Winokur et al., 1995). Further, Kendler et al. (1993) found increased risk for schizophrenia among the relatives of individuals with psychotic affective illness.

Table 3.2 summarizes nominal significance levels for statistical analysis of marker genotypes located in a ~10 cM region of chromosome 18p11. Results are presented for a narrow phenotypic definition in which only BPI was affected (Nothen et al., 1999; Knowles et al., 1998) or for a broader definition (Berrettini et al., 1997; Stine et al., 1995; Schwab et al., 1998).

If the locus described by Berrettini et al. (1994, 1997) increases risk for BP disorder by a factor of ~2, simulations indicate that ~200 affected sibling pairs are required to have >90% power to

**TABLE 3.2**
**Linkage Results (p Values) for Bipolar Disorder and 18p11 DNA Markers**

| | DNA Marker (Distance in cM) | | | | | |
|---|---|---|---|---|---|---|
| Study | D18S53 (2) | S37 (2) | S453 (2) | S40 (3) | S45 | MP† |
| Berrettini et al., 1997 | 0.04 | 0.01 | 0.06 | 0.0046 | 0.002 | 0.00008 |
| Stine et al., 1995 | 0.02 | 0.0003 | NR* | 0.02 | NS** | NR |
| Nothen et al., 1998‡ | 0.03 | 0.005 | 0.002 | 0.005 | NR | 0.0004 |
| Knowles et al., 1998 | 0.001 | 0.08 | 0.50 | NR | 0.0003 | NS |
| Schwab et al., 1998° | 0.0001 | 0.0056 | 0.078 | 0.038 | NR | 0.0006 |

† MP = multipoint; ‡results for paternal kindreds; * NR = not reported; NS ** = not significant; ° schizophrenia probands.

detect it (Hauser et al., 1996) at a significance level (LOD > 1.2 or p < 0.01) adequate for confirmation (Lander and Kruglyak, 1995). Kalsi et al. (1997), De Bryn et al. (1997), Coon et al. (1994), Pauls et al. (1995), Bowen et al. (1998), McMahon et al. (1997), and Kelsoe et al. (1995) studied samples from European, Icelandic, and North American populations and found no evidence for confirmation of linkage on 18p. But these sample sizes did not exceed 100 affected sibling pairs in any one study. However, the 18p BP locus has not been confirmed in the NIMH Collaborative Study (Detera-Wadleigh et al., 1997) in which an adequate sample size was evaluated.

Genetic Analysis Workshop 10 (Goldin et al., 1997) allowed statistical geneticists to analyze data from Berrettini et al. (1997), Nothen et al. (1998), Stine et al. (1995), Knowles et al. (1998), and Kalsi et al. (1997). Results of several different analyses were consistent with the existence of a BP susceptibility gene. For example, Lin and Bale (1997) analyzed the entire data set of 382 affected sibling pairs (defined under a broad affection status model) using a multipoint non-parametric method. At D18S37, for 382 affected sibling pairs excess allele sharing (58%) was evident, with $p = 2.8 \times 10^{-8}$. Stine et al. (1995) also reported evidence for linkage to a distinct and separate region, 18q21-2. This 18q linkage was supported by the LOD score method (LOD is 3.51 for D18S41) and the ASP method (0.00002 at D18S41) in paternal pedigrees. In an extension of this work, McMahon et al. (1997) provided additional evidence for linkage to 18q21-2 in 30 new BP kindreds. This locus may have been detected by Freimer et al. (1996) and McInnes et al. (1996) who studied Costa Rican BP kindreds. McInnes et al. (1996) described evidence for increased allele sharing at some of the same markers identified by McMahon et al. (1997). For example, at D18S55, McMahon et al. (1997) reported a non-parametric LOD score of 2.2, while McInnes et al. (1996) at this same marker reports a maximum likelihood estimate of the LOD score as 1.67.

Straub et al. (1994) described linkage of BP illness to 21q21, near the phosphofructokinase locus. An extended BP pedigree with a LOD score of 3.41 was reported from a series of 57 BP kindreds; further, the APM method yielded evidence for linkage (P < 0.0003 for PFKL). A confirmatory report has been described from a two-locus analysis of genotypic data from 21q21 and 11p15.5 in a study by Gurling et al. (1995). This 21q21 BP susceptibility locus has been confirmed by Detera-Wadleigh et al. (1996), who employed multipoint ASP analyses (p < 0.001). Confirmation has been recorded by the NIMH Genetics Initiative collaborative study of BP disorder (Detera-Wadleigh et al., 1997). Thus, there are three independent confirmatory studies of this BP susceptibility locus.

Xq26, including the coagulation factor IX (F9) locus, is a third region of interest regarding BP susceptibility loci. The F9 locus was identified as a region of interest by Mendlewicz et al. (1987). A number of supportive reports followed (Lucotte et al., 1991; Gill et al., 1992; Craddock and Owen, 1992; Jefferies et al., 1993). However, these reports involved either a single or a few DNA markers with low polymorphism content or clinically assessed F9 deficiency as markers in single kindreds.

Pekkarinen et al. (1995) reported evidence for BP linkage (a LOD score of 3.54 at DXS994) by using multiple microsatellite DNA markers in the region near HPRT, which is ~10 cM centromeric to F9, in a single large Finnish pedigree. This finding probably represents a confirmation of the previous reported F9 linkage. Confirmatory affected sibling pair data have also been published for Xq26 markers (Stine et al., 1997) in an analysis of affected sisters.

Blackwood et al. (1996) reported on a single large Scottish kindred that showed linkage (LOD 4.1 at D4S394) to 4p DNA markers, near the alpha 2c adrenergic and D5 dopaminergic receptor genes. They found weakly positive LOD scores in several smaller kindreds of the same ethnic origins. They found no mutations in the dopamine receptor gene. Confirmation of the 4p locus has been noted by Nothen et al. (1997), in which increased allele-sharing was shown at D4S394 (p = 0.0009).

Ginns et al. (1996) conducted a genomic scan of multiple kindreds from the Old Order Amish community near Lancaster, PA. This group reports modest evidence for BP susceptibility loci on chromosomes 6 (LOD 2.5 at D6S7), 13 (at D13S1), and 15 (at D15S45). Confirmation of these loci has not been reported.

Kelsoe et al. (1996) reported some evidence for a BP susceptibility locus on chromosome 5p15.5, near the dopamine transporter locus, in North American and Icelandic kindreds. In an affected sibling pair analysis, at D5S392, p = 0.0008. This report, which did not reach statistical criteria for significant linkage (Lander and Kruglyak, 1995) requires confirmation.

Ewald et al. (1995) reported evidence for a BP susceptibility locus on 16p13 in two Danish kindreds. Assuming a recessive mode of inheritance, a two-point LOD score of 2.52 was found for marker D16S510, and a three-point LOD score of 2.65. Support for this 16p13 locus has been described, in a preliminary publication by Edenberg et al. (1997), but the original report (Ewald et al., 1995) did not describe evidence for significant linkage (Lander and Kruglyak, 1995). Thus, this locus must be studied in greater detail.

Lachman et al. (1996) described limited evidence for a BP susceptibility locus on chromosome 22, near the velo-cardiofacial syndrome locus. This region has been implicated in risk for schizo-phrenia (Pulver et al., 1994; Wildenauer et al., 1996) and modest supportive evidence for linkage to BP disorder has been reported by Edenberg et al. (1997). This region deserves further study.

Anticipation is the term used to define an observation that a familial disorder occurs with earlier age-at-onset and/or increasing severity among younger generations, compared to older generations. Anticipation occurs in several neurodegenerative diseases, including Huntington's Disease, Fragile X, myotonic dystrophy, spinocerebellar ataxias, and others. The molecular explanation for antici-pation in these disorders involves unstable intragenic tri-nucleotide repeats, which expand in subsequent generations, giving rise to increasing levels of gene disruption and thus to earlier age-at-onset and increasingly severe phenotype in younger generations (for review, see Petronis and Kennedy, 1995).

Evidence for anticipation has been reported in several family studies of BP illness (McInnis et al., 1993; Lipp et al., 1995; Nylander et al., 1994; Gershon et al., 1987), but some authorities suggest that there is intractable ascertainment bias (Penrose, 1948; Hodge and Wickramaratne, 1995). Individuals with earlier age-at-onset BP disorder may have reduced capacity to reproduce, so parents with such early-onset disorders may be under-represented in the general population. Individuals with familial BP disorder may come to treatment earlier than those with sporadic disease, such that less severe mood disorder episodes are detected medically, and an earlier age-at-onset is defined. Such individuals (by virtue of their familiarity with mood disorder symptoms) may be more likely to report minor mood disturbance in terms of "diagnosable syndromes." Some evidence for anticipation in BP disorder comes from extensive studies of multiplex BP families for linkage studies. These linkage studies select for earlier age-at-onset cases, because preference is given to densely-affected kindreds. Among broader cultural factors possibly underlying the evidence for anticipation, if stigma concerning mood disorders is less among younger affected persons (compared to older individuals), then younger cohorts might describe their experiences more easily in terms of a diagnosable mood disorder, since denial (due to stigma) is less prevalent among the younger cohorts. These potential confounding factors make detection of anticipation in BP disorder difficult.

The hypothesis that anticipation in BP disorder reflects causative expanding trinucleotide CTG repeat sequences has generated genomic searches for such sequences (O'Donovan et al., 1995 and 1996; Lindblad et al., 1995; Oruc et al., 1997a), using the repeat expansion detection method (Schalling et al., 1993). These three groups have noted increased lengths of CTG repeats in BP disorders, especially among those with familial disease. However, not all studies have reported this difference (Vincent et al., 1996) and no report shows transmission of an expanding repeat within BP families, the definitive evidence. Furthermore, greater than 90% of the expanded CTG repeats detected by the method of Schalling et al. (1993) are from two apparently non-pathogenic unstable CTG repeats on 17q and 18q21 (Sidransky et al., 1998). The hypothesis that unstable trinucleotide repeats represent BP susceptibility factors deserves continued study.

## SUMMARY

Linkage studies have defined at least five BP susceptibility loci that meet suggested guidelines (Lander and Kruglyak, 1995) for initial identification and subsequent confirmation. These loci, found on 18p11, 18q22, 21q21, 4p16, and Xq26, are targets for BP candidate gene investigations. Molecular dissection of expressed sequences for these regions is likely to yield specific BP susceptibility alleles in most cases. In all probability, these BP susceptibility alleles will be common in the general population and, individually, will be neither necessary nor sufficient for manifestation of the syndrome. Additive or multiplicative oligogenic models involving several susceptibility loci appear most reasonable at present. It is hoped that these BP susceptibility genes will increase understanding of many mysteries surrounding these disorders, including drug response, cycling patterns, age-of-onset, and modes of transmission.

## REFERENCES

Allen, M.G., Cohen, S., Pollin, W., and Greenspan, S.I., Affective illness in veteran twins: a diagnostic review, *Am. J. Psychiat.,* 131, 1234, 1974.

Andrews G., Stewart G., Allen, R., and Henderson, A.S., The genetics of six neurotic disorders: a twin study, *J. Affective Disord.,* 19, 23, 1990.

Angst, J., Frey, R., Lohmeyer, R., and Zerben-Rubin, E., Bipolar Manic depressive psychoses: results of a genetic investigation, *Hum. Gen.,* 55, 237, 1980.

Baron, M., Linkage between an X-chromosome marker (deutan color blindness) and bipolar affective illness, *Arch. Gen. Psychiatry,* 34, 721, 1977.

Baron, M., Gruen, R., Anis, L., and Kane, J., Schizoaffective illness, schizophrenia and affective disorders: morbidity risk and genetic transmission, *Acta. Psychiatr. Scand.,* 65, 253, 1983.

Baron, M., Freimer, N.F., Risch, N., Lerer, B., Alexander, J.R., Straub, R.E., Asokan, S., Das, K., Peterson, A., Amos, J., Endicott, J., Ott, J., and Gilliam, C., Diminished support for linkage between manic depressive illness and X-chromosome markers in three Israeli pedigrees, *Nat. Genet.,* 3, 49, 1993.

Baron, M., Risch, N., Hamburger, R., Mandel, B., Kushner, S., Newman, M., Drumer, D., and Belmaker, R.H., Genetic linkage between X-chromosome markers and bipolar affective illness, *Nature,* 326, 289, 1987.

Berrettini, W.H., Goldin, L.R., Gelernter, J., Gejman, P.V., Gershon, E.S., and Detera-Wadleigh, S., X-chromosome markers and manic-depressive illness: rejection of linkage to Xq28 in nine bipolar pedigrees, *Arch. Gen. Psychiatr.,* 47, 336, 1990.

Berrettini, W., Ferraro, T., Goldin, L., Weeks, D., Detera-Wadleigh, S.D., Nurnberger, Jr., J.I., and Gershon, E., Chromosome 18 DNA markers and manic depressive illness: evidence for a susceptibility gene, *PNAS (USA),* 91, 5918, 1994.

Berrettini, W., Ferraro, T., Choi, H., Goldin, L., Detera-Wadleigh, S.D., Muniec, D., Hsieh, W.-T., Hoehe, M., Guroff, J., Kazuba, D., Nurnberger, Jr., J.I., and Gershon, E., Linkage studies of bipolar illness, *Arch. Gen. Psychiatr.,* 54, 32, 1997.

Bertelsen, A., Harvald, B., and Hauge, M., A Danish twin study of manic-depressive disorders, *Br. J. Psychiatr.,* 130, 330, 1977.

Blackwood, D.H.R., He, L., Morris, S.W., McLean, A., Whitton, C., Thomson, M., Walker, M.T., Woodburn, K., Sharp, C.M., Wright, A.F., Shibasaki, Y., St. Clair, D.M., Porteous, D.J., and Muir, W.J., A locus for bipolar affective disorder on chromosome 4p, *Nat. Genet.,* 12, 427, 1996.

Bowen, T., Kirov, G., Gill, M., Spurlock, G., Vallada, H.P., Murray, R.M., McGuffin, P., Collier, D.A., Own, M.J., and Craddock, N., Linkage studies of bipolar disorder with chromosome 18 markers, *Neuropsychiatr. Genet.,* in press.

Cadoret, R.J., Evidence for genetic inheritance of primary affective disorder in adoptees, *Am. J. Psychiatr.,* 135(4), 463, 1978.

Cavazzoni, P., Alda, M., Turecki, G., Rouleau, G., Grof, E., Martin, R. et al., Lithium-responsive affective disordeers: no associations with the tyrosine hydroxylase gene, *Psychiatr. Res.,* 64, 91, 1996.

Clerget-Darpoux, F., Bonaiti-Pellie, C., and Hochez, J., Effects of misspecifying genetic parameters in lod score analysis, *Biometrics,* 42, 245, 1986.

Coon, H., Hoff, M., Holik, J., Hadley, D., Fang, N., Reimherr, F., Wender, P., Leppert, M., and Byerley, W., Analysis of chromosome 18 DNA markers in multiplex pedigrees with manic depression, *Biol. Psychiatr.*, 39, 689, 1996.

Coon, H., Jensen, S., Hoff, M., Holik, J., Plaetke, R., Reimherr, F., Wender, P., Leppert, M., and Byerley, W., A genome-wide search for genes predisposing to manic depression assuming autosomal dominant inheritance, *Am. J. Hum. Genet.*, 52, 619, 1993.

Craddock, N. and Owen, M., Christmas disease and major affective disorder, *Br. J. Psychiatr.*, 160, 715, 1992.

Curtis, D., Sherrington, R., Brett, P., Holmes, D.S., Kalsi, G., Brynjolfsson, J., Petursson, H., Rifkin, L., Murphy, P., Moloney, E., Melmer, G., and Gurling, H.M.D., Genetic linkage analysis of manic depression in Iceland, *J. R. Soc. Med.*, 86, 506, 1993.

Davies, J.L., Kawaguchi, Y., Bennett, S.T., Copeman, J.B., Cordell, H.J., Pritchard, L.E., Reed, P.W., Gough, S.C.L., Jenkins, S.C., Palmer, S.M., Balfour, K.M., Rowe, B.R., Farrall, M., Barnett, A.H., Bain, S.C., and Todd, J.A., A genome-wide search for human type 1 diabetes susceptibility genes, *Nature*, 371, 130, 1994.

De bruyn, A., Souery, D., Mendelbaum, K., Mendlewicz, J., Van and Broekhoven, C., Linkage analysis of families with bipolar illness and chromosome 18 markers, *Biol. Psychiatr.*, 39, 679, 1996.

DeCastro, I.P., Santos, S., Torres, P., Visedo, G., Saiz-Ruiz, J., Llinares, D. et al., A weak association between TH and DRD2 genes and bipolar affective disorder in a Spanish sample, *J. Med. Genet.*, 32, 131, 1995.

Del Zompo, M., Bochetta, A., Goldin, L.R., and Corsini, G.U., Linkage between X-chromosome markers and manic-depressive illness: two Sardinian pedigrees, *Acta. Psychiatr. Scand.*, 70, 282, 1984.

Del Zompo, M., Bocchetta, A., Ruiu, S., Goldin, L.R., Berrettini, W.H., Association and linkage studies of affective disorders, in *Biological Psychiatry,* Vol. 2, Racagni, G., Brunello, N., Fukuda, T., Eds., Elsevier Science Pub Co., Inc., New York, 1991, 446.

Detera-Wadleigh, S.D., Badner, J.A., Goldin, L.R., Berrettini, W.H., Sanders, A.R., Rollins, D.Y., Turner, G., Moses, T., Haerian, H., Muniec, D., Nurnberger, J.I., Jr., and Gershon, E.S., Affected sib-pair analyses reveal support of prior evidence for a susceptibility locus for bipolar disorder on 21q, *Am. J. Hum. Genet.*, 58, 1279, 1996.

Detera-Wadleigh, S.D., Berrettini, W.H., Goldin, L.R., Boorman, D., Anderson, S., and Gershon, E.S., Close linkage of c-Harvey-ras-1 and the insulin gene to affective disorder is ruled out in three North American pedigrees, *Nature*, 325, 806, 1987.

Detera-Wadleigh, S.D., Hsieh, W.-T., Berrettini, W.H., Goldin, L.R., Rollins, D.Y., Muniec, D., Grewal, R., Guroff, J.J., Turner, G., Coffman, D., Barrick, J., Mills, K., Murray, J., Donohue, S.J., Klein, D.C., Sanders, J., Nurnberger, Jr., J.I., and Gershon, E.S., Genetic linkage mapping for a susceptibility locus to bipolar illness: chromosomes 2, 3, 4, 7, 9, 10p, 11p, 22, and Xpter, *Neuropsychiatr. Genet.*, 54, 206, 1994.

Dib, C., Faure, S., Fizames, C., Samson, D., Drouot, N., Vignal, A., Millasseau, P., Marc, S., Hazan, J., Seboun, E., Latrhop, M., Gyapay, G., Morissette, J., and Weissenbach, J., A comprehensive genetic map of the human genome based on 5264 microsatellites, *Nature*, 380, 152, 1996.

Edenberg, H., Foroud, T., Conneally, M., Sorbel, J., Carr, K., Crose, C., Willig, C., Zhao, J., Miller, M., Bowman, E., Mayeda, A., Rau, L., Smiley, C., Goate, A., Reich, T., Stine, C., McMahon, F., DePaulo, R., Meyers, D., Detera-Wadleigh, S., Goldin, L., Gershon, E., Blehar, M., and Nurnberger, Jr. J.I., Initial genomic scan of the NIMH genetics initiative bipolar pedigrees: chromosomes 3, 5, 15, 16, 17, and 22, *Neuropsychiatr. Genet.*, 74, 238, 1997.

Egeland, J.A., Gerhard, D.S., Pauls, D.L., Sussex, J.N., Kidd, K.K., Allen, C.R., Hostetter, A.M., and Housman, D.E., Bipolar affective disorder linked to DNA markers on chromosome 11, *Nature*, 325, 783, 1987.

Ewald, H., Mors, O., Koed, K., Eiberg, H., and Kruse, T.A., A possible locus for manic depressive illness on chromosome 16p13, *Psychiatr. Genet.*, 5, 71, 1995.

Freimer, N.B., Reus, V.I., Escamilla, M.A., McInnes, L.A., Spesny, M., Leon, P., Service, S.K., Smith, L.B., Silva, S., Rojas, E., Gallegos, A., Meza, I., Fournier, E., Baharloo, S., Blankenship, K., Tyler, D.J., Batki, S., Vinogradov, S., Weissenbach, J., Barondes, S.H., and Sandkuijl, L.A., Genetic maping using haplotype, association and linkage methods suggests a locus for severe bipolar disorder (BP) at 18q22-q23, *Nat. Genet.*, 12, 436, 1996.

Gershon, E.S., DeLisi, L.E., Hamovit, J., Nurnberger, Jr., J.I., Maxwell, M.E., Schreiber, J., Dauphinais, D. et al., A controlled family study of chronic psychoses, *Arch. Gen. Pscyhiatr.*, 45, 328, 1988.

Gershon, E.S., Badner, J.A., Ferraro, T.N., Detera-Wadleigh, S., and Berrettini, W.H., Maternal inheritance and chromosome 18 allele sharing in unilineal bipolar illness pedigrees, *Neuropsychiatr. Genet.*, 67, 1, 1996.

Gershon, E.S., Hamovit, J., Guroff, J.J., Dibble, E., Leckman, J.F., Sceery, W., Targum, S.D., Nurnberger Jr., J.I., Goldin, L.R., and Bunney, Jr., W.E., A family study of schizoaffective, bipolar I, bipolar II, unipolar, and normal control probands, *Arch. Gen. Psychiatr.,* 39, 1157, 1982.

Gershon, E.S., Hamovit, J.H., Guroff, J.J., and Nurnberger, Jr., J.I., Birth-cohort changes in manic and depressive disorders in relatives of bipolar and schizoaffective patients, *Arch. Gen. Psychiatr.,* 440, 314, 1987.

Gill, M., McKeon, P., and Humphries, P., Linkage analysis of manic depression in an Irish family using H-ras 1 and INS DNA markers, *J. Med. Genet.,* 25, 634, 1988.

Gill, M., Castle, D., Hunt, N., Clements, A., Sham, P., Murray, R.M., Tyrosine hydroxylase polymorphisms and bipolar affective disorder, *J. Psychiatr. Res.,* 25, 179, 1991.

Gill, M., Castle, D., and Duggan, C., Cosegregation of Christmas disease and major affective disorder in a pedigree, *Br. J. Psychiatr.,* 160, 112, 1992.

Ginns, E.I., Ott, J., Egeland, J.A., Allen, C.R., Fann, C.S.J., Pauls, D.L., Weissenbach, J., Carulli, J.P., Falls, K.M., Keith, T.P., and Paul, S.M., A genome-wide search for chromosomal loci linked to bipolar affective disorder in the Old Order Amish, *Nat. Genet.,* 12, 431, 1996.

Goldgar, D.E., Lewis, C.U., and Gholami, K., Analysis of discrete phenotypes using a multipoint identity by descent method: application to Alzheimer's disease, *Genet. Epidemiol.,* 10, 383, 1993.

Goldin, L.R., Gershon, E.S., Berrettini, W.H., Stine, O.C., DePaulo, R., McMahon, F., Meyers, D., Nothen, M., Propping, P., Cichon, S., Fimmers, R., Baur, M., Albus, M., Franzek, E., Kreiner, R., Maier, W., Rietschel, M., Baron, M., Knowles, J., Gilliam, C., Endicott, J., Gurling, H., Curtis, D., Smyth, C., and Kelsoe, J., Description of the Genetic Analysis Workshop 10 bipolar disorder linkage data sets, *Genet. Epidemiol.,* 14, 563, 1997.

Gurling, H., Smyth, C., Kalsi, G., Moloney, E., Rifkin, L., O'Neill, J., and Murphy, P., Linkage findings in bipolar disorder, *Nat. Genet.,* 10, 8, 1995.

Gyapay, G., Morissette, J., Vignal, A., Deib, C., Fizames, C., Millasseua, P., Marc, S., Bernardi, G., Lathrop, M., and Weissenbach, J., 1993-4 Genethon human genetic linkage map, *Nat. Genet.,* 7, 246, 1994.

Hauser, E.R. and Boehnke, M., Confirmation of linkage results in affected sib-pair linkage analysis for complex genetic traits, *Am. J. Hum. Genet.,* 61 (Suppl., p. A278), Abstr. 1621, 1997.

Hauser, E.R., Boehnke, M., Guo, S.-W., and Risch, N., Affected sib pair interval mapping and exclusion for complex genetic traits, *Genet. Epidemiol.,* 13, 117, 1996.

Harvald, B. and Hauge, M., in *Genetics and the Epidemiology of Chronic Diseases,* Neal, J.V., Shaw, M.W., and Shull, W.J., Eds., PHS Publication No. 1163, Washington, D.C., US DHEW, 61, 1975.

Helzer, J.E. and Winokur, G., A family interview study of male manic-depressives, *Arch. Gen. Psychiatr.,* 31, 73, 1974.

Hodge, S.E. and Wickramaratne, P., Statistical pitfalls in detecting age-at-onset anticipation: the role of correlation in studying anticipation and detecting ascertainment bias, *Psychiatr. Genet.,* 5, 43, 1995.

Hodgkinson, S., Sherrington, R., Gurling, H., Marchbanks, R., Reeders, S., Mallet, J., McInnis, M., Petursson, H., Brynjolfsson, J., Molecular genetic evidence for heterogeneity in manic depression, *Nature,* 325, 805, 1987.

Inayama, Y., Yoneda, H., Sakai, T., Ishida, T., Kobayashi, S., Nonomura, Y., Kono, Y., Koh, J., and Asaba, H., Lack of association between bipolar affective disorder and tyrosine hydroxylase DNA marker, *Am. J. Med. Genet.,* 48, 87, 1993.

James, N.M. and Chapman, C.J., A genetic study of bipolar affective disorder, *Br. J. Psychiatr.,* 126, 449, 1975.

Johnson, G.F.S. and Leeman, M.M., Analysis of familial factors in bipolar affective illness, *Arch. Gen. Psychiatr.,* 34, 1074, 1977.

Joyce, P.R., Oakley-Brown, M.A., Wells, J.E., Bushnell, J.A., and Hornblow, A.R., Birth cohort trends in major depression: increasing rates and earlier onset in New Zealand, *J. Affect. Disord.,* 18, 83, 1990.

Jeffries, F.M., Reiss, A.L., Brown, W.T., Meyers, D.A., Glicksman, A.C., and Bandyopadhyay, S., Bipolar spectrum disorder and fragile X syndrome: a family study, *Biol. Psychiatr.,* 33, 213, 1993.

Kallman, F., in *Depression,* Hoch, P.H. and Zubin, J., Eds., Grune & Stratton, New York, 1954, 1.

Kalsi, G., Smyth, C., Brynjolfsson, J., Sherrington, R.S., O'Neill, J., Curtis, D., Rifkin, L., Murphy, P., Petursson, H., and Gurling, H.M.D., Linkage analysis of manic depression (bipolar affective disorder) in Icelandic and British kindreds using markers on the short arm of chromosome 18, *Hum. Heredity,* 47, 268, 1997.

Kelsoe, J., Ginns, E., Egeland, J., Gerhard, D., Goldstein, A., Bale, S., Pauls, D., Long, R., Kidd, K., Conte, G., Housman, D., and Paul, S., Re-evaluation of the linkage relationship between chromosome 11p loci and the gene for bipolar affective disorder in the Old Order Amish, *Nature,* 342, 238, 1989.

Kelsoe, J., Sadovick, A., Kristbjarnarson, H., Bergesch, P., Mroczkowski-Parker, Z., Flodman, P., Rapaport, M., Mirow, A., Egeland, J., Spence, M., Remick, R., Genetic linkage studies of bipolar disorder and chromosome 18 markers in North American, Icelandic and Amish pedigrees, *Psychiatr. Genet.*, 5, S17, 1995.

Kelsoe, J.R., Sadovnick, A.D., Kristbjarnarson, H., Bergesch, P., Mroczkowski-Parker, Z., Drennan, M.D., Rapaport, M.H., Flodman, P., Spence, M.A., and Remick, R.A., Possible locus for bipolar disorder near the dopamine transporter on chromosome 5, *Am. J. Med. Genet.*, 67, 533, 1996.

Kendler, K.S., Neale, M.C., Kessler, R.C., Heath, A.C., and Eaves, L.J., A population based twin study of major depression in women: the impact of varying definitions of illness, *Arch. Gen. Psychiatr.*, 49, 257, 1992.

Kendler, K.S., Petersen, N., Johnson, L., Neale, M.C., Mathe, A.A., A pilot Swedish twin study of affective illness, including hospital and population ascertained subsamples, *Arch. Gen. Psychiatr.*, 50, 699, 1993.

Kendler, K.S., McGuire, M., Gruenberg, A.M., O'Hare, A., Spellman, M., and Walsh, D., The Roscommon family study, *Arch. Gen. Psychiatr.*, 50, 527, 1993.

Kennedy, J.L., Sidenberg, D.G., Macciardi, F.M., Joffe, R.T., Genetic association study of tyrosine hydroxylase and D4 receptor variants in bipolar I patients, *Psychiatr. Genet.*, 3, 120, 1993.

Knowles, J.A., Rao, P.A., Cox-Matise, T. et al., No evidence for significant linkage between bipolar affective disorder and chromosome 18 pericentromeric markers in a large series of multiplex entended pedigrees, *Am. J. Hum. Genet.*, 62, 916, 1998.

Klerman, G.L., Clinical epidemiology of suicide, *J. Clin. Psychiatr.*, Suppl. 48, 33, 1987.

Klerman, G.L., Lavori, P.W., Rice, J., Reich, T., Endicott, J., Andreasen, N.C., Keller, M.B., and Hirschfield, P.M.A., Birth-cohort trends in rates of major depressive disorder among relatives of patients with affective disorder, *Arch. Gen. Psychiatr.*, 42, 689, 1985.

Klerman, G.L. and Weissman, M.M., Increasing rates of depression, *JAMA*, 261, 2229, 1989.

Korner, J., Fritze, J., and Propping, P., RFLP alleles at the tyrosine hydroxylase locus: no assoiation found to affective disorders, *Psychiatr. Res.*, 32, 275, 1990.

Korner, J., Rietschel, M., Hunt, N., Castle, D., Gill, M., Nothen, M., Craddock, N., Daniels, J., Owen, M., Frimmers, R., Fritze, J., Moller, H.-J., Propping, P., Association and haplotype analysis at the tyrosine hydroxylase locus in a combined German-British sample of manic-depressive patients and controls, *Psychiatr. Genet.*, 4, 167, 1994.

Kruglyak, L., Daly, M.J., Reeve-Daly, M.P., and Lander, E.S., Parametric and nonparametric linkage analysis: a unified multipoint approach, *Am. J. Hum. Genet.*, 58, 1347, 1996.

Kunugi, H., Kawada, Y., Tatsumi, M., Sasaki, T., Nanko, S., Manic-depressive illness and tyrosine hydroxylase gene, *Lancet*, 349, 336, 1996.

Lachman, H.M., Kelsoe, J.R., Remick, R.A., Sadovnick, A.D., Rapaport, M.H., Lin, M., Pazur, B.A., Roe, A.A., Saito, T., and Papolos, D.F., Linkage studies support a possible locus for bipolar disorder in the velocardiofacial syndrome region on chromosome 22, *Am. J. Med. Genet.*, 74, 121, 1996.

Lander, E. and Kruglyak, L., Genetic dissection of complex traits: guidelines for interpreting and reporting linkage results, *Nat. Genet.*, 11, 241, 1995.

Lander, E. and Schork, N., Genetic dissection of complex traits, *Science*, 265, 2037, 1994.

Leboyer, M., Malafosse, A., Boularand, S., Campion, D., Gheysen, F., Samolyk, D., Henriksson, B., Denise, E., des Lauriers, A., and Lepine, J.P., Tyrosine hydroxylase polymorphisms associated with manic-depressive illness, *Lancet*, 335, 1219, 1990.

Lin, J.P. and Bale, S.J., Parental transmission and D18S37 allele sharing in bipolar affective disorder, *Genet. Epidemiol.*, 14, 665, 1997.

Lindblad, K., Nylander, P.-O., De Bruyn, A., et al., Detection of expanded CAG repeats in bipolar affective disorder using the repeat expansion detection (RED) method, *Neurobiol. Dis.*, 2, 55, 1998.

Lipp, O., Souery, D., Mahieu, B., De Bruyn, A., Van Broeckhoven, C., and Mendlewicz, J., Anticipation in affective disorders, *Psychiatr. Genet.*, 5, S8, 1995.

Lucotte, G., Landoulsi, A., Berriche, S., David, F., and Babron, M.C., Manic depressive illnessis linked to factor IX in a French pedigree, *Ann. Genet.*, 35, 93, 1992.

Luxenberger, H., Psychiatrisch-neurologische Zwillings pathologie, *Zentralblatt fur diagesamte Neurologie and Psychiatrie*, 14, 56, 145, 1930.

Maier, W., Lichtermann, D., Minges, J., Hallmayer, J., Heun, R., Benkert, O., and Levinson, D.F., Continuity and discontinuity of affective disorders and schizophrenia. Results of a controlled family study, *Arch. Gen. Psychiatr.*, 50, 871, 1993.

McGuffin, P., Katz, R., Watkins, S., and Rutherford, J., A hospital-based twin register of the heritability of DSM-IV unipolar depression, *Arch. Gen. Psychiatr.*, 53, 129, 1996.

McInnes, L.A., Escamilla, M.A., Service, S.K. et al., A complete genome screen for genes predisposing to severe bipolar disorder in two Costa Rican pedigrees, *PNAS*, 93, 13060, 1996.

McInnis, M.G., McMahon, F.J., Chase, G.A., Simpson, S.G., Ross, C.A., Anticipation in bipolar affective disorder, *Am. J. Hum. Genet.*, 53, 385, 1993.

McMahon, F.J., Hopkins, P.J., Xu, J., McInnis, M.G., Shaw, S., Cardon, L., Simpson, S.G., MacKinnon, D.F., Stine, O.C., Sherrington, R., Meyers, D.A., and DePaulo, J.R., Linkage of bipolar affective disorder to chromosome 18 markers in a new pedigree series, *Am. J. Hum. Genet.*, 61, 1397, 1997.

Meloni, R., Leboyer, M., Campion, D., Savoye, C., Poirier, M.-F., Samolyk, D., Malafosse, A., Mallet, J., Association of manic depressive illness with the TH locus using a microsatellite marker localized in the tyrosine hydroxylase gene, *Psychiatr. Genet.*, 3, 121, 1993.

Meloni, R., Leboyer, M., Bellivier, F., Barbe, B., Samolyk, D., Allilaire, J.F. et al., Association of manic-depressive illness with tyrosine hydroxylase microsatellite marker, *Lancet*, 345, 932, 1995.

Mendlewicz, J., Linkowski, P., Guroff, J.J., and Van Praag, H.M., Color blindness linkage to manic-depressive illness, *Arch. Gen. Psychiatr.*, 36, 1442, 1979.

Mendlewicz, J. and Fleiss, J.L., Linkage studies with X-chromosome markers in bipolar (manic-depressive) and unipolar (depressive) illnesses, *Biol. Psychiatr.*, 9, 261, 1974.

Mendlewicz, J., Linkowski, P., and Wilmotte, J., Linkage between glucose-6-phosphate dehyrogenase deficiency and manic-depressive psychosis, *Brit. J. Psychiatr.*, 137, 337, 1980.

Mendlewicz, J. and Rainer, J.D., Adoption study supporting genetic transmission in manic-depressive illness, *Nature*, 368(5618), 327, 1977.

Mendlewicz, J., Simon, P., Sevy, S. et al., Polymorphic DNA marker on X-chromosome and manic depression, *Lancet*, 1, 1230, 1987.

Mendlewicz, J., Leboyer, M., De Bruyn, A., Malafosse, A., Sevy, S., Hirsch, D., Van Broeckhoven, C., Mallet, J., Absence of linkagebetween chromosome 11p15 markers and manic-depressive illness in a Belgian pedigree, *Am. J. Psychiatr.*, 148, 1683, 1991.

Mirow, A.L., Kristbjnarson, H., Egeland, J.A., Helgason, T., Gillin, J.C., Hirsch, S., and Kelsoe, J.R., A linkage study of distal chromosome 5q and bipolar disorder, *Biol. Psychiatr.*, 36, 223, 1994.

Mitchell, P., Waters, B., Morrison, N., Shine, J., Donald, J., Eissman, J., Close linkage of bipolar disorder to chromosome 11 markers is excluded in two large Australian pedigrees, *J. Aff. Disord.*, 20, 23, 1991.

Nothen, M.M., Cichon, S., Rohleder, H., Hemmer, S., Franzek, E., Fritze, J., Albus, M., Borrmann-Hassenbach, M., Kreiner, R., Weigelt, B., Minges, J., Lichtermann, D., Maier, W., Craddock, N., Fimmers, R., Holler, T., Bauer, M.P., Rietschel, M., and Propping, R., Evaluation of linkage of bipolar affective disorder to chromosome 18 in a sample of 57 German families, *Mol. Psychiatr.*, 4, 76, 1999.

Nothen, M.M., Cichon, S., Franzek, E., Albus, M., Bormann, M., Rietschel, R., Lichtermann, D., Weigelt, B., Maier, W., Fimmers, R., and Propping, R., Systematic search for susceptibility genes in bipolar affective disorder — evidence for disease loci at 18p and 4 p, *Am. J. Hum. Genet.*, 61(S), A288, 1997.

Nothen, M., Korner, J., Lanczik, M., Fritze, J., and Propping, P., Tyrosine hydroxylase polymorphisms and manic-depressive illness, *Lancet*, 336, 575, 1990.

Nylander, P.O., Engstrom, C., Shotai, J., Wahlstrom, J., and Adolfsson, R., Anticipation in Swedish families with bipolar affective disorder, *J. Med. Genet.*, 31, 686, 1994.

O'Donovan, M.C., Guy, C., Craddock, N., Murphy, K.C., Cardno, A.G., Jones, L.A., Owen, M.J., and McGuffin, P., Expanded CAG repeats in schizophrenia and bipolar disorder, *Nat. Genet.*, 10, 380, 1995.

O'Donovan, M.C., Guy, C., Craddock, N., Bown, T., McKeon, P., Macedo, A., Owen, M.J., and McGuffin, P., Confirmation of association between expanded CAG/CTG repeats and both schizophrenia and bipolar disorder, *Psychol. Med.*, 26, 1145, 1996.

Oruc, L., Lindblad, K., Verheyen, G.R., Ahlberg, S., Jakovljevic, M., Ivezic, S., Raeymaikers, P., Van Broeckhoven, C., and Schalling, M., CAG repeat expansions in bipolar and unipolar disorders, *Am. J. Hum. Genet.*, 60, 730, 1997a.

Oruc, L., Verheyen, G.R., Furac, I., Jakovljevic, M., Ivezic, S., Raeymaikers, P., Van Broeckhoven, C., Analysis of the tyrosine hydroxylase and dopamine D4 receptor genes in a Croatian sample of bipolar I and unipolar patients, *Am. J. Med. Genet.*, 74, 176, 1997b.

Pauls, D.L., Ott, J., Paul, S.M., Allen, C.R., Fann, C.S.J., Carulli, J.P., Falls, K.M., Bouthillier, C.A., Gravius, T.C., Keith, T.P., Egeland, J.A., and Ginns, E.I., Linkage analyses of chromosome 18 markers do not identify a major susceptibility locus for Bipolar affective disorder in the Old Order Amish, *Am. J. Hum. Genet.*, 57, 636, 1995.

Pekkarinen, P., Terwilliger, J., Bredbacka, P.-E., Lonnquist, J., and Peltonen, L., Evidence of a predisposing locus to bipolar disorder on Xq24-q27.1 in an extended Finnish pedigree, *Genome Res.*, 5, 105, 1995.

Penrose, L.S., The problem of anticipation in pedigrees of dystrophia myotonica, *Ann. Eugenics*, 14, 125, 1948.

Petronis, A. and Kennedy, J.L., Unstable genes — unstable mind? *Am. J. Psychiatr.*, 152, 64, 1995.

Pulver, A.E., Karayiorgou, M., Wolyniec, P.S., Lasseter, V.K., Kasch, L., Nestadt, G., Antonarakis, S., Housman, D., Kazazian, H.H., and Meyers, D., Sequential strategy to identify a susceptibility gene for schizophrenia: report of potential linkage on chromosome 22q12-q13.1: Part 1, *Am. J. Med. Genet.*, 54, 36, 1994.

Rice, J., Goate, A., Williams, J., Bierut, L., Dorr, D., Wu, W., Shears, S., Gopalakrishnan, G., Edenberg, H., Foroud, T., Nurnberger, Jr., J.I., Gershon, E., Detera-Wadleigh, S., Goldin, L., Guroff, J., McMahon, F., Simpson, S., MacKinnon, D., McInnis, M., Stine, C., DePaulo, R., Blehar, M., and Reich, T., Initial genome scan of the NIMH genetics initiative bipolar pedigrees: chromosomes 1,6,8,10 and 12, *Neuropsychiatr. Genet.*, 74, 247, 1997.

Rietschel, M., Nothen, M.M., Maier, W., Albus, M., Frianzek, E., and Propping, P., Tyrosine hydroxylase gene and manic depressive illness, *Lancet*, 345, 1368.

Risch, N., Linkage strategies for genetically complex traits. II. The power of affected relative pairs, *Am. J. Hum. Genet.*, 46, 229, 1990a.

Rosanoff, A.J., Handy, L., and Plesset, I.R., The etiology of manic-depressive syndromes with special reference to their occurrence in twins, *Am. J. Psychiatr.*, 91, 725, 1935.

Sandbrink, R., Hartmann, T., Masters, C.L., and Beyreuther, K., Genes contributing to Alzheimer's disease, *Mol. Psychiatr.*, 1, 27, 1996.

Schalling, M., Hudson, T.J., Buetow, K.H., and Housman, D.E., Direct detection of novel expanded trinucleotide repeat in the human genome, *Nat. Genet.*, 4, 135, 1993.

Schwab, S.G., Hallmayer, J., Lerer, B., Albus, M., Borrmann, M., Honig, S., Strauss, M., Segman, R., Lichtermann, D., Knapp, M., Trixler, M., Maier, W., and Wildenauer, D.B., Support for a chromosome 18p locus conferring susceptibility to functional psychoses in families with schizophrenia, by association and linkage analysis, *Am. J. Hum. Genet.*, 63, 1139, 1998.

Serretti, A., Macciardi, F., Cusin, C., Verga, M., Pedrini, S., and Smeraldi, E., Tyrosine hydroxylase gene in linkage disequilibrium with mood disorders, *Mol. Psychiatr.*, 3, 169, 1998.

Sidransky, E., Burgess, C., Ikeuchi, T., Lindblad, K., Long, R.T., Philbert, R.A., Rapoport, J., Schalling, M., Tsuji, S., and Ginns, E.I., A triplet repeat on 17q accounts for most expansions detected by the repeat-expansion detection technique, *Am. J. Hum. Genet.*, 62, 1548, 1998.

Slater, E., Psychotic and neurotic illness in twins, Medical Research Council Special Report Series No. 278, Her Majesty's Stationery Office, London, 1953.

Smyth, C., Kalsi, G., Brynjolfsson, J., O'Neill, J., Curtis, D., Rifkin, L., Moloney, E., Murphy, P., Sherrington, R., Petursson, H., and Gurling, H., Further tests for linkage of bipolar affective disorder to the tyrosine hydroxylase gene locus on chromosome 11p15 in a new series of multiplex British affective disorder pedigrees, *Am. J. Psychiatr.*, 153, 271, 1996.

Smyth, C., Kalsi, G., Brynjolfsson, J., O'Neill, J., Curtis, D., Rifkin, L., Moloney, E., Murphy, P., Petursson, H., and Gurling, H., Test of Xq26.3-28 linkage in bipolar and unipolar affective disorder in families selected for absence of male to male transmission, *Br. J. Psychiatr.*, 171, 578, 1997.

Souza, F.G.M. and Goodwin, F.K., Lithium treatment and prophylaxis in unipolar depression: a meta-analysis, *Br. J. Psychiatr.*, 158, 666, 1991.

Stine, O.C., Xu, J., Koskela, R., McMahon, F.J., Gschwend, M., Friddle, C., Clark, C.D., McInnis, M.G., Simpson, S.G., Breschel, T.S., Vishio, E., Riskin, K., Feilotter, H., Chen, E., Shen, S., Folstein, S., Meyers, D.A., Botstein, D., Marr, T.G., and DePaulo, J.R., Evidence for linkage of bipolar disorder to chromosome 18 with a parent-of-origin effect, *Am. J. Hum. Genet.*, 57, 1384, 1995.

Stine, C., McMahon, F., Chen, L., Xu, J., Meyers, D., MacKinnon, D., Simpson, S., McInnis, M., Rice, J., Goate, A., Reich, T., Edenberg, H., Foroud, T., Nurnberger, Jr., J.I., Detera-Wadleigh, S., Goldin, L., Guroff, J., Gershon, E., Blehar, M., and DePaulo, R., Initial genome screen for bipolar disorder in the NIMH genetics initiative pedigrees: chromosomes 2,11,13,14, and X, *Neuropsychiatr. Genet.,* 74, 263, 1997.

Straub, R.E., Lehner, T., Luo, Y., Loth, J.E., Shao, W., Sharpe, L., Alexander, J.R., Das, K., Simon, R., Fieve, R.R., Lerer, B., Endicott, J., Ott, J., Gilliam, C.T., and Baron, M., A possible vulnerability locus for bipolar affective disorder on chromosome 21q22.3, *Nat. Genet.,* 8, 291, 1994.

Suarez, B., Harpe, C.L., and Van Eerdeweugh, P., Problems of replicating linkage claims in psychiatry, in Genetic Approaches to Mental Disorders. Proceedings of the 82nd Annual Meeting of the American Psychopathological Association, Gershon, E.S. and Cloninger, C.R., Eds., American Psychiatric Press, Washington, D.C., 23, 1994.

Thompson, G., Identification complex disease genes: progress and paradigms, *Nat. Genet.,* 8, 108, 1994.

Todd, R.D., Lobos, E.A., Parisian, A., Simpson, S., and DePaulo, J.R., Manic-depressive illness and tyrosine hydroxylase markers, *Lancet,* 347, 1634, 1996.

Verga, M., Marino, C., Petronis, A., Cavallini, M.C., Cauli, G., Smeralki, E., Kennedy, J.L., and Macciardi, F., Association of tyrosine hydoxylase gene and neuropsychitric disorders, *Psychiatr. Genet.,* 3, 168, 1993.

Vincent, J.B., Klempan, T., Parikh, S.S., Sasaki, T., Meltzer, H.Y., Sirugo, G., Cola, P., Petronis, A., and Kennedy, J.L., Frequency analysis of large CAG/CTG trinucleotide repeats in schizophrenia and bipolar affective disorder, *Mol. Psychiatr.,* 1, 141, 1996.

Von Knorring, A.L., Cloninger, C.R., Bohman, M., and Sigvardsson, A., An adoption study of depressive disorders and substance abuse, *Arch. Gen. Psychiatr.,* 40, 943, 1983.

Weber, J.L. and May, P.E., Abundant class of human DNA polymorphisms which can be typed using the polymerase chain reaction, *Am. J. Hum. Genet.,* 44, 388, 1989.

Weeks, D.E. and Lange, K., The affected-pedigree-member method of linkage analysis, *Am. J. Hum. Genet.,* 42, 315, 1988.

Weeks, D.E. and Lange, K., A multilocus extension of the affected-pedigree-member method of linkage analysis, *Am. J. Hum. Genet.,* 50, 859, 1992.

Weissman, M.M., Advances in psychiatric epidemiology: rates and risks for major depression, *Am. J. Public. Health.,* 77, 445, 1987.

Weissman, M.M., Gershon, E.S., Kidd, K.K., Prusoff, B.A., Leckman, J.F., Dibble, E., Hamovit, J., Thompson, W.D., Pauls, D.L., and Guroff, J.J., Psychiatric disorders in the relatives of probands with affective disorder, *Arch. Gen. Psychiatr.,* 41, 13, 1984.

Weissman, M.M. and Myers, J.K., Affective disorders in a US urban community: the use of research diagnostic criteria in an epidemiological survey, *Arch. Gen. Psychiatr.,* 35, 1304, 1978.

Wender, H., Kety, S.S., Rosenthal, D., Schulsinger, F., Ortmann, J., and Lunde, I., Psychiatric disorders in the biological and adoptive families of adopted individuals with affective disorders, *Arch. Gen. Psychiatr.,* 43, 923, 1986.

Wildenauer, D.B., Albus, M., Lerer, B., and Maier, W., Searching for susceptibility genes in schizophrenia by genetic linkage analysis, in *Function and Dysfunction in the Nervous System,* Cold Spring Harbor laboratory Press, Cold Spring Harbor, NY, 1996.

Winokur, G., Coryell, W., Keller, M., Endicott, J., and Leon, A., A family study of manic-depressive (bipolar I) disease. Is it a distinct illness separable from primary unipolar depression?, *Arch. Gen. Psychiatr.,* 52(5), 367, 1995.

Winokur, G., Clayton, P.J., and Reich, T., *Manic-depressive Illness,* C.V. Mosby Co., St. Louis, 1969, 112.

Winokur, G., Tsuang, M.T., and Crowe, R.R., The Iowa 500: affective disorder in relatives of manic and depressed patients, *Am. J. Psychiatr.,* 139, 209, 1982

# 4 Genetics of Schizophrenia

*Douglas F. Levinson and Bryan J. Mowry*

## CONTENTS

## DIAGNOSTIC ISSUES

Schizophrenia is a syndrome of unknown pathophysiology characterized by psychotic symptoms such as hallucinations, delusional ideas, disordered speech and thinking, as well as deficits in emotional and social behavior. The modern definition began with Kraepelin[1] who differentiated dementia praecox as a syndrome of psychotic symptoms and chronic deterioration of personality (but not cognitive decline), from manic-depressive or circular insanity with remission between periods of mood disturbance. Eugen Bleuler[2] coined the term schizophrenia ("split between thinking and perception") and emphasized blunted or unusual emotional expressions, abnormal association of ideas, and impaired volition. Schneider[3] noted certain symptoms that continue to be considered suggestive (although not pathognomonic) of schizophrenia, such as beliefs that one's thoughts are being controlled, stolen, or heard by others and hallucinations of people conversing about or commenting on oneself or of one's thoughts heard aloud. Crow,[4] applying a classical neurological concept, suggested that schizophrenia's "negative" symptoms such as blunted emotions and loss of motivation might be due to primary neurological deficit, while "positive" symptoms such as hallucinations and delusions might be due to compensatory mechanisms. Most factor analytic studies[5-7] suggest that there are at least three symptom factors (positive, negative, and disorganized) that could have distinct neural mechanisms, but there is no evidence of as yet that they define etiologic or familial subtypes.

Interpretation of genetic research on schizophrenia is complicated by historical changes in diagnostic practices. Prior to the 1960s, genetic studies[8] were based on descriptive and somewhat subjective diagnostic definitions. Particularly problematic has been the differentiation of schizophrenia from the psychotic forms of mood ("affective") disorders, which usually include "mood congruent" psychotic ideas — e.g., grandiose delusions during mania or delusions of disease during depression — but can include mood incongruent psychosis with irrational thought and behavior resembling schizophrenia. A landmark study[9] trained research teams to make consistent diagnoses and then demonstrated that psychiatrists in the U.S. (but not the U.K.) diagnosed 70–80% of psychotic mood disorders as schizophrenia, prompting the development of more specific diagnostic criteria to increase reliability. The St. Louis criteria[10] were the precursors of the Research Diagnostic Criteria, or RDC,[11] which in turn evolved into the DSM-III,[12] the first comprehensive set of specified diagnostic criteria for clinical use. DSM-IIIR[13] and DSM-IV[14] criteria incorporated further modifications, while the European ICD-10 criteria[15] are similar but not identical. Most modifications have been attempts to deal with three central issues:

1. **Duration.** The RDC require only two weeks of symptoms. From DSM-III through DSM-IV, schizophrenia includes six months of active, prodromal, and residual symptoms, or two years for chronic schizophrenia. Schizophreniform disorder is defined as a schizophrenic syndrome lasting less than six months. The European ICD-10 system, on the other hand, requires that active symptoms should have been present for at least one month and that the course — e.g., continuous, episodic, or in remission — should be specified only after the illness has been observed for at least 12 months.

2. **Core syndrome.** The RDC require two of a list of "core" features of schizophrenia during active periods of illness. DSM-III broadens the diagnosis, e.g., two features such as a combination of any hallucinations and any delusions qualify as the core syndrome (whereas RDC further qualify the required types of psychotic symptoms), and certain single core symptoms are considered sufficient (such as voices conversing with each other or "bizarre," i.e., completely implausible, delusions). DSM-IV further broadens the definition to include, for example, delusions and prominent negative symptoms. ICD-10 resembles DSM-IV in requiring at least one core symptom or two or more other symptoms, and it revives the older concept of simple schizophrenia, defined as the insidious development of negative symptoms and social decline in the absence of active psychosis.

3. **Mood disturbance.** Kraepelin[1] described patients with both dementia praecox and mood disorder, and Kasanin[16] suggested the term schizoaffective disorder. The RDC define schizoaffective disorder as co-occurrence of manic and/or depressive syndromes with core features of schizophrenia. Studies of treatment response and familial patterns of illness suggested that the RDC "mainly affective" subtype, with simultaneous remission of psychosis and mood disorder, resembles primary mood psychoses, and DSM-III categorizes most such cases as mood disorders with mood incongruent psychosis. The RDC define a "mainly schizophrenic" subtype with persistence of psychosis after remission of mood syndrome. It resembles schizophrenia in treatment and family studies and was the basis for the DSM-IIIR and DSM-IV criteria for schizoaffective disorder, which require persistence of psychosis for at least two weeks after remission of the mood syndrome. DSM-IIIR requires that mood syndrome be "not brief" in relation to total duration of active psychosis, while DSM-IV require that mood syndrome be present for a "substantial" portion of the total duration of illness (not just of acute exacerbations). ICD-10 also requires at least a two-week co-occurrence of prominent acute schizophrenia symptoms and mood syndrome. In general, the proportion of schizoaffective patients is highest by RDC and lowest by DSM-IV. It has proven difficult even for highly trained research clinicians to distinguish reliably between schizophrenia and schizoaffective disorder.[17,18] This places the genetics researcher in a dilemma: These two similar disorders can be combined as a phenotypic category, although they have differences as well as similarities, or they can be separated, although there is no reliable way to accomplish

this. Alternatively, schizophrenia-like and mood features could be measured separately using a more dimensional approach, but this has not been tried in any large-scale linkage investigation.

Finally, modern criteria for schizotypal and paranoid personality disorders (STPD and PPD) were developed on the basis of case histories of biological relatives of adoptees who developed schizophrenia as adults.[19] Essentially, these individuals resemble people with schizophrenia but without overt psychotic symptoms, with STPD characterized by oddities of emotion, speech, thinking, and emotional and social behavior, and PPD by a striking degree of suspiciousness and guardedness. Other diagnostic issues include the differentiation of psychosis induced by substance abuse or by neurological disorder.[20] Most molecular genetic studies combine schizophrenia and schizoaffective disorder into a "core" phenotype. Some studies also include "other non-affective psychoses," including schizophreniform disorder, delusional disorder (non-bizarre delusions without other features of schizophrenia) and psychoses not quite meeting criteria for schizophrenia (atypical psychosis or "psychosis not otherwise specified"). A few studies have included a separate analysis, including relatives with STPD with or without PPD, mood disorders with mood incongruent psychosis, or other psychiatric disorders less clearly related to schizophrenia.

## GENETIC EPIDEMIOLOGY

A series of twin, adoption, and family studies suggest that genetic factors confer most of the susceptibility to schizophrenia. Older studies have been reviewed elsewhere,[20-22] and this summary will focus on more recent, larger studies using modern diagnostic criteria. Twin studies most specifically implicate genetic factors, but there have been few modern studies. Older studies demonstrated MZ concordance of about 50% and DZ concordance of 10–15%.[8] McGuffin and colleagues[23] re-analyzed the case material from the Maudsley twin sample using DSM-IIIR criteria and reported an estimated broad heritability of over 0.8 for either RDC or DSM-IIIR schizophrenia. The ratio of MZ to DZ concordance was slightly increased by including a broader set of psychoses as a single phenotype.[24] A small new twin study[25] reported that concordance for DSM-IIIR schizophrenia was 85.7% for 22 MZ and 25% for 23 DZ twin pairs, higher estimates than in other studies. Taking an entirely different approach, Cannon et al.[26] determined registry-based clinical diagnoses for all twins born in Finland between 1940 and 1957 and estimated heritability of 0.83 for schizophrenia, identical to that reported in the Maudsley study. In general, methodological criticisms of twin studies have not been consistently supported by evidence. For example, Klaning et al.[27] reported that twins in Denmark have a higher incidence of schizophrenia than the general population, but Kendler et al.[28] found no such increase for any psychotic or affective disorder in Swedish twins. A small study suggested that MZ concordance was higher for monochorionic than dichorionic MZ twin pairs, which could implicate non-genetic factors such as infection,[29] but this has not been replicated.

Two Danish adoption studies played a pivotal role in directing the attention of modern psychiatry from psychoanalytic to genetic theories of schizophrenia. Kety et al.[30] studied the biological relatives of schizophrenic vs. control adoptees and found that the former had an increased risk of "schizophrenia spectrum" disorders, including non-affective psychoses and what are now termed STPD and PPD. Rosenthal et al.[31] reported that adopted-away offspring of a schizophrenic parent had an increased risk compared with adopted-away offspring of well parents. The original case material from each of these studies was subsequently subjected to re-diagnosis by DSM-III criteria and the conclusions were confirmed.[32,33] Tienari et al.[34] replicated the latter cross-fostering design in Finland using modern criteria and reported that 9% of index adoptees and 1.1% of control adoptees developed schizophrenia. He also reported that severe disorganization in the family environment was associated with development of overt schizophrenia in those at genetic risk, a finding that has not been tested elsewhere.

Large samples of relatives of different proband groups have been possible in family studies in which first-degree relatives of probands are given semi-structured research interviews linked to diagnostic criteria. Gottesman and Shields[8] reviewed studies prior to the use of specified diagnostic criteria and found a 10–13% lifetime risk of schizophrenia in first-degree relatives vs. about 1% in the general population, with similar risks in offspring, parents, and siblings. The risk to second-degree relatives was 2–3%. Studies using RDC and DSM-III have reported a 3–8% risk in first-degree relatives of schizophrenia probands vs. 0.2–0.6% in control relatives,[35-39] with one study finding a risk of 16.9%.[40] These studies also demonstrated an increased risk of schizophrenia in relatives of schizoaffective probands (about 3–8%), primarily those with RDC mainly schizophrenic or chronic subtypes. A greater risk of mood disorders was also observed, suggesting that schizoaffective disorder is more heterogeneous.

Table 4.1 shows the results of the three large family studies completed in the last decade, each with a large sample of relatives interviewed for each proband group, and with most relatives interviewed rather than diagnosed by records or others' reports. Again, lifetime risk of schizophrenia was 3–8% in first-degree relatives of schizophrenia probands and 0.5–1.1% in relatives of control probands. Also shown are results of two recent reports of follow-up studies of offspring of a schizophrenic parent, and again increased risk is confirmed. Several other patterns of results are noteworthy:

1. None of the studies found increased bipolar disorder in relatives of schizophrenia probands or increased schizophrenia in relatives of bipolar probands (data not shown), suggesting that these two disorders do not lie on a single genetic continuum.
2. Psychotic disorders resembling, but not quite meeting, criteria for schizophrenia (variously termed probable schizophrenia, other non-affective psychoses, or atypical psychoses) co-aggregate with schizophrenia in families.[20,24,30,32,39,41,43]
3. Schizophrenic and schizoaffective disorders co-aggregate in families in all studies (indeed, risks of both disorders are similarly elevated, regardless of which disorder is observed in the proband). Although it appeared from earlier studies that this increase is limited to probands with RDC mainly schizophrenic subtype (similar to DSM-IIIR schizoaffective disorder), Maier et al.[46] did not find this to be the case.
4. Two studies (Gershon[41] and Maier[46]) reported that relatives of schizophrenic probands are at increased risk of unipolar major depressive disorder. This was not observed in previous studies using RDC or DSM-III, nor by Kendler et al.[44]
5. Results regarding familial co-segregation of schizophrenia and psychotic mood disorders are also unclear. An increased risk for psychotic mood disorders was reported in relatives of schizophrenic probands by Kendler et al. but not by Maier et al., and not in the offspring of a schizophrenic parent.[49] Kendler et al.[43] also reported that DSM-IIIR schizoaffective disorder was increased in relatives of probands with psychotic mood disorders, particularly bipolar disorders. Thus, the relationships among schizophrenic, schizoaffective, and mood disorders have yet to be fully clarified, but they appear to be more complex than either the dichotomy proposed by Kraepelin or a simple continuum model.

Studies using DSM-III/IIIR criteria and in-person semi-structured interviews suggest that first-degree relatives of schizophrenia probands have an increased risk of STPD and of PPD[37] or of the combined incidence of these two disorders,[50] while studies in which few relatives were interviewed in person had both positive[41] and negative[51] results. These findings confirmed the results of the two early adoption studies, which were updated by re-analyzing data with DSM-III criteria.[32,33] Table 4.2 summarizes the results of two recent large family studies in which most relatives were personally interviewed. Kendler et al.[52] reported an increase in STPD and PPD in the relatives of both schizophrenic and (to a lesser degree) schizoaffective probands, while Maier et al.[53] reported an increase of STPD in both groups but an increase in PPD only in the relatives of depressed probands.

**TABLE 4.1**
**Recent Family Studies of Schizophrenic and Schizoaffective Disorders**

| First Author, Year | Diagnostic Criteria | MR of SCZ in Relatives of: | | | | MR of SA in Relatives of: | | | | MR of MDD in Relatives of: | | | |
|---|---|---|---|---|---|---|---|---|---|---|---|---|---|
| | | SCZ | SA | MDD | NL | SCZ | SA | MDD | NL | SCZ | SA | MDD | NL |
| *Recent Large Family Studies* | | | | | | | | | | | | | |
| Gershon, 1988[41 a] | RDC | 3.1* | 2.5* | — | 0.6 | 5.0* | 2.5* | — | 0.6 | 16.0* | 18.1* | — | 7.0 |
| Kendler, 1993[42-44], 1995[45 b] | DSM-IIIR | 8.0*† | 0.6 | 1.1 | 3.0† | 2.4 | 2.4 | 1.1 | 21.1 | 44.9‡ | 24.5 | 21.4 | |
| Maier, 1993[46 c] | RDC | 5.2† | 4.0† | 0.6 | 0.5 | 5.2† | 8.0† | 0.6 | 0.5 | 20.7† | 20.3† | 21.6† | 10.6 |
| *Recent Studies of At-Risk Offspring* | | | | | | | | | | | | | |
| Parnas, 1993[47] | DSM-IIIR | 16.2* | 4.6* | — | 1.9 | — | — | — | — | — | — | — | — |
| Erlenmeyer-Kimling, 1997[48 d] | Unclear | 13.1*† | — | 0 | 0 | 1.2 | — | 6.0 | 0.7 | 27.4 | — | 29.9 | 27.2 |

*Note:* Numbers represent morbid risks (%) to first-degree relatives (family studies) or offspring of probands in each group. RDC = Research Diagnostic Criteria, SA = schizoaffective disorder, SA-S-RDC = mainly schizophrenic subtype, SA-A-RDC = mainly affective subtype, SCZ = schizophrenia, MDD = major depressive disorder, NL = normal.

MRS that are statistically significantly elevated *vs.* another group of relatives (p < 0.5) are indicated as follows: * = Relatives of NL probands, † = Relatives of AD probands, ‡ = Relatives of SCZ probands.

[a] Compared across groups was the risk of SCZ + SA + other psychoses + other hospitalized disorders.

[b] Risk of SCZ in relatives of MDD probands is for all non-psychotic affective illness probands, and risk of MDD for MDD probands is non-psychotic non-bipolar disorder in non-psychotic affective disorder probands. (MDD is not reported as a separate category.)

[c] Totals for SA estimated from data for unipolar and bipolar SA.

[d] Proband group listed here as MDD actually includes all mothers with affective illness; risks given for MDD include psychotic and non-psychotic MDD.

**TABLE 4.2**
**Recent Family Studies of DSM-IIIR Schizotypal and Paranoid Personality Disorders**

| First Author, Year | Prev. of STPD in Relatives of: | | | | Prev. of PPD in Relatives of: | | | |
|---|---|---|---|---|---|---|---|---|
| | SCZ | SA | AD | Cont | SCZ | SA | AD | Cont |
| Kendler, 1993[52] | 6.9* | 2.8* | 2.4 | 1.4 | 1.4* | 2.9* | 0.8 | 0.4 |
| Maier, 1994[53] | 2.1* | 2.0* | 0.7 | 0.3 | 1.7 | 2.0 | 2.9* | 0.9 |

*Note:* Prev = lifetime prevalence (shown in %). STPD = schizotypal personality disorder; PPD = paranoid personality disorder; AD = affective disorder (all AD for Kendler, MDD for Maier); Cont = controls. Other abbreviations as in Table 4.1.

* Prevalence for this group is > for relatives of control probands (p < .05).

Erlenmeyer-Kimling et al.[49] found an increase in STPD, PPD, and schizoid personality disorders combined in the offspring of schizophrenic as well as affectively ill mothers.

Non-genetic factors appear to play some role in the etiology of schizophrenia, but it is unclear which factors are significant, whether they interact with genetic factors, or whether there is a subset of non-genetic cases. A preponderance of studies suggests an excess of winter births among individuals with schizophrenia, although the effect is seen for major mood disorders as well.[54] This finding is generally interpreted as supporting the hypothesis that viral infection early in pregnancy, or possibly immunological response to such infection, could play an etiologic role. Pulver et al.[55] have suggested that winter birth interacts with familial loading and that infectious or immunological factors may interact with familial-genetic mechanisms. Numerous studies have examined whether maternal exposure to the influenza virus increases risk of schizophrenia in offspring, but results have been contradictory,[56,57] and the one study in which exposure of individuals was documented (rather than exposure of populations) found no effect.[58] Numerous studies also suggest an increase in history of birth complications in subjects with schizophrenia, possibly interacting with genetic factors,[59] but one recent multicenter analysis that supported a role of birth complications in schizophrenia and earlier age of onset failed to find any relationship with family history.[60] One recent meta-analysis of published reports found very modest effects of birth complications on risk of schizophrenia, with an increased effect in offspring of mothers with schizophrenia,[61] and another meta-analysis found evidence for publication bias inflating the reported effects.[62] An interaction of genetic and unknown environmental factors comes from studies of African-Caribbean immigrants to the United Kingdom that report a significantly higher morbid risk for schizophrenia in siblings of psychotic probands among second-generation African-Caribbeans in the United Kingdom probands, compared with relatives of first-generation African-Caribbeans or whites.[63]

Thus, although significant issues remain to be resolved, the preponderance of evidence suggests that schizophrenia and schizoaffective disorder are genetically related disorders and that a spectrum of related psychotic and personality disorders are also more common in relatives of schizophrenic probands.

## MODE OF TRANSMISSION

No modern segregation analysis has been able to resolve the mode of transmission of schizophrenia, although a polygenic-multifactorial model is most plausible.[21] Single major locus models cannot be supported; to account for the modestly increased risk to siblings would require an estimate of penetrance that greatly underestimates the risk to MZ twins.[64] Similarly, the highly observed MZ:DZ concordance ratio is most consistent with polygenic inheritance.[24] Risch[64] applied the James[65] model of the relationship of mode of transmission to the relative risk of first-, second-, and third-degree relatives (i.e., the risk of illness in each group vs. the risk in the general population), and fit models of multigenic transmission to the older European data summarized by Gottesman and Shields,[8] which is the only source of data for more distant relatives. The best-fitting model included 3–5 interacting genetic loci, none of which increased risk of illness in siblings by more than two- to three-fold, and with allowance for additional minor loci. Linkage studies were viewed as the most direct test of the hypothesis that there might be major locus effects in multiply-affected families.[66] The failure of multiple genome scans to detect replicable major locus effects (as discussed below) supports the conclusion from genetic modeling that genetic susceptibility to schizophrenia is usually due to multigenic transmission. And if major locus effects exist, they are too rare to detect in current samples.

Molecular studies would be facilitated by the identification of subtypes or dimensions of illness that could be analyzed separately. Although there are clinical subtypes such as paranoid, disorganized, and deficit (negative) symptom schizophrenia that can be distinguished by clinical course, none has proven to be familial,[67] nor does age of onset predict familial risk.[68] Sibling pairs show modest correlations in symptoms such as affective features[69] and ratings of positive, negative, and affective symptoms.[69] Kendler et al.[70] recently presented a latent class analysis of symptom ratings for schizophrenic and mood disorder probands in their large Irish family study, and they identified several symptom clusters. One syndrome ("hebephrenic," an old term for a disorganized form of schizophrenia), found in about 5% of probands, was associated with a greatly increased familial risk of schizophrenia (16%) compared with all other types. This syndrome was characterized by severe core features of schizophrenia as well as mania-like symptoms, which would lead to a schizoaffective diagnosis in some cases by current criteria. Interestingly, Liang and Pulver[71] had previously reported that the greatest familial risk of schizophrenia (by family history report) was observed for schizophrenic probands with some evidence of a manic syndrome by interview or records. However, it remains to be determined whether this finding can be replicated. A further issue is whether the genetic loci that contribute to multigenic transmission of schizophrenia vary across populations with different genetic histories. Terwilliger,[72] reviewing the work of Weiss,[73] has argued that genetic loci underlying complex traits frequently vary across populations and that the same is likely to be true for psychiatric disorders. Some studies of schizophrenia have been initiated in relatively isolated populations in part for this reason. On the other hand, Risch[74] has pointed out that in the absence of familial clinical subtypes or of known population differences in incidence of schizophrenia, it is quite possible that the most common multigenic mechanism(s) for schizophrenia will be found in most populations.

It is beyond the scope of this chapter to review hypotheses concerning possible component psychological or physiological traits that might be more closely linked than a schizophrenia diagnosis to individual genetic loci. An example of this approach is described subsequently.

## CANDIDATE GENE STUDIES

The nomination of "candidate" genes is on shaky grounds when the pathophysiology of a disorder is completely unknown. Nevertheless, many investigators have used case-control or haplotype relative risk (transmission disequilibrium) methods to study the association between schizophrenia and loci related to monoamine receptors or metabolic pathways, CNS-related peptides, other neuropsychiatric disorders, and trinucleotide repeat expansion mechanisms. These studies are fraught with all of the problems common to association studies. Sample sizes are typically small, although some larger studies are beginning to appear. Interpretation of statistical significance is difficult: One can calculate the statistical effect of searching for linkage throughout the genome (because linkage results are non-independent across extended distances),[75] but for assocation studies in heterogeneous or old populations, all but the most tightly linked loci are essentially independent so that many very low p values will be observed by chance given the thousands of loci that may need to be studied.[76-78] Yet, investigators who study many loci in relation to a particular disease tend to report one or a few loci at a time and interpret very modest p values (0.05 or 0.01) as "significant." Further, as is the case for linkage, the field as a whole can be expected to study all plausible candidate loci repeatedly, so frequent false-positive results can be expected. More realistic discussion of statistical interpretation would be useful in this area. Finally, population stratification can produce false-positive linkage results within samples. Kidd and colleagues[79] have adopted a particularly elegant approach to studying the distribution of tightly linked haplotypes around candidate genes for psychiatric disorders in populations throughout the world, which facilitates the identification of false-positive results based on patterns of association that are inconsistent with the known linked haplotypes. In general, then, "significant" association results from small samples should be viewed with caution, and findings should be replicated across multiple samples or subjected to testing in large or collaborative samples.

## LOCI RELATED TO DOPAMINERGIC NEUROTRANSMISSION

There are at least five human dopamine (DA) receptors. DRD1 has been mapped to chromosome 5q34-35,[80] DRD2 to 11q22-23,[81] DRD3 to 3q13.1,[82] DRD4 to 11p15.5 (pter near Hras),[83] and DRD5 to 4p16.1.[84] Classical or "typical" antipsychotic drugs act by blocking D2 receptors post-synaptically, so that this receptor has received considerable attention. However, D2 blocking agents are hardly specific to schizophrenia, are partially or minimally effective in a large proportion of cases, and probably have more reliable therapeutic effects for psychosis caused by apparently unrelated disorders such as acute mania, Alzheimer's disease, and acute delirium associated with medical disease. The newer "atypical" neuroleptics appear to have increased efficacy for schizo-phrenia and have blocking activity at other DA receptors as well as serotonergic receptors. Thus, all of these receptors and their regulatory regions have been considered candidate genes.

DRD3 is the only DA receptor for which there is substantial positive evidence for an association with schizoprenia. Crocq et al.[85] described biallelic *Bal* I and *Msc* I RFLPs within this locus and reported that schizophrenia was significantly associated with homozygosity at these sites. This group[86] later further characterized these two sites, which are 40,000 base pairs apart, and reported a weak association only with the *Bal* I site. Of the case-control studies that have produced some form of positive evidence for association at the *Bal* I site, the observed effects have been incon-sistent: Associations have reported between schizophrenia and DRD3 *Bal* I homozygosity (geno-types 1-1 or 2-2 vs. 1-2);[85-87] 1-1 homozygosity;[88] 2-2 homozygosity;[89] excess allele 1 but, only in the subgroup with positive family history;[90] excess allele 1, but only in one of two case samples and in comparison with one of two control groups;[91] excess allele 1, but only in adult-onset and not childhood-onset cases;[92] and excess 2-2 in cases with tardive dyskinesia.[93] There have also been reports of weak association between the *Msc* I site and schizophrenia[94] or delusional disorder.[95] There have been at least 14 negative studies.[96-109] A possible explanation for this pattern of results would be that the positive reports are false-positives resulting from multiple testing of diverse hypotheses and clinical subgroups.

However, two recent large analyses do lend more substantial support to an association at the *Bal* I site. Spurlock et al.[87] reported that analyses of 311 patients and 306 controls from the large European Multicentre Association Study of Schizophrenia (EMASS) supported an excess of any homozygosity (p = 0.003) and of the 1-1 genotype (p = 0.004). Williams et al.[110] performed a meta-analysis of published case-control studies totalling 5,351 individuals and found an excess of any homozygosity (p = 0.0009) with an odds ratio of 1.21 and no evidence of cross-sample heterogeneity or of a publication bias. These same authors reported an analysis of 57 parent/proband trios using the transmission disequilibrium test, which showed an excess of any homozygosity (p = 0.004) and of the 1-1 genotype (p = 0.02). Confirmation of an effect in two multicenter analyses, plus a TDT analysis (which is less prone to artifact than case-control analyses), provide support for the interpretation that the inconsistent results across studies are due to a weak association that cannot be detected in every sample. Alternatively, the results could represent a false-positive given the modest level of significance in relation to the multiple tests performed in every analysis and the multiple loci that are being tested in the field.

At the DRD2 locus, Itokawa et al.[111] reported a missense mutation resulting in a Ser311→Cys mutation within the receptor which was non-significantly more common in 50 schizophrenic patients vs. 110 controls (frequency of 0.04 vs. 0.023). This group later reported that in 291 schizophrenic, 78 mood disorder, and 579 control subjects, the variant was more common in schizophrenic patients without negative symptoms ((17.1%, p < 0.00001) and mood-incongruent psychotic mood disorders (33.3%, p < 0.0001), but not in schizophrenics with negative symptoms (5.7%, p = 0.46) compared with controls (4.1%).[112] A family was also reported in which a 1q:11q translocation was associated with schizophrenia, with the 11q breakpoint near the DRD2 region.[113] There have been numerous negative studies including association studies of the Ser311C mutation and linkage analyses of this mutation or of other nearby chromosome 11 markers.[87,114-133] It therefore seems unlikely that there is a pathogenic locus in this region.

There have been a number of negative association and linkage studies of the DRD1 receptor region.[134-136] DRD4 has attracted interest because the effective atypical antipsychotic drug clozapine binds with highest affinity to this receptor.[137] A number of expressed polymorphisms have been reported, but association and linkage studies have been negative for schizophrenia[138-150] except for one report of an excess of the seven-repeat allele of a 48-base-pair polymorphism[138] in schizophrenic, schizoaffective, and mood disorder subjects vs. controls.[151] A drug with D4 antagonist activity appeared to worsen schizophrenia, although a full range of dosages was not studied.[152] At DRD5, Williams et al.[153] reported an excess of an allele of a microsatellite marker in 97 schizophrenic vs. 97 control subjects, but the modest effect (p = .019) is probably best explained by multiple testing. Two other studies of the D5 region have been negative.[154,155] The **tyrosine hydroxylase** locus is in the same region as DRD4 and thus was studied in many of the negative reports of that region cited above. Meloni et al.[156] reported that a rare 10-tetranucleotide-repeat allele of the TH01 microsatellite within the TH gene was found in 5–9% of schizophrenic patients in two samples and no controls. Thibaut et al.[157] reported that this allele was associated with reduced monoamine metabolite levels. Burgert et al.[158] failed to find any association between this allele and schizophrenia. Negative findings have been reported at the **dopamine transporter**[159-166] and **dopamine beta hydroxylase**[167] loci.

## Loci Related to Serotonergic Transmission

Inayama et al.[168] reported a positive association between schizophrenia and a T102C polymorphism in the **5HT2A** (serotonin type 2A) receptor locus on chromosome 13q14.3. A negative study followed,[169] and negative linkage results had previously been reported in this region in a large multiplex pedigree.[170] Williams et al.[171] reported a positive association with allele 2 of this polymorphism (p = 0.003, odds ratio 1.3) in a multicenter case-control study of 571 schizophrenic and 639 control subjects. Spurlock et al.[172] reported positive evidence (p = 0.006) in 63 parent/proband trios. Williams et al.[173] reported a meta-analysis of published studies and found weak evidence for an association between schizophrenia and allele 2 (p < 0.001, odds ratio 1.16). However, this is a silent polymorphism, and efforts to identify mutations in this region associated with schizophrenia have so far failed.[174-176] A number of other serotonin-related loci have also been studied with generally negative results.[177-180]

## Other Candidate Genes

No consistent pattern of association has been demonstrated between schizophrenia and a number of loci related to **monoamine metabolism,**[181-186] **excitatory amino acid** neurotransmitters,[187-191] **phospholipase,**[192,193] classical **blood markers**[194] including **HLA** alleles,[195-201] **amyloid precursor protein**[202-210] or **apolipoprotein E,**[211] or **interleukin.**[212-214] A null mutation in the **ciliary neu-**

**rotrophic factor** (CNTF) gene was found to be associated with schizophrenia in one of two samples,[215] but evidence was negative in a subsequent study[216] and association was not observed for the **neural cell adhesion molecule** locus.[217] A positive association was observed at the **neurotrophin-3** locus,[218] but subsequent studies were negative.[219,220] There is one positive report for the **histamine-2** receptor gene.[221] Several studies have reported weak evidence of differences in **mitochondrial DNA,** but these studies appear to be in a very early stage.[222,223] A family was reported with co-segregation of schizophrenia and **Marfan's syndrome,**[224] but no linkage was found in the region.[225] There are reports of families with apparent co-segregation of **Becker's**[226] and **Duchenne's**[227] **muscular dystrophy**, but no mutations were observed in the **dystrophin gene** in schizophrenic subjects.[228] There was a report of linkage disequilibrium between schizophrenia and the **spinocerebellar ataxia-1** locus,[229] which is in a region of chromosome 6p that is of interest as discussed below, and another report of a mutation in this locus in an affected sibling pair and an additional affected individual,[230] but no association was observed in a subsequent study.[231] And this locus has not been the site of reported positive evidence for linkage on chromosome 6p. Weak evidence for linkage at the **androgen receptor** locus was reported in one study[232] but not in another.[233] No linkage was observed at the locus for **Darier's disease**.[234] A negative association has been reported between schizophrenia and **rheumatoid arthritis** in numerous older studies,[235,236] although a recent study found no decrease in risk of schizophrenia in a large sample of subjects with rheumatoid arthritis.[237] Wright et al.[238] reported decreased frequency of the HLA DRB1*04 allele (which is positively associated with rheumatoid arthritis) in subjects with schizophrenia and in mothers of a separate sample of schizophrenia patients. An association was also recently reported between **Rh incompatibility** (in second- or later-born offspring) and risk of schizophrenia in a large sample.[239]

## ANTICIPATION AND TRINUCLEOTIDE REPEAT EXPANSIONS

A number of investigators have searched for trinucleotide repeat expansions that might contribute to susceptibility to schizophrenia, stimulated by the discovery that several degenerative neuropsychiatric disorders are caused by this type of mutation.[240] One clinical characteristic of these disorders has been termed "anticipation," i.e., one observes increasing severity of illness and declining age of onset in successive generations as the unstable site expands further, typically with a more robust effect for paternal meioses. Evidence concerning anticipation in schizophrenia has been reviewed.[241,242] Several authors have presented evidence for anticipation for age of onset alone,[243-248] for age of onset and severity,[249] for an age of onset effect associated with paternal transmission,[250] and for anticipation of age of onset and a paternal transmission effect for severity.[251] Negative data have also been reported,[252,253] and many authors have presented concerns that the apparent age of onset effect could be due to ascertainment bias.[248,250,253] For example, early-onset schizophrenia tends to be a severe form of disease and is presumably therefore more likely to be ascertained, but an ill parent of such a subject is likely to have had a later onset because in very early onset cases the social relatedness that leads to procreation is less likely to be present. Anticipation has been reported to be strongest for early-onset offspring[250] and to be absent for early-onset parents.[253] The patterns of age at onset and familial risk are complex and might not be related to genetic factors directly determining disease liability.[254] Methods of correcting for bias in analyses of anticipation were discussed by Huang and Vieland,[255] but they were not found to be entirely satisfactory, i.e., it is difficult to separate the effects of anticipation and ascertainment bias for schizophrenia.

Methods for detecting long triplet repeat sequences[256] have been applied to schizophrenia, and positive studies have been reported,[257,258] although other studies using similar methods[259] or probing for specific repeat loci[260-262] were negative, as were studies searching specifically for very long expansions in schizophrenia.[263,264] Sidransky et al.[265] recently reported that up to 94% of the trinucleotide repeat expansions observed in humans represented common expansions, not associated with psychiatric disease, at two specific loci on chromosomes 17q and 18q with characteristic repeat lengths. Thus, future studies will have to focus on determining whether there are other loci at which expansions are associated with schizophrenia.

## CONCLUSIONS

Generally, candidate gene studies have failed to yield consistently replicable findings. Multicenter studies and meta-analyses lend some support to findings of association between schizophrenia and polymorphisms within the DRD3 and 5HT2A loci, although the magnitude of the observed effects is not sufficiently robust to be accepted as proof of association given the large number of loci that are being studied.

# LINKAGE STUDIES

Over the past ten years, many research groups have collected samples of families with multiple cases of schizophrenia-related disorders and initiated genetic linkage studies. Many of these projects began with the goal of completing a genome scan, i.e., of genotyping each subject using a set of DNA marker loci distributed relatively evenly on every chromosome, to permit linkage analyses to identify one or more regions likely to contain a susceptibility locus. Remarkably, it has taken most of the decade to complete this first generation of studies. Many factors contributed to the slow rate of progress, including the scarcity of eligible families, the resistance of many subjects in such families to participate in research (which appears to be a greater obstacle for schizophrenia than for most psychiatric or medical disorders), the labor-intensive nature of the recruitment and assessment process (similar to other major psychiatric disorders), and the many changes in laboratory and statistical methods that, while increasing the power of the studies, have often required time-consuming adjustments. These studies are just now reaching fruition.

Most of the early schizophrenia genome scan projects were designed to test the hypothesis that major locus effects were most likely to be observed in multiply-affected families. Planned sample sizes were considered adequate to detect a locus of major effect segregating in most or at least a substantial minority of families. The failure of these studies to detect a major locus effect places schizophrenia in the same position as many genetically complex disorders: There is a growing number of suggestively positive linkage findings whose interpretation is unclear, none of which can be considered clear and replicated proof of linkage. It would be valuable to have much larger samples to detect loci that produce small increases in risk of developing schizophrenia and to narrow the regions of interest sufficiently to facilitate identification of the relevant genes. Pragmatically it will require some years of collaborative effort to collect such samples. Meanwhile, a substantial body of schizophrenia linkage data has been collected, and it is quite possible that regions containing important pathogenic loci have already been identified. However, it remains possible that all regions identified so far are false-positives. Here we review the most promising findings and then discuss their implications for future research.

## MAJOR SCHIZOPHRENIA LINKAGE STUDIES

It may be useful to identify the major ongoing studies to assist in following the literature for new results. Many of these projects are only now beginning to report their genome scan results. At the time of this writing, nine completed genome scans had been published or were in press,[266-275] results from an almost-completed scan were discussed at an international meeting,[276] a group has presented positive findings from an ongoing scan,[277] a partial scan of a single large pedigree has been published[278] (although data were considered essentially negative), and two additional scans have been presented, but details of linkage analyses are not yet available.[279,280] Some characteristics of eleven of these projects are summarized in Table 4.3, omitting the three for which comparable data are not yet available.[278-280]

**TABLE 4.3**
**Schizophrenia Genome Scan Projects**

| Group | Final Report | # of Pedigrees | Diagnostic Models Tested | # of Tests per Model | 2p | 5q | 6p | 6q | 8p | 9q | 10p | 10q | 13q | 22g | Other Positive Regions |
|---|---|---|---|---|---|---|---|---|---|---|---|---|---|---|---|
| Utah[266] | Y | 9 | Scz/Sa | 3 | X | | | | | | | | | x | |
| Kiel[267] | Y | 5 (+f/u in 65[a]) | Scz/Sa | 1 | | | x | | | x | | x | | | 2q, 4q |
| NIMH/Millennium | | | | | | | | | | | | | | | |
| Eur-Amer[269] | Y | 43 | Scz/Sa-D | 1 | | | | x | | | X | x | | | 2q |
| Afr-Amer[270] | Y | 30 | Scz/Sa-D | 1 | | | | | | | | | | | 15q |
| Johns Hopkins[271] | Y | 54 {+f/u in 51[b]) | Scz/Sa | 1+2[g] | | | | | x | x | | | XX | x | 7q, 14p |
| SUNY/AxyS[272] | Y | 70 | Scz; +Sa; +Stpd/Ppd/Onap | 3 | x | | | | X | x | | x | x | x | 4p, 1g, 11p, 5p, 16q |
| Utah/Palau[273] | Y | (isolate) | Scz/Sa | 2 | | | | | | | | | | | |
| Univ Wales[274] | N | 100 (+f/u in +100[c]) | Scz/Sa | 1 | | | | | | | | x | | | 18p |
| Med Coll Virginia[275] | N | 265 (in thirds[d]) | Scz/Sa; 3 Broader models[f] | 4 | | X | XX | | x | | x | | | | |
| Bonn[276] | N | 72 | Scz/Sa | 2 | | x | x | | | | x | | | x | 18p |
| NIMH/U Chicago[277] | N | 53 (+f/u[e]) | Scz/Sa | 1 | | | | X | | | | | | | |

*Note:* Shown are the chromosomal regions where each study observed the most positive linkage results. XX = result with nominal p < 0.00002 (without correction for multiple tests); X = result with nominal p < 0.001; x = most positive results in genome scan (above a threshold identified by authors, or selected from a longer list of "positives").

a. Five moderate-size Icelandic pedigrees studied with genome map; 65 additional pedigrees from multiple datasets studied with 10 most positive markers.

b. Genome scan completed in 54 pedigrees; most positive results followed up in a separate sample of 51 pedigrees.

c. Genome scan (20 cM) completed in 100 affected sib pairs (all possible pairs), and follow up of regions with nominal p < .05 in an additional 100 pairs.

d. Genome scan completed by typing each marker using one-third of pedigrees, so that each one-third was scanned every 30 cM; largest positives studied in all pedigrees.

e. Genome scane progressing in 53 independent affected sib pairs, and positives (such as 6q) studied in an additional 87 sib pairs from the NIMH Genetics Initiative dataset, and in 54 sib pairs (43 pedigrees) from the US/Australia dataset.

## REVIEW OF LINKAGE FINDINGS

Results are presented first for proximal chromosome 5q, the first reported schizophrenia linkage that was not replicated, and chromosome 22q, the first region with suggestive evidence for linkage for which some corroborative evidence has been observed. Next, the strongest linkage findings to date, on chromosomes 6p, 8p, 13q, and 10p, are presented in approximate historical order to give a sense of the progress of efforts in this field. Finally, the other regions shown in Table 4.3 are discussed, i.e., those for which there is at least one report with p < 0.001, plus another positive report, or those with three or more modestly positive reports.

**Proximal chromosome 5q.** Following up on a report of a balanced translocation associated with schizophrenia in one family, Sherrington et al.[281] reported a peak LOD score of 4.33 at D5S76 on chromosome 5q11-13 in a series of Icelandic pedigrees, with the score rising to above 6.0 when the phenotype was extended to include most psychiatric disorders. Many investigators subsequently reported that the finding could not be replicated, as reviewed by McGuffin et al.,[282] and this region has not been implicated by more recent studies except for one report by Silverman et al.[283] of a multipoint LOD score of 4.37 on chromosome 5p11-13 (within 10-15 cM of the original 5q finding) in one family from rural Puerto Rico using a schizophrenia spectrum model, although with negative results in another 23 families. It seems unlikely that any common locus for schizophrenia exists in this region.

**Chromosome 22q.** Part way through a preliminary genome scan in 39 multiply-affected small pedigrees, Pulver et al.[284] reported that their most positive LOD score was at IL2RB (22q13), with a Zmax of 1.54 in a screening analysis and 2.82 after maximization of the LOD score over several parameters. They later reported that the maximum evidence for linkage in an expanded sample of 57 pedigrees was observed at CRYB2 (about 11 cM centromeric to IL2RB) with p = .009 (Sibpal).[285] Two large multicenter studies were organized to study this region. The first studied a combined sample of 256 pedigrees with negative results at three markers (D22S268, IL2RB, and D22S307).[286] Gill et al.[287] then organized a collaboration of eleven groups and reported a maximum LOD score (MLS) of 1.79 (p = .004) in 620 affected sibling pairs at D22S278. Both multicenter analyses included the original Pulver sample; in the Gill et al. study, in the six samples that had not studied this region previously, a chi-square analysis of proportion of alleles shared IBD produced p = 0.03. Positive results were reported separately in two of the samples that were included in the latter collaboration: a sib-pair analysis with p = 0.019 at D22S278 in 105 pedigrees,[289] a Zmax of 1.45 (one of the largest scores observed in a genome scan of 9 pedigrees),[266] and a Zmax of 2.09 at D22S276 using a denser map of the region,[281] both slightly distal to D22S278. There were two positive tests for linkage disequilibrium at D22S278[290,291] and one negative report.[292] Linkage and LD data for chromosome 22q have been reviewed elsewhere.[293]

The initial linkage results prompted Pulver to focus on chromosome 22 abnormalities that were associated with psychosis. Velo-cardio-facial syndrome (VCFS) is a syndrome of characteristic facies, cleft palate, cardiac anomalies, and learning disability, sometimes accompanied by psychotic or mood disorders.[294] VCFS and DiGeorge syndrome are variants with similar deletion regions of about 3 mB on chromosome 22q11.2,[295] at least 20 cM centromeric from D22S278 but less than 10 cM from CRYB2. Several series of VCFS patients have been studied for psychiatric disorders. Pulver et al.[296] found that 4 of 14 had schizophrenia or schizoaffective disorder. Murphy and Owen[297] reported that 10 of 40 (25%) met DSM-IV criteria for schizophrenia or schizoaffective disorder, 13% for major depression, and 3% for rapid-cycling bipolar disorder. Bassett et al.[298] have described 10 patients with schizophrenia or schizoaffective disorder with confirmed 22q11 deletions, of whom 8 were identified from 15 psychiatric patients referred because they also had clinical features suggesting VCFS. These latter authors also provide a useful review of the literature in this area. Karayiorgou et al.[299] found microdeletions in the VCFS region in 3 of 207 consecutively ascertained schizophrenia probands, and in 0 of 312 controls. To complicate matters, cycling mood disorders were the most common psychiatric finding in one sample of VCFS patients, although the deletion regions did not differ between patients and without psychiatric disorders.[300] There are two reports of positive evidence for linkage of bipolar disorder on chromosome 22q.[301,302] Note that a second VCFS deletion region has been reported on chromosome 10p[303] which may be implicated in schizophrenia (see Table 4.3 and subsequent material).

It is not yet known whether the weakly positive linkage results reported by several groups in this region are in fact related to a locus within the VCFS deletion region, whether there is a second schizophrenia-related locus on distal chromosome 22q, or whether these were false-positive results that simply directed the attention of the field to this syndrome. No psychosis-related mutations have been described in genes within the VCFS deletion region. However, it seems likely that deletions in this region represent an under-recognized cause of schizophrenic symptoms, that some gene or combination of genes within this region is causally implicated in schizophrenia in a small proportion of patients, and that discovery of these genes and the mechanism of their association with psychosis could play an important role in this field.

**Chromosome 6p.** Preliminary evidence for linkage on chromosome 6p was originally observed in the Medical College of Virginia Irish High-Density Family Study,[304] and Straub et al.[305] subsequently presented a full analysis of 265 families demonstrating a peak heterogeneity LOD score of 3.51 at D6S296 on chromosome 6p24-22 using a schizophrenia spectrum model that included non-affective psychotic disorders and schizotypal personality disorder. While a heterogeneity LOD score above 3.3 would traditionally be considered statistically significant evidence for linkage, this was 1 of 16 analyses performed (4 diagnostic models, 4 genetic models for each), so that although these analyses cannot be considered entirely independent, the statistical interpretation of the result is not straightforward. As shown in Table 4.3, only two of the many published genome scans have reported that this region yielded some of their most positive evidence for linkage. However, an international collaboration formed to study regions of chromosomes 3, 6, and 8[306] did detect suggestive evidence for linkage on chromosome 6p: maximum LOD scores (MLS) were 2.19 in 463 affected sib pairs not including the MCV sample (p = .001) and 2.68 in all 687 pairs (p =.0004). Other positive results (in samples overlapping with the large collaborative study) include p = 0.005 (Weighted Ranked Pairs Correlation test) in 65 pedigrees in a two-stage genome scan,[267] and MLS of 2.2 in 54 pedigrees, both at D6S274.[307] Linkage studies in this region have been reviewed elsewhere.[308,309] Note that the peak evidence for linkage in these studies and in the collaborative analysis is observed about 10 cM centromeric to that in the MCV sample. In a different type of analysis, Brzustowicz et al.[310] reported positive evidence on chromosome 6p for linkage of positive symptom scores as a quantitative trait, but not schizophrenia as a diagnostic category.

**Chromosome 8p.** Pulver et al.[311] had reported maximum heterogeneity LOD scores of 2.35 (dominant) and 2.20 (recessive) at D8S136 on chromosome 8p22-21, with a peak of 66.7% shared alleles in 57 pedigrees using a narrow diagnostic model. Recently, this group reported that in the 54 informative pedigrees from this sample, a denser map revealed a maximum Non-Parametric Linkage (NPL) score of 3.64 (p = 0.0001) at D8S1771. And in a sample of 51 additional pedigrees, the maximum NPL score was 1.95 (p = 0.023) at D8S1752 in the same region.[271] The large international collaborative study[306] found a maximum heterogeneity LOD score of 2.22 (p = .0014) in 384 pedigrees, not including the Pulver sample, at D8S261, and 3.06 (p = 0.00018) in all 429 pedigrees at the same locus. The peak multipoint result was near D8S133 in 677 affected sib pairs and was quite consistent in the Pulver and other samples. The MCV group, which participated in this collaboration, separately reported heterogeneity LOD scores between 2 and 2.5 in this region, depending on the diagnostic model,[312] and the NIMH Genetics Initiative group reported that in their 30 African-American pedigrees, an NPL score of 2.27 (p = 0.013) was observed at D8S1819 on chromosome 8p23,[270] although negative results were observed in 43 European-American pedigrees.[269]

**Chromosome 13q.** Lin et al.[313] reported a modestly positive heterogeneity LOD score (1.61) at the marker D13S144 and later reported[314] that in an expanded European sample the maximum multipoint LOD score across the 13q14.1-32 region was 2.58 near D13S122 and D13S128, although the region was negative in Asian families in the same study. Pulver et al.[271] subsequently reported that the only significant result in their genome scan was in this region. This group used a single diagnostic model (schizophrenia and schizoaffective disorder) and considered multipoint NPL analysis the primary analytic method for the completed genome scan. After saturating the region, the maximum NPL score for 54 pedigrees was 4.18 near D13S174, 8-10 cM centromeric from the Lin et al. result. The associated p value was 0.00002, a value that was shown in one study to be a conservative threshold for a 5% genome-wide significance level[75] (i.e., this value would be expected about once in every 20 very dense genome scans using a single multipoint test). Given that NPL scores reflect only the deviation of marker allele sharing from chance, rather than any one genetic model, and are known to be rather conservative generally, this is an impressive result. Dominant and recessive heterogeneity LOD score tests were also applied to give an indication of mode of inheritance. Under the dominant model, the maximum LOD score was 4.54. These authors also studied a second data set of 51 families and observed an NPL score of 2.36 (p = 0.007) in the same region. In the SUNY/AxyS study, a maximum NPL score of 1.83 was observed at D13S170 (13q31), considering schizophrenia and schizoaffective disorder as affected.[272]

**Chromosome 10p/10q.** The National Institute of Mental Health sponsored a three-center Genetics Initiative to collect a sample of multiplex schizophrenia pedigrees for which DNA and diagnostic data are available to qualified scientists for studies of schizophrenia (http://www-grb.nimh.nih.gov/gi.html). The collaboration recently published a description of the effort[315] and genome scan results with separate analyses of 43 pedigrees of European-American ancestry[269] as well as 30 pedigrees of African-American ancestry.[270] The most positive results were seen in the European-American sample on chromosome 10p, with a maximum NPL score of 3.36 (p = 0.0004) at D10S1423. Two other groups also reported positive evidence for linkage in exactly the same region: The Medical College of Virginia group[316] observed a maximum NPL score of 1.88 (p = 0.03) and a maximum multipoint heterogeneity LOD score of 1.91 (p = 0.006). This was the fourth most positive region in their genome scan (after chromosome 6p and 8p as discussed above, and chromosome 5q as discussed below). The Bonn group[317] reported a maximum lod score of 2.13 by multipoint affected sib pair analysis.

The strongly positive region in the Genetics Initiative European-American sample extended about 60 cM (D10S189/10p15 to D10S1225/10q21) with a somewhat distinct second peak on 10q21 with a maximum NPL score of 2.79 (p = 0.003) at D10S604 (10q11.2).[269] On the CHLC map, D10S604 is located at about 48 cM from pter, and the edge of the positive region (D10S1225) at about 82 cM. The most positive scores in several other scans were grouped farther down chromosome 10q. In the SUNY/AxyS group, there were NPL scores between 2.01 and 2.18 at five markers, one of which was D10S677 (10q22-q24, NPL = 2.08, p = 0.01), at about 126 cM on the CHLC map.[272] The US/Australia study observed one of its two highest scores (NPL = 2.02 , p < 0.01) at D10S1239 (10q23) at about 132 cM. Finally, a preliminary report by the University of Wales group on its first stage (20 cM) scan of 100 affected sib pairs[274] found one of its highest scores (MLS = 2.31) at D10S217, at about 138 cM on the Southamptom map, which is not exactly comparable to CHLC. These three regions (10p, proximal 10q, distal 10q) are too far apart to represent a single linkage.

**Distal chromosome 5q.** The Medical College of Virginia group observed a multipoint heterogeneity LOD score of 3.35 (p = 0.0002) on chromosome 5q22-31 at D5S804, under a narrow diagnostic model and a recessive genetic model, with an NPL score of 2.84 (p = 0.002).[318] The Bonn group simultaneously reported more modestly positive results in the same region, with maximum LOD scores of 1.8 at marker IL9 in 14 pedigrees and 1.27 at D5S399 in an additional 40 pedigrees.[319] This region is 70-80 cM distal to the proximal 5q region discussed previously.

**Chromosome 6q.** Gejman and colleagues at the NIMH intramural program genotyped markers on chromosome 6 in the initial stages of a genome scan of 53 pedigrees containing 61 independent affected sib pairs.[320] No evidence for linkage was observed on chromosome 6p, although these pedigrees were included in the large multicenter study that yielded somewhat positive results.[306] However, on chromosome 6q, affected sib pair analysis produced a p value of 0.00018, following which 87 independent sib pairs from the NIMH Genetics Initiative were typed and a p value of 0.00095 was observed.[320] Further analyses (Gejman, personal communication, paper submitted) were reported on these two samples and a third collaborating sample (the U.S./Australia sample listed on Table 4.3, 43 pedigrees, 54 independent ASPs). The latter data set produced slightly positive evidence for linkage (p = 0.013). The two replication datasets combined (total of 141 independent affected sib pairs) produced an overall p value of 0.000014 (multipoint maximum LOD score = 3.82). In the NIMH Intramural data set, there were positive scores across a broad region with a maximum at D6S474, while in both of the other two datasets maximum scores were observed 20 cM toward the centromere at D6S424.

**V.B.9. Chromosome 9q.** Moises completed a genome scan in 5 Icelandic pedigrees with about 5 affected members each, analyzed the data with a single non-parametric test (Weighted Pairs Ranked Correlation), and then collaborated with other groups to type 10 markers in the most positive regions in 65 additional pedigrees.[267] In the complete scan, p values less than 0.05 were observed near 4 markers on chromosome 9q. And at one of these markers on proximal 9q21 (D9S175), a p value less than 0.01 was also observed in the replication sample. The U.S./Australia group observed an NPL score of 1.54 (p < 0.05) at D9S257 (9q22),[268] and NIMH Genetics Initiative group observed NPL scores with p values less than 0.05 at several markers on 9q32-34 in African-American pedigrees, with a maximum score of 2.17 (p = 0.017) at D9S1830.

**Chromosome 2p.** The Utah group collected schizophrenia pedigrees on the South Pacific island nation of Palau, a very old genetic isolate, and completed a genome scan on the largest pedigree with 14 affected members. A maximum LOD score of 2.17 was observed near D2S441 (2p13-14).[273] In all 17 pedigrees, the maximum LOD score in this region was 1.69. The SUNY/AxyS group observed one of its highest scores in this region at D2S1337 (NPL = 2.13 and p = 0.01 with a broad diagnostic model; and NPL = 1.97 and p = 0.02, heterogeneity LOD score = 2.19, with a narrow model).[272]

**Other regions.** Freedman et al.[321] reported strong evidence for linkage between a physiological trait associated with schizophrenia and DNA markers on chromosome 15q14, with a maximum LOD score of 5.3 at D15S1360. The trait, observed in schizophrenic patients and some of their well relatives, is deficient inhibition of the P50 event-related brain potential when a test auditory stimulus is preceded by another stimulus. These authors did not observe linkage to schizophrenia as a diagnosis, and other scans have generally not observed evidence for linkage in this region. Leonard et al.[322] typed three 15q markers in 26 sib pairs and a small number of additional affected relatives from 20 families from the

Genetics Initiative data set selected to exclude all schizoaffective diagnoses and pairs without two available parents. The maximum LOD score was 1.46 at D15S1360, with 58% of alleles shared (p = 0.0024).

Table 4.3 lists the other "most positive" regions in the reported genome scans. Several have been reported more than once but with only very modestly positive scores. The one region that has been subjected to a multistage analysis is chromosome 20q. Moises et al.[267] observed a p value of 0.009 at D20S40 in their genome scan of 5 pedigrees and a p value less than 0.001 in 65 additional pedigrees. Another region of interest has been the X chromosome, which DeLisi and Crow have hypothesized might be involved in schizophrenia because of the pattern of sex differences (male excess and excess of same-sex pairs) sometimes observed.[323] Several studies have produced modestly positive evidence for linkage within the pseudoautosomal region where recombination is observed between the X and Y chromosomes,[324-329] including one report of positive scores at the androgen receptor.[330] However, there have been a number of negative reports,[331-336] and this region has not produced among the highest scores in the reported genome scans to date.

## DISCUSSION

The existing schizophrenia linkage data are difficult to interpret. Do we have converging evidence for a number of chromosomal regions that contain schizophrenia susceptibility loci or a pattern of false-positive results whose sheer number ensures that some regions will be positive two or three times by chance alone? The same issue applies to all linkage studies of genetically complex disorders.

Lander and Kruglyak[75] have presented one framework: a theoretical approach to determining the number of false positives to be expected in complete genome scans. And they suggest the terms "suggestive," "significant," and "highly significant" linkage for results expected by chance once, 0.05, or 0.001 times per genome scan respectively. The first two categories correspond to LOD scores of 1.9 (pointwise p = 0.0017) and 3.3 (p = 0.000049), and to maximum LOD scores in sib pair studies of 2.2 (p = 0.00074) and 3.6 (p = 0.000022). They further argue that replication of a significant linkage requires a p value less than 0.01. Note, however, that their model assumes a single multipoint analysis of a very dense map. Most studies have computed multiple partially correlated two-point and multipoint statistical analyses, and sometimes with multiple diagnostic models. Thus many reported significance levels are likely to be inflated. On the other hand, most findings were generated with 10 cM or sparser maps, and simulation data suggest that the thresholds proposed by Lander and Kruglyak may be overly conservative in this situation.[337] Therefore, it is difficult to compare results across studies.

Nevertheless, in terms of the proposed model, the only reported finding that formally meets the criteria for replicated linkage is the data of the Johns Hopkins group on chromosome 13q,[271] given that a primary multipoint analysis method resulted in a p value of 0.00002, with a value of 0.007 in a second dataset, and positive results in two other datasets. This finding has yet to be subjected to multicenter analysis (one is planned). The other strongest finding to data is that of the Medical College of Virginia group on chromosome 6p. The robust maximum LOD score of 3.51 must be considered in the context of this group's approach of conducting 16 partially correlated tests (four genetic tests of each of four diagnostic models). As noted by Lander and Kruglyak,[75] this finding is probably at least rather close to a genome-wide significance of 0.05 and has been supported by positive scores in two single samples and in a large multicenter analysis. The other major findings discussed above (5q, 6q, 8p, 9q, 10p) also must be considered promising in that each has produced among the most-observed results in two or three independent genome scans. A more pessimistic view would be that one clearly "significant" result would be expected by chance in 13 separate genome scan analyses to date,[266-278] and with multiple testing taken into account, the current pattern of results could in fact represent all false-positives.

Another vantage point on these data is suggested by the work of Suarez et al.[338] They demonstrated that for a set of loci of small effect that interact to cause a disease, there will be random fluctuations in the observed strength of the linkage between nearby marker loci and the disease, especially in smaller samples, such that the effect may be exaggerated in some samples (leading to initial "detection") and greatly underestimated in others (leading to failure to replicate). Under these circumstances, one would not expect true linkages to be observed in most samples, and true positives (if any exist) would be more likely than false-positives to produce substantial evidence for linkage in multiple scans. This would suggest that the regions producing the most positive results in multiple scans should be considered strong candidate regions regardless of the precise p values in each study.

Our view is that both the consideration of appropriate statistical thresholds and the application of a pragmatic "rank order and replicate" strategy are necessary in this field.[339] The former keeps us honest: We should not fool ourselves into believing prematurely that we have an answer. But the latter is the *de facto* manner in which the field progresses: Highly significant findings cannot realistically be expected, given the myriad of uncertainties in defining both the phenotype(s) and the mode of transmission, and all highly suggestive leads must be pursued vigorously because of the devastating nature of this common disease. Multicenter studies represent one way to evaluate the strength of findings, but it should be kept in mind that weak susceptibility loci may never produce significant linkage results[75] and that multiple sampling and analytic and molecular methods are likely to be needed to confirm each finding and identify the relevant genes.

## CONCLUSIONS

Molecular genetic investigations of schizophrenia have progressed rapidly in the last decade. The development of robust methods of non-parametric linkage analysis has been stimulated in part by studies of schizophrenia and other psychiatric disorders, the technology of high throughput genotyping of complete maps has been harnessed, larger samples are being collected, and reasonably consistent diagnostic models applied. It is now possible to assess the results of multiple independent genome scans, with large multicenter analyses beginning to emerge of both linkage and association findings. Although interpretation of association studies is particularly difficult because of the problem of multiple testing, multicenter analyses are somewhat supportive of associations between schizophrenia and polymorphisms in both the DRD3 and 5HT2A loci. Robust linkage findings have been reported on chromosomes 13q and 6p, and a number of other candidate regions have emerged on the basis of observing the most positive results across multiple genome scans. Although it is statistically possible that all of these results represent false positives, it appears more likely that some represent true linkages to loci that produce small increases in risk in many families, that some produce larger increases in risk in fewer families, or that some are associated with clinical subtypes that are not well identified by current clinical methods. The challenge for the future is determining which linkages are real, identifying the pathogenic loci, and beginning to understand the molecular biology that underlies susceptibility to schizophrenia.

## REFERENCES

1. Kraepelin, E., *Dementia praecox and paraphrenia*, facsimile 1919 edition, Robertson, G. M., Ed., Krieger, Huntington, NY, 1971.
2. Bleuler, E., *Dementia praecox; or, the group of schizophrenias,* International Universities Press, New York, 1950.
3. Schneider, K., *Clinical psychopathology,* Grune & Stratton, New York, 1959.
4. Crow, T.J., Molecular pathology of schizophrenia: more than one disease process? *Br. Med. J.,* 280, 66, 1980.

5. Peralta, V. and Cuesta, M. J., Factor structure and clinical validity of competing models of positive symptoms in schizophrenia, *Biol. Psychiatr.*, 44, 107, 1998.

6. Lindenmayer, J. P., Grochowski, S., and Hyman, R. B., Five factor model of schizophrenia: replication across samples, *Schizophr. Res.*, 14, 229, 1995.

7. Andreasen, N. C., Arndt, S., Alliger, R., Miller, D., and Flaum, M., Symptoms of schizophrenia. Methods, meanings, and mechanisms, *Arch. Gen. Psychiatr.*, 52, 341, 1995.

8. Gottesman, I. I. and Shields, J., *Schizophrenia: the epigenetic puzzle,* Cambridge University Press, Cambridge, 1982.

9. Cooper, J. E., Kendell, R. E., Gurland, B. J., Sharpe, L., Copeland, J. R. M., and Simon, R., *Psychiatric Diagnosis in New York and London: A Comparative Study of Mental Hospital Admissions,* Oxford University Press, London, 1972.

10. Feighner, J. P., Robins, E., Guze, S. B. et al., Diagnostic criteria for use in psychiatric research, *Arch. Gen. Psychiatr.,* 26, 57, 1972.

11. Spitzer, R. L., Endicott, J., and Robins, E., *Research Diagnostic Criteria (RDC) for a Selected Group of Functional Disorders,* New York State Psychiatric Institute, Biometrics Research, New York, 1981.

12. American Psychiatric Association, *Diagnostic and Statistical Manual of Mental Disorders,* 3rd ed., American Psychiatric Association, Washington, DC, 1980.

13. American Psychiatric Association, *Diagnostic and Statistical Manual,* 3rd ed. (Revised), American Psychiatric Association, Washington, DC, 1987.

14. American Psychiatric Association, *Diagnostic and Statistical Manual,* 4th ed., American Psychiatric Association, Washington, DC, 1994

15. World Health Organization, The ICD-10 classification of mental and behavioural disorders: diagnostic criteria for research, Geneva, 1992.

16. Kasanin, J., The acute schizoaffective psychoses, *Am. J. Psychiatr.*, 13, 97, 1933.

17. Spitzer, R. L., Endicott, J., and Robins, E., Research diagnostic criteria: rationale and reliability, *Arch. Gen. Psychiatr.*, 35, 773, 1978.

18. Nurnberger, J. I., Jr., Blehar, M. C., Kaufmann, C. A., York-Cooler, C., Simpson, S. G., Harkavy-Friedman, J., Severe, J. B., Malaspina, D., and Reich, T., Diagnostic interview for genetic studies. Rationale, unique features, and training, *Arch. Gen. Psychiatr.,* 51, 849, 1994.

19. Spitzer, R. L., Endicott, J., and Gibbon, M., Crossing the border into borderline personality and borderline schizophrenia, the development of criteria, *Arch. Gen. Psychiatr.*, 36, 17, 1979.

20. Levinson, D. F. and Mowry, B. J., Defining the schizophrenia spectrum: issues for genetic linkage studies, *Schizophr. Bull.*, 17, 491, 1991.

21. Tsuang, M. T., Gilbertson, M. W., and Faraone, S. V., The genetics of schizophrenia: current knowledge and future directions, *Schizophr. Res.,*, 4, 157, 1991.

22. Kendler, K. S. and Diehl, S. R., The genetics of schizophrenia: a current, genetic-epidemiologic perspective, *Schizophr. Bull.*, 19, 261, 1993.

23. McGuffin, P., Farmer, A. E., Gottesman, I. I., Murray, R. M., and Reveley, A. M., Twin concordance for operationally defined schizophrenia, *Arch. Gen. Psychiatr.*, 41, 541, 1984.

24. Farmer, A. E., McGuffin, P., and Gottesman, I. I., Twin concordance for DSM-III schizophrenia. Scrutinizing the validity of the definition, *Arch. Gen. Psychiatr.,* 44, 634, 1987.

25. Franzek, E. and Beckmann, H., Different genetic background of schizophrenia spectrum psychoses: a twin study, *Am. J. Psychiatr.*, 155, 76, 1998.

26. Cannon, T. D., Kaprio, J., Lonnqvist, J., Huttunen, M., and Koskenvuo, M., The genetic epidemiology of schizophrenia in a Finnish twin cohort. A population-based modeling study, *Arch. Gen. Psychiatr.*, 55, 67, 1998

27. Klaning, U., Mortensen, P. B., and Kyvik, K. O., Increased occurrence of schizophrenia and other psychiatric illnesses among twins, *Br. J. Psychiatr.*, 168, 688, 1996.

28. Kendler, K. S., Pedersen, N. L., Farahmand, B. Y., and Persson, P. G., The treated incidence of psychotic and affective illness in twins compared with population expectation: a study in the Swedish Twin and Psychiatric Registries, *Psychol. Med.*, 26, 1135, 1996.

29. Davis, J. O., Phelps, J. A., and Bracha, H. S., Prenatal development of monozygotic twins and concordance for schizophrenia, *Schizophr. Bull.*, 21, 357, 1995.

30. Kety, S. S., Rosenthal, D., Wender, P. H., Schulsinger, F., and Jacobsen, B., The biologic and adoptive families of adopted individuals who became schizophrenic: prevalence of mental illness and other characteristics, in *The Nature of Schizophrenia: New Approaches to Research and Treatment,* Wynne, L.C., Cromwell, R.L., and Matthysse, S., Eds., John Wiley & Sons Inc., New York, 1978, 25.

31. Rosenthal, D., Wender, P. H., Kety, S. S., Welner, J., and Schulsinger, F., The adopted-away offspring of schizophrenia, *Am. J. Psychiatr.,* 128, 307, 1971.

32. Kendler, K. S. and Gruenberg, A. M., An independent analysis of the Danish Adoption Study of Schizophrenia. VI. The relationship between psychiatric disorders as defined by DSM-III in the relatives and adoptees, *Arch. Gen. Psychiatr.,* 41, 555, 1984.

33. Lowing, P. A., Mirsky, A. F., and Pereira, R. The inheritance of schizophrenia spectrum disorders: a reanalysis of the Danish adoptee study data, *Am. J. Psychiatr.,* 140, 1167, 1983.

34. Tienari, P., Interaction between genetic vulnerability and family environment: the Finnish adoptive family study of schizophrenia, *Acta. Psychiatr. Scand.,* 84, 460, 1991.

35. Scharfetter, C. and Nusperli, M., The group of schizophrenias, schizoaffective psychoses, and affective disorders, *Schizophr. Bull.,* 6, 586, 1980.

36. Baron, M., Gruen, R., Asnis, L., and Kane, J., Schizoaffective illness, schizophrenia and affective disorders: morbidity risk and genetic transmission, *Acta. Psychiatr. Scand.,* 65, 253, 1982.

37. Baron, M., Gruen, R., Rainer, J. D., Kane, J., Asnis, L., and Lord, S., A family study of schizophrenic and normal control probands: implications for the spectrum concept of schizophrenia, *Am. J. Psychiatr.,* 142, 447, 1985.

38. Kendler, K. S., Gruenberg, A. M., and Tsuang, M. T., Psychiatric illness in first-degree relatives of schizophrenic and surgical control patients. A family study using DSM-III criteria, *Arch. Gen. Psychiatr.,* 42, 770, 1985.

39. Kendler, K. S., Gruenberg, A. M., and Tsuang, M. T., A DSM-III family study of the nonschizophrenic psychotic disorders, *Am. J. Psychiatr.,* 143, 1098, 1986.

40. Mendlewicz, J., Linkowski, P., Wilmotte, J., Relationship between schizo-affective illness and affective disorders or schizophrenia. Morbidity risk and genetic transmission, *J. Aff. Disorders,* 2, 289, 1980.

41. Gershon, E. S., DeLisi, L.E., Hamovit, J., Nurnberger, J.I. Jr., Maxwell, M.E., Schreiber, J., Dauphinais, D., Dingman, C.W., II, and Guroff, J.J, A controlled family study of chronic psychoses, *Arch. Gen. Psychiatr.,* 45, 328, 1988.

42. Kendler, K. S., McGuire, M., Gruenberg, A. M., O'Hare, A., Spellman, M., and Walsh, D., The Roscommon family study. I. Methods, diagnosis of probands, and risk of schizophrenia in relatives, *Arch. Gen. Psychiatr.,* 50, 527, 1993.

43. Kendler, K. S., McGuire, M., Gruenberg, A. M., Spellman, M., O'Hare, A., and Walsh, D., The Roscommon family study. II. The risk of nonschizophrenic nonaffective psychoses in relatives, *Arch. Gen. Psychiatr.,* 50, 645, 1993.

44. Kendler, K. S., McGuire, M., Gruenberg, A. M., O'Hare, A., Spellman, M., and Walsh, D., The Roscommon family study. IV. Affective illness, anxiety disorders, and alcoholism in relatives, *Arch. Gen. Psychiatr.,* 50, 952, 1993.

45. Kendler, K. S., McGuire, M., Gruenberg, A. M., and Walsh, D., Examining the validity of DSM-III-R schizoaffective disorder and its putative subtypes in the Roscommon family study, *Am. J. Psychiatr.,* 152, 755, 1995.

46. Maier, W., Lichtermann, D., Minges, J., Hallmayer, J., Heun, R., Benkert, O., and Levinson, D. F., Continuity and discontinuity of affective disorders and schizophrenia. Results of a controlled family study, *Arch. Gen. Psychiatr.,* 50, 871, 1993.

47. Parnas, J., Cannon, T. D., Jacobsen, B., Schulsinger, H., Schulsinger, F., and Mednick, S. A., Lifetime DSM-III-R diagnostic outcomes in the offspring of schizophrenic mothers. Results from the Copenhagen high-risk study, *Arch. Gen. Psychiatr.,* 50, 707, 1993.

48. Erlenmeyer-Kimling, L., Adamo, U. H., Rock, D., Roberts, S. A., Bassett, A. S., Squires-Wheeler, E., Cornblatt, B. A., Endicott, J., Pape, S., and Gottesman, I. I., The New York high-risk project. Prevalence and comorbidity of axis I disorders in offspring of schizophrenic parents at 25-year follow-up, *Arch. Gen. Psychiatr.,* 54, 1096, 1997.

49. Erlenmeyer-Kimling, L., Squires-Wheeler, E., Adamo, U. H., Bassett, A. S., Cornblatt, B. A., Kestenbaum, C. J., Rock, D., Roberts, S. A., and Gottesman, I. I., The New York high-risk project. Psychoses and cluster A personality disorders in offspring of schizophrenic parents at 23 years of follow-up, *Arch. Gen. Psychiatr.,* 52, 857, 1995.

50. Frangos, E., Athanassenas, G., Tsitourides, S., Katsanou, N., and Alexandrakou, P., Prevalence of DSM III schizophrenia among the first-degree relatives of schizophrenic probands, *Acta. Psychiatr. Scand.*, 72, 382, 1985.

51. Coryell, W. and Zimmerman, M., The heritability of schizophrenia and schizoaffective disorder: a family study, *Arch. Gen. Psychiatr.*, 45, 323, 1988.

52. Kendler, K. S., McGuire, M., Gruenberg, A. M., O'Hare, A., Spellman, M., and Walsh, D., The Roscommon family study. III. Schizophrenia-related personality disorders in relatives, *Arch. Gen. Psychiatr.*, 50, 781, 1993.

53. Maier, W., Lichtermann, D., Minges, J., and Heun R., Personality disorders among the relatives of schizophrenia patients, *Schizophr. Bull.*, 20, 481, 1994.

54. Torrey, E. F., Miller, J., Rawlings, R., and Yolken, R. H., Seasonality of births in schizophrenia and bipolar disorder: a review of the literature, *Schizophr. Res.*, 28, 1, 1997.

55. Pulver, A. E., Liang, K. Y., Brown, C. H., Wolyniec, P., McGrath, J., Adler, L., Tam, D., Carpenter, W. T., and Childs, B., Risk factors in schizophrenia. Season of birth, gender, and familial risk, *Br. J. Psychiatr.*, 160, 65, 1992.

56. McGrath, J. and Castle, D., Does influenza cause schizophrenia? A five year review, *Aust. N. Z. J. Psychiatr.*, 29, 23, 1995.

57. Crow, T. J., Prenatal exposure to influenza as a cause of schizophrenia. There are inconsistencies and contradictions in the evidence, *Br. J. Psychiatr.*, 164, 588, 1994.

58. Cannon, M., Cotter, D., Coffey, V. P., Sham, P. C., Takei, N., Larkin, C., Murray, R. M., and O'Callaghan, E., Prenatal exposure to the 1957 influenza epidemic and adult schizophrenia: a follow-up study, *Br. J. Psychiatr.*, 168, 368, 1996.

59. Cannon, T. D., Abnormalities of brain structure and function in schizophrenia: implications for aetiology and pathophysiology, *Ann. Med.*, 28, 533, 1996.

60. Verdoux, H., Geddes, J. R., Takei, N., Lawrie, S. M., Bovet, P., Eagles, J. M., Heun, R., McCreadie, R. G., McNeil, T. F., O'Callaghan, E., Stober, G., Willinger, M. U., Wright, P., and Murray, R. M., Obstetric complications and age at onset in schizophrenia: an international collaborative meta-analysis of individual patient data, *Am. J. Psychiatr.*, 154, 1220, 1997.

61. Sacker, A., Done, D. J., and Crow, T. J., Obstetric complications in children born to parents with schizophrenia:a meta-analysis of case-control studies, *Psychol. Med.*, 26, 279, 1996.

62. Geddes, J. R., Lawrie, S. M., Obstetric complications and schizophrenia: a meta-analysis, *Br. J. Psychiatr.*, 167, 786, 1995.

63. Hutchinson, G., Takei, N., Fahy, T. A., Bhugra, D., Gilvarry, C., Moran, P., Mallett, R., Sham, P., Leff, J., and Murray, R. M., Morbid risk of schizophrenia in first-degree relatives of white and African-Caribbean patients with psychosis, *Br. J. Psychiatr.*, 169, 776, 1996.

64. Risch, N., Linkage strategies for genetically complex traits I. Multilocus models, *Am. J. Hum. Genet.*, 46, 222, 1990.

65. James, J. W., Frequency in relatives for an all-or-none trait, *Ann. Hum. Genet.*, 35, 47, 1971.

66. Kidd, K. K., Searching for major genes for psychiatric disorder, in *Molecular Genetic Approaches to Human Polygenic Disease*, Ciba Foundation Symposium 130, Bock, G. and Collins, G. M., Eds., John Wiley & Sons Inc., Chichester, England, 1987, 193.

67. Kendler, K. S., McGuire, M., Gruenberg, A. M., and Walsh, D., Clinical heterogeneity in schizophrenia and the pattern of psychopathology in relatives: results from an epidemiologically based family study, *Acta. Psychiatr. Scand.*, 89, 294, 1994.

68. Kendler, K. S., Karkowski-Shuman, L., and Walsh, D., Age at onset in schizophrenia and risk of illness in relatives. Results from the Roscommon family study, *Br. J. Psychiatr.*, 169, 213, 1996.

68. DeLisi, L. E., Goldin, L. R,, Maxwell, E., Kazuba, D. M., and Gershon, E. S., Clinical features of illness in siblings with schizophrenia or schizoaffective disorder, *Arch. Gen. Psychiatr.*, 44, 891, 1987.

69. Kendler, K. S., Karkowski-Shuman, L., O'Neill, F. A., Straub, R. E., MacLean, C. J., and Walsh, D., Resemblance of psychotic symptoms and syndromes in affected sibling pairs from the Irish study of high-density schizophrenia families: evidence for possible etiologic heterogeneity, *Am. J. Psychiatr.*, 154, 191, 1997.

70. Kendler, K. S. and Walsh, D., The structure of psychosis: syndromes and dimensions, *Arch. Gen. Psychiatr.*, 55, 508, 1998.

71. Liang, K. Y. and Pulver, A. E., Analysis of case-control/family sampling design, *Genet. Epidemiol.*, 13, 253, 1996.

72. Terwilliger, J. D., book review of *Genetic Variation and Human Disease: Principles and Evolutionary Approaches, Am. J. Hum. Genet.*, Weiss, K. M., 60, 1565, 1997.

73. Weiss, K. M., *Genetic Variation and Human Disease: Principles and Evolutionary Approaches,* Cambridge University Press, Cambridge, 1993.

74. Risch, N., Genetic linkage and complex diseases, with special reference to psychiatric disorders, *Genet. Epidemiol.*, 7, 3, 1990.

75. Lander, E. and Kruglyak, L., Genetic dissection of complex traits: guidelines for interpreting and reporting linkage findings, *Nature Genet.*, 11, 241, 1995.

76. Hodge, S. E., An oligogenic disease displaying weak marker associations: a summary of contributions to problem 1 of GAW9, *Genet. Epidemiol.*, 12, 545, 1995.

77. Kruglyak, L. and Lander, E. S., High-resolution genetic mapping of complex traits, *Am. J. Hum. Genet.*, 56, 1212, 1995.

78. Risch, N. and Merikangas, K., The future of genetic studies of complex human diseases, *Science*, 273, 1516, 1996.

79. Cubells, J.F., Kobayashi, K., Nagatsu, T., Kidd, K. K., Kidd, J. R., Calafell, F., Kranzler, H. R., Ichinose, H., and Gelernter, J., Population genetics of a functional variant of the dopamine beta-hydroxylase gene (DBH), *Am. J. Med. Genet.*, 74, 374, 1997.

80. Kostrzewa, M., Grady, D. L., Moyzis, R. K., Floter, L., and Muller, U., Integration of four genes, a pseudogene, thirty-one STSs, and a highly polymorphic STRP into the 7-10 Mb YAC contig of 5q34-q35, *Hum. Genet.*, 97, 399, 1996.

81. Grandy, D. K., Litt, M., Allen, L., Bunzow, J. R., Marchionni, M., Makam, H., Reed, L., Magenis, R. E., and Civelli, O., The human dopamine D2 receptor gene is located on chromosome 11 at q22-q23 and identifies a TaqI RFLP, *Am. J. Hum. Genet.*, 45,77, 1989.

82. Le Coniat, M., Sokoloff, P., Hillion, J., Martres, M. P., Giros, B., Pilon, C., Schwartz, J. C., Berger, R., Chromosomal localization of the human D3 dopamine receptor gene, *Hum. Genet.*, 87, 618, 1991.

83. Gelernter, J., Kennedy, J., L., van Tol, H. H., Civelli, O., and Kidd, K. K., The D4 dopamine receptor (DRD4) maps to distal 11p close to HRAS, *Genomics,* 13, 208, 1992.

84. Sherrington, R., Mankoo, B., Attwood, J., Kalsi, G., Curtis, D., Buetow, K., Povey, S., and Gurling, H., Cloning of the human dopamine D5 receptor gene and identification of a highly polymorphic microsatellite for the DRD5 locus that shows tight linkage to the chromosome 4p reference marker RAF1P1, *Genomics*, 18, 423, 1993.

85. Crocq, M. A., Mant, R., Asherson, P., Williams, J., Hode, Y., Mayerova, A., Collier D., Lannfelt, L., Sokoloff, P., Schwartz, J. C. et al., Association between schizophrenia and homozygosity at the dopamine D3 receptor gene, *J. Med. Genet.*, 29, 858, 1992.

86. Griffon, N., Crocq, M. A., Pilon, C., Martres, M. P., Mayerova, A., Uyanik, G., Burgert E, Duval, F., Macher, J. P., Javoy-Agid, F., Tamminga, C. A., Schwartz J. C., and Sokoloff, P., Dopamine D3 receptor gene: organization, transcript variants, and polymorphism associated with schizophrenia, *Am. J. Med. Genet.*, 67, 63, 1996.

87. Spurlock, G., Williams, J., McGuffin, P., Aschauer, H. N., Lenzinger, E., Fuchs, K., Sieghart, W. C., Meszaros, K., Fathi, N., Laurent, C., Mallet, J., Macciardi, F., Pedrini, S., Gill, M., Hawi, Z., Gibson, S., Jazin, E. E., Yang, H. T., Adolfsson R., Pato, C. N., Dourado, A. M., and Owen, M. J., European Multicentre Association study of schizophrenia: a study of the DRD2 Ser311Cys and DRD3 Ser9Gly polymorphisms, *Am. J. Med. Genet.*, 81, 24, 1998.

88. Asherson, P., Mant, R., Holmans, P., Williams, J., Cardno, A., Murphy, K., Jones, L., Collier, D., McGuffin, P., and Owen, M. J., Linkage, association and mutational analysis of the dopamine D3 receptor gene in schizophrenia, *Mol. Psychiatr.*, 1, 125, 1996.

89. Ebstein, R. P., Macciardi, F., Heresco-Levi, U., Serretti, A., Blaine, D., Verga, M., Nebamov, L., Gur, E., Belmaker, R. H., Avnon, M., and Lerer, B., Evidence for an association between the dopamine D3 receptor gene DRD3 and schizophrenia, *Hum. Hered.*, 47, 6-16, 1997.

90. Nimgaonkar, V. L., Zhang, X. R., Caldwell, J. G., Ganguli, R., and Chakravarti, A., Association study of schizophrenia with dopamine D3 receptor gene polymorphisms: probable effects of family history of schizophrenia? *Am. J. Med. Genet.*, 48, 214, 1993.

91. Nimgaonkar, V. L., Sanders, A. R., Ganguli, R., Zhang, X. R., Brar, J., Hogge, W., Fann, W. E., Patel, P. I., and Chakravarti, A., Association study of schizophrenia and the dopamine D3 receptor gene locus in two independent samples, *Am. J. Med. Genet.*, 67, 505, 1996.

92. Maziade, M., Martinez, M., Rodrigue, C., Gauthier, B., Tremblay, G., Fournier, C., Bissonnette, L., Simard, C., Roy, M. A., Rouillard, E., and Merette, C., Childhood/early adolescence-onset and adult-onset schizophrenia. Heterogeneity at the dopamine D3 receptor gene, *Br. J. Psychiatr.*, 170, 27, 1997.

93. Steen, V. M., Lovlie, R., MacEwan, T., and McCreadie, R. G., Dopamine D3.ceptor gene variant and susceptibility to tardive dyskinesia in schizophrenic patients, *Mol. Psychiatr.*, 2, 139.5, 1997.

94. Kennedy, J. L., Billett, E. A., Macciardi, F. M., Verga, M., Parsons, T. J., Meltzer, H. Y., Lieberman, J., and Buchanan, J. A., Association study of dopamine D3 receptor gene and schizophrenia, *Am. J. Med. Genet.*, 60, 558.2, 1995.

95. DiBella, B. D., Catalano, M., Strukel, A., Nobile, M., Novelli, E., and Smeraldi, E., Distribution of the MscI polymorphism of the dopamine D3 receptor in an Italian psychotic population, *Psychiatr. Genet.*, 4, 39, 1994.

96. Hawi, Z., McCabe, U., Straub, R. E., O'Neill, A., Kendler, K. S., Walsh, D., and Gill, M., Examination of new and reported data of the DRD3/MscI polymorphism: no support for the proposed association with schizophrenia, *Mol. Psychiatr.*, 3, 150, 1998.

97. Malhotra, A. K., Goldman, D., Buchanan, R. W., Rooney, W., Clifton, A., Kosmidis M. H., Breier, A., and Pickar, D., The dopamine D3 receptor (DRD3) Ser9Gly polymorphism and schizophrenia: a haplotype relative risk study and association with clozapine response, *Mol. Psychiatr.*, 3, 72, 1998.

98. Chen, C. H., Liu, M. Y., Wei, F. C., Koong, F. J., Hwu, H. G., and Hsiao, K. J., Further evidence of no association between Ser9Gly polymorphism of dopamine D3 receptor gene and schizophrenia, *Am. J. Med. Genet.*, 74, 40, 1997.

99. Tanaka, T., Igarashi, S., Onodera, O., Tanaka, H., Takahashi, M., Maeda, M., Kameda K., Tsuji, S., and Ihda, S., Association study between schizophrenia and dopamine D3 receptor gene polymorphism, *Am. J. Med. Genet.*, 67, 366.8, 1996.

100. Rothschild, L. G., Badner, J., Cravchik, A., Gershon, E. S., and Gejman, P. V., No association detected between a D3 receptor gene-pressed variant and schizophrenia, *Am. J. Med. Genet.*, 67, 232.4, 1996.

101. Inada, T., Sugita, T., Dobashi, I., Inagaki, A., Kitao, Y., Matsuda, G., Kato, S., Takano, T., Yagi, G., and Asai, M., Dopamine D3 receptor gene polymorphism and the psychiatric symptoms seen in first-break schizophrenic patients, *Psychiatr. Genet.*, 5, 113, 1995.

102. Macciardi, F., Verga, M., Kennedy, J. L., Petronis, A., Bersani, G., Pancheri, P., and Smeraldi, E., An association study between schizophrenia and the dopamine receptor genes DRD3 and DRD4 using haplotype relative risk, *Hum. Hered.*, 44, 328.6, 1994.

103. Rietschel, M., Nothen, M. M., Albus, M., Maier, W., Minges, J., Bondy, B., Korner J., Hemmer, S., Fimmers, R., Moller, H. J., Wildenauer, D., and Propping, P., Dopamine D3 receptor Gly9/Ser9 polymorphism and schizophrenia: no increased frequency of homozygosity in German familial cases, *Schizophr. Res.*, 20, 181, 1996.

104. Wiese, C., Lannfelt, L., Kristbjarnarson, H., Yang, L., Zoega, T., Sokoloff P., Ivarsson O, Schwartz, J. C., Moises, H. W., and Helgason, T., No evidence of linkage between schizophrenia and D3 dopamine receptor gene locus in Icelandic pedigrees, *Psychiatr. Res.*, 46, 69, 1993.

105. Sabate, O., Campion, D., d'Amato, T., Martres, M. P., Sokoloff, P., Giros, B., Leboyer, M., Jay, M., Guedj, F., Thibaut, F. et al., Failure to find evidence for linkage or association between the dopamine D3 receptor gene and schizophrenia, *Am. J. Psychiatr.*, 151, 107, 1994.

106. Nanko, S., Fukuda, R., Hattori, M., Sasaki, T., Dai, X. Y., Yamaguchi, K., and Kazamatsuri, H., Further evidence of no linkage between schizophrenia and the dopamine D3 receptor gene locus, *Am. J. Med. Genet.*, 54, 264, 1994.

107. Yang, L., Li, T., Wiese, C., Lannfelt, L., Sokoloff, P., Xu, C. T., Zeng, Z., Schwartz, J. C., Liu, X., and Moises, H. W., No association between schizophrenia and homozygosity at the D3 dopamine receptor gene, *Am. J. Med. Genet.*, 48, 83, 1993.

108. Nanko, S., Sasaki, T., Fukuda, R., Hattori, M., Dai, X. Y., Kazamatsuri, H., Kuwata S., Juji, T., and Gill, M., A study of the association between schizophrenia and the dopamine D3receptor gene, *Hum. Genet.*, 92, 336, 1993.

109. Gaitonde, E. J., Morris, A., Sivagnanasundaram, S., McKenna, P. J., Hunt, D., M., and Mollon, J. D., Assessment of association of D3 dopamine receptor MscI polymorphism with schizophrenia: analysis of symptom ratings, family history, age atonset, and movement disorders, *Am. J. Med. Genet.*, 67, 455, 1996.

110. Williams, J., Spurlock, G., Holmans, P., Mant, R., Murphy, K., Jones, L. et al., A meta-analysis and transmission disequilibrium study of association between the dopamine D3 receptor gene and schizophrenia, *Mol. Psychiatr.*, 3, 141.9, 1998.

111. Itokawa, M., Arinami, T., Futamura, N., Hamaguchi, H., and Toru, M., A structural polymorphism of human dopamine D2 receptor, D2(Ser311→Cys), *Biochem. Biophys. Res. Commun.*, 196, 1369, 1993.

112. Arinami, T. and Itokawa, M., Analysis of the dopamine D2 receptor gene, *Jpn. J. Psychopharm.*, 14, 129, 1994.

113. Muir, W. J., Gosden, C. M., Brookes, A. J., Fantes, J., Evans, K. L., Maguire S. M., Stevenson, B., Boyle, S., Blackwood, D. H., St. Clair, D. M., Porteous, D. J., and Weith, A., Direct microdissection and microcloning of a translocation breakpoint region, t(1;11)(q42.2;q21), associated with schizophrenia, *Cytogenet. Cell. Genet.*, 70, 35, 1995.

114. Catalano, M., Nobile, M., Novelli, E., and Smeraldi, E., Use of polymerase chain reaction and denaturing gradient gel electrophoresis to identify polymorphisms in three exons of dopamine D2 receptor gene in schizophrenic and delusional patients, *Neuropsychobiology*, 26, 1, 1992.

115. Coon, H., Byerley, W., Holik, J., Hoff, M., Myles-Worsley, M., Lannfelt, L., Sokoloff P., Schwartz, J. C., Waldo, M., Freedman, R. et al., Linkage analysis of schizophrenia with five dopamine receptor genes in nine pedigrees, *Am. J. Hum. Genet.*, 52, 327, 1993.

116. Crawford, F., Hoyne, J., Cai, X., Osborne, A., Poston, D., Zaglul, J., Dajani, N., Walsh, S., Bradley, R., Solomon, R., and Mullan, M., Dopamine DRD2/Cys311 is not associated with chronic schizophrenia, *Am. J. Med. Genet.*, 67. 483, 1996.

117. Dollfus, S., Campion, D., Vasse, T., Preterre, P., Laurent, C., d'Amato, T., Thibaut F., Mallet, J., and Petit, M., Association study between dopamine D1, D2, D3, and D4 receptor genes and schizophrenia defined by several diagnostic systems, *Biol. Psychiatr.*, 40, 419, 1996.

118. Gejman, P. V., Ram, A., Gelernter, J., Friedman, E., Cao, Q., Pickar, D., Blum K., Noble, E. P., Kranzler, H. R., O'Malley, S. et al., No structural mutation in the dopamine D2 receptor gene in alcoholism or schizophrenia, Analysis using denaturing gradient gel electrophoresis, *JAMA*, 271, 204, 1994.

119. Gill, M., McGuffin, P., Parfitt, E., Mant, R., Asherson, P., Collier, D., Vallada H., Powell, J., Shaikh, S., Taylor, C. et al., A linkage study of schizophrenia with DNA markers from the long arm of chromosome 11, *Psychol. Med.*, 23, 27, 1993.

120. Grassi, E., Mortilla, M., Amaducci, L., Pallanti, S., Pazzagli, A., Galassi, F., Guarnieri B. M., Petruzzi, C., Bolino, F., Ortenzi, L., Nistico, R., De Cataldo, S., Rossi, A., and Sorbi, S., No evidence of linkage between schizophrenia and D2 dopamine receptor gene locus in Italian pedigrees, *Neurosci. Lett.*, 206, 196, 1996.

121. Hallmayer, J., Maier, W., Schwab, S., Ertl, M. A., Minges, J., Ackenheil, M., Lichtermann, D., and Wildenauer, D. B., No evidence of linkage between the dopamine D2 receptor gene and schizophrenia, *Psychiatr. Res.*, 53, 203, 1994.

122. Hattori, M., Nanko, S., Dai, X. Y., Fukuda, R., and Kazamatsuri, H., Mismatch PCR RFLP detection of DRD2 Ser311Cys polymorphism and schizophrenia, *Biochem. Biophys. Res. Commun.*, 202, 757, 1994.

123. Kalsi, G., Mankoo, B. S., Curtis, D., Brynjolfsson, J., Read, T., Sharma, T., Murphy, P., Petursson, H., and Gurling, H. M., Exclusion of linkage of schizophrenia to the gene for the dopamine D2 receptor (DRD2) and chromosome 11q translocation sites, *Psychol. Med.*, 25, 531, 1995.

124. Kaneshima, M., Higa, T., Nakamoto, H., and Nagamine, M., An association study between the Cys311 variant of dopamine D2 receptor gene and schizophrenia in the Okinawan population, *Psychiatry Clin. Neurosci.*, 51, 379, 1997.

125. Maziade, M., Raymond, V., Cliche, D., Fournier, J. P., Caron, C., Garneau Y. Nicole L., Marcotte, P., Couture, C., Simard, C. et al., Linkage results on 11Q21-22, in Eastern Quebec pedigrees densely affected by schizophrenia, *Am. J. Med. Genet.*, 60, 522, 1995.

126. Moises, H. W., Gelernter, J., Giuffra, L. A., Zarcone, V., Wetterberg, L., Civelli O., Kidd KK, Cavalli-Sforza, L. L., Grandy, D. K., Kennedy, J. L. et al., No linkage between D2 dopamine receptor gene region and schizophrenia, *Arch. Gen. Psychiatr.*, 48, 643, 1991.

127. Mulcrone, J., Whatley, S. A., Marchbanks, R., Wildenauer, D., Altmark, D., Daoud H., Gur E, Ebstein, R. P., and Lerer, B., Genetic linkage analysis of schizophrenia using chromosome 11q13. markers in Israeli pedigrees, *Am. J. Med. Genet.*, 60, 103, 1995.

128. Nanko, S., Gill, M., Owen, M., Takazawa, N., Moridaira, J., and Kazamatsuri, H., Linkage study of schizophrenia with markers on chromosome 11 in two Japanese pedigrees, *Jpn. J. Psychiatr. Neurol.*, 46, 155, 1992.

129. Sarkar, G., Kapelner, S., Grandy, D. K., Marchionni, M., Civelli, O., Sobell, J., Heston, L., and Sommer, S. S., Direct sequencing of the dopamine D2 receptor (DRD2) in schizophrenics reveals three polymorphisms but no structural change in the receptor, *Genomics*, 11, 8, 1991.

130. Seeman, P., Ohara, K., Ulpian, C., Seeman, M. V., Jellinger, K., Van, TH. H., and Niznik, H. B., Schizophrenia: normal sequence in the dopamine D2 receptor region that couples to G-proteins. DNA polymorphisms in D2, *Neuropsychopharmacology*, 8, 137-42, 1993.

131. Su, Y., Burke, J., O'Neill, F. A., Murphy, B., Nie, L., Kipps, B., Bray J., Shinkwin R., Ni Nuallain, M., MacLean, C. J. et al., Exclusion of linkage between schizophrenia and the D2 dopamine receptor gene region of chromosome 11q in 112 Irish multiplex families, *Arch. Gen. Psychiatr.*, 50, 205, 1993.

132. Verga, M., Macciardi, F., Pedrini, S., Cohen, S., and Smeraldi, E., No association of the Ser/Cys311 DRD2 molecular variant with schizophrenia using a classical case control study and the haplotype relative risk, *Schizophr. Res.*, 25, 117, 1997.

133. Wang, Z. W., Black, D., Andreasen, N. C., and Crowe, R. R., A linkage study of chromosome 11q in schizophrenia, *Arch. Gen. Psychiatr.*, 50, 212, 1993.

134. Cichon, S., Nothen, M. M., Stober, G., Schroers, R., Albus, M., Maier, W., Rietschel M., Korner, J., Weigelt, B., Franzek, E., Wildenauer, D., Fimmers, R., and Propping, P., Systematic screening for mutations in the 5′-regulatory region of the human dopamine D1 receptor (DRD1) gene in patients with schizophrenia and bipolar affective disorder, *Am. J. Med. Genet.*, 67, 424, 1996.

135. Jensen, S., Plaetke, R., Holik, J., Hoff, M., Myles-Worsley, M., Leppert, M., Coon, H., Vest, K., Freedman, R., Waldo, M. et al., Linkage analysis of schizophrenia: the D1 dopamine receptor gene and several flanking DNA markers, *Hum. Hered.*, 43, 58, 1993.

136. Ohara, K., Ulpian, C., Seeman, P., Sunahara, R. K., Van, T. H., and Niznik, H. B., Schizophrenia: dopamine D1 receptor sequence is normal but has DNA polymorphisms, *Neuropsychopharmacology*, 8, 131, 1993.

137. Van Tol, H. H., Wu, C. M., Guan, H. C., Ohara, K., Bunzow, J. R., Civelli, O., Kennedy, J., Seeman, P., Niznik, H. B., and Jovanovic, V., Multiple dopamine D4 receptor variants in the human population, *Nature*, 358, 149, 1992.

138. Sommer, S. S., Lind, T. J., Heston, L. L., and Sobell, J. L., Dopamine D4 receptor variants in unrelated schizophrenic cases and controls, *Am. J. Med. Genet.*, 48, 90, 1993.

139. Ohara, K., Nakamura, Y., Xie, D. W., Ishigaki, T., Deng, Z. L., Tani, K., Zhang H. Y., Kondo, N., Liu, J. C., Miyasato, K., and Ohara, K., Polymorphisms of dopamine D2-like (D2, D3, and D4) receptors in schizophrenia, *Biol. Psychiatr.*, 40, 1209, 1996.

140. Petronis, A., Macciardi, F., Athanassiades, A., Paterson, A. D., Verga M., Meltzer H. Y., Cola, P., Buchanan, J. A., Van Tol, H. H., and Kennedy, J. L., Association study between the dopamine D4 receptor gene and schizophrenia, *Am. J. Med. Genet.*, 60, 452, 1995.

141. Catalano, M., Nobile, M., Novelli, E., Nothen, M. M., and Smeraldi, E., Distribution of a novel mutation in the first exon of the human dopamine D4 receptor gene in psychotic patients, *Biol. Psychiatr.*, 34, 459, 1993.

142. Tani, M., Koide, S., Onishi, H., Nakamura, Y., and Yamagami, S., Association analysis of the SmaI polymorphism of the dopamine D4 receptor in Japanese schizophrenics, *Psychiatr. Genet.*, 7, 137, 1997.

143. Hong, C. J., Chiu, H. J., Chang, Y. S., and Sim, C. B., Twelve-nucleotide repeat polymorphism of D4 dopamine receptor gene in Chinese familial schizophrenic patients, *Biol. Psychiatr.*, 43, 432, 1998.

144. Kohn, Y., Ebstein, R. P., Heresco-Levi, U., Shapira, B., Nemanov, L., Gritsenko, I., Avnon, M., and Lerer, B., Dopamine D4 receptor gene polymorphisms: relation to ethnicity, no association with schizophrenia and response to clozapine in Israeli subjects, *Eur. Neuropsychopharmacol.*, 7, 39, 1997.

145. Seeman, P., Ulpian, C., Chouinard, G., Van Tol, H. H., Dwosh, H., Lieberman, J. A., Siminovitch, K., Liu, I. S., Waye, J., Voruganti, P. et al., Dopamine D4 receptor variant, D4GLYCINE194, in Africans, but not in Caucasians: no association with schizophrenia, *Am. J. Med. Genet.*, 54, 384, 1994.

146. Daniels, J., Williams, J., Mant, R., Asherson, P., McGuffin, P., and Owen, M. J., Repeat length variation in the dopamine D4 receptor gene shows no evidence of association with schizophrenia, *Am. J. Med. Genet.*, 54, 256, 1994.

147. Maier, W., Schwab, S., Hallmayer, J., Ertl, M. A., Minges, J., Ackenheil M., Lichtermann, D., and Wildenauer, D., Absence of linkage between schizophrenia and the dopamine D4 receptor gene, *Psychiatr. Res.*, 53, 77, 1994.

148. Macciardi, F., Petronis, A., Van Tol, H. H., Marino, C., Cavallini, M. C., Smeraldi E., and Kennedy, J. L., Analysis of the D4 dopamine receptor gene variant in an Italian schizophrenia kindred, *Arch. Gen. Psychiatr.*, 51, 288, 1994.

149. Shaikh, S., Gill, M., Owen, M., Asherson, P., McGuffin, P., Nanko, S., Murray, R. M., and Collier, D. A., Failure to find linkage between a functional polymorphism in the dopamine D4 receptor gene and schizophrenia, *Am. J. Med. Genet.*, 54, 8, 1994.

150. Barr, C. L., Kennedy, J. L., Lichter, J. B., Van Tol, H. H., Wetterberg, L., Livak K. J., and Kidd, K. K., Alleles at the dopamine D4 receptor locus do not contribute to the genetic susceptibility to schizophrenia in a large Swedish kindred, *Am. J. Med. Genet.*, 48, 218, 1993.

151. Weiss, J., Magert, H. J., Cieslak, A., and Forssmann, W. G., Association between different psychotic disorders and the DRD4 polymorphism, but no differences in the main ligand binding region ofthe DRD4 receptor protein compared to controls, *Eur. J. Med. Res.*, 1, 439, 1996.

152. Kramer, M. S., Last, B., Getson, A., and Reines, S. A., The effects of a selective D4 dopamine receptor antagonist (L-745,870) in acutely psychotic inpatients with schizophrenia, D4 Dopamine Antagonist Group, *Arch. Gen. Psychiatr.*, 54, 567, 1997.

153. Williams, J., Spurlock, G., McGuffin, P., Mallet, J., Nothen, M. M., Gill, M., Aschauer, H., Nylander, P. O., Macciardi, F., and Owen, M. J., Association between schizophrenia and T102C polymorphism of the 5-hydroxytryptamine type 2a-receptor gene. European Multicentre Association Study of Schizophrenia (EMASS) Group, *Lancet*, 347, 1294, 1996.

154. Kalsi, G., Sherrington, R., Mankoo, B. S., Brynjolfsson, J., Sigmundsson, T., Curtis D., Read, T., Murphy, P., Butler, R., Petursson, H., and Gurling, H. M., Linkage study of the D5 dopamine receptor gene (DRD5) in multiplex Icelandic and English schizophrenia pedigrees, *Am. J. Psychiatr.*, 153, 107, 1996.

155. Ravindranathan, A., Coon, H., DeLisi, L., Holik, J., Hoff, M., Brown, A., Shields G., Crow, T., and Byerley, W., Linkage analysis between schizophrenia and a microsatellite polymorphism for the D5 dopamine receptor gene, *Psychiatr. Genet.*, 4, 77, 1994.

156. Meloni, R., Laurent, C., Campion, D., Hadjali, B. B., Thibaut, F., Dollfus, S. et al., A rare allele of a microsatellite located in the tyrosine hydroxylase gene found in schizophrenia patients, *C. R. Acad. Sci. III*, 318, 803, 1995

157. Thibaut, F., Ribeyre, J. M., Dourmap, N., Meloni, R., Laurent, C., Campion, D., Menard, J. F., Dollfus, S., Mallet, J., and Petit, M., Association of DNA polymorphism in the first intron of the tyrosine hydroxylase gene with disturbances of the catecholaminergic system in schizophrenia, *Schizophr. Res.*, 23, 259, 1997.

158. Burgert, E., Crocq M.-A., Bausch, E., Macher, J.-P., and Morris-Rosendahl, D. J., No association between the tyrosine hydroxylase microsatellite marker HUMTH01 and schizophrenia or bipolar I disorder, *Psychiatr. Genet.*, 8, 45, 1998.

159. Persico, A. M. and Macciardi, F., Genotypic association between dopamine transporter gene polymorphisms and schizophrenia, *Am. J. Med. Genet.*, 74, 53, 1997.

160. Inada, T., Sugita, T., Dobashi, I., Inagaki, A., Kitao, Y., Matsuda, G., Kato, S., Takano, T., Yagi, G., and Asai, M., Dopamine transporter gene polymorphism and psychiatric symptoms seen in schizophrenic patients at their first episode, *Am. J. Med. Genet.*, 67, 406, 1996.

161. Maier, W., Minges, J., Eckstein, N., Brodski, C., Albus, M., Lerer B., Hallmayer, J., Fimmers, R., Ackenheil, M., Ebstein, R. E., Borrmann M., Lichtermann, D., and Wildenauer, D. B., Genetic relationship between dopamine transporter gene and schizophrenia: linkage and association, *Schizophr. Res.*, 20, 175, 1996.

162. Bodeau-Pean, S., Laurent, C., Campion, D., Jay, M., Thibaut, F., Dollfus S., Petit, M., Samolyk, D., d'Amato, T., Martinez, M. et al., No evidence for linkage or association between the dopamine transporter gene and schizophrenia in a French population, *Psychiatr. Res.*, 59, 1, 1995.

163. Persico, A. M., Wang, Z. W., Black, D. W., Andreasen, N. C., Uhl, G. R., and Crowe, R. R., Exclusion of close linkage of the dopamine transporter gene with schizophrenia spectrum disorders, *Am. J. Psychiatr.*, 152, 134, 1995.

164. Li, T., Yang, L., Wiese, C., Xu, C. T., Zeng, Z., Giros, B., Caron, M. G., Moises, H. W., and Liu, X., No association between alleles or genotypes at the dopamine transporter gene and schizophrenia, *Psychiatr. Res.*, 52, 17, 1994.

165. Byerley, W., Coon, H., Hoff, M., Holik, J., Waldo, M., Freedman, R., Caron M. G., and Giros, B., Human dopamine transporter gene not linked to schizophrenia in multigenerational pedigrees, *Hum. Hered.*, 43, 319, 1993.

166. Persico, A. M., Wang, Z. W., Black, D. W., Andreasen, N. C., Uhl, G. R., and Crowe, R. R., Exclusion of close linkage between the synaptic vesicular monoamine transporter locus and schizophrenia spectrum disorders, *Am. J. Med. Genet.*, 60, 563, 1995.

167. Meszaros, K., Lenzinger, E., Fureder, T., Hornik, K., Willinger, U., Stompe, T., Heiden, A. M., Resinger, E., Fathi, N., Gerhard, E., Fuchs, K., Miller-Reiter, E., Pfersmann, V., Sieghart, W., Aschauer, H. N., and Kasper, S., Schizophrenia and the dopamine-beta-hydroxylase gene: results of a linkage and association study, *Psychiatr. Genet.* 6, 17, 1996.

168. Inayama, Y., Yoneda, H., Sakai, T., Ishida, T., Nonomura, Y., Kono, Y., Takahata, R., Koh, J., Sakai, J., Takai, A., Inada, Y., and Asaba, H., Positive association between a DNA sequence variant in the serotonin 2A receptor gene and schizophrenia, *Am. J. Med. Genet.*, 67, 103, 1996.

169. Verga, M., Macciardi, F., Cohen, S., Pedrini, S., and Smeraldi, E., No association between schizophrenia and the serotonin receptor 5HTR2a in an Italian population, *Am. J. Med. Genet.*, 74, 21, 1997.

170. Hallmayer, J., Kennedy, J. L., Wetterberg, L., Sjogren, B., Kidd, K. K., and Cavalli-Sforza, L. L., Exclusion of linkage between the serotonin2 receptor and schizophrenia in a large Swedish kindred, *Arch. Gen. Psychiatr.*, 49, 216, 1992.

171. Williams, J., Spurlock, G., McGuffin, P., Mallet, J., Nothen, M. M., Gill, M., Aschauer H, Nylander, P. O., Macciardi, F., and Owen, M. J., Association between schizophrenia and T102C polymorphism of the 5-hydroxytryptamine type 2a-receptor gene. European Multicentre Association Study of Schizophrenia (EMASS) Group, *Lancet*, 347, 1294, 1996.

172. Spurlock, G., Heils, A., Holmans, P., Williams, J., D'Souza, U. M., Cardno, A., Murphy, K. C., Jones, L., Buckland, P. R., McGuffin, P., Lesch, K. P., and Owen, M. J., A family-based association study of T102C polymorphism in 5HT2A and schizophrenia, plus identification of new polymorphisms in the promoter, *Mol. Psychiatr.*, 3, 42, 1998.

173. Williams, J., McGuffin, P., Nothen, M., and Owen, M. J., Meta-analysis of association between the 5.2a receptor T102C polymorphism and schizophrenia, EMASS Collaborative Group, *Lancet,* 349(9060), 1221, 1997.

174. Ohara, K., Ino, A., Ishigaki, T., Tani, K., Tsukamoto, T., Nakamura, Y., and Ohara, K., Analysis of the 5'-flanking promoter region of the 5HT-2A receptor gene in schizophrenia, *Neuropsychopharmacology*, 17, 274, 1997.

175. Ishigaki, T., Intact 5.2A receptor exons and the adjoining intron regions in schizophrenia, *Neuropsychopharmacology*, 15, 323, 1996.

176. Erdmann, J., Shimron-Abarbanell, D., Rietschel, M., Albus, M., Maier, W., Korner, J., Bondy, B., Chen, K., Shih, J. C., Knapp, M., Propping, P., and Nothen, M. M., Systematic screening for mutations in the human serotonin. (5-HT2A) receptor gene: identification of two naturally occurring receptor variants and association analysis in schizophrenia, *Hum. Genet.*, 97, 614, 1996.

177. Bonnet-Brilhault, F., Laurent, C., Thibaut, F., Campion, D., Chavand, O., Samolyk, D., Martinez, M., Petit, M., and Mallet, J., Serotonin transporter gene polymorphism and schizophrenia: an association study, *Biol. Psychiatr.*, 42, 634, 1997.

178. Shimron-Abarbanell, D., Harms, H., Erdmann, J., Albus, M., Maier, W., Rietschel, M., Korner J, Weigelt, B., Franzek, E., Sander, T., Knapp, M., Propping, P., and Nothen, M. M., Systematic screening for mutations in the human serotonin 1F receptor gene in patients with bipolar affective disorder and schizophrenia, *Am. J. Med. Genet.*, 67, 225, 1996.

179. Erdmann, J., Nothen, M. M., Shimron-Abarbanell, D., Rietschel, M., Albus, M., Borrmann, M., Maier, W., Franzek, E., Korner, J., Weigelt, B., Fimmers, R., and Propping, P., The human serotonin 7 (5-HT7) receptor gene: genomic organization and systematic mutation screening in schizophrenia and bipolar affective disorder, *Mol. Psychiatr.*, 1, 392, 1996.

180. Sidenberg, D. G., Bassett, A. S., Demchyshyn, L., Niznik, H. B., Macciardi F., Kamble, A. B., Honer, W. G., and Kennedy, J. L., New polymorphism for the human serotonin 1D receptor variant (5-HT1D beta) not linked to schizophrenia in five Canadian pedigrees, *Hum. Hered.*, 43, 315, 1993.

181. Sobell, J. L., Lind, T. J., Hebrink, D. D., Heston, L. L., and Sommer, S. S., Screening the monoamine oxidase B gene in 100 male patients with schizophrenia: a cluster of polymorphisms in African-Americans but lack of functionally significant sequence changes, *Am. J. Med. Genet.*, 74, 44, 1997.

182. Coron, B., Campion, D., Thibaut, F., Dollfus, S., Preterre, P., Langlois, S., Vasse T., Moreau, V., Martin, C., Charbonnier, F., Laurent, C., Mallet, J. P., and Frebourg, T., Association study between schizophrenia and monoamine oxidase A and BDNA polymorphisms, *Psychiatr. Res.*, 62, 221, 1996.

183. Stober, G., Nothen, M. M., Porzgen, P., Bruss, M., Bonisch, H., Knapp, M., Beckmann H., and Propping, P., Systematic search for variation in the human norepinephrine transporter gene: identification of five naturally occurring missense mutations and study of association with major psychiatric disorders, *Am. J. Med. Genet.*, 67, 523, 1996.

184. Hayakawa, T., Ishiguro, H., Toru, M., Hamaguchi, H., and Arinami, T., Systematic search for mutations in the 14-3-3-beta chain gene on chromosome 22 in schizophrenics, *Psychiatr. Genet.*, 8, 33, 1998.

185. Chen, C. H., Lee, Y. R., Liu, M. Y., Wei, F. C., Koong, F. J., Hwu, H. G., and Hsiao, K. J., Identification of a BglI polymorphism of catechol-O-methyltransferase (COMT) gene, and association study with schizophrenia, *Am. J. Med. Genet.*, 67, 556, 1996.

186. Strous, R. D., Bark, N., Woerner, M., and Lachman, H. M., Lack of association of a functional catechol-O-methyltransferase gene polymorphism in schizophrenia, *Biol. Psychiatr.*, 41, 493, 1997.

187. Pariseau, C., Gregor, P., Myles-Worsley, M., Holik, J., Hoff, M., Waldo, M., Freedman R., Coon, H., Byerley, W., Schizophrenia and glutamate receptor genes, *Psychiatr. Genet.*, 4, 161, 1994.

188. Chen, A. C., Kalsi, G., Brynjolfsson, J., Sigmundsson, T., Curtis, D., Butler R., Read T, Murphy, P., Petursson, H., Barnard, E. A., and Gurling, H. M., Lack of evidence for close linkage of the glutamate GluR6 receptor gene with schizophrenia, *Am. J. Psychiatr.*, 153, 1634, 1996.

189. Chen, A. C., Kalsi, G., Brynjolfsson, J., Sigmundsson, T., Curtis, D., Butler R., Read T., Murphy, P., Barnard, E. A., Petursson, H., and Gurling, H. M., Exclusion of linkage between schizophrenia and the gene encoding a neutral amino acid glutamate/aspartate transporter, SLC1A5, *Am. J. Med. Genet.*, 74, 50, 1997.

190. Chen, A. C., Kalsi, G., Brynjolfsson, J., Sigmundsson, T., Curtis, D., Butler R., Read T., Murphy, P., Petursson, H., Barnard, E. A., and Gurling, H. M., Exclusion of linkage of schizophrenia to the gene for the glutamate GluR5 receptor, *Biol. Psychiatr.*, 41, 243, 1997.

191. Riley, B. P., Tahir, E., Rajagopalan, S., Mogudi-Carter, M., Faure, S., Weissenbach, J., Jenkins, T., and Williamson, R., A linkage study of the N-methyl-D-aspartate receptor subunit gene loci and schizophrenia in southern African Bantu-speaking families, *Psychiatr. Genet.*, 7, 57, 1997.

192. Bell, R., Collier, D. A., Rice, S. Q., Roberts, G. W., MacPhee, C. H., Kerwin R. W., Price, J., and Gloger, I. S., Systematic screening of the LDL-PLA2 gene for polymorphic variants and case-control analysis in schizophrenia, *Biochem. Biophys. Res. Commun.*, 241, 630, 1997.

193. Price, S. A., Fox, H., St. Clair, C. D., and Shaw, D. J., Lack of association between schizophrenia and a polymorphism close to the cytosolic phospholipase A2 gene, *Psychiatr. Genet.*, 7, 111, 1997.

194. Andrew, B., Watt, D. C., Gillespie, C., and Chapel, H., A study of genetic linkage in schizophrenia, *Psychol. Med.*, 17, 363, 1987.

195. McGuffin, P. and Sturt, E., Genetic markers in schizophrenia, *Hum. Hered.*, 36, 65, 1986.

196. Chadda, R., Kulhara, P., Singh, T., and Sehgal, S., HLA antigens in schizophrenia: a family study, *Br. J. Psychiatry*, 149, 612, 1986.

197. Goldin, L. R., DeLisi, L. E., and Gershon, E. S., Relationship of HLA to schizophrenia in 10 nuclear families, *Psychiatr. Res.*, 20, 69, 1987.

198. Alexander, R. C., Coggiano, M., Daniel, D. G., and Wyatt, R. J., HLA antigens in schizophrenia, *Psychiatr. Res.*, 31, 221, 1990.

199. Owen, M. J. and McGuffin, P., DNA and classical genetic markers in schizophrenia, *Eur. Arch. Psychiatr. Clin. Neurosci.*, 240, 193, 1991.

200. Campion, D., Leboyer, M., Hillaire, D., Halle, L., Gorwood, P., Cavelier B., Soufflet, M. F., d'Amato, T., Muller, B., Kaplan, C. et al., Relationship of HLA to schizophrenia not supported in multiplex families, *Psychiatr. Res.*, 41, 99, 1992.

201. Blackwood, D. H., Muir, W. J., Stephenson, A., Wentzel, J., Ad'hiah, A., Walker, M. J., Papiha S. S., St. Clair, D. M., and Roberts, D. F., Reduced expression of HLA.5 in schizophrenia, *Psychiatr. Genet.*, 6, 51, 1996.

202. Jones, C. T., Morris, S., Yates, C. M., Moffoot, A., Sharpe, C., Brock, D. J., and St. Clair, D., Mutation in codon 713 of the beta amyloid precursor protein gene presenting with schizophrenia, *Nat. Genet.*, 1, 306, 1992.

203. Fukuda, R., Hattori, M., Sasaki, T., Kazamatsuri, H., Kuwata, S., Shibata, Y., and Nanko, S., No evidence for a point mutation at codon 713 and 717 of amyloid precursor protein gene in Japanese schizophrenics, *Jpn. J. Hum. Genet.*, 38, 407, 1993.

204. Arnholt, J. C., Sobell, J. L., Heston, L. L., and Sommer, S. S., APP mutations and schizophrenia, *Biol. Psychiatr.*, 34, 739, 1993.

205. Nothen, M. M., Erdmann, J., Propping, P., Lanczik, M., Rietschel, M., Korner, J., Maier W., Albus, M., Ertl, M. A., and Wildenauer, D. B., Mutation in the beta amyloid precursor protein gene and schizophrenia, *Biol. Psychiatr.*, 34, 502, 1993.

206. Coon, H., Hoff, M., Holik, J., Delisi, L. E., Crowe, T., Freedman, R., Shields, G., Boccio, A. M., Lerman, M., Gershon, E. S. et al., C to T nucleotide substitution in codon 713 of amyloid precursor-protein gene not found in 86 unrelated schizophrenics from multiplexfamilies, *Am. J. Med. Genet.*, 48, 36, 1993.

207. Asherson, P., Mant, R., Taylor, C., Sargeant, M., Collier, D., Clements, A. et al., Failure to find linkage between schizophrenia and genetic markers on chromosome 21, *Am. J. Med. Genet.*, 48, 161, 1993.

208. Carter, D., Campion, D., d'Amato, T., Jay, M., Brice, A., Bellis, M., Mallet, J., and Agid, Y., No mutation in codon 713 of the amyloid precursor gene in schizophrenic patients, *Hum. Mol. Genet.*, 2, 321, 1993.

209. Mortilla, M., Amaducci, L., Bruni, A., Montesi, M. P., Trubnikov, A., De Cataldo S., Pallanti, S., Pazzagli, A., Grecu, L., Servi, P. et al., Absence of APP713 mutation in Italian and Russian families with schizophrenia, *Neurosci. Lett.*, 165, 45, 1994.

210. Karayiorgou, M., Kasch, L., Lasseter, V. K., Hwang, J., Elango, R., Bernardini, D. J., Kimberland, M., Babb, R., Francomano, C. A., Wolyniec, P. S. et al., Report from the Maryland epidemiology schizophrenia linkage study: no evidence for linkage between schizophrenia and a number of candidates and other genomic regions using a complex dominant model, *Am. J. Med. Genet.*, 54, 345, 1994.

211. Zhu, S., Nothen, M. M., Uhlhaas, S., Rietschel, M., Korner, J., Lanczik, M., Fimmers R., Propping, P., Apolipoprotein E genotype distribution in schizophrenia, *Psychiatr. Genet.*, 6, 75, 1996.

212. Nimgaonkar, V. L., Yang, Z. W., Zhang, X. R., Brar, J. S., Chakravarti, A., and Ganguli R., Association study of schizophrenia and the IL-2 receptor beta chain gene, *Am. J. Med. Genet.*, 60, 448, 1995.

213. Tatsumi, M., Sasaki, T., Sakai, T., Kamijima, K., Fukuda, R., Kunugi, H., Hattori, M., and Nanko, S., Genes for interleukin-2 receptor beta chain, interleukin-1 beta, and schizophrenia: no evidence for the association or linkage, *Am. J. Med. Genet.*, 74, 338, 1997.

214. Laurent, C., Thibaut, F., Ravassard, P., Campion, D., Samolyk, D., Lafargue C., Petit, M., Martinez, M., and Mallet, J., Detection of two new polymorphic sites in the human interleukin-1 beta gene: lack of association with schizophrenia in a French population, *Psychiatr. Genet.*, 7, 103, 1997.

215. Thome, J., Durany, N., Harsanyi, A., Foley, P., Palomo, A., Kornhuber, J., Weijers, H. G., Baumer, A., Rosler, M., Cruz-Sanchez, F. F., Beckmann, H., and Riederer, P., A null mutation allele in the CNTF gene and schizophrenic psychoses, *Neuroreport*, 7, 1413, 1996.

216. Sakai, T., Sasaki, T., Tatsumi, M., Kunugi, H., Hattori, M., and Nanko, S., Schizophrenia and the ciliary neurotrophic factor (CNTF) gene: no evidence for association, *Psychiatr. Res.*, 71, 7, 1997.

217. Vicente, A. M., Macciardi, F., Verga, M., Bassett, A. S., Honer, W. G., Bean, G., and Kennedy, J. L., NCAM and schizophrenia: genetic studies, *Mol. Psychiatr.*, 2, 65, 1997.

218. Nanko, S., Hattori, M., Kuwata, S., Sasaki, T., Fukuda, R., Dai, X. Y., Yamaguchi K., Shibata, Y., and Kazamatsuri, H., Neurotrophin-3 gene polymorphism associated with schizophrenia, *Acta. Psychiatr. Scand.*, 89, 390, 1994.

219. Gill, M., Hawi, Z., O'Neill, F. A., Walsh, D., Straub, R. E., and Kendler, K. S., Neurotrophin-3 gene polymorphisms and schizophrenia: no evidence for linkage or association, *Psychiatr. Genet.*, 6, 183, 1996.

220. Jonsson, E., Brene, S., Zhang, X. R., Nimgaonkar, V. L., Tylec, A., Schalling, M., and Sedvall, G., Schizophrenia and neurotrophin-3 alleles, *Acta. Psychiatr. Scand.*, 95, 414-9, 1997.

221 Orange, P. R., Heath, P. R., Wright, S. R., Ramchand, C. N., Kolkeiwicz, L., and Pearson, R. C., Individuals with schizophrenia have an increased incidence of the H2R649G allele for the histamine H2 receptor gene, *Mol. Psychiatr.*, 1, 466, 1996.

222. Whatley, S. A., Curti, D., and Marchbanks, R. M., Mitochondrial involvement in schizophrenia and other functional psychoses, *Neurochem. Res.*, 21: 995, 1996.

223. Lindholm, E., Cavelier, L., Howell, W. M., Eriksson, I., Jalonen, P., Adolfsson, R., Blackwood, D. H., Muir, W. J., Brookes, A. J., Gyllensten, U., and Jazin, E. E., Mitochondrial sequence variants in patients with schizophrenia, *Eur. J. Hum. Genet.*, 5, 406, 1997.

224. Sirota, P., Frydman, M., and Sirota, L., Schizophrenia and Marfan syndrome, *Br. J. Psychiatr.*, 157, 433, 1990.

225. Kalsi, G., Mankoo, B. S., Brynjolfsson, J., Curtis, D., Read, T., Murphy, P., Sharma T., Petursson, H., and Gurling, H. M., The Marfan syndrome gene locus as a favoured locus for susceptibility to schizophrenia, *Psychiatr. Genet.*, 4, 219, 1994.

226. Zatz, M., Vallada, H., Melo, M. S., Passos-Bueno, M. R., Vieira, A. H., Vainzof, M., Gill, M., and Gentil, V., Cosegregation of schizophrenia with Becker muscular dystrophy: susceptibility locus for schizophrenia at Xp21 or an effect of the dystrophin gene in the brain?, *J. Med. Genet.*, 30, 131, 1993.

227. Melo, M., Vieira, A. H., Passos-Bueno, M. R., and Zaty, M., Association of schizophrenia and Duchenne muscular dystrophy, *Br. J. Psychiatr.*, 162, 711, 1993.

228. Lindor, N. M., Sobell, J. L., Heston, L. L., Thibodeau, S. N., and Sommer, S. S., Screening the dystrophin gene suggests a high rate of polymorphism in general but no exonic deletions in schizophrenics, *Am. J. Med. Genet.*, 54, 1, 1994.

229. Wang, S., Detera-Wadleigh, S. D., Coon, H., Sun, C. E., Goldin, L. R., Duffy D. L., Byerley, W. F., Gershon, E. S., and Diehl, S. R., Evidence of linkage disequilibrium between schizophrenia and the SCA1 CAG repeat on chromosome 6p23, *Am. J. Hum. Genet.*, 59, 731, 1996.

230. Pujana, M. A., Martorell, L., Volpini, V., Valero, J., Labad, A., Vilella, E., and Estivill, X., Analysis of amino-acid and nucleotide variants in the spinocerebellar ataxia type 1 (SCA1) gene in schizophrenic patients, *Hum. Genet.*, 99, 772, 1997.

231. Morris-Rosendahl, D. J., Burgert, E., Uyanik, G., Mayerova, A., Duval, F., Macher, J. P., and Crocq, M. A., Analysis of the CAG repeats in the SCA1 and B37 genes in schizophrenic and bipolar I disorder patients: tentative association between B37 and schizophrenia, *Am. J. Med. Genet.*, 74, 324, 1997.

232. Crow, T. J., Poulter, M., Lofthouse, R., Chen, G., Shah, T., Bass, N., Morganti, C., Vita, A., Smith, C., Boccio-Smith, A. et al., Male siblings with schizophrenia share alleles at the androgen receptor above chance expectation, *Am. J. Med. Genet.*, 48, 159, 1993.

233. Arranz, M., Sharma, T., Sham, P., Kerwin, R., Nanko, S., Owen, M., Gill, M., and Collier, D., Schizophrenia and the androgen receptor gene: report of a sibship showing co-segregation with Reifenstein syndrome but no evidence for linkage in 23 multiply affected families, *Am. J. Med. Genet.*, 60, 377, 1995.

234. O'Malley, M. P., Bassett, A. S., Honer, W. G., Kennedy, J. L., King, N., and Berg, D., Linkage analysis between schizophrenia and the Darier's disease region on 12q, *Psychiatr. Genet.*, 6, 187, 1996.

235. Eaton, W. W., Hayward, C., and Ram, R., Schizophrenia and rheumatoid arthritis: a review, *Schizophr. Res.*, 6, 181, 1992.

236. Vinogradov, S., Gottesman, I. I., Moises, H. W., and Nicol, S., Negative association between schizophrenia and rheumatoid arthritis, *Schizophr. Bull.*, 17, 669, 1991.

237. Lauerma, H., Lehtinen, V., Joukamaa, M., Jarvelin, M. R., Helenius, H., and Isohanni, M., Schizophrenia among patients treated for rheumatoid arthritis and appendicitis, *Schizophr. Res.*, 29, 255, 1998.

238. Wright, P., Donaldson, P. T., Underhill, J. A., Choudhuri, K., Doherty, D. G., and Murray, R. M., Genetic association of the HLA DRB1 gene locus on chromosome 6p21.3 with schizophrenia, *Am. J. Psychiatr.*, 153, 1530, 1996.

239. Hollister, J. M., Laing, P., and Mednick, S. A., Rhesus incompatibility as a risk factor for schizophrenia in male adults, *Arch. Gen. Psychiatr.*, 53, 19, 1996.

240. La Spada, A. R., Trinucleotide repeat instability: genetic features and molecular mechanisms, *Brain Pathol.*, 7, 943, 1997.

241. Petronis, A. and Kennedy, J. L., Unstable genes — unstable mind?, *Am. J. Psychiatry*, 152, 164, 1995.

242. Petronis, A., Sherrington, R. P., Paterson, A. D., and Kennedy, J. L., Genetic anticipation in schizophrenia: pro and con, *Clin. Neurosci.*, 3, 76, 1995.

243. Chotai, J., Engstrom, C., Ekholm, B., son Berg, M. L., Adolfsson, R., and Nylander, P. O., Anticipation in Swedish families with schizophrenia, *Psychiatr. Genet.*, 5, 181, 1995.

244. Gorwood, P., Leboyer, M., Falissard, B., Jay, M., Rouillon, F., and Feingold, J., Anticipation in schizophrenia: new light on a controversial problem, *Am. J. Psychiatr.*, 153, 1173, 1996.

245. Stober, G., Franzek, E., Lesch, K. P., and Beckmann, H., Periodic catatonia: a schizophrenic subtype with major gene effect and anticipation, *Eur. Arch. Psychiatr. Clin. Neurosci.*, 245, 135, 1995.

246. Thibaut, F., Martinez, M., Petit, M., Jay, M., and Campion, D., Further evidence for anticipation in schizophrenia, *Psychiatr. Res.*, 59, 25, 1995.

247. Valero, J., Martorell, L., Marine, J., Vilella, E., and Labad, A., Anticipation and imprinting in Spanish families with schizophrenia, *Acta. Psychiatr. Scand.*, 97, 343, 1998.

248. Johnson, J. E., Cleary, J., Ahsan, H., Harkavy Friedman, J., Malaspina, D., Cloninger, C. R., Faraone, S. V., Tsuang, M. T., and Kaufmann, C. A., Anticipation in schizophrenia: biology or bias?, *Am. J. Med. Genet.*, 74, 275, 1997.

249. Bassett, A. S. and Honer, W. G., Evidence for anticipation in schizophrenia, *Am. J. Hum. Genet.*, 54, 864, 1994.

250. Husted, J., Scutt, L. E., and Bassett, A. S., Paternal transmission and anticipation in schizophrenia, *Am. J. Med. Genet.*, 81, 156, 1998.

251. Ohara, K., Xu, H. D., Mori, N., Suzuki, Y., Xu, D. S., Ohara, K., and Wang, Z. C., Anticipation and imprinting in schizophrenia, *Biol. Psychiatr.*, 42, 760, 1997.

252. Asherson, P., Walsh, C., Williams, J., Sargeant, M., Taylor, C., Clements, A., Gill, M., Owen, M., and McGuffin, P., Imprinting and anticipation. Are they relevant to genetic studies of schizophrenia?, *Br. J. Psychiatry*, 164, 619, 1994.

253. Yaw, J., Myles-Worsley, M., Hoff, M., Holik, J., Freedman, R., Byerley, W., and Coon, H., Anticipation in multiplex schizophrenia pedigrees, *Psychiatr. Genet.*, 6, 7, 1996.

254. Kendler, K. S. and MacLean, C. J., Estimating familial effects on age at onset and liability to schizophrenia. I. Results of a large sample family study, *Genet. Epidemiol.*, 7, 409, 1990.

255. Huang, J. and Vieland, V., A new statistical test for age-of-onset anticipation: application to bipolar disorder, *Genet. Epidemiol.*, 14, 1091, 1997.

256. Schalling, M., Hudson, T. J., Buetow, K. H., and Housman, D. E., Direct detection of novel expanded trinucleotide repeats in the human genome, *Nat. Genet.*, 4, 135, 1993.

257. Morris, A. G., Gaitonde, E., McKenna, P. J., Mollon, J. D., and Hunt, D. M., CAG repeat expansions and schizophrenia: association with disease in females and with early age-at-onset, *Hum. Mol. Genet.*, 4, 1957, 1995.

258. O'Donovan, M. C., Guy, C., Craddock, N., Bowen, T., McKeon, P., Macedo, A., Maier, W., Wildenauer, D., Aschauer, H. N., Sorbi, S., Feldman, E., Mynett-Johnson, L., Claffey, E., Nacmias, B., Valente, J., Dourado, A., Grassi, E., Lenzinger, E., Heiden, A. M., Moorhead, S., Harrison, D., Williams, J. M. P., and Owen, M. J., Confirmation of association between expanded CAG/CTG repeats and both schizophrenia and bipolar disorder, *Psychol. Med.*, 26, 1145, 1996.

259. Laurent, C., Zander, C., Thibaut, F., Bonnet-Brilhault, F., Chavand, O., Jay, M., Samolyk, D., Petit, M., Martinez, M., Campion, D., Neri, C., Mallet, J., and Cann, H., Anticipation in schizophrenia: no evidence of expanded CAG/CTG repeat sequences in French families and sporadic cases, *Am. J. Med. Genet.*, 81, 342, 1998.

260. Ohara, K., Tani, K., Tsukamoto, T., Ino, A., Nagai, M., Suzuki, Y., Xu, H. D., Xu, D. S., Wang, Z. C., and Ohara, K., Three CAG trinucleotide repeats on chromosome 6 (D6S1014, D6S1015, and D6S1058) are not expanded in 30 families with schizophrenia, *Neuropsychopharmacology*, 17, 279, 1997.

261. Ohara, K., Tani, K., Tsukamoto, T., Suzuki, Y., Xu, H. D., Xu, D. S., Wang, Z. C., and Ohara, K., Exclusion of five trinucleotide repeat (CAG and CCG) expansions in 17 families with schizophrenia, *Biol. Psychiatr.*, 42, 756, 1997.

262. Lesch, K. P., Stober, G., Balling, U., Franzek, E., Li, S. H., Ross, C. A., Newman, M., Beckmann, H., and Riederer, P., Triplet repeats in clinical subtypes of schizophrenia: variation at the DRPLA (B 37 CAG repeat) locus is not associated with periodic catatonia, *J. Neural Transm. Gen. Sect.*, 98, 153, 1994.

263. Petronis, A., Bassett, A. S., Honer, W. G., Vincent, J. B., Tatuch, Y., Sasaki, T., Ying, D. J., Klempan, T. A., and Kennedy, J. L., Search for unstable DNA in schizophrenia families with evidence for genetic anticipation, *Am. J. Hum. Genet.*, 59, 905, 1996.

264. Schurhoff, F., Stevanin, G., Trottier, Y., Bellivier, F., Mouren-Simeoni, M. C., Brice, A., and Leboyer, M., A preliminary study on early onset schizophrenia and bipolar disorder: large polyglutamine expansions are not involved, *Psychiatr. Res.*, 72, 141, 1997.

265. Sidransky, E., Burgess, C., Ikeuchi, T., Lindblad, K., Long, R. T., Philibert, R. A., Rapoport, J., Schalling, M., Tsuji, S., and Ginns, E. I., A triplet repeat on 17q accounts for most expansions detected by the repeat-expansion-detection technique, *Am. J. Hum. Genet.*, 62, 1548, 1998.

266. Coon, H., Jensen, S., Holik, J., Hoff, M., Myles-Worsley, M., Reimherr, F., Wender, P., Waldo, M., Freedman, R., Leppert, M., and Byerley, W., Genomic scan for genes predisposing to schizophrenia, *Am. J. Med. Genet.*, 54, 59, 1994.

267. Moises, H. W., Yang, L., Kristbjarnarson, H., Wiese, C., Byerley, W., Macciardi, F., Arolt, V., Blackwood, D., Liu, X., Sjogren, B., Aschauer, H. N., Hwu, H.-G., Jang, K., Livesley, W. J., Kennedy, J. L., Zoega, T., Ivarsson, O., Bui, M.-T., Yu, M.-H., Havsteen, B., Commenges, D., Weissenbach, J., Schwinger, E., Gottesman, I. I., Pakstis, A. J., Wetterberg, L., Kidd, K. K., and Helgason, T., An international two-stage genome-wide search for schizophrenia susceptibility genes, *Nat. Genet.*, 11, 321, 1995.

268. Levinson, D. F., Mahtani, M. M., Nancarrow, D. J., Brown, D. M., Kruglyak, L., Kirby A, Hayward, N. K., Crowe, R. R., Andreasen, N. C., Black, D. W., Silverman J. M., Endicott, J., Sharpe, L., Mohs, R. C., Siever, L. J., Walters, M. K., Lennon D. P., Jones, H. L., Nertney, D. A., Daly, M. J., Gladis, M., and Mowry, B. J., Genome scan of schizophrenia, *Am. J. Psychiatr.*, 155, 741, 1998.

269. Faraone, S. V., Matise, T., Svrakic, D., Pepple, J., Malaspina, D., Suarez, B., Hampe, C., Zambuto, C. T., Schmitt, K., Meyer, J., Markel, P., Lee, H., Harkevy-Friedman, J., Kaufmann, C. A., Cloninger, C. R., and Tsuang, M. T., A genome scan of the European-American schizophrenia pedigrees of the NIMH Genetics Initiative, *Am. J. Med. Genet.*, 81, 290-295, 1998.

270. Kaufmann, C. A., Suarez, B., Malaspina, D., Pepple, J., Svrakic, D., Markel, P. D., Meyer, J., Zambuto, C. T., Schmitt, K., Matise, T. C., Harkavy-Friedman, J. M., Hampe, C., Lee, H., Shore, D., Wynne, D., Faraone, S. V., Tsuang, M. T., and Cloninger, C. R., The NIMH genetics initiative Millennium schizophrenia consortium: linkage analysis of African-American pedigrees, *Am. J. Med. Genet.*, 81, 282, 1998.

271. Blouin, J.-L., Dombrowski, B., Nath, S. K., Lasseter, V., Wolyniec, P. S., Nestadt, G., Thornquist, M., Ullrich, G., McGrath, J., Kasch, L., Lamacz, M., Thomas, M. G., Gehrig, C., Radhakrishna, U., Snyder, S. E., Balk, K. G., Neufeld, K., DeMarchi, N., Papadimitriou, G. N., Childs. B., Housman, D. E., Kazazian, H. H., Antonarakis, S. E., and Pulver, A. E., Schizophrenia susceptibility loci identified on chromosomes 13q32 and 8p21, *Nat. Genet.*, 20, 70, 1998.

272. Shaw, S. H., Kelly M., Smith, A. B., Shields, G., Hopkins, P. J., Loftus, J., Laval, S. H., Vita, A., De Hert, M., Cardon, L., Crow, T. J., Sherrington, R., and DeLisi, L. E., A genome-side search for schizophrenia susceptibility genes, *Am. J. Med. Genet.*, 81, 364, 1998.

273. Coon, H., Myles-Worsley, M., Tiobech, J., Hoff, M., Rosenthal, J., Bennett, P., Reimherr, F., Wender, P., Dale, P., Polloi, A., and Byerley, W., Evidence for a chromosome 2p13-14 schizophrenia susceptibility locus in families from Palau, Micronesia, *Mol. Psychiatr.*, 3, 521, 1998.

274. Williams, N. M., Rees, M. I., Holmans, P., Daniels, J., Fenton, I., Cardno, A. G., Murphy, K. C., Jones, L. A., Asherson, P., McGuffin, P., and Owen, M. J., Genome search for schizophrenia susceptibility genes using a two-stage sib-pair approach, *Am. J. Med. Genet.*,74 (Abst.), 559, 1997.

275. Straub, R. E., MacLean, C. J., O'Neill, F. A., Walsh, D., and Kendler, K. S., Genome scan for schizophrenia genes: a detailed progress report in an Irish cohort, *Am. J. Med. Genet.*, 74 (Abstr.), 559, 1997.

276. Wildenauer, D. B., Albus, M., Schwab, S. G., Hallmayer, J., Hanses, C., Eckstein, G. N., Zill, P., Hönig, S., Lerer, B., Ebstein, R., Lichtermann, D., Trixler, M., Borrmann, M., and Maier, W., Searching for susceptibility genes in schizophrenia by affected sib-pair analysis (Germany), *Am. J. Med. Genet.*, 74 (Abstr.), 558, 1997.

277. Cao, Q., Martinez, M., Zhang, J., Sanders, A. R., Badner, J. A., Cravchik, A., Markey, C. J., Beshah, E., Guroff, J. J., Maxwell, M. E., Kazuba, D. M., Whiten, R., Goldin, L. R., Gershon, E. S., and Gejman, P. V., Suggestive evidence for a schizophrenia susceptibility locus on chromosome 6q and a confirmation in an independent series of pedigrees, *Genomics*, 43, 1, 1997.

278. Barr, C. L., Kennedy, J. L., Pakstis, A. J., Wetterberg, L., Sjogren, B., Bierut, L., Wadelius, C., Wahlstrom, J., Martinsson, T., Giuffra, L. et al., Progress in a genome scan for linkage in schizophrenia in a large Swedish kindred, *Am. J. Med. Genet.*, 54, 51, 1994.

279. Mallet, J., Molecular neurobiology of mental illness, a European consortium, *Am. J. Med. Genet.*, 74 (Abstr.), 557, 1997.

280. Hovatta, I., Varilo, T., Suvisaari, J., Vaisanen, Terwilliger, J. D., Lonnqvist, J., and Peltonen, L., A genome-wide search for schizophrenia genes in an internal isolate of Finland, *Am. J. Med. Genet.*, 74 (Abstr.), 559, 1997.

281. Sherrington, R., Brynjolfsson, J., Petursson, H., Potter, M., Dudleston, K., Barraclough B, Wasmuth, J., Dobbs, M., and Gurling, H., Localization of a susceptibility locus for schizophrenia on chromosome 5, *Nature*, 336, 164, 1988.

282. McGuffin, P., Sargeant, M., Hetti, G., Tidmarsh, S., Whatley, S., and Marchbanks, R. M., Exclusion of a schizophrenia susceptibility gene from the chromosome 5q11.3 region: new data and a reanalysis of previous reports, *Am. J. Hum. Genet.*, 47, 524-35, 1990.

283. Silverman, J. M., Greenberg, D. A., Altstiel, L. D., Siever, L. J., Mohs, R. C., Smith, C. J., Zhou, G., Hollander, T. E., Yang, X. P., Kedache, M., Li, G., Zaccario, M. L., and Davis, K. L., Evidence of a locus for schizophrenia and related disorders on the short arm of chromosome 5 in a large pedigree, *Am. J. Med. Genet.*, 67, 162, 1996.

284. Pulver, A. E., Karayiorgou, M., Wolyniec, P. S., Lasseter, V. K., Kasch, L., Nestadt, G., Antonarakis S., Housman, D., Kazazian, H. H., Meyers, D., Ott, J., Lamacz, M., Liang, K.-Y., Hanfelt, J., Ullrich, G., DeMarchi, N., Ramu, E., McHugh, P. R., Adler, L., Thomas, M., Carpenter, W. T., Manschreck, T., Gordon, C. T., Kimberland, M., Babb, R., Puck, J., and Childs, B., Sequential strategy to identify a susceptibility gene for schizophrenia: report of potential linkage on chromosome 22q12-q13.1: Part 1, *Am. J. Med. Genet.*, 54, 36, 1994.

285. Lasseter, V. K., Pulver, A. E., Wolyniec, P. S., Nestadt, G., Meyers, D., Karayiorgou, M., Housman, D., Antonarakis, S., Kazazian, H., Kasch, L. et al., Follow-up report of potential linkage for schizophrenia on chromosome 22q: Part 3, *Am. J. Med. Genet.*, 60, 172, 1995.

286. Pulver, A. E., Karayiorgou, M., Lasseter, V. K., Wolyniec, P., Kasch, L., Antonarakis, S., Housman, D., Kazazian, H. H., Meyers, D., Nestadt, G. et al., Follow-up of a report of a potential linkage for schizophrenia on chromosome 22q12-q13.1: Part 2, *Am. J. Med. Genet.*, 54, 44, 1994.

287. Gill, M., Vallada, H., Collier, D., Sham, P., Holmans, P., Murray, R., McGuffin, P., Nanko, S., Owen, M., Antonarakis, S., Housman, D., Kazazian, H., Nestadt, G., Pulver, A. E., Straub, R. E., MacLean, C. J., Walsh, D., Kendler, K. S., DeLisi, L., Polymeropoulos, M., Coon, H., Byerley, W., Lofthouse, R., Gershon, E. S. et al., A combined analysis of D22S278 marker alleles in affected sib-pairs: support for a susceptibility locus for schizophrenia at chromosome 22q12. Schizophrenia Collaborative Linkage Group (Chromosome 22), *Am. J. Med. Genet.*, 67, 40, 1996.

288. Coon, H., Holik, J., Hoff, M., Reimherr, F., Wender, P., Myles-Worsley, M., Waldo, M., Freedman, R., and Byerley, W., Analysis of chromosome 22 markers in nine schizophrenia pedigrees, *Am. J. Med. Genet.*, 54, 72, 1994.

289. Polymeropoulos, M. H., Coon, H., Byerley, W., Gershon, E. S., Goldin, L., Crow, T. J., Rubenstein, J., Hoff, M., Holik, J., Smith, A. M. et al., Search for a schizophrenia susceptibility locus on human chromosome 22, *Am. J. Med. Genet.*, 54, 93, 1994.

290. Moises, H. W., Yang, L., Li, T., Havsteen, B., Fimmers, R., Baur, M. P., Liu X., and Gottesman, I. I., Potential linkage disequilibrium between schizophrenia and locus D22S278 on the long arm of chromosome 22, *Am. J. Med. Genet.*, 60, 465, 1995.

291. Vallada, H., Curtis, D., Sham, P. C., Murray, R. M., McGuffin, P., Nanko, S., Gill M., Owen, M., and Collier, D. A., Chromosome 22 markers demonstrate transmission disequilibrium with schizophrenia, *Psychiatr. Genet.*, 5, 127, 1995.

292. Williams, N. M., Jones, L. A., Murphy, K. C., Cardno, A. G., Asherson, P., Williams, J., McGuffin, P., and Owen, M. J., No evidence for an allelic association between schizophrenia and markers D22S278 and D22S283, *Am. J. Med. Genet.*, 74, 37, 1997.

293. Levinson, D. F. and Coon, H., Chromosome 22 workshop, *Psychiatr. Genet.*, 8, 115, 1998.

294. Goldberg, R., Motzkin, B., Marion, R., Scambler, P. J., and Shprintzen, R. J., Velo-cardio-facial syndrome: a review of 120 patients, *Am. J. Med. Genet.*, 45, 313, 1993.

295. Gottlieb, S., Emanuel, B. S., Driscoll, D. A., Sellinger, B., Wang, Z., Roe, B., and Budarf, M. L., The DiGeorge syndrome minimal critical region contains a goosecoid-like (GSCL) homeobox gene that is expressed early in human development, *Am. J. Hum. Genet.*, 60, 1194, 1997.

296. Pulver, A. E., Nestadt, G., Goldberg, R., Shprintzen, R. J., Lamacz, M., Wolyniec, P. S., Morrow, B., Karayiorgou, M., Antonarakis, S. E., Housman, D. et al., Psychotic illness in patients diagnosed with velo-cardio-facial syndrome and their relatives, *J. Nerv. Ment. Dis.*, 182, 476, 1994.

297. Murphy, K. C. and Owen, M. J., The behavioral phenotype in velo-cardio-facial syndrome, *Am. J. Med. Genet.*, 74 (Abstr.), 660, 1997.

298. Bassett, A. S., Hodgkinson, K., Chow, E. W. C., Correia, S., Scutt, L. E., and Weksberg, R., 22q11 deletion syndrome in adults with schizophrenia, *Am. J. Med. Genet.*, 81, 328, 1998.

299. Karayiorgou, M., Morris, M. A., Morrow, B., Shprintzen, R. J., Goldberg, R., Borrow, J,. Gos, A., Nestadt, G., Wolyniec, P. S., Lasseter, V. K. et al., Schizophrenia susceptibility associated with interstitial deletions of chromosome 22q11, *Proc. Natl. Acad. Sci. U.S.A.*, 92, 7612, 1995.

300. Carlson, C., Papolos, D., Pandita, R. K., Faedda, G. L., Veit, S., Goldberg, R., Shprintzen, R., Kucherlapati, R., and Morrow, B., Molecular analysis of velo-cardio-facial syndrome patients with psychiatric disorders, *Am. J. Hum. Genet.*, 60, 851, 1997.

301. Lachman, H. M., Kelsoe, J. R., Remick, R. A., Sadovnick, A. D., Rapaport, M. H., Lin, M., Pazur, B. A., Roe, A. M., Saito, T., and Papolos, D. F., Linkage studies suggest a possible locus for bipolar disorder near the velo-cardio-facial syndrome region on chromosome 22, *Am. J. Med. Genet.*, 74, 121,1997.

302. Edenberg, H. J., Foroud, T., Conneally, P. M., Sorbel, J. J., Carr, K., Crose, C., Willig, C., Zhao, J., Miller, M., Bowman, E., Mayeda, A., Rau, N. L., Smiley, C., Rice, J. P., Goate, A., Reich, T., Stine, O. C., McMahon, F., DePaulo, J. R., Meyers, D., Detera-Wadleigh, S. D., Goldin, L. R., Gershon, E. S., Blehar, M. C., and Nurnberger, J. I. Jr., Initial genomic scan of the NIMH genetics initiative bipolar pedigrees: chromosomes 3, 5, 15, 16, 17, and 22, *Am. J. Med. Genet.*, 74, 238, 1997.

303. Daw, S. C., Taylor, C., Kraman, M., Call, K., Mao, J., Schuffenhauer, S., Meitinger, T., Lipson, T., Goodship, J., and Scambler, P., A common region of 10p deleted in DiGeorge and velocardiofacial syndromes, *Nat. Genet.*, 13, 458, 1996.

304. Wang, S., Sun, C. E., Walczak, C. A., Ziegle, J. S., Kipps, B. R., Goldin, L. R., and Diehl, S. R., Evidence for a susceptibility locus for schizophrenia on chromosome 6pter-p22, *Nat. Genet.*, 10, 41, 1995.

305. Straub, R. E., MacLean, C. J., O'Neill, F. A., Burke, J., Murphy, B., Duke, F., Shinkwin, R., Webb, B. T., Zhang, J., Walsh, D. et al., A potential vulnerability locus for schizophrenia on chromosome 6p24- 22: evidence for genetic heterogeneity, *Nat. Genet.*, 11, 287, 1995.

306. Schizophrenia Linkage Collaborative Group for Chromosomes 3, 6 and 8, Additional support for schizophrenia linkage on chromosomes 6 and 8: a multicenter study, *Am. J. Med. Genet.*, 67, 580, 1996.

307. Schwab, S. G., Albus, M., Hallmayer, J., Honig, S., Borrmann, M., Lichtermann, D., Ebstein, R. P., Ackenheil, M., Lerer, B., Risch, N. et al., Evaluation of a susceptibility gene for schizophrenia on chromosome 6p by multipoint affected sib-pair linkage analysis, *Nat. Genet.*, 11, 325, 1995.

308. Mowry, B. J. and Levinson, D. F., Genetic linkage and schizophrenia: methods, recent findings and future directions, *Aust. N. Z. J. Psychiatr.*, 27, 200, 1993.

309. Straub, R. E., MacLean, C. J., and Kendler, K. S., The putative schizophrenia locus on chromosome 6p: a brief overview of the linkage studies, *Mol. Psychiatr.*, 1, 89, 1996.

310. Brzustowicz, L. M., Honer, W. G., Chow, E. W., Hogan, J., Hodgkinson, K., and Bassett, A. S., Use of a quantitative trait to map a locus associated with severity of positive symptoms in familial schizophrenia to chromosome 6p, *Am. J. Hum. Genet.*, 61, 1388, 1997.

311. Pulver, A. E., Lasseter, V. K., Kasch, L., Wolyniec, P., Nestadt, G., Blouin, J. L., Kimberland, M., Babb, R., Vourlis, S., Chen, H. et al., Schizophrenia: a genome scan targets chromosomes 3p and 8p as potential sites of susceptibility genes, *Am. J. Med. Genet.*, 60, 252, 1995.

312. Kendler, K. S., MacLean, C. J., O'Neill, F. A., Burke, J., Murphy, B., Duke, F., Shinkwin, R., Easter, S. M., Webb, B. T., Zhang, J., Walsh, D., and Straub, R. E., Evidence for a schizophrenia vulnerability locus on chromosome 8p in the Irish Study of High-Density Schizophrenia Families, *Am. J. Psychiatry*, 153, 1534, 1996.

313. Lin, M. W., Curtis, D., Williams, N., Arranz, M., Nanko, S., Collier, D., McGuffin, P., Murray, R., Owen, M., Gill, M. et al., Suggestive evidence for linkage of schizophrenia to markers on chromosome 13q14.1-q32, *Psychiatr. Genet.*, 5, 117, 1995.

314. Lin, M. W., Sham, P., Hwu, H. G., Collier, D., Murray, R., and Powell, J. F., Suggestive evidence for linkage of schizophrenia to markers on chromosome 13 in Caucasian but not Oriental populations, *Hum. Genet.*, 99, 417, 1997.

315. Cloninger, C. R., Kaufmann, C. A., Faraone, S. V., Malaspina, D., Svrakic, D. M., Harkavy-Friedman, J., Suarez, B. K., Matise, T. C., Shore, D., Lee, H., Hampe, C. L., Wynne, D., Drain, C., Markel, P. D., Zambuto, C. T., Schmitt, K., and Tsuang, M. T., A genome-wide search for schizophrenia susceptibility loci: the NIMH genetics initiative & Millennium consortium, *Am. J. Med. Genet.*, 81, 275, 1998.

316. Straub, R. E., MacLean, C. J., Martin, R. B., Myakishev, M. V., Harris-Kern, C., O'Neill, F. A., Walsh, D., and Kendler, K. S., A schizophrenia locus may be located in region 10p15.1, *Am. J. Med. Genet.*, 81, 296, 1998.

317. Schwab, S. G., Hallmayer, J., Albus, M., Lerer, B., Hanses, C., Kanyas, K., Segman, R., Borrman, M., Dreikorn, B., Lichtermann, D., Rietschel, M., Trixler, M., Maier, W., and Wildenauer, D. B., Further evidence for a susceptibility locus on chromosome 10p14-p11 in 72 families with schizophrenia by nonparametric linkage analysis, *Am. J. Med. Genet.*, 81, 302, 1998.

318. Straub, R. E., MacLean, C. J., O'Neill, F. A., Walsh, D., and Kendler, K. S., Support for a possible schizophrenia vulnerability locus in region 5q22-31 in Irish families, *Mol. Psychiatr.*, 2, 148, 1997.

319. Schwab, S. G., Eckstein, G. N., Hallmayer, J., Lerer, B., Albus, M., Borrmann, M., Lichtermann, D., Ertl, M. A., Maier, W., and Wildenauer, D. B., Evidence suggestive of a locus on chromosome 5q31 contributing to susceptibility for schizophrenia in German and Israeli families by multipoint affected sib-pair linkage analysis, *Mol. Psychiatr.*, 2, 156, 1997.

320. Cao, Q., Martinez, M., Zhang, J., Sanders, A. R., Badner, J. A., Cravchik, A., Markey, C. J., Beshah, E., Guroff, J. J., Maxwell, M. E., Kazuba, D. M., Whiten, R., Goldin, L. R., Gershon, E. S., and Gejman, P. V., Suggestive evidence for a schizophrenia susceptibility locus on chromosome 6q and a confirmation in an independent series of pedigrees, *Genomics*, 43, 1, 1997.

321. Freedman, R., Coon, H., Myles-Worsley, M., Orr-Urtreger, A., Olincy, A., Davis, A., Polymeropoulos, M., Holik, J., Hopkins, J., Hoff, M., Rosenthal, J., Waldo, M. C., Reimherr, F., Wender, P., Yaw, J., Young, D. A., Breese, C. R., Adams, C., Patterson, D., Adler, L. E., Kruglyak, L. L. S., and Byerley, W., Linkage of a neurophysiological deficit in schizophrenia to a chromosome 15 locus, *Proc. Natl. Acad. Sci. U.S.A.*, 94, 587, 1997.

322. Leonard, S., Gault, J., Moore, T., Hopkins, J., Robinson, M., Olincy, A., Adler, L. E., Cloninger, C. R., Kaufmann, C. A., Tsuang, M. T., Faraone, S. V., Malaspina, D., Svrakic, D. M., and Freedman, R., Further investigation of a chromosome 15 locus in schizophrenia: analysis of affected sibpairs from the NIMH Genetics Initiative, *Am. J. Med. Genet.*, 81, 308, 1998.

323. DeLisi, L. E., Friedrich, U., Wahlstrom, J., Boccio-Smith, A., Forsman A., Eklund, K., and Crow, T. J., Schizophrenia and sex chromosome anomalies, *Schiz. Bull.*, 20, 495, 1994.

324. Crow, T. J., Delisi, L. E., Lofthouse, R., Poulter, M., Lehner, T., Bass, N., Shah, T., Walsh, C., Boccio-Smith, A., Shields, G. et al., An examination of linkage of schizophrenia and schizoaffective disorder to the pseudoautosomal region (Xp22.3), *Br. J. Psychiatr.*, 164, 159, 1994.

325. Collinge, J., Delisi, L. E., Boccio, A., Johnstone, E. C., Lane, A., Larkin, C., Leach, M., Lofthouse, R., Owen, F., Poulter, M., et al., Evidence for a pseudo-autosomal locus for schizophrenia using the method of affected sibling pairs, *Br. J. Psychiatr.*, 158, 624, 1991.

326. d'Amato, T., Waksman, G., Martinez, M., Laurent, C., Gorwood, P., Campion, D., Jay, M., Petit, C., Savoye, C., Bastard, C. et al., Pseudoautosomal region in schizophrenia: linkage analysis of seven loci by sib-pair and lod-score methods, *Psychiatr. Res.*, 52, 135, 1994.

327. d'Amato, T., Waksman, G., Martinez, M., Laurent, C., Gorwood, P., Campion D., Jay M., Petit, C., Savoye, C., Bastard, C. et al., Pseudoautosomal region in schizophrenia: linkage analysis of seven loci by sib-pair and lod-score methods, *Psychiatr. Res.*, 52, 135, 1994.

328. Dann, J., DeLisi, L. E., Devoto, M., Laval, S., Nancarrow, D. J., Shields, G., Smith, A., Loftus, J., Peterson, P., Vita, A., Comazzi, M., Invernizzi, G., Levinson, D. F., Wildenauer, D., Mowry, B. J., Collier, D., Powell, J., Crowe, R. R., Andreasen, N. C., Silverman, J. M., Mohs, R. C., Murray, R. M., Walters, M. K., Lennon, D. P., Crow, T. J. et al., A linkage study of schizophrenia to markers within Xp11 near the MAOB gene, *Psychiatr. Res.*, 70, 131, 1997.

329. DeLisi, L. E., Devoto, M., Lofthouse, R., Poulter, M., Smith, A., Shields, G., Bass, N., Chen, G., Vita, A., Morganti, C. et al., Search for linkage to schizophrenia on the X and Y chromosomes, *Am. J. Med. Genet.*, 54, 113, 1994.

330. Crow, T. J., Poulter, M., Lofthouse, R., Chen, G., Shah, T., Bass, N., Morganti, C., Vita, A., Smith, C., Boccio-Smith, A. et al., Male siblings with schizophrenia share alleles at the androgen receptor above chance expectation, *Am. J. Med. Genet.*, 48, 159, 1993.

331. Arranz, M., Sharma, T., Sham, P., Kerwin, R., Nanko, S., Owen, M., Gill, M., and Collier, D., Schizophrenia and the androgen receptor gene: report of a sibship showing co-segregation with Reifenstein syndrome but no evidence for linkage in 23 multiply affected families, *Am. J. Med. Genet.*, 60, 377, 1995.

332. Delisi, L. E., Crow, T. J., Davies, K. E., Terwilliger, J. D., Ott, J., Ram, R., Flint, T., and Boccio, A., No genetic linkage detected for schizophrenia to Xq27-q28, *Br. J. Psychiatr.*, 158, 630, 1991.

333. Maier, W., Schmidt, F., Schwab, S. G., Hallmayer, J., Minges, J., Ackenheil, M., Lichtermann, D., and Wildenauer, D. B., Lack of linkage between schizophrenia and markers at the telomeric end of the pseudoautosomal region of the sex chromosomes, *Biol. Psychiatr.*, 37, 344, 1995.

334. Okoro, C., Bell, R., Sham, P., Nanko, S., Asherson, P., Owen, M., Gill, M., McGuffin, P., Murray, R. M., and Collier, D., No evidence for linkage between the X-chromosome marker DXS7 and schizophrenia, *Am. J. Med. Genet.*, 60, 461, 1995.

335. Wang, Z. W., Black, D. W., Andreasen, N., and Crowe, R. R., No evidence of a schizophrenia locus in a second pseudoautosomal region, *Arch. Gen. Psychiatr.*, 51, 427, 1994.

336. Asherson, P., Parfitt, E., Sargeant, M., Tidmarsh, S., Buckland, P., Taylor, C., Clements A, Gill, M., McGuffin, P., and Owen, M., No evidence for a pseudoautosomal locus for schizophrenia. Linkage analysis of multiply affected families, *Br. J. Psychiatr.*, 161, 63, 1992.

337. Holmans, P. and Craddock, N., Efficient strategies for genome scanning using maximum likelihood affected-sib-pair analysis, *Am. J. Hum. Genet.*, 60, 657, 1997.

338. Suarez, B., Hampe, C. L., and van Eerdewegh, P., Problems of replicating linkage claims in psychiatry, in *New Genetic Approaches to Mental Disorders,* Gershon, E. S. and Cloninger, C. R., Eds., Washington, DC, American Psychiatric Press, 1995.

339. Levinson, D. F., Pragmatics and statistics in psychiatric genetics, *Am. J. Med. Genet.*, 74, 220, 1997.

# 5 The Genetic Basis and Neurobiological Characteristics of Obsessive-Compulsive Disorder

*Christina Sobin and Maria Karayiorgou*

## CONTENTS

## INTRODUCTION

Once believed to be a rare mental illness, obsessive-compulsive disorder (OCD) has been increasingly recognized as a relatively common neurobiologic disease with an obvious genetic component. A differential concordance in mono- and dizygotic twins has been demonstrated, though in monozygotes transmission is not complete. Thus environmental factors are presumed to contribute to the manifestation of OCD. Although patients sometimes attribute their disorder to situational precipitants,[1] few significant associations have been found between the onset of OCD and negative events or child-rearing practices.[2] On the other hand, environmental factors such as exposure to birth complications and subsequent physiologic vulnerability may have special significance in OCD.

Patients afflicted with OCD experience intrusive, disturbing, repetitive thoughts (obsessions) and the uncontrollable urge to repeatedly enact stereotypic behaviors or rituals (compulsions), thereby reducing the psychic anxiety produced by the obsessional process.[3] Unlike normal worrying or ruminations, specific obsessional thoughts persist and recur. Compulsions can include checking, handwashing, tapping, or counting and are often repeated several times in a particular pattern that, if interrupted, must be started again. When severe, the completion of obsessive-compulsive cycles requires hours each day. OCD causes chronic inner turmoil because sufferers know their thoughts and behaviors are unreasonable, but they are powerless to resist or control them. The consequences of their symptoms can be devastating. Children with OCD can suffer social isolation, academic impairment, and familial rejection. Spouses and family members cannot understand why the adult patient is unable to stop behaviors he knows to be excessive and absurd, and employers do not understand why simple tasks are never completed.

## EPIDEMIOLOGY

Since the mid-1980s, OCD has been increasingly recognized as a relatively common psychiatric disorder. The first U.S.-based epidemiologic survey of psychiatric disorders to be completed (National Epidemiologic Catchment Area Survey, ECA)[4,5] included approximately 21,000 adults from five distributed sites and exploited the use of newly developed diagnostic criteria[6] and standardized diagnostic instrumentation (Diagnostic Interview Schedule, DIS).[7] Six-month point prevalence and lifetime prevalence rates of OCD were found to be 1.6% and 2.5%, respectively. Though later studies have criticized the stability and thus validity of the ECA diagnoses,[8] these rates have been approximated in subsequent U.S. studies[9] as well as in studies of OCD in an array of cultures including Africa,[10] Canada,[111-13] Finland,[14] Israel,[15] Korea, New Zealand, Puerto Rico,[13] and Zurich.[16] Current estimates of the lifetime prevalence of OCD are generally 1 to 2%.

## DEMOGRAPHICS

The results of demographic studies of OCD have been strikingly consistent over the past 30 years. One of the first estimates of the gender distribution was based on a compilation of inpatient and outpatient studies of OCD patients published prior to 1974[4] and reported a male to female sex ratio of 1:1.1. This ratio has since been closely approximated in subsequent clinical,[5] epidemiologic,[6] and meta-analytic[7] studies. With regard to age of onset, in one of the largest clinical studies to date,[1] 250 consecutive OCD outpatient admissions were evaluated, and the results substantiated those of smaller previous studies. Mean age of onset for 200 patients was reported to be 19.8 ± 9.6, while the mean age for treatment seeking was 25.6 ± 10.7. Approximately 80% of patients reported the onset of first symptoms between the ages of 10 and 24, suggesting that although full diagnostic criteria was not met for many years, the onset of partial symptomatology occurred in childhood for most OCD patients. For a majority of patients (72%) disease course was continual, with waxing and waning of symptoms; 16% reported a continuous course, 9% had a deteriorative course, and only 3% reported an episodic course, with distinct periods of complete symptom remission.

The clinical manifestation of OCD is remarkably consistent across cultures, with a relatively limited number of types of obsessions and compulsions described.[5,8,9] In the large majority of cases, the focus of obsessions includes fears of contamination and illness, pathologic self-doubt, need for symmetry, and aggressive and sexual thoughts. Compulsions include checking, cleaning/washing, counting, asking and confessing, obtaining symmetry and precision, and hoarding.[1] With regard to comorbidity, it has long been reported that approximately 80% of OCD patients experience comorbid major depressive disorder during the course of their OCD illness.[10-12]

Although the symptoms and course of OCD are nearly identical in children and adults, the male to female ratio of childhood OCD has been estimated to be approximately 3:1.[13-17] Interestingly, two epidemiologic[18,19] studies of adolescents reported a male to female sex ratio that more closely approximated that of adults, suggesting that the shift in sex ratio among OCD patients occurred between childhood and adolescence. Among clinic-referred child OCD patients, mean ages of onset are typically reported to be between 9 and 10 years of age,[20-21] while a large epidemiologic study of adolescents estimated approximately 13 years of age.[18] As in adults, over 80% of childhood onset cases have other comorbid psychiatric diagnoses, in particular major depressive disorder.

It should not be surprising that many sufferers attempt to hide or deny their symptoms. Approximately 60% of anxiety patients never see a mental health professional,[22] and OCD patients are most likely to seek treatment from specialists such as dermatologists for secondary symptoms that occur as a consequence of their compulsions.[23] This complicates the unbiased ascertainment of clinical cases in practically all types of studies. As recognition of the disorder has grown, so has our insight regarding the identification, diagnosis, and heterogeneity of this disease, and our knowledge of its multiple biologic and neurobiologic concomitants. More than ever before, we possess the knowledge with which to explore OCD from a genetic perspective.

## TWIN STUDIES

Few twin studies of OCD have been completed, and the hidden nature of OCD has complicated the unbiased ascertainment of affected probands. A large study of systematically ascertained OCD twins has not yet been conducted, and, unsurprisingly, small studies have produced inconclusive results. Inconsistencies in ascertainment methodology, including the identification and diagnosis of probands and family members, have often been cited as the primary source of discrepant findings.[36] On the other hand, positive findings have been reported more often than not, and methodologic issues appear to account for virtually all of the negative findings.

A review of anecdotal, predominantly single-case reports found 63% (32/51) concordance of OCD among monozygotic twins,[37] and studies comparing the rates of OCD in mono- and dizygotic twins extend these findings. Disease concordance was examined in 10 monozygotic and 4 dizygotic OCD index cases identified in private practice.[38] Zygosity was assessed by laboratory method, and all twins were interviewed directly and followed during and after treatment. The ratio of concordance was 1.6:1. Similar results were achieved[39] in a study that ascertained 15 mono- and 15 dizygotic twin pairs through the consecutive hospital admissions of affected singletons. Approximately 50% of the monozygotic and 20% of the dizygotic co-twins of patients hospitalized for OCD had sought psychiatric treatment for obsessions. Concordance rates for full-syndromal OCD did not differ. However, the inclusion of clinically significant subsyndromal obsessive-compulsive symptoms (OCS) — those that do not cause significant impairment or interference — yielded a concordance ratio of 1.6:1.

A later study found no concordance in a same-sex twin sample;[40] however, only 2 of the 12 OCD pairs were male. Robust gender differences in clinical and genetic findings between male and female OCD patients (see subsequent material) may explain these negative results. Recently, OCD was identified in 18 mono- and 30 dizygotic twin pairs and, again, no concordance differences were found.[41] Again, however, the sample was gender-biased, with females representing 72% mono- and 70% dizygotic twin pairs. In addition, the extensive ascertainment and enrollment requirements of this particular study might have eliminated severely affected individuals. Indeed, of the nearly 5967 twin pairs originally contacted for participation in a research registry, only 446 (7%) participated in this study.

## FAMILY STUDIES

A review of family studies conducted over the past 30 years (see Table 5.1)[36,42-51] revealed that estimates of familial risk have fluctuated depending upon the methodology employed. All of the studies reviewed applied a criteria-driven system of diagnosis. The studies varied greatly, however, with regard to the number of probands and relatives directly interviewed and the percentage of participating relatives. Our current knowledge suggests that these factors could significantly influence the estimated familial rates of OCD.

**TABLE 5.1**
**Family Studies of Obsessive-Compulsive Disorder**

| Source | Sample — Probands (Males) | (Females) | Sample — 1° Relatives (fa) | (mo) | (sibs) | Rates of OCD Syndromes* in 1° Relatives (syndrome) | (fa) | (mo) | (sibs) | Comments |
|---|---|---|---|---|---|---|---|---|---|---|
| Kringlen[42] (1965) | 91 adults / 38 | 53 | 91 | 91 | 182 | OCS | 0.7.7% | 12.1% | — | – all probands diagnosed by hospital admission records<br>– 100% of relatives diagnosed by direct interview |
| Lo[43] (1967) | 88 adults / 64 | 24 | 88 | 88 | 309 | OCS | 8.6% |  | 4.6% | – all probands diagnosed by direct interview<br>– all relatives diagnosed by family member report<br>– one-third of probands reported childhood onset |
| Rosenberg[44] (1967) | 144 adults / 52 | 92 | 144 | 144 | 259 | OCS |  | 0.04% |  | – two-thirds of probands diagnosed by medical record<br>– 91% of relatives diagnosed by proband report alone |
| Insel et al.[45] (1983) | 27 adults / 17 | 10 | 27 | 27 | 54 | OCD |  | 0% |  | – all probands diagnosed by direct interview<br>– two-thirds of relatives diagnosed by proband report alone<br>– 50% of probands did not manifest compulsions ("pure obsessionals")<br>– 81% of probands reported childhood onset |
| Rasmussen and Tsuang[46] (1986) | 44 adults / 16 | 28 | 44 | 44 | 88 | OCS | 09.0% | 14.0% | — | – all probands diagnosed by direct interview<br>– all relatives diagnosed on the basis of proband report alone |
| McKeon and Murray[47] (1987) | 50 adults | | 61 | 88* | 149 | OCD rates in relatives of proband and comparison groups did not differ | | | | – all probands diagnosed by direct interview<br>– participating half of the parents diagnosed by direct interview |

| Study | Probands | Relatives | Diagnosis | | Rates | | Proband Diagnosis | | Notes |
|---|---|---|---|---|---|---|---|---|---|
| | | | | | | | OCD | Controls | |
| Bellodi et al.[48] (1992) | 92 adults (46, 46) | 184 / 336, 152 | OCD | | 03.2% | | | 04.5% | – all probands diagnosed by direct interview<br>– one-third of relatives diagnosed by family history report |
| Black et al.[49] (1992) | 42 adults (23, 19) | 52 ma, 120, 68 fem | OCD + OCS | Parents<br>Siblings | 07.9%<br>10.3% | | 15.6%<br>20.4% | 2.9%<br>19.0% | Comparison Analysis of Rates in Parents<br>– all probands diagnosed by direct interview<br>– 72% of relatives diagnosed by direct interview |
| Pauls et al.[36] (1995) | 108 adults (39, 69) | 239 ma, 466, 227 fem | OCS<br>OCD | Parents<br>Siblings | 13.0%<br>25.0% | | —<br>— | —<br>— | – all probands diagnosed by direct interview<br>– 61% of relatives' diagnosed by proband and family member report<br>– diagnoses of all probands and relatives based on expert consensus procedure |
| Lenane et al.[50] (1990) | 46 child/adol (29, 17) | 46, 44, 56 / 146 | OCS<br>OCD | | 13.0%<br>09.0% | | 13.0%<br>25.0% | 03.0%<br>05.0% | – all probands diagnosed by direct interview<br>– 96% of parents and 98% of siblings participated<br>– probands had severe primary childhood onset OCD |
| Riddle et al.[51] (1990) | 20 child/adol (9, 11) | 20, 20 / 40, — | OCS<br>OCD | | 25.0%<br>10.0% | | 30.0%<br>10.0% | —<br>— | – all probands diagnosed by direct interview<br>– 80% of parents diagnosed by direct interview |

\* OCD indicates rates of OCD diagnosis by eith DSM-III, DSM-III-R, RDC, or ICD; OCS indicates rates of clinically significant o-c symptoms that may not have met diagnostic criteria.

\** Sample includes siblings and children over age 16.

Studies in which all probands and a minimum of 70% of family members were directly interviewed[42,49-51] found markedly increased rates of familial risk for OCD and OCS: 10 to 25% for full-syndromal OCD, and 7.7 to 30% for OCS. These rates represent a 4- to 15-fold risk increase for family members of patients with OCD. Furthermore, a more genetically influenced subtype has been suggested by the consistently higher rates of familial OCD among parents of probands with primary childhood onset.[50] It is interesting to note that in the study by Black et al.,[49] the significant risk was carried primarily by the mothers (21.1%), while rates among fathers were less than that of controls, though not significantly so (7.7%). Although reporting bias could explain this differential, a gender-based transmission preference may be suggested.

Three other studies that interviewed less than 30% of family members also reported significant, though less marked, risk increase in first degree relatives.[43,46,48] These significant results, despite the methodologic weakness, could have occurred for several reasons. In the Lo study,[43] one-third of the population reported onset in childhood, suggesting a severe and perhaps more familial form of the disorder. For patients in the Rasmussen and Tsuang study,[46] 70% had an onset between the ages of 10 and 23, with bi-modal distribution peaks at 10 to 12 years and 20 to 22 years, suggesting that at least one-third of their sample endorsed onset in childhood or adolescence. While one study with negative results[45] reported a sample consisting of 81% of childhood onset cases, over 50% of this same sample included patients with no history of compulsions. Recent reports indicate that over 90% of patients have OCD manifest compulsions,[52] suggesting that this sample may not be representative of the majority of OCD patients.

Thus, while results of twin and family studies have been somewhat variable, only some studies have employed the optimal methods for accurately ascertaining affectedness status in family members. Those that have have found markedly increased familial risk. The high risk among families of children with OCD suggest that the childhood onset form of this disorder may represent a more familial variety.

## GENETIC STUDIES

While twin and family studies have suggested the presence of a genetic component in the etiology of OCD, its mode of inheritance is unknown. And it is likely to be complex, involving both genes and environmental influences. Over the past few years, highly polymorphic markers, more or less evenly spaced throughout the genome, have become increasingly available for genome-wide searches for OCD susceptibility loci. Linkage analysis, using DNA markers in families with multiply affected individuals, has proven useful in the detection of monogenic or oligogenic inheritance where one or a small number of genes have a major effect. However, linkage analyses are based on assumptions that are unlikely to apply to OCD — that a gene of major effect segregates in at least a proportion of the families, that within-family homogeneity exists, and that the mode of inheritance can be at least approximately inferred.

On the other hand, nonparametric approaches involving analysis of pairs of affected relatives (that are less atypical than multiplex families) have also been developed.[53-55] In these analyses, evidence for linkage is inferred if affected pairs share more alleles at a given marker locus than would be expected by chance. Because the mode of inheritance need not be specified, these approaches may be more appropriate for the detection of genes of moderate, rather than major, effect. The simplest model is the affected sibling pair method, which has already been used successfully in other complex genetic diseases, such as type 1 diabetes mellitus.[56] In nonparametric methods, however, the power of the method is low. Therefore, large sample sizes are required.

Yet another approach for detecting genes of weak effect are association studies.[57] Association studies are based on the assumption that complex psychiatric disorders are likely to be associated with low penetrance but common functional variations in a number of susceptibility genes (alleles of weak or moderate effect). However, individual susceptibility alleles, necessary to confer increased risk, are not sufficient to cause the disease. Two alternative experimental designs are used more frequently. In case-control studies, the distribution of a specific allele of a polymorphic locus is tested in affected individuals compared to controls who have been matched by ethnicity, sex, and age. Problems with this design include stratification effects, where a section of the population contains a disease and a marker allele more commonly than expected without a causal relationship between them. At the same time, population stratification might be less concerning if association is identified in distinct subsets of patients. Another design, the Transmission Disequilibrium Test (TDT),[58,59] tests the transmission of a particular allele from a parent to the affected individual using the other untransmitted allele from the parent as the "control." Because both the experimental and the control alleles are drawn from the same gene pool, this test eliminates the need for matched controls. In both methodologies, differences in the distribution or transmission of alleles may suggest that the marker may have some direct influence on susceptibility to the disease, or, alternatively, that the marker allele may be very closely linked to the disease allele ("linkage disequilibrium").

Association studies have several practical advantages over linkage studies. Families with multiple affected individuals, which are relatively scarce, are not required, and no assumptions are made about the mode of inheritance of the disease. Unlike linkage studies, association studies have considerable statistical power to detect genes of weak effect. For example, if an allele that occurs naturally in 10% of the population doubles an individual's risk of disease, more than 5000 affected sib-pairs would be needed to detect linkage at LOD > 3. Using an association study design, however, the same effect could be detected with a sample of approximately 700 family trios (affected individual and both biological parents).[57] A limitation of association studies is that functional polymorphisms within candidate genes, or at least polymorphisms in strong disequilibrium, must be identified before the test can be performed. The reason is that in heterogeneous populations, departure from the expected distribution is detectable only very close to or at the "associated" polymorphism. As a potential solution to this problem, genomic association schemes have been recently proposed[57,60] and are likely to benefit genetic research in OCD in the immediate future. According to these schemes, a large number of genes and polymorphisms (preferentially ones that create alterations in derived proteins or their expression) must first be identified (partly as a byproduct of the Human Genome Project), and an extremely large number of bi-allelic polymorphisms need to be tested on several hundred families. It has been estimated that systematic mapping by association requires a marker map of density at least 100-fold greater than that for linkage mapping. Risch and Merikangas[57] analyzed this requirement in some detail and concluded that these limitations are imposed by current technological capabilities and not because of insufficient number of families with the disease or inadequate statistical power. For that reason, more efficient methods of screening large numbers of polymorphisms such as sample pooling or high density DNA arrays[61] must be optimized and applied. It is likely that the required number of markers and families needed for initial screening could be reduced if one allows for linkage disequilibrium, or if the screening is performed in a more homogeneous population (genetic isolates).

As more genes of interest are sequenced, association studies are becoming increasingly feasible although, for now, there are few clear pathophysiological hypotheses upon which this approach can be based. Current models are primarily based on the partial efficacy of pharmacological agents such as selective serotonin reuptake inhibitors (SSRIs), which has led to the hypothesis that OCD may be associated with dysregulation of serotonergic neurotransmission.[62] However, a substantial proportion (30 to 40%) of OCD patients do not respond to SSRIs.[63,64] Treatment augmentation with dopamine receptor blockers has proven useful among some of these patients, thus implicating involvement of dopaminergic pathways as well. The list of genes involved in both pathways (serotonergic and dopaminergic) that have been tested so far is quite small and includes (1) the DRD2, DRD3, and 5HT2A receptor genes (with negative results);[65] (2) the dopamine D4 receptor gene, with mildly positive results for an association between the seven-repeat variant of the receptor and the subset of OCD patients who also have tics,[66] and negative results for an association between a null mutation (13bp deletion) in the first exon of the gene and OCD;[67] (3) the gene for the serotonin transporter (5-HTT), which mediates the reuptake of serotonin into the presynaptic neuron (action site of the SSRIs). A 44-bp insertion/deletion polymorphism in the promoter region of the gene, which alters expression of the transporter protein,[68] has been used in association studies with negative results as well.[69] In addition, mutational screening and sequencing of the coding sequence of the 5-HTT gene did not reveal any mutations associated with OCD.[70,71] And (4) the COMT gene, which besides being a key modulator of dopaminergic and noradrenergic neurotransmission, is also a strong "positional candidate" gene, since it maps in the q11 band of human chromosome 22.[72] Two independent studies raised the possibility that this chromosomal region may harbor genes predisposing to OCD. These studies reported increased rates of comorbid OCD or OCS among schizophrenic patients with the 22q11 microdeletion[73] and similarly increased rates of anxiety, and OCS and OCD in children and adolescents with the 22q11 microdeletion in the absence of schizophrenia.[74] A common functional polymorphism in the COMT gene is associated with a three-to-four-fold variation in COMT enzyme activity. This variation in activity is due to a G→A transition at codon 158 of the COMT gene that results in a valine (high activity allele, COMT*H) to methionine (low activity allele, COMT*L).[75-78]

A significant association has been described between the COMT*L allele and susceptibility to OCD, particularly in males in a case/control design.[79] The COMT*L/COMT*L genotype appeared to be a risk factor for OCD with an approximate relative risk of 5.91 (95% C.I.: 2.40, 14.53) estimated for genotype COMT*L/COMT*L vs. non-COMT*L/COMT*L. The mechanism underlying this sex-selective association awaits definition and may include a sexual dimorphism in COMT activity, although close linkage with a nearby disease susceptibility locus could not be excluded.

It should be emphasized that although pharmacotherapy-based hypotheses for genetic association studies represent a reasonable start, they can, in principle, be flawed because it is rather risky to deduce etiology from our knowledge of pharmacotherapy. This is particularly important to consider when analyzing candidate genes encoding proteins that serve as targets of treatment. In addition, there is no reason to assume that the OCD syndrome includes only one disorder. Indeed, it is very likely that the careful dissection of subtypes will prove increasingly important for understanding the manifestations of the disease. Genetic studies can be crucial in this regard: When a sufficient number of patients are available, the transmission of alleles and genotypes among patient cases can also be examined in relation to several variables such as gender, family history, age at disease onset, presence of co-morbid conditions, response to treatment, and symptom type. It is likely that this kind of analysis will eventually provide objective criteria to define etiologic subtypes of this disorder.

## EVIDENCE OF THE NEUROBIOLOGIC SUBSTRATES OF OCD

In order to further explore the genetic basis of OCD and develop new hypotheses regarding candidate regions and genes, it is useful to consider its possible neurobiologic underpinnings.

### GENDER

For several years, researchers have reported that male OCD patients have a significantly earlier onset age than females. This result has been most strikingly demonstrated in OCD onset age distributions. Recent studies have found that more males report illness onset between the ages of 5 to 15, and more females report onset between the ages of 26 to 35.[80-81] On the basis of these results, it has been suggested that earlier age of onset represented a more severe form of OCD specific to males.[9] Furthermore, markedly different male to female illness ratios among child (3:1) vs. adult (approximately 1:1) OCD patients, and earlier onset ages of male and female children vs. adolescents, have suggested that OCD onset may result from an interaction of neurodevelopmental factors and gender-specific gonadal hormones.[52,82,83] Lensi et al.[80] explored this hypothesis by comparing the incidence of perinatal trauma in their male and female subjects. Dystocic delivery requiring the use of forceps, breech presentation, or prolonged hypoxia were found to be present in 22.4% males vs. 12.9% females. They noted that such organic insults were more common among males in a variety of pediatric onset disorders, perhaps suggesting a greater sensitivity to neurophysiologic, immunologic, and endocrinologic insult in males.

The relation of gender to the onset of OCD becomes more obvious when the influence of gonadal hormones on the function of major neurotransmitters — in particular the serotonergic system — is considered. This has special significance in the case of OCD, since serotonin re-uptake blockades such as clomipramine and selective serotonin re-uptake inhibitors (SSRIs) moderate its symptoms.[84] (Investigators have identified several types of serotonergic dysfunction in OCD patients[85,86] and a series of studies has shown marked differences in the serotonergic function of OCD patients and normal controls, although the direction of effect has been variable.[87-97] The significant effects of both gender and age on serotonergic function[98-104] were not always controlled in these studies and could have contributed to the variable outcomes.) A recent report indicated that a marked cortisol response to d-fenfluramine may be specific to female OCD patients,[97] suggesting a differential role for the 5-HT system in female vs. male OCD patients. Future research may benefit from the integration of animal study results demonstrating that estrogen and progesterone influence CNS mediated serotonin activity[105-107] and that gonadal hormones profoundly influence many aspects of brain development, including total brain volume, neuronal structures, and synapse characteristics.[108]

It is also significant that fluctuations in gonadal hormones have been reported to trigger or alter OCD symptoms[109,110] in some female patients, though very few studies in this area have been completed. Two survey studies of outpatients[111,112] reported the onset of OCD during pregnancy in 15 to 39% of referred women, with 21% reporting symptom onset during puerperium. One of the only controlled studies of the influence of premenstruum, pregnancy, and puerperium gives the strongest evidence to date of the influence of female hormonal fluctuations on OCD.[113] A 5-year retrospective outpatient hospital chart review yielded 72 females with DSM-III-R OCD, 57 of whom participated in the study. Forty-two percent reported premenstrual worsening of OCD. Of the 31 (54%) women who had delivered at least 1 child, 5 reported the onset of OCD during pregnancy (16% of women carrying to term). Onset of OCD during puerperium was not reported by any patient. Fifty-one percent of the sample (29) had an onset of OCD prior to becoming pregnant. Among these women, 31% (9) experienced either marked worsening or improvement of OCD symptoms during pregnancy, and 5 women experienced worsening during puerperium.

Thus, the intriguing relationship between gender-based differences and OCD is far from fully understood. Differences in the rates of birth complications in males and females, studies demonstrating the influence of gonadal hormones on the serotonergic systems, and perhaps developing brain structure suggest a complex interaction. Although earlier age of onset in males implicates a range of factors that could contribute to symptom manifestation, it seems unlikely that gender alone defines a distinct subtype of OCD.

## NEUROLOGIC DISEASES OF THE BASAL GANGLIA

The manifestation of obsessions and compulsions in patients with neurologic disease — specifically, those of the basal ganglia — has provided important clues regarding the areas of brain anomalies that may contribute to idiopathic OCD. In the original description of nine cases of patients with involuntary vocal and motor tics, Gilles de la Tourette also described unceasing repetitive thoughts.[114] Long presumed to be a basal-ganglia-based disorder, recent structural[115,116] and functional[117] brain imaging studies have begun to confirm this assumption and have reported specific volumetric asymmetries in basal ganglia structures, particularly the caudate nucleus, as well as perfusion abnormalities, in right basal ganglia structures. While vocal (involuntary shouting, echolalia, and coprolalia) and simple and complex motor tics are the cardinal symptoms of Tourette's disorder, a review of studies conducted over the past 20 years suggested that 30 to 40% of adult

Tourette's patients manifest obsessive-compulsive symptoms.[118] Moreover, OCD is over-represented among parents of Tourette's patients,[119,120] additionally supporting the neurobiologic link between this basal ganglia disease and OCD. Further substantiating the etiologic link between these diseases, motor tic syndromes have been identified in 20% of children with OCD.[1]

Sydenham's chorea is associated with acute rheumatic fever in children. It is characterized by jerky, irregular, and involuntary movements of the limbs, neck, and face. The motor pathology of this disorder has been attributed in part to anti-neuronal antibodies targeting basal ganglia structures, in particular, the caudate nucleus and subthalamic nuclei.[121] Studies of OCD and OCS in children with Sydenham's chorea have revealed rates of 70 to 80%.[122,123]

Postencephalitic Parkinsonism struck over 800,000 survivors of Von Economo's encephalitis during the pandemic of 1917 to 1928.[124] This number represented 80% of the victims, and a key feature of this syndrome was the onset of compulsions during oculogyric crises.[125] Current post-mortem neuropathologic studies of these patients have reported degenerative lesions in the substantia nigra in 60% (3/5) of cases[126] and near complete loss of the dopaminergic neurons of the nigrostriatal and mesolimbic system in 100% (6/6) of cases.[127]

Huntington's disease[128] consists of degeneration and neuronal cell loss in the neostriatum, with the most profound losses noted in the caudate and putamen and more diffuse damage in the ventral striatum.[129] Through the course of the illness, pathological changes extend from the basal ganglia and affect occipital, temporal, and frontal regions[130,131] and include neuronal loss in the prefrontal cortex.[132,133] Although no large-scale studies have been completed, case reports[134] and preliminary data[135] have suggested that rates of OCD and OCS are significantly increased in this population. Case reports have also described obsessions and compulsions in patients with bilateral basal ganglia lesions observed with MRI and CT scan.[136,137]

## AUTOIMMUNE RESPONSE

In addition to investigations of brain structure, neurobiologic mechanisms have been considered, and it has been postulated that the post-streptococcal autoimmunity response may be strongly associated with the onset of OCD in certain populations.[138,139] In addition to the high rates of OCD among children with Sydenham's chorea,[139] high rates of antineuronal antibodies have been found in children with OCD and Tourette's syndrome.[140] Interestingly, studies of Tourette's patients have revealed two subtypes of neuropsychiatrically ill children. Patients with pre-existing OCD experience a dramatic exacerbation of their OC symptoms and/or tics in response to group A β-hemolytic streptococcal infections. Patients who were without OC symptoms and tics prior to β-hemolytic streptococcal infections experience an abrupt onset of both Tourette's Disorder and OCD.[141]

PANDAS (pediatric autoimmune neuropsychiatric disorders associated with streptococcal infections) and Sydenham's chorea appear to represent two distinct disease entities in which the response to β-hemolytic streptococcal infections differs. This response could be influenced by several factors, not the least of which is the specificity of the host-microbe interaction. It is likely that in Sydenham's chorea, cross-reactive antibodies recognize a different population of cells from those involved in the pathogenesis of OCD, OCS, or tics.[142,143] However, antibodies raised against invading bacteria might cross-react with basal ganglia structures, resulting in exacerbation of OCD or tic disorders. Indeed, acute basal ganglia enlargement, as shown by serial MRI scans, accompanied by exacerbation of OC symptoms and treatment with plasmapheresis has been described in one adolescent boy.[144] Furthermore, plasmapheresis and other immunological treatments (i.e., intravenous immunoglobulin and immunosuppression) proved useful in treating four adolescent boys with abrupt post-streptococcal onset of OC symptoms or tic disorder.[145]

Monoclonal antibody D8/17 identifies a B lymphocyte antigen and is a trait marker of rheumatic fever susceptibility,[146] which appears to be a genetically determined trait. Interestingly, two recent reports described increased frequency of the marker D8/17 among the PANDAS children[141] and children with OCD and/or Tourette's syndrome or chronic tic disorder.[147] None of the children had rheumatic fever or Sydenham's chorea. Thus, there is a possibility that an infection-triggered, autoimmune subtype of pediatric OCD and Tourette's syndrome may exist that is episodic in nature and manifests with abrupt onset or exacerbation of symptoms following post-streptococcal infections. Clearly, this is a developing story and elucidation of the physiological substrate that represent the targets of the immune response might provide valuable insights in the pathogenesis of the disease and, possibly, additional genetically-testable hypotheses.

## STRUCTURAL AND FUNCTIONAL BRAIN DIFFERENCES

Structural and functional brain-imaging studies of patients with OCD have begun to explain the prevalence rates of OCD among neurologic patients with basal ganglia pathology. Over the past 15 years, Computed Tomography (CT), Magnetic Resonant Imaging (MRI), and Positron Emission Tomography (PET) have been used in a variety of paradigms to assess the structure and function of brain areas in OCD patients (see Table 5.2).[148-163]

In all of these studies, patients were diagnosed according to DSM criteria, ratings were conducted "blind" to diagnostic status, controls were free of psychiatric and neurologic illness, and all studies reviewed employed standard imaging methods. However, imaging methods, as well as statistical analytic approaches, varied across studies. Some investigators greatly increased the likelihood of Type I error by analyzing many brain regions, as opposed to just a few. Male and female patients were nearly always mixed, differed in their distribution, and drug status was often not controlled. Furthermore, all of the studies reviewed were "resting state" studies. Patients are likely to vary greatly in their ability to achieve a true resting state, and in the future, more consistent results might be achieved through the addition of a simple standardized mental task. These factors likely contributed to the inconsistency of the findings.

On the other hand, the studies do not completely diverge and several findings are replicated. Overall, positive findings have outweighed negative findings 3 to 1. With regard to the analyses of structure (CT, MRI),[148-159] caudate abnormalities are most evident. It is important to note that the ratio of males to females in the studies with positive caudate findings was 1.3:1, and approximately 30% of the patient sample consisted of patients with childhood onset OCD. In structure studies with negative caudate findings, the male to female ratio was 0.7:1, and none of the samples selected patients for childhood disease onset. The significance of gender and childhood onset in the neural characteristics of OCD could not be evaluated from these data.

Interestingly, all 4 studies of brain metabolism (PET) reviewed[160-163] found hyperfunction in various brain regions of OCD patients, including the basal ganglia, thalamus, orbital gyri, sensorimotor, cingulate, and orbital frontal cortex. Using an atypical approach, Horwitz et al.[162] examined activity throughout the brain by correlating the metabolism rates of paired brain regions. The total number of correlations that differentiated OCD and controls exceeded chance estimates, though both groups had an approximately equal distribution of large vs. small correlations. It was suggested that reorganization of neural pathways, rather than loss of integrative function or hypermetabolism, may best characterize brain anomalies in OCD.

PET studies that have imaged patients pre- and post-treatment, and during exposure to symptom-eliciting stimuli, lend support to the positive findings noted previously. Several studies have demonstrated that metabolism resolves to normal levels following symptom remission in the cingulate cortex[163] and the caudate nuclei.[161,164,165] Metabolism and blood flow in OCD patients have also been compared in rest and exposure-to-symptom-producing-stimuli (e.g., a dirty object) conditions. In these studies, similar activation shifts have been reported. Blood flow, measured via oxygen 15-labeled carbon dioxide tracer, increased significantly in symptomatic vs. resting states in the right caudate nucleus, left anterior cingulate cortex, and bilateral orbitofrontal cortex.[166] In an F18 PET study, the metabolism rates of patients exposed to a hierarchy of noxious stimuli were positively

**TABLE 5.2**
**Brain Imaging Studies of Patients with OCD**

| Source | Sample | Target Structure/Regions | Findings |
|---|---|---|---|
| | | **Computed Tomography (CT) Studies of Brain Structure** | |
| Behar et al.[148] (1984) | 16 OCD adolescents (14 m/3 f, age 9–18) 16 age-sex matched controls | ventricular system | – ventricular/brain ratio in OCD > controls |
| Luxenberg[149] (1988) | 10 OCD adults (10 m, age 17–24) w/ childhood onset 10 age-sex matched controls | third and lateral ventricles, caudate nucleus, lenticular nucleus | – volume of bilateral caudate nuclei in OCD < controls |
| Stein et al.[150] (1993) | 16 OCD adults (5 m/3 f, mean age 33.5 ± 9.8) w/ high/low "soft sign" scores 8 controls (3 m/8 f, mean age 34.4 ± 12.7) | ventricular system, caudate nucleus, lenticular nucleus | – ventricular volume of high "soft sign" OCD > low "soft sign" OCD and controls |
| Insel et al.[151] (1983) | 10 OCD adults (6 m/4 f, age 18–65) 10 age-sex matched controls | ventricular system and hemisphere volume | – no differences found |
| | | **Magnetic Resonance Imaging (MRI) Studies of Brain Structure** | |
| Scarone[152] (1992) | 20 OCD adults (8 m/12 f, mean age 32.5 ± 10.1) 16 controls (10 m/6 f, mean age 31.1 ± 6.5) | caudate nucleus head | – right caudate head volume in OCD > controls |
| Calabrese[153] (1993) | 20 OCD adults (8 m/12 f, mean age 32.5 ± 10.1) 14 controls (8 m/6 f, 31.2 ± 6.9) | caudate nucleus volume and density | – right caudate nucleus volume in OCD > controls – assymetry in caudate density in OCD not found in controls |
| Robinson[154] (1995) | 26 OCD adults (14 m/12 f, age 19–44) 26 controls (16 m/10 f, age 20–45) | prefrontal cortex, caudate nucleus, lateral and third ventricles | – caudate nucleus volumes in OCD < controls |
| Jenike et al.[155] (1996) | 10 OCD adults (10 f, mean age 31.6 ± 8.6) 10 age-sex matched controls | cerebral cortex, diencephalon, caudate, putamen, globus pallidus, hippocampus, amygdala, third/fourth ventricles, corpus callosum, operculum, cerebellum, brain stem | – total white matter in OCD < normal controls – total cortex and operculur volumes in OCD > controls – severity of OCD correlated with opercular volume |
| Rosenberg[156] (1997) | 19 OCD adolescents (13 m/6 f, age 7–18) 19 age-sex matched controls | prefrontal cortex, striatum (caudate and putamen), lateral and third ventricles, intracranial volume | – striatal volumes in OCD patients < controls – third ventral volumes in OCD patients > controls |

| Study | Sample | Brain region examined | Findings |
|---|---|---|---|
| Kellner[157] (1991) | 12 OCD adults (5 m/7 f, age 23–66), 12 age-sex matched controls | head of caudate, cingulate gyrus thickness, corpus callosum | – no differences found |
| Aylward[158] (1996) | 24 OCD adults (17 m/7 f, mean age 34.4 ± 9.8) w/ adult onset, 21 control (14 m/7 f, mean age 31.3 ± 10.4) | caudate nucleus and putamen | – no differenced found |
| Stein[159] (1997) | 13 OCD adults (13 f), 12 normal controls | caudate and ventricular system | – no differences found |

**Positron Emission Tomography (PET) Studies of Brain Function**

| Study | Sample | Brain region examined | Findings |
|---|---|---|---|
| Swedo et al.[160] (1989) | 18 OCD adults (9 m/9 f, age 20–46) w/ childhood onset, 18 age-sex matched controls | whole brain | – metabolism of left orbital frontal, right sensorimotor, bilateral prefrontal, anterior cingulate regions in OCD > controls<br>– state/trait measures of OCD patients correlated with metabolism rates |
| Baxter et al.[161] (1987) | 14 OCD adults (9 m/5 f, age 22–62), 14 controls (7 m/7 f, age 25–39) | hemispheres, prefrontal cortex, temporal cortex, hippocampal complex, cingulate, caudate, putamen, thalamus | – metabolism of cerebral hemispheres, head of caudate, orbital gyri in OCD > controls |
| Horwitz et al.[162] (1991) | 18 OCD adults (9 m/9 f, age 20–46) w/ childhood onset, 18 age-sex matched controls | investigated functional associations between metabolism in paired brain regions | – regional metabolic correlations of left hemisphere superior parietal and left anterior medial temporal (inclu amygdala) in OCD > controls<br>– correlations between metabolism in anterior limbic/paralimbic regions and frontal areas in OCD > controls |
| Perani et al.[163] (1995) | 11 OCD adults (3 m/8 f, age 18–36), 15 age-matched controls (11 m/4 f) | frontal, parietal, temporal, cingulate cortex, occipital, basal ganglia, thalamus, cerebellum | – metabolism of cingulate cortex, thalamus, pallidum/putamen in OCD > controls |

correlated with symptom intensity in the right inferior frontal gyrus, caudate nucleus, putamen, globus pallidus, and thalamus, left hippocampus, and posterior cingulate gyrus.[167] A functional MRI study using this same paradigm reported similar results.[168]

Thus, while comparison studies of OCD patients and normal controls require further replication, pre-/post-treatment and provocation studies within OCD patient groups suggest brain areas that are likely involved in OCD symptom manifestation, including the basal ganglia, cingulate cortex, orbital gyri, left hemisphere superior parietal, and anterior medial temporal lobes. Whether these brain variations in themselves represent heritable differences or are the consequence of some more fundamental neurobiologic abnormality awaits further investigation.

Given the evidence regarding the high rates of OCD among patients with basal ganglia disease and the suggestion from imaging studies that, among other regions, the basal ganglia appear to be involved in the manifestation of OCD symptomatology, it may seem contradictory that the most effective biologic treatment for OCD to date has been selective serotonin re-uptake inhibitors (SSRIs).[169] While more than 80% of the dopamine in the brain is produced in the basal ganglia,[170] it is well established that the striatum receives a primary serotonergic projection from the dorsal raphe nucleus,[171] and a high concentration of serotonin and serotonin receptors in the basal ganglia has been demonstrated.[172,173] Although the cumulative evidence suggests a critical role for abnormal function of 5-hydroxytryptamine (5-HT) in OCD,[169] the complex sequencing of neurotransmitter interactions in response to these agents is unknown. The action of 5-HT in obsessive and compulsive symptoms could be far "downstream" of a more primary functional deficit and likely participates in a complex feed-forward/feed-back interaction with the noradrenergic system. It has been noted that symptom response follows re-uptake inhibition by as much as several weeks, and an average of only 50% of patients respond to these agents, typically with only partial symptom remission.[169]

## SUMMARY

OCD is a relatively common and highly debilitating neurobiologic psychiatric disorder that strikes 1 to 2% of the population. Twin and family studies have suggested that genetic factors contribute to the onset of the disorder; however, the mode of inheritance is likely to be complex. Because drugs affecting both the serotonergic as well as the dopaminergic systems have been used to moderate effect in OCD patients, the genes believed to influence these pathways were among the first to be explored. At the same time, pursuing only those genes believed to be associated with neurotransmitter activity may be unnecessarily limiting, and it is likely that there are many other neurobiologic mechanisms worthy of consideration. For example, the over-representation of OCD among patients with basal ganglia disease may offer important clues as to its neurobiologic underpinnings. In addition, several studies have suggested that OCD can emerge following an anti-neuronal autoimmune response to streptococcal infection in vulnerable individuals. The greatly increased interest in OCD over the past decade has provided the necessary clinical groundwork from which to further explore the genetics of OCD.

## REFERENCES

1. Rasmussen, S. A. and Eisen J. L., Phenomenology of OCD: clinical subtypes, heterogeneity and coexistence, in *The Psychobiology of Obsessive-Compulsive Disorder,* Zohar, J., Insel, T., and Rasmussen, S., Ed., Springer Publishing, New York, 13, 1991.
2. Swedo, S., Rapoport J., Leonard, H., Lenane, M., and Cheslow, D., Obsessive-compulsive disorder in children and adolescents: clinical phenomenology of 70 consecutive cases, *Arch. Gen. Psychiatr.,* 46, 335, 1989.
3. American Psychiatric Association, *Diagnostic and Statistical Manual of Mental Disorders,* 4th ed., American Psychiatric Press Inc., Washington, DC, 1994.

4. Myers, J. K., Weissman, M. M., Tischler, G. L., Holzer, C. E. 3d, Leaf, P. J., Orvaschel, H., Anthony, J. C., Boyd, J. H., Burke, J. D., Jr., and Kramer M., Six-month prevalence of psychiatric disorders in three communities 1980 to 1982, *Arch. Gen. Psychiatr.,* 41, 959, 1984.
5. Robins, L. N., Helzer, J. E., Weissman, M. M., Orvaschel, H., Gruenberg E., Burke, J. D., Jr., and Regier, D. A., Lifetime prevalence of specific psychiatric disorders in three sites, *Arch. Gen. Psychiatr.,* 41, 949, 1984.
6. American Psychiatric Association, *Diagnostic and Statistical Manual of Mental Disorders,* 3rd ed., American Psychiatric Press Inc., Washington, DC, 1980.
7. Robins, L. N., Helzer, J. E., Croughan, J., and Ratcliff, K. S., The NIMH diagnostic interview schedule: its history, characteristics, and validity, *Arch. Gen. Psychiatr.,* 46, 1006, 1989.
8. Nelson, E. and Rice, J., Stability of diagnosis of obsessive-compulsive disorder in the epidemiologic catchment area study, *Am. J. Psychiatr.,* 154, 826, 1997.
9. Flament, M. F., Whitaker, A., Rapoport, J. L. et al., Obsessive-compulsive disorder in adolescence: an epidemiologic study, *J. Am. Acad. Child Adol. Psychiatr.,* 27, 764, 1988.
10. Orley, J. and Wing, J. K., Psychiatric disorders in two African villages, *Arch. Gen. Psychiatr.,* 36, 513, 1979.
11. Bland, R. C., Newman, S. C., and Orn, H., Lifetime prevalence of psychiatric disorder in Edmonton, *Acta. Psychiatr. Scand.,* 77(Suppl. 338), 24, 1988.
12. Bland, R. C., Newman, S. C., and Orn, H., Period prevalence of psychiatric disorder in Edmonton, *Acta. Psychiatr. Scand.,* 77(Suppl. 338), 33, 1988.
13. Weissman, M. M., Bland, R. C., Canino, G. J., Greenwald, S., Hwu H. G., Lee, C. K., Newman, S. C., Oakley-Browne, M. A., Rubio-Stipec, M., Wickramaratne, P. J. et al., The cross-national epidemiology of obsessive-compulsive disorder, *J. Clin. Psychiatr.,* 55(Suppl.), 5, 1994.
14. Vaisaner, E., Psychiatric disorders in Finland, *Acta. Psychiatr. Scand.,* 27(Suppl. 263), 1975.
15. Zohar, A. H., Ratzoni, G., Pauls, D. L. et al., An epidemiological study of obsessive-compulsive disorder and related disorders in Israeli adolescents, *J. Am. Acad. Child Adol. Psychiatr.,* 31, 1057, 1992.
16. Degonda, M., Wyss, M., and Angst, J., The Zurich Study. XVIII. Obsessive-compulsive disorders and syndromes in the general population, *Eur. Arch. Psychiatr. Clini. Neurosci.,* 243, 16, 1993.
17. Black, A., The natural history of obsessional neurosis, in *Obsessional States,* Beech, H. R., Ed., Methuen Co., London, 1, 1974.
18. Rassmussen, S. A. and Eisen J. L., Epidemiology and clinical features of obsessive-compulsive disorder, in *Obsessive-Compulsive Disorders: Theory and Management, 2nd ed.,* Jenike, M. A., Baer, L., and Minichiello, W. E., Eds., Year Book, Chicago, IL, 1990.
19. Karno, M. and Golding, J. M., Obsessive-compulsive disorder, in *Psychiatric disorders in America, The Epidemiologic Catchment Area Study,* Robins, L. N. and Regier, D. A., Eds., Macmillan, London, 204, 1991.
20. Rasmussen, S. A. and Tsuang, M., Epidemiology and clinical features of obsessive-compulsive disorder, in *Obsessive-Compulsive Disorders,* Jenike, M. A., Baer, L., and Minichiello, W. E., Eds., PSG Publishing Co., Littleton, MA, 23, 1986.
21. Okasha, A., Kamel, M., and Hassan, A. H. Preliminary psychiatric observations in Egypt, *Br. J. Psychiatr.,* 114, 949, 1968.
22. Ahktar, S., Wig, N. N., Varma, V. K. et al., A phenomenological analysis of symptoms in obsessive-compulsive neurosis, *Br. J. Psychiatr.,* 127, 342, 1975.
23. Lewis, A. J., Problems of obsessional illness, *Proc. R. Soc. Med.,* 29, 325, 1936.
24. Goodwin, D. W., Guze S. B., and Robins, E., Follow-up studies in obsessional neurosis, *Arch. Gen. Psychiatr.,* 20, 182, 1969.
25. Rasmussen, S. A. and Eisen, J. L., Clinical and epidemiologic findings of significance to neuropharmacologic trials in OCD, *Psychopharmacol. Bull.,* 24, 466, 1988.
26. Hollingsworth, C., Tanguay, P., Grossman, L. et al., Long-term outcome of obsessive-compulsive disorder in childhood, *J. Am. Acad. Child Adol. Psychiatr.,* 19,134, 1980.
27. DeVeaugh-Geiss, J., Moroz, G., Biederman, J. et al., Clomipramine hydrochloride in childhood and adolescent obsessive-compulsive disorder: a multicenter trial, *J. Am. Acad. Child Adol. Psychiatr.,* 31, 45, 1992.
28. Last, C. G. and Strauss, C. C., Obsessive-compulsive disorder in childhood, *J. Anxiety Disorders,* 3, 295, 1989.

29. Swedo, S. E., Rapoport, J. L., Leonard, H., Lenane, M., and Cheslow, D., Obsessive-compulsive disorder in children and adolescents, *Arch. Gen. Psychiatr.,* 46, 335, 1989.

30. Thomsen, P. H., Obsessive-compulsive symptoms in children and adolescents: a phenomenological analysis of 61 Danish cases, *Psychopathology,* 24, 12, 1991.

31. Zohar, A. H., Ratzoni, G., Pauls, D. L. et al, Obsessive-compulsive disorder in adolescence: an epidemiologic study, *J. Am. Acad. Child Adol. Psychiatr.,* 27, 764, 1988.

32. Rapoport, J. L., *Obsessive-compulsive disorder in children and adolescents,* American Psychiatric Press, Washington, DC, 1989.

33. Hanna, G. L., Demographic and clinical features of obsessive-compulsive disorder in children and adolescents, *J. Am. Acad. Child Adol. Psychiatr.,* 34, 19, 1995.

34. Shapiro S., Skinner, E. A., Kessler, L. G. et al., Utilization of health and mental health services, *Arch. Gen. Psychiatr.,* 41, 971, 1984.

35. Rasmussen, S. A., Obsessive-compulsive disorder in dermatologic practice, *J. Am. Acad. Dermatol.,* 13, 965, 1986.

36. Pauls, D. L., Alsobrook, J. P., Goodman, W., Rasmussen, S., and Leckman, J., A family study of obsessive-compulsive disorder, *Am. J. Psychiatr.,* 152, 76, 1995.

37. Rasmussen, S. A. and Tswang, M. T., The epidemiology of obsessive-compulsive disorders, *J. Clin. Psychiatr.,* 45, 11, 1984.

38. Inouye, E., Similar and dissimilar manifestations of obsessive-compulsive neurosis in monozygotic twins, *Am. J. Psychiatr.,* 121, 1171, 1965.

39. Carey, G. and Gottesman, J. J., Twin and family studies of anxiety, phobic, and obsessive disorders, in *Anxiety: New Research and Changing Concepts,* Klein, D. F. and Rabkin, J., Eds., Raven Press, New York, 117, 1981.

40. Torgersen, S., Genetic factors in anxiety disorder, *Arch. Gen. Psychiatr.,* 40, 1085, 1983.

41. Andrews, G., Stewart, G., Allen, R., and Henderson, A. S., The genetics of six neurotic disorders: a twin study, *J. Affect. Disord.,* 19, 23, 1990.

42. Kringlen, E., Obsessional neurotics: a long-term follow-up, *Br. J. Psychiatr.,* 111, 709, 1965.

43. Lo, W. H., A follow-up study of obsessional-neurotics in Hong Kong Chinese, *Br. J. Psychiatr.,* 113, 823, 1967.

44. Rosenberg, C. M., Familial aspects of obsessional neurosis, *Arch. Gen. Psychiatr.,* 113, 405, 1967.

45. Insel, T. R., Hoover, C., and Murphy, D. L., Parents of patients with obsessive-compulsive disorder, *Psycholog. Med.,* 13, 807, 1983.

46. Rasmussen, S. A. and Tsuang, M. T., Clinical characteristics and family history in DSM-III obsessive-compulsive disorder, *Am. J. Psychiatr.,* 143, 317, 1986.

47. McKeon, P. and Murray, R., Familial aspects of obsessive-compulsive neurosis, *Br. J. Psychiatr.,* 151, 528, 1987.

48. Bellodi, L., Sciuto, G., Diaferia, G., Ronchi, P., and Smeraldi, E., Psychiatric disorders in the families of patients with obsessive-compulsive disorder, *Psychiatr. Res.,* 42, 111, 1992.

49. Black, D. W., Noyes, R., Jr., Goldstein, R. B., and Blum, N., A family study of obsessive-compulsive disorder, *Arch. Gen. Psychiatr.,* 49, 362, 1992.

50. Lenane, M. C., Swedo, S. E., Leonard, H., Pauls, D. L., Sceery, W., and Rapoport, J. L., Psychiatric disorders in first degree relatives of children and adolescents with obsessive-compulsive disorder, *J. Am. Acad. Child Adol. Psychiatr.,* 29, 407, 1990.

51. Riddle, M. A., Scahill, L., King, R., Hardin, M. T., Towbin, K. E., Ort, S. I., Leckman, J. F., and Cohen, D. J., Obsessive-compulsive disorder in children and adolescents: phenomenology and family history, *J. Am. Acad. Child Adol. Psychiatr.,* 29, 766, 1990.

52. Noshirvani, H. F., Kasvikis, Y., Marks, I. M., Tsakiris, F., and Monteiro, W. O., Gender-divergent aetiological factors in obsessive-compulsive disorder, *Br. J. Psychiatr.,* 158, 260, 1991.

53. Cloninger, C. R., Turning point in the design of linkage studies of schizophrenia, *Am. J. Med. Genet.,* 54, 83, 1994.

54. Shah, S. and Green, J. R., Disease susceptibility genes and the sib-pair method: a review of recent methodology, *Ann. Hum. Genet.,* 58, 381, 1994.

55. Curtis, D. and Sham, P. C., Using risk calculation to implement an extended relative pair analysis, *Ann. Hum. Genet.,* 58, 151, 1994.

56. Davies, J. L., Kawaguchi, Y., Bennett, S. T., Copeman, J. B., Cordell, H. J., Pritchard, L. E., Reed, P. W., Gough, S. C. L., Jenkins, S. C., Palmer, S. M., Balfour, K. M., Rowe, B. R., Farrall, M., Barnett, A. H., Bain, S. C., and Todd, J. A., A genome-wide search for human type 1 diabetes susceptibility genes, *Nature*, 371, 130, 1994.

57. Risch, N. and Merikangas, K., The future of genetic studies of complex human diseases, *Science*, 275, 1327, 1997.

58. Terwilliger, J. and Ott, J., A haplotype-based "haplotype relative risk" approach to detecting allelic associations, *Hum. Hered.*, 42, 337, 1992.

59. Spielman, R. S., McGinnis, R. E., and Ewens, W. J., Transmission test for linkage disequilibrium: the insulin gene region and insulin-dependent diabetes mellitus (IDDM), *Am. J. Hum. Genet.*, 52, 506, 1993.

60. Collins, F. S., Guyer, M. S., and Charkravarti, A., Variations on a theme: cataloging human DNA sequence variation, *Science*, 278, 1580, 1997.

61. Chee, M., Yang, R., Hubbell, E., Berno, A., Huang, X. C., Stern, D., Winkler, J., Lockhart, D. J., Morris, M. S., and Fodor, S. P. A., Accessing genetic information with high-density DNA arrays, *Science*, 274, 610, 1996.

62. Murphy, D. L., Greenberg, B., Altemus, M., Benjamin, J., Grady, T., and Pigott, T., The neuropharmacology and neurobiology of obsessive-compulsive disorder: an update on the serotonin hypothesis, in Advances in the *Neurobiology of Anxiety Disorders,* Westenberg, H. G. M. and Boer, J. A. D., Eds., Wiley Publications, 1995.

63. McDougle, C. J., Goodman, W. K., Leckman, J. F., Lee, N. C., Heninger, G. R., and Price, L. H., Haloperidol addition in fluvoxamine-refractory obsessive-compulsive disorder: a double-blind, placebo-controlled study in patients with and without tics, *Arch. Gen. Psychiatr.*, 51, 302, 1994.

64. Greist, J. H., Jefferson, J. W., Kobak, K. A., Katzelnick, D. J., and Serlin, R. C., Efficacy and tolerability of serotonin transport inhibitors in obsessive-compulsive disorder, *Arch. Gen. Psychiatr.*, 52, 53, 1995.

65. Nicolini, H., Cruz, C., Camarena, B., Orozco, B., Kennedy, J. L., King, N., Weissbecker, K., de la Fuente, J. R., and Sidenberg, D., DRD2, DRD3 and 5HT2A receptor genes polymorphisms in obsessive-compulsive disorder, *Mol. Psychiatr.*, 1, 461, 1996.

66. Cruz, C., Camarena, B., King, N., Paez, F., Sidenberg, D., de la Fuente, J. R., and Nicolini, H., Increased prevalence of the seven-repeat variant of the dopamine D4 receptor gene in patients with obsessive-compulsive disorder with tics, *Neurosci. Lett.*, 231, 1, 1997.

67. Di Bella, D., Catalano, M., Cichon, S., and Nothen, M. M., Association study of a null mutation in the dopamine D4 receptor gene in Italian patients with obsessive-compulsive disorder, bipolar mood disorder and schizophrenia, *Psychiatr. Genet.*, 6, 119, 1996a.

68. Lesch, K. P., Bengel, D., Heils, A., Sabol, S. Z., Greenberg, B. D., Petri, S., Benjamin, J., Muller, C. R., Hamer, D. H., and Murphy, D. L., Association of anxiety-related traits with a polymorphism in the serotonin transporter gene regulatory region, *Science*, 274, 1527, 1996.

69. Billett, E. A., Richter, M. A., King, N., Heils, A., Lesch, K. P., and Kennedy, J. L., Obsessive-compulsive disorder, response to serotonin reuptake inhibitors and the serotonin transporter gene, *Mol. Psychiatr.*, 2, 403, 1997.

70. Di Bella, D., Catalano, M., Balling, U., Smeraldi, E., and Lesch, K. P., Systematic screening for mutations in the coding region of the human serotonin transporter (5-HTT) gene using PCR and DGGE, *Am. J. Med. Genet.*, 67, 541, 1996b.

71. Altemus, M., Murphy, D. L., Greenberg, B., and Lesch, K. P., Intact coding region of the serotonin transporter gene in obsessive-compulsive disorder, *Am. J. Med. Genet.*, 67, 409, 1996.

72. Karayiorgou, M., Morris, M. A., Morrow, B. et al., Schizophrenia susceptibility associated with interstitial deletions of chromosome 22q11, *Proc. Ntl. Acad. Sci. U.S.A.*, 92, 7612, 1995.

73. Pulver, A. E., Nestadt, G., Goldberg, R. et al., Psychotic illness in patients diagnosed with velo-cardio-facial syndrome and their relatives, *J. Nerv. Ment. Dis.*, 182, 476, 1994.

74. Papolos, D. F., Faedda, G. L., Veitm, S. et al., Bipolar spectrum disorders in patients diagnosed with velo-cardio-facial syndrome: does a hemizygous deletion of chromosome 22q11 result in bipolar affective disorder?, *Am. J. Psychiatr.*, 153, 1541, 1996.

75. Grossman, M. H., Littrell, J. B., Weinstein, R., Szumlanski, C., and Weinshilboum, R. M., Identification of the possible basis for inherited differences in human catechol-*O*-methyltransferase, *Trans. Neurosci. Soc.*, 18, 70, 1992.

76. Lotta, T., Vidgren, J., Tilgmann, C. et al., Kinetics of human soluble and membrane bound catechol-O-methyltransferase: a revised mechanism and description of the thermolabile variant of the enzyme, *Biochemistry*, 34, 4202, 1995.

77. Lachman, H. M., Papolos, D. F., Saito, T., Yu, Y.-M., Szumlanski, C. L., and Weinshilboum, R. M., Human catechol-O-methyltransferase pharmacogenetics: description of a functional polymorphism and its potential application to neuropsychiatric disorders, *Pharmacogenetics*, 6, 243, 1996.

78. Karayiorgou, M., Gogos, J. A., Galke, B. L. et al., Identification of sequence variants and analysis of the role of the COMT gene in schizophrenia susceptibility, *Biolog. Psychiatr.*, 43, 425, 1998.

79. Karayiorgou, M., Altemus, M., Galke, B. L., Goldman, D., Murphy, D. L., Ott, J., and Gogos, J. A., Genotype determining low catechol-O-methyltransferase activity as a risk factor for obsessive-compulsive disorder, *Proc. Natl. Acad. Sci. U.S.A.*, 94, 4572, 1997.

80. Lensi, P., Cassano, G. B., Correddu, G., Ravagli, S., Kunovac, J. L., and Akiskal, H. S., Obsessive-compulsive disorder: familial-developmental history, symptomatology, comorbidity and course with special reference to gender-related differences, *Br. J. Psychiatr.*, 169, 101, 1996.

81. Castle, D. J., Deale, A., and Marks, I. A., Gender differences in obsessive-compulsive disorder, *Aust. N. Z. J. Psychiatr.*, 29, 114, 1195.

82. Hanna, G. L., Demographic and clinical features of obsessive-compulsive disorder in children and adolescents, *J. Am. Acad. Child Adol. Psychiatr.*, 34, 19, 1995.

83. Apter, A. and Tyano, S., Obsessive-compulsive disorders in adolescence, *J. Adol.*, 11, 183, 1988.

84. Piccinelli, M., Pini, S., Bellantuono, C. et al., Efficacy of drug treatment in obsessive-compulsive disorder: a meta-analytic review, *Br. J. Psychiatr.*, 166, 424, 1995.

85. Lewis, D. A. and Sherman, B. M., Serotonergic regulation of prolactin and growth hormone secretion in humans, *Acta. Endocrinol.*, 110, 152, 1985.

86. Gorard, D. A., Taylor, T. M., Medbak, S. H., Perry, L. A., Libby, G. W., and Farthing, M. J. G., Plasma prolactin, adrenocoritcoptrophic hormone and cortisol after administration of d-fenfluramine or placebo to healthy subjects, *Int. Clin. Psychopharmacol.*, 8, 123, 1993.

87. Hewlett W. A., Vinogradov, S., Berman, S., Cxernansky, J. G., and Agars, W. S., Fenfluramine stimulation of prolactin in obsessive-compulsive disorder, *Psychiatr. Res.*, 42, 81–92, 1992.

88. Hollander, E., DeCaria, C. M., Nitesai, A., Gully, R., Suckow, R. F., Cooper, T. B., Gorman, J. M., Klein, D. F., and Liebowitz, M. R., Serotonergic function in obsessive-compulsive disorder: behavioral and neruoendocrine responses to oral m-chlorophenylpiperazine and fenfluramine in patients and normal volunteers, *Arch. Gen. Psychiatr.*, 49, 21, 1992.

89. Lucey, J., Butcher, G., Clare, A., and Dinan, T., Buspirone-induced prolactin responses in obsessive-compulsive disorder: is OCD a 5-HT2 receptor disorder?, *Int. Clin. Psychopharmacol.*, 7, 45, 1992.

90. Fineberg, N. A., Bullock, T., Montgomery, D. B., and Montgomery, S. A., Serotonin re-uptake inhibitors are the treatment of choice in obsessive-compulsive disorder, *Int. Clin. Psychopharmacol.*, 7, 43, 1994.

91. Lopez-Ibor, J. J., Siaz, J., and Vinas, R., Obsessive-compulsive disorder and depression, in *Psychopharmacology of Depression*, Montgomery, S. A., and Corn, Eds., Oxford University Press, Oxford, 1994, 185.

92. McBride, P. A., DeMeo, M. D., Sweeney, J. A., Halper, J., Mann, J. J., and Shear, M. K., Neuroendocrine and behavioral responses to challenge with the indirect serotonin agonist dl-fenfluramine in adults with obsessive-compulsive disorder, *Biol. Psychiatr.*, 31, 19, 1992.

93. Piggot, T. A., Hill J. L., Grady, T. A., L'Heaveux, F., Bernstein, S., Rubenstein, S., and Murphy, D., A comparison of the behavioral effects of oral versus intravenous mCPP administration in OCD patients and the effect of metergoline prior to IV mCPP, *Biol. Psychiatr.*, 33, 3, 1993.

94. Charney, D. S., Goodman, W. K., Price, C. H., Woods, S. W., Rasmussen, S. A., and Heninger, G. R., Serotonin function in obsessive-compulsive disorder, *Arch. Gen. Psychiatr.*, 45, 177, 1988.

95. Zohar, J. and Insel, T. R., Obsessive-compulsive disorder: psychobiological approaches to diagnosis, treatment, and pathophysiology, *Biol. Psychiatr.*, 22, 667, 1987.

96. Fineberg, N. A., Roberts, A., Montgomery, S. A., and Cowen, P.J., Brain 5-HT function in obsessive-compulsive disorder: prolactin responses to d-fenfluramine, *Br. J. Psychiatr.*, 171, 280, 1997.

97. Monteleone, P., Catapano, F., Tortorella, A., and Maj, M., Cortisol response to d-fenfluramine in patients with obsessive-compulsive disorder and in healthy subjects: evidence for a gender-related effect, *Neuropsychobiology*, 36, 8 1997.

98. Lepage, P. and Steiner, M., Gender and serotonergic dysregulation: implications for late luteal phase dysphoric disorder, in *Royal Society of Medicine Services International Congress and Symposium Series No. 165,* Cassano, G. B. and Akiskal, H. S., Eds., Royal Society of Medicine Services Limited, Oxford, 131, 1991.

99. Halbreich, U. and Tworek, H., Altered serotonergic activity in women with dysphoric premenstrual syndromes, *Int. J. Psychiatr. Med.,* 23, 1, 1993.

100. Halbreich, U., Rojansky, N., Zander, K. J., and Barkai, A., Influence of age, sex and diurnal variability on imipramine receptor binding and serotonin uptake in platelets of normal subjects, *J. Psychiatr. Res.,* 25, 7, 1991.

101. Marazziti, D., Giusti, P., Rotondo, A., Placidi, G. F., and Pacifici, G. M., Imipramine receptors in human platelets: effect on age, *J. Clin. Pharmacol. Res.,* 7, 145, 1987.

102. Arora, R. C. and Meltzer, H. Y., [³H] Imipramine binding in the frontal cortex of suicides, *Psychiatr. Res.,* 30, 125, 1989.

103. Arato, M., Frecska, E., Tekes, K., and MacCrimmon, D. J., Serotonergic interhemispheric asymmetry: gender difference in the orbital cortex, *Acta. Psychiatr. Scand.,* 84, 110, 1991.

104. Halbreich, U., Asnis, G. M., Zumoff, B., Nathan, R. S., and Shindledecker, R., Effect of age and sex on cortisol secretion in depressives and normals, *Psychiatr. Res.,* 13, 221, 1984.

105. McEwen B. S., Davis, P. G., Parsons, B. et al., The brain as a target for steroid hormone action, *Ann. Rev. Neurosci.,* 2, 65, 1979.

106. Luine V. N., Khylchevskaya, R. I., and McEwen, B. S., Effects of gonadal steroids on activities of monoamine oxidase and choline acetylase in rat brain, *Brain Res.,* 86, 293, 1975.

107. Biegon A., Reches, A., Snyder L. et al., Serotonergic and noradrenergic receptors in the rate brain: modulation by chronic exposure to ovarian hormones, *Life Sci.,* 17, 2015, 1983.

108. Halbreich, U. and Lumley, L. A., The multiple interactional biological processes that might lead to depression and gender differences in its appearance, *J. Affect. Disord.,* 29, 159, 1993.

109. Weiss M., Baerg, E., Wisebord, S., and Temple, J., The influence of gonadal hormones on periodicity of obsessive-compulsive disorder, *Can. J. Psychiatr.,* 40, 205, 1995.

110. Sichel, D. A., Cohen, L. S., Dimmock, J. A. et al., Postpartum obsessive-compulsive disorder: a case series, *J. Clin. Psychiatr.,* 54, 156, 1993.

111. Neziroglu, F., Anemone, R., and Yaryura-Tobias, J. A., Onset of obsessive-compulsive disorder in pregnancy, *Am. J. Psychiatr.,* 149, 950, 1992.

112. Buttolph, M. L. and Holland, A., Obsessive-compulsive disorders in pregnancy and childbirth, in *Obsessive-Compulsive Disorders, Theory and Management,* Jenike, M., Baer, L., and Minichiello, W. E., Eds., Yearbook Medical Publishers, Chicago, IL, 1990.

113. Williams, K. R. and Koran, L. M., Obsessive-compulsive disorder in pregnancy, the puerperium, and the premenstruum, *J. Clin. Psychiatr.,* 58, 330, 1997.

114. de la Tourette, G., Etude sur une affection nerveuse caracterisee par de l'incoordination mortirice accompagnee d'echolalie et de copralalie, *Arch. Neurol.,* 9,158, 1885.

115. Hyde T. M., Stacey, M. E., Coppola, R., Handel, S. F., Rickler, K. C., and Weinberger, D., Cerebral morphometric abnormalities in Tourette's syndrome: a quantitative MRI study of monozygotic twins, *Neurology,* 45, 1176, 1995.

116. Peterson, B. S., Gore, J.C., Riddle, M.A., Cohen, D. J., and Leckman, J. F., Abnormal magnetic resonance imaging T2 relaxation time asymmetries in Tourette's syndrome, *Psychiatr. Res.,* 55, 205, 1994.

117. Klieger, P. S., Fett, K. A., Dimitsopulos, T., and Karlan, R., Asymmetry of basal ganglia perfusion in Tourette's syndrome shown by technetium-99m-HMPAO SPECT, *J. Nucl. Med.,* 38, 188, 1997.

118. Leckman, J. F., Tourette's disorder, in *Obsessive-Compulsive Related Disorders,* Hollander, E., Ed., American Psychiatric Press, Washington, DC, 120, 1993.

119. Frankel, M., Cummings, J., Robertson, M., Trimble, M., Hill, M., and Benson, D., Obsessions and compulsions in Gilles de la Tourette syndrome, *Neurology,* 36, 378, 1986.

120. Pauls, D., Towbin, K., Leckman, J. et al., Gilles de la Tourette syndrome and obsessive-compulsive disorder: evidence supporting a genetic relationship, *Arch. Gen. Psychiatr.,* 43, 1180, 1986.

121. Husby, G., van de Rijn, I., Zabriskie, J. B., Abdin, Z. H., and Williams, R. C., Jr., Antibodies reacting with cytoplasm of subthalamic and caudate nuclei neurons in chorea and acute rheumatic fever, *J. Exp. Med.,* 144, 1094, 1976.

122. Swedo, S. E., Rapoport, J. L., Cheslow, D. L., Leonard, H. L., Ayoub, E. M., Hosier, D. M., and Wald, E. R., High prevalence of obsessive-compulsive symptoms in patients with Sydenham's chorea, *Am. J. Psychiatr.,* 146, 246, 1989.

123. Swedo, S. E., Leonard, H. L., Schapiro, M.B., Casey, B.J., Mannheim, G.B., Lenane, M.C., and Rettew, D. C., Sydenham's chorea: physical and psychological symptoms of St. Vitus dance, *Pediatrics,* 91, 706, 1993.

124. Cheyette, S. R. and Cummings, J. L., Encephalitis lethargica: lessons for contemporary neuropsychiatry, *J. Neuropsychiatr. Clin. Neurosci.,* 7, 125–134, 1995.

125. von Economo, C., *Encephalitis lethargica: Its sequellae and treatment,* Oxford University Press, London, 1931.

126. Mizukami, K., Sasaki, M., Shiraishi, H., Ikeda, K., and Kosaka, K., A neuropathologic study of long-term, Economo-type postencephalitic parkinsonism with a prolonged clinical course, *Psychiatr. Clin. Neurosci.,* 50, 79, 1996.

127. Bogerts, B., Hantsch, J., and Herzer, M., A morphometric study of the dopamine-containing cell groups in the mesencephalon of normals, Parkinson patients, and schizophrenics, *Biol. Psychiatr.,* 18, 951, 1983.

128. Huntington, G., On chorea, *Adv. Neurol.,* 1, 33, 1872.

129. Bruyn, G. W., Huntington's chorea: historical, clinical, and laboratory synopsis, in *Handbook of Clinical Neurology,* Vinken, P. J. and Bruyn, G. W., Eds., Elsevier, Amsterdam, 298, 1968.

130. Lange, H. W., Quantitative changes of telencephalon, diencephalon, and mesencephalon in Huntington's chorea, post-encephalitic and idiopathic Parkinsonism, *Verhandlunge Anatomische Gesellschaft,* 75, 923, 1981.

131. De La Monte, S. M., Vonsattel, J. P., and Richardons, E. P., Morphometric demonstration of atrophic changes in the cerebral cortex, white matter and neostriatum in Huntington's disease, *J. Neuropathol. Exp. Neurol.,* 47, 516, 1988.

132. Sotrel, A., Paskevich, P. A., Kiely, D. K. et al., Morphometric analysis of the prefrontal cortex in Huntington's disease, *Neurology,* 41, 1117, 1991.

133. Hedreen, J. C., Peyser, C. E., Folstein, S. E. et al., Neuronal loss in lamina VI of cerebral cortex in Huntington's disease, *Neuroscience,* 133, 257, 1991.

134. Cummings, J. L. and Cunningham, K., Obsessive-compulsive disorder in Huntington's disease, *Biolog. Psychiatr.,* 31, 263, 1992.

135. Rapoport, J. L., Recent advances in obsessive-compulsive disorder, *Neuropsychopharmacology,* 5, 1, 1991.

136. Laplane, D., Levasseur, M., Pilloni, B., Dubois, B., Baulac, M., Mazoyer, B., Tranbinh, S., Settle, G., Danze, F., and Baron, J., Obsessive-compulsive and other behavioral changes with bilateral basal ganglia lesions, *Brain,* 112, 649, 1989.

137. Tonkonogy, J. and Barreira, P., Obsessive-compulsive disorder and caudate-frontal lesion, *Neuropsychiatr., Neuropsychol. Behav. Neurol.,* 2, 203, 1989.

138. Swedo, S. E., Sydenham's chorea. A model for childhood autoimmune neuropsychiatric disorders [clinical conference], *JAMA,* 272, 1788, 1994a.

139. Swedo, S. E., Leonard, H. L., and Kiessling, L. S., Speculations on antineuronal antibody-mediated neuropsychiatric disorders of childhood, *Pediatrics,* 93, 323, 1994b.

140. Leonard, H. L., Swedo, S. E., Lenane, M. C., Rettew, D. C., Hamburger, S. D., Bartko, J. J., and Rapoport, J. L., A 2- to 7-year follow-up study of 54 obsessive-compulsive children and adolescents, *Arch. Gen. Psychiatr.,* 50, 429, 1993.

141. Swedo, S. E., Leonard, H. L., Mittleman, B. B., Allen, A. J., Rapoport, J. L., Dow, S. P., Kanter, M. E., Chapman, F., and Zabriskie, J., Identification of children with pediatric autoimmune neuropsychiatric disorders associated with streptococcal infections by a marker associated with rheumatic fever, *Am. J. Psychiatr.,* 154, 110, 1997.

142. Behar, S. M. and Porcelli, S. A., Mechanisms of autoimmune disease induction. The role of the immune response to microbial pathogens, *Arthr. Rheum.,* 38, 458, 1995.

143. Bronze, M. S. and Dale, J. B., The reemergence of serious group A streptococcal infections and acute rheumatic fever, *Am. J. Med. Sci.,* 311, 41, 1996.

144. Giedd, J. N., Rapoport, J. L., Leonard, H. L., Richter, D., and Swedo, S. E., Case study: acute basal ganglia enlargement and obsessive-compulsive symptoms in an adolescent boy, *J. Am. Acad. Child Adol. Psychiatr.,* 35, 913, 1996.

145. Allen, A. J., Leonard, H. L., and Swedo, S. E., Case study: a new infection-triggered, autoimmune subtype of pediatric OCD and Tourette's syndrome, *J. Am. Acad. Child Adol. Psychiatr.*, 34, 307, 1995.

146. Khanna, A. K., Buskirk, D. R., Williams, R. C., Jr., Gibofsky, A., Crow, M. K., Menon, A., Fotino, M., Reid, H. M., Poon-King, T., Rubinstein, P., and Zabriskie, J. B., Presence of a non-HLA B cell antigen in rheumatic fever patients and their families as defined by a monoclonal antibody, *J. Clin. Invest.*, 83, 1710, 1989.

147. Murphy, T. K., Goodman, W. K., Fudge, M. W., Williams, R. C., Jr., Ayoub, E. M., Dalal, M., Lewis, M. H., and Zabriskie, J. B., B lymphocyte antigen D8/17: a peripheral marker for childhood-onset obsessive-compulsive disorder and Tourette's syndrome?, *Am. J. Psychiatr.*, 154, 402, 1997.

148. Behar, D., Rapoport, J. L., Berg, C. J., Denckla, M. B., Mann L., Cox, C., Fedio, P., Szahn, T., and Wolfman, M. G., Computerized tomography and neuropsychological test measures in adolescents with obsessive-compulsive disorder, *Am. J. Psychiatr.*, 141, 363, 1984.

149. Luxenberg, J., Swedo, S., Flament, M., Friedland, R., Rapoport, J., and Rapoport, S. I., Neuroanatomical abnormalities in obsessive-compulsive disorder detected with quantitative X-ray computed tomography, *Am. J. Psychiatr.*, 145, 1089, 1988.

150. Stein, D. J., Hollander, E., Chan, S., DeCaria, C. M., Hilal, S., Liebowitz, M. R., and Klein, D. F., Computed tomography and neurological soft signs in obsessive-compulsive disorder, *Psychiatr. Res.: Neuroimaging*, 50, 143, 1993.

151. Insel, T. R., Donnelly, E. F., Lalakea, M. L., Alterman, I. S., and Murphy, D. L., Neurological and neuropsychological studies of patients with obsessive-compulsive disorder, *Biol. Psychiatr.*, 18, 741, 1983.

152. Scarone, S., Colombo, C., Livian, S., Abbruzzes, M., Ronchi, P., Locatelli, M., Scotti, G., and Smeraldi, E., Increased right caudate nucleus size in obsessive-compulsive disorder: detection with magnetic resonance imaging, *Psychiatr. Res.: Neuroimaging*, 45, 115, 1992.

153. Calabrese, G., Colombo, C. Bonfanti, A., Scotti, G., and Scarone, S., Caudate nucleus abnormalities in obsessive-compulsive disorder: measurements of MRI signal intensity, *Psychiatr. Res.*, 50, 89, 1993.

154. Robinson, D., Wu, H., Munne, R. A. et al., Reduced caudate nucleus volume in obsessive-compulsive disorder, *Arch. Gen. Psychiatr.*, 52, 393, 1995.

155. Jenike, M. A., Breiter, H. C., Baer, L. et al. Cerebral structural abnormalities in obsessive-compulsive disorder, *Arch. Gen. Psychiatr.*, 53, 625, 1996.

156. Rosenberg, D. R., Keshavan, M. S., O'Hearn, K. M., Dick, E. L., Bagwell, W. W., Seymour, A. B., Montrose, D. M., Pierri, J. N., and Birmaher, B., Frontostriatal measurement in treatment-naive children with obsessive-compulsive disorder, *Arch. Gen. Psychiatr.*, 54, 824, 1997.

157. Kellner, C. H., Jolley, R. R., Holgate, R. C., Austin, L., Lydiard, R. B., Laraia, M., and Ballenger, J. C., Brain MRI in obsessive-compulsive disorder, *Psychiatr. Res.*, 36, 45, 1991.

158. Aylward, E. H., Harris, G. J., Hoehn-Saric, R., Barta, P. E., Machlin, S. R., and Pearlson, G. D., Normal caudate nucleus in obsessive-compulsive disorder assessed by quantitative neuroimaging, *Arch. Gen. Psychiatr.*, 53, 577, 1996.

159. Stein, D. J., Coetzer, R., Lee, M., Davids, B., and Bouwer, C., Magnetic resonance brain imaging in women with obsessive-compulsive disorder and trichotillomania, *Psychiatr. Res.*, 74, 177, 1997.

160. Swedo, S. E., Pietrini, P., Leonard, H. L. et al., Cerebral glucose metabolism in childhood-onset obsessive-compulsive disorder, *Arch. Gen. Psychiatr.*, 49, 690, 1992.

161. Baxter, L. R., Jr., Thompson, J. M., Schwartz, J. M., Guze, B. H., Phelps, M. E., Mazziotta, J. C., Selin, C. E., and Moss, L., Trazodone treatment response in obsessive-compulsive disorder — correlated with shifts in glucose metabolism in the caudate nuclei, *Psychopathology*, 20 (Suppl. 1), 114, 1987.

162. Horwitz, B., Swedo, S. E., Grady, C. L., Pietrini, P., Schapiro, M. B., Rapoport, J. L., and Rapoport S. I., Cerebral metabolic pattern in obsessive-compulsive disorder: altered intercorrelations between regional rates of glucose utilization, *Psychiatr. Res.*, 40, 221, 1991.

163. Perani, D., Colombo, C., Bressi, S. et al., [18F] FDG PET study in obsessive-compulsive disorder: a clinical/metabolic correlation study after treatment, *Br. J. Psychiatr.*, 166, 244, 1995.

164. Baxter, L. R., Schwartz, J. M., Bergman, K. S. et al., Caudate glucose metabolic rate changes with both drug and behavior therapy for obsessive-compulsive disorder, *Arch. Gen. Psychiatr.*, 49, 681, 1992.

165. Schwartz, J. M., Stoessel, P. W., Baxter, L. R. et al., Systematic changes in cerebral glucose metabolic rate after successful behavior modification treatment of obsessive-compulsive disorder, *Arch. Gen. Psychiatr.*, 53, 109, 1996.

166. Rauch, S. L., Jenike, M. A., Alpert, N. M. et al., Regional cerebral blood flow measured during symptom provocation in obsessive-compulsive disorder using oxygen 15-labeled carbon dioxide and positron emission tomography, *Arch. Gen. Psychiatr.*, 51, 62, 1994.

167. McGuire, P. K., Bench C. J., Frith, C. D., Marks, I. M., Frackowiak, R. S., and Dolan R. J., Functional anatomy of obsessive-compulsive phenomena, *Br. J. Psychiatr.*, 164, 459, 1994.

168. Breiter, H. C., Rauch, L. S., Kwong K. K. et al., Functional magnetic resonance imaging of symptom provocation in obsessive-compulsive disorder, *Arch. Gen. Psychiatr.*, 53, 595, 1996.

169. Goodman, W. K., Ward, H., Kablinger, A., and Murphy, T., Fluvoxamine in the treatment of obsessive-compulsive disorder and related conditions, *J. Clin. Psychiatr.*, 58 (Suppl. 5), 32, 1997.

170. Carllson, A., The occurrence, distribution and physiological role of catecholamines in the nervous system, *Pharmacol. Rev.*, 11, 490, 1959.

171. Martin, J., *Neuroanatomy Text and Atlas*, Appleton & Lange, Norwalk, CT, 1989.

172. Pazos A. and Palacios, J., Quantitative autoradiographic mapping of serotonin receptors in rat brain, I. Serotonin I receptors, *Brain Res.*, 346, 205, 1986.

173. Stuart A., Slater, J., Unwin H., and Crossman, A., A semi-quantitative atlas of 5-hydroxytryptamine-1 receptors in the primate brain, *Neuroscience*, 18, 619, 1986.

# 6 Panic Disorder: A Genetic Perspective

*D. J. Niehaus,\* J. A. Knowles, and D. J. Stein\**

## CONTENTS

## INTRODUCTION

Panic disorder is a highly prevalent psychiatric disorder with significant associated morbidity. The disorder often begins with spontaneous panic attacks, characterized by sympathetic discharge. Agoraphobia and other avoidance behavior may follow these attacks. Ultimately, significant distress and marked impairment of social and occupational function may be seen.

Organisms ranging from protozoa to mammals have evolved defensive responses to external stimuli. Anxiety responses, such as the classical "flight-or-fight" phenomenon, have clear neurobiological underpinnings and presumably a genetic substrate.[1] While certain kinds of anxiety disorder may reflect a low threshold for normal anxiety responses, panic disorder arguably represents a specific kind of false alarm rather than a general anxiety response.

Indeed, a range of evidence suggests that panic disorder may have a specific genetic basis. In this chapter we review early family and twin studies and then discuss more recent searches for candidate genes.[2] The search for a genetic basis for panic disorder may ultimately lead to improved understanding of and treatment for this disorder.

## DIAGNOSTIC CERTAINTY — A PREREQUISITE FOR GENETIC STUDIES

Studies of the neurobiology of panic disorder received significant impetus from the provision of specific diagnostic criteria for panic disorder in DSM-III. The new international classifications of

\* Dr. Neihaus and Dr. Stein are supported by the MRC Research Unit on Anxiety and Stress Disorders (South Africa).

mental disorders (ICD and DSM) and the use of diagnostic criteria have allowed a marked improvement in the diagnostic process. Structured interviews have been developed based on these classification systems. The Diagnostic Interview for Genetic Studies (DIGS), the Structured Clinical Interview for Psychiatric Disorders (SCID), and the Schedule for Affective Disorders and Schizophrenia (SADS) are some examples. These interviews have good test-retest reliability and inter-rater reliability.[3,4] Diagnostic precision is also necessary when assessing panic disorder in family members of patients with panic disorder. When family members are not directly interviewed, there may be an overestimation of the diagnosis.[5,6] Thus, the tendency for panic disorder to "run in families" may be overestimated.[6]

## EPIDEMIOLOGY

Lifetime rates of panic disorder measured in the general population range from 1.2 to 2.4%. This is much lower than the reported 7.7 to 20.5% lifetime rate in first-degree relatives of panic probands.[7] There is a large sex effect on the prevalence, as females are affected twice as often as males.[8]

Panic disorder has wide range in the age of onset, with a mean age of around 25.[8] In a number of psychiatric disorders, early age of onset has been associated with increased familial risk, severity, and distinctive patterns of comorbidity. In panic disorder, family history of panic disorder with agoraphobia (thought to be a more severe variant of the disorder) influenced age at onset of panic disorder.[9,10] The age at onset of the disorder may be useful in differentiating familial subtypes of panic disorder. In one study of 838 adult first-degree relatives of 152 probands and controls, the rate of panic disorder in adult first-degree relatives of probands with panic disorder onset at or before 20, or after 20 years of age, were increased 17-fold and 6-fold, respectively, as compared to the rate in the relatives of the controls.[11] The age of onset did not appear to be specifically transmitted within families in this study.

An age of onset effect raises the possibility that genetic anticipation may be occurring. Given the finding of expanding trinucleotide DNA sequences as the basis of a number of neuropsychiatric disorders, interest in anticipation has revived.[12] Anticipation can involve an earlier age at onset, greater severity, and/or a higher number of affected individuals (increased penetrance) in successive generations within a family. Unfortunately, studies of anticipation are difficult to interpret because of multiple methodological problems. There is some controversy concerning detection of age-at-onset anticipation, due to problems of sampling bias. Differential age at interview between parent and child is one source of bias.[13] Parents may have passed through more of the risk period than their offspring. Results show that the timing of diagnostic assessment can produce a severely biased sample exhibiting apparent anticipation. There also seems to be a reduction in agreement between probands and family members over time of the age at onset.[3] There was a tendency for older respondents to increase the age of onset of their depression across two interviews, although the increase was only a few years.[3] In the face of these difficulties, the two studies of anticipation in panic disorder came to different conclusions.[13,14]

## FAMILY, RISK, AND PANIC DISORDER

Panic disorder appears to be one of the more familial psychiatric diseases. Two-thirds of panic disorder probands have relatives affected with the same condition, and the estimated risk to first degree relatives is approximately 3 to 6 times the rate of the general population (5.7% to 18% compared to 1.8% to 3% in control groups).[17-19] Some authors argue that given a lifetime cumulative risk of 1.9% for males and 4.7% for females, models predict an approximately five times increased risk of developing panic disorder if an affected parent or sibling is present.[8]

A family history study of second-degree relatives of 19 patients with panic disorder and 19 controls showed a morbidity risk of 9.5%[20] for relatives of affecteds. This was significantly higher than the 1.4% among the control group. These risks were approximately half those found among first-degree relatives. Female relatives were at higher risk for panic disorder, probably in keeping with prevalence differences.[20]

Given the prevalence data, family aggregation in panic disorder was examined in 636 individuals at the Austin Hospital in Victoria, Australia. The findings show consistence with the idea of family aggregation. The data alone were not sufficient to draw conclusions about the cause of the aggregation.[21]

In summary, studies indicate that first- and second-degree family members of affected probands are at greater risk for developing panic disorder and family aggregation does occur.

## THE ROLE OF TWIN STUDIES IN PANIC DISORDER

Twin studies of panic anxiety disorders, although limited in number, suggest that panic disorder has a genetic component.[21] One recent study reported a 30 to 73% concordance among monozygotic (MZ) twins, against 0 to 4% among dizygotic (DZ) twins.[23] In a study of 32 MZ and 53 DZ adult same-sexed twin pairs, anxiety disorders with panic attacks were more than five times as frequent in MZ as in DZ co-twins in a combined group of probands with panic disorders and agoraphobia with panic attacks.[24] Another study of anxiety disorders by the same group focused on a sample of 20 MZ and 29 DZ co-twins of anxiety disorder probands and a comparison group of co-twins of 12 MZ and 20 DZ twin probands with other nonpsychotic mental disorders.[25] Panic disorder was significantly more prevalent in co-twins of panic probands than in controls. In both studies, the MZ:DZ concordance ratio in panic disorder was more than 2:1.[25]

In the largest twin study of panic disorder, 1,033 female twin pairs from a population-based twin registry were directly interviewed by trained interviewers. This study estimated that additive genetic factors account for 30 to 40% of the variance to develop panic disorder using a multifactorial-threshold model.[15] One interpretation of this result is that panic disorder in the general population is only a moderately heritable condition. The studies described previously were based on clinically ascertained samples. One possibility is that these findings could be due to sampling differences. However, a study addressed this aspect and found that selecting probands from treatment clinics rather than from the general population does not necessarily lead to greater estimates of familial aggregation of panic disorder and therefore might not account for the higher observed heritability in the clinic samples.[16]

The genetics of panic disorder may differ from that of sporadic panic attacks or panic symptoms. Sporadic panic attacks (SPAs) in 120 twins recruited from the general population showed a concordance ratio in MZ to DZ twins (57%:43% = 1.32) that was lower than that seen in panic disorder twins.[23] This suggests that genetic factors may not be crucial for the development of sporadic panic attacks.[23] A large self-report study of panic symptoms, the "Virginia 30,000" twin-family study suggests that genetic factors significantly influence these symptoms while familial and/or environmental factors play little or no etiologic role.[26] In an extension of this work, data from 2,903 adult same-sex twin pairs suggest that much of the genetic variation influencing the physical symptoms associated with panic is nonadditive, perhaps due to dominance or epistasis. In both sexes, these nonadditive genetic effects on physical symptoms influence the reporting of "feelings of panic."[27]

## COMORBIDITY: SHARED GENETIC FACTORS?

Systematic community and clinical sampling has found that anxiety and affective disorders frequently coexist in the same individual. The National Comorbidity Survey, for example, found that comborbity is high for major depression and all anxiety disorders.[28] Comorbidity between anxiety and affective disorders suggests the need to investigate the possibility of shared genetic factors. In addition, the etiological factors underlying panic disorder may overlap with those of alcoholism and those of unipolar major depression.[18]

Comorbidity of panic disorder with agoraphobia may also shed light on the genetics of panic disorder. Agoraphobia is closely linked to panic disorder and segregation analysis showed that agoraphobia segregated predominantly among female relatives of agoraphobic probands. Family studies of agoraphobia, panic disorder, and nonanxious controls showed the morbidity risk for all anxiety disorders to be 32% among first-degree relatives of agoraphobics, 33% among relatives of patients with panic disorder, and 15% among relatives of controls.[29] The morbidity risk for panic disorder was increased among the relatives of agoraphobics (8.3%). The morbidity risk for agoraphobia is not increased among the relatives of panic disorder patients (1.9%).[19,30] Probands and relatives with agoraphobia reported an earlier onset of illness, more persistent and disabling symptoms, more frequent complications, and a less favorable outcome than probands and relatives with panic disorder. The findings suggest that agoraphobia might be a more severe variant of panic disorder.[29]

On the other hand, the genetic overlap between panic disorder and the other anxiety disorders may differ. Multiple family studies that have compared panic disorder and GAD in their pattern of their symptom, age and type of onset, personality characteristics, course of illness, and outcome support the hypothesis that the disorders are discrete.[32-34] Further support comes from a study of 1033 female-female twin pairs from a population-based registry.[31] The results suggest that GAD is a moderately familial disorder, as heritability was estimated around 30%.[31] Further study into the overlap of these genetic factors with those that contribute to panic disorder in this twin sample suggested that they might be different.[69] These studies suggest the validity of generalized anxiety disorder and panic disorder as discrete diagnostic entities in both family and twin studies.[7,17,32-34]

There is an increased prevalence (as high as 27% in one study) of obsessive-compulsive symptoms in patients with panic disorder. Some authors suggests that a subgroup of patients with panic disorder along with obsessive-compulsive symptoms have an earlier onset of illness, were more likely to have personal and family histories of major depression and substance abuse, and showed a poorer outcome after treatment.[35]

An average of 50% of subjects with social phobia in the community reported a concomitant anxiety disorder including another phobic disorder, generalized anxiety, or panic disorder.[36] However, social phobia in individuals who subsequently develop panic disorder show a different form of familial transmission from social phobia without lifetime anxiety comorbidity. Thus, social phobia may be nonfamilial and/or causally related to panic disorder.[37]

"Behavioral inhibition to the unfamiliar," as defined by the Harvard Infant Study program, is viewed as an early temperamental characteristic of children at risk for adult panic disorder and agoraphobia.[38] The possibility of an early marker for panic disorder may help the search for genetic mechanisms. A three-generation family's presentation of separation anxiety disorder manifested itself as school phobia in childhood, which evolves toward panic disorder with agoraphobia in young adulthood and diffuse anxiety with hypochondriacal features in later life. The information from this family would fit into a model that describes a spectrum of anxiety disorders presenting with "different manifestations of overt psychopathology throughout the life cycle."[39]

Panic disorder in the parents conferred more than a threefold increased risk of separation anxiety in the children. Other factors that increased the risk to children were degree of familial loading for psychiatric illness, parental assortative mating, and parental recurrent depression.[40] On the other hand, a retrospective twin study revealed a substantial genetic contribution to separation anxiety in females, but not in males, with unique environmental influences being important in both gender groups.[41] Further study is needed to examine the interface between these disorders.

The co-occurrence of alcoholism and anxiety disorders in epidemiological and clinical samples is well established. The lifetime rate for panic disorder was significantly higher in alcoholics (4.2%) and their close biological family (3.4%) than in controls (1.0%).[42,43] In a family study of

113 families of probands with either panic disorder or alcoholism or both, and 80 families of healthy controls, an elevated risk of alcoholism in relatives of probands with panic disorder as compared to the relatives of controls was observed.[44] Self-medication of anxiety disorder probands with the anxiolytic substance alcohol may be one reason for this association. Alternatively, common genetic susceptibility factors for both disorders are a possible explanation, but it is not supported by a twin study.[69]

Findings on the relationship between anxiety and affective disorders are inconsistent. A controlled family study found that both unipolar depression and panic disorder were aggregating in families.[53] The comorbid condition did not represent a distinct subtype in terms of familial aggregation. A sharing of familial factors of etiological relevance between panic disorder and unipolar depression might explain a limited proportion of comorbid cases. A case of identical twins concordant for attention deficit disorder and somnambulism but discordant for major depression and agoraphobia with panic attacks has been reported.[47] Major depression plus panic disorder in probands was associated with a marked increase in risk in relatives for major depression, panic disorder, phobia, and/or alcoholism than the relatives of probands with major depression without any anxiety disorder.[46-48] These findings are consistent with the hypothesis that panic disorder plus agoraphobia and panic disorder-agoraphobia with major depression (MDE) share common familial etiologic factors while MDE alone is an independent disorder.[49] In 646 female twins from a population-based register with a lifetime diagnosis of MDE, panic disorder was one of the four variables that uniquely predicted an increased risk for MDE in the co-twin.[50] Although these and other studies indicate that there is a substantial comorbidity in individuals between panic disorder and major depression, other studies suggest that these two disorders are separate conditions that are independently and specifically transmitted within families.[7,30,37,51,52] However, the major proportion of comorbidity between panic disorder and unipolar depression may still be due to nonfamilial factors.[53]

The possible intersection between bipolar mood disorder and panic disorder in families has also been the subject of a number of studies. In one study, probands with both bipolar mood disorder (BMD) and panic disorder had a significantly higher prevalence of panic disorder, bipolar II, cyclothymia, and dysthymia among their first-degree relatives, but they had lower prevalence of substance abuse than the relatives of the bipolar probands without panic disorder.[54] The authors suggested the possibility that panic disorder might be a marker for heterogeneity in bipolar mood disorder.

Comorbidity of several disorders appears to be independent of panic disorder. In one family study, the risk for schizophrenia was not elevated in relatives of patients with panic disorder (0%) in comparison to controls (0.3%, P > 0.05).[55] Likewise, a case-control study suggests that somatization disorder is not simply a form of panic disorder and that the two disorders may coexist in the same subject without sharing a common genetic diathesis.[56]

Analyses undertaken to examine whether a wide range of psychiatric disorders, including anxiety disorders, represent variant manifestations or expressions of Tourette's syndrome (TS) could not be proven in one study.[57] However, two studies by another group did purport to find a relationship. This group found a high rate of panic attacks in patients with TS (33 to 55%).[58] They have also described a series of 11 TS pedigrees with a high rate of panic disorder.[59]

In summary, these studies suggest that there may be a genetic relationship between panic disorder, major depression, and/or bipolar depression. Additional studies in larger epidemiological samples may be required to assess generalizability of these findings.[17,37]

## THE ROLE OF PANIC SUBTYPES

In the search for the most suitable phenotype for genetic studies, there is a tendency to either lump or split panic disorder symptoms or comorbid conditions into new subtypes. Both lumpers and splitters believe that their approach will be justified when the final word has been spoken on the genetics of panic disorder. The subgroupers believe that a narrower spectrum will lead to a more pronounced gene effect.

As with many psychiatric mental disorders, two predisposing factors have been found to operate: a genetic element, as shown by family and twin studies, and an element relating to environmental psychoaffective development (traumatic events in childhood, separation). As described previously, panic disorder is commonly associated with other anxiety disorders, depression, alcoholism, and drug abuse. These associations not only represent poor prognostic factors that are associated with chronicity, attempted suicide, and social handicap, but they may also be significant from a genetic perspective.

In the future, validation of subcategories of panic disorder may be obtained from molecular genetics studies. On the other hand, subtyping of panic disorder may be useful in helping the genetic studies find the relevant genes. Thus, ongoing interaction between clinical and molecular science is necessary to resolve questions of subtyping.[3]

Two subtype panic disorder researchers have employed various techniques. For example, one hypothesis suggests that an abnormal threshold for suffocation alarm underlies panic disorder.[60] In support of this hypothesis, a number of experiments indicate that hypersensitivity to carbon dioxide may act as a disease-specific trait marker. Because this disorder is highly familial, evidence of an abnormal suffocation threshold is expected in high-risk individuals before they develop clinical illness. Single inhalation of 35% $CO_2$ or air is used for evaluation. In one study, 45.5% of individuals were at high risk for panic but none of the controls showed a positive panic response.[61] Another study of 203 patients with panic disorder yielded a significantly higher genetic morbidity risk of 14.4% for patients who reacted positively to the carbon dioxide challenge than for non-responders (morbidity risk = 3.9%).[62] The response to $CO_2$ can be modified by a one-week treatment with the serotonergic drugs, clomipramine and fluvoxamine, in $CO_2$-sensitive patients[63] and appears to be highly heritable in a recent twin study.[64] Thus, hypersensitivity to $CO_2$ might be associated with a subtype of panic disorder specifically related to a greater familial loading.

The triggering effect of lactate infusion in predisposed individuals is also well-known. Subjects with lactate-induced panic attacks seem to show a high prevalence of anxiety disorders among first-degree relatives, while there was no significant difference in the prevalence of mood disorders and substance abuse between first-degree relatives of subjects with (N = 45) and without (N =115) lactate-induced panic attacks. The results suggest that individuals with a family history of anxiety disorders may be vulnerable to lactate-induced panic attacks.[65]

In the search for genetic causes of illness, an association with other medical abnormalities sometimes leads to possible candidate genes or mechanisms. The association with mitral valve prolapse suggests that perhaps 38% of patients presenting with symptoms of panic disorders have mitral valve prolapse on echocardiography. One explanation for the association between mitral valve prolapse (MVP) and panic attacks in clinic populations is that panic attacks represent a set of symptoms caused by mitral prolapse. Since mitral prolapse is often familial, it predicts a higher incidence of panic attacks in relatives of persons with prolapse than the general population and the incidence of panic attacks in the relatives should be independent of panic attacks in the proband. In one study, the incidence of panic attacks in first degree relatives of MVP probands was 4.5%.[66] This is comparable to control rates reported in the authors' previous studies. The study also found that the incidence of panic attacks in relatives of MVP probands with panic disorder was more than ten times greater than the rate found among relatives of probands without panic attacks. These findings are consistent with the hypothesis that mitral prolapse and panic attacks are segregating independently in these families.[66] The prevalence of 12.5% of panic disorder in a sample of cardiac outpatients from the Toronto area is in keeping with previous studies.[67] Finally, panic disorder was significantly more prevalent among the first-degree relatives of probands with normal coronary arteries diagnosed with panic disorder or panic attacks than among the family members of probands with normal coronary arteries without panic disorder or attacks (17.4% vs. 15.7% vs. 4.0%).[68]

## MODE OF TRANSMISSION

The genetic mode of transmission of panic disorder remains unclear. The first segregation analysis examined 19 kindreds of panic disorder and suggested that this disorder is transmitted as an autosomal dominant trait.[71] This pedigree sample was increased to 41 pedigrees and a second segregation analysis found the best fit to a single major locus that was roughly dominant in action; however, a multifactorial model could not be excluded.[17] The same group also found that the number of observed unilateral pairs (two ancestral cases from one side of the kindred) was significantly greater than the number expected in a sample of 15 kindreds, supporting dominant transmission.[72] The observation of panic disorder in four generations of a family also supports an autosomal dominant mode of inheritance.[73]

The possibility that the transmission of the disorder is not dominant is raised by the findings of other studies, however. In a subgroup of the Yale Family Study, 28 families in which the proband had both depression and panic disorder, a major gene mode of inheritance was not supported.[74] In another study of 30 pedigrees of panic disorder without comorbid major depression, there was approximately equal support for the best-fit recessive and dominant models.[75] In addition, fully penetrant models with no phenocopies for both dominant and recessive modes of transmission could be excluded. In the last study, comparable support for dominant and recessive models with nonzero phenocopy rates was also observed.[70] This sample consisted of 126 families of probands with panic disorder (some with comorbid depression). In this study, the nongenetic transmission model was rejected when compared to the recessive model. Given these nonspecific and contradictory findings, it is safe to state that the mode of transmission remains unclear but that segregation analysis indicates that panic disorder may be compatible with either autosomal dominant, autosomal recessive, or a more complex mode of inheritance.

## CANDIDATE GENES AND GENOME SEARCHES

The previously outlined studies indicate that it is likely that genetic factors contribute to an individual's predisposition to develop panic disorder. Advances in molecular genetics over the last two decades have made it possible to study genes in human disease in ways not previously possible. Both candidate gene and genome-wide searches have been used in the search for the "panic" gene or genes.

An early study indicated a possible association between panic disorder and the major histocompatibility gene complex (HLA). Members of a family with high morbidity of panic disorder were tested for the multiple antigens (11 HLA-A, 16 HLA-B, and 5 HLA-Cw). The results suggested an HLA component for panic disorder, based on the presence of the same haplotype (A3B18) in the six affected members (panic disorder with agoraphobia) of the family.[78]

The involvement of pro-opiomelanocortin (POMC) gene has also been postulated. This gene is responsible for the synthesis of a large precursor peptide from which ACTH is derived. The ACTH gene may be relevant to panic disorder, since post-dexamethasone suppression of cortisol is abnormal in some cases of panic disorder. A linkage study of 16 family members (6 affected in 4 generations) in one large family excluded this locus, but clearly this gene needs to be tested in a larger sample.[79]

Noradrenergic dysfunction has also been hypothesized as a mechanism in panic disorder. Stimulation of the nucleus ceruleus in monkeys increases fear-like behavior, while ablation reduces the behavior. In humans, the noradrenergic system reaches maximal activity at the age when the incidence of panic disorder peaks. This fits with the symptoms of panic attacks and findings of increased noradrenergic function in panic disorder.[80] Five polymorphic DNA markers near the various adrenergic receptor loci were genotyped in 154 individuals from 14 families (68 affecteds), but no evidence for linkage was detected.[79] The Cys311 DRD2 variant described in a pedigree with multiple affecteds with panic disorder also seems not to be causative.[80] These findings do not rule out a noradrenergic mechanism, and further studies may still unearth different findings.

Although inconsistent, many studies have implicated abnormalities in the serotonergic system in patients with panic disorder.[83] This prompted studies to determine whether there is an abnormal gene encoding for 5HT1d2 and/or 5HT1db receptors in patients with panic disorder. Genomic DNA from panic disorder patients was sequenced for each receptor gene. Two DNA polymorphisms that did not change the protein sequence of the receptors were observed, making it unlikely that they had a role in the etiology of panic disorder.[80] The serotonin transporter gene is the action site of selective serotonin reuptake inhibitors, widely used in the treatment of panic disorder. The short allele of a polymorphism in the promoter of this gene has been associated with higher anxiety levels in a general population sample.[81] However, two associations studies of this polymorphism and panic disorder have failed to find any evidence of linkage disequilibrium and disorder.[82,83]

The cholecystokinin B (CCK-B) receptor gene has also been investigated as a candidate gene. Cholecystokinin tetrapeptide (a CCK-B agonist) is known to induce panic attacks more easily in patients with panic disorder than in controls. The CCK-B receptor was screened for mutations in 22 probands with panic disorder, using single stranded conformation polymorphism (SSCP) analysis. Two polymorphisms were found, one of which altered the protein coding sequence. The frequency of this polymorphism was found to be 8.8% in panic disorder patients and 4.4% in controls, but it did not segregate with disease in the two families that had the polymorphism.[84] The CCK gene has also been scanned for polymorphisms and a C to T transition in the 5' promoter, in a SP1 binding site was found.[85] The "T" allele was observed in 16% of individuals with panic disorder and 8% of controls. The TDT, a test of association in the presence of linkage was not significant when panic disorder was considered "affected," but it was suggestive when individuals with panic attacks were added ($p < 0.05$).[85] The role of this polymorphism awaits follow-up studies.

The GABA receptor, with its known 13 subunits, has also been implicated in the etiology of panic disorder. Chronic administration of imipramine (tricyclic antidepressant) and phenelzine (a monoamine oxidase inhibitor) lead to increased levels of GABA receptor alpha 1, beta 2, and gamma 2 subunits in a rat brainstem.[84] The anti-panic properties of benzodiazepines also support a modulating role for the GABA receptors in panic disorder. Functional benzodiazepine receptors are composed of multiple subunits. A distinct gene that is differentially expressed in the brain encodes each subunit. Polymorphisms have been found in 8 of 13 subunit receptor genes. No evidence for linkage between panic disorder and the 8 polymorphisms was found.[87,88]

A possible association between panic disorder and the low voltage electroencephalogram trait (LVEEG) has also been observed.[89] Some cases with LVEEG (approximately one third) have been linked to chromosome 20 in close proximity to the gene for neuronal nicotine acetylcholine receptor alpha4 subunit (CHRNA4). Therefore, the hypothesis that polymorphisms in the CHRNA4 gene are associated with panic disorder was tested. The allelic frequencies of three different CHRNA4 polymorphisms in patients with panic disorder and in healthy controls were examined. No significant difference in allele frequencies were noted and thus an association between panic disorder and the CHRNA4 gene is unlikely.[90,91]

Linkage studies of psychiatric disorders can yield informative results by identifying tentative locations that merit further investigation and by excluding regions of the genome from future linkage searches. In the first linkage study of panic disorder, 29 polymorphic blood antigens distributed over 10 chromosomes were typed in 26 families.[91] The LOD score for alpha-haptoglobin (16q22) was suggestive of linkage (2.27 at a recombination fraction of 0.0),[91] but a follow-up study using an additional 10 families and a DNA marker yielded markedly decreased support for linkage.[92]

Two genome-wide scans for anxiety genes in humans have been published. Knowles et al. completed the first genome-wide genetic screen for panic disorder on 23 multiplex families with panic disorder.[93] The first-pass genomic screen with 540 microsattelite DNA markers covered the genome with an average marker density of 11 cM. Only two markers gave a LOD score above 1.0 under a dominant model (chromosomes 1p and 20p). Under a recessive model, four markers gave

LOD scores in excess of 1.0 (chromosomes7p, 20q and X/Y). In an affected sib-pair analysis loci on chromosomes 2, 7p, 8p, 8q, 9p, 11q, 12q, 16p, 20p, and 20q exceeded a LOD score of 1.0. If panic disorder was transmitted as a homogenous single gene dominant disorder, this study had adequate power to detect such a locus. Given the absence of detection of a locus with the dominant model it is likely that the genetic predisposition to panic disorder is due to a recessive locus or multiple genetic loci. A genome-wide scan for the personality trait of harm avoidance (as measured by the TPQ) in 177 families detected linkage genetic linkage to a locus on chromosome 8p21-23 that explained 38% of variance of the trait.[94] Other epistatic loci were observed on chromosomes 18p, 20p, and 21q.[94] Some of these loci are in regions that had LOD scores over 1.0 in the panic disorder genome screen, raising the possibility that the genes that affect the personality trait of harm avoidance may overlap with those for panic disorder.

## CONCLUSION

There is good evidence that genetic factors play an important role in panic disorder. Diagnostic accuracy and reliable pedigree information are essential for future genetic studies in this genetic work. Neurobiologically relevant parameters like $CO_2$ sensitivity may further support these developments. Comorbidity studies may also provide additional ways to subtype panic disorder and address possible shared etiological factors. The apparent lack of definite positive results in candidate and genome-wide searches makes further studies with larger sample sizes necessary.

## REFERENCES

1. Marks, I.M., Genetics of fear and anxiety disorders, *Br. J. Psychiatr.,* 149, 406, 1986.
2. Berg, K., Mullican, C., Maestri, N. et al., Psychiatric genetic research at the National Institute of Mental Health, *Am. J. Med. Genet.,* 54(4), 295, 1994.
3. Prusoff, B.A., Merikangas, K.R., and Weissman, M.M., Lifetime prevalence and age of onset of psychiatric disorders: recall 4 years later, *J. Psychiatr. Res.,* 22(2), 107, 1988.
4. Skre, I., Onstad, S., Torgersen, S. et al., High interrater reliability for the Structured Clinical Interview for DSM-III-R Axis I (SCID-I), *Acta. Psychiatr. Scand.,* 84(2), 167, 1991.
5. Brown, T.A., Familial aggregation of panic in nonclinical panickers, *Behav. Res. Ther.,* 32(2), 233, 1994.
6. Chapman, T.F., Mannuzza, S., Klein, D.F. et al., Effects of informant mental disorder on psychiatric family history data, *Am. J. Psychiatr.,* 151(4), 574, 1994.
7. Weissman, M.M., Family genetic studies of panic disorder, *J. Psychiatr. Res.,* 27 (Suppl. 1), 69, 1993.
8. Hopper, J.L., Judd, F.K., Derrick, P.L. et al., A family study of panic disorder, *Genet. Epidemiol.,* 4(1), 33, 1987.
9. Battaglia, M., Bertella, S., Politi, E. et al., Age at onset of panic disorder: influence of familial liability to the disease and of childhood separation anxiety disorder, *Am. J. Psychiatr.,* 152(9), 1362, 1995.
10. Gruppo Italiano Disturbi d'Ansia, Familial analysis of panic disorder and agoraphobia, *J. Affect. Disord.,* 17(1), 1, 1989.
11. Goldstein, R.B., Wickramaratne, P.J., Horwath, E. et al., Familial aggregation and phenomenology of 'early'-onset (at or before age 20 years) panic disorder, *Arch. Gen. Psychiatr.,* 54(3), 271, 1997.
12. Petronis, A. and Kennedy, J.L., Unstable genes — unstable mind?, *Am. J. Psychiatr.,* 152, 164, 1995.
13. Heiman, G.A., Hodge, S.E., and Wickramaratne, P., Age-at-interview bias in anticipation studies: computer simulations and an example with panic disorder, *Psychiatr. Genet.,* 6, 61, 1996.
14. Battaglia, M., Bertella, S., Bajo, S. et al., Anticipation of age at onset in panic disorder, *Am. J. Psychiatr.,* 155, 590, 1998.
15. Kendler, K.S., Neale, M.C., Kessler, R.C. et al., Panic disorder in women: a population-based twin study, *Psychol. Med.,* 23(2), 397, 1993.
16. Wickramaratne, P.J., Weissman, M.M., Horwath, E. et al., The familial aggregation of panic disorder by source of proband ascertainment, *Psychiatr. Genet.,* 4(3), 125, 1994.

17. Crowe, R.R., Noyes, R., Pauls, D.L. et al., A family study of panic disorder, *Arch. Gen. Psychiatr.,* 40(10), 1065, 1983.

18. Maier, W., Lichtermann, D., Minges, J. et al., A controlled family study in panic disorder, *J. Psychiatr. Res.,* 27 (Suppl 1), 79, 1993.

19. Leboyer, M. and Lepine, J.P., [Is anxiety hereditary?]'anxiete est-elle hereditaire?, *Encephale.,* 14(2), 49, 1988.

20. Pauls, D.L., Noyes, R., Jr., and Crowe, R.R., The familial prevalence in second-degree relatives of patients with anxiety neurosis (panic disorder), *J. Affect. Disord.,* 1(4), 279, 1979.

21. Burrows, G.D., Judd, F.K., and Hopper, J.L., The biology of panic-genetic evidence, *Int. J. Clin. Pharmacol. Res.,* 9(2), 147, 1989.

22. Weissman, M.M. and Merikangas, K.R., The epidemiology of anxiety and panic disorders: an update, *J. Clin. Psychiatr.,* 47, 11, 1986.

23. Perna, G., Caldirola, D., Arancio, C. et al., Panic attacks: a twin study, *Psychiatr. Res.,* 66(1), 69, 1997.

24. Torgersen, D., Genetic factors in anxiety disorders, *Arch. Gen. Psychiatr.,* 40(10), 1085, 1983.

25. Skre, I., Onstad, S., Torgersen, S. et al., A twin study of DSM-III-R anxiety disorders, *Acta. Psychiatr. Scand.,* 88(2), 85, 1993.

26. Kendler, K.S., Walters, E.E., Truett, K.R. et al., A twin-family study of self-report symptoms of panic-phobia and somatization, *Behav. Genet.,* 25(6), 499, 1995.

27. Martin, N.G., Jardine, R., Andrews, G. et al., Anxiety disorders and neuroticism: are there genetic factors specific to panic?, *Acta. Psychiatr. Scand.,* 77(6), 698, 1988.

28. Lepine, J.P., Comorbidite des troubles anxieux et depressifs: perspectives epidemiologiques, *Encephale*, 20 (Spec. 4), 683, 1994.

29. Harris, E.L., Noyes, R., Jr., Crowe, R.R. et al., Family study of agoraphobia. Report of a pilot study, *Arch. Gen. Psychiatr.,* 40(10), 1061, 1983.

30. Noyes, R., Jr., Crowe, R.R., Harris, E.L. et al., Relationship between panic disorder and agoraphobia. A family study, *Arch. Gen. Psychiatr.,* 43(3), 227, 1986.

31. Kendler, K.S., Neale, M.C., Kessler, R.C. et al., Generalized anxiety disorder in women. A population-based twin study [see comments], *Arch. Gen. Psychiatr.,* 49(4), 267, 1992.

32. Raskin, M., Peeke, H.V., Dickman, W. et al., Panic and generalized anxiety disorders. Developmental antecedents and precipitants, *Arch. Gen. Psychiatr.,* 39(6), 687, 1982.

33. Noyes, R., Jr., Clarkson, C., Crowe, R.R. et al., A family study of generalized anxiety disorder, *Am. J. Psychiatr.,* 144(8), 1019, 1987.

34. Anderson, D.J., Noyes, R., Jr., and Crowe, R.R., A comparison of panic disorder and generalized anxiety disorder, *Am. J. Psychiatr.,* 141(4), 572, 1984.

35. Mellman, T.A. and Uhde, T.W., Obsessive-compulsive symptoms in panic disorder, *Am. J. Psychiatr.,* 144(12), 1573, 1987.

36. Merikangas, K.R. and Angst, J., Comorbidity and social phobia: evidence from clinical, epidemiologic, and genetic studies, *Eur. Arch. Psychiatr. Clin. Neurosci.,* 244(6), 297, 1995.

37. Crowe, R.R., Wang, Z., Noyes, R., Jr. et al., Candidate gene study of eight GABAA receptor subunits in panic disorder, *Am. J. Psychiatr.,* 154(8), 1096, 1997.

38. Rosenbaum, J.F., Biederman, J., Gersten, M. et al., Behavioral inhibition in children of parents with panic disorder and agoraphobia. A controlled study, *Arch. Gen. Psychiatr.,* 45(5), 463, 1988.

39. Deltito, J.A. and Hahn, R.A., Three-generational presentation of separation anxiety in childhood with agoraphobia in adulthood, *Psychopharmacol. Bull.,* 29(2), 189, 1993.

40. Weissman, M.M., Leckman, J.F., merikangas, K.R. et al., Depression and anxiety disorders in parents and children. Results from the Yale family study, *Arch. Gen. Psychiatr.,* 41(9), 845, 1984.

41. Silove, D., Manicavasagar, V., O'Connell, D. et al., Genetic factors in early separation anxiety: implications for the genesis of adult anxiety disorders, *Acta. Psychiatr. Scand.,* 92(1), 17, 1995.

42. Schuckit, M.A., Tipp, J.E., Bucholz, K.K. et al., The lifetime rates of three major mood disorders and four major anxiety disorders in alcoholics and controls, *Addiction*, 92(10), 1289, 1997.

43. Schuckit, M.A., Heeselbrock, V.M., Tipp, J. et al., The prevalence of major anxiety disorders in relatives of alcohol dependent men and women, *J. Stud. Alcohol.,* 56(3), 309, 1995.

44. Maier, W., Minges, J., and Lichtermann, D., Alcoholism and panic disorder: co-occurrence and co-transmission in families, *Eur. Arch. Psychiatr. Clin. Neurosci.,* 243(3–4), 205, 1993.

45. Cohen, L.S. and Biederman, J., Further evidence for an association between affective disorders and anxiety disorders: review and case reports, *J. Clin. Psychiatr.,* 49(8), 313, 1988.

46. Leckman, J.F., Weissman, M.M., Merikangas, K.R. et al., Panic disorder and major depression. Increased risk of depression, alcoholism, panic, and phobic disorders in families of depressed probands with panic disorder, *Arch. Gen. Psychiatr.,* 40(10), 1055, 1983.

47. Mannuzza, S., Champan, T.F., Kelin, D.F. et al., Familial transmission of panic disorder: effect of major depression comorbidity, *Anxiety,* 1(4), 180, 1994–5.

48. Goldstein, R.B., Weissman, M.M., Adams, P.B. et al., Psychiatric disorders in relatives of probands with panic disorder and/or major depression, *Arch. Gen. Psychiatr.,* 51(5), 383, 1994.

49. Biederman, J., Rosenbaum, J.F., Bolduc, E.A. et al., A high risk study of young children of parents with panic disorder and agoraphobia with and without comorbid major depression, P*sychiatr. Res.,* 37(2), 333, 1991.

50. Kendler, K.S., Neale, M.C., Kessler, R.C. et al., The clinical characteristics of major depression as indices of the familial risk to illness, *Br. J. Psychiatr.,* 165(2), 66, 1994.

51. Skre, I., Onstad, S., Edvardsen, J.A. et al., Family study of anxiety disorders: familial transmission and relationship to mood disorder and psychoactive substance use disorder, *Acta. Psychiatr. Scand.,* 90(5), 366, 1994.

52. Weissman, M.M., Wickramaratne, P., Adams, P.B. et al.,The relationship between panic disorder and major depression. A new family study, *Arch. Gen. Psychiatr.,* 50(10), 767, 1993.

53. Maier, W., Minges, J., and Lichtermann, D., The familial relationship between panic disorder and unipolar depression, *J. Psychiatr. Res.,* 29(5), 375, 1995.

54. MacKinnon, D.F., McMahon, F.J., Simpson, S.G. et al., Panic disorder with familial bipolar disorder, *Biol. Psychiatr.,* 42(2), 90, 1997.

55. Heun, R. and Maier, W., Relation of schizophrenia and panic disorder: evidence from a controlled family study, *Am. J. Med. Genet.,* 60(2), 127, 1995.

56. Battaglia, M., Bernardeschi, L., Politi, E. et al., Comorbidity of panic and somatization disorder: a genetic-epidemiological approach, *Compr. Psychiatr.,* 36(6), 411, 1995.

57. Pauls, D.L., Leckman, J.F., and Cohen, D.J., Evidence against a genetic relationship between Tourette's syndrome and anxiety, depression, panic and phobic disorders, *Br. J. Psychiatr.,* 164(2), 215, 1994.

58. Comings, D.E. and Comings, B.G., A controlled study of Tourette syndrome. III. Phobias and panic attacks, *Am. J. Hum. Genet.,* 41(5), 761, 1987.

59. Comings, D.E. and Comings, B.G., Hereditary agoraphobia and obsessive-compulsive behaviour in relatives of patients with Gilles de la Tourett'e syndrome, *Br. J. Psychiatr.,* 151, 195, 1987.

60. Klein, D.F., False suffocation alarms, spontaneous panics, and related conditions. An integrative hypothesis, *Arch. Gen. Psychiatr.,* 50, 306, 1993.

61. Coryell, W., Hypersensitivity to carbon dioxide as a disease-specific trait marker, *Biol. Psychiatr.,* 41(3), 259, 1997.

62. Perna, G., Bertani, A., Caldirola, D. et al., Family history of panic disorder and hypersensitivity to $CO_2$ in patients with panic disorder, *Am. J. Psychiatr.,* 153(8), 1060, 1996.

63. Perna, G., Bertani, A., Gabriele, A., Politi, E., and Bellodi, L., Modification of 35% carbon dioxide hypersensitivity across one week of treatment with clomipramine and fluvoxamine: a double-blind, randomized, placebo-controlled study, *J. Clin. Psychopharmacol.,* 17(3), 173, 1997.

64. Bellodi, L., Perna, G., Caldirola, D., Arancio, C., Bertani, A., and Di Bella, D., $CO_2$-induced panic attacks: a twin study, *Am. J. Psychiatr.,* 155(9), 1184, 1998.

65. Balon, R., Jordan, M., Pohl, R. et al., Family history of anxiety disorders in control subjects with lactate-induced panic attacks, *Am. J. Psychiatr.,* 146(10), 1304, 1989.

66. Crowe, R.R., Gaffney, G., and Kerber, R., Panic attacks in families of patients with mitral valve prolapse, *J. Affect. Disord.,* 4(2), 121, 1982.

67. Morris, A., Baker, B., Defins, G.M. et al., Prevalence of panic disorder in cardiac outpatients, *Can. J. Psychiatr.,* 42(2), 185, 1997.

68. Kushner, M.G., Thomas, A.M., Bartels, K.M. et al., Panic disorder history in the families of patients with angiographically normal coronary arteries, *Am. J. Psychiatr.,* 149(11), 1563, 1992.

69. Kendler, K.S., Walters, E.E., and Neale, M.C., The structure of the genetic and environmental risk factors for six major psychiatric disorders in women. Phobia, generalized anxiety disorder, panic disorder, bulimia, major depression, and alcoholism, *Arch. Gen. Psychiatr.,* 52(5), 374, 1995.

70. Vieland, V.J., Goodman, D.W., and Chapman, T., New segregation analysis of panic disorder, *Am. J. Med. Genet.,* 67(2), 147, 1996.

71. Pauls, D.L., Bucher, K.D., and Crowe, R.R., A gentic study of panic disorder pedigrees, *Am. J. Hum. Genet.,* 32(5), 639, 1980.

72. Pauls, D.L., Crowe, R.R., and Noyes, R., Jr., Distribution of ancestral secondary cases in anxiety neurosis (panic disorder), *J. Affect. Disord.,* 1(14), 387, 1979.

73. Van Winter, J.T. and Stickler, G.B., Panic attack syndrome, *J. Pediatr.,* 105(4), 661, 1984.

74. Price, R.A., Kidd, K.K., and Weissman, M.M., Early onset (under age 30 years) and panic disorder as markers for etiologic homogeneity in major depression, *Arch. Gen. Psychiatr.,* 44(5), 434, 1987.

75. Vieland, V.J., Hodge, S.E., Lish, J.D. et al., Segregation analysis of panic disorder, *Psychiatr. Genet.,* 3, 63, 1993.

76. Ayuso Gutierrez, J.L., Llorente-Perez, L.J., and Ponce de Leon, C., [Genetic studies of panic disorder] Estudios geneticos del trastorno de panico, *Arch. Neurobiol. Madr.,* 54(3), 104, 1991.

77. Crowe, R.R., Noyes, R., Jr., and Persico, A.M., Pro-opiomelanocortin (POMC) gene excluded as a cause of panic disorder in a large family, *J. Affect. Disord.,* 12(1), 23, 1987.

78. Wang, Z.W., Crowe, R.R., and Noyes, R., Jr., Adrenergic receptor genes as candidate genes for panic disorder: a linkage study, *Am. J. Psychiatr.,* 149(4), 470, 1992.

79. Crawford, F., Hoyne, J., Diaz, P. et al., Occurrence of the Cys311 DRD2 variant in a pedigree multiply affected with panic disorder, *Am. J. Med. Genet.,* 60(4), 332, 1995.

80. Ohara, K., Xie, D.W., Ishigaki, T. et al., The genes encoding the 5HT1D alpha and 5HT1D beta receptors are unchanged in patients with panic disorder, *Biol. Psychiatr.,* 39(1), 5, 1996.

81. Lesch, K.P., Bengel, D., Heils, A. et al., Association of anxiety-related traits with a polymorphism in the serotonin transporter gene regulatory region, *Science,* 274(5292), 1527, 1996.

82. Deckert, J., Catalano, M., Heils, A. et al., Functional promoter polymorphism of the human serotonin transporter: lack of association with panic disorder, *Psychiatr. Genet.,* 7(1), 45, 1997.

83. Ishiguro, H., Arinami, T., Yamada, K. et al., An association study between a transcriptional polymorphism in the serotonin transporter gene in panic disorder in a Japanese population, *Psychiatr. Clin. Neurosci.,* 51, 333, 1997.

84. Kato, T., Wang, Z.W., Zoega, T. et al., Missense mutation of cholecystokinin B receptor gene: lack of association with panic disorder, *Am. J. Med. Genet.,* 67(4), 401, 1996.

85. Wang, Z., Valdes, J., Noyes, R. et al., Possible association of a cholecystokinin promotor polymorphism (CCK-36CT) with panic disorder, *Am. J. Med. Genet.,* 81(3), 228, 1998.

86. Tanay, V.A., Glencorse, T.A., Greenshaw, A.J. et al., Chronic administration of antipanic drugs alters rat brainstem CABA A receptor subunit mRNA levels, *Neuropharmacology,* 35(9–10), 1475, 1996.

87. Crowe, R.R., Wang, Z., Noyes, R., Jr. et al., Candidate gene study of eight GABA A receptor subunits in panic disorder, *Am. J. Psychiatr.,* 154(8), 1096, 1997.

88. Schmidt, S.M., Zoega, T., and Crowe, R.R., Excluding linkage between panic disorder and the gamma-aminobutyric acid beta 1 receptor locus in five Icelandic pedigrees, *Acta. Psychiatr. Scand.,* 88(4), 225, 1993.

89. Enoch, M.A., Rohrbaugh, J.W., Davis, E.Z. et al., Relationship of genetically transmitted alpha EEG traits to anxiety disorders and alcoholism, *Am. J. Med. Genet.,* 60(5), 400, 1995.

90. Steinlein, O.K., Deckert, J., nothen, M.M. et al., Neuronal nicotine acetylcholine receptor alpha 4 subunit (CHRNA4) and panic disorder: an associative study, *Am. J. Med. Genet.,* 74(2), 199, 1997.

91. Crowe, R.R., Noyes, R., Jr., and Wilson, A.F., A linkage study of panic disorder, *Arch. Gen. Psychiatr.,* 44(11), 933, 1987.

92. Crowe, R.R., Noyes, R., Jr., Samuelson, S. et al., Close linkage between panic disorder and alpha-haptoglobin excluded in 10 families, *Arch. Gen. Psychiatr.,* 47(4), 377, 1990.

93. Knowles, J.A., Fyer, A.J., Vieland, V.J. et al., Results of a genome-wide genetic screen for panic disorder, *Am. J. Med. Genet.,* 81, 139, 1998.

94. Cloninger, C.R., Van Eerdewegh, P., Goate, A. et al., Anxiety proneness linked to epistatic loci in genome scan of human personality traits, *Am. J. Med. Genet.,* 81(4), 313, 1998.

# 7 The Genetics of Epilepsy: Mouse and Human Studies

*Thomas N. Ferraro and Russell J. Buono*

## CONTENTS

## INTRODUCTION

Epilepsy is characterized by uncontrolled, abnormal firing of neurons in the central nervous system (CNS). This phenomenon of neuronal hyperexcitability is recurrent and frequently associated with behavioral manifestations referred to as seizures. In specific instances it is known that seizures may be caused by developmental anomalies, enzyme deficiencies, or traumatic brain injury. But more often they are idiopathic. In general, seizure disorders are a heterogeneous group with diverse etiologies.

The molecular and cellular mechanisms underlying the initiation and propagation of seizure activity are not well understood. From a systems perspective, seizures result from a loss of function (disinhibition) or gain of function (excitation), and these alterations may occur in discrete brain regions, especially neocortical and limbic structures. Feedback control of neuronal circuitry and the ability of the seizure focus to spread by recruiting additional neurons may also play a role. Each of these factors is influenced by genetic variation, and a number of studies have now revealed genetic contributions to specific seizure phenotypes in humans.

The clinical heterogeneity associated with epilepsy is reflected by a broad range of phenotypes. This may be matched by a high degree of genetic variability, as several seizure syndromes have now been characterized as having more than one genetic susceptibility factor. The many different forms of epilepsy that exist and the multiple genetic influences that may cause or influence any particular form combine to make it difficult to collect the appropriate numbers of well-defined patients with similar phenotypes required to carry out rigorous genetic studies. Thus, whereas a large fraction of all idiopathic human epilepsies are believed to have genetic components, only a few rare forms (nearly all inherited in Mendelian fashion) have yielded positive linkage results and fewer have been defined at the gene level.

The significant clinical and genetic obstacles that hinder the search for human epilepsy genes provide impetus for studying simpler systems, and of those available, mouse models offer significant advantages and promise. A variety of genetic mouse models of human seizure syndromes have been discovered or developed. Both at the level of phenotype and genotype, these models have utility in circumventing some limitations of carrying out genetic studies in humans. The greatest advantages include the ability to reproducibly measure seizure sensitivity or threshold and the opportunity to carry out genetic engineering and directed breeding in genetically well-defined populations. These factors combine to facilitate the identification of genes related to neuronal excitability and seizures in mice. A serious limitation, however, is the uncertainty with which mouse seizure-related genes can be associated with human epilepsy. Even so, such studies can provide starting points for genetic studies of epilepsy in humans by suggesting candidate genes or genomic regions or even by revealing new mechanisms by which epilepsy may be triggered. Thus, at the present time, seizure models and genetic resources that involve mice have great potential for providing insight into human epilepsy.

The purpose of this chapter is to review the evidence for genetic involvement in human epilepsies and in mouse models of such disorders. The chapter is divided into two parts. The first part describes relevant studies in mice, including naturally occurring mutants, transgenic and knockout mice, and seizure-related quantitative trait loci. The second part of the chapter reviews studies in humans with a focus on reports of linkage or association. Ultimately, as genetic maps and regions of synteny between man and mouse become more refined, an integrated map of seizure-related loci will provide a foundation for determining future research directions by way of a translational approach that moves compelling animal data to human studies.

## GENETIC MOUSE MODELS OF EPILEPSY

One strategy for studying genetic influences on human predisposition to epilepsy is based upon prior identification of susceptibility alleles in animal models of seizure syndromes. By identifying alterations in genes that result in experimental seizure phenotypes, testable hypotheses are readily generated for subsequent human studies. Thus, in this way, not only does the identification of seizure susceptibility genes in animals provide a direct entry point for studying the genetics of human epilepsy through gene cloning and sequencing efforts but it also provides an indirect entry point by suggesting alternative genes that are involved in homologous seizure mechanisms. For example, the recent characterization of a seizure phenotype in a serotonin receptor (5HT2c) knockout mouse[1] provides impetus not only for studying the potential role of this specific receptor in forms of human epilepsy, but it also provides a rationale for studying genes involved in other aspects of serotonin neurotransmission, including genes for other serotonin receptor subtypes, metabolic enzymes and uptake proteins, as well as genes involved in the function of other neurons in circuits with those bearing 5HT2c receptors. Overall, identification of seizure-related genes in experimental models can provide information relevant to unraveling complex genetic mechanisms of human epilepsy.

Genetic animal models of epilepsy have been described in various species and among these, studies involving mice are prominent. In fact, it could be argued that as a result of the genetic tools and resources that are currently available, mice represent the most useful species for developing genetic models of human disease.[2] Strategies employed previously involve naturally occurring mutants, inbred and recombinant inbred strains, and, more recently, transgenic and knockout mice. These resources all have strengths and weaknesses that influence their utility. Some of these issues are subsequently discussed.

**TABLE 7.1**
**Spontaneous Mouse Mutants with a Phenotype That Includes Behavioral Seizure Activity**

| Mutant | Locus | Position (cM) | Seizure Phenotype | Ref. |
|--------|-------|---------------|-------------------|------|
| Dilute (d) | Myo5a | chr. 9 (42) | generalized seizures; opisthotonus | 22, 23 |
| Ducky (du) | ? | chr. 9 (60.0) | seizures (not well described) | 24 |
| Frings (mass1) | ? | chr. 13 (40.0) | generalized seizures (audiogenic) | 30, 31 |
| Jimpy (jp) | Plp | chr. X (56) | generalized seizures | 59, 90 |
| Jittery (ji) | ? | chr. 10 (43.0) | tonic/clonic seizures | 26, 91 |
| Lethargic (lh) | Cacnb4 | chr. 2 (33.9) | absence & partial motor seizures | 6, 92 |
| Quaking (qk) | Star? | chr. 17 (5.9) | absence(?) & generalized seizures | 90, 93 |
| Shiverer (shi) | Mbp | chr. 18 (55.0) | generalized seizures | 94, 95 |
| Slow Wave (swe) | Slc9a1 | chr. 4 (64.6) | absence & generalized seizures | 13 |
| Stargazer (stg) | Cacng2 | chr. 15 (45.2) | absence seizures | 14, 16 |
| Teetering (tn) | ? | chr. 11 (79.0) | "running fits" | 28 |
| Tippy (tip) | ? | chr. 9 (63.0) | sudden bursts of activity | 25 |
| Tottering (tg) | Cacna1a | chr. 8 (37.5) | absence & partial motor seizures | 8, 10 |
| Trembler (tr) | Pmp22 | chr. 11 (34.4) | transient juvenile seizures | 96–98 |
| Trembly-like (tyl) | ? | chr. X | generalized seizures | 27 |
| Weaver (wv) | Kcnj6 | chr. 16 (67.6) | tonic/clonic seizures | 17, 99 |

## SINGLE GENE MUTANTS

Seizure-prone mice that carry mutations of single genes have been the focus of considerable study.[3,4] Arising spontaneously in large breeding colonies, these unique animals are identified by behavioral or other phenotypic traits and subsequently perpetuated for study under controlled breeding conditions. Table 7.1 lists a number of such mutant mouse strains. In general, these mice exhibit seizures of various kinds as part of a broader, complex neurological syndrome. Such pleiotropy is a factor that hinders interpretation of these models, especially in their relation to human epilepsy, since many of the most common seizure disorders in humans do not present with other neurological symptoms.

Neurological mutant mouse strains that do exhibit seizures as a major feature of the abnormal phenotype include lethargic, stargazer, slow wave epilepsy and tottering. Each of these strains is characterized by a defect in a gene that codes for a protein involved in ion flux (see Table 7.1). Indeed, a significant increase in understanding the basic mechanisms of neuronal hyperexcitability may come about with the recent identification of causative genes in these mutants. For example, the lethargic mouse, exhibiting a phenotype that begins with hypoactivity and ataxia, develops both spontaneous focal motor seizures as well as absence-like seizures with generalized cortical spike-wave discharges.[5] This mutant has been found to result from a defect in the calcium channel β-subunit gene, Cacnb4, in which a 4-nucleotide insertion into a splice donor site causes exon skipping, translational frameshift, and a truncated protein devoid of the α1-binding site.[6] Neuron densities appear normal in the brains of lethargic mice although functional changes, including decreased conduction velocity and prolonged distal latency have been documented in the peripheral nervous system.[7] The difficulty now comes in determining how a defective calcium channel results in a paroxysmal syndrome such as epilepsy.

Another mutant, tottering, exhibits a severe seizure disorder which develops at about three weeks of age.[8] The first signs of abnormal phenotype in these mice are often related to an ataxic or wobbly gait. Early seizure symptoms include absence-like attacks characterized by spike-wave discharges and behavioral arrest. Then, at about four weeks of age, these mice develop partial motor seizures. Interestingly, roller and leaner mutants, which are allelic with tottering,[9] do not exhibit motor seizures but exhibit more severe forms of ataxia compared to tottering.[8] The tottering gene, *Cacn1a4*, has recently been cloned and found to encode a voltage dependent P/Q-type calcium channel in which a single base pair mutation results in an amino acid substitution in the second "subunit motif" of the channel protein.[10] Reported histopathology in tottering mice includes increased density of noradrenergic fibers in terminal fields of the locus ceruleus, including the hippocampus.[11] This finding, plus the observation that ataxic and seizure symptoms of tottering are abolished following chemical lesion of the locus ceruleus in neonatal mice,[12] suggests that the influence of *Cacn1a4* may be most pronounced on noradrenergic neurons and that this pathway is important in the control of seizure activity.

The slow wave epilepsy mouse is another interesting neurological mutant. This strain exhibits both absence-like as well as generalized tonic-clonic seizures, and recent work on the gene underlying this mutant has revealed it to be *Nhe1*, a gene for a $Na^+/H^+$ exchange protein that contains a single A to T transversion in a putative stop codon.[13] This defect likely accounts for the very low levels (<10% of wild type) of full-length *Nhe1* mRNA in the brain of slow wave epilepsy mutants.[13] Selective neuronal degeneration is observed in deep cerebellar, vestibular, and cochlear nuclei, further reinforcing a relationship between cerebellum, ataxia, and seizure activity. The discovery that a molecule involved in the maintenance of cellular $p$H and osmotic pressure can also be involved in the generation of seizure activity provides a new perspective on understanding epilepsy and developing new treatments. This is consistent with the idea that proteins involved with ion flux contribute in a major way to the regulation of neuronal excitability. Nonetheless, it is still a major challenge to integrate this genetic information with mechanisms of epileptogenesis.

Stargazer is described primarily as a seizure mutant, but again, the phenotype is a broader one that begins with ataxia and progresses to include an unusual head-tossing behavior.[14] The abnormal movements become more severe with age, yet there is no documented cell loss in the central nervous system. Seizures involve mainly a behavioral arrest and neurophysiological correlates reveal generalized spike-wave discharges in the cerebral cortex. The stargazer gene is mapped to chromosome 15[14,15] and has recently been shown to encode a γ-subunit of a voltage-sensitive calcium channel.[16] The gene, *Cacng2*, is disrupted by an early transposon insertion in stargazer mice and this mutation results in appreciably lower levels of full-length *Cacng2* transcripts.[16] γ-subunits had not previously been found in the brain; however, the demonstration of functional electrophysiologic effects of *Cacng2* in artificially expressed calcium channels[16] provides support for the involvement of this gene in neuronal hyperexcitability disorders.

Weaver is another example of a seizure-prone mutant mouse with a defect in ion flux properties, since the defective gene in this mutant is a voltage-gated potassium channel, *Kcnj6*.[17] Like myelin-deficient mutants (discussed subsequently), weaver mice are best known for their phenotype of ataxia and hypotonia. Unlike the previous examples of ion-flux gene mutants, seizures are a rare feature of the weaver phenotype, but nonetheless mice carrying the weaver mutation are known to be seizure-sensitive.[18,19] Degeneration of cerebellar granule cells is a well-documented weaver trait, again emphasizing a possible link between cerebellar defects and seizures. However, loss of pigmented nuclei in the brainstem of weaver mice, particularly in the substantia nigra, occurs later and may have a closer temporal association with the seizures than the cerebellar pathology. The involvement of basal ganglia in seizure disorders has been documented previously with studies in mice showing that lesion of substantia nigra neurons has significant influences upon seizure sensitivity.[20]

In Table 7.1, four different mutants, jimpy, quaking, shiverer, and trembler, are all characterized by defects in genes that are involved in the process of myelination. The primary phenotypic trait of dysmyelinating mutants is a body tremor that precedes and/or accompanies locomotion. In general, seizures are not a major feature of these mutant phenotypes but rather occur at various developmental timepoints, often being reported as a cause of early death. Seizures in myelin-deficient mutant mice are usually described as being absence-like, with mice remaining immobile throughout the ictal interval. Actual "convulsions" also have been reported, but these are much less common. The relationship between the global loss of axonal myelination and the development of seizures in these mice is not clear. For example, peripheral myelin protein-22, the product of the *Pmp22* gene that is defective in trembler mice, may not even be produced in the central nervous system.[21] So, while such models can provide insight regarding basic mechanisms of epilepsy by identifying individual genetic factors that are sufficient in and of themselves to cause seizures, they are fraught with interpretational difficulties.

Several other spontaneous mutant mouse strains with a phenotype that includes seizures have been discovered (see Table 7.1); again, the seizure syndrome is not the main feature of the phenotype and has not been well characterized. For example, the mouse dilute mutation, best known for its effect upon melanocyte morphology and coat color, is also associated with seizures.[22] The dilute gene, *Myo5a*, is located on chromosome 9 and codes for an unconventional myosin-like protein.[23] The neurological phenotype in dilute mice involves generalized convulsive seizures or a convulsive arching of the back called opisthotonus; however, not all allelic variants of the dilute gene are associated with seizures.[22] The mouse mutant ducky is another model that presents primarily with ataxia. The central nervous system of this mouse shows severe developmental abnormalities, especially in the hindbrain and spinal cord, and there is selective demyelination.[24] In addition to deficits in voluntary motor activity, these mice are subject to occasional seizures,[24] although again these are not well described. The ducky gene has not been identified but has been mapped to chromosome 9 and is linked to *Myo5a*, the dilute gene.[25] Careful review of the literature reveals that seizures are also occasionally seen in the jittery mutant[26] and the trembly-like mutant,[27] two mouse strains that are defined primarily by ataxia and tremors. Still other mutants are said to exhibit behaviors that are merely suggestive of seizure activity such as "running fits" in the teetering mutant[28,29] and sudden bursts of activity in the tippy mutant.[25] These models require further study and verification of electroencephalographic seizure activity before genetic correlates of their behavior will be relevant to human epilepsy. An interesting mutant with a predominantly seizure-related phenotype is the frings mouse, which exhibits high susceptibility to generalized audiogenic seizures throughout its lifespan.[30] The gene responsible for the frings mutation, *mass1*, has recently been mapped to chromosome 13[31] but is as yet unidentified.

Overall, several observations are worth emphasizing regarding seizure-prone mouse mutants. In general, these mice have abnormal neurological phenotypes that go beyond seizure susceptibility. In fact, each known mutant also presents with ataxia or other involuntary movements. These signs are frequently the first indication of nervous system dysfunction and often remain the major abnormal feature of the phenotype. Although seizures occur in humans as secondary features of larger syndromes, the majority and most common human epilepsies do not present with neurological symptoms other than seizures. Even though the pleiotropic effects of genes identified in mouse seizure mutants may hinder interpretation of the phenotype, it does not obviate the possibility of directly examining the human orthologs of these genes in patients with epilepsy.

## KNOCKOUT AND TRANSGENIC MICE

A number of genetically modified mice have been shown to exhibit seizures, either spontaneous or experimentally-induced; these are listed in Table 7.2. In many cases, targeted genes have been candidates for regulating neuronal excitability such as genes involved in the function of GABA and glutamate, the two main fast transmitters in the central nervous system. GABAergic mechanisms have long been associated with seizure activity, and a number of clinically useful anticonvulsant drugs work by enhancing the function of GABA synapses.[32] GABAergic deficits have been induced in mice by knocking out *Gad2*, a gene for an isoform of glutamic acid decarboxylase (GAD), the synthetic enzyme for GABA, and this has resulted in a seizure-susceptible phenotype.[33,34] Another gene, *Gad1*, codes for a different GAD isoform, and deletion of this gene leads to a severely abnormal developmental phenotype that includes cleft palate and death in the early post-natal

**TABLE 7.2**
**Knockout and Transgenic Mice with a Phenotype That Includes Spontaneous Seizures or Increased Seizure Susceptibility**

| Gene | Gene Product | Type | Location (cM) | Ref. |
|------|--------------|------|---------------|------|
| *Gad1* | glutamic acid decarboxylase | ko | chr. 2 (43.0) | 35 |
| *Gad2* | glutamic acid decarboxylase | ko | chr.2 (9.0) | 33 |
| *Gad2* | glutamic acid decarboxylase | ko | chr.2 (9.0) | 34 |
| *Gabrb3* | $GABA_A$ receptor β-subunit | ko | chr. 7 (28.6) | 38 |
| *Grik2* | glutamate receptor β2 subunit | ko | chr. 10 (27.0) | 39 |
| *Gria2* | glutamate receptor α2 subunit | ko | chr. 3 (33.6) | 40 |
| *Slc1a2* | glutamate transporter | ko | chr. 2 (54.0) | 41 |
| *Npy* | neuropeptide Y | ko | chr. 6 (22.0) | 44 |
| *Htr2c* | serotonin 2C receptor | ko | chr. X (66.1) | 1 |
| *Akp2* | alkaline phosphatase 2 | ko | chr. 4 (70.2) | 36 |
| *Syn1, Syn2* | synapsin I, II | ko | chr. X, 6 (6.2, 49.0) | 45 |
| *Kcnj6* | K-channel (inwardly rectifying) | ko | chr.16 (67.6) | 42 |
| *Kcna1* | K-channel (voltage-gated) | ko | chr.6 (60.0) | 43 |
| *Camk2a* | calmodulin kinase II α subunit | ko | chr. 18 (33.0) | 46 |
| *Itpr1* | inositol triphosphate receptor | ko | chr. 6 (48.2) | 47 |
| *Fyn* | tyrosine kinase receptor | ko | chr. 10 (25.0) | 100 |
| *Plat* | tissue plasminogen activator | ko | chr. 8 (9.0) | 48 |
| *Tnfr1, Tnfr2* | tumor necrosis factor receptor 1,2 | ko | chr. 6, 4 (57.1,75.5) | 49 |
| *Jrk* | centromere binding protein | ko/transgene | chr. 15 (42.8) | 50 |
| *Otx1* | orthodenticle (homeobox) | ko | chr. 11 (13.0) | 51 |
| *Pcmt1* | protein iso-Asp O-methyltrferase | ko | chr. 10 (7.0) | 52 |
| *Pcmt1* | protein iso-Asp O-methyltrferase | ko | chr. 10 (7.0) | 53 |
| *Cdk5r* | cyclin dependent kinase 5, regulatory subunit (p35) | ko | chr. 11 (46.5) | 54 |
| *Bdnf* | brain-derived ntropic factor | ko | chr. 2 (62.0) | 55 |
| *Gap43* | growth accentuating protein | overexpress | chr. 16 (29.5) | 56 |
| *Sap* | sphingolipid activating protein | ko | ? | 57 |
| *Hexa, Hexb* | β-hexosaminidase A | double ko | chr. 9, 13 (29.0, 46.0) | 58 |
| *Plp* | myelin proteolipoprotein | overexpress | chr. X (56.0) | 60 |
| *Hdh* | Huntingtin disease gene homolog | CAG repeat | chr. 5 (20.0) | 61 |
| *Hprt* | hypoxanthine guanine phosphoribosyltransferase | CAG repeat | chr. X (17.0) | 62 |

period.[35] The existence of several distinct GAD genes clouds the relationship between GABA synthesis and seizures in GAD knockout mice, but this relationship is better elucidated in *Akp2* knockout mice. *Akp2* is the gene that codes for tissue nonspecific alkaline phosphatase, an enzyme involved in the production of vitamin B6, a required co-factor for all GAD activity. Mice with a targeted deletion of the *Akp2* gene were shown to have significantly reduced brain GABA levels and a severe seizure phenotype that could be "rescued" by treatment with pyridoxal.[36] Another way that GABA transmission has been genetically altered in mice has been through deletion of genes coding for various subunits of the $GABA_A$ receptor. So far, studies have been reported on the knockout of the α6[37] and β3[38] subunit genes. Like *Gad2* knockouts, these mice also have been shown to exhibit seizure disorders. The β3 knockout mouse has been suggested to be a model of Angelman's syndrome (epilepsy and mental retardation) due to synteny between mouse chromosome 7, the location of several GABA subunit genes such as β3, and human 15, the location of the Angelman defect.[37]

As noted, alteration of glutamate transmission also can be manipulated to influence seizure activity, and several knockout mice have been generated that directly affect function of glutamatergic synapses (see Table 7.2). Each of these mice exhibits a phenotype that involves seizure activity without other gross phenotypic or behavioral abnormalities. Knocking out *Grik2*, the gene for a subunit of the kainate subtype of glutamate receptor, yields a mouse that is resistant to kainate-induced seizures,[39] whereas mice engineered to harbor an editing deficient *Gria2* allele produce AMPA-type glutamate receptors that are abnormally permeable to calcium and exhibit a severe seizure disorder and early death.[40] Deletion of *Slc1a2*, the gene encoding an astroglial glutamate transporter responsible for clearing extracellular glutamate following synaptic release, also results in a seizure phenotype.[41] Both glutamate and GABA ultimately affect the movement of ions through activated ligand-gated channels, and in that sense, the knockout mice described above all have disturbances of ion conductance, either anionic in the case of GABA, or cationic in the case of glutamate. This is consistent with the concept that altered neuronal excitability characteristic of epilepsy is due to alteration of normal synaptic ion regulation. Further support for this notion is provided by studies of mice in which potassium channels have been knocked out. In one case, knockout of the weaver mutant potassium channel gene GIRK2 (*Kcnj6*) yields mice that have spontaneous seizures as well as increased sensitivity to pentylenetetrazol-induced seizures.[17,42] In another case, knockout of the voltage-gated potassium ion channel $K_v1.1$ (*Kcna1*) yields a mouse with subtle defects in CA3 pyramidal cell physiology and a seizure syndrome with both limbic and generalized features.[43] In all, there is compelling evidence from knockout strategies that genes that regulate transmembrane ion flux can, when altered, lead to neuronal hyperexcitability and behavioral seizure activity.

A number of other seizure-prone knockout mice have been generated by targeting genes for neurotransmitters that are less obvious candidates for participating in the genesis of seizure activity. These include knockouts of *Htr2c*, a serotonin receptor gene[1] and the gene for neuropeptide Y.[44] Also, knockouts of genes involved in more generic synaptic mechanisms such as the neurotransmitter release-associated synapsins[45] and signal transduction-related molecules, including calmodulin kinase II α subunit[46] and inositol triphosphate receptor I,[47] provide evidence that genetic influences on chemical synaptic transmission and signal transduction are profound and occur at many different levels of nervous system function.

Seizure-prone knockout mice have also been created by targeting genes with effects related more to synaptic development and plasticity rather than to effects on short-term signal transduction mechanisms. Thus, deletion of genes such as *Plat*, the tissue plasminogen activator gene;[48] *Tnfr1* and *Tnfr2*, genes for tumor necrosis factor receptors;[49] *jerky*, the gene for a centromere-binding-like protein;[50] *Otx1*, a gene involved in brain development;[51] *Pcmt1*, the protein isoaspartate-O-methyltransferase gene;[52,53] *Cdk5r*, a gene encoding a neuronal specific activator of cyclin-dependent kinase 5;[54] and *Bdnf*, the gene encoding brain derived neurotrophic factor[55] all result in mice with spontaneous seizures or lowered seizure threshold. Compared to genes for neurotransmitter or second messenger system components, these genes code for protein products that provide longer-term regulation of CNS structure and function. Another gene that likely has longer-term regulatory effects in the CNS is *Gap43*, a gene coding for the growth accentuating protein, which, when overexpressed, also results in the development of seizure-prone mice.[56] The roles that these genes fulfill in nervous system development and plasticity provide a link between mouse seizure models and human epilepsy syndromes associated with developmental defects. Characterization of altered neurophysiological effects induced by manipulation of these genes will facilitate delineation of the molecular mechanisms of neuronal hyperexcitability related to seizure activity.

Targeted deletion of genes involved in lipid metabolism have also yielded seizure-sensitive mice. This is consistent with reports of seizures in humans with lipid storage diseases. Thus, deletion of either the sphingolipid activator protein (*sap*) gene[57] or the genes for hexosaminidase A and B[58] cause ataxia and seizures in mice. These results suggest alternative mechanisms by which biochemical defects in the nervous system initiate seizure activity and emphasize a need for further investigation into the relationship between lipid metabolism and neuronal excitability. This concept is simultaneously reinforced and complicated by the observation that overexpression of *Plp*, the same gene that is defective in jimpy mice,[59] leads to an abnormal phenotype marked by ataxia, tremors, and seizures.[60]

In summary, the occurrence of seizure- and non-seizure-related phenotypic abnormalities in many knockout mice raises questions regarding pleiotropism and model interpretation, making it a challenge to fully understand the relationship between specific gene mutations and the ultimate expression of seizure activity. Indeed, transgenic mice carrying a CAG/polyglutamine expansion in the huntingtin gene present with a complex phenotype involving a broad range of motor abnormalities including seizures.[61] Also, the severe progressive neurological phenotype that includes seizures in transgenic mice bearing CAG repeats within the gene for hypoxanthine phosphoribosyltransferase (*Hprt*) indicates that neurotoxic effects can be induced by CAG expansion within genes that regulate classical metabolic pathways.[62] Although complex, knockout models provide further evidence that genetic contributions to seizure activity are significant and can facilitate studies of human genetic variation that predispose to epilepsy.

## MULTIGENIC MOUSE SEIZURE MODELS

In contrast to single-gene models of epilepsy, multifactorial models result from the combined effects of a variety of natural factors and are by definition more difficult to dissect relative to single-gene models. Seizure susceptibility in a multigenic model is more likely to exhibit homology with common heritable forms of human epilepsy that almost always show complicated modes of inheritance and variable phenotypes within families.[63] The wide range of clinical characteristics exhibited within epilepsy families, together with the complex patterns of heritability, suggest that seizure disorders result from the interaction of multiple genes and the environment. Thus, it is possible that natural variation in specific gene sequences may mediate seizure susceptible phenotypes either directly by contributing to lower seizure threshold or indirectly by interacting with other genes of major effect and with environmental factors such as toxins, trauma, or stress.

In spite of their complex nature, multigenic or multifactorial traits are now beginning to yield to newer more sophisticated means of genetic investigation. Unlike single gene models, multifactorial models of epilepsy in the mouse have not yet resulted in the identification of specific genetic influences on seizure susceptibility; however, a number of promising leads are being uncovered for future investigation using an approach known as quantitative trait loci (QTL) mapping.[64] Quantitative traits exhibit a continuous range of values across a given population reflecting the influences of multiple genes interacting with each other and with the environment. QTL analysis provides an opportunity to detect individual components of quantitative or complex traits, facilitating their approximate genomic localization and also allowing an estimation of the relative effect size for individual factors. However, together with the promise of QTL studies come pitfalls. Since complex traits are polygenic by definition, a likely possibility is that there will be many genetic factors operative in the phenotype, with most having such a small overall effect that detecting them with certainty above chance findings is beyond the practical limit of sensitivity of current methods of analysis. It is now recognized that linkages may often occur by chance, and guidelines have been proposed to indicate the relative level of confidence for a particular finding.[65] This will help to standardize the reporting of results from different laboratories and guide efforts in pursuing important findings. Additionally, the use of complementary statistical mapping procedures for the analysis of the same phenotype can provide an added degree of confidence if the results from the different tests support one another.[66] Nonetheless, successful QTL mapping relies on the utilization of appropriate statistical tests and models, and these are still the subject of significant discussion. In Table 7.3, QTLs are categorized according to the statistical guidelines of Lander and Kruglyak.[65]

The first multigenic mouse model of epilepsy studied successfully with a QTL approach involved the audiogenic seizure susceptibility of the immature DBA mouse.[67] In a series of exper-

**TABLE 7.3**
**Seizure-Related Quantitative Trait Loci (QTLs) in Mice**

| QTL | Status[1] | Location (cM) | Reference |
|------|-----------|---------------|-----------|
| *Asp1* | suggestive | chr. 12 (20.5) | 68-70 |
| *Asp2* | suggestive | chr. 4 (45.1) | 68-70 |
| *Asp3* | suggestive | chr. 7 (35.0) | 68,69 |
| *Bis1* | suggestive | chr. 4 (80.1) | 77 |
| *Bis2* | significant | chr. 13 (8.0) | 77 |
| *El1* | significant | chr. 9 (55.0) | 74,101 |
| *El2* | significant | chr. 2 (57.0) | 74,101 |
| *El3* | suggestive | chr. 10 (46.0) | 74 |
| *El4* | suggestive | chr. 9 (23.0) | 74 |
| *El5* | significant | chr. 14 (28.5) | 74 |
| *El6* | suggestive | chr. 11 (68.0) | 74 |
| EtOH-sz(1) | significant | chr. 4 (41.3) | 87 |
| EtOH-sz(2) | significant | chr. 1 (95.8) | 87 |
| EtOh-sz(3) | suggestive | chr. 11 (17.0) | 87 |
| EtOH-sz(4) | suggestive | chr. 2 (37.0) | 87 |
| EtOH-sz(5) | suggestive | chr. 2 (69.0) | 87 |
| *Szf1* | significant | chr. 7 (28.0) | 102 |
| *Szs1* | significant | chr. 1 (90.0) | 66 |
| *Szs2* | suggestive | chr. 4 (44.5) | 66 |
| *Szs3* | suggestive | chr. 11 (61.5) | 66 |
| *Szs4* | suggestive | chr. 18 (50.0) | 66 |
| *Szv1* | suggestive | chr. 5 (26.0) | 66 |
| *Szv2* | suggestive | chr. 7 (57.5) | 66 |
| *Szv3* | suggestive | chr. 15 (11.4) | 66 |

[1] Based on statistical criteria set forth in Lander and Kruglyak.[65]

iments, the exquisite sensitivity of young DBA mice to sound-induced tonic/clonic seizures was shown to be a result of several different genetic influences, including as yet unidentified genes on chromosomes 12 (*Asp1*), 4 (*Asp2*), and 7 (*Asp3*).[68-70] So far, none of the genes underlying *Asp* QTLs has been identified. Another multigenic model of epilepsy that is in the process of being analyzed using QTL strategies is represented by the EL mouse strain, which exhibits a seizure syndrome with parallels to human temporal lobe epilepsy.[71-73] Recent work on this model has documented the effects of multiple genetic factors localizing to chromosomes 9 (*El1*, *El4*), 2 (*El2*), 10 (*El3*), 14 (*El5*), and 11 (*El6*).[74] Although several candidate genes have been proposed for various *El* QTLs, none has yet been confirmed, and congenic strains are being developed to characterize these influences further.[75]

Other complex genetic influences upon seizure susceptibility are being dissected with QTL strategies by studying seizure models based upon strain differences in seizure response to various chemoconvulsants. Preliminary identification of genetic influences upon cocaine-induced seizures has been carried out using BXD RI strains,[76] although these results are not compelling in terms of statistical significance and do not reach recently proposed standards for declaring linkage.[65] An independent confirmation of these loci with an F2 QTL strategy seems warranted. Another recent chemoconvulsant model involving differential strain sensitivity to β-carboline induced seizures identified two significant genetic factors, *Bis1* and *Bis2* (chromosomes 4 and 13, respectively). These QTLs are specific for β-carboline-induced seizures and are not involved in general seizure susceptibility of the strains studied.[77] Other more recent work is based upon the well-documented significant difference in seizure sensitivity between DBA/2J (D2) mice, which are relatively seizure sensitive, and C57BL/6J (B6) mice, which are relatively seizure resistant.[78-84] The strategy involves determining genetic factors that are involved in the general seizure sensitivity of these strains by localizing QTLs that are in common between different convulsant paradigms. So far, a set of QTLs has been characterized that is involved in regulation of kainic acid seizure sensitivity differences between these strains with a locus of moderate to large effect identified on chromosome 1 (*Szs1*) and a series of additional loci of minor to moderate effect identified on six other chromosomes.[66] Additionally, a different chemoconvulsant, pentylenetetrazol, has been used to confirm a seizure QTL of large effect on chromosome 1 near *Szs1*.[85] Taken together, these results suggest that the area on chromosome 1 near markers D1Mit30 and D1Mit16 contains a gene(s) important for controlling the difference between seizure expression in these two common inbred mouse strains. Further support for this conclusion derives from a recent report of QTL mapping in F2 progeny from B6 and D2 mice related to severity of withdrawal from high dose ethanol which also detected a significant QTL on distal chromosome 1.[86] Although several other putative ethanol-withdrawal loci were detected in that study (see Table 7.3), it is possible that the chromosome 1 locus may not be related specifically to ethanol intoxication but rather may reflect a generally lower seizure threshold in D2 compared to B6 mice.

Although QTL mapping has made great strides in the past few years, moving from QTL region to the underlying gene(s) remains a major obstacle. Even if a QTL of large effect can be determined unambiguously, the resolution of current mapping techniques is not sufficient to undertake positional cloning experiments. As it stands, the multifactorial nature of complex traits results in a relatively poor relationship between genotype and phenotype for any one individual in a first generation backcross or intercross.[2] At best, QTL mapping can localize genetic influences to chromosomal regions of about 20 cM, although newer computational approaches to mapping can decrease the size of these intervals somewhat further.[87,88] Nonetheless, these regions are still too large to consider positional cloning strategies for gene identification, since this latter approach requires that the location of the gene (QTL) be narrowed to intervals of 1 cM or less. Cloning efforts, therefore, require additional fine mapping of QTLs. One favored approach to such fine mapping is the generation of congenic strains,[89] which are also useful for determining the effects of genetic background and epistatic interactions.[75] Alternatively, it is possible to avoid positional cloning if a candidate gene approach is able to identify the structural basis of the strain difference. However, even if positional cloning experiments are successful or candidate gene analysis indicates structural variants which translate into functional differences, this can only prove that the gene explains phenotypic differences between the specific strains of mice being examined. Thus, while generalization of results to other strains of mice, let alone humans, is not possible *a priori*, the value of QTL studies lies in their ability to generate candidate genes or genomic intervals for evaluation in alternate seizure models and also directly in humans with seizure disorders.

## GENETICS OF HUMAN EPILEPSY

The clinical presentation of epilepsy in humans is extremely heterogeneous. The International League Against Epilepsy has categorized over 40 different human disorders in which seizures are the primary symptom.[103] These disorders are classified as *bona fide* epilepsy based on criteria that include characteristic electroencephalograph (EEG) recordings and stereotypical behavioral and motor signs. There are over 40 additional human syndromes or diseases in which seizures comprise a major component of a larger constellation of neurological and other defects. Additionally, many case reports have documented individuals suffering from seizures coincident with congenital enzyme deficiencies, developmental anomalies, chromosomal aberrations and traumatic brain injuries. This section reviews the genetic linkage and association studies performed on those disorders which are classified as epilepsy. Human syndromes which are heritable and include seizures as a major component of a more complex phenotype are then discussed. Finally, congenital diseases that result from enzyme defects leading to neural degeneration and/or developmental anomalies that are associated with seizures are briefly discussed. The search strategy employed to identify these heritable conditions utilized the Online Mendelian Inheritance in Man (OMIM) database (http//www.ncbi.nlm.nih.gov/omim/) in conjunction with the Medline database. Searching the OMIM database using the terms "epilepsy," "seizure," and "convulsion" retrieved over 150 entities. Careful examination of these citations revealed that congenital diseases such as Gauchers[104] and Tay Sachs,[105] in which seizures do occur in some percentage of patients, were not retrieved in the search. Thus, the review of genetic data regarding those conditions classified as *bona fide* epilepsy may be considered comprehensive, whereas the review of additional entities in which seizures occur is less so.

## GENETIC LINKAGE AND HUMAN EPILEPSY

Currently there are over 15 disorders classified as epilepsy in which a genomic region or gene has been implicated as a seizure susceptibility factor. Table 7.4 summarizes these conditions and the genomic regions or genes that demonstrate linkage. One of the most common forms of primary generalized epilepsy, juvenile myoclonic epilepsy (JME), has been intensively studied with respect to genetic linkage. Initial reports demonstrated linkage on chromosome 6p near the location of the human leukocyte antigen (HLA).[106] Subsequent reports reproduced this finding,[107-108] while others were unable to confirm 6p linkage in separate families.[109-110] Elmslie et al. have recently reported positive linkage to markers on chromosome 15q for a JME susceptibility factor.[111] These data provide evidence that multiple loci are acting to increase risk for JME and demonstrate the multigenic nature of this complex trait. Adding support to this hypothesis are recent reports of positive association between polymorphisms in genes for acetylcholine[112] and glutamate receptor subunits[113] and subtypes of primary generalized epilepsy, including JME, juvenile absence epilepsy (JAE), and childhood absence epilepsy (CAE).

Linkage analysis and positional cloning have successfully identified causal genes in several rare types of epilepsy. In cases of benign familial neonatal convulsions (BFNC), a rare epilepsy inherited in a Mendelian fashion, linkage was demonstrated on two separate chromosomes in two different families. Linkage on chromosomes 20q[114] and 8q[115] were subsequently confirmed, and the potassium ion channel genes KCNQ2[116] and KCNQ3,[117] respectively, were identified as harboring mutations causal to BFNC. These data demonstrate the genetic heterogeneity that exists such that even in cases of monogenic Mendelian inheritance with high penetrance, different loci when mutated, can resolve into the same seizure phenotype. In other rare epilepsies with monogenic inheritance patterns, polymorphisms in specific genes have been discovered that are causal to the disorder. This is the case for Progressive Myoclonic Epilepsy of the Unverricht-Lundborg type (EPM1).[118] Mutations in the gene encoding the cystatin b protease inhibitor have been discovered, and these are linked to EPM1 disease.[119-120] Mutations in the gene that codes for an acetylcholine receptor subunit (ChRNA4) have been linked to a rare generalized epilepsy subtype, autosomal dominant nocturnal frontal lobe epilepsy (ADNFLE).[121-123] Finally, recent reports document a mutation in a sodium channel subunit (SCN1B) in a newly recognized clinical form of generalized epilepsy designated GEFS for generalized epilepsy with febrile seizures.[124]

There are several epilepsy subtypes that are known to have an inherited component, but the regions of the genome involved have not been identified. These include photogenic epilepsy,[125-127] reading epilepsy,[128] retinal degeneration with epilepsy,[129] photogenic epilepsy with spastic diplegia and mental retardation,[130] spastic paraplegia with myoclonus epilepsy,[131] Rolandic epilepsy and speech dyspraxia,[132] benign occipital epilepsy,[133-134] benign adult familial myoclonus epilepsy,[135] Hartung myoclonus epilepsy,[136] and febrile seizures.[137] The prevalence of some of these subtypes is relatively high and heritability has been well documented, yet only segregation data exist thus far.

**TABLE 7.4**
**Summary of Positive Linkage Between Genes or Chromosomal Regions and Human Epilepsy Subtypes**

| Epilepsy Type | Chromosome/Gene | Ref. |
|---|---|---|
| Juvenile Myoclonic Epilepsy | 6p | 106–108 |
| Juvenile Myoclonic Epilepsy | 15q.14 | 111 |
| Juvenile Absence Epilepsy | 21q.22/GRIKI | 212 |
| Progressive Myoclonic-Lafora | 6q24/EPM2 | 213, 276, 277 |
| Progressive Myoclonic-Unverricht/Lunborg | 21q22.3/cystatin b | 118–120 |
| Autosomal Dominant Nocturnal | | |
|    Frontal Lobe Epilepsy | 20q.13.2/ChRNA4 | 121–123 |
| | 15q24 | 279 |
| Benign Familial Neonatal Convulsions: | | |
|    BFNC1 | 20q.13.3/KCNQ2 | 114, 116 |
|    BFNC2 | 8q24/KCNQ3 | 115, 117 |
| Familial Febrile Convulsions: | | |
|    FEB1 | 8q13-q2 | 214 |
|    FEB2 | 19p13.3 | 215 |
| Benign Familial Infantile Convulsion | 19q | 216 |
| Partial Epilepsy | 10q | 217 |
|    Familial Partial Epilepsy with Variable Foci | 2q33-37 | 280 |
| Idiopathic Generalized: | | |
| Childhood Absence, JAE, JME | 8q | 218 |
|    Childhood Absence | 8q24 | 278 |
| Generalized Epilepsy with | | |
| Febrile Seizures (GEFS) | 19q13.1/SCN1B | 124 |
| Familial Infantile Convulsions with | | |
| Paroxysmal Choreoathetosis (ICCA) | 16p12-q12 | 177 |
| Epilepsy with Mental Retardation | | |
|    (EPMR-northern) | Xq22 | 219–220 |
| Mental Retardation/X linked 14 | | |
|    MRX14 | Xp11.3q13.3 | 221 |
| Epilepsy-Female Restricted with | | |
|    Mental Retardation (EFMR) | Xq22 | 222 |
| Benign Rolandic Epilepsy | 1q ?? | 146 |
| | 15q14 | 281 |

These studies typify the current thinking with respect to the etiology of common idiopathic epilepsies. The collective data suggest that an inherited component exists, which, in the proper environmental context, predisposes specific individuals toward disease. In some cases, a single gene defect can result in a seizure disorder; however, for the more common forms of epilepsy, these studies demonstrate that there are multiple susceptibility genes and that subtle alterations in combinations of these genes can increase the risk for disease. It is clear that the majority of gene polymorphisms thus far associated with epilepsy are directly involved with ion homeostasis. These include mutations in ion channel subunits or neurotransmitter receptors that directly or indirectly gate ions across excitable membranes. Whereas genetic factors are being discovered which increase the risk for disease, the environmental components which act to influence seizure susceptibility have not been systematically evaluated. In some instances, fatigue, drugs, and/or diet have been noted to influence seizure phenotype,[138] but the elucidation of the environmental factors that contribute to epilepsy will likely be at least as difficult to discern as genetic factors given the current strategies and study designs employed.

## CHROMOSOMAL ABERRATIONS AND EPILEPSY

Evidence for chromosomal aberrations being associated with epilepsy has also been documented. Trisomy of chromosome 9p has been associated with epilepsy in several case reports and includes a constellation of comorbid conditions such as cerebellar degeneration; limb, craniofacial, and dermatoglyphic anomalies; short stature; and mental retardation.[139] Trisomy of 12p[140] and 15p[141] have also been associated with generalized epilepsy and myoclonic seizures. Chromosomal translocations (2q:14q),[142] duplication,[143] and inversion (15q)[144] have been associated with seizures as well. Additionally, a case report of a combination of duplication (14q) and deletion (15q) in an individual was accompanied by myoclonic epilepsy.[145] There are over 30 case reports in which deletions of distal chromosome 1q have been associated with abnormal EEG and behavioral seizures very similar to those observed in benign Rolandic epilepsy.[146] These data do little to address specific loci that act as susceptibility factors; they do, however, point to regions of the genome that harbor genes necessary for normal neuronal function.

## SYNDROMES AND DISEASES WITH SEIZURES AS A PROMINENT COMPONENT

Over 40 human syndromes or diseases have been described in which seizures comprise a major component of a larger constellation of neurological and other defects. Table 7.5 summarizes those syndromes that are frequently accompanied by seizures, but it is by no means an exhaustive list.

The vast majority of syndromes and diseases that are accompanied by seizures have documented neural degeneration as a common feature. This neural degeneration is mostly confined to the cerebrum and cerebellum. Degeneration in the cerebrum often leads to mental retardation. Degeneration of the cerebrum has been documented in syndromes and diseases such as Angelman,[147-149] Gurrieri,[150-151] Landau-Kleffner,[152-153] West,[154] Delleman,[155] Borjeson,[156-157] Aicardi,[158-160] and many others (see Table 7.5). Cerebellar degeneration is a common theme among several syndromes as well, including Hermann,[161] Rett,[162-164] Pheo,[165-166] Paine,[167] Alpers,[168-169] Flynn-Aird,[170] Ramsay-Hunt,[171,172] ataxia with myoclonus and presenile dementia,[173] and familial myoclonus with deafness,[174] which all include seizures as a prominent component. This is interesting with respect to homology between human syndromes and mouse models of epilepsy, since many of the mouse mutants that are spontaneously seizure-prone were first discovered because of pronounced ataxia, presumably caused by a cerebellar defect. In addition, degeneration of basal ganglia resulting in choreoathetosis occurs in some diseases together with seizures, most notably dentatorubral-pallidoluysian atrophy (DRPLA)[175-176] and a familial infantile convulsive syndrome, ICCA[177] (see Table 7.4).

Pathology involving neurectodermal structures such as the retina or organ of Corti are also common among syndromes or diseases that include seizures. CHIME,[178-179] DOOR,[180-181] Flynn-Aird,[170] Juberg-Marsidi,[182] Rud,[183-184] Aicardi,[158-160] and Hermann[161] are just some examples of syndromes and diseases in which deafness or retinopathy accompany seizure phenotype. Pathology of cutaneous structures such as skin and hair follicles is also commonly associated with seizure phenotypes. These include Moynahan's alopecia,[185-186] Shokeir,[187] Rud,[183-184] Darrier-White,[188-189] neurofibromatosis,[190] nevus sebaceus of Jadassohn,[191] and tuberous sclerosis.[192-196] Renal (Hermann[161]), hepatic (Alpers[168,169]), and cardiac (Schinzel-Giedion[197,198]) pathology accompany some syndromes, and diabetes can also be a complicating comorbid condition. Skeletal dysplasia, dysmorphic features, periodontal and gingival disease, and growth retardation occur in a number of these syndromes and diseases as well.

Developmental anomalies in which cysts, tumors, vascular malformation, or lissencephaly occur can also lead to seizure phenotype. Schizencephaly can occur from mutations in developmental loci such as the homeobox gene EMX2.[199] In this condition, large regions of the cerebrum are absent and replaced by ventricular space and CSF. Heterotopic development of neocortex can also

**TABLE 7.5**
**Summary of Human Syndromes and Diseases in which Seizures Occur**

| Syndrome/Disease | Major Pathology | Segregation/Linkage | Ref. |
|---|---|---|---|
| CHIME | neuroectodermal | segregation | 178, 179 |
| DOOR | neuroectodermal | segregation | 180, 181 |
| Kohlschutter | neural/periodontal | segregation | 223, 224 |
| West | neural | segregation | 154 |
| Landau-Kleffner | neural | segregation | 152, 153 |
| Kifafa | neural | segregation | 225, 226 |
| Hallervorden-Spatz | neural | linkage (20p13) | 227, 228 |
| Pheo | neural | segregation | 165, 166 |
| Angelman | neural/branchial arch | linkage (15q11) | 147, 149 |
| Gurrieri | neural/skeletal dysplasia | segregation | 150, 151 |
| Alpers | neural/hepatic | segregation | 168, 169 |
| Celiac | intestinal/cerebral calcifications | linkage (6p) | 229 |
| Ramon | neural/skeletal/gingival | segregation | 230, 231 |
| Juberg-Marsidi | neural/neuroectodermal growth retardation | linkage (Xq13) | 182 |
| Flynn-Aird | neuroectodermal/cutaneous dental/skeletal | segregation | 170 |
| Familial sleep apnea plus | neural/muscle | segregation | 232 |
| Worster-Drought | neural/muscle | segregation | 233 |
| Cohen | neural/dental/dysmorphic | linkage (8q22) | 234, 236 |
| Delleman | neural/cutaneous | segregation | 155 |
| Borjeson | neural/cutaneous/metabolic dysmorphic | linkage (Xq26) | 156, 157 |
| Rud | neural/neuroectodermal cutaneous | segregation | 183, 184 |
| Aicardi | neural/neuroectodermal skeletal | linkage (Xp22) | 158, 160 |
| Paine | neural | segregation | 167 |
| Rasmussen encephalitis | neural | link. (Xq25/GLUR3) | 237, 238 |
| Darrier-White | neural/cutaneous | linkage (12q23) | 188, 189 |
| Rett | neural | linkage (Xp ??) | 162, 164 |
| Hermann | neural/neuroectodermal diabetes/nephropathy | segregation | 161 |
| Ramsay-Hunt "Feigenbaum" | neural | segregation | 171, 172 |
|  | atherosclerosis/photomyoclonic epilepsy/ deafness, nephropathy/diabetes/neural | segregation | 239 |
| Familial myoclonus with congenital deafness | neural/neuroectodermal | segregation | 174 |
| Ataxia with Myoclonus and Presenile dementia | neural | segregation | 173 |
| Moynahan alopecia | neural/cutaneous | segregation | 185, 186 |
| Shokeir | neural/cutaneous/periodontal | segregation | 187 |
| Alopecia, mental retardation w/convulsion, hyo-gonadism | neural/cutaneous/metabolic | segregation | 240 |
| Seitelberger | neural | segregation | 241 |
| DRLPA | neural | linkage (12p13.3) | 175, 176 |
| Schizencephaly | neural/developmental | linkage (10q26) | 199 |
| Schinzel-Giedion | skeletal/renal/cardiac dysmorphic/growth retardation | segregation | 197, 198 |
| Familial frontonasal cysts | cutaneous | segregation | 242 |
| Congenital perisylvian | neural | segregation | 243, 244 |
| Doublecortin | neural/development | linkage (Xq23/DCX) | 200, 201 |
| Cerebral cavernous malformation | vascular | linkage (7q11.2) | 245 |

**TABLE 7.5** *(continued)*
**Summary of Human Syndromes and Diseases in which Seizures Occur**

| Syndrome/Disease | Major Pathology | Segregation/Linkage | Ref. |
|---|---|---|---|
| Familial nodular heterotopia | neural/development | linkage (Xq28) | 202, 203 |
| Familial occipital calcification with epilepsy | neural/development | segregation | 246 |
| Nevus sebaceous Jadassohn | cutaneous | segregation | 191 |
| Neurofibromatosis | cutaneous | linkage — 17q11.2 | 190 |
| Tuberous sclerosis | neural/cutaneous/periodontal | linkage (9q34/tsc1) | 196 |
| | | linkage (16p.13/tsc2) | 193, 195 |

lead to seizure activity as is the case for mutations in the Doublecortin gene[200-201] and in the syndrome designated as familial nodular heterotopia.[202-203]

In summary, a wide range of human diseases and syndromes include seizure as a prominent phenotype. Neural degeneration is the most common underlying cause; however, the exact location of cell loss is variable. In the majority of these conditions, genomic regions have not been linked or associated with the illness, as most were identified as being heritable from familial segregation studies.

## ENZYME OR MITOCHONDRIAL DEFECTS WHICH LEAD TO SEIZURE PHENOTYPE

Defective enzymes involved in lysosomal storage, lipid catabolism, post-translational modification, and glycolytic and oxidative metabolism can lead to seizure phenotypes in specific instances. Table 7.6 lists some of these disorders and the underlying enzyme defect causal to the disease or syndrome. Several of these disorders result from defects in neuraminidase deficiency or lead to gangliosidosis or lipofuscinosis. Others perturb mitochondrial function and result in myoclonus and muscle pathology. Myoclonic epilepsy with ragged red fibers (MERRF) was one of the first, and now best known, of the mitochondrial mutations that lead to a seizure phenotype. Mutations in the transfer RNA molecules for lysine[204] and leucine[205] can cause MERRF. A separate entity known as mitochondrial myopathy, encephalopathy, lactic acidosis, and stroke-like episodes (MELAS) is also caused by mutations in the mitochondrial tRNAs for leucine[205] and serine[206] and includes seizures as part of the phenotype. Defects in amino acid synthesis and degradation can also lead to seizure phenotypes, as is the case for arginemia,[207-208] hyperglycinemia,[209] and phenylketonuria.[210]

Thus, congenital enzyme defects affect many different tissues, and a significant proportion of individuals afflicted exhibit seizures. While gross defects in these enzymes are sufficient to cause seizure phenotypes in the context of a larger syndrome of deficits, the genes encoding these enzymes, if altered in a subtle way, may be good candidates for investigations designed to elucidate the multigenic nature of the common forms of human epilepsy.

## SUMMARY

Identification of epilepsy susceptibility genes in humans is progressing rapidly. For rare seizure disorders inherited in a Mendelian fashion, traditional family linkage studies have identified mutations in specific genes causal to the disease. However, the most promising way to identify genetic factors contributing to the etiology of common types of epilepsy may be association studies with candidate genes. Animal models and human case reports are generating candidate genes for study in haplotype relative risk analyses in order to identify seizure susceptibility factors for the most common forms of the disease. In the next several years, an increasing number of candidate genes

**TABLE 7.6**
**Enzyme Deficiencies or Mitochondrial Mutations Associated with Seizures**

| Disorder | Major Pathology | Segregation/Linkage | Ref. |
|---|---|---|---|
| Phenylketonuria | neural | linkage (12q24.1) | 210 |
| Congenital lactic acidosis | neural/renal | segregation | 247 |
| GAD deficiency | neural | linkage (2q31) | 248, 249 |
| Phosphoglycerate kinase 1 | neural/blood/muscle | linkage (Xq13) | 250, 251 |
| Phosphoglycerate dehydrog. | neural/dysmorphic | segregation | 252 |
| Pyruvate dehdrogenase | lactic acidosis | linkage (Xp22.2) | 253 |
| Guanidinoacetate methyl transferase | neural/hepatic | segregation | 254 |
| Adenylosuccinate lyase | neural/muscle | linkage (22q.13) | 255-257 |
| D2 hydroxyglutaric aciduria | neural | segregation | 258 |
| L2 hydroxyglutaric aciduria | neural | segregation | 259 |
| Hyperglycinemia | neural | linkage — 9p22 | 209 |
| Arginemia | neural | linkage — 6q22 | 208 |
| Cytochrome Oxidase | neural/hepatic/renal | segregation | 260, 261 |
| NARP Syndrome | neural/muscle/neuroectodermal | mitochondrial | 262 |
| ATPsynthase 6 | neural/neuroectodermal | mitochondrial | 263 |
| MELAS | neural/muscle | mitochondrial | 205, 206 |
| MERRF | neural/muscle | mitochondrial | 204, 205 |
| Gaucher (glucocerebrosidase) | neural | linkage (1q23) | 104 |
| Tay Sachs (gangliosidosis Gm2) | neural/resp/neuroectodermal | linkage (15q23) | 105 |
| Gangliosidosis-(GM1) | neural | segregation | 264 |
| Neimann-Pick (sphingomyelin lipidosis) | neural/resp/skeletal | linkage (11p15.4) | 265, 266 |
| Neuronal ceroid lipofuscinosis 3 (NCL) Batten Disease | neural | linkage (16p12.1) | 267 |
| NCL-5 | neural | linkage (13q.21) | 268, 269 |
| Palmityol protein thioesterase NCL | neural | linkage (1q32) | 270 |
| Neuraminidase (Cherry red spot and Myoclonus) | neural/skeletal neuroectodermal | linkage (6p21.3) | 271 |
| Goldberg (neuraminidase and beta galactosidase) | neural/skeletal/neurectodermal | linkage (10q, 20q13.1) | 272, 273 |

will likely be found harboring mutations that will be linked to specific seizure phenotypes. The identification of these genetic factors will improve the diagnostic criteria for the separate epilepsy entities as well as provide specific targets for therapeutic drug design. Ultimately, it may be feasible to consider gene therapy for some of the most debilitating and untreatable seizure disorders.[211]

## UPDATE

Since preparing this review several new studies have been published which deserve mention. Regarding mouse models, a targeted knockout of the gene responsible for human PME of the Unverricht-Lundborg Type[120] results in a mouse with myoclonic seizures and ataxia.[274] Also, with regard to polygenic models, the utility of QTL studies has been enhanced by a newly described strategy for high resolution mapping of quatitative traits.[275] New human studies include identification of a novel protein tyrosine phosphatase gene (EPM2/EPM2A) as the gene responsible for PME of the Lafora type.[276-277] Also, in agreement with earlier studies on a broad idiopathic generalized epilepsy phenotype,[218] a new locus has been identified for childhood absence epilepsy at 8q24.[278] In addition, a new locus has been identified for ADNFLE at 15q24.[279] In the realm of partial epilepsy, a new syndrome has been identified called familial partial epilepsy with variable foci

(FPEVF) which may be linked to markers on chromosome 2q33-37.[280] Finally, an additional linkage has been detected for benign Rolandic epilepsy on chromosome 15q14.[281] (*Note:* For references pertaining to these new studies, see Ref. 274–281.)

## REFERENCES

1. Tecott, L.H., Sun, L.M., Akara, S.F., Strack, A.M., Lowenstein, D.H., Dallman, M.F., and Julius, D., Eating disorder and epilepsy in mice lacking 5-Ht2c serotonin receptors, *Nature,* 374, 542, 1995.
2. Frankel, W.N., Taking stock of complex trait genetics in mice, *Trends Genet.,* 11, 471, 1995.
3. Noebels, J.L., Targeting epilepsy genes, *Neuron,* 16, 241, 1996.
4. Buchhalter, J.R., Animal models of inherited epilepsy, *Epilepsia,* 34(Suppl. 3), S31, 1993.
5. Noebels, J.L. and Sidman, R.L., Three mutations causing spike-wave seizuers in the mouse, *Neuroscience,* (Suppl) 7, 159, 1982.
6. Burgess, D.L., Jones, J.M., Meisler, M.H., and Noebels, J.L., Mutation of the Ca2+ channel b subunit gene Cchb4 is associated with ataxia and seizures in the lethargic (lh) mouse, *Cell,* 88, 385, 1997.
7. Herring, J.M., Dung, H.C., Yoo, J.H., and Yu, J., Chronological studies of peripheral motor nerve conduction in lethargic mice, *Electromyogr. Clin. Neurophysiol.,* 21, 121, 1981.
8. Green, M.C. and Sidman, R.L., Tottering — a neuromuscular mutation in the mouse, *J. Hered.,* 53, 23, 1962.
9. Tsuji, S. and Meier, H., Evidence for allelism of leaner and tottering in the mouse, *Genet. Res.,* 17, 83, 1971.
10. Fletcher, C.F., Lutz, C.M., O'Sullivan, T.N., Shaughnessy, J.D., Hawkes, R., Frankel, W.N., Copeland, N.G., and Jenkins, N.A., Absence epilepsy in tottering mutant mice is associated with calcium channel defects, *Cell,* 87, 607, 1996.
11. Levitt, P. and Noebels, J.L., Mutant mouse tottering: selective increase of locus ceruleus axonsin a defined single-locus mutation, *Proc Natl. Acad. Sci.,* 78, 4630, 1981.
12. Noebels, J.L., A single gene error of noradrenergic axon growth synchronizes central neurones, *Nature,* 310, 409, 1984.
13. Cox, G.A., Lutz, C.M., Yang, C.L., Biemesderfer, D., Bronson, R.T., Fu, A., Aronson, P.S., Noebels, J.L., and Frankel, W.N., Sodium/hydrogen exchanger gene defect in slow-wave epilepsy mutant mice, *Cell,* 3, 139, 1997.
14. Noebels, J.L., Qiao, X., Bronson, R.T., Spencer, C., and Davisson, M.T., Stargazer: a new neurological mutant on chromosome 15 in the mouse with prolonged cortical seizures, *Epilepsy Res.,* 7(2), 129, 1990.
15. Letts, V.A., Valenzuela, A., Kirley, J.P., Sweet, H.O., Davisson, M.T., and Frankel, W.N., Genetic and physical maps of the stargazer locus on mouse chromosome 15, *Genomics,* 43, 62, 1997.
16. Letts, V.A., Felix, R., Biddlecome, G.H., Arikkath, J., Mahaffey, C.L., Valenzuela, A., Barlett, F.S., Mori, Y., Campbell, K.P., and Frankel, W.N., The mouse stargazer gene encodes a neuronal Ca2+-channel g subunit, *Nat. Genet.,* 19, 340, 1998.
17. Patil, N., Cox, D.R., Bhat, D., Faham, M., Myers, R.M., and Peterson, A.S., A potassium channel mutation in weaver mice implicates membrane excitability in granule cell differentiation, *Nat. Genet.,* 11, 126, 1995.
18. Goldowitz, D. and Koch, J., Performance of normal and neurological mutant mice on radial arm maze and active avoidance tasks, *Behav. Neural Biol.,* 46, 216, 1986.
19. Eisenberg, B. and Messer, A., Tonic/clonic seizures in a mutant mouse carrying the weaver gene, *Neurosci. Lett.,* 96, 168, 1989.
20. Bonuccelli, U., Garant, D.S., Maggio, R., and Fariello, R.G., Motor expression of kainic acid seizures is attenuated by dopamine depletion in mice, *Brain Res.,* 657, 269, 1994.
21. Snipes, G.J., Suter, U., Welcher, A.A., and Shooter, E.M., Characterization of a novel peripheral nervous sytem myelin protein (PMP-22/SR13), *J. Cell Biol.,* 117, 225, 1992.
22. Searle, A.G., A lethal allele of dilute in the house mouse, *Heredity,* 6, 395, 1952.
23. Mercer, J.A., Seperack, P.K., Strobel, M.C., Copeland, N.G., and Jenkins, N.A., Novel myosin heavy chain encoded by murine dilute coat colour locus, *Nature,* 349, 709, 1991.
24. Snell, G.D., Ducky, a new second chromosome mutation in the mouse, *J. Hered.,* 46, 27, 1955.

25. Holz, A., Frank, M., Copeland, N.G., Gilbert, D.J., Jenkins, N.A., and Schwab, M.E., Chromosomal localization of the myelin-associated oligodendrocytic basic protein and expression in the genetically linked neurological mouse mutants ducky and tippy, *J. Neurochem.*, 69, 1801, 1997.
26. DeOme, K.B., A new recessive lethal mutation in mice, *Univ. Calif. Publ. Zool.*, 53, 41, 1945.
27. Sweet, H.O., Bronson, R.T., Harris, B.S., and Davisson, M.T., Trembly-like (Tyl) — a new X-linked mutation in the mouse, *Mouse Genome,* 87, 112, 1990.
28. Lane, P.W. and Green, M.C., Teetering (tn), *Mouse News Lett.,* 27, 38, 1962.
29. Meier, H., The neuropathy of teetering, a neurological mutation in the mouse, *Arch. Neurol.,* 16, 59, 1967.
30. Frings, H., Frings, M., and Hamilton, M., Experiments with albino mice from stocks selected for predictable susceptibilites to audiogenic seizures, *Int. J. Behav. Res.,* 9, 44, 1956.
31. Skradski, S.L., White, S.H., and Ptacek, L.J., Genetic mapping of a locus mass1 causing audiogenic seizures in mice, *Genomics*, 49, 188, 1998.
32. Loscher, W., Basic aspects of epilepsy, *Curr. Opin. Neurol. Neurosurg.,* 6, 223.
33. Kash, S.F., Johnson, R.S., Tecott, L.H., Noebels, J.L., Mayfield, R.D., Hanaman, D., and Baekkeskov, S., Epilepsy in mice deficient in the 65-kDa isoform of glutamic acid decarboxylase, *Proc. Nat. Acad. Sci.,* 94, 14060, 1997.
34. Asada, H., Kawamura, Y., Maruyama, K., Hume, H., Ding, R.-G., Ji, F.Y., Kanbara, N., Kuzume, H., Sanbo, M., Yagi, T., and Obata, K., Mice lacking the 65 kDa isoform of glutamic acid decarboxylase (GAD65) maintain normal levels of GAD67 and GABA in their brains but are not susceptible to seizures, *Biochem. Biophys. Res. Commun.,* 229, 891, 1996.
35. Condie, B.G., Bain, G., Gottlieb, D.I., and Capecchi, M.R., Cleft palate in mice with a targeted mutation in the gamma-aminobutyric acid-producing enzyme glutamic acid decarboxylase 67, *Proc. Natl. Acad. Sci.,* 94, 11451, 1997.
36. Waymire, K.G., Mahuren, D., Jaje, J.M., Guilarte, T.R., Coburn, S.P., and MacGregor, G.R., Mice lacking tissue non-specific alkaline phosphatase die from seizures due to defective metabolism of vitamin B6, *Nat. Genet.,* 11, 45, 1995.
37. Homanics, G.E., Ferguson, C., Quinlan, J.J., Daggett, J., Synder, K., Lagenaur, C., Mi, Z.P., Wang, X.H., Grayson, D.R., and Firestone, L.L., Gene knockout of the alpha 6 subunit of the gamma-aminobutyric acid type A receptor: lack of gene effect on responses to ethanol, pentobarbital and general anesthetics, *Mol. Pharmacol.*, 51, 588, 1997.
38. Homanics, G.E., DeLorey, T.M., Firestone, L.L., Quinlan, J.L., Handforth, A., Harrison, N.L., Krasowski, M.D., Rick, C.E., Korpi, E.A., Makela, R., Brilliant, M.H., Hagiwara, N., Ferguson, C., Snyder, K., and Olsen, R.W., Mice devoid of g-aminobutyrate type A receptor b3 subunit have epilepsy, cleft palate and hypersensitvive behavior, *Proc. Natl. Acad. Sci. U.S.A.,* 94, 4143, 1997b.
39. Mulle, C., Sailer, A., Perez-Otano, I., Dickinson-Anson, H., Castilo, P., Bureau, I., Maron, C., Gage, F.H., Mann, J.R., Bettler, B., and Heinemann, S., Altered synaptic physiology and reduced susceptibility to kainate-induced seizures in GluR6-deficient mice, *Nature,* 392, 601, 1998.
40. Brusa, R., Zimmerman, F., Koh, D.S., Feldmeyer, D., Gass, P., Sakmann, B., Seeberg, P.H., and Sprengel, R., Early-onset epilepsy and postnatal lethality associated with an editing deficient GluR-B allele in mice, *Science,* 270, 1677, 1995.
41. Tanaka, K., Watase, K., Manabe, T., Yamada, K., Watanabe, M., Takahashi, K., Iwama, H., Nishikawa, T., Ichihara, N., Kikuchi, T., Okuyama, S., Kawashima, N., Hori, S., Takimoto, M., and Wada, K., Epilepsy and exacerbation of brain injury in mice lacking the glutamate transporter GLT-1, *Science,* 276, 1699, 1997
42. Signorini, S., Liao, Y.J., Duncan, S.A., Jan, L.Y., and Stoffel, M., Normal cerebellar development but susceptibility to seizures in mice lacking G protein-coupled, inwardly rectifying K+ channel GIRK2, *Proc. Natl. Acad. Sci. U.S.A.,* 94(3), 923, 1997
43. Smart, S.L., Lopantsev, V., Zhang, C.L., Robbins, C.A., Wang, H., Chiu, S.Y., Schwartzkroin, P.A., Messing, A., and Tempel, B.L., Deletion of the K(V)1.1 potassium channel causes epilepsy in mice, *Neuron,* 20(4), 809, 1998.
44. Erickson, J.C., Clegg, K.E., and Palmiter, R.D., Sensitivity to leptin and susceptibility to seizures of mice lacking neuropeptide Y, *Nature,* 381, 415, 1996.
45. Rosahl, T.W., Spillane, D., Missler, M., Herz, J., Selig, D.K., Wolff, J.R., Hammer, R.E., Malenka, R.C., and Sudhof, T.C., Essential functions of synapsins I and II in synaptic vesicle regulation, *Nature,* 375, 488, 1995.

46. Butler, L.S., Silva, A.J., Abeliovich, A., Watanabe, Y., Tonegawa, S., and McNamara, J.O., Limbic epilepsy in transgenic mice carrying a Ca2+/calmodulin-dependent kinase II alpha-subunit mutation, *Proc. Natl. Acad. Sci. U.S.A.,* 92(15), 6852, 1995.

47. Matsumoto, M., Nakagawa, T., Inoue, T., Nagata, E., Tanaka, K., Takano, H., Minowa, O., Kuno, J., Sakakibara, S., Yamada, M., Yoneshima, H., Miyawaki, A., Fukuuchi, Y., Furuichi, T., Okano, H., Mikoshiba, K., and Noda, T., Ataxia and epileptic seizures in mice lacking type 1 inositol 1,4,5-trisphosphate receptor, *Nature,* 379(6561), 168, 1996.

48. Tsirka, S.E., Gualandris, A., Amaral, D.G., and Strickland, S., Excitotoxin-induced neuronal degeneration and seizure are mediated by tissue plasminogen activator, *Nature,* 377(6547), 340, 1995.

49. Bruce, A.J., Boling, W., Kindy, M.S., Peschon, J., Kraemer, P.J., Carpenter, M.K., Holtsberg, F.W., and Mattson, M.P., Altered neuronal and microglial responses to excitotoxic and ischemic brain injury in mice lacking TNF receptors, *Nat. Med.,* 2(7), 788, 1996.

50. Toth, M., Grimsby, J., Buzsaki, G., and Donovan, G.P., Epileptic seizures caused by inactivation of a novel gene, jerky, related to centromere binding protein-B in transgenic mice, *Nat. Gen.,* 11(1), 71, 1995.

51. Acampora, D., Mazan, S., Avantaggiato, V., Barone, P., Tuorto, F., Lallemand, Y., Brulet, P., and Simeone, A., Epilepsy and brain abnormalities in mice lacking the Otx1 gene, *Nat. Genet.,* 14(2), 218, 1996.

52. Yamamoto, A., Takagi, H., Kitamura, D., Tatsuoka, H., Nakano, H., Kawano, H., Kuroyanagi, H., Yahagi, Y., Kobayashi, S., Koizumi, K., Sakai, T., Saito, K., Chiba, T., Kawamura, K., Suzuki, K., Watanabe, T., Mori, H., and Shirasawa, T., Deficiency in protein L-isoaspartyl methyltransferase results in a fatal progressive epilepsy, *J. Neurosci.,* 18(6), 2063, 1998.

53. Kim, E., Lowenson, J.D., MacLaren, D.C., Clarke, S., and Young, S.G., Deficiency of a protein-repair enzyme results in the accumulation of altered proteins, retardation of growth, and fatal seizures in mice, *Proc. Natl. Acad. Sci. U.S.A.,* 94(12), 6132, 1997.

54. Chae, T., Kwon, Y.T., Bronson, R., Dikkes, P., Li, E., and Tsai, L.H., Mice lacking p35, a neuronal specific activator of Cdk5, display cortical lamination defects, seizures, and adult lethality, *Neuron,* 18(1), 29, 1997.

55. Kokaia, M., Ernfors, P., Kokaia, Z., Elmer, E., Jaenisch, R., and Lindvall, O., Suppressed epileptogenesis in BDNF mutant mice, *Exp. Neurol.,* 133(2), 215, 1995.

56. Aigner, L., Arber, S., Kapfhammer, J.P., Laux, T., Schneider, C., Botteri, F., Brenner, H.R., and Caroni, P., Overexpression of the neural growth-associated protein GAP-43 induces nerve sprouting in the adult nervous system of transgenic mice, *Cell,* 83(2), 269, 1995.

57. Fujita, N., Suzuki, K., Vanier, M.T., Popko, B., Maeda, N., Klein, A., Sandhoff, K., Nakayasu, H., and Suzuki, K., Genetic sphingolipid activator protein (sap) deficiencies in human and a "knockout" mouse, *J. Neurochem.,* 66, S1, 1996.

58. Sango, K., McDonald, M.P., Crawley, J.N., Mack, M.L., Tifft, C.J., Skop, E., Starr, C.M., Hoffmann, A., Sandhoff, K., Suzuki, K., and Proia, R.L., Mice lacking both subunits of lysosomal beta-hexosaminidase display gangliosidosis and mucopolysaccharidosis, *Nat. Genet.,* 14(3), 348, 1996.

59. Nave, K.A., Lai, C., Bloom, F.E., and Milner, R.J., Jimpy mutant mouse: a 74-base deletion in the mRNA for myelin proteolipid protein and evidence for a primary defect in RNA splicing, *Proc Natl. Acad. Sci.,* 83, 9264, 1986.

60. Kagawa, T., Ikenaka, K., Inoue, Y., Kuriyama, S., Tsujii, T., Nakao, J., Nakajima, K., Aruga, J., Okano, H., and Mikoshiba, K., Glial cell degeneration and hypomyelination caused by overexpression of myelin proteolipid protein gene, *Neuron,* 13(2), 427, 1994.

61. Mangiarini, L., Sathasivam, K., Seller, M., Cozens, B., Harper, A., Hetherington, C., Lawton, M., Trottier, Y., Lehrach, H., Davies, S.W., and Bates, G.P., Exon 1 of the HD gene with an expanded CAG repeat is sufficient to cause a progressive neurological phenotype in transgenic mice, *Cell,* 87(3), 493, 1996.

62. Ordway, J.M., Tallaksen-Greene, S., Gutekunst, C.A., Bernstein, E.M., Cearley, J.A., Wiener, H.W., Dure, L.S., 4th, Lindsey, R., Hersch, S.M., Jope, R.S., Albin, R.L., and Detloff, P.J., Ectopically expressed CAG repeats cause intranuclear inclusions and a progressive late onset neurological phenotype in the mouse, *Cell,* 91(6), 753, 1997.

63. Sander, T., The genetics of idiopathic epilepsy: implications for the understanding of its aetiology, *Mol. Med. Today,* 8, 173, 1996.

64. Lander, E.S. and Botstein, D., Mapping Mendelian factors underlying quantitative traits using RFLP linkage maps, *Genetics*, 121, 185, 1989.

65. Lander, E.S. and Kruglyak, L., Genetic dissection of complex traits: guidelines for interpreting and reporting linkage results, *Nat. Genet.*, 11, 241, 1995.

66. Ferraro, T.N., Golden, G.T., Smith, G.G., Schork, N.J., St. Jean, P., Ballas, C., Choi, H., and Berrettini, W.H., Mapping murine loci for seizure response to kainic acid, *Mammalian Genome*, 7, 200, 1997.

67. Hall, C.S., Genetic differences in fatal audiogenic seizures between two inbred strains of house mice, *J. Hered.*, 38, 2, 1947.

68. Neumann, P.E. and Collins, R.L., Confirmation of the influence of a chromosome 7 locus on susceptibility to audiogenic seizures, *Mammalian Genome*, 3, 250, 1992.

69. Neumann, P.E. and Collins, R.L., Genetic dissection of susceptibility to audiogenic seizures in inbred mice, *Proc. Natl. Acad. Sci. U.S.A.*, 88, 5408, 1991.

70. Neumann, P.E. and Seyfried, T.N., Mapping of two genes that influence susceptibility to audiogenic seizures in crosses of C57BL/6J and DBA/2J mice, *Behav. Genet.*, 20, 307, 1990.

71. Seyfried, T.N. and Glaser, G.H., A review of mouse mutants as genetic models of epilepsy, *Epilepsia*, 26, 143, 1985.

72. Suzuki, J., Paroxysmal changes in the electroencephalogram of the El mouse, *Experientia*, 32, 336, 1976.

73. Suzuki, J. and Nakamoto, Y., Seizure patterns and electroencephalograms of El mouse, *Electroencephalogr. Clin. Neurophysiol.*, 43, 299, 1977.

74. Frankel, W.N., Valenzuela, A., Lutz, C.M., Johnson, E.W., Dietrich, W.F., and Coffin, J.M., New seizure frequency QTL and the complex genetics of epilepsy in EL mice, *Mammalian Genome*, 6, 830, 1995.

75. Frankel, W.N., Johnson, E.W., and Lutz, C.M., Congenic strains reveal effects of the epilepsy quantitative trait locus, El2, separate from other El loci, *Mammalian Genome*, 6, 839, 1995.

76. Miner, L.L. and Marley, R.J., Chromosomal mapping of loci influencing sensitivity to cocaine-induced seizures in BXD recombinant inbred strains of mice, *Psychopharmacology*, 117, 62, 1995.

77. Martin, B., Clement, Y., Venault, P., and Chapouthier, G., Mouse chromosomes 4 and 13 are involved in β-carboline-induced seizures, *J. Hered.*, 86, 274, 1995.

78. Ferraro, T.N., Golden, G.T., Smith, G.G., and Berrettini, W.H., Differential susceptibility to seizures induced by systemic kainic acid treatment in mature DBA/2J and C57BL/6J mice, *Epilepsia*, 36, 301, 1995.

79. Kosobud, A.E. and Crabbe, J.C., Genetic correlations among inbred strain sensitivities to convulsions induced by 9 convulsant drugs, *Brain Res.*, 526, 8, 1990.

80. Engstrom, F. and Woodbury, D.M., Seizure susceptibility in DBA and C57 mice: the effects of various convulsants, *Epilepsia*, 29, 389, 1988.

81. Freund, R.K., Marley, R.J., and Wehner, J.M., Differential sensitivity to bicuculline in three inbred mouse strains, *Brain Res. Bull.*, 18, 657, 1987.

82. Marley, R.J., Gaffney, D., and Wehner, J.M., Genetic influences on GABA-related seizures, *Pharmacol. Biochem Behav.*, 24, 665, 1986.

83. Taylor, B.A., Genetic analysis of susceptibility to isoniazid-induced seizures in mice, *Genetics*, 83, 373, 1976.

84. Schlesinger, K., Boggan, W.O., and Griek, B.J., Pharmacogenetic correlates of pentylenetetrazole and electroconvulsive seizure thresholds in mice, *Psychopharmacololy*, (Berl.), 13, 181, 1968.

85. Ferraro, T.N. and Berrettini, W.H., Quantitative trait loci mapping in mouse models of complex behavior, in *Function and Dysfunction in the Nervous System, Proceedings of the Cold Spring Harbor Symposium on Quantitative Biology*, Stillman, B., Ed., 1996. Cold Spring Harbor Laboratory Press, Cold Spring Harbor, NY.

86. Buck, K.J., Metten, P., Belknap, J.K., and Crabbe, J.C., Quantitative trait loci involved in genetic predisposition to acute alcohol withdrawal in mice, *J. Neurosci.*, 17(10), 3946, 1997.

87. Zeng, Z.-B., Precision mapping of quantitative trait loci, *Genetics*, 136, 1457, 1994.

88. Zeng, Z.-B., Theoretical basis of separation of multiple linked gene effects on mapping quantitative trait loci, *Proc. Natl. Acad. Sci. U.S.A.*, 90, 10972, 1993.

89. Snell, G.D., Methods for the study of histocompatibility genes, *J. Genet.*, 49, 87, 1948.

90. Sidman, R.L., Dickie, M.M., and Appel, S.H., Mutant mice (quaking and jimpy) with deficient myelination in the central nervous system, *Science*, 144, 309, 1964.

91. Kapfhamer, D., Sweet, H.O., Sufalko, D., Warren, S., Johnson, K.R., and Burmeister, M., The neurological mouse mutations jittery and hesitant are allelic and map to the region of mouse chromosome 10 homologous to 19p13.3, *Genomics*, 35(3), 533, 1996.

92. Dickie, M.M., Lethargic (lh), *Mouse News Lett.*, 30, 31, 1964.

93. Ebersole, T.A., Chen, Q., Justice, M.J., and Artzt, K., The quaking gene product necessary in embryogenesis and myelination combines features of RNA binding and signal transduction proteins, *Nat. Genet.*, 12, 260, 1996.

94. Roach, A., Takahashi, N., Pravtcheva, D., Ruddle, F., and Hood, L., Chromosomal mapping of mouse myelin basic protein gene and structure and transcription of of the partially deleted gene in shiverer mutant mice, *Cell*, 42, 149, 1985.

95. Chernoff, G.F., Shiverer: an autosomal recessive mutant mouse with myelin deficiency, *J. Hered.*, 72, 128, 1981.

96. Popko, B., Puckett, C., and Hood, L., A novel mutation in myelin-deficient mice results in unstable myelin basic protein gene transcripts, *Neuron*, 1, 221, 1988.

97. Falconer, D.S., Two new mutants, "trembler" and "reeler," with neurological actions in the house mouse, *J. Genet.*, 50, 192, 1951.

98. Suter, U., Welcher, A.A., Ozcelik, T., Snipes, G.J., Kosaras, B., Francke, U., Billings-Gagliardi, S., and Sidman, R.L., Shooter EM: Trembler mouse carries a point mutation in a myelin gene, *Nature*, 356, 241, 1992.

99. Lane, P.W. and Weaver, W.V., *Mouse News Lett.*, 30, 32, 1964.

100. Cain, D.P., Grant, S.G., Saucier, D., Hargreaves, E.L., and Kandel, E.R., Fyn tyrosine kinase is required for normal amygdala kindling, *Epilepsy Res.*, 22, 107, 1995.

101. Rise, M.L., Frankel, W.N., Coffin, J.M., and Seyfried, T.N., Genes for epilepsy mapped in the mouse, *Science*, 253, 669, 1991.

102. Frankel, W.N., Taylor, B.A., Noebels, J.L., and Lutz, C.M., Genetic epilepsy model derived from common inbred mouse strains, *Genetics*, 138, 481, 1994.

103. Anon., Proposal for revised classification of epilepsies and epileptic syndromes. Commission on Classification and Terminology of the International League Against Epilepsy, *Epilepsia*, 30(4), 389, 1989.

104. Beutler, E., Gauchers Disease, *N. Engl. J. Med.*, 325, 1354, 1991.

105. Myerowitz, R., Tay Sachs disease causing mutations and neutral polymorphisms in the HexA gene, *Hum. Mut.*, 9, 195, 1997.

106. Greenberg, D.A., Delgado-Escueta, A.V., Widelitz, H., Sparkes, R.S., Treiman, L., Maldonado, H.M., Park, M.S., and Terasaki, P.I., Juvenile myoclonic epilepsy (JME) may be linked to the BF and HLA loci on human chromosome 6, *Am. J. Med. Genet.*, 31(1), 185, 1988.

107. Liu, A.W., Delgado-Escueta, A.V., Serratosa, J.M., Alonso, M.E., Medina, M.T., Gee, M.N., Cordova, S., Zhao, H.Z., Spellman, J.M., Peek, J.R. et al., Juvenile myoclonic epilepsy locus in chromosome 6p21.2-p11: linkage to convulsions and electroencephalography trait, *Am. J. Hum. Genet.*, 57(2), 368, 1995.

108. Weissbecker, K. A., Durner, M., Janz, D., Scaramelli, A., Sparkes, R. S., and Spence, M.A., Confirmation of linkage between juvenile myoclonic epilepsy locus and the HLA region of chromosome 6, *Am. J. Med. Genet.*, 38, 32, 1991.

109. Whitehouse, W.P., Rees, M.. Curtis, D., Sundqvist, A., Parker, K., Chung, E., Baralle, D., and Gardiner, R.M., Linkage analysis of idiopathic generalized epilepsy (IGE) and marker loci on chromosome 6p in families of patients with juvenile myoclonic epilepsy: no evidence for an epilepsy locus in the HLA region, *Am. J. Hum. Genet.*, 53(3), 652, 1993.

110. Whitehouse, W., Diebold, U., Rees, M., Parker, K., Doose, H., and Gardiner, R.M., Exclusion of linkage of genetic focal sharp waves to the HLA region on chromosome 6p in families with benign partial epilepsy with centrotemporal sharp waves, *Neuropediatrics*, 24(4), 208, 1993.

111. Elmslie, F.V., Rees, M., Williamson, M.P., Kerr, M., Kjeldsen, M.J., Pang, K.A., Sundqvist, A., Friis, M.L., Chadwick, D., Richens, A., Covanis, A., Santos, M., Arzimanoglou, A., Panayiotopoulos, C.P., Curtis, D., Whitehouse, W.P., and Gardiner, R.M., Genetic mapping of a major susceptibility locus for juvenile myoclonic epilepsy on chromosome 15q, *Hum. Mol. Genet.*, 6(8), 1329, 1997.

112. Steinlein, O., Sander, T., Stoodt, J., Kretz, R., Janz, D., and Propping, P., Possible association of a silent polymorphism in the neuronal nicotinic acetylcholine receptor subunit alpha4 with common idiopathic generalized epilepsies, *Am. J. Med. Genet.*, 74(4), 445, 1997.

113. Sander, T., Hildmann, T., Kretz, R., Furst, R., Sailer, U., Bauer, G., Schmitz, B., Beck-Mannagetta, G., Wienker, T.F., and Janz, D., Allelic association of juvenile absence epilepsy with a GluR5 kainate receptor gene (GRIK1) polymorphism, *Am. J. Med. Genet.*, 74(4), 416, 1997.

114. Leppert, M., Anderson, V.E., Quattlebaum, T., Stauffer, D., O'Connell, P., Nakamura, Y., Lalouel, J.-M., and White, R., Benign familial neonatal convulsions linked to genetic markers on chromosome 20, *Nature*, 337, 647, 1989.

115. Lewis, T.B., Leach, R.J., Ward, K., O'Connell, P., and Ryan, S.G., Genetic heterogeneity in benign familial neonatal convulsions: identification of a new locus on chromosome 8q, *Am. J. Hum. Genet.*, 53, 670, 1993.

116. Singh, N.A., Charlier, C., Stauffer, D., DuPont, B.R., Leach, R.J., Melis, R., Ronen, G.M., Bjerre, I., Quattlebaum, T., Murphy, J.V., McHarg, M.L., Gagnon, D., Rosales, T.O., Peiffer, A., Anderson, V.E., and Leppert, M., A novel potassium channel gene, KCNQ2, is mutated in an inherited epilepsy of newborns, *Nat. Genet.*, 18(1), 25, 1998.

117. Charlier, C., Singh, N. A., Ryan, S. G., Lewis, T. B., Reus, B. E., Leach, R. J., and Leppert, M., A pore mutation in a novel KQT-like potassium channel gene in an idiopathic epilepsy family, *Nat. Genet.*, 18, 53, 1998.

118. Lehesjoki, A.E., Koskiniemi, M., Norio, R., Tirrito, S., Sistonen, P., Lander, E., and de la Chapelle, A., Localization of the EPM1 gene for progressive myoclonus epilepsy on chromosome 21: linkage disequilibrium allows high resolution mapping, *Hum. Mol. Genet.*, 2(8), 1229, 1993.

119. Lafreniere, R.G., Rochefort, D.L., Chretien, N., Rommens, J.M., Cochius, J.I., Kalviainen, R., Nousiainen, U., Patry, G., Farrell, K, Soderfeldt, B., Federico, A., Hale, B.R., Cossio, O.H., Sorensen, T., Pouliot, M.A., Kmiec, T., Uldall, P., Janszky, J., Pranzatelli, M.R., Andermann, F., Andermann, E., and Rouleau, G.A., Unstable insertion in the 5' flanking region of the cystatin B gene is the most common mutation in progressive myoclonus epilepsy type 1, EPM1, *Nat. Genet.*, 15(3), 298, 1997.

120. Pennacchio, L.A., Lehesjoki, A.E., Stone, N.E., Willour, V.L., Virtaneva, K., Miao, J., D'Amato, E., Ramirez, L., Faham, M., Koskiniemi, M., Warrington, J.A., Norio, R., de la Chapelle, A., Cox, D.R., and Myers, R.M., Mutations in the gene encoding cystatin B in progressive myoclonus epilepsy (EPM1), *Science*, 271(5256), 1731, 1996.

121. Phillips, H.A., Scheffer, I.E., Berkovic, S.F., Hollway, G.E., Sutherland, G.R., and Mulley, J.C., Localization of a gene for autosomal dominant nocturnal frontal lobe epilepsy to chromosome 20q 13.2, *Nat. Genet.*, 10(1), 117, 1995.

122. Steinlein, O.K., Mulley, J.C., Propping, P., Wallace, R.H., Phillips, H.A., Sutherland, G.R., Scheffer, I.E., and Berkovic, S.F., A missense mutation in the neuronal nicotinic acetylcholine receptor alpha 4 subunit is associated with autosomal dominant nocturnal frontal lobe epilepsy, *Nat. Genet.*, 11(2), 201, 1995.

123. Steinlein, O.K., Magnusson, A., Stoodt, J., Bertrand, S., Weiland, S., Berkovic, S.F., Nakken, K.O., Propping, P., and Bertrand, D., An insertion mutation of the CHRNA4 gene in a family with autosomal dominant nocturnal frontal lobe epilepsy, *Hum. Mol. Genet.*, 6(6), 943, 1997.

124. Wallace, R.H., Wang, D.W., Singh, R., Scheffer, I.E., George, A.L., Phillips, H.A., Saar, K., Reis, A., Johnson, E.W., Sutherland, G.R., Berkovic, S.F., and Mulley, J.C., Febrile Seizures and generalized epilepsy associated with a mutation in the Na+-channel beta 1 subunit gene SCN1B, *Nat. Genet.*, 19, 366, 1998.

125. Davidson, S. and Watson, C.W., Hereditary ligh sensitive epilepsy, *Neurology*, 6, 231, 1956.

126. Gerken, H., Doose, H., Volzke, E., Voltz, C., and Hein-Volpel, K.F., Genetics of childhood epilepsy with photic sensitivity, *Lancet*, 1, 1377, 1968.

127. Doose, H. and Waltz, S., Photosensitivity — genetics and clinical significance, *Neuropediatrics*, 24(5), 249, 1993.

128. Daly, R.F. and Forster, F.M., Inheritance of reading epilepsy, *Neurology*, 25(11), 1051, 1975.

129. Cohan, S.L., Kattah, J.C., and Limaye, S.R., Familial tapetoretinal degeneration and epilepsy, *Arch. Neurol.*, 36(9), 544, 1979.

130. Daly, D., Siekert, R. G., and Burke, E. C., A variety of familial light sensitive epilepsy, *Electroenceph. Clin. Neurophysiol.*, 11, 141, 1959.

131. Sommerfelt, K., Kyllerman, M., and Sanner, G., Hereditary spastic paraplegia with epileptic myoclonus, *Acta. Neurolog. Scand.*, 84(2), 157, 1991.

132. Scheffer, I.E., Jones, L., Pozzebon, M., Howell, R.A., Saling, M.M., and Berkovic, S.F., Autosomal dominant rolandic epilepsy and speech dyspraxia: a new syndrome with anticipation, *Ann. Neurol.*, 38(4), 633, 1995.

133. Kuzniecky, R. and Rosenblatt, B., Benign occipital epilepsy: a family study, *Epilepsia*, 28(4), 346, 1987.

134. Gastaut, H., A new type of epilepsy: benign partial epilepsy of childhood with occipital spike-waves, *Clin. Electroencephalogr.*, 13(1), 13, 1982.

135. Kuwano, A., Takakubo, F., Morimoto, Y., Uyama, E., Uchino, M., Ando, M., Yasuda, T., Terao, A., Hayama, T., Kobayashi, R., and Kondo, I., Benign adult familial myoclonus epilepsy (BAFME): an autosomal dominant form not linked to the dentatorubral pallidoluysian atrophy (DRPLA) gene, *J. Med. Genet.*, 33(1), 80, 1996.

136. Hartung, E., Zwei Faelle von Paramyoclonus multiplex mit Epilepsie, *Z. Ges. Neurol. Psychiat.*, 56, 150, 1920.

137. Johnson, W.G., Kugler, S.L., Stenroos, E.S., Meulener, M.C., Rangwalla, I., Johnson, T.W., and Mandelbaum, D.E., Pedigree analysis in families with febrile seizures, *Am. J. Med. Genet.*, 61(4), 345, 1996.

138. Aird, R.B., The importance of seizure-inducing factors in the control of refractory forms ofepilepsy, *Epilepsia*, 24(5), 567, 1983.

139. Stern, J.M., The epilepsy of trisomy 9p, *Neurology*, 47(3), 821, 1996.

140. Guerrini, R., Bureau, M., Mattei, M.G., Battaglia, A., Galland, M.C., and Roger, J., Trisomy 12p syndrome: achromosomal disorder associated with generalized 3-Hz spike and wave discharges, *Epilepsia*, 31(5), 557, 1990.

141. Taysi, K., Devivo, D.C., and Sekhon, G.S., Partial trisomy 15 and intractable seizures, *Acta. Paediatrica. Scand.*, 68(3), 445, 1979.

142. Carpentier, S., Lebon, P., Bonis, A., and Lhermitte, F., Brain cell chromosome translocation (2q;14q) associated with continuous partial epilepsy, *Lancet*, 1(8287), 1473, 1982.

143. Bundey, S., Hardy, C., Vickers, S., Kilpatrick, M.W., and Corbett, J.A., Duplication of the 15q11-13 region in a patient with autism, epilepsy and ataxia, *Dev. Med. Child Neurol.*, 36(8), 736, 1994.

144. Bingham, P.M., Spinner, N.B., Sovinsky, L., Zackai, E.H., and Chance, P.F., Infantile spasms associated with proximal duplication of chromosome 15q, *Pediatr. Neurol.*, 15(2), 163, 1996.

145. Iannetti, P., Spalice, A., Mingarelli, R., Raucci, U., Novelli, A., and Dallapiccola, B., Myoclonic epilepsy, neuroblast migration disorders, and maternally derived partial duplication 14q/deletion 15q, *Ann. Genet.*, 39(1), 26, 1996.

146. Vaughn, B.V., Greenwood, R.S., Aylsworth, A.S., and Tennison, M.B., Similarities of EEG and seizures in del(1q) and benign rolandic epilepsy, *Pediatr. Neurol.*, 15(3), 261, 1996.

147. Kishino, T., Lalande, M., and Wagstaff, J., UBE3A/E6-AP mutations cause Angelman syndrome, *Nat. Genet.*, 15(1), 70, 1997.

148. Kuwano, A., Mutirangura, A., Dittrich, B., Buiting, K., Horsthemke, B., Saitoh, S., Niikawa, N., Ledbetter, S.A., Greenberg, F., Chinault, A.C. et al., Molecular dissection of the Prader-Willi/Angelman syndrome region (15q11-13) by YAC cloning and FISH analysis, *Hum. Mol. Genet.*, 1(6), 417, 1992.

149. Viani, F., Romeo, A., Viri, M., Mastrangelo, M., Lalatta, F., Selicorni, A., Gobbi, G., Lanzi, G., Bettio, D., Briscioli, V. et al., Seizure and EEG patterns in Angelman's syndrome, *J. Child Neurol.*, 10(6), 467, 1995.

150 Battaglia, A., Orsitto, E., and Gibilisco, G., Mental retardation, epilepsy, short stature, and skeletal dysplasia: confirmation of the Gurrieri syndrome, *Am. J. Med. Genet.*, 62(3), 230, 1996.

151. Gurrieri, F., Sammito, V., Bellussi, A., and Neri, G., New autosomal recessive syndrome of mental retardation, epilepsy, short stature, and skeletal dysplasia, *Am. J. Med. Genet.*, 44(3), 315, 1992.

152. Ansink, B.J., Sarphatie, H., and van Dongen, H.R., The Landau-Kleffner syndrome — case report and theoretical considerations, *Neuropediatrics*, 20(3), 170, 1989.

153. Feekery, C.J., Parry-Fielder, B., and Hopkins, I.J., Landau-Kleffner syndrome: six patients including discordant monozygotic twins, *Pediatr. Neurol.*, 9(1), 49, 1993.

154. Pavone, L., Mollica, F., Incorpora, G., and Pampiglione, G., Infantile spasms syndrome in monozygotic twins. A 7-year follow-up, *Ital. J. Neurolog. Sci.*, 6(4), 503, 1985.

155. Moog, U., de Die-Smulders, C., Systermans, J.M., and Cobben, J.M., Oculocerebrocutaneous syndrome: report of three additional cases and aetiological considerations, *Clin. Genet.*, 52(4), 219, 1997.

156. Borjeson, M., Forssman, H., and Lehmann, O., An X-linked, recessively inherited syndrome characterized by grave mental deficiency, epilepsy, and endocrine disorder, *Acta. Med. Scand.*, 171, 13, 1962.

157. Turner, G., Gedeon, A., Mulley, J., Sutherland, G., Rae, J., Power, K., and Arthur, L., The Borjeson-Forssman-Lehmann syndrome:clinical manifestations and gene localization to Xq26-27, *Am. J. Med. Genet.*, 24, 463, 1989.

158. Aicardi, J., Chevrie, J. J., and Rousselie, F., Le syndrome spasmes en flexion, agenesic calleuse, anomalies chorio-retiniennes, *Arch. Franc. Pediat.*, 26, 1103, 1969.

159. Neidich, J.A., Nussbaum, R.L., Packer, R.J., Emanuel, B.S., and Puck, J.M., Heterogeneity of clinical severity and molecular lesions in Aicardi syndrome, *J. Pediatr.*, 116(6), 911, 1990.

160. Ropers, H.H., Zuffardi, O., Bianchi, E., and Tiepolo, L., Agenesis of corpus callosum, ocular, and skeletal anomalies (X-linked dominant Aicardi's syndrome) in a girl with balanced X/3 translocation, *Hum. Genet.*, 61(4), 364, 1982.

161. Herrmann, C., Jr., Aguilar, M.J., and Sacks, O.W., Hereditary photomyoclonus associated with diabetes mellitus, deafness, nephropathy, and cerebral dysfunction, *Neurology*, 14, 212, 1964.

162. Rett, A., Cerebral atrophy associated with hyperammonaemia, in *Handbook of Clinical Neurology*, Vinken, P.J. and Bruyn, G.W., Eds., North Holland, Amsterdam, 1977, 305.

163. Miyamoto, A., Yamamoto, M., Takahashi, S., and Oki, J., Classical Rett syndrome in sisters: variability of clinical expression, *Brain Dev.*, 19(7), 492, 1997.

164. Migeon, B.R., Dunn, M.A., Thomas, G., Schmeckpeper, B.J., and Naidu, S., Studies of X inactivation and isodisomy in twins provide further evidence that the X chromosome is not involved in Rett syndrome, *Am. J. Hum. Genet.*, 56(3), 647, 1995.

165. Chitty, L.S., Robb, S., Berry, C., Silver, D., and Baraitser, M., PEHO or PEHO-like syndrome?, *Clin. Dysmorphol.*, 5(2), 143, 1996.

166. Haltia, M. and Somer, M., Infantile cerebello-optic atrophy. Neuropathology of the progressive encephalopathy syndrome with edema, hypsarrhythmia and optic atrophy (the PEHO syndrome), *Acta. Neuropathol.*, 85(3), 241, 1993.

167. Paine, R.S., Evaluation of familial biochemically determined mental retardation in children, with special reference to aminoaciduria, *N. Engl. J. Med.*, 262, 658, 1960.

168. Alpers, B.J., Diffuse Progressive degeneration of gray matter of cerebrum, *Arch. Neurol. Psychiat.*, 25, 469, 1931.

169. Frydman, M., Jager-Roman, E., de Vries, L., Stoltenburg-Didinger, G., Nussinovitch, M., and Sirota, L., Alpers progressive infantile neuronal poliodystrophy: an acute neonatal form with findings of the fetal akinesia syndrome, *Am. J. Med. Genet.*, 47(1), 31, 1993.

170. Flynn, P. and Aird, R.B., A neuroectodermal syndrome of dominant inheritance, *J. Neurolog. Sci.*, 2(2), 161, 1965.

171. Andermann, F., Berkovic, S., Carpenter, S., and Andermann, E., The Ramsay Hunt syndrome is no longer a useful diagnostic category, *Move. Disord.*, 4(1), 13, 1989.

172. Marsdan, C.D. and Obeso, J.A., Viewpoints on the Ramsay Hunt syndrome: 1. The Ramsay Hunt syndrome is a useful clinical entity, *Move. Disord.*, 4, 6, 1989.

173. Skre, H. and Loken, A.C., Myoclonus epilepsy and subacute presenile dementia in heredo-ataxia, *Acta. Neurolog. Scand.*, 46(1), 18, 1970.

174. Latham, A.D. and Munro, T.A., Familial myoclonus epilepsy associated with deaf-mutism in a family showing other psychobiological abnormalities, *Ann. Eugen.*, 8, 166, 1937.

175. Koide, R., Ikeuchi, T., Onodera, O., Tanaka, H., Igarashi, S., Endo, K., Takahashi, H., Kondo, R., Ishikawa, A., Hayashi, T. et al., Unstable expansion of CAG repeat in hereditary dentatorubral-pallidoluysian atrophy (DRPLA), *Nat. Genet.*, 6(1), 9, 1994.

176. Takano, T., Yamanouchi, Y., Nagafuchi, S., and Yamada, M., Assignment of the dentatorubral and pallidoluysian atrophy (DRPLA) gene to 12p 13.31 by fluorescence in situ hybridization, *Genomics*, 32(1), 171, 1996.

177. Szepetowski, P., Rochette, J., Berquin, P., Piussan, C., Lathrop, G.M., and Monaco, A.P., Familial infantile convulsions and paroxysmal choreoathetosis: a new neurological syndrome linked to the pericentromeric region of human chromosome 16, *Am. J. Hum. Genet.*, 61(4), 889, 1997.

178. Shashi, V., Zunich, J., Kelly, T.E., and Fryburg, J.S., Neuroectodermal (CHIME) syndrome: an additional case with long term follow up of all reported cases, *J. Med. Genet.,* 32(6), 465, 1995.

179. Zunich, J. and Kaye, C.I., New syndrome of congenital ichthyosis with neurologic abnormalities, *Am. J. Med. Genet.,* 15(2), 331, 335, 1983.

180. Cantwell, R.J., Congenital sensori-neural deafness associated with onycho-osteo dystrophy and mental retardation (DOOR syndrome), *Humangenetik,* 26(3), 261, 1975.

181. Winter, R.M., Eronen syndrome identical with DOOR syndrome?, *Clin. Genet.,* 43(3), 167, 1993.

182. Juberg, R. C. and Marsidi, I., A new form of X-linked mental retardation with growth retardation, deafness, and microgenitalism, *Am. J. Hum. Genet.,* 32, 714, 1980.

183. Munke, M., Kruse, K., Goos, M., Ropers, H.H., and Tolksdorf, M., Genetic heterogeneity of the ichthyosis, hypogonadism, mental retardation, and epilepsy syndrome. Clinical and biochemical investigations on two patients with Rud syndrome and review of the literature, *Eur. J. Pediatr.,* 141(1), 8, 1983.

184. Wisniewski, K., Levis, A. R., and Shanske, A. L., X-linked inheritance of the Rud syndrome (Abst.), *Am. J. Hum. Genet.,* 37, A83, 1985.

185. Moynahan, E.J., Familial congenital alopecia, epilepsy, mental retardation with unusual electroencephalograms, *Proc. Roy. Soc. Med.,* 55, 411, 1962.

186. van Haeringen, A., Hurst, J.A., Savidge, R., and Baraitser, M., A familial syndrome of microcephaly, sparse hair, mental retardation, and seizures, *J. Med. Genet.,* 27(2), 127, 1990.

187. Timar, L., Czeizel, A.E., and Koszo, P., Association of Shokeir syndrome (congenital universal alopecia, epilepsy, mental subnormality and pyorrhea) and giant pigmented nevus, *Clin. Genet.,* 44(2), 76, 1993.

188. Hitch, J.M., Callaway, J.L., and Moseley, V., Familiar Darier's disease (keratosis follicularis), *Sth. Med. J.,* 34, 578, 1941.

189. Monk, S., Sakuntabhai, A., Carter, S.A., Bryce, S.D., Cox, R., Harrington, L., Levy, E., Ruiz-Perez, V.L., Katsantoni, E., Kodvawala, A., Munro, C.S., Burge, S., Larregue, M., Nagy, G., Rees, J.L., Lathrop, M., Monaco, A.P., Strachan, T., 'Hovnanian, A., Refined genetic mapping of the darier locus to a <1-cM region of chromosome 12q24.1, and construction of a complete, high-resolution P1 artificial chromosome/bacterial artificial chromosome contig of the critical region, *Am. J. Hum. Genet.,* 62(4), 890, 1998.

190. Friedman, J.M. and Birch, P.H., Type 1 neurofibromatosis: a descriptive analysis of the disorder in 1,728 patients, *Am. J. Med. Genet.,* 70(2), 138, 1997.

191 Sahl, W.J., Jr., Familial nevus sebaceus of Jadassohn: occurrence in three generations, *J. Am. Acad. Dermatol.,* 22(5 Pt 1), 853, 1990.

192. Gunther, M. and Penrose, L.S., Gentics of epiloia, *J. Genet.,* 31, 413, 1935.

193. Jones, A.C., Daniells, C.E., Snell, R.G., Tachataki, M., Idziaszczyk, S.A., Krawczak, M., Sampson, J.R., and Cheadle, J.P., Molecular genetic and phenotypic analysis reveals differences between TSC1 and TSC2 associated familial and sporadic tuberous sclerosis, *Hum. Mol. Genet.,* 6(12), 2155, 1997.

194. Flanagan, N., O'Connor, W.J., McCartan, B., Miller, S., McMenamin, J., and Watson, R., Developmental enamel defects in tuberous sclerosis: a clinical genetic marker?, *J. Med. Genet.,* 34(8), 637, 1997.

195. van Slegtenhorst, M., Nellist, M., Nagelkerken, B., Cheadle, J., Snell, R., van den Ouweland, A., Reuser, A., Sampson, J., Halley, D., and van der Sluijs, P., Interaction between hamartin and tuberin, the TSC1 and TSC2 gene products, *Hum. Mol. Genet.,* 7(6), 1053, 1998.

196. van Slegtenhorst, M., de Hoogt, R., Hermans, C., Nellist, M., Janssen, B., Verhoef, S., Lindhout, D., van den Ouweland, A., Halley, D., Young, J., Burley, M., Jeremiah, S., Woodward, K., Nahmias, J., Fox, M., Ekong, R., Osborne, J., Wolfe, J., Povey, S., Snell, R.G,. Cheadle, J.P., Jones, A.C., Tachataki, M., Ravine, D., Kwiatkowski, D.J. et al., Identification of the tuberous sclerosis gene TSC1 on chromosome 9q34, *Science,* 277(5327), 805, 1997.

197. al-Gazali, L.I., Farndon, P., Burn, J., Flannery, D.B., Davison, C., and Mueller, R.F., The Schinzel-Giedion syndrome, *J. Med. Genet.,* 27(1), 42, 1990.

198. Schinzel, A. and Giedion, A., A syndrome of severe midface retraction, multiple skull anomalies, clubfeet, and cardiac and renal malformations in sibs, *Am. J. Med. Genet.,* 1(4), 361, 1978.

199. Brunelli, S., Faiella, A., Capra, V., Nigro, V., Simeone, A., Cama, A., and Boncinelli, E., Germline mutations in the homeobox gene EMX2 in patients with severe schizencephaly, *Nat. Genet.,* 12(1):94-6, 1996.

200. des Portes, V., Pinard, J.M., Billuart, P., Vinet, M.C., Koulakoff, A., Carrie, A., Gelot, A., Dupuis, E., Motte, J., Berwald-Netter, Y., Catala, M., Kahn, A., Beldjord, C., and Chelly, J., A novel CNS gene required for neuronal migration and involved in X-linked subcortical laminar heterotopia and lissencephaly syndrome, *Cell*, 92(1), 51, 1998.

201. Gleeson, J.G., Allen, K.M., Fox, J.W., Lamperti, E.D., Berkovic, S., Scheffer, I., Cooper, E.C., Dobyns, W.B., Minnerath, S.R., Ross, M.E., and Walsh, C.A., Doublecortin, a brain-specific gene mutated in human X-linked lissencephaly and double cortex syndrome, encodes a putative signaling protein, *Cell*, 92(1), 63, 1998.

202. Fink, J.M., Dobyns, W.B., Guerrini, R., and Hirsch, B.A., Identification of a duplication of Xq28 associated with bilateral periventricular nodular heterotopia, *Am. J. Hum. Genet.*, 61(2), 379, 1997.

203. Jardine, P.E., Clarke, M.A., and Super, M., Familial bilateral periventricular nodular heterotopia mimics tuberous sclerosis, *Arch. Dis. Childhood*, 74(3), 244, 1996.

204. Shoffner, J.M., Lott, M.T., Lezza, A.M., Seibel, P., Ballinger, S.W., and Wallace, D.C., Myoclonic epilepsy and ragged-red fiber disease (MERRF) is associated with a mitochondrial DNA tRNA(Lys) mutation, *Cell*, 61(6), 931, 1990.

205. Kobayashi, Y., Momoi, M.Y., Tominaga, K., Momoi, T., Nihei, K., Yanagisawa, M., Kagawa, Y., and Ohta, S., A point mutation in the mitochondrial tRNA(Leu)(UUR) gene in MELAS (mitochondrial myopathy, encephalopathy, lactic acidosis and stroke-like episodes), *Biochem. Biophysic. Res. Comm.*, 173(3), 816, 1990

206. Nakamura, M., Nakano, S., Goto, Y., Ozawa, M., Nagahama, Y., Fukuyama, H., Akiguchi, I., Kaji, R., and Kimura, J., A novel point mutation in the mitochondrial tRNA(Ser(UCN)) gene detected in a family with MERRF/MELAS overlap syndrome, *Biochem. Biophys. Res. Comm.*, 214(1), 86, 1995.

207. Terheggen, H.G., Schwenk, A., Lowenthal, A., Van Sande, M., and Colombo, J.P., Argininaemia with arginase deficiency, *Lancet*, II, 748, 1969.

208. Sparkes, R. S., Dizikes, G. J., Klisak, I., Grody, W. W., Mohandas, T., Heinzmann, C., Zollman, S., Lusis, A. J., and Cederbaum, S. D., The gene for human liver arginase (ARG1) is assigned to chromosome band 6q23, *Am. J. Hum. Genet.*, 39, 186, 1986.

209. Isobe, M., Koyata, H., Sakakibara, T., Momoi-Isobe, K., and Hiraga, K., Assignment of the true and processed genes for human glycine decarboxylase to 9p23-24 and 4q12, *Biochem. Biophys. Res. Commun.*, 20: 1483-1487,1994.

210. Kalaydjieva, L., Dworniczak, B., Kucinskas, V., Yurgeliavicius, V., Kunert, E., and Horst, J., Geographical distribution gradients of the major PKU mutations and the linked haplotypes, *Hum. Genet.*, 86(4), 411, 1991.

211. Serratosa, J.M. and Delgado-Escueta, A.V., Mapping Human Epilepsy Genes: Implications for the Treatment of Epilepsy, *CNS Drugs*, 5(3), 155, 1996.

212. Sander, T., Hildmann, T., Kretz, R., Furst, R., Sailer, U., Bauer, G., Schmitz, B., Beck-Mannagetta, G., Wienker, T.F., and Janz, D., Allelic association of juvenile absence epilepsy with a GluR5 kainate receptor gene (GRIK1) polymorphism, *Am. J. Med. Genet.*, 74(4), 416, 1997.

213. Sainz, J., Minassian, B.A., Serratosa, J.M., Gee, M.N., Sakamoto, L.M., Iranmanesh, R., Bohlega, S., Baumann, R.J., Ryan, S., Sparkes, R.S., and Delgado-Escueta, A.V., Lafora progressive myoclonus epilepsy: narrowing the chromosome 6q24 locus by recombinations and homozygosities, *Am. J. Hum. Genet.*, 61(5), 1205, 1997.

214. Wallace, R.H., Berkovic, S.F., Howell, R.A., Sutherland, G.R., and Mulley, J.C., Suggestion of a major gene for familial febrile convulsions mapping to 8q13-21, *J. Med. Genet.*, 33(4), 308, 1996.

215. Johnson, E.W., Dubovsky, J., Rich, S.S., O'Donovan, C.A., Orr, H.T., Anderson, V.E., Gil-Nagel, A., Ahmann, P., Dokken, C.G., Schneider, D.T., and Weber, J.L., Evidence for a novel gene for familial febrile convulsions, FEB2, linked to chromosome 19p in an extended family from the Midwest, *Hum. Mol. Genet.*, 7(1), 63, 1998.

216. Guipponi, M., Rivier, F., Vigevano, F., Beck, C., Crespel, A., Echenne, B., Lucchini, P., Sebastianelli, R., Baldy-Moulinier, M., and Malafosse, A., Linkage mapping of benign familial infantile convulsions (BFIC) to chromosome 19q, *Hum. Mol. Genet.*, 6(3), 473, 1997.

217. Ottman, R., Risch, N., Hauser, W.A., Pedley, T.A., Lee, J.H., Barker-Cummings, C., Lustenberger, A., Nagle, K.J., Lee, K.S., Scheuer, M.L. et al., Localization of a gene for partial epilepsy to chromosome 10q, *Nat. Genet.*, 10(1), 56, 1995.

218. Zara, F., Bianchi, A., Avanzini, G., Di Donato, S., Castellotti, B., Patel, P.I., and Pandolfo, M., Mapping of genes predisposing to idiopathic generalized epilepsy, *Hum. Mol. Genet.,* 4(7), 1201, 1995.

219. Tahvanainen, E., Ranta, S., Hirvasniemi, A., Karila, E., Leisti, J., Sistonen, P., Weissenbach, J., Lehesjoki, A.E., and de la Chapelle, A., The gene for a recessively inherited human childhood progressive epilepsy with mental retardation maps to the distal short arm of chromosome 8, *Proc. Natl. Acad. Sci. U.S.A.,* 91(15), 7267, 1994.

220. Ranta, S., Lehesjoki, A.E., Hirvasniemi, A., Weissenbach, J., Ross, B., Leal, S.M., de la Chapelle, A., and Gilliam, T.C., Genetic and physical mapping of the progressive epilepsy with mental retardation (EPMR) locus on chromosome 8p, *Genome Res.,* 6(5), 351, 1996.

221. Gendrot, C., Ronce, N., Toutain, A., Moizard, M.P., Muh, J.P., Raynaud, M., Dourlens, J., Briault, S., and Moraine, C., X-linked mental retardation exhibiting linkage to DXS255 and PGKP1: a new MRX family (MRX14) with localization in the pericentromeric region, *Clin. Genet.,* 45(3), 145, 1994.

222. Ryan, S.G., Chance, P.F., Zou, C.H., Spinner, N.B., Golden, J.A., and Smietana, S., Epilepsy and mental retardation limited to females: an X-linked dominant disorder with male sparing, *Nat. Genet.,* 17(1), 92, 1997.

223. Kohlschutter, A., Chappuis, D., Meier, C., Tonz, O., Vassella, F., and Herschkowitz, N., Familial epilepsy and yellow teeth — a disease of the CNS associated with enamel hypoplasia, *Helvetica Paediatrica. Acta.,* 29(4):283-94, 1974.

224. Petermoller, M., Kunze, J., and Gross-Selbeck, G., Kohlschutter syndrome: syndrome of epilepsy — dementia — amelogenesis imperfecta, *Neuropediatrics,* 24(6), 337, 1993.

225. Jilek-Aall, L., Jilek, W., and Miller, J.R., Clinical and genetic aspects of seizure disorders prevalent in an isolated African population, *Epilepsia,* 20(6), 613, 1979.

226. Neuman, R.J., Kwon, J.M., Jilek-Aall, L., Rwiza, H.T., Rice, J.P., and Goodfellow, P.J., Genetic analysis of kifafa, a complex familial seizure disorder, *Am. J. Hum. Genet.,* 57(4), 902, 1995.

227. Casteels, I., Spileers, W., Swinnen, T., Demaerel, P., Silberstein, J., Casaer, P., and Missotten, L., Optic atrophy as the presenting sign in Hallervorden-Spatz syndrome, *Neuropediatrics,* 25(5), 265, 1994.

228. Taylor, T.D., Litt, M., Kramer, P., Pandolfo, M., Angelini, L., Nardocci, N., Davis, S., Pineda, M., Hattori, H., Flett, P.J., Cilio, M.R., Bertini, E., and Hayflick, S.J., Homozygosity mapping of Hallervorden-Spatz syndrome to chromosome 20p12.3-p13, *Nat. Genet.,* 14(4), 479, 1996.

229. Gobbi, G., Bouquet, F., Greco, L., Lambertini, A., Tassinari, C.A., Ventura, A., and Zaniboni, M.G., Coeliac disease, epilepsy, and cerebral calcifications. The Italian Working Group on Coeliac Disease and Epilepsy, *Lancet,* 340(8817), 439, 1992.

230. de Pina-Neto, J.M., de Souza, N.V., Velludo, M.A., Perosa, G.B., de Freitas, M.M., and Colafemina, J.F., Retinal changes and tumorigenesis in Ramon syndrome: follow-up of a Brazilian family, *Am. J. Med. Genet.,* 77(1), 43, 1998.

231. Ramon, Y., Berman, W., and Bubis, J.J., Gingival fibromatosis combined with cherubism, *Oral Surg., Oral Med., Oral Pathol.,* 24(4), 435, 1967.

232. Manon-Espaillat, R., Gothe, B., Adams, N., Newman, C., and Ruff, R., Familial 'sleep apnea plus' syndrome: report of a family *Neurology,* 38(2), 190, 1988.

233. Patton, M.A., Baraitser, M., and Brett, E.M., A family with congenital suprabulbar paresis (Worster-Drought syndrome), *Clin. Genet.,* 29(2), 147, 1986.

234. Tahvanainen, E., Norio, R., Karila, E., Ranta, S., Weissenbach, J., Sistonen, P., and de la Chapelle, A., Cohen syndrome gene assigned to the long arm of chromosome 8 by linkage analysis, *Nat. Genet.,* 7(2), 201, 1994.

235. North, C., Patton, M.A., Baraitser, M., and Winter, R.M., The clinical features of the Cohen syndrome: further case reports, *J. Med. Genet.,* 22(2), 131, 1985.

236. Kolehmainen, J., Norio, R., Kivitie-Kallio, S., Tahvanainen, E., de la Chapelle, A., and Lehesjoki, A.E., Refined mapping of the Cohen syndrome gene by linkage disequilibrium, *Eur. J. Hum. Genet.,* 5(4), 206, 1997.

237. Rasmussen, T., Olszewiski, J., and Lloyd-Smith, D., Focal seizures due to chronic localized encephalitis, *Neurology,* 8, 435, 1958.

238. Rogers, S.W., Andrews, P.I., Gahring, L.C., Whisenand, T., Cauley, K., Crain, B., Hughes, T.E., Heinemann, S.F., and McNamara, J.O., Autoantibodies to glutamate receptor GluR3 in Rasmussen's encephalitis, *Science,* 265(5172), 648, 1994.

239. Feigenbaum, A., Bergeron, C., Richardson, R., Wherrett, J., Robinson, B., and Weksberg, R., Premature atherosclerosis with photomyoclonic epilepsy, deafness, diabetes mellitus, nephropathy, and neurodegenerative disorder in two brothers: a new syndrome?, *Am. J. Med. Genet.*, 49(1), 118, 1994.

240. Devriendt, K., Van den Berghe, H., and Fryns, J.P., Alopecia-mental retardation syndrome associated with convulsions and hypergonadotropic hypogonadism, *Clin. Genet.*, 49(1), 6, 1996.

241. Scheithauer, B.W., Forno, L.S., Dorfman, L.J., and Kane, C.A., Neuroaxonal dystrophy (Seitelberger's disease) with late onset, protracted course and myoclonic epilepsy, *J. Neurolog. Sci.*, 36(2), 247, 1978.

242. Plewes, J.L. and Jacobson, I., Familial frontonasal dermoid cysts. Report of four cases, *J. Neurosurg.*, 34(5), 683, 1971.

243. Hattori, H., Higuchi, Y., Maihara, T., Jung, E.Y., Furusho, K., and Asato, R., Congenital bilateral perisylvian syndrome: first report in a Japanese patient, *Jpn. J. Hum. Genet.*, 41(1), 189, 1996.

244. Kuzniecky, R., Andermann, F., and Guerrini, R., Congenital bilateral perisylvian syndrome: study of 31 patients. The CBPS Multicenter Collaborative Study, *Lancet*, 341(8845), 608, 1993.

245. Dubovsky, J., Zabramski, J.M., Kurth, J., Spetzler, R.F., Rich, S.S., Orr, H.T., and Weber, J.L., A gene responsible for cavernous malformations of the brain maps to chromosome 7q, *Hum. Mol. Genet.*, 4(3), 453, 1995.

246. Tortorella, G., Magaudda, A., Mercuri, E., Longo, M., and Guzzetta, F., Familial unilateral and bilateral occipital calcifications and epilepsy, *Neuropediatrics*, 24(6), 341, 1993.

247. Goodyer, P.R. and Lancaster, G.A., Inherited lactic acidosis: correction of the defect in cultured fibroblasts, *Pediatr. Res.*, 18(11), 1144, 1984.

248. Bejsovec, M., Kulenda, Z., and Ponca, E., Familial intrauterine convulsions in pyridoxine dependency, *Arch. Dis. Childhood*, 42(222), 201, 1967.

249. Goutieres, F. and Aicardi, J., Atypical presentations of pyridoxine-dependent seizures: a treatable cause of intractable epilepsy in infants, *Ann. Neurol.*, 17(2), 117, 1985.

250. Krietsch, W.K., Krietsch, H., Kaiser, W., Dunnwald, M., Kuntz, G.W., Duhm, J., and Bucher, T., Hereditary deficiency of phosphoglycerate kinase: a new variant in erythrocytes and leucocytes, not associated with haemolytic anaemia, *Eur. J. Clin. Invest.*, 7(5), 427, 1977.

251. Sugie, H., Sugie, Y., Ito, M., and Fukuda, T., A novel missense mutation (837T→C) in the phospho-glycerate kinase gene of a patient with a myopathic form of phosphoglycerate kinase deficiency, *J. Child Neurol.*, 13(2), 95, 1998.

252. Jaeken, J., Detheux, M., Van Maldergem, L., Frijns, J.P., Alliet, P., Foulon, M., Carchon, H., and Van Schaftingen, E., 3-Phosphoglycerate dehydrogenase deficiency and 3-phosphoserine phosphatase deficiency: inborn errors of serine biosynthesis, *J. Inher. Metab. Dis.*, 19(2), 223, 1996.

253. Otero, L.J., Brown, G.K., Silver, K., Arnold, D.L., and Matthews, P.M., Association of cerebral dysgenesis and lactic acidemia with X-linked PDH E1 alpha subunit mutations in females, *Pediatr. Neurol.*, 13(4), 327, 1995.

254. Stockler, S., Isbrandt, D., Hanefeld, F., Schmidt, B., and von Figura, K., Guanidinoacetate methyl-transferase deficiency: the first inborn error of creatine metabolism in man, *Am. J. Hum. Genet.*, 58(5), 914, 1996.

255. Jaeken, J., Van den Bergh, F., Vincent, M.F., Casaer, P., and Van den Berghe, G., Adenylosuccinase deficiency: a newly recognized variant, *J. Inher. Metab. Dis.*, 15(3), 416, 1992.

256. Fon, E.A., Demczuk, S., Delattre, O., Thomas, G., and Rouleau, G.A., Mapping of the human adenylosuccinate lyase (ADSL) gene to chromosome 22q13.1→q13.2, *Cytogenet. Cell Genet.*, 64(3–4), 201, 1993.

257. Maaswinkel-Mooij, P.D., Laan, L.A., Onkenhout, W., Brouwer, O.F., Jaeken, J., and Poorthuis, B.J., Adenylosuccinase deficiency presenting with epilepsy in early infancy, *J. Inher. Metab. Dis.*, 20(4), 606, 1997.

258. Nyhan, W.L., Shelton, G.D., Jakobs, C., Holmes, B., Bowe, C., Curry, C.J., Vance, C., Duran, M., and Sweetman, L., D-2-hydroxyglutaric aciduria, *J. Child Neurol.*, 10(2), 137, 1995.

259. Barth, P.G., Hoffmann, G.F., Jaeken, J., Lehnert, W., Hanefeld, F., van Gennip, A.H., Duran, M., Valk, J., Schutgens, R.B., Trefz, F.K. et al., L-2-hydroxyglutaric acidemia: a novel inherited neurometabolic disease, *Ann. Neurol.*, 32(1), 66, 1992.

260. Chabrol, B., Mancini, J., Chretien, D., Rustin, P., Munnich, A., and Pinsard, N., Valproate-induced hepatic failure in a case of cytochrome c oxidase deficiency, *Eur. J. Pediatr.*, 153(2), 133, 1994.

261. Parfait, B., Percheron, A., Chretien, D., Rustin, P., Munnich, A., and Rotig, A., No mitochondrial cytochrome oxidase (COX) gene mutations in 18 cases of COX deficiency, *Hum. Genet.,* 101, 247, 1997.

262. Holt, I.J., Harding, A.E., Petty, R.K., and Morgan-Hughes, J.A., A new mitochondrial disease associated with mitochondrial DNA heteroplasmy, *Am. J. Hum. Genet.,* 46(3), 428, 1990.

263. de Coo, I.F., Smeets, H.J., Gabreels, F.J., Arts, N., van and Oost, B.A., Isolated case of mental retardation and ataxia due to a de novo mitochondrial T8993G mutation, *Am. J. Hum. Genet.,* 58(3), 636, 1996.

264. Yamamoto, A., Adachi, S., Kawamura, S., Takahashi, M., and Kitani, T., Localized beta-galactosidase deficiency. Occurrence in cerebellar ataxia with myoclonus epilepsy and macular cherry-red spot — a new variant of GM1-gangliosidosis?, *Arch. Intern. Med.,* 134(4), 627, 1974.

265. Pereira, L. V., Desnick, R. J., Adler, D. A., Disteche, C. M., and Schuchman, E. H., Regional assignment of the human acid sphingomyelinase gene (SMPD1) by PCR analysis of somatic cell hybrids and in situ hybridization to11p15.1-p15.4, *Genomics,* 9, 229, 1991.

266. Takahashi, T., Suchi, M., Desnick, R. J., Takada, G., and Schuchman, E. H., Identification and expression of five mutations in the human acid sphingomyelinase gene causing types A and B Niemann-Pick disease:molecular evidence for genetic heterogeneity in the neuronopathic and non-neuronopathic forms, *J. Biol. Chem.,* 267, 12552, 1992.

267. Callen, D.F., Baker, E., Lane, S., Nancarrow, J., Thompson, A., Whitmore, S.A., MacLennan, D.H., Berger, R,. Cherif, D., Jarvela, I. et al., Regional mapping of the Batten disease locus (CLN3) to human chromosome 16p12, *Am. J. Hum. Genet.,* 49(6), 1372, 1991.

268. Savukoski, M., Kestila, M., Williams, R., Jarvela, I., Sharp, J., Harris, J., Santavuori, P., Gardiner, M., and Peltonen, L., Defined chromosomal assignment of CLN5 demonstrates that at least four genetic loci are involved in the pathogenesis of human ceroid lipofuscinoses, *Am. J. Hum. Genet.,* 55(4), 695, 1994.

269. Savukoski, M., Klockars, T., Holmberg, V., Santavuori, P., Lander, E.S., and Peltonen, L., CLN5, a novel gene encoding a putative transmembrane protein mutated in Finnish variant late infantile neuronal ceroid lipofuscinosis, *Nat. Genet.,* 19(3), 286, 1998.

270. Mitchison, H.M., Hofmann, S.L., Becerra, C.H., Munroe, P.B., Lake, B.D., Crow, Y.J., Stephenson, J.B., Williams, R.E., Hofman, I.L., Taschner P.E.M., Martin, J.J., Philippart, M., Andermann, E., Andermann, F., Mole, S.E., Gardiner, R.M., and O'Rawe, A.M., Mutations in the palmitoyl-protein thioesterase gene (PPT; CLN1) causing juvenile neuronal ceroid lipofuscinosis with granular osmiophilic deposits, *Hum. Mol. Genet.,* 7(2), 291, 1998.

271. Bonten, E., van der Spoel, A., Fornerod, M., Grosveld, G., and d'Azzo, A., Characterization of human lysosomal neuraminidase defines the molecular basis of the metabolic storage disorder sialidosis, *Genes Dev.,* 10(24), 3156, 1996.

272. Goldberg, M.F., Cotlier, E., Fichenscher, L.G., Kenyon, K., Enat, R., and Borowsky, S.A., Macular cherry-red spot, corneal clouding, and -galactosidase deficiency. Clinical, biochemical, and electron microscopic study of a new autosomal recessive storage disease, *Arch. Intern. Med.,* 128(3), 387, 1971.

273. Mueller, O.T., Henry, W.M., Haley, L.L., Byers, M.G., and Eddy, R.L., Shows TB. Sialidosis and galactosialidosis: chromosomal assignment of two genes associated with neuraminidase-deficiency disorders, *Proc. Natl. Acad. Sci. U.S.A.,* 83(6), 1817, 1986.

274. Pennacchio, L.A., Bouley, D.M., Higgins, K.M., Scott, M.P., Noebels, J.L., and Myers, R.M., Progressive ataxia, myoclonic epilepsy and cerebellar apoptosis in cystatin B-deficient mice, *Nat. Genet.,* 20(3), 251, 1998.

275. Talbot, C.J., Nicod, A., Cherny, S.S., Fulker, D.W., Collins, A.C., and Flint, J., High resolution mapping if quantitative trait loci in outbred mice, *Nat. Genet.,* 21, 305, 1999.

276. Minassian, B.A., Lee, J.R., Herbrick, J.A., Huizenga, J., Soder, S., Mungall, A.J., Dunham, I., Gardner, R., Fong, C.Y., Carpenter, S., Jardim, L., Satishchandra, P., Andermann, E., Snead, O.C., 3rd, Lopes-Cendes, I., Tsui, L.C., Delgado-Escueta, A.V., Rouleau, G.A., and Scherer, S.W., Mutations in a gene encoding a novel protein tyrosine phosphatase cause progressive myoclonus epilepsy, *Nat. Genet.,* 20(2), 171, 1988.

277. Serratosa, J.M., Gomez-Garre, P., Gallardo, M.E., Anta, B., de Bernabe, D.B., Lindhout, D., Augustijn, P.B., Tassinari, C.A., Malafosse, R.M., Topcu, M., Grid, D., Dravet, C., Berkovic, S.F., and De Cardoba, S.R., A novel protein tyrosine phosphatase gene in mutated in progressive myoclonus epilepsy of the Lafora type (EPM2), *Hum. Mol. Genet.,* 8(2), 345, 1999.

278. Fong, G.C., Shah, P.U., Gee, M.N., Serratosa, J.M., Castroveijo, I.P., Khan, S., Ravat, S.H., Mani, J., Huang, Y., Zhao, H.Z., Medina, M.T., Treiman, L.J., Pineda, G., and Delgado-Escueta, A.V., Childhood absense epilepsy with tonic-clonic seizures and electroencephalogram 3-4-Hz spike and multispike-slow wave complexes: linkage to chromosome 8q24, *Am. J. Hum. Genet.,* 63(4), 1117, 1998.

279. Phillips, H.A., Scheffer, I.E., Crossland, K.M., Bhatia, K.P., Fish, D.R., Marsden, C.D., Howell, S.J., Stephenson, J.B., Tolmie, J., Plazzi, G., Eeg-Olofsson, O., Singh, R., Lopes-Cendes, I., Andermann, E., Andermann, F., Berkovic, S.F., and Mulley, J.C., Autosomal dominant nocturnal frontal-lobe epilepsy: genetic heterogeneity and evidence for a second locus at 15q24, *Am. J. Hum. Genet.,* 63(4), 1108, 1998.

280. Scheffer, I.E., Phillips, H.A., O'Brien, C.E., Saling, M.M., Wrennall, J.A., Wallace, R.H., Mulley, J.C., and Berkovic, S.F., Familial partial epilepsy with variable foci: a new partial epilepsy syndrome with suggestion of linkage to chromosome 2, *Ann. Neurol.,* 44(6), 890, 1998.

281. Neubauer, B.A., Fiedler, B., Himmelein, B., Kampfer, F., Lassker, U., Schwabe, G., Spanier, I., Tams, D., Bretscher, C., Moldenhauer, K., Kurlemann, G., Weise, S., Tedroff, K., Eeg-Olofsson, O., Wadelius, C., and Stephani, U., Centrotemporal spikes in familes with rolandic epilepsy: linkage to chromosome 15q14, *Neurology,* 51(6), 1608, 1998.

# 8 Genetics of Alcoholism

*Mary-Anne Enoch and David Goldman*

## CONTENTS

## INTRODUCTION

Alcohol consumption has been a pleasurable pastime for humans from antiquity to the present day. Remarkable ingenuity has been employed in the choice of indigenous raw materials for the fermentation of alcohol: from yak milk in the Caucasus to the agave in Mexico, from rice in Japan to palm sap in Papua New Guinea. There are few regions in the world without some form of alcoholic drink. Most people drink in a controlled fashion and derive social benefit from alcohol consumption; however, a certain number of individuals become addicted with often disasterous physical and social consequences. Approximately 90% of the adult population of the United States consumes alcohol at some time, whereas the lifetime prevalence of alcohol dependence, a severe form of alcoholism, is 8 to 14%.[1] Alcohol consumption is generally higher in Western countries than Southeast Asia and is markedly higher in several ancient cultures, including Native American in whom the lifetime prevalence of alcohol dependence in males can be as high as 85%.

Alcoholism is a complex multifactorial disease. However, advances in the understanding of the genetics and neurobiology of alcoholism are leading to the development of specific pharmacotherapies and in time, by using molecular diagnostic approaches, will hopefully produce more accurate targeting of therapies. Twin studies have established that the role of genetic variation in alcoholism is profound; however, twins with the same genotype often do not share the same behavior, suggesting that environmental interactions are involved. Cross-cultural and cross-generational differences in prevalence have confirmed the importance of availability and custom of use in alcoholism. Advances in the understanding of the neurobiology of addiction and relevant behaviors, such as anxiety and impulsivity, have identified multiple points where genetic variation could influence vulnerability to alcoholism. Over the past several decades, much work has been done to identify these genetic components, but the difficulties are great and progress has been slow in identifying roles in vulnerability for genes expressed in the brain. Recently, linkage analyses in humans and rodents have pointed to genomic regions harboring genes that influence alcoholism and associated behaviors. Risk-influencing variants of alcohol metabolic genes have been found by the case-control association approach.

## COMORBIDITY AND SHARED NEUROBIOLOGY — IS ALCOHOLISM VULNERABILITY A SUBSET OF ADDICTIVE VULNERABILITY?

Alcoholism is frequently comorbid with psychostimulant and opioid abuse disorders and with nicotine dependency.[2] The greater the severity of alcoholism, the more likely it is that other substance dependencies are present. More than 80% of alcoholics smoke cigarettes, and 70% are heavy smokers compared with 30% of the general population who smoke and 10% who smoke heavily.[3] This suggests that a portion of the genetic vulnerability to alcoholism may be in common with other addictive disorders with which the neurobiology is shared. Although ethanol and other addictive disorders affect a range of neurotransmitter systems in different regions of the brain, a pathway that appears to be crucial to the action of all addictive drugs is the mesolimbic dopamine system and certain other neurotransmitters and neurotransmitter systems, e.g., the serotonin, $GABA_A$, and opioid peptide systems. The mesolimbic dopamine pathway extends from the ventral tegmentum to the nucleus accumbens, with projections to the limbic system and the orbitofrontal cortex,[4] and is associated with the ability to feel pleasure. Ethanol appears to act at a variety of sites within the cell membrane probably inducing effects on neurotransmitter membrane receptors and ion channels as well as modulating neurotransmitter release.[5] Enhanced gamma-aminobutyric acid (GABA), glutamate (NMDA receptors), dopaminergic, opioid peptide, and serotoninergic neurotransmission have been associated with acute ethanol administration and potentially mediate some of alcohol's reinforcing effects.[6] The development of tolerance may be related to ethanol-induced adaptive changes in the $GABA_A$ receptor system, and changes in the NMDA receptor system may underlie withdrawal symptoms.[7] $GABA_A$ receptors are the primary site of action of benzodiazepines and barbiturates. The success of the opioid antagonist naltrexone in modifying drinking behavior, controlling craving, and preventing relapse in alcoholics indicates that opioid receptor-mediated mechanisms are activated by alcohol.[5] Opioid receptors are, of course, the primary site of action of opiate drugs. Selective serotonin re-uptake inhibitors play a role in modifying craving for alcohol and modify other behaviors, such as depression and anxiety, that appear to play a wider role in vulnerability to drug abuse.

## GENETIC INFLUENCES ON ALCOHOLISM: HERITABILITY STUDIES

Alcoholism tends to run in families. Having an alcoholic parent is a significant risk factor for the development of the disease: Children of alcoholics are five times more likely to develop alcohol-related problems than children of nonalcoholics.[8,9] There is a tendency for alcoholics to marry amongst themselves rather than with nonalcoholics (assortative mating),[10] suggesting that shared or mutually attracting behavioral traits or common environment are important. Studies of heritability, a measure of the genetic component of variance in interindividual vulnerability, indicate that genetic influences are substantially responsible for the observed patterns of familiality. Large, well-constructed twin studies[11-14] and adoption studies[13,15] demonstrate that genetic factors are important in determining vulnerability to alcoholism, particularly in the more severe forms of the disease.[16] In a study of 1030 female twin pairs ascertained from the Virginia Twin Registry, Kendler et al.[12] demonstrated that whether using narrow, intermediate, or broad definitions, the concordance for alcoholism was consistently higher in monozygotic than dizygotic twin pairs, and that the heritability of alcoholism was 0.5 to 0.6. In a study of 114 male twin pairs and 55 female twin pairs, Pickens et al.[11] found that the overall heritability was 0.35 in males and 0.24 in females. However, heritability was considerably higher for the more severe form of alcoholism studied (dependence): approximately 0.50 to 0.60 in both men and women. These heritability values were recently replicated in a study of 2685 twin pairs from the Australian twin registry.[14] Adoption studies have shown that alcoholism in the biological parents predicts alcoholism in their male children even when they are reared by unrelated adoptive parents.[15]

## GENETIC HETEROGENEITY

Individual differences in the clinical presentation of alcoholism suggest variation in mechanisms of vulnerability. Alcoholics vary in their drinking patterns, the severity of their symptoms, and the behavioral, physical, and psychiatric sequelae. Vulnerability factors found in some, but by no means all, alcoholics include attentional deficits reflected by the P300 event-related potential trait, anxiety reflected by the low voltage alpha EEG trait, and diminished subjective response to alcohol. Alcoholism may be present alone or may be comorbid with other drug abuse, major depression, bulimia, or anti-social personality disorder (ASPD), identifying subgroups on a clinical basis. Severe alcoholism, suicidality, and impulsivity often co-occur.[17]

### Clinical Subgroups

Because of the heterogeneity of alcoholism, careful delineation of phenotypes is a key to the piecemeal analysis of the genetic influences on this disease. Attempts have been made to classify alcoholics into more homogenous clinical groups by severity (dependence or abuse), withdrawal signs, tolerance, medical sequelae, or latent class phenotypes. For example, latent class analysis of the large COGA (National Institute on Alcohol Abuse and Alcoholism's Collaborative Study on Genetics of Alcoholism) family data set did reveal, in the three-class and four-class solutions, a group of alcoholics distinguished by a high frequency of withdrawal symptoms.[18] Cloninger[19] divided alcoholics on a clinical and genetic epidemiological basis into Type 1 (milieu-limited, later onset) and Type 2 (early onset, male-dominated, associated with ASPD). This influential classification linked dimensionally defined pre-morbid personality with alcoholism vulnerability and identified an alcoholism subtype, Type 2, with stronger inheritance. Alcoholics with impulsive behavior, and to some extent with ASPD, have lower levels of the serotonin metabolite 5-HIAA, in cerebrospinal fluid,[20] relating clinical phenotype to neurochemistry.

Genetic heterogeneity (the influence of different genetic factors in different individuals) is likely in alcoholism. In addition, polygenicity, a requirement for the combined action of variants at different genes, may be important. Polygenic models of alcoholism are supported by the observation that a simple Mendelian mode of inheritance is not observable in families, and by classic genetic analyses in rodents showing that alcohol-related phenotypes (e.g., alcohol sensitivity) are polygenic. On the other hand, twin data showing an approximately 2:1 monozygotic/dizygotic ratio of concordance for alcoholism and the high recurrence rate in first-degree relatives of alcoholics are less compatible with polygenic models. It is also important to recognize that the triggering of addiction may be influenced by interactions with multiple environmental components. Finally, some subtypes of disease may have a larger genetic component than others, like alcoholism associated with ASPD, for example.

In order to dissect the multiple genetic influences on alcoholism vulnerability, it may be necessary to consider several phenotypes representing different aspects of the disease. Valuable information can, and has been, gained by studying traits associated with alcoholism that may be markers for alcoholism genes.

## Electrophysiological Markers

Event-related potentials (ERPs) are electrical signals that are elicited in response to specific visual or auditory stimuli and are sensitive measures of cognitive brain activity. A robust and extensively replicated finding is that the P300 ERP is of lower amplitude in alcoholics and in young, nondrinking sons of alcoholic men[21] compared to controls. The low amplitude P300 may be co-transmitted with alcoholism in families.[22] Low voltage alpha (LVA) is a heritable variant of the resting EEG in which the alpha rhythm is absent or occurs infrequently with low amplitude. It is several times more common in alcoholics than in nonalcoholics and is particularly abundant in alcoholics with anxiety disorders.[23]

## Alcohol Sensitivity

In a series of studies, Schuckit[24] has measured subjective (feelings of intoxication) and nonsubjective (hormonal response and instability) responsiveness to alcohol in young, nonalcoholic men who drank little or no alcohol. These men were later re-interviewed in a long-term follow-up study. Lower response to modest doses of alcohol were found in sons of alcoholics. Also, the level of low response to alcohol predicted a fourfold increase in the risk of future alcoholism in young men, irrespective of family history.

## Neurochemical Markers

### Monoamine Oxidase (MAO)

It has been suggested by some researchers that the level of platelet MAO activity may be a marker for alcoholism vulnerability. MAO catalyses the oxidation of serotonin to 5-HIAA and dopamine to HVA. Many studies have shown that low platelet MAO-B activity is associated with alcoholism, particularly early onset, severe alcoholism in men. However, other well-executed studies have failed to replicate this finding.[25] Also, platelet MAO is influenced by many other factors, both intrinsic (e.g., gender, medical and psychiatric illness, personality traits) and extrinsic (e.g., nicotine and other drug use, recent alcohol intake). Therefore, it appears that platelet MAO is more of a state marker for behavior rather than a genetic marker for psychopathology underlying alcoholism.

### Adenylate Cyclase (AC)

AC is an intracellular enzyme involved in signal transduction and is therefore a potential site for genetic variation in neurotransmitter sensitivity. AC is modulated by G proteins, and after activation it produces cyclic adenosine monophosphate (cAMP), which serves as a second messenger within the cell. Drug-mediated reinforcement requires the interaction of neuronal dopamine receptors in the nucleus accumbens with the AC system. Several studies have found lower AC activity in the platelets and lymphocytes of abstinent alcoholics and in postmortem brains of alcoholics,[25] but it is not yet clear whether this is a trait rather than a state effect.

## DETERMINING THE GENETIC BASIS OF VULNERABILITY TO ALCOHOLISM

The analysis of complex traits is a piecemeal process, as the population-wide effects of variation at any single gene are likely to be small. To detect subtle genetic effects, large samples are needed. The four methods[26] most widely used are (a) linkage analysis: the elucidation of the inheritance pattern of phenotypes and genotypes in pedigrees; (b) allele sharing methods: affected relatives are compared to detect excess genotype sharing; (c) case-control association studies: unrelated affected and unaffected individuals are compared; (d) analysis of inbred, transgenic, and gene-knockout animals (mainly mice and rats).

## GENETIC STUDIES OF ALCOHOLISM IN HUMANS

### ALCOHOL METABOLISM AND THE FLUSHING RESPONSE

At the present time, the only genes for alcoholism that have been identified are protective, although it could be argued that in each case while one gene variant (allele) is a protective allele, the other allele is a vulnerability allele. Alcohol dehydrogenase (ADH) metabolizes ethanol to acetaldehyde, which is, in turn, converted to acetate by aldehyde dehydrogenase (ALDH). Approximately half the population of southeast Asian countries such as China, Japan, and Korea (which together make up a quarter of the world's population) have functional polymorphisms at three different genes: ADH2, ALDH1, and ALDH2. The most important variants are ALDH2*2 ($Glu_{487} - Lys_{487}$) which dominantly inactivates mitochondrial ALDH2, the enzyme responsible for most acetaldehyde metabolism in cells, and ADH2*2 ($Arg_{47} - His_{47}$), a superactive variant. These genetic variants, either by increasing the rate of synthesis of acetaldehyde or by decreasing its rate of metabolism, or by interacting together and doing both, raise the levels of this toxic intermediary. The result is that ingestion of even small amounts of ethanol produces an unpleasant reaction characterised by flushing, tachycardia, nausea, and headache.[27] This flushing reaction, severe in homozygotes but milder in heterozygotes, deters individuals with the protective alleles from becoming alcoholic. For example, the ALDH2*2 variant, which renders this enzyme inactive, has an allele frequency of about 0.30 in the Japanese and Chinese. Across populations, the ALDH2*2 variant appears on a similar genetic background (haplotype) and thus has probably had the same evolutionary origin.[28] Due to the dominant action of the $Lys_{487}$ subunit in the aldehyde dehydrogenase tetramer, about one in two Japanese and Chinese experiences flushing after alcohol consumption. Their risk of alcoholism is reduced approximately 4 to 10 fold. Approximately 10% of Japanese are ALDH2*2/ALDH2*2 homozygotes, but not one has yet been observed to be alcoholic across a series of studies in which several hundred alcoholics have been genotyped. The ADH2 genotype has been shown to be an independent factor contributing to the risk of alcoholism[29] and acts additively with the ALDH2*2 variant. A recent pilot study has shown that the ADH2*2 allele accounts for 20 to 30% of the alcohol intake variance between two groups of light drinking and heavy drinking Israeli Jews and suggests that the relatively high frequency of the ADH2*2 allele might contribute to the generally perceived lower levels of alcohol consumption and increased sensitivity to alcohol observed among Jews.[30] Although Monteiro et al.[31] observed that Jews showed a greater sensitivity to the intoxicating effects of alcohol, it should be pointed out that this finding may not be directly related to an increased rate of flushing.

The genetic influence of ALDH2 variants on alcohol consumption may be modified by environment. Tu and Israel[32] found that acculturation accounted for a significant proportion (7 to 11%) of the variance in alcohol consumption for southeast Asian males born in North America, although the ALDH2 polymorphism predicted two-thirds of the alcohol consumption and excessive alcohol use. Also, there are large differences in the prevalence of alcohol dependence in populations that have similar ALDH2 allele frequencies. The prevalence of alcoholism is 2.9% in Taiwan and 17.2% in Korea, suggesting that there may be potential interactions between the metabolic gene variants and other genetic and environmental factors.[22]

Disulfiram, used in the treatment of alcoholism, and some antiprotozoal drugs, such as metronidazole, inhibit ALDH2, thereby causing a flushing reaction after alcohol consumption. Therefore, the protective effect of ALDH2 genotypes can be regarded as directly analogous to protection with disulfiram. Finally, the observed lower risk of alcoholism in women may be due to a reduced rate of ethanol metabolism by Class II ADH in the stomach.[33] However, this difference in enzyme activity could also explain the tendency for women to experience more severe sequelae of alcohol use after an equivalent lifetime intake of alcohol.

## PROP TASTER STATUS AND ALCOHOLISM

There is a bimodal distribution for the ability to register propylthiouracil (PROP) as a bitter taste. Inability to taste PROP is a single locus recessive trait with a frequency of 30% amongst U.S. Caucasians; however, there is a great deal of variability amongst taster and nontaster groups.[34] The results of some studies suggest that the inability to taste PROP (i.e., bitterness) is a vulnerability factor for alcoholism, as pure alcohol is perceived as bitter.[34] However, other studies have found no such association.[35] It would be easier to determine the validity of such an association if studies had careful comparisons between taster status, alcoholism, and the nature of alcoholic beverages consumed; taster status may only be protective against alcoholism in societies in which no bitterness-disguised alcoholic drinks (e.g., wines) are available.

## CANDIDATE GENE APPROACH: CASE-CONTROL ASSOCIATION STUDIES

One starting point for understanding alcoholism vulnerability is to look for variants in genes involved in neurotransmitter metabolism. Genes for the metabolising enzymes, transporters, and receptors are good candidates. Because of the complexity of causation of alcoholism, any genetic determinants of vulnerability to alcoholism and differing sensitivity to alcohol's effects are likely to be subtle.

### Serotonin

Serotonin is involved in behavioral inhibition and is a target for the pharmacological treatment of alcoholism. Therefore, genetic variation of serotonin genes has been studied in humans. Pathologically low levels of serotonin may contribute to impulsivity; for example, a group of criminal, alcoholic Finns was shown to have low CSF 5-HIAA, the lowest levels being found in those who had committed impulsive crimes.[20] These are the alcoholics who would be most likely to have a structural serotonin gene variant. A functional polymorphism, 5-HTTLPR, has been found in the serotonin transporter promoter region and associated with anxiety.[36] The higher repeat number yallele is associated with reduced sensitivity to alcohol, i.e., with individuals who may be more vulnerable to developing alcoholism.[37] A tryptophan hydroxylase (TPH) intron variant that affects splicing is associated with reduced 5-HIAA and suicidality in impulsive alcoholics.[38,39] Associations with alcoholism have been found for two serotonin receptors: A synonomous substitution in 5-HT1B was significantly increased in Finnish criminal alcoholics as was a nonconservative amino acid substitution in 5-HT2C.[40]

### Dopamine

Dopamine is involved in arousal, reward, and motivation. Structural variants, some altering function or level of expression of gene product, have been found in the dopamine transporter and in several dopamine receptor genes (DRD2, DRD3, and DRD4). At the present time, no role for variation in dopamine-related genes in alcoholism has been consistently demonstrated. The controversial association of a DRD2 dopamine receptor polymorphism with alcoholism has been replicated in some but not in numerous other case-control studies, nor was it supported in two recent and very large family studies. These family studies were not subject to the potential problems of ethnic stratification inherent in some of the DRD2 case-control association studies.[41]

## Opioid Receptors

Human and animal studies implicate the opioid system, particularly the mu opioid receptor, in both initial sensitivity or response to alcohol and in the reinforcing effects of alcohol. However in humans, no association has been found between genetic variation at the mu opioid receptor gene and alcohol dependence.[42]

## WHOLE GENOME LINKAGE SCANS

The power of the genetic linkage analysis approach for alcoholism has been greatly improved by the recent collection of very large family and population data sets such as that of COGA and the NIAAA Southwestern Indian sample. Two recently published studies utilizing these data sets have detected evidence for linkage of alcoholism to several chromosomal regions, with some convergent findings. In the Southwestern Native American population isolate, suggestive evidence was found for linkage of alcohol dependence to the ADH region on chromosome 4q and to two regions harboring neurogenetic candidate genes. Those locations were chromosome 11p, in close proximity to the DRD4 dopamine receptor and tyrosine hydroxylase gene (the rate limiting enzyme in dopamine biosynthesis), and chromosome 4p, near a GABA receptor gene cluster (GABA receptors are allosterically modulated by ethanol).[43] In the COGA families, which derive from the cosmopolitan, diverse population of the United States, modest evidence was also found for linkage to the ADH region on 4q. There was also evidence of linkage to chromosomes 1 and 7, and to chromosome 2 at the location of the opioid receptor gene.[44]

## GENETIC STUDIES OF ALCOHOLISM IN ANIMALS

### NONHUMAN PRIMATE MODELS FOR ALCOHOLISM

The drinking behavior of two nonhuman primates, the vervet monkey[45] and the rhesus macaque monkey,[46] has been used as an animal model of alcoholism. In both vervet and rhesus macaque monkeys, individual monkeys vary with respect to their alcohol consumption and some will consume large amounts of alcohol willingly and on a long-term basis. These monkeys suffer symptoms of intoxication and withdrawal. The mating, rearing, and environmental conditions of these monkeys can be manipulated in order to study gene-gene and gene-environment interactions in alcoholism.[46] Studies suggest that both genetic and environmental factors are important in high alcohol consumption in nonhuman primates and provide some support for multiple origins of alcoholism.[47]

### RODENT MODELS: QUANTITATIVE TRAIT LOCUS ANALYSES

Rodent strains are inbred to produce large numbers of genetically identical animals that can be maintained under controlled environmental conditions. Genetically influenced traits underlying the vulnerability to alcoholism are called quantitative traits as they are continuously distributed in a population. Each of the multiple genes responsible for alcoholism can be regarded as a quantitative trait locus (QTL). Several QTLs may influence one trait; one QTL may affect several traits and these QTLs can be individually mapped, with the ultimate goal being to identify genes which play a role in human addiction to alcohol. Furthermore, the knockout of an individual gene in the mouse can reveal a potential role for the equivalent (homologous) gene in the human.

Several QTLs for alcohol-associated behaviors have been identified in mice by using recombinant inbred strains that differ widely with respect to many alcohol-related traits. The behaviors for which QTLs have been mapped include alcohol acute and chronic withdrawal sensitivity, alcohol consumption, and alcohol-associated hypothermia. Buck et al.[48] have shown that 68% of the genetic variability for genes influencing alcohol withdrawal severity can be assigned to QTLs on mouse chromosomes 1, 4, and 11 (the latter being near the gene for $GABA_A$ receptors). QTLs for alcohol-induced hypothermia, alcohol consumption, and certain responses to amphetamine and morphine are located on chromosome 1[49] and on chromosome 9 near the serotonin 5-HT1B receptor and dopamine D2 receptor genes.[50] These genetic inter-relationships between different phenotypes indicate that the same genes influence different alcohol-associated behaviors, that several genes may affect one phenotype, and that some loci for drug abuse are not drug-specific. This is also evident in studies of mice in which specific genes that map to the QTL regions or candidate genes located elsewhere in the mouse genome have been knocked out. 5-HT1B knockout mice drink twice as much ethanol, are less intoxicated, and show enhanced aggression compared with normal mice.[51] On chronic exposure to alcohol they show less evidence of tolerance. These mice also work harder to self-administer cocaine and show an increased locomoter response, behaving as if already sensitized to the drug.[52]

Two dopamine-related genes that have been knocked out are the DRD4 dopamine receptor, which is located at the site of one of the alcohol QTLs, the D1 dopamine receptor, and the dopamine transporter (DAT). The DRD4 knockout mice appear to be supersensitive to ethanol, cocaine, and amphetamines.[53]

For morphine preference, three loci identified on murine chromosomes 1, 6, and, 10 are apparently responsible for nearly 85% of the genetic variance in this trait.[54] The mu opioid receptor gene is located at the site of the largest QTL.

QTL analysis in rodents has limitations. Frequently the result is a large genomic region of interest rather than a gene. Mice and humans may not share the same functional variants at the same allele; for example, the ALDH2*2 allele is not even present in all human populations, and is not known in mice. Another problem is that behaviors in mice cannot be freely extrapolated to humans; for example, alcohol preference in mice may well be taste preference.[55] However, QTL analyses in mice are useful for the identification of candidate genes and gene regions.

## DISCUSSION

There is no longer any doubt that the influence of genotype in alcoholism vulnerability is frequently profound. Also, the influence of genotype tends to be greatest in individuals with the most severe form of the disease. Addictive disorders tend to co-occur within the same individual and share in part an underlying neurobiology. It seems increasingly likely that many addicted individuals, including alcoholics, share a genetic vulnerability, perhaps relating to shared behavioral traits. However, it is also ever more apparent that alcoholism is etiologically very complex both genetically and environmentally and that some of the vulnerability factors are specific to it, both genetic factors (e.g., ALDH2) and environmental factors (e.g., alcohol availability). Recent twin and family studies have shown that a substantial proportion of the inheritance of alcohol and certain other drugs, e.g., opiods, is substance-specific.[56]

The focus of genetic research in alcoholism has shifted from detecting genetic influence to identifying neurogenetic factors that, like the ALDH2 variant, affect an individual's response to drinking alcohol and vulnerability to alcoholism. Intensive studies in large data sets such as the National Institute on Alcohol Abuse and Alcoholism's (NIAAA) Collaborative Study on Genetics of Alcoholism (COGA) and intramural NIAAA studies on isolates (e.g., Native Americans), are under way to identify such factors using linkage analysis and other methods. QTL analysis in mice is beginning to identify candidate genes for exploration in humans. Case-control association studies have identified the role of the alcohol metabolizing genes in the prevention of alcoholism and are being used to investigate candidate genes involved in brain function. Nonhuman primate models for alcoholism appear to be promising tools for manipulating and understanding gene-gene and gene-environment interactions.

Widespread recognition of the genetic component for alcoholism has advanced our concept of alcoholism as a disease. Future progress in understanding the neurogenetics of alcoholism should mean that treatment and prevention will benefit from the tools of molecular diagnosis identifying alcoholic subtypes who are more responsive to particular forms of therapy.

## REFERENCES

1. American Psychiatric Association, *Diagnostic and Statistical Manual of Mental Disorders,* 4th ed., Washington, DC, American Psychiatric Press, 1994.
2. Kreek, M.J., Cocaine, dopamine and the endogenous opioid system, *J. Addict. Dis.,* 1996; 15(4), 73, 1996.
3. NIAAA, Alcohol and tobacco. Alcohol Alert, NIAAA publications, 39, 1, 1998.
4. Koob, G.F. and Le Moal, M., Drug abuse: hedonic homeostatic dysregulation, *Science,* 278, 52, 1997.
5. Herz, A., Endogenous opioid systems and alcohol addiction, *Psychopharmacology,* 129, 99, 1997.
6. Markou, A., Kosten, T.R., and Koob, G.F., Neurobiological similarities in depression and drug dependence: a self medication hypothesis, *Neuropsychopharmacology,* 18(3), 135, 1998.
7. Tabakoff, B. and Hoffman, P.L., Alcohol addiction: an enigma among us, *Neuron,* 16, 909, 1996.
8. Winokur, G. and Clayton, P.J., Family history studies, IV: comparison of male and female alcoholics, *Q. J. Stud. Alcohol.,* 29, 885, 1968.
9. Midanik, L., Familial alcoholism and problem drinking in a national drinking survey, *Addict. Behav.,* 8, 133, 1983.
10. Tambs, K. and Vaglum, P., Alcohol consumption in parents and offspring: a study of the family correlation structure in a general population, *Acta. Psychiatr. Scand.,* 82, 145, 1990.
11. Pickens, R.W., Svikis, D.S., McGue, M., Lykken, D.T., Heston, L.L., and Clayton, P.J., Heterogeneity in the inheritance of alcoholism, *Arch. Gen. Psychiatr.,* 48, 19, 1991.
12. Kendler, K.S., Heath, A.C., Neale, M.C., Kessler, R.C., and Eaves, L.J., A population-based twin study of alcoholism in women, *JAMA,* 268 (14), 1877, 1992.
13. Heath, A.C., Genetic influences on alcoholism risk. A review on adoption and twin studies, *Alcohol Health Res. World,* 19(3), 166, 1995.
14. Heath, A.C., Bucholz, K.K., Madden, P.A.F., Dinwiddie, S.H., Slutske, W.S., Beirut, D.J., Statham, D.J., Dunne, M.P., Whitfield, J.B., and Martin, N.G., Genetic and environmental contributions to alcohol dependence risk in a national twin sample: consistency of findings in women and men, *Psychol. Med.,* 27, 1381, 1997.
15. Goodwin, D.W., Alcoholism and Heredity, *Arch. Gen. Psychiatr.,* 36, 57, 1979.
16. Kaij, L., *Alcoholism in twins,* Almqvist & Wiksell, Stockholm, 1960.
17. Brown, G.L., Kline, W.J., Goyer, P.F., Minichiello, M.D., Krusei, M.J.P., and Goodwin, F.K., Relationship of childhood characteristics to cerebrospinal fluid 5-hydroxyindoleacetic acid in aggressive adults, in *Biological Psychiatry,* Chagass, C., Ed., Elsevier, New York, 1985, 177.
18. Hesselbrock, V., The collaborative study on the genetics of alcoholism: clinical assessment, *Alcohol Health Res. World, NIAAA,* 19(3), 230, 1995.
19. Cloninger, C.R., Bohman, M., and Sigvardsson, S., Inheritance of alcohol abuse: cross-fostering analysis of adopted men, *Arch. Gen. Psychiatr.,* 38, 861, 1981.
20. Linnoila, M., Virkkunen, M., and Roy, A., Biochemical aspects of aggression in man, *Clin. Neuropharmacol.,* 9(Suppl. 4), 377, 1986.
21. Begleiter, H., Porjesz, B., Bihari, B., and Kissin, B., Event-related brain potentials in boys at risk for alcoholism, *Science,* 225, 1493, 1980.
22. Goldman, D., Genetic transmission, in *Recent Developments in Alcoholism, Vol. 11: Ten years of Progress,* Galanter, M., Ed., Plenum Press, New York, 14, 231, 1993.
23. Enoch, M.-A., Rohrbaugh, J.W., Davis, E.Z., Harris, C.R., Ellingson, R.J., Andreason, P., Moore, V., Varner, J.L., Brown, G.L., Eckardt, M.J., and Goldman, D., Relationship of genetically transmitted alpha EEG traits to anxiety disorders and alcoholism, *Am. J. Med. Genet. (Neuropsychiatr. Genet.),* 60, 400, 1995.

24. Schuckit, M.A., Low level of response to alcohol as a predictor of future alcoholism, *Am. J. Psychiatr.,* 151(2), 184, 1994.

25. Anthenelli, R.M. and Tabakoff, B., The genetics of alcoholism: the search for biochemical markers, *Alcohol Health Res. World,* 19(3), 176, 1995.

26. Lander, E.S. and Schork, N.J., Genetic dissection of complex traits, *Science,* 265, 2037, 1994.

27. Harada, S., Agarwal, D.P., and Goedde, H.W., Aldehyde dehydrogenase deficiency as cause of facial flushing reaction to alcohol in Japanese, *Lancet,* ii(8253), 982, 1982.

28. Peterson, R.J., Goldman, D., and Long, J.C., Nucleotide sequence diversity in non-coding regions of ALDH2 as revealed by restriction enzyme and SSCP analysis, *Hum. Genet.,* 104, 177, 1999.

29. Thomasson, H.R., Edenburg, H.J., Crabb, D.W., Mai, X.L., Jerome, R.E., Li, T.K., Wang, S.P., Lin, Y.T., Lu, R.B., and Yin, S.J., Alcohol and aldehyde dehydrogenase genotypes and alcoholism in Chinese men, *Am. J. Hum. Genet.,* 48, 677, 1991.

30. Neumark, Y.D., Friedlander, Y., Thomasson, H.R., and Li, T.K., Association of the ADH2*2 allele with reduced ethanol consumption in Jewish men in Israel: a pilot study, *J. Stud. Alcohol.,* 59(2), 133, 1998.

31. Monteiro, M.G., Klein, J.L., and Schuckit, M.A., High levels of sensitivity to alcohol in young adult Jewish men: a pilot study, *J. Stud. Alcohol.,* 52(5), 464, 1991.

32. Tu, G.C. and Israel, Y., Alcohol consumption by orientals in North America is predicted largely by a single gene, *Behav. Genet.,* 25(1), 59, 1995.

33. Frezza, M., di Padora, C., Pozzata, G. et al., High blood alcohol levels in women, *N. Engl. J. Med.,* 322, 95, 1990.

34. Pelchat, M.L. and Danowski, S., A possible genetic association between PROP-tasting and alcoholism, *Physiol. Behav.,* 51, 1261, 1992.

35. Kranzler, H.R., Moore, P.J., and Hesselbrock, V.M., No association of PROP taster status and paternal history of alcohol dependence, *Alcohol. Clin. Exp. Res.,* 20(8), 1496, 1996.

36. Lesch, K.P., Bengel, D., Heils, A., Sabol, S.Z., Greenberg, B.D., Petri, S., Benjamin, J., Muller, C.R., Hamer, D.H., and Murphy, D.L., Association of anxiety-related traits with a polymorphism in the serotonin transporter gene regulatory region, *Science,* 274(5292), 1527, 1996.

37. Schuckit, M.A., Mazzanti, C.M., Smith, T.L., Ahmed, U., Radel, M., Iwata, N., and Goldman, D., Selective genotyping for the role of 5-HT2A, 5-HT2C and GABA$_{A6}$ receptors and the serotonin transporter in the level of response to alcohol — a pilot study, *Biol. Psychiatr.,* in press.

38. Nielsen, D.A., Goldman, D., Virkkunen, M., Tokola, R., Rawlings, R., and Linnoila, M., Suicidality and 5-hydroxyindole acetic acid concentration associated with a tryptophan hydroxylase polymorphism, *Arch. Gen. Psychiatr.,* 51(1), 34, 1994.

39. Nielsen, D.A., Virkkunen, M., Lappalainen, J., Eggert, M., Brown, G.L., Long, J.C., Goldman, D., and Linnoila, M., A tryptophan hydroxylase gene marker for suicidality and alcoholism, *Arch. Gen. Psychiatr.,* 55(7), 593, 1998.

40. Goldman, D., Candidate genes in alcoholism, *Clin. Neurosci.,* 3, 174, 1995.

41. Goldman, D., Urbanek, M., Guenther, D., Robin, R., and Long, J.C., A functionally deficient DRD2 variant [Ser311Cys] is not linked to alcoholism and substance abuse, *Alcohol,* 16(1), 47, 1998.

42. Bergen, A.W., Kokoszka, J., Peterson, R., Long, J.C., Virkkunen, M., Linnoila, M., and Goldman, D., Mu opioid receptor gene variants: lack of association with alcohol dependence, *Mol. Psychiatr.,* 2, 490, 1997.

43. Long, J.C., Knowler, W.C., Hanson, R.L., Robin, R.W., Urbanek, M., Moore, E., Bennett, P.H., and Goldman, D., Evidence for genetic linkage to alcohol dependence on chromosomes 4 and 11 from an autosome-wide scan in an American Indian population, *Am. J. Med. Genet. (Neuropsych. Genet.),* 81, 216, 1998.

44. Reich, T., Edenberg, H.J., Goate, A., Williams, J.T., Rice, J.P., Van Eerdewegh, P., Foroud, T., Hesselbrock, V., Schuckit, M.A., Bucholz, K., and Begleiter, H., Genome-wide search for genes affecting the risk for alcohol dependence, *Am. J. Med. Genet.,* 81(3), 207, 1998.

45. Ervin, F.R., Palmour, R.M., Young, S.N., Guzman-Flores, C., and Juarez, J., Voluntary consumption of beverage alcohol by vervet monkeys: population screening, descriptive behavior and biochemical measures, *Pharmacol. Biochem. Behav.,* 36, 367, 1990.

46. Higley, J.D., Hasert, M.F., Suomi, S.J., and Linnoila, M., Nonhuman primate model of alcohol abuse: effects of early experience, personality and stress on alcohol consumption, *Proc. Natl. Acad. Sci. U.S.A.*, 88, 7261, 1991.

47. Higley, J.D., Primates in alcohol research, *Alcohol Health Res. World, NIAAA*, 19(3), 213, 1995.

48. Buck, K.J., Metten, P., Belknap, J.K., and Crabbe, J.C., Quantitative trait loci involved in genetic predisposition to acute alcohol withdrawal in mice, *J. Neurosci.*, 17(10), 3946, 1997.

49. Crabbe, J.C., Belknap, J.K., Mitchell, S.R., and Crawshaw, L.I., Quantitative trait loci mapping of genes that influence the sensitivity and tolerance to ethanol-induced hypothermia in BXD recombinant inbred mice, *J. Pharmacol. Exp. Therapeut.*, 269(1), 184, 1994.

50. Crabbe, J.C., Buck, K.J., Metten, P., and Belknap, J.K., Strategies for identifying genes underlying drug abuse susceptibility, in *Molecular Approaches to Drug Abuse Research, Vol III*, Lee, T.N.H., Ed., NIDA Research Monograph, 161, 201, 1996.

51. Crabbe, J.C., Phillips, T.J., Feller, D.J., Hen, R., Wenger, C.D., Lessov, C.N., and Schafer, G.L., Elevated alcohol consumption in null mutant mice lacking 5-HT1B serotonin receptors, *Nat. Genet.*, 14(1), 98, 1996.

52. Rocha, B.A., Scearce-Levie, K., and Hen, R., Increased vulnerability to cocaine in mice lacking the serotonin-1B receptor, *Nature*, 393(6681), 175, 1998.

53. Rubinstein, M., Phillips, T.J., Bunzow, J.R., Falzone, T.L., Dziewczapolski, G., Zhang, G., Fang, Y., Larson, J.L., McDougall, J.A., Chester, J.A., Saez, C., Pugsley, T.A., Gershanik, O., Low, M.J., and Grandy, D.K., Mice lacking dopamine D4 receptors are supersensitive to ethanol, cocaine and methamphetamine, *Cell*, 90, 991, 1997.

54. Berrettini, W.H., Ferraro, T.N., Alexander, R.C., Buchberg, A.M., and Vogel, W.H., Quantitative trait loci mapping of three loci controlling morphine preference using inbred mouse strains, *Nat. Genet.*, 7, 54, 1994.

55. Grisel, J.E., Metten, P., Hughes, A.M., Lowe, D., and Crabbe, J.C., DBA/2J Mice like beer. RSA Abstr. 52, *Alcohol. Clin. Exp. Res.*, Suppl. 22(3), 12A, 1998.

56. Goldman, D. and Bergen, A., General and specific inheritance of substance abuse and alcoholism, *Arch. Gen. Psychiatr.*, 55(11), 964, 1998.

# 9 Alcohol and Other Abused Drugs

*Kari J. Buck, John C. Crabbe, and John K. Belknap*

## CONTENTS

0-8493-2688-5/00/$0.00+$.50

## DRUGS AND BEHAVIOR

Many of the most highly prevalent human diseases have multiple behavioral signs and symptoms that contribute to (or entirely constitute) their detrimental effects. Behavioral responses are almost invariably oligogenically or polygenically controlled. Hence, the effects of any single gene on individual differences are likely to be small and difficult to detect. Described here are genetic approaches that have been used successfully to identify some of the genes underlying complex behavioral responses.

### Effects of Drugs on Behavior

For several reasons, drug effects on behavior offer ideal model systems for ascertaining gene-behavior correlations. Drugs often have reasonably well-defined mechanisms of action, which can focus subsequent searches for relevant genes. Drug responses are generally dose-dependent, and the relative roles of tissue sensitivity (pharmacodynamic effects) can readily be distinguished from those of drug availability (e.g., absorption, distribution, or metabolism; collectively, pharmacokinetic effects) in determining individual differences in drug response. Pharmacokinetics is the predominant domain of what has traditionally been termed "pharmacogenetics." Historically, this field has concentrated on identifying and studying genetically determined variants in enzyme function that lead to large individual differences in patterns of drug metabolism. Reviews of pharmacokinetic factors can be found elsewhere[1,2] and will not be our focus here. However, most studies that analyze genetic aspects of pharmacodynamics do not actually measure drug concentrations at the target tissue (e.g., brain drug levels). For example, a strain showing relatively greater behavioral response to a given drug may do so simply because more drug reaches its brain after a fixed dose. If the goal is to understand genetic differences in pharmacodynamic responses to drugs, understanding drug pharmacokinetics is an important control issue that is all too frequently neglected.

### Sensitivity, Tolerance, Dependence, and Withdrawal

Drug effects are not monolithic. A naive organism responds characteristically when first challenged with a drug, and individual differences in drug *sensitivity* comprise one domain for genetic analysis. When a drug is repeatedly administered, the initial response is often attenuated (termed *tolerance*), although in some cases (primarily drug-stimulated responses), responsiveness increases with repetition, which is termed *sensitization*. For many classes of drugs, including alcohol, other sedative-hypnotics such as barbiturates, minor tranquilizers, and opioid drugs such as heroin or codeine, a state of *dependence* can be induced, which is inferred from the waxing and waning appearance of characteristic signs of *withdrawal* after the drug is discontinued. All these various aspects have a role in determining individual differences in susceptibility to chronic abuse of drugs and have been independently modeled in animals (see section titled "Genetic Animal Models Employed").

## Reinforcement and Drug-Seeking Behaviors

In the previous section, we reviewed different aspects of characterizing responses to drugs of abuse. Using a different conceptual framework, it is important to note that drugs of abuse have reinforcing effects. That is, their effects can be seen to have rewarding or punishing effects on behavior. Human abusers clearly engage in drug-seeking behaviors, and animal models for drug-seeking behaviors have been employed for experimental studies of genetic controls.

## Genetic Animal Models Employed

Many genetic animal models have been employed to examine behavioral responses to drugs and to identify the relevant genetic influences. More details regarding the relative strengths and weaknesses of these models can be found in reviews elsewhere.[3,4] Studies have made it clear that genetic determinants of one drug effect (e.g., sensitivity to alcohol-induced hypothermia) may be completely unrelated to the determinants of another effect of the drug (e.g., alcohol preference drinking). Conversely, some genes may influence a variety of effects of multiple drugs.[5,6]

### Selective Breeding

Perhaps the oldest genetic model involves selective breeding. In much the same way that multiple breeds of dogs have been bred from the ancestral wolf, mice or rats have been bred for sensitivity or tolerance to, dependence on, or self-administration of many drugs of abuse. These studies have been systematically reviewed.[7] Selection studies employ a starting population of genetically heterogeneous animals, i.e., one in which there is allelic variation at many genes. Successful selective breeding demonstrates that high and low behavioral response is genetically determined and not solely due to environmental factors. Moreover, as lines of animals are selected for high (or low) behavioral response, the frequencies of all alleles affecting the trait are systematically altered: alleles favoring high (or low) response increase in frequency, and alleles opposing the direction of selection become more rare and are often eliminated from the population. The great advantage of selection is its ability to capture genes with very small effects on the drug response selected. However, the principal limitation is that the genes captured are entirely anonymous. Thus, comparison of high- and low-response populations can elucidate general mechanisms underlying drug responses, but other techniques must then be used to identify the specific genes responsible.

### Inbred Strains

An inbred strain is one in which close genetic relatives (usually siblings) have been mated for 20-plus generations. The effect of this inbreeding is to force all genes to a state of homozygosity (fixation) within each strain, rendering all same-sex individuals virtually genetically identical, like monozygotic twins. Individual differences in drug response within an inbred strain are by definition due to non-genetic (i.e., environmental) causes. When several inbred strains are compared, differences among strains are due to the influence of genes, while those within each strain are nongenetic.[4] Thus, the effect of inbreeding differs from that seen in selected lines in that the particular allelic form of each gene that is fixed is due to chance and is not systematically related to any particular drug response. Inbred strains may be compared on multiple drug responses, and their similarities in responsiveness suggest the influence of some common genes on multiple drug responses. As with selective breeding, the particular genes influencing drug responses are initially anonymous.

### Transgenics and Knockouts

Recent advances in molecular biology have allowed the development of mouse models in which a specific gene is altered and then introduced into the genome of a host mouse strain. Such transgenic mice may have over- or under-expression of the transgene, and their function may be attenuated or even entirely deleted. Hundreds of specific genes with potential relevance for influence on drug abuse-related traits have been examined using transgenic mice. A few direct examples are discussed (see section titled "Targeted (Candidate) Gene Manipulation and Behavior"). Potential pitfalls of interpreting the results of such studies have also been discussed.[8] Results of such experiments can identify genes relevant for drug abuse, and the effects of such genes can then be studied using other genetic animal models such as inbred strains and selectively bred lines.

### Recombinant Inbreds, QTL Mapping, and Congenics

A specialized type of inbred strains, recombinant inbred (RI) strains, can be used to begin to specify the particular genes affecting drug response.[9,10] When two inbred strains are crossed to form an F1 generation, which is then intercrossed to form an F2 population, the genetic variability made available by segregation and independent assortment can be isolated into new inbred strains, after 20 generations of re-inbreeding. These RI strains represent more or less random samples of the available genetic variability. Because they have been genotyped for many hundreds of marker genes whose positions in the mouse genome are known, (genetic) differences among RI strains for a drug response can be compared with the specific allele each strain possesses at each of those marker genes. Whenever drug response covaries with alleles at a particular group of markers, a quantitative trait locus (QTL) is roughly mapped. This greatly narrows the range of genetic material through which one must look to identify the relevant genes.

Once a region of interest has been identified, it is possible to produce *congenic* strains by backcrossing mice containing the desired chromosomal segment onto a background strain with the opposite drug response. The resulting congenic strain thus offers a background of low drug response, for example, against which the effect of the introduced segment to increase response can easily be tested. Multiple congenics are typically produced, because each has a slightly different introduced segment of DNA, and identifying which congenics still possess the gene of interest can reduce the QTL region considerably, facilitating high-resolution mapping and making such strategies as positional cloning of the relevant gene more feasible.

## TARGETED (CANDIDATE) GENE MANIPULATION AND BEHAVIOR

The recent development of transgenic mice that lack or overexpress specific genes offers a powerful opportunity to evaluate the roles of their gene products in a variety of behaviors mediated by alcohol and other drugs of abuse. Antisense oligodeoxynucleotide (ODN) strategies represent a complementary approach to correlate the properties of cloned genes with their *in vivo* pharmacological effects. This technique entails synthesis of an ODN sequence designed to hybridize with the native mRNA coding for synthesis of a target protein. Injection of an antisense ODN into the brain can often lead to a blockade of production of the targeted gene product. A puzzling feature of the antisense ODN strategy is that it does not appear to block expression of particular genes, but the reasons for such systematic failures remain unclear. In this section, we provide an overview of studies employing transgenic and antisense ODN strategies that specifically study the effects of drugs of abuse (e.g., psychostimulants, opioids, ethanol) on behaviors thought to be involved in drug abuse (e.g., conditioned preference and aversion, tolerance, sensitization, withdrawal). We do not review many other studies employing antisense ODN or transgenic approaches to examine genetic effects on behaviors other than drug responses, such as testing responses to opioid receptor ligands that are not drugs of abuse or those studying other behaviors.

### PSYCHOSTIMULANTS

The mesolimbic dopamine system is involved in behaviors linked with motivation and reward,[11] and imbalance in dopamine systems is thought to play an important role in drug addiction.[12,13] Here, we discuss studies in which specific genes involved in dopaminergic transmission have been disrupted or overexpressed in order to evaluate the roles of these genes in the actions of psychostimulants and in some dopamine-mediated behaviors.

## Dopamine and Vesicular Monoamine Transporters

Inhibition of dopamine reuptake is thought to be the primary or initial mechanism mediating self-administration of cocaine and related psychostimulants.[12] Disruption of the mouse dopamine transporter gene (*Dat1*) by homologous recombination results in spontaneous hyperlocomotion.[14] Dopamine persisted ~100 times longer in the extracellular space in null mutant mice and may explain the biochemical basis of the hyperdopaminergic phenotype, although major adaptive decreases in neurotransmitter and receptor levels may also contribute to the mutant phenotype.[14] Studies using the null mutant mice establish the central importance of the dopamine transporter as an obligatory target for the behavioral and biochemical actions of amphetamines and cocaine. Intraperitoneal injections of high doses of cocaine (40 mg/kg) and d-amphetamine (10 mg/kg) produced no significant locomotor effects or stereotyped behaviors in *Dat1* null mutant mice, whereas treatment with either cocaine or d-amphetamine produced a six- to eight-fold increase in locomotor activity in wild-type or heterozygous mice, which was elevated to the same level as the spontaneous activity in the null mutant mice. These doses of cocaine and d-amphetamine also produced stereotyped behaviors in wild-type mice, but not in mice lacking the dopamine transporter.

Knockout of the vesicular monoamine transporter 2 (*Vmat2*), the predominant form expressed in the brain, indicates that the vesicular transporter also plays a crucial role in the effects of psychostimulants and other drugs of abuse.[15,16] Homozygous null mutants die within a few days of birth, but heterozygotes are viable into adulthood. Takahaski et al.[16] showed that in heterozygotes, amphetamine produced enhanced locomotion but diminished reward as measured by amphetamine conditioned place preference. Wang et al.[15] showed that heterozygous adult mice are supersensitive to the locomotor effect of the dopamine agonist apomorphine, psychostimulants such as amphetamines and cocaine, and ethanol. These studies suggest that intact synaptic vesicle function contributes to psychostimulant-induced locomotion and to amphetamine-conditioned reward.

## D$_1$ Dopamine Receptor

The D$_1$ dopamine receptor plays a key role in dopaminergic transmission and is thought to be involved in some actions of psychostimulants. Ablation of the D$_1$ receptor gene (*Drd1a*) by homologous recombination produced mice that show less amphetamine-induced locomotor activity than similarly treated control mice.[17] Both null mutant and control mice exhibited behavioral sensitization, although the sensitized response was less pronounced in the D$_1$ receptor deficient mutants.[17] Null mutant mice treated with cocaine showed no change[19] or decreased[18] locomotor activity levels, whereas increased activity was noted in wild-type and heterozygous animals. Null mutant mice retained cocaine-conditioned place preference and exhibited increases in preference for drug-paired environments similar to that for wild-type and heterozygous mice.[19] These results are consistent with the involvement of the D$_1$ receptor in the locomotor stimulant effects of cocaine, but they suggest little role in the rewarding and reinforcing effects of cocaine. However, extracellular single unit recording of dopamine-sensitive nucleus accumbens neurons revealed a marked reduction in the inhibitory effects of cocaine and dopamine on the generation of action potentials in null mutant mice compared to wild-type controls, suggesting that D$_1$ receptors play a fundamental role in cocaine- and dopamine-mediated neurophysiological effects within the nucleus accumbens.[18,20]

Repeated administration of psychostimulants such as cocaine and amphetamine often results in long-lasting behavioral alterations, including sensitization and dependence. The postsynaptic effects of indirect dopamine receptor agonists, such as cocaine and amphetamines, include altered gene expression, which may underlie some of the behavioral changes produced by these drugs.[21] In contrast to wild-type mice, cocaine failed to induce the immediate-early genes *c-fos* and *zif 268* in mutant mice lacking the D$_1$ receptor, whereas substance P expression was abnormally increased.[22] These data indicate that some effects of cocaine on gene regulation are mediated via D$_1$ receptor-dependent mechanisms.

### D$_2$-Like Dopamine Receptors

Antisense ODNs directed against the D$_2$ dopamine receptor have been used to examine its role in cocaine- and dopamine-mediated behaviors in mice and rats.[23-25] Unilateral administration of antisense ODN, via intracerebral cannula, into the substantia nigra of rats caused dramatic contralateral rotational behavior in rats in response to a subcutaneous injection of cocaine.[25] Without cocaine, no spontaneous rotational behavior was observed. The robustness of cocaine-induced rotation and the impaired ability of sulpiride to enhance dopamine release from slices suggested that nigrostriatal D$_2$ autoreceptors play a direct role in reducing the motor response to cocaine administration. Mice with targeted genetic disruption of the D$_2$ receptor gene (Drd2) have been generated, but their responses to psychostimulant drugs of abuse are as yet unknown. However, D$_2$ receptor knockout mice have been used to examine the role of this gene in responses to morphine.[26]

The D$_3$ dopamine receptor is expressed in brain regions thought to influence motivational and motor functions. Xu et al.[27] showed that D$_3$ receptor null mutant mice were transiently more active than wild-type mice in a novel environment and exhibited enhanced behavioral sensitivity to cocaine, amphetamines, and to combined injections of D$_1$ and D$_2$ class agonists. These data suggest that the D$_3$ receptor may modulate these behaviors by inhibiting the cooperative effects of postsynaptic D$_1$ and other D$_2$ class receptors at the systems level.

Mice with targeted genetic disruption of the D$_4$ dopamine receptor gene (Drd4) displayed enhanced sensitivity to the locomotor stimulant effects of cocaine, methamphetamine, and ethanol.[28] Null mutant mice were also less active in open field tests than wild-type mice, but they outperformed wild-type mice on the rotorod. Biochemical analyses showed that dopamine synthesis and its conversion to DOPAC were elevated in the dorsal striatum of null mutant mice.

### 5-HT$_{1B}$ Receptor

Serotonergic transmission has also been suggested to modulate the effects of cocaine. Mice lacking the presynaptic 5-hydroxytryptamine$_{1B}$ (5-HT$_{1B}$) receptor have recently been used to test the possible contribution of this receptor subtype to intravenous cocaine self-administration.[29] Both mutant and wild-type mice had similar dose responses for cocaine, but mutant mice presented a significantly shorter latency to meet intravenous cocaine self-administration. These data implicate the 5-HT$_{1B}$ receptor in the propensity to self-administer cocaine, whereas other mechanisms might be involved in the maintenance of cocaine self-administration. 5-HT$_{1B}$ knockout mice also display a markedly attenuated induction of c-fos by cocaine in different brain regions, most notably the striatum.[30] Administration of RU24969, a 5-HT$_{1B}$ receptor antagonist, to wild-type mice results in striatal induction of c-fos expression very similar to that induced by cocaine in its time course, cellular and anatomical distribution, and pharmacology. These converging lines of evidence indicate that cocaine may act as an indirect agonist at 5-HT$_{1B}$ receptors in vivo.

### c-fos

Amphetamines can produce long-lived changes in behavior, including sensitization and dependence, but the neural substrates of these drugs' effects are unknown. Based on their prolonged time course, they are thought to involve drug-induced alterations in gene expression. Systemic administration of cocaine and amphetamines induces c-fos expression in the rat striatum, including the nucleus accumbens that are implicated in reward.[31] The functional significance of c-fos induction is unknown, but it has been suggested as a possible mechanism for delayed effects of central stimulants.[32] Antisense ODN approaches were used to examine the role of this immediate early gene in responses to psychostimulants,[33-35] as well as the role in arginine vasopressin maintenance of ethanol tolerance[36] (discussed later) and GABA transmission.[37] Null mutation of the c-fos proto-oncogene has also been shown to have pleiotropic effects.[38]

Bilateral administration of an antisense ODN against *c-fos* in the nucleus accumbens blocked cocaine induced locomotor stimulation without affecting spontaneous exploratory activity.[32] A control-sense ODN was inactive. Another study showed that unilateral attenuation of *c-fos* expression by the direct infusion of antisense ODNs into the left or right striatum had behavioral consequences, e.g., induction of amphetamine-stimulated rotational behavior.[34] These studies support a role for c-*fos* in mediating locomotor effects of cocaine and amphetamines.

## cAMP Response Element-Binding Protein (CREB)

Cyclic AMP (cAMP) response elements (CREs) and CREB protein are involved in the transcriptional regulation of many genes, including immediate-early genes such as *c-fos*,[39,40] and several neuropeptide genes, such as proenkephalin, dynorphin, somatostatin, and vasoactive intestinal polypeptide.[41-44] Konradi et al.[45] have shown that amphetamines induce phosphorylation of transcription factor CREB in rat striatum *in vivo* and that dopamine $D_1$ receptor stimulation also induces phosphorylation of CREB within specific complexes bound to cAMP regulatory elements. Antisense ODN injection showed that CREB is necessary for *c-fos* induction by amphetamines *in vivo*.[45] These studies indicate that CREB phosphorylation is an important early nuclear event mediating long-term consequences of amphetamine administration and dopamine $D_1$ receptor stimulation in striatal neurons.

## CuZn-Superoxide Dismutase

Methamphetamine-induced neurotoxicity is thought to involve release of dopamine and increased formation of oxygen-based free radicals. Transgenic mice that express the human CuZn-superoxide dismutase enzyme show gene dosage-dependent attenuation of methamphetamine-induced depletion of striatal dopamine and dihydroxyphenylacetic acid (DOPAC).[46,47] Similarly, 1-methyl-1,2,3,6-tetrahydropyridine (MPTP) caused marked depletion of striatal dopamine and DOPAC in nontransgenic, but not in transgenic mice. These studies suggest that CuZn-superoxide dismutase may protect dopaminergic neurons against neurotoxic effects of methamphetamine and MPTP.

## Cocaine- and Amphetamine-Regulated Transcript (CART)

Messenger RNA (mRNA) differential display screening has recently been used to identify a rat brain mRNA that is transcriptionally regulated in the striatum following acute administration of cocaine and amphetamines.[48] This gene appears to encode a novel neuroendocrine signaling molecule which is expressed in diverse rat and human brain structures, as well as endocrine tissues. The chromosomal location of this gene has not been determined in rats, but PCR/Southern blot analysis of DNA from human/rodent somatic cell hybrid panels localized the human homologue, CART (Cocaine- and Amphetamine-Regulated Transcript), to human chromosome 5.[49] The role of the CART gene in behavioral responses to cocaine and amphetamines remains to be determined. These studies show that mRNA differential display represents a powerful approach for identifying novel genes that are transcriptionally regulated by drugs of abuse for future study.

## G-Protein β1 Subunit ($G_{β1}$)

Uhl and colleagues[50] have also used subtractive differential display to identify genes whose expression is regulated by cocaine and amphetamines. In rats, $G_{β1}$ was upregulated by cocaine or amphetamine treatments in neurons of the nucleus accumbens shell region. Antisense oligodeoxynucleotide treatments directed against $G_{β1}$ attenuated its expression in regions including the nucleus accumbens and abolished the development of behavioral sensitization when administered during repeated cocaine exposures that establish sensitization. However, antisense treatment failed to alter the acute cocaine response and did not block the expression of cocaine sensitization.

## OPIOIDS

"Opiate" is used to designate drugs derived from opium, such as morphine and codeine. After the development of totally synthetic entities with morphine-like actions, the word "opioid" was introduced to refer in a generic sense to all drugs, natural and synthetic, with morphine-like actions. Opioids share some of the properties of certain naturally occurring peptides — the enkephalins, the endorphins, and the dynorphins — and exert their pharmacological actions through three opioid-receptor classes, $\mu$, $\delta$, and $\kappa$, whose genes have been cloned.[51] These receptor subtypes are expressed in circuits that can modulate nociception and receive inputs from endogenous opioid neuropeptide ligands, but the roles played by each receptor subtype in nociceptive processing and drug reward are only beginning to be elucidated. Here, we discuss studies using transgenic and antisense ODN approaches directed against specific genes involved in opioid transmission and gene expression in order to evaluate their roles in behaviors mediated by morphine and morphine-like opioid compounds.

### Disruption of $\mu$-Opioid Receptor Expression

Endogenous brain systems mediating analgesia and reward employ enkephalin recognition by $\mu$-opioid receptors as major neurochemical transmission pathways. Disruption of the $\mu$-opioid receptor by homologous recombination,[52,53] and by using antisense ODNs directed against the $\mu$-receptor gene,[54-58] have been used to delineate the contribution of $\mu$-receptors in opioid function *in vivo*. Investigation of the behavioral effects of morphine reveals that lack of $\mu$-receptors in null mutant mice abolishes the analgesic effect of morphine in the tail-immersion and hot-plate tests.[52,53] In the case of the hot-plate test, this is not surprising, as the response is mainly controlled by $\mu$-mediated supraspinal mechanisms.[59] But lack of morphine response in the tail-flick test was unexpected because the latter primarily involves spinal mechanisms with participation of $\mu$, $\delta$, and $\kappa$ receptors.[60] However, the $\mu$-opioid receptor appears to be necessary for analgesia produced by the $\delta$-opioid receptor agonist [D-Pen$^2$, D-Pen$^5$] enkephalin (DPDPE) in the tail-flick and hot-plate tests, suggesting that $\delta$-opioid receptor mediated analgesia may be masked when the $\mu$-opioid receptor has been knocked out.[61] Previous studies showed that administration of an antisense ODN targeting the 5'-untranslated region of the cloned $\mu$-receptor gene administered into the periaqueductal gray blocked morphine analgesia in rats examined using the tail-flick test.[54] Additional studies have confirmed the activity of $\mu$-receptor antisense ODNs following intracerebroventricular injection into rats.[56,57]

$\mu$-opioid receptor expression in regions containing dopaminergic neurons and their terminals has been postulated to play a significant role in opioid-induced reward, whereas expression on locus coeruleus neurons using norepinephrine has been associated with several of the autonomic signs of the opioid withdrawal syndrome.[62,63] Null mutant mice lacked morphine-conditioned place preference and physical dependence, as evidence by naloxone-induced withdrawal.[52] Although the high doses of morphine and naloxone used are presumably able to interact with all opioid receptors, morphine physical dependence was totally suppressed in the absence of the $\mu$-receptor. Wild-type mice, but not null-mutant animals, showed classical signs of chronically morphine-treated mice, such as Straub-tail reflex and increased locomotor activity. These studies suggest that the $\mu$-opioid receptor gene product is the molecular target of morphine *in vivo* and that it is a mandatory component of the opioid system for morphine action regarding several morphine effects.

Notably, null mutant mice showed no overt behavioral abnormalities or major compensatory changes within the opioid system. *In situ* hybridization analysis showed no modification in the expression pattern of proenkephalin, prodynorphin, or proopiomelanocortin messenger RNAs, suggesting that $\mu$-receptor expression does not exert regulatory control on the transcription of genes encoding endogenous ligands.[52] Scatchard analysis and autoradiographic mapping using selective $\delta$ and $\kappa$ radioligands showed similar plots for wild-type and mutant mice, suggesting that the absence of the μ-receptor did not alter the number or distribution of δ and κ opioid receptors in brain regions where binding was detected in wild-type mice.[52,53] In contrast, saturation binding with the μ-selective ligand [³H]DAMGO showed a 54 to 60% reduction of μ-receptors in whole brains of heterozygous mice and total loss in homozygous mutants. These data provide supporting evidence that both putative μ₁ and μ₂ receptor sites arise from the mouse *Oprm* gene.

## μ-Opioid Receptor Overexpression

The 4.8 kb tyrosine hydroxylase 5'-flanking promoter region was used to drive expression of the rat μ-opioid receptor cDNA in dopaminergic and noradrenergic neurons in transgenic mice.[64] Expression of the transgene mRNA was confirmed by analyses of RNase-protected fragments specific for the rat and endogenous mouse *Oprm* gene. However, transgenic animals did not differ from age- and sex-matched littermate controls in size, development, apparent fertility, nociceptive thresholds, or locomotion in novel environments. Transgenic mice displayed marked reductions in the number of withdrawal signs when abstinence was precipitated by naloxone following three days of morphine pellet implantation. This result is consistent with the postulated role of noradrenergic cells in opioid withdrawal. The transgenic mice were also significantly different from controls in morphine effects in the hot-plate test, but not for the tail-flick test. μ-opioid receptor overexpression in dopamine cells did not yield heightened conditioned place preference when transgenic mice were conditioned using 5 mg/kg morphine doses.[64]

## δ- and κ-Opioid Receptors

Gene-targeted disruption of the κ-opioid receptor indicates that this receptor participates in the expression of morphine abstinence but does not contribute to morphine analgesia or reward.[65] Mice lacking the κ-opioid receptor were also insensitive to the hypolocomotor, analgesic and aversive actions of the prototypic κ-agonist U-50,488H.[65] Transgenic mutants for the δ-opioid receptors have not yet been produced, but studies using antisense ODNs directed against the δ opioid receptor have been used to examine the contribution of this receptor in opioid function *in vivo*. Injection of an antisense ODN directed against the δ-opioid receptor, but not a mismatch ODN, blocked[66] or attenuated[67] the development of acute morphine dependence and blocked chronic morphine tolerance[66] in rats. Studies using drugs with relative preferences for δ- and κ-opioid receptors provide supporting evidence that these receptor subtypes play significant roles in the development of opioid physical dependence[68-70] and suggest a role for these receptor subtypes in analgesic responses induced by morphine-like drugs.[71,72] However, studies using μ-receptor null mutant mice suggested that δ and κ receptors did not mediate any of the major biological actions of morphine in the absence of the μ-receptor and were generally consistent with studies showing that antisense ODNs directed against the δ- and κ₁-opioid receptors selectively block the analgesic actions of their respective ligands but did not block morphine analgesia.[73,74] One study reported that a δ-opioid receptor ODN attenuated analgesia produced by (D-Ala²) deltorphin II (a δ-opioid receptor agonist), and to a lesser extent by morphine, but not by β-endorphin or U50,488H.[75] These studies suggest that cooperativity between opioid receptors may be critical. It has been suggested that cooperativity might take place at the molecular level through receptor allosteric interactions or second messenger systems or occur at a functional level on separate neurons.[76]

## β-Endorphin

A physiological role for β-endorphin in endogenous pain inhibition has been investigated by targeted mutagenesis of the proopiomelanocortin gene (*Pomc1*) in mouse embryonic stem cells.[77] Because β-endorphin is post-translationally processed from a larger multifunctional precursor, a point mutation of the *Pomc1* gene was introduced that translates to a truncated prohormone lacking the entire C-terminal amino acid region encoding β-endorphin. The resulting transgenic mice displayed no overt developmental or behavioral alterations and have a normally functioning hypothalamic-pituitary-adrenal axis. Null mutant mice exhibited normal analgesia in response to morphine. However, these mice lacked the opioid (naloxone reversible) analgesia induced by mild swim stress. Null mutant mice also displayed greater nonopioid analgesia in response to the more severe cold water swim stress compared with controls. These data suggest that loss of β-endorphin does not alter response to morphine, but it may produce compensatory upregulation of alternative pain inhibitory mechanisms.

## G-Protein Transduction of Opioid Responses

Supraspinal morphine-induced antinociception is mediated through multiple second-messenger pathways, including inhibitory $G_i$ proteins sensitive to ADP-ribosylation by pertussis toxin. The effect of the $G_i1\alpha$, $G_i2\alpha$, $G_i3\alpha$, and $G_s\alpha$ subunits on the antinociceptive signal induced by supraspinal administration of μ-opioid agonists was examined using antisense ODNs directed against each of these G-protein subunit mRNAs.[78] Intraventricular injection of the antisense ODN directed against $G_i2\alpha$, but not the other subunits examined, reduced morphine-induced antinociception by 75% in the tail-flick test in rats. Additional studies showed that microinjection of an antisense ODN directed against $G_i2\alpha$ blocked morphine analgesia but was inactive against analgesia induced by the potent morphine metabolite, morphine-6β-glucuronide.[79] Conversely, an antisense ODN against $G_i1\alpha$ inhibited morphine-6β-glucuronide analgesia without affecting morphine analgesia. An antisense ODN against $G_o\alpha$ was ineffective against both compounds. These results indicate that morphine analgesia is associated with $G_i2\alpha$ and suggest that morphine-6β-glucuronide acts through a different effector. None of these ODNs altered naloxone-precipitated jumping (acute dependence) or morphine-induced constipation.[80]

In a separate study, five consecutive days of repeated i.c.v. administration of antisense ODNs directed against Gα subunit mRNAs was used to impair the function of mouse $G_i1\alpha$, $G_i2\alpha$, $G_i3\alpha$, and $G_{x/z}\alpha$ regulatory proteins.[81] Decreases of 20 to 60% on the Gα-like immunoreactivity were observed in neuronal structures of mouse brain. This effect was not produced by a random-sequence ODN used as a control. In mice injected with the ODN to $G_i2\alpha$, the antinociceptive activity of all the opioids tested was greatly impaired. The ODN to $G_i3\alpha$ reduced the effects of the selective δ-opioid receptor agonists, [D-Pen$^{2,5}$]enkephalin and [D-Ala$^2$] deltorphin II. Conversely, the analgesia evoked by opioids binding μ-opioid receptors, morphine and [D-Ala$^2$, N-MePhe$^4$, Gly-ol$^5$]-enkephalin, was consistently and significantly attenuated in mice injected with the ODN to $G_{x/z}\alpha$. The effect of the neuropeptide β-endorphin(1-31) agonist at μ- and δ-opioid receptors was also reduced by ODNs to $G_i3\alpha$ or $G_{x/z}\alpha$. The ODN to $G_i1\alpha$ lacked effect on opioid-evoked analgesia for all of the opioids tested. Injection of antibodies directed to these Gα subunits antagonized opioid-induced analgesia with a pattern similar to that observed for the ODNs.[81]

## cAMP Response Element-Binding Protein (CREB)

Chronic opioid administration regulates protein components of the cAMP signaling pathway, specifically in the nucleus accumbens, a brain region implicated in the reinforcing properties of opioids. It has been suggested that such adaptations may contribute to changes in reinforcement mechanisms that characterize addiction.[21] Widnell et al.[82] examined a possible role for the transcription factor CREB in mediating these long-term effects of opioids. Chronic, but not acute, morphine administration decreased levels of CREB immunoreactivity in the nucleus accumbens, an effect not seen in other brain regions studied. The functional significance of the CREB down-regulation was examined using an antisense ODN directed against CREB, which produced a specific and sustained decrease in CREB levels in the nucleus accumbens, without detectable toxicity. Antisense ODN reduction in CREB levels mimicked the effect of morphine on certain, but not all, cAMP pathway proteins in this brain region, whereas a large number of other signal transduction proteins tested were unaffected by ODN treatment. These findings support a role for CREB in autoregulation of the cAMP pathway in the nervous system, as well as in mediating some of the effects of morphine on this signaling pathway in the nucleus accumbens.

The nucleus locus coeruleus is also implicated in the expression of opioid physical dependence and withdrawal. Infusion of an antisense oligodeoxynucleotide, directed against CREB directly into the locus coeruleus for five days, reduced CREB immunoreactivity and completely blocked morphine-induced upregulation of type VIII adenylyl cyclase, but not of protein kinase-A.[83] Antisense infusion also blocked morphine-induced upregulation of tyrosine hydroxylase but not of $G_i\alpha$, two other proteins induced in the locus coeruleus by chronic morphine treatment. These data provide direct evidence for a role for CREB in mediating opioid-induced upregulation of the cAMP pathway.

In mice with targeted genetic disruption of the $\alpha$ and $\delta$ isoforms of CREB, the main symptoms of morphine withdrawal were strongly attenuated.[84] No change in opioid binding sites or in morphine-induced analgesia was observed in the mutant mice, and the increase of adenylyl cyclase activity and immediate early gene expression after morphine withdrawal was normal. These data suggest that CREB-dependent gene transcription is a factor in the onset of behavioral manifestations of opioid withdrawal. In the brain, CREB has been implicated in transcriptional activation of the proenkephalin gene,[43] which may play a role in the onset of opioid dependence.

## $D_2$ Dopamine Receptor

Investigation of adaptive responses to repeated morphine administration in mice with targeted genetic disruption of the $D_2$ receptor gene (*Drd2*) shows that expression of morphine withdrawal is unchanged in mice lacking $D_2$ receptors, whereas a total suppression of morphine rewarding properties was observed in a place preference test.[26] These data suggest that $D_2$ receptors may play a crucial role in the motivational component of morphine dependence.

## ETHANOL, BENZODIAZEPINES, AND BARBITURATES

Transgenic mice and antisense ODN studies have helped define the roles of some candidate genes and their protein products in ethanol-related behaviors, including drinking, sensitivity and tolerance in the grid test for ataxia, and hypnotic sensitivity and tolerance. The relationships between a variety of ethanol-related behaviors and specific genes are discussed in the next few sections. Ethanol and sedative-hypnotic benzodiazepines and barbiturates have extensive pharmacological similarities, and selective breeding studies suggest the influence of common genes, but the specific genes involved remain to be elucidated. The role of specific $GABA_A$ receptor subunit genes and the $PKC\gamma$ gene on ethanol, benzodiazepine, and barbiturate responses is discussed in the section titled "$GABA_A$ Receptors and Modulation by $PKC\gamma$."

## $5-HT_{1B}$ Receptor

Lowered central serotonin (5-HT) neurotransmission has been found in a subgroup of alcoholics, possibly those with more aggressive, assaultive tendencies. Substantial evidence also links ethanol drinking and 5-HT functioning in animals. Null mutant mice lacking the $5-HT_{1B}$ receptor display enhanced aggression and altered 5-HT release in slice preparations from some, but not all, brain areas.[85-87] Null mutant mice also drink twice as much ethanol as wild-type mice and voluntarily ingest solutions containing up to 20% ethanol in water.[88] Null mutants drank 10 g/kg/day on average and consumed as much as 16 to 28 g/kg in the last 24 hours, whereas wild-type mice drank 4 g/kg/day on average. Null mutant intake of food and water and sucrose, saccharin, and quinine solutions, was normal. Null mutant mice were also less sensitive than wild-types in a test of ethanol-induced ataxia. And with repeated drug administration, they tended to develop tolerance more slowly in the grid test. In tests of ethanol withdrawal and metabolism, mutants and wild-type mice showed equivalent responses.[88] These results suggest that the $5\text{-HT}_{1B}$ receptor participates in the regulation of ethanol drinking, and they demonstrate that serotonergic manipulations lead to reduced responsiveness to certain ataxic effects of ethanol without affecting physical dependence.

The motivational effects of ethanol have also been examined by studying the acquisition of ethanol-induced conditioned taste aversion and ethanol-induced conditioned place preference.[89] In the taste conditioning procedure, mice received access to 0.2 M NaCl solution, followed by i.p. injection of 0 to 4 g/kg ethanol. Ethanol produced dose-dependent conditioned taste aversion that was the same in both genotypes. In the place conditioning procedure, null mutant and wild-type mice received six pairings of a tactile stimulus with 2 g/kg ethanol, whereas a different tactile stimulus was paired with saline. Ethanol produced increases in locomotor activity, with wild-type mice showing higher levels of ethanol-stimulated activity than null mutant mice during conditioning trials five and six. Wild-type mice demonstrated conditioned place preference for the ethanol-paired stimulus, whereas null mutant mice showed no evidence of place conditioning. These analyses tend to suggest a role for $5\text{-HT}_{1B}$ receptors in the rewarding effects of ethanol, but not for the aversive effects of ethanol.

## GABA$_A$ Receptors and Modulation by PKCγ

Numerous studies using diverse paradigms have implicated GABA$_A$ receptors in many behavioral effects of ethanol, benzodiazepines, and barbiturates. Antisense ODNs directed against particular GABA$_A$ receptor subunits have been used to examine the role of these subunits *in vitro* for modulation by barbiturates, benzodiazepines, and ethanol.[90] Antisense ODNs against the $\alpha_1$, $\beta_1$, $\gamma_1$, $\gamma_{2S+2L}$, $\gamma_{2L}$, or $\gamma_3$ subunits did not alter GABA action or enhancement by pentobarbital. Diazepam modulation was prevented by the antisense ODN to $\gamma_{2S+2L}$ and reduced by the antisense ODN to $\gamma_{2L}$ but was not affected by antisense ODNs directed against the other receptor subunits. Ethanol enhancement of GABA action was prevented only by an antisense ODN to $\gamma_{2L}$, a subunit that differs from $\gamma_{2S}$ only by the addition of eight amino acids bearing a consensus site for phosphorylation by protein kinase C. Site-directed mutagenesis of key residues in this phosphorylation site of the $\gamma_{2L}$ subunit prevents ethanol modulation of receptors containing the mutant $\gamma_{2L}$ subunit.[91] Null mutant mice that lack the γ isoform of protein kinase C show reduced sensitivity to ethanol on loss of righting reflex and hypothermia, but they show normal responses to flunitrazepam or pentobarbital.[92] Likewise, GABA$_A$ receptor function of isolated brain membranes showed that the null mutation abolished modulation by ethanol but did not alter the actions of flunitrazepam or pentobarbital.[92] These studies suggest that phosphorylation of a site on the $\gamma_{2L}$ subunit is involved in loss of righting reflex and hypothermic responses to ethanol. Antisense ODNs directed against the GABA$_A$ receptor $\gamma_2$ subunit have also been used to examine the role of this subunit *in vivo* using benzodiazepine receptor binding in rat brain[93] and convulsive threshold for benzodiazepine inverse agonists.[94]

Targeted disruption of specific subunits of multimeric GABA$_A$ receptors has only recently been reported.[95,96] Interestingly, inactivation of the $\alpha_6$ gene was shown to abolish δ subunit expression, suggesting that perturbation of one subunit can have profound affects on the expression of other GABA$_A$ receptor subunits or can influence GABA$_A$ receptor assembly.[95] Deletion of the $\alpha_6$ subunit gene did not change GABA$_A$ receptor density in mice, but it markedly reduced affinity for muscimol. Mutant mice did not differ from wild-type mice in sensitivity to loss of righting reflex after pentobarbital or ethanol or to volatile anesthetics.[96]

## Transforming Growth Factor-α (TGFα)

Male transgenic mice that overexpress the human TGFα[97] maintain highly elevated aggressive behavior following alcohol administration,[98] suggesting that male TGFα mice may serve as a model to examine the biological basis of aggression exhibited under the influence of ethanol. The transgenic mice exhibited elevated plasma levels of 17β-estradiol,[99] which has been implicated in aggression and alcohol intake in male animals. Castration abolished the difference in aggressive behavior between transgenic and nontransgenic CD-1 mice by reducing aggression, and alcohol did not increase aggressive behavior in these mice.[100] Treatment with pellets releasing 17β-estradiol over a 60-day period increased aggression in the castrated male TGFα mice and nontransgenic controls to the levels seen in intact male transgenic mice. Alcohol did not significantly alter aggressive behavior in the 17β-estradiol treated castrated mice.[100] These results support a role for high plasma 17β-estradiol in aggressive behavior that is maintained by ethanol.

Overexpression of human TGFα changes neurotransmitter systems, e.g., brain monoamines and $GABA_A$ receptors, that are associated with behaviors altered in the transgenic mice. The 5-hydroxyindoleacetic acid (5-HIAA)/5-HT ratio was significantly reduced in the brain stem of the male TGFα mice and in the frontal cortex of female TGFα mice. Female, but not male, transgenic mice showed elevated levels of norepinephrine in the hypothalamus and 5-HT in the cortex and brain stem when compared with nontransgenic mice.[101] TBPS binding to $GABA_A$ receptors in the forebrain of transgenic mice was less sensitive to blockade of GABA agonist sites by bicuculline or SR95531 as compared to nontransgenic mice. These data suggest either altered endogenous GABA concentrations or a change in $GABA_A$ receptor populations in TGFα transgenic mice, which may contribute to aggressive behavior following alcohol administration.

### Insulin-Like Growth Factor-I (IGF-I) and IGF Binding Protein-1

Transgenic mice that overexpress IGF-I are less sensitive to the hypnotic effect of ethanol as compared to nontransgenic littermate controls.[102] A similar trend, though not significant, was observed for ethanol-induced hypothermia. Transgenic and nontransgenic mice did not differ in ethanol-induced ataxia. IGF-I transgenic mice did not acquire tolerance to either the hypnotic or hypothermic effects of ethanol following seven days of treatment. In contrast, ectopic expression of IGF binding protein-1, a protein not normally expressed in the brain, produced tolerance that was significantly more pronounced than in control mice.[102] No significant differences were observed in blood ethanol concentrations.

### c-fos

Administration of arginine vasopressin reduces the rate of dissipation of tolerance to the hypnotic effect of ethanol in mice, as determined as the duration of loss of righting reflex after i.p. injection of a challenge dose of 3.2 g/kg ethanol.[103,104] Vasopressin, at a dose that maintained ethanol tolerance, acts through $V_1$ receptors in the mouse septum to produce increased c-fos mRNA levels,[105] suggesting a mechanism by which short-lived vasopressin-related neuropeptides could have relatively long-term effects on ethanol tolerance. Intracerebroventricular injection of antisense ODN against c-fos completely blocked the ability of vasopressin to maintain ethanol tolerance and attenuated the increase in septal c-fos mRNA levels caused by vasopressin, whereas a missense ODN was without effect.[35] These studies suggest a role of c-fos expression in the maintenance of ethanol tolerance by vasopressin.

## QUANTITATIVE TRAIT LOCUS (QTL) MAPPING

In the section titled "Recombinant Inbreds, QTL Mapping, and Congenics," the use of RI strains for QTL mapping was introduced. Mapping can also be carried out in F2 crosses, backcrosses, selected lines, congenics, and inbred strains. For a thorough review of QTL mapping methods, see Tanksley.[106] A great deal of effort is currently invested in QTL mapping of drug response genes.[6,107]

## ALCOHOL

More QTL mapping work has been carried out for alcohol-related responses than for other drugs of abuse. This doubtlessly reflects the pharmacogenetic literature in general, which shows a similar preponderance of work with alcohol (see Crabbe and Harris, 1991).[7]

## Withdrawal

Using a two-step genetic mapping strategy, we have mapped loci to regions of mouse chromosomes 1, 4, and 11 that contain genes that influence acute alcohol withdrawal severity. These three risk markers accounted for 68% of the aggregate genetic variability in alcohol withdrawal.[108] In addition, suggestive linkages were indicated for two loci on mouse chromosome 2, one near *Gad1* (encoding glutamic acid decarboxylase) and the other near the *El2* locus (which influences the seizure phenotype in the epilepsy model strain El). The QTL analyses were conducted using the C57BL/6J (B6) and DBA/2J (D2) inbred strains of mice, their BXD RI strains, a B6D2 F2 intercross, and mice selectively bred for high vs. low alcohol withdrawal liability [High (HAW) and Low (LAW) Alcohol Withdrawal].[109] The D2 strain is well-known to display severe withdrawal convulsions, whereas the B6 strain has mild withdrawal reactions.[110]

Genes encoding the $\alpha_1$, $\alpha_6$, and $\gamma_2$ subunits of type-A receptors for the inhibitory neurotransmitter, GABA, map in close proximity to the QTL identified on chromosome 11. Buck and Hood[111] have identified a polymorphism in the coding region of the $\gamma_2$ subunit gene that is genetically correlated with alcohol withdrawal severity. This is the first report that QTL mapping for an alcohol-related trait has successfully led to the identification of a polymorphic candidate gene product that is genetically associated with the trait. Syntenic conservation between human and mouse chromosomes suggests that human homologues of genes that increase risk for physiological dependence may localize to 1q21-q32, 2q24-q37/11p13, 9p21-p23/1p22-32, and 5q32-q35.[108]

## Loss of Righting Reflex Sensitivity

Starting with mouse lines selectively bred for sensitivity to (Long-Sleep, or LS) or resistance to (Short-Sleep, or SS) ethanol-induced loss or righting reflex, Markel et al.[112] have successfully mapped several QTLs influencing the hypnotic response to ethanol. QTLs provisionally identified in RI strains from the LSXSSF2 cross were verified in subsequent analyses with large F2 populations from the LS X SS cross. Five significant QTLs were identified on chromosomes 1, 2, 8, 11, and 15, with LOD scores ranging from 3.4 to 6.9. Together, they account for about 53% of the total genetic variance in the trait. Several potential candidate genes lie within the QTL regions mapped in this study, including *Ntsr*, coding for the high-affinity neurotensin receptor (Chromosome 2), and two acetylcholine subunit genes, *Acrg* and *Acrd*, coding for γ and δ subunits, respectively, near the QTL on chromosome 1. QTL mapping studies in the LSXSS RI strains have also provisionally mapped a QTL for neurotensin high-affinity receptor density near *Ntsr* on chromosome 2.[113,114]

## Reinforcement

Several studies by three groups have published QTL analyses of ethanol preference drinking and acceptance of ethanol solutions, using the B6, D2, and BXD RI strains, short-term selective breeding from the B6D2F2, and other specific genetic crosses.[110,115-119] Significant QTLs have been reported on mouse chromosomes 1, 2, 4, 9, and 11, and rat chromosome 4. The QTL on chromosome 9 is near genes for both the dopamine D2 and serotonin 5-HT$_{1B}$ receptor genes, *Drd2* and *Htr1b*, respectively. Several suggestive QTLs have been provisionally identified on chromosomes 3, 4, 5, 7, 10, 13 and 15. The current picture awaits further clarification for several reasons — different crosses, sexes, strains, and mapping strategies were employed in these studies. Finally, three different measures of preference-related drinking were employed.

Three other analyses of ethanol responses related to reinforcement in BXD RI studies have provisionally nominated QTL regions for ethanol conditioned place preference,[120] conditioned taste aversion, [121] locomotor stimulation[120,122] and sensitization.[122] Confirmatory studies are under way for these traits.

## Other Responses to Alcohol

Other responses to alcohol that have been analyzed for QTLs using the BXD RI strains include hypothermia, locomotor ataxia, and dowel balancing, as well as tolerance to all three of these phenotypes,[123-126] and corticosterone response to acute ethanol injection.[127] All these studies require verification in other crosses to achieve statistically reliable evidence for linkage.[128]

## Opioids

It has been known since the 1970s that large inbred strain differences in morphine sensitivity exist, suggesting sizeable genetic determination. The strains that have been best characterized are C57BL/6 (B6) and DBA/2 (D2), which show large differences on many measures of morphine sensitivity, and the CXBK recombinant inbred strain, which is deficient in μ-opioid receptors and in many morphine responses. This extensive literature has been thoroughly reviewed.[129-131] This makes crosses among these diverse morphine-response strains particularly attractive as QTL mapping populations. Populations derived from B6 and D2, in particular, have been used in all of the reported QTL studies to date.

Voluntary morphine drinking was studied in a large (N = 606) $F_2$ population derived from the B6 and D2 inbred strains (i.e., B6D2F2).[132] Saccharin was used to partially mask the bitter taste of morphine. The mice were presented with a choice between morphine/saccharin and quinine/saccharin (quinine approximates the bitter taste of morphine). The progenitor B6 and D2 strains were known to differ markedly for this behavior,[133-135] which insured that this trait would be highly heritable in the $F_2$ population. The extreme ends of the $F_2$ trait distribution (93 mice) were genotyped for 157 microsatellite markers distributed throughout the genome. They found three QTLs with LOD scores >3.0, which collectively accounted for an estimated 85% of the genetic variance. These QTLs were mapped to chromosomes 1 (~90 cM), 6 (~60 cM) and 10 (~5 cM).

The chromosome 10 QTL was the largest detected and mapped to the same proximal region of chromosome 10 as does the μ-opioid receptor locus (Oprm). Since μ-opioid receptors are known to mediate most effects of morphine, the possibility is raised that the QTL and Oprm may be one and the same. Several morphine sensitivity traits have been studied in the BXD RI set, including morphine drinking (morphine/saccharin vs. water), activity, thermal responses, and hot plate-assessed analgesia.[134,135] For thermal responses (hypothermia) and analgesia, an $F_2$ population derived from B6 and D2 was also tested, and a QTL mapped to the proximal region of chromosome 10 (~5 cM) with LOD scores of about 4.0 for each trait.[6,136] Overall, these findings suggest that Orpm is a plausible candidate gene in a variety of morphine-induced behaviors, and this will be tested further in higher resolution mapping and functional studies. This candidate gene is further supported by the finding that D2 mice show higher receptor density (Bmax) values for [³H]naloxone binding to opioid receptors in whole brain homogenates than B6 mice.[136,137] An even larger difference is seen with the μ-opioid receptor preferring ligand, [³H]DAMGO.[138] In the BXD set, [³H]naloxone Bmax is correlated with morphine analgesia and thermal responses (p < .05) in the expected direction and with allelic variation of chromosome 10 markers in this same region.[136]

## Psychostimulants and Other Drugs

Sizeable inbred mouse strain differences have been reported for several cocaine and amphetamine effects; these have been thoroughly reviewed.[138-143] Several BXD RI QTL studies have been reported for cocaine activity and seizures,[144-146] amphetamine-induced activity,[147,148] stereotypy, and thermal responses.[148] However, none of these provisional QTLs has as yet been tested in an independent population to disconfirm false positives. The only exception known to us is the BXD/B6D2F2 QTL study presently ongoing in our laboratories for cocaine-induced seizures using a timed i.v. infusion, which allows a seizure threshold to be determined for each mouse. Thus far, using combined BXD and B6D2F2 data (N = 400), QTLs influencing cocaine seizures have been mapped to the mid-portion of chromosome 9 (LOD > 5), the mid-portion of chromosome 14 (LOD > 4) and the mid-region of chromosome 15 (LOD > 4).[6] Studies of the BXD RI strains for phencyclidine-induced locomotor activity,[147] nitrous oxide antinociception,[149] and withdrawal[111] have also been conducted, but the provisional QTLs identified await follow-up verification tests.

## RELEVANCE TO HUMAN QTL MAPPING

Most mammals share a number of genetic similarities; these have been best characterized in comparisons of mice and humans, the two best studied mammalian species. For example, the physical length of the genome appears to be similar in mice and humans — about 3 billion base pairs (bp) in both species, and the number of estimated genes is also similar, about 100,000.[150] However, the genetic (linkage) map length is about 3300 cM in humans and 1500 cM in mice, which is largely due to the higher rate of recombination (crossing-over) per generation seen in humans, thus expanding the linkage map. The question of the human relevance of mouse QTL data depends on the degree of homology that exists at primarily three levels-gene homology, linkage homology and trait homology. Molecular cloning has shown that almost all genes in mice or rats have homologues in humans, and vice versa.[150] Gene homology exists when similarity in base pair sequence can be shown between the species, which often implies similarity in function for the homologous gene.

Linkage homology exists when loci located close together on a chromosome in one species are also located close together in another species. This has been most thoroughly studied comparing the mouse to human genomes. This form of homology implies that sizeable segments of chromosomes have remained relatively intact (syntenic conservation) since a common ancestor existed some 70 million years ago.[150] Over 1000 homologous genes have been mapped in both mice and humans, allowing the identification of over 100 chromosomal regions in the mouse genome showing linkage homology with portions of the human genome, each averaging about 9 cM in length. It has been estimated that 80% of the mouse genome shows linkage homology with portions of the human genome.[151] This greatly increases the probability that a QTL mapping result in the mouse will immediately suggest a map site in the human genome, and vice versa. For example, the μ-opioid receptor gene (*Oprm*) was recently mapped in the mouse to proximal chromosome 10.[152] Based on linkage homology, the human μ-opioid receptor gene (*OPRM*) should map to human chromosome 6q24-25. This is indeed where human *OPRM* is located. Since QTL mapping is much more straightforward in the mouse, this genetically well-studied laboratory species is likely to become an important tool in mapping human QTLs. This is a major reason why sequencing the mouse genome is an important objective of the Human Genome Project.[153] The identification of genes underlying QTLs should also be much easier in the mouse, since environmental effects on behavior are more readily controlled and minimized, thus providing a valuable testing ground for candidate genes prior to testing in human populations.

Trait homology implies that a trait measured in two species will have some similarity in its determinants and function. For many cellular and physiological traits, trait homology can be demonstrated, e.g., regulation of blood pressure and gonadal hormone output, and homologous genes can be identified that appear to contribute to the trait homology. For behavioral traits, those showing prominent biological determinants, such as consummatory behavior, sexual behavior, stress reactions, seizure activity, drug withdrawal syndromes, and some aspects of drug reward, appear to show sufficient trait homology to make mouse models useful as a testing ground for genetic hypotheses relevant to human behavior. However, the lack of knowledge concerning the determinants and function of many behavioral traits in *either* species makes trait homology difficult to assess. Thus, we are often forced to rely on mouse behaviors where the degree of trait homology is not well known.

## IDENTIFICATION OF GENES UNDERLYING A QTL

Knowing the approximate map location of a QTL represents an important step toward identifying the specific gene underlying a QTL. This is because of the large number of mapped genes of obvious neurochemical importance in the mouse, making it likely that QTL map sites emerging from genome searches can immediately suggest plausible candidate genes previously mapped to the same region. A good example (see the recent section titled "Opioids") is the mapping of a QTL influencing morphine-induced analgesia, hypothermia, and morphine drinking, to the same proximal chromosome 10 region as the μ-opioid receptor gene (*Oprm*). Candidate genes can be examined initially by screening for variants (polymorphisms) between the two progenitor strains in or around *Oprm*, and by gene expression studies. If, on the other hand, no candidate gene emerges in the chromosomal region of a known QTL, a long-range strategy when there are no candidate genes suitable for testing is positional cloning based on map position.

## NARROWING THE REGION OF A QTL: FINE MAPPING

If a QTL is initially identified using a full genome scan in 1000 $F_2$ mice, for example, the mapping resolution, expressed as the 95% confidence interval, is estimated to be ~25 cM for QTLs accounting for ~3% of the phenotypic variance, and 7 cM for QTLs accounting for 10% of the phenotypic variance.[154,155] This range of QTL effect sizes (3 to 10%) is fairly typical of those attaining Lander and Kruglyak[128] significance levels in published studies with mice and rats. There are undoubtedly even larger numbers of QTLs smaller than this range, but they would be beyond reliable detection for the sample sizes now in common use (200 to 1000 $F_2$ mice). A chromosomal region of 7 cM would involve about 14 million base pairs (Mb) in size (1 cM equals on average ~2 Mb and contains ~65 genes in the mouse).[150] With an estimated 100,000 genes in the mouse genome, we can expect roughly 400 genes to be found in a 7 cM region, only one (or few) of which is responsible for a QTL. Clearly, much higher mapping resolution would be invaluable in the search to identify the gene responsible for a QTL, or for positional cloning.

## CONCLUSIONS AND FUTURE DIRECTIONS

Studies using animal models (e.g., selectively bred lines) clearly demonstrate that acute and chronic responses to drugs of abuse are determined by genetic factors as well as environmental influences. Earlier we focused on studies using transgenic and antisense ODN strategies to examine the roles of targeted genes *in vivo* on behaviors thought to be related to drug abuse. In many ways traditional transgenic and antisense ODN approaches are complementary in their efficacies and limitations.[157] Traditional transgenic null mutant mice experience global gene knockout throughout their development, which can have profound compensatory effects. In contrast, antisense ODN strategies have traditionally been used to produce more discrete gene knock-down, such as at a particular stage of development or within a general region of the brain or spinal cord. However, improvements in transgenic technology (e.g., conditional knockouts) will soon allow rapid induction (or suppression) of transgenes, allowing rapid manipulation of transgene expression at any stage of development. Other advances will allow transgene expression in specific cell types or within discrete brain regions. These methodological improvements will address some of the most serious limitations of transgenic studies using traditional null mutant animals.

Transgenic technology will also make it possible to examine the behavioral effects of substituting one allele for another variant of the same gene. Many investigators consider phenotype switch (e.g., from drug preference to drug aversion) following allele substitution as the gold standard for demonstrating that a specific candidate gene and a QTL are one and the same. Congenic strain strategies can also be used for allele substitution within a discrete chromosome region. They are extremely valuable for higher resolution mapping of a QTL in efforts to identify a candidate gene or as a first step toward positional cloning of the underlying gene.

Since the first mouse genes were mapped about 80 years ago, the number of mapped genes increased by only one or two per year in the first two-thirds of this century. The advent of recombinant molecular biology has led to a meteoric increase in mapped genes to over 5000,[157] largely due to the development of techniques to detect polymorphisms, which can serve as markers and used to detect linkage with newly discovered genes. Most of these genes can be described as single locus traits determined *in vitro*, such as RFLPs based on previously cloned genes.[151] However, the same marker technologies have opened the door to the "last frontier" of gene mapping, the detection and mapping of loci influencing quantitative traits *in vivo*. Less than 2% of all mapped genes are known QTLs for a quantititive trait, but this number is sure to increase in the near future. Given that quantitative traits are far more numerous than qualitative (single locus) traits when studying living organisms,[106] the traits of most interest to behavioral scientists are now amenable to gene mapping, the all-important first step toward identifying the genes involved and their modes of action.[6,107] This promises to be a very lively and exciting field of research in the years to come.

## REFERENCES

1. Vesell, E. S., Advances in pharmacogenetics, *Prog. Med. Genet.*, 9, 291, 1973.
2. Meyer, U. A., Molecular genetics and the future of pharmacogenetics, *Pharmacol. Ther.*, 46, 349, 1990.
3. Crabbe, J. C. and Li, T.-K., Genetic strategies in preclinical substance abuse research, in *Psychopharmacology: A Fourth Generation of Progress*, Bloom, F. E. and Kupfer, D. J., Eds., Raven Press, New York, 1995, 799.
4. McClearn, G. E., The tools of pharmacogenetics, in *The Genetic Basis of Alcohol and Drug Actions*, Crabbe, J. C. and Harris, R. A., Eds., Plenum Publishing Corporation, New York, 1991, 1.
5. Crabbe, J. C., Belknap, J. K., and Buck, K. J., Genetic animal models of alcohol and drug abuse, *Science*, 264, 1715, 1994.
6. Crabbe, J. C., Phillips, T. J., Buck, K. J., Cunningham, C. L., and Belknap, J. K., Identifying genes for alcohol and drug sensitivity: recent progress and future directions, *Trends Neurosci.*, 22, 173, 1999.
7. Crabbe, J. C. and Harris, R. A., Eds., *The Genetic Basis of Alcohol and Drug Actions*, Plenum Publishing Corporation, New York, 1991.
8. Gerlai, R., Gene-targeting studies of mammalian behavior: is it the mutation or the background genotype?, *Trends Neurosci.*, 19, 177, 1996.
9. Plomin, R., McClearn, G. E., and Gora-Maslak, G., Quantitative trait loci and psychopharmacology, *J. Psychopharmacol.*, 5, 1, 1991.
10. Gora-Maslak, G., McClearn, G. E., Crabbe, J. C., Phillips, T. J., Belknap, J. K., and Plomin, R., Use of recombinant inbred strains to identify quantitative trait loci in pharmacogenetic research, *Psychopharmacology*, 104, 413, 1991.
11. Robbins, T. W., Cador, M., Taylor, J. R., and Everitt, B. J., Limbic-striatal interactions in reward-related processes, *Neurosci. Biobehav. Rev.*, 13, 155, 1989.
12. Ritz, M. C., Lamb, R. J., Goldberg, S. R., and Kuhar, M. J., Cocaine receptors on dopamine transporters are related to self-administration of cocaine, *Science*, 237, 1219, 1987.
13. Koob G. F. and Bloom, F. E., Cellular and molecular mechanisms of drug dependence, *Science*, 242, 715, 1988.
14. Giros, B., Jaber, M., Jones, S. R., Wightman, R. M., and Caron, M. G., Hyperlocomotion and indifference to cocaine and amphetamine in mice lacking the dopamine transporter, *Nature*, 379, 606, 1996.
15. Wang, Y. M., Gainetdinov, R. R., Fumagalli, F., Xu, F., Jones, S. R., Bock, C. B., Miller, G. W., Wightman, R. M., and Caron, M. G., Knockout of the vesicular monoamine transporter 2 gene results in neonatal death and supersensitivity to cocaine and amphetamine, *Neuron*, 19, 1285, 1997.

16. Takahaski, N., Miner, L. L., Sora, I., Ujike, H., Revay, R. S., Kostic, V., Jackson-Lewis, V., Przedborski, S., and Uhl, G. R., *VMAT2* knockout mice: heterozygotes display reduced amphetamine-conditioned reward, enhanced amphetamine locomotion, and enhanced MPTP toxicity, *Proc. Natl. Acad. Sci. U.S.A.*, 94, 9938, 1997.

17. Crawford, C. A., Drago, J., Watson, J. B., and Levine, M. S., Effects of repeated amphetamine treatment on the locomotor activity of the dopamine $D_{1A}$-deficient mouse, *Neuroreport*, 8, 2523, 1997.

18. Xu, M., Moratella, R., Gold, L. H., Hiroi, N., Koob, G. F., Graybiel, A. M., and Tonegawa, S., Dopamine $D_1$ receptor mutant mice are deficient in striatal expression of dynorphin and in dopamine-mediated behavioral responses, *Cell*, 79, 729, 1994.

19. Miner, L. L., Drago, J. Chamberlain, P. M., Donovan, D., and Uhl, G. R., Retained cocaine conditioned place preference in $D_1$ receptor deficient mice, *Neuroreport*, 6, 2314, 1995.

20. Xu, M., Hu, X. T., Cooper, D. C., White, F. J., and Tonegawa, S., A genetic approach to study mechanisms of cocaine action, *Ann. N.Y. Acad. Sci.*, 801, 51, 1996.

21. Self, D. W. and Nestler, E. J., Molecular mechanisms of drug reinforcement and addiction, *Annu. Rev. Neurosci.*, 18, 463, 1995.

22. Drago, J., Gerfen, C. R., Westphal, H., and Steiner, H., $D_1$ dopamine receptor-deficient mouse: cocaine-induced regulation of immediate-early gene and substance P expression in the striatum, *Neuroscience*, 74, 813, 1996.

23. Weiss, B., Zhou, L.-W., Shang, S.-P., and Qin Z.-H., Antisense oligodeoxynucleotide inhibits D2 dopamine receptor-mediated behavior and $D_2$ messenger RNA, *Neuroscience*, 55, 607, 1993.

24. Zhang, M. and Creese, I., Antisense oligodeoxynucleotide reduces brain dopamine $D_2$ receptors: behavioral correlates, *Neurosci. Lett.*, 161, 223, 1993.

25. Silvia, C. P., King, G. R., Lee, T. H., Xue, Z. Y., Caron, M. G., and Ellinwood, E. H., Intranigral administration of $D_2$ dopamine receptor antisense oligodeoxynucleotides establishes a role for nigrostriatal $D_2$ autoreceptors in the motor actions of cocaine, *Mol. Pharmacol.*, 46, 51, 1994.

26. Maldonado, R., Saiardi, A., Valverde, O., Samad, T. A., Roques, B. P., and Borrelli, E., Absence of opiate rewarding effects in mice lacking dopamine $D_2$ receptors, *Nature,* 388, 586, 1997.

27. Xu, M., Koeltzow, T. E., Santiago, G. T., Moratalla, R., Cooper, D. C., Hu, X. T., White, N. M., Graybiel, A. M., White, F. J., and Tonegawa, S., Dopamine $D_3$ receptor mutant mice exhibit increased behavioral sensitivity to concurrent stimulation of $D_1$ and $D_2$ receptors, *Neuron,* 19, 837, 1997.

28. Rubinstein, M., Phillips, T. J., Bunzow, J. R., Falzone, T. L., Dziewczapolski, G., Zhang, G., Fang, Y., Larson, J. L., McDougall, J. A., Chester, J. A., Saez, C., Pugsley, T. A., Gershanik, O., Low, M. J., and Grandy, D. K., Mice lacking dopamine $D_4$ receptors are supersensitive to ethanol, cocaine and methamphetamine, *Cell*, 90, 991, 1997.

29. Rocha, B. A., Ator, R., Emmett-Oglesby, M. W., and Hen, R., Intravenous cocaine self-administration in mice lacking 5-HT$_{1B}$ receptors, *Pharmacol. Biochem. Behav.*, 57, 407, 1997.

30. Lucas, J. J., Segu, L., and Hen, R., 5-Hydroxytryptamine$_{1B}$ receptors modulate the effect of cocaine on *c-fos* expression: converging evidence using 5-Hydroxytryptamine$_{1B}$ knockout mice and the 5-Hydroxytryptamine$_{1B/1D}$ antagonist GR127935, *Mol. Pharmacol.*, 51, 755, 1997.

31. Koob, G. F., Drugs of abuse: anatomy, pharmacology and function of reward pathways, *Trends Pharmacol. Sci.*, 13, 177, 1992.

32. Graybiel, A. M., Moratalla, R., and Robertson, H. A., Amphetamine and cocaine induce drug-specific activation of the *c-fos* gene in striosome-matrix compartments and limbic subdivisions of the striatum, *Proc. Natl. Acad. Sci. U.S.A.*, 87, 6912, 1990.

33. Heilig, M., Engel, J. A., and Soderpalm, B., *c-fos* antisense in the nucleus accumbens blocks the locomotor stimulant action of cocaine, *Eur. J. Pharmacol.*, 236, 339, 1993.

34. Sommer, W., Bjelke, B., Ganten, D., and Fuxe, K., Antisense oligonucleotide to *c-fos* induces ipsilateral rotational behavior to D-amphetamine, *Neuroreport*, 5, 277, 1993.

35. Hooper, M. L., Chiasson, B. J., and Robertson, H. A., Infusion into the brain of an antisense oligonucleotide to the immediate-early gene *c-fos* suppresses production of Fos and produces a behavioral effect, *Neuroscience*, 63, 917, 1994.

36. Szabo, G., Nunley, K. R., and Hoffman, P. L., Antisense oligonucleotide to *c-fos* blocks the ability of arginine vasopressin to maintain ethanol tolerance, *Eur. J. Pharmacol.*, 306, 67, 1996.

37. Sommer, W., Rimondini, R., O'Connor, W., Hansson, A. C., Ungerstedt, U., and Fuxe, K., Intrastriatally injected *c-fos* antisense oligonucleotide interferes with striatonigral but not striatopallidal γ-aminobutyric acid transmission in the conscious rat, *Proc. Natl. Acad. Sci. U.S.A.*, 93, 14134, 1996.

38. Johnson, R. S., Spiegelman, B. M., and Papaioannou, V., Pleiotropic effects of a null mutation in the *c-fos* proto-oncogene, *Cell*, 71, 577, 1992.

39. Sassone-Corsi, P., Visvader, J., Ferland, L., Mellon, P. L., and Verma, I. M., Induction of proto-oncogene *fos* transcription through the adenylate cyclase pathway: characterization of a cAMP-responsive element, *Genes Dev.*, 2, 1529, 1988.

40. Sheng, M., McFadden, G., and Greenberg, M., Membrane depolarization and calcium induce *c-fos* transcription via phosphorylation of transcription factor CREB, *Neuron*, 4, 571, 1990.

41. Fink, J. S., Verhave, M., Kasper, S., Tsukada, T., Mandel, G., and Goodman, R. H., The CGTCA sequence motif is essential for biological activity of the vasoactive intestinal peptide gene cAMP-regulated enhancer, *Proc. Natl. Acad. Sci. U.S.A.*, 85, 6662, 1988.

42. Gonzalez, G. A. and Montminy, M. R., Cyclic AMP stimulates somatostatin gene transcription by phosphorylation of CREB at serine 133, *Cell*, 59, 675, 1989.

43. Montminy, M. R., Gonzalez, G. A., and Yamamoto, K. K., Regulation of cAMP-inducible genes by CREB, *Trends Neurosci.*, 13, 184, 1990.

44. Konradi, C., Kobierski, L., Nguyen, T. V., Heckers, S., and Hyman, S. E., The cAMP-response-element-binding protein interacts, but fos protein does not interact, with the proenkephalin enhancer in rat striatum, *Proc. Natl. Acad. Sci. U.S.A.*, 90, 7005, 1993.

45. Konradi, C., Cole, R. L., Heckers, S., and Hyman, S. E., Amphetamine regulates gene expression in rat striatum via transcription factor CREB, *J. Neurosci.*, 14, 5623, 1994.

46. Cadet, J. L., Ali, S. F., Rothman, R. B., and Epstein, C. J., Neurotoxicity, drugs and abuse, and the CuZn-superoxide dismutase transgenic mice, *Mol. Neurobiol.*, 11, 155, 1995.

47. Hirata, H., Ladenheim, B., Carlson, E., Epstein, C., and Cadet, J. L., Autoradiographic evidence for methamphetamine-induced striatal dopaminergic loss in mouse brain: attenuation in CuZn-superoxide dismutase transgenic mice, *Brain Res.*, 714, 95, 1996.

48. Douglass, J. and Daoud, S., Characterization of the human cDNA and genomic DNA encoding CART: a cocaine- and amphetamine-regulated transcript, *Gene*, 169, 241, 1996.

49. Douglass, J., McKinzie, A. A., and Couceyro, P., PCR differential display identifies a rat brain mRNA that is transcriptionally regulated by cocaine and amphetamine, *J. Neurosci.*, 15, 2471, 1995.

50. Wang, X. B., Funada, M., Imai, Y., Revay, R. S., Ujike, H., Vandenbergh, D. J., and Uhl, G. R., *rGbeta1*: a psychostimulant-regulated gene essential for establishing cocaine sensitization, *J. Neurosci.*, 17, 5993, 1997.

51. Kieffer, B. L., Recent advances in molecular recognition and signal transduction of active peptides: receptors for opioid peptides, *Cell. Mol. Neurobiol.*, 15, 615, 1995.

52. Matthes, H. W. D., Maldonado, R., Simonin, F., Valverde, O., Slowe, S., Kitchen, I., Befort, K., Dierich, A., Le Meur, M., Dolle, P., Tzavara, E., Hanoune, J., Roques, B. P., and Kieffer, B. L., Loss of morphine-induced analgesia, reward effect and withdrawal symptoms in mice lacking the µ-opioid receptor gene, *Nature*, 383, 819, 1996.

53. Sora, I., Takahashi, N., Funada, M., Ujike, H., Revay, R. S., Donovan, D. M., Miner, L. L., and Uhl, G. R., Opiate receptor knockout mice define µ receptor roles in endogenous nociceptive responses and morphine-induced analgesia, *Proc. Natl. Acad. Sci. U.S.A.*, 94, 1544, 1997.

54. Rossi, G., Pan, Y.-X., Cheng, J., and Pasternak, G. W., Blockade of morphine analgesia by an antisense oligodeoxynucleotide against the µ-receptor, *Life Sci.*, 54, PL-375, 1994.

55. Sanchez-Blazquez, P., Garcia-Espana, A., and Garzon, J., Antisense oligodeoxynucleotides to opioid µ and δ receptors reduced morphine dependence in mice: role of $\delta_2$ opioid receptors, *J. Pharmacol. Exp. Ther.*, 280, 1423, 1997.

56. Chen, X.-H., Adams, J. U., Geller, E. B., De Riel, J. K., Adler, M. W., and Liu-Chen, L.-Y., An antisense oligodeoxynucleotide to µ-opioid receptors inhibits µ-opioid receptor agonist-induced analgesia in rats, *Eur. J. Pharmacol.*, 275, 105, 1995.

57. Chen, X.-H., Liu-Chen, L.-Y., Tallarida, R. J., Geller, E. B., De Riel, J. K., and Adler, M. W., Use of a µ-antisense oligodeoxynucleotide as a µ-opioid receptor noncompetitive antagonist *in vivo*, *Neurochem. Res.*, 21, 1363, 1996.

58. Idanpaan-Heikkila, J. J., Pauhala, P., and Mannisto, P. T., μ- and δ-opioid receptor antisense oligodeoxynucleotides antagonize morphine-induced growth hormone secretion in rats, *Eur. J. Pharmacol.*, 284, 227, 1995.

59. Baamond A., Dauge, V., Gacel, C., and Rogues, B. P. J., Systemic administration of Tyr-D-Ser(O-tert-butyl)-Gly-Phe-Leu-Thr(O-tert-butyl), a highly selective δ-opioid agonist, induces μ-receptor-mediated analgesia in mice, *J. Pharmacol. Exp. Ther.*, 257, 767, 1991.

60. Dickenson, A. H., Mechanisms of the analgesic actions of opiates and opioids, *Br. Med. Bull.*, 47, 690, 1991.

61. Sora, I., Funada, M., and Uhl, G. R., The μ-opioid receptor is necessary for [D-Pen$^2$, D-Pen$^5$]enkephalin-induced analgesia, *Eur. J. Pharmacol.*, 324, R1, 1997.

62. Nestler, E. J., Alreja, M., and Aghajanian, G. K., Molecular and cellular mechanisms of opiate action: studies in the rat locus coeruleus, *Brain Res. Bull.*, 35, 521, 1994.

63. Rasmussen, K., Batner-Johnson, D., Krystal, J. H., Aghajanian, G. K., and Nestler, E. J., Opiate withdrawal and the rat locus coeruleus. Behavioral, electrophysiological and biochemical correlates, *J. Neurosci.*, 10, 2308, 1990.

64. Donovan, D. M., Miner, L. L., Sharpe, L., Cho, H. J., Takemura, M., Unterwald, E., Nyberg, F., Kreek, M. J., and Uhl, G. R., Transgenic mice: preproenkephalin and the μ-opiate receptor, *Ann. N.Y. Acad. Sci.*, 780, 19, 1996.

65. Simonin, F., Valverde, O., Smadja, C., Slowe, S., Kitchen, I., Dierich, A., LeMeur, M., Roques, B. P., Maldonado, R., and Kieffer, B. L., Disruption of the κ-opioid receptor gene in mice enhances sensitivity to chemical visceral pain, impairs pharmacological actions of the selective κ-agonist U-50,488H and attenuates morphine withdrawal, *EMBO J.*, 17, 886, 1998.

66. Kest, B., Lee, C. E., McLemore, G. L., and Inturrisi, C. E., An antisense oligodeoxynucleotide to the δ-opioid receptor *(DOR-1)* inhibits morphine tolerance and acute dependence in mice, *Brain Res. Bull.*, 39, 185, 1996.

67. Suzuki, T., Ikeda, H., Tsuji, M., Misawa, M., Narita, M., and Tsen, L. F., antisense oligodeoxynucleotide to δ-opioid receptors attenuates morphine dependence in mice, *Life Sci.*, 61, PL 165, 1997.

68. Maldonado, R., Negus, S., and Koob, G. F., Precipitation of morphine withdrawal syndrome in rats by administration of μ-, δ- and κ-selective opioid antagonists, *Neuropharmacology*, 31, 1231, 1992.

69. Wei, E. T., Enkephalin analogs and physical dependence, *J. Pharmacol. Exp. Ther.*, 216, 12, 1981.

70. Cowan, A., Zhu, X. Z., Mosberg, H. I., Omnass, J. R., and Porreca, F., Direct dependence studies in rats with agents selective for different types of opioid receptor, *J. Pharmacol. Exp. Ther.*, 246, 950, 1988.

71. Uphouse, L. A., Welch, S. P., Ward, C. R., Ellis, E. F., and Embrey, J. P., Antinociceptive activity of intrathecal ketorolac is blocked by the κ-opioid receptor antagonist, nor-binaltorphimine, *Eur. J. Pharmacol.*, 242, 53, 1993.

72. Herz, A., *Opiates*, 2nd ed., Springer, Berlin, 1993, 21.

73. Standifer, K. M., Chien, C.-C., Wahlestedt, C., Brown, G. P., and Pasternak, G. W., Selective loss of δ opioid analgesia and binding by antisense oligonucleotides to a δ opioid receptor, *Neuron*, 12, 805, 1994.

74. Chien, C.-C., Brown, G., Pan, Y.-X., and Pasternak, G. W., Blockade of U50,488H analgesia by antisense oligodeoxynucleotides to a κ-opioid receptor, *Eur. J. Pharmacol.*, 253, R7, 1994.

75. Wang, H.Q., Kampine, J.P., and Tseng, L.F., Antisense oligodeoxynucleotide to a δ-opioid receptor messenger RNA selectively blocks the antinociception induced by intracerebroventricularly administered δ-, but not μ-, ε- or κ-opioid receptor agonists in the mouse, *Neuroscience*, 72, 2, 1996

76. Traynor, J. R. and Elliot, J., δ-Opioid receptor subtypes and cross-talk with μ-receptors, *Trends Pharmacol. Sci.*, 14, 84, 1993.

77. Rubinstein, M., Mogil, J. S., Japon, M., Chan, E. C., Allen, R. G., and Low, M. J., Absence of opioid stress-induced analgesia in mice lacking β-endorphin by site-directed mutagenesis, *Proc. Natl. Acad. Sci. U.S.A.*, 93, 3995, 1996.

78. Raffa, R. B., Martinez, R. P., and Connelly, C. D., G-protein antisense oligodeoxy-nucleotides and μ-opioid supraspinal antinociception, *Eur. J. Pharmacol.*, 258, R5, 1994.

79. Rossi, G., Standifer, K. M., and Pasternak, G. W., Differential blockade of morphine and morphine-6β-glucuronide analgesia by antisense oligodeoxynucleotides directed against *MOR-1* and G-protein α subunits in rats, *Neurosci. Lett.*, 198, 99, 1995.

80. Raffa, R. B., Goode, T. L., Martinez, R. P., and Jacoby, H. I., A $G_i2\alpha$ antisense oligonucleotide differentiates morphine antinociception, constipation and acute dependence in mice, *Life Sci.*, 58, PL73, 1996.

81. Sanchez-Blazquez, P., Garcia-Espana, A., and Garzon, J., *In vivo* injection of antisense oligodeoxynucleotides to G$\alpha$ subunits and supraspinal analgesia evoked by $\mu$ and $\delta$ opioid agonists, *J. Pharmacol. Exp. Ther.*, 275, 1590, 1995.

82. Widnell, K. L., Self, D. W., Lane, S. B., Russell, D. S., Vaidya, V. A., Miserendino, M. J. D., Rubin, C. S., Duman, R. S., and Nestler, E. J., Regulation of *CREB* expression: *In vivo* evidence for a functional role in morphine action in the nucleus accumbens, *J. Pharmacol. Exp. Ther.*, 276, 306, 1995.

83. Lane-Ladd, S. B., Pineda, J., Boundy, V. A., Pfeuffer, I., Krupinski, J., Aghajanian, G. K., and Nestler, E. J., CREB (cAMP response element-binding protein) in the locus coeruleus: biochemical, physiological, and behavioral evidence for a role in opiate dependence, *J. Neurosci.*, 17, 7890, 1997.

84. Maldonado, R., Blendy, J. A., Tzavara, E., Gass, P., Roques, B. P., Hanoune, J., and Schutz, G., Reduction of morphine abstinence in mice with a mutation in the gene encoding CREB, *Science*, 273, 657, 1996.

85. Saudou, F., Amara, D. A., Dierich, A., LeMeur, M., Ramboz, S., Segu, L., Buhot, M. C., and Hen, R., Enhanced aggressive behavior in mice lacking 5-HT$_{1B}$ receptor, *Science*, 265, 1875, 1994.

86. Ramboz, S., Saudou, F., Ait Amara, D., Belzung, C., Segu, L., Misslin, R., and Buhot, M. C., Hen, R., 5-HT$_{1B}$ receptor knock out — behavioral consequences, *Behav. Brain Res.*, 73, 305, 1996.

87. Piñeyro, G., Castanon, N., Hen, R., and Blier, P., Regulation of [$^3$H]5-HT release in raphe, frontal cortex and hippocampus of 5-HT$_{1B}$ knock-out mice, *Neuroreport*, 7, 353, 1995.

88. Crabbe, J. C., Phillips, T. J., Feller, D. J., Hen, R., Wenger, C. D., Lessov, C. N., and Schafer, G. L., Elevated alcohol consumption in null mutant mice lacking 5-HT$_{1B}$ serotonin receptors, *Nat. Genet.*, 14, 98, 1996.

89. Risinger, F. O., Bormann, N. M., and Oakes, R. A., Reduced sensitivity to ethanol reward, but not ethanol aversion, in mice lacking 5-HT$_{1B}$ receptors, *Alc. Clin. Exp. Res.*, 20, 1401, 1996.

90. Wafford, K. A., Burnett, D. M., Leidenheimer, N. J., Burt, D. R., Wang, J. B., Kofuji, P., Dunwiddie, T. V., Harris, R. A., Sikela, J.M., Ethanol sensitivity of the GABA$_A$ receptor expressed in *Xenopus* oocytes requires 8 amino acids contained in the $\gamma_{2L}$ subunit, *Neuron*, 7, 27, 1991.

91. Wafford, K. A., Burnett, D., Harris, R. A., and Whiting, P. J., GABA$_A$ receptor subunit expression and sensitivity to ethanol, *Alc. Alcoholism*, 2, 327, 1993.

92. Harris, R. A., McQuilkin, S. J., Paylor, R., Abeliovich, A., Tonegawa, S., and Wehner, J. M., Mutant mice lacking the $\gamma$ isoform of protein kinase C show decreased behavioral actions of ethanol and altered function of $\gamma$-aminobutyrate type-A receptors, *Proc. Natl. Acad. Sci. U.S.A.*, 92, 3658, 1995.

93. Karle, J. and Nielsen, M., Modest reduction of benzodiazepine binding in rat brain *in vivo* induced by antisense oligonucleotide to GABA$_A$ receptor $\gamma$2 subunit subtype, *Eur. J. Pharmacol.*, 291, 439, 1995.

94. Karle, J., Witt, M. R., and Nielsen, M., Antisense oligonucleotide to GABA$_A$ receptor $\gamma$2 subunit induces loss of neurons in rat hippocampus, *Neurosci. Lett.*, 202, 97, 1995.

95. Zhao, T. J., Rosenberg, H. C., and Chiu, T. H., Treatment with an antisense oligodeoxy-nucleotide to the GABA$_A$ receptor $\gamma$2 subunit increases convulsive threshold for $\beta$-CCM, a benzodiazepine inverse agonist in rats, *Eur. J. Pharmacol.*, 306, 61, 1996.

96. Jones, A., Korpi, E. R., McKernan, R. M., Pelz, R., Nusser, Z., Makela, R., Mellor, J. R., Pollard, S., Bahn, S., Stephenson, F. A., Randall, A. D., Sieghart, W., Somogyi, P., Smith, A. J. H., and Wisden, W., Ligand-gated ion channel subunit partnerships: GABA$_A$ receptor $\alpha_6$ subunit gene inactivation inhibits $\delta$ subunit expression, *J. Neurosci.*, 17, 1350, 1997.

97. Jhappan, C., Stahle, C., Harkins, R. N., Fausto, N., Smith, G. H., and Merlino, G. T., TGF$\alpha$ overexpression in transgenic mice induces liver neoplasia and abnormal development of the mammary gland and pancreas, *Cell*, 61, 1137, 1990.

98. Hilakivi-Clarke, L. A. and Goldberg, R., The effects of alcohol on elevated levels of aggressive behavior in transgenic mice overexpressing transforming growth factor-$\alpha$, *Neuroreport*, 4, 155, 1993.

99. Hilakivi-Clarke, L. A., Arora, P. K., Sabol, M. B., Clarke, R., Dickson, R. B., and Lippman, M. E., Alterations in behavior, steroid hormones and natural killer cell activity in male transgenic TGF$\alpha$ mice, *Brain Res.*, 588, 97, 1992.

100. Hilakivi-Clarke, L. A. and Goldberg, R., Gonadal hormones and aggression-maintaining effect of alcohol in male transgenic transforming growth factor-$\alpha$ mice, *Alc. Clin. Exp. Ther.*, 19, 708, 1995.

101. Hilakivi-Clarke, L. A., Corduban, T. D., Taira, T., Hitri, A., Deutsch, S., Korpi, E. R., Goldberg, R., and Kellar, K. J., Alterations in brain monoamines and $GABA_A$ receptors in transgenic mice overexpressing TGFα, *Pharmacol. Biochem. Behav.*, 50, 593, 1995.

102. Pucilowski, O., Ayensu, W. K., and D'Ercole, A. J., Insulin-like growth factor I expression alters acute sensitivity and tolerance to ethanol in transgenic mice, *Eur. J. Pharmacol.*, 305, 57, 1996.

103. Hoffman, P. L., Ritzmann, R. F., Walter, R., and Tabakoff, B., Arginine vasopressin maintains ethanol tolerance, *Nature*, 276, 614, 1978.

104. Lê, A.-D., Kalant, H., and Khanna, J. M., Interaction between des-glycinamide$^9$[Arg$^8$]-vasopressin and serotonin on ethanol tolerance, *Eur. J. Pharmacol.*, 80, 337, 1982.

105. Giri, P. R., Dave, B., Tabakoff, B., and Hoffman, P. L., Arginine vasopressin induces the expression of *c-fos* in the mouse septum and hippocampus, *Mol. Brain Res.*, 7, 131, 1990.

106. Tanksley, S. D., Mapping polygenes, *Ann. Rev. Genet.*, 27, 205, 1993.

107. Belknap, J. K., Dubay, C., Crabbe, J. C., and Buck, K. J., Mapping quantitative trait loci for behavioral traits in the mouse, in *Handbook of Psychiatric Genetics*, Blum, K. and Noble, E. P., Eds., CRC Press LLC, New York, 435, 1997.

108. Buck, K. J., Metten, P., Belknap, J. K., and Crabbe, J. C., Quantitative trait loci involved in genetic predisposition to acute alcohol withdrawal in mice, *J. Neurosci.*, 17, 3946, 1997.

109. Metten, P., Belknap, J. K., and Crabbe, J. C., Drug withdrawal convulsions and susceptibility to convulsants after short-term selective breeding for acute alcohol withdrawal, *Behav. Brain Res.*, 95, 113, 1998.

110. Crabbe, J. C., Kosobud, A., Young, E., and Janowsky, J., Polygenic and single-gene determination of response to ethanol in BXD/Ty recombinant inbred mouse strains, *Neurobehav. Toxicol. Terat.*, 5, 181, 1983.

111. Buck, K. J. and Hood, H. M., Genetic association of a $GABA_A$ receptor $\gamma_2$ subunit variant with acute physiological dependence on alcohol, *Mammalian Genome*, 9, 975, 1998.

112. Markel, P. D., Bennett, B., Beeson, M., Gordon, L., and Johnson, T. E., Confirmation of quantitative trait loci for ethanol sensitivity in long-sleep and short-sleep mice, *Genome Res.*, 7, 92, 1997.

113. Erwin, V. G., Markel, P. D., Johnson, T. E., Gehle, V. M., and Jones, B. C., Common quantitative trait loci for alcohol-related behaviors and central nervous system neurotensin measures: hypnotic and hypothermic effects, *J. Pharmacol. Exper. Ther.*, 280, 911, 1997.

114. Erwin, V. G., Radcliffe, R. A., Gehle, V. M., and Jones, B. C., Common quantitative trait loci for alcohol-related behaviors and central nervous system neurotensin measures: locomotor activation, *J. Pharmacol. Exper. Ther.*, 280, 919, 1997.

115. Phillips, T. J., Belknap, J. K., Buck, K. J., and Cunningham, C. L., Genes on mouse chromosomes 2 and 9 determine variation in ethanol consumption, *Mammalian Genome*, 9, 936, 1998.

116. Tarantino, L. M., McClearn, G. E., Rodriguez, L. A., and Plomin, R., Confirmation of quantitative trait loci for alcohol preference in mice, *Alc. Clin. Exp. Res.*, 22, 1099, 1998.

117. Melo, J. A., Shendure, J., Pociask, K., and Silver, L. M., Identification of sex-specific quantitative trait loci controlling alcohol preference in C57BL/6 mice, *Nat. Genet.*, 13, 147, 1996.

118. Gill, K., Desaulniers, N., Desjardins, P., and Lake, K., Alcohol preference in AXB/BXA recombinant inbred mice: gender differences and gender-specific quantitative trait loci, *Mammalian Genome*, 9, 929, 1998.

119. Bice, P., Foroud, T., Bo, R., Castelluccio, P., Lumeng, L., Li, T.-K., and Carr, L., Genomic screen for GTLs underlying alcohol consumption in the P and NP rat lines, *Mammalian Genome*, 9, 949, 1998.

120. Cunningham, C. L., Localization of genes influencing ethanol-induced conditioned place preference and locomotor activity in BXD recombinant inbred mice, *Psychopharm.*, 120, 28, 1995.

121. Risinger, F. O. and Cunningham, C. L., Identification of genetic markers associated with sensitivity to ethanol-induced conditioned taste aversion, *Alc. Clin. Exp. Res.*, 18, 451, 1994.

122. Phillips, T. J., Huson, M., Gwiazdon, C., Burkhart-Kasch, S., and Shen, E. H., Effects of acute and repeated ethanol exposures on the locomotor activity of BXD recombinant inbred mice, *Alc. Clin. Exp. Res.*, 19, 269, 1995.

123. Gallaher, E. J., Jones, G. E., Belknap, J. K., and Crabbe, J. C., Identification of genetic markers for initial sensitivity and rapid tolerance to ethanol-induced ataxia using quantitative trait locus analysis in BXD recombinant inbred mice, *J. Pharm. Exp. Ther.*, 227, 604, 1996.

124. Crabbe, J. C., Belknap, J. K., Mitchell, S. R., and Crawshaw, L. I., Quantitative trait loci mapping of genes that influence the sensitivity and tolerance to ethanol-induced hypothermia in BXD recombinant inbred mice, *J. Pharmacol. Exp. Ther.*, 269, 184, 1994.

125. Crabbe, J. C., Phillips, T. J., Gallaher, E. J., Crawshaw, L. I., and Mitchell, S. R., Common genetic determinants of the ataxic and hypothermic effects of ethanol in BXD/Ty recombinant inbred mice: genetic correlations and quantitative trait loci, *J. Pharmacol. Exp. Ther.*, 277, 624, 1996.

126. Phillips, T. J., Lessov, C. N., Harland, R. D., and Mitchell, S. R., Evaluation of potential genetic associations between ethanol tolerance and sensitization in BXD/Ty recombinant inbred mice, *J. Pharmacol. Exp. Ther.*, 277, 613, 1996.

127. Roberts, A. J., Phillips, T. J., Belknap, J. K., Finn, D. A., and Keith, L. D., Genetic analysis of the corticosterone response to ethanol in BXD recombinant inbred mice, *Behav. Neurosci.*, 109, 1199, 1995.

128. Lander, E. S., and Kruglyak, L., Genetic dissection of complex traits: guidelines for interpreting and reporting linkage results, *Nature Genet.*, 11, 241, 1995.

129. Shuster, L., Genetic markers of drug abuse in mouse models, in *Genetic and Biological Markers in Drug Abuse and Alcoholism,* NIDA Res. Monogr. 66, Braude, M. C. and Chao, H. M., Eds., USGPO, Washington, DC, 71, 1986.

130. Belknap, J. K. and O'Toole, L. A., Studies of genetic differences in responses to opioid drugs, in *The Genetic Basis of Alcohol and Drug Actions*, Crabbe, J. C. and Harris, R. A., Eds., Plenum Press Corporation, New York, 225, 1991.

131. Mogil, J. S., Sternberg, W. F., Marek, P., Sadowski, B., Belknap, J. K., and Liebeskind, J. C., The genetics of pain and pain inhibition, *Proc. Nat. Acad. Sci.*, 93, 3048, 1996.

132. Berrettini, W. H., Ferraro, T. N., Alexander, R. C., Buchberg, A. M., and Vogel, W. H., Quantitative trait loci mapping of three loci controlling morphine preference using inbred mouse strains, *Nat. Genet.*, 7, 54, 1994.

133. Belknap, J. K., Physical dependence induced by the voluntary consumption of morphine in inbred mice, *Pharmacol. Biochem. Behav.*, 35, 311, 1990.

134. Belknap, J. K. and Crabbe, J. C., Chromosome mapping of gene loci affecting morphine and amphetamine responses in BXD recombinant inbred mice, in *The Neurobiology of Alcohol and Drug Addiction, Annals of the N.Y. Academy of Sciences*, Kalivas, P. and Samson, H., Eds., 654, 311, 1992.

135. Phillips, T. J., Belknap, J. K., and Crabbe, J. C., Use of voluntary morphine consumption in recombinant inbred strains to assess vulnerability to drug abuse at the genetic level, *J. Add. Dis.*, 10, 73, 1991.

136. Belknap, J. K., Mogil, J. S., Helms, M. L., Richards, S. P., O'Toole, L. A., Bergeson, S. E., and Buck, K. J., Localization to chromosome 10 of a locus influencing morphine-induced analgesia in crosses derived from C57BL/6 and DBA/2 mice, *Life Sci. (Pharmacol. Lett.)*, 57, PL117, 1995.

137. Belknap, J.K., Noordewier, B., and Lamè, M., Genetic dissociation of multiple morphine effects among C57BL/6J, DBA/2J and C3H/HeJ inbred mouse strains, *Physiol. Behav.*, 46, 69, 1989.

138. Broadhurst, P., *Drugs and the Inheritance of Behavior*, Plenum Press Corporation, New York, 1978.

139. Seale, T. W., Genetic differences in response to cocaine and stimulant drugs, in *The Genetic Basis of Alcohol and Drug Actions*, Crabbe, J. C. and Harris, R. A., Eds., Plenum Press Corporation, New York, 279, 1991.

140. Seale, T. W. and Carney, J. M., Genetic determinants of susceptibility to the rewarding and other behavioral actions of cocaine, *J. Add. Dis.*, 10, 141, 1991.

141. Marley, R. J., Elmer, G. I., and Goldberg, S. R., The use of pharmacogenetic techniques in drug abuse research, *Pharmacol. Ther.*, 53, 217, 1992.

142. Morse, A. C., Erwin, V. G., and Jones, B. C., Pharmacogenetics of cocaine: a critical review, *Pharmacogenetics*, 5, 183, 1995.

143. Uhl, G. R., Elmer, G. I., LaBuda, M. C., and Pickens, R. W., Genetic influences in drug abuse, in *Psychopharmacology: The Fourth Generation of Progress*, Bloom, F. E. and Kupfer, D. J., Eds., Raven Press, New York, 1793, 1995.

144. Miner, L. L. and Marley, R. J., Chromosomal mapping of loci influencing sensitivity to cocaine-induced seizures in BXD recombinant inbred strains of mice, *Psychopharmogy*, 117, 62, 1995.

145. Miner, L. L. and Marley, R. J., Chromosomal mapping of the psychomotor stimulant effects of cocaine in BXD recombinant inbred mice, *Psychopharmogy*, 122, 209, 1995.

146. Tolliver, B. K., Belknap, J. K., Woods, W. E., and Carney, J. M., Genetic analysis of sensitization and tolerance to cocaine, *J. Pharm. Exp. Ther.*, 270, 1230, 1994.

147. Alexander, R. C., Wright, R., and Freed, W., Quantitative trait loci contributing to phencyclidine-induced and amphetamine-induced locomotor behavior in inbred mice, *Neuropsychopharmacology*, 15, 484, 1996.

148. Grisel, J. E., Belknap, J. K, O'Toole, L. A., Helms, M. L., Wenger, C. D., and Crabbe, J. C., Quantitative trait loci affecting methamphetamine responses in BXD recombinant inbred mouse strains, *J. Neurosci.*, 17, 745, 1997.

149. Quock, R. M., Mueller, J. L., Vaughn, L. K., and Belknap, J. K., Nitrous oxide antinociception in BXD recombinant mouse strains and identification of quantitative trait loci, *Brain Res.*, 725, 1996.

150. Silver, L. M., *Mouse Genetics: Concepts and Applications,* Oxford Press, Oxford, 1995.

151. Copeland, N. G., Jenkins, N. A., Gilbert, D. J., Eppig, J. T., Maltais, L. J., Miller, J. C., Dietrich, W. F., Weaver, A., Lincoln, S. E., Steen, R. G., Stein, L. D., Nadeau, J. H., and Lander, E. S., A genetic linkage map of the mouse: current applications and future prospects, *Science*, 262, 57, 1993.

152. Kozak, C. A., Filie, J., Adamson, M. C., Chen, Y., and Yu, L., Murine chromosomal location of the $\mu$- and $\kappa$-opioid receptor genes, *Genomics*, 21, 659, 1994.

153. Collins, F. and Galas, D., A new five-year plan for the U. S. human genome project, *Science*, 262, 43, 1993.

154. Darvasi, A. and Soller, M., Advanced intercross lines, an experimental population for fine genetic mapping, *Genetics*, 141, 1199, 1995.

155. Darvasi, A. and Soller, M., A simple method to calculate resolving power and confidence interval of QTL map location, *Behav. Genet.*, 27, 125, 1997.

156. Wehner, J. M. and Bowers, B., Use of transgenics, null mutants, and antisense approaches to study ethanol's actions, *Alc. Clin. Exp. Res.*, 19, 811, 1995.

157. Silver, L. M. and Nadeau, J. H., Encyclopedia of the Mouse Genome VI, *Mammalian Genome*, 7, S1 (Special issue), 1997.

# 10 Molecular Genetics of Primary Tumors of the Central Nervous System

*Frank J. Coufal, H.-J. Su Huang, and Webster K. Cavenee*

## CONTENTS

## INTRODUCTION

During the development of the central nervous system (CNS), a careful equilibrium is maintained between the processes of cell proliferation, differentiation, and apoptosis. In contrast, tumorigenesis is a complex, multistep sequence of cellular transformations resulting from the accumulation of mutations in the very genes that govern these three balanced processes. According to the currently accepted paradigm for tumorigenesis, these mutations can manifest by either of two mechanisms: activation of a dominantly acting oncogene or inactivation of a recessive tumor suppressor gene. Particularly exciting in recent years has been the elucidation of not only the diversity of mutations in these genes, but also the interplay that occurs between them in the genesis/progression of specific CNS tumor types. As a consequence of this improved molecular genetic understanding, it is now possible to make sense of the extraordinary karyotypic changes observed in tumor cells and thereby gain insight into the pathogenesis, progression, and prognosis of primary CNS tumors. The following chapter reviews the emerging knowledge of the molecular mechanisms involved in the initiation and progression of primary CNS tumors, with particular emphasis on the malignant lesions whose prognosis still remains grim.

## GENETICS OF GLIAL TUMORS

### DIFFUSE ASTROCYTOMAS

### Low-Grade Astrocytoma

The diffuse astrocytic tumors exhibit the highest incidence among primary CNS tumors and consequently have been the focus of intense clinical and histopathologic study. This has led to the successful development of a classification scheme that reliably correlates the microscopic appearance of these tumors with their clinical behavior. This scheme, proposed by the WHO, is four-tiered in its design. Whereas grade I applies to circumscribed pilocytic astrocytomas exclusively, grades II-IV describe progressive degrees of malignancy of the diffuse astrocytomas (Kleihues, 1992).

WHO grade II astrocytomas are interchangeably referred to as "low-grade" tumors and comprise 20% of all gliomas (Nelson, 1993). They most often occur in the third and fourth decades of life, with symptoms referrable to their more common location in the cerebral hemispheres. Although they are typically slow-growing tumors with a reported median survival of 8.2 years (Vertosick, 1991), they nonetheless possess features belying their low-grade designation. One such feature is the ability to diffusely infiltrate surrounding brain, thereby preventing curative resection. More ominous, though, is the marked potential of these tumors for undergoing malignant progression.

The most frequently observed genetic change in grade II astrocytomas is allelic loss of chromosome 17p, occurring in nearly half of all grade II cases (von Deimling, 1992). This chromosomal site notably includes the p53 tumor suppressor gene, which correspondingly displays frequent mutations in the remaining 17p allele (Fults, 1992; Sidransky, 1992; Van Meir, 1994), this being the genetic behavior for recessive tumor suppressor gene inactivation (Cavenee, 1983; Cavenee, 1985). The p53 notably encodes a transcription factor that not only acts to arrest the cell in $G_1$ phase but also initiates DNA repair by induction of the GADD (growth arrest and DNA damage inducible) DNA repair enzymes, or alternatively instigates apoptosis if DNA damage is irreparable (Louis, 1994). Mutations that occur in this gene are primarily missense alterations involving the exon 5-8 conserved domains, with hotspots noted at codons 175, 248, and 273 (Bogler, 1995). Also detected in a variety of other human tumors (Hollstein, 1991), mutation of p53 apparently exerts its profound effect by altering protein residues that are necessary for DNA-binding and transcriptional transactivation (Bargonetti, 1993). As a consequence, a number of p53-mediated cellular processes are deregulated; in some instances the result is a loss of suppressor gene function, while in others it is the product of a dominant gain of function.

Because p53 mutations are also commonly found in anaplastic astrocytomas and glioblastomas, their involvement would initially appear less related to malignant astrocytic progression than to the early stages of tumor formation (Louis, 1994; Sidransky, 1992). Indeed, experimental support for the functional involvement of p53 in astrocytoma tumorigenesis comes from the finding that introduction and expression of an exogenous wild-type p53 gene in glioblastoma cell lines suppresses their growth (Mercer, 1990; Van Meir, 1995). The implication of p53 mutation as an early event is also supported by the finding of p53 germline mutations in a series of patients with multifocal glioma and family history (Kyritsis, 1994) in conjunction with studies of Li-Fraumeni patients (Malkin, 1990). Other studies, however, have reported findings that may nonetheless implicate p53 mutation in the progression of at least a subset of astrocytomas. First, progression from low-grade to high-grade astrocytoma appears to result from the clonal expansion of cells with p53 mutation (Sidransky, 1992). Higher-grade tumors exhibiting p53 mutation, however, appear to be unique in terms of accumulated genetic changes as well as clinicopathologic behavior. In this regard, younger astrocytoma patients not only had higher incidences of p53 mutation/17p loss (Rasheed, 1994) but also survived longer than those without mutations (von Deimling, 1993). Therefore, glioblastomas may develop by either of two distinct pathways: one that requires p53 inactivation and one that does not.

Whether alternative genetic loci may substitute for p53 inactivation in those astrocytomas lacking p53 mutations is not yet clear. One possibility that has been suggested is that wild-type p53 may be bypassed by mutations in associated effector genes. The MDM2 oncogene, for instance, has in rare cases displayed amplification (Reifenberger, 1994). As an upstream inhibitor of p53 function, it would thus be a candidate for such a bypass gene. Similarly, the WAF1/CIP1 gene that is induced by p53 to express the p21 protein could also provide a bypass target. However, investigation of astrocytic as well as other human brain tumors have so far revealed neither mutations in this gene nor associated loss of heterozygosity for the chromosome 6p region carrying the WAF1/CIP1 gene (von Deimling, 1995). Finally, wild-type p53 has been found to stimulate transcription from the human TGFα promoter, thus indicating a dual growth-stimulatory role (Shin, 1995). In the absence of detectable p53 mutation, TGFα overexpression might then similarly bypass the wild-type p53 signal.

Involvement of p53 effector genes may yet be revealed by closer inspection of those tumors lacking p53 mutations but having elevated levels of wild-type protein (Rubio, 1993). Initially, this atypical prolongation of p53 protein half-life was suspected to reflect mutations at nonconserved domains. However, analysis of exons 4, 9, and 10 in one study did not reveal any mutations (Rubio, 1993). Since this protein accumulation is more common in higher grade tumors (Louis, 1994; Rubio, 1993), a consensus view has formed that it may merely reflect a physiological response by p53 to the accumulation of DNA damage that attends progression to higher grade tumors (Louis and Gusella, 1995). Alternatively, this protein accumulation may be more intimately associated with the regulatory mechanisms for apoptosis. Such an association has been revealed by showing that astrocytomas accumulating wild-type p53 also express more of the apoptosis inhibitor, bcl-2 (Alderson, 1995).

The product of the bcl-2 oncogene has been shown to inhibit Fas/APO-1 induced apoptosis and prolong cell survival by arresting glioblastoma cells in the $G_0/G_1$ phase of the cell cycle (Weller, 1995). A potential mechanism by which p53 induces apoptotic cell death was suggested when a temperature-sensitive p53 induced decreases in bcl-2 expression at the permissive temperature while simultaneously increasing expression of bax (a dominant inhibitor of bcl-2) (Miyashita, 1994). In the situation of wild-type p53 overexpression, this mechanism could allow for the abrogation of p53-mediated apoptosis by bcl-2 overexpression. Alternatively, a common selective pressure such as hypoxia might independantly result in the increased expression of both the p53 and bcl-2 proteins (Graeber, 1996). Though seemingly important, attempts at showing a correlation between bcl-2 protein expression and degree of malignancy have been unsuccessful (Krishna, 1995); instead, a predilection for bcl-2 expression in low-grade glial tumors and reactive astrocytes has been found, suggesting that resistance to apoptosis may be non-specific.

In addition to its role in cell cycle progression and apoptosis, p53 has also been implicated in tumor angiogenesis (Hanahan and Folkman, 1996). Loss of wild-type p53 function in glioblastoma cell lines has been correlated with down-regulation of a secreted angiogenesis inhibitory factor and a corresponding increase in rat corneal neovascularization (Van Meir, 1994). Similar to this effect, mutant p53 has also been shown to reverse wild type p53-mediated repression of the bFGF transcriptional promoter (Ueba, 1994). This release, in turn, results in the propagation of a bFGF autocrine loop that has been linked to astrocytoma neovascularization (Takahashi, 1992). Finally, mutant p53 has been found to potentiate protein kinase C induction of VEGF, an endothelial cell mitogen and angiogenesis inducer (Kieser, 1994).

In summary, the early p53 mutations in astrocytoma cells likely impact on the biologic choices made during progression. This view certainly is supported by recent work examining p53 status and astrocytoma resistance to cancer therapy. In particular, p53 mutations can predispose cells to decreased killing by chemotherapy or gamma-irradiation (Iwadate, 1996; Lowe, 1994). Unfortunately, such useful prognostication of the response of a tumor to treatment does not apply as well to the more accessible technique of p53 immunohistochemistry. Indeed, p53 immunopositivity has been found more frequently in astrocytomas with 17p loss and aneuploidy, but it has exhibited no association with survival time (Danks, 1995; Cunningham, 1997). This has been particularly frustrating, since p53 protein accumulation correlates with prognosis for patients with many other types of malignancy (Lane, 1994).

Though p53 inactivation is the most prevalent alteration in WHO grade II astrocytomas, activation of the platelet-derived growth factor (PDGF) system and 22q allelic loss also appear to be important. Regarding the PDGF system, studies of both glioma cell lines and astrocytoma specimens have shown a differential over-expression of the PDGF A-chain and its corresponding PDGFα receptor (Hermanson, 1992; Westermark, 1995). As with p53 mutation, this up-regulated autocrine stimulatory loop is detected equally in all histologic grades of astrocytoma, thereby defining its occurrence as an early event. Still unresolved, however, is the mechanism by which this up-regulation occurs. One possibility would be PDGFα receptor amplification, but this event has only rarely been observed in astrocytomas (Fleming, 1992). Alternatively, the finding that 17p loss/p53 gene inactivation is correlated with PDGFα over-expression suggests that these two genes may be functionally interdependant (Hermanson, 1996). This notion has been supported by experimental evidence demonstrating cooperation between p53 inactivation and specific growth factors in the transformation of primary cortical astrocytes (Bogler, 1995).

Chromosome 22q allelic loss is yet another frequent association with low-grade tumors. Occurring in 20 to 30% of astrocytomas of all grades, it has led to speculation that another tumor suppressor gene is involved in early events (James, 1988; Fults, 1990). Though there was initial excitement that the 22q-associated NF2 gene might be this putative tumor suppressor gene, subsequent work has excluded this possibility. Specifically, sequence examination of the NF2 gene in various grades of astrocytoma did not detect any mutations (Rubio, 1994). Moreover, other cytogenetic/LOH studies have pinpointed 22q loss to regions telomeric to NF2 (Ransom, 1992).

## Anaplastic Astrocytoma

Marked changes in histologic appearance accompany the progression from low-grade to anaplastic (WHO grade III) astrocytoma. Typically, such tumors display considerably more cellularity, nuclear and cellular pleomorphism, and frequent mitotic figures (Burger and Vogel, 1982). Although many patients with low-grade astrocytomas survive beyond five years, patients with anaplastic astrocytomas often die within three years and frequently show progression to glioblastoma.

The transition from low-grade to anaplastic astrocytoma is accompanied by allelic losses on chromosomes 9p, 13q, and 19q, and less frequently, 12q amplification. Interestingly, these abnormalities now appear to converge on one critical cell cycle regulatory complex, as has been the subject of several reviews (Furnari, 1996; Louis and Cavenee, 1995; Louis and Gusella, 1995; von Deimling, 1995; Westermark, 1995). Simply stated, the p15 (MTS2) and p16 (MTS1) genes code for cyclin-dependant kinase 4 (CDK4)/cyclin D1 complex inhibitors, with the RB gene providing the substrate for CDK4-mediated phosphorylation. Thus, inactivation of CDK4/cyclin

D1 inhibitors results in the hindrance of pRb function as a negative regulator of $G_1$-S cell cycle transition (Serrano, 1993).

These CDK4 inhibitory genes have been mapped to the chromosome 9p (region 9p21), a site that is significantly associated with interstitial and homozygous deletions in high-grade astrocytomas and two-thirds of glioma cell lines (Olopade, 1992; James, 1991). While there remains debate whether p15 and p16 represent the tumor suppressor gene targeted by these structural abnormalities, there is experimental evidence to support such a role. In particular, transfection into and expression of the p16 gene in glioblastoma cell lines lacking the gene has been shown to result in growth suppression (Arap, 1995). Arguing against p15/p16 functional primacy, however, has been the results of mutational analyses of these genes showing few mutations in primary astrocytomas with 9p allelic loss (Ueki, 1994). This is in contrast to the high frequency seen in corresponding cell lines, raising the question of whether these changes result from *in vitro* selection. Indeed, closer inspection of several genetic analyses of primary tumor tissue has shown disproportionately high homozygous deletion rates (Giani, 1994; Jen, 1994; Schmidt, 1994). Moreover, there are several mechanisms other than mutation/deletion by which p15/p16 function can be overcome. One such mechanism involves amplification of the CDK4 gene, a finding that is frequent in primary malignant tumor tissue and cell lines possessing normal p16 expression (Nishikawa, 1994). Alternatively, hypermethylation/inactivation of this genetic locus has been suggested as an explanation for malignant astrocytomas having reduced p16 expression without p16 loss/mutation (Costello, 1996).

The 9p homozygous deletions that appear so important to progression invariably include the class I interferon (IFN) gene cluster in addition to p15/p16 (James, 1993; Olopade, 1992; James, 1991). Although the involvement of IFN may merely be the consequence of the proximity between this cluster and the biologically significant p16 locus (Kamb, 1994; Nobori, 1994), the antiproliferative, differentiative, and immune-modulatory properties of the class I IFNs have raised the counterargument that IFN inactivation might confer a growth advantage on tumor cells. In this regard, IFNα gene transfer and constitutive expression in glioblastoma cells lacking the IFN cluster caused growth suppression in soft agar and a reduction of tumorigenicity (He, 1996). Consequently, loss of IFN genes, when it occurs, may play an additional role in the progression of astrocytic tumors.

Other investigations of anaplastic tumors have also revealed frequent allelic loss of chromosome 13q (James, 1988; Venter, 1991), which occurs in one-third to one-half of cases and has been shown to result in RB gene inactivation (Henson, 1994). Interestingly, RB and p15/p16 alterations in high-grade astrocytomas display an inverse correlation, rarely occurring together in the same tumor (Ueki, 1996). Thus, interruption of the function of the p16-cdk4/cyclin D1-Rb regulatory complex is a common end result manifested by mutually exclusive genetic mutations for each component of the complex.

The final chromosomal site implicated in progression to be discussed here is 19q, with allelic loss reported in 40% of high-grade tumors (von Deimling, 1994). A tumor suppressor gene is likely unmasked by this, but in this case the identity of the biologically significant gene remains to be identified. Deletion mapping using LOH studies has narrowed the candidate region for this putative suppressor gene to the 19q13.3 region between the APOC2 and HRC genes (Yong, 1995; Rubio, 1994). Interestingly, several candidate genes reside within this common area of deletion, including the DNA repair genes ERCC1, ERCC2, and LIG1, as well as BAX and a serine-threonine phosphatase gene (Liang, 1995; Chou, 1996; Yong, 1995). Unfortunately, there is no experimental evidence showing mutation, copy number change, or altered expression for these genes in tumors exhibiting loss of the other copy of 19q. However, cytogenetic comparisons of the three types of diffuse gliomas (astrocytoma, oligodendroglioma, and mixed glioma), showed that chromosome 19q was the only region in which loss was common to all (Yong, 1995). Moreover, 19q loss has not been described in a consistent fashion in other human tumors (Seizinger, 1991). Together, these findings suggest that this 19q region may harbor a glioma-specific tumor suppressor gene.

## Glioblastoma Multiforme

The transition from anaplastic astrocytoma to glioblastoma (WHO grade IV) is unmistakably apparent by the histopathologic features of endothelial cell proliferation and pseudopalisading pattern of necrosis (Russell and Rubinstein, 1989). In addition, the extent of local invasiveness is remarkably different. Whereas lower-grade astrocytomas are able to invade adjacent normal brain, glioblastomas are generally observed to have more distant spread of tumors cells in tandem with a greater variety of secondary structure patterns (Scherer, 1940; Russell and Rubinstein, 1989).

In terms of their clinical presentation, glioblastomas are the most common as well as most malignant glial tumors, and usually occur in the fifth and sixth decades of life. While the majority of these tumors appear to develop without a documented history of a previous astrocytic tumor, they may also develop in patients with a previously diagnosed lower-grade tumor. Interestingly, glioblastomas with a history of an antecedent lower-grade tumor appear to have a better prognosis than those whose tumors arose *de novo* (Winger, 1989). Moreover, younger age at the time of diagnosis of glioblastoma has been correlated with a longer survival (Burger and Green, 1987). Taken together, the results of these outcome analyses suggest that glioblastoma can be subdivided such that younger patients with an antecedent lower-grade tumor represent a more favorable subgroup.

Equally intringuing has been the question of whether these tumors can be subtyped on a genetic basis, and how such information might shed light on the origin of glioblastoma. In this regard, there is now a large body of cytogenetic and molecular genetic analysis from which conclusions can be drawn. For example, numerous cytogenetic studies of glioblastoma have demonstrated a stereotypical pattern of non-random numerical and structural abnormalities. Of these, chromosome 10 monosomy and amplification of portions of chromosome 7 as double minutes are most frequently detected (Bigner, 1988; Hecht, 1995; James, 1988; Kim, 1995;). Less commonly seen have been losses of 1q (Li, 1995), Y (Hecht, 1995), 6q, 9p, and 14q (Debiec-Rychter, 1995; Kim, 1995), as well as translocations involving the 17q11.2 and 12q13 break points (Hecht, 1995; McKeever, 1996). In particular, chromosome 6 and 14 losses have a strong association with recurrent tumors, suggesting that these changes play a role in either tumor recurrence or resistance to therapy (McKeever, 1996; Kim, 1995).

Specific patterns emerge when these glioblastoma-specific chromosomal abnormalities are examined for linkage with lower-grade-specific abnormalities. For instance, chromosome 17p LOH and chromosome 7 amplification have not often been found together, thus possibly allowing the separation of glioblastoma into at least two genetic subsets (von Deimling, 1993). Chromosome 10 loss, however, has been found in virtually all glioblastomas and in association with both chromosome 7 gain and 17p loss (James, 1988; Louis, 1996). Therefore, 10 loss appears to be a common genetic endpoint for glioblastoma cells, whose presence or absence cannot be used to further subclassify these tumors.

Attempts to map putative tumor suppressor genes on chromosome 10 have been hampered by the fact that in most cases the entire chromosome is lost. However, a number of deletion mapping studies have suggested that this chromosome may possess as many as three discrete suppressor genes, mapping to 10p (Karlbom, 1993; von Deimling, 1992), 10q25 (Fults, 1993; Karlbom, 1993; Rasheed, 1992), and 10q near the centromere (Karlbom, 1993). Interestingly, there is now data from an LOH study of high-grade tumors implicating a specific gene with one of these sites (Albarosa, 1995). This gene, MXI1, has been localized to the 10q25 region and encodes a protein that interacts with Max, a regulatory factor of the MYC oncogene. Although MYC amplification is an infrequent finding in astrocytic tumors (Collins, 1995), MXI1 could act as a tumor suppressor gene by competing with Max for proteins other than Myc. Though work is still needed to confirm such candidate genes, there is little doubt that chromosome 10 loci are functionally linked to glioma progression. This supposition has been borne out by experimental data showing that introduction of chromosome 10 into glioblastoma cells completely suppresses tumor formation when the resulting hybrid cells were injected into nude mice (Pershouse, 1993).

The majority of gene amplification events in high-grade astrocytomas have been shown to involve the gene for the receptor tyrosine kinase, epidermal growth factor receptor (EGFR) (Bigner, 1990a; Libermann, 1985; Wong, 1987). This receptor has been associated with cell growth and proliferation in several *in vitro* systems, but more important, it can transform cells in culture in a ligand-dependant fashion when expressed at high levels (DiFiore, 1987; Riedel, 1988). This ligand-dependancy is particularly important in light of the finding that the two endogenous ligands for EGFR, TGFα, and EGF are uniformly expressed in glioblastoma specimens (Ekstrand, 1991). Thus, co-expression of receptor and ligands may act as an autocrine or paracrine stimulating mechanism.

An alternate mechanism for activation of the EGFR-mediated growth stimulatory pathway also appears to be integral to glioblastoma formation. Approximately half of those tumors with EGFR amplification have been found to also harbor EGFR gene rearrangements, with the majority encoding a truncated constitutively active mutant similar to the viral erbB1 oncogene (Wong, 1992). Specifically, this mutation results in the deletion of exons 2-7 in the corresponding mRNA, which in turn leads to an in-frame truncation of a portion of the extracellular domain of the protein (Ekstrand, 1992; Humphrey, 1990; Sugawa, 1990). A growing body of experimental evidence has now yielded an understanding of the functional impact of this mutation. Retrovirally mediated transfer of the mutant receptor (ΔEGFR) to glioblastoma cells caused a significant growth advantage in vivo (Nishikawa, 1994). This ΔEGFR-mediated growth advantage has further been defined as an effect governed by two mechanisms. Not only does ΔEGFR increase the number of cells expressing proliferation markers, but it also reduces the apoptosis rate by augmenting Bcl-X$_L$ expression (Nagane, 1996). What is not yet known, however, is whether the mutant receptor interacts with signal-transducing molecules in a manner aberrant from the ligand-activated wild-type receptor. Such aberrant interactions might in turn influence more complex processes such as tumor cell invasiveness or neovascularization. The MAP kinase and Ras-Shc-Grb2 pathways have been implicated in ΔEGFR signal transduction but with as yet unclear ramifications (Montgomery, 1995; Prigent, 1996). From a clinical perspective, the EGFR gene appears to have importance as a prognostic marker that mirrors its biologic impact in functional studies. Indeed, several studies have reported a correlation between EGFR gene amplification and a shorter interval to relapse/shorter survival (Hurtt, 1992; Schlegel, 1994). In contrast to this prognostic importance for EGFR, a group of less commonly amplified oncogenes probably does not play a generalized role in glioblastoma growth. This group includes N-myc, c-myc (Fujimoto, 1988), gli (Fuller, 1992; Collins, 1993), PDGFα (Westermark, 1995), c-src, H-ras, or N-ras (Takenaka, 1985), SAS and MDM2 (Reifenberger, 1994), and MET (Fischer, 1995).

However, another group of recently characterized genes appears to be differentially regulated within the specific microenvironment of a glioblastoma *in situ*. Particularly important features include the circumferential parenchymal interface and regions of necrosis/hypoxia; sites where the complex processes of invasion and angiogenesis are enacted, respectively. For example, vascular endothelial growth factor (VEGF) and its receptors have been shown to play a major role in glioblastoma neovascularization (Strawn, 1996; Cheng, 1996; Plate, 1994; Hanahan and Folkman, 1996; Plate, 1992; Millauer, 1994), with VEGF appearing to be preferentially upregulated by tumor cells surrounding areas of necrosis. Based on experimental evidence, hypoxia likely induces VEGF expression and secretion by tumor cells which then binds in a paracrine fashion to its cognate receptors (flk-1, flt-1) on endothelial cells (Shweiki, 1992; Plate, 1993; Goldberg, 1994; Minchenko, 1994).

Similarly, the invasive margin of glioblastomas has been demonstrated to be a site of upregulation of a variety of factors. Included in this group are MT-MMP1 (Yamamoto, 1996), bFGF (Finkelstein, 1994), uPA (Yamamoto, 1994), and members of the cathepsin family (Sivaparvathi, 1996). In addition, a second group of factors has been described whose expression level correlates with various *in vitro* indices of invasiveness (for cultured cells) or tumor grade (for tissue). This group has not demonstrated selective *in situ* expression at the invasive margin, but nonetheless it may be important in the facilitating processes of matrix degradation and tumor cell migration.

These include NCAM (Edvardsen, 1994), L1 (Izumoto, 1996), CD44 (Kaaijk, 1995; Eibl, 1995; Okada, 1996), versican (Paulus, 1996), GFAP (Rutka, 1994), TIMP-1 (Matsuzawa, 1996), MMP-9 (Rao, 1996), and selected integrin molecules (Paulus, 1993; Chintala, 1996; Friedlander, 1996). Significantly, the histologic grade of an astrocytic tumor does not seem to be strictly correlated with the degree of local invasion, as low-grade tumors may show extensive infiltration of local brain (Guthrie and Laws, 1990). However, as stated previously, the variety of substrates that an invasive tumor cell interacts with does increase with grade. Consequently, a better understanding of these factors implicated in the invasiveness of glioblastoma cells may help explain features of the behavior of this tumor type, such as the rare occurrence of extra-CNS metastasis.

Thus, genetic analysis of glioblastoma has contributed to a more precise understanding of glioma progression. Based on the accumulated evidence, it appears that the end-stage of glioblastoma can be arrived at by any of at least three pathways. It may arise from either a lower-grade astrocytoma via p53 gene inactivation, de novo with EGFR amplification, or from oligodendroglial tumors through chromosome 19q and 1p-related mechanisms. More important, the genotypes associated with the first and second pathways are linked to unique clinical behaviors for their respective tumors. Whereas patients with tumors harboring p53 mutation/17p loss tend to be younger and have longer courses, patients with de novo tumors and EGFR amplification tend to be older and succumb more rapidly (von Deimling, 1993). The demonstration of such clinical correlates provides credibility to the large efforts being made to provide a genetic subclassification of glioblastoma.

## OTHER ASTROCYTIC TUMOR TYPES

### Juvenile Pilocytic Astrocytoma

In contrast to the diffuse fibrillary astrocytic tumors, pilocytic astrocytomas are typically circumscribed tumors that arise primarily in the cerebellum, peri-chiasmatic region, and temporal lobe. These tumors usually present in the second decade of life and are associated with a notably favorable course following resection (Garcia, 1985), despite histologic features such as marked cellular atypia, endothelial proliferation, and leptomeningeal invasion (Russell and Rubenstein, 1989). Despite these latter features, de novo malignant transformation and postirradiation malignant changes are rare. DNA ploidy analysis has shown these tumors to be diploid with the exception of the de novo and postirradiation tumors, which were aneuploid and tetraploid (Tomlinson, 1992). Cytogenetic analysis of these tumors so far has not revealed a consistent genetic abnormality (Ransom, 1994), although one series suggested an association of pilocytic tumors with partial chromosome 1p and 17q deletions (Bello, 1995).

Because pilocytic astrocytomas frequently occur in patients with type 1 neurofibromatosis (NF1) and the NF1 gene has been mapped to 17q11.2, there has been particular interest in 17q allelic status. Consistent with the clinical association, allelic loss on chromosome 17q has been detected in 25% of cases, including both sporadic and NF1-associated tumors (von Deimling, 1993), suggesting the presence of a tumor suppressor gene on 17q. The candidacy of the NF1 gene has been tested by quantitatively and qualitatively assaying NF1 gene expression in six sporadic tumors (Platten, 1996). Surprisingly, NF1 gene expression was not reduced, rather, it was overexpressed when compared to a normal brain. Furthermore, screening of portions of the NF1 coding sequence and immunohistochemical evaluation of its protein product (neurofibromin) strongly suggests that NF1 is not the target gene and the actual target has yet to be identified.

The overexpression of wild-type NF1/neurofibromin in pilocytic tumors may be a part of a reactive physiologic response to uncontrolled proliferation. Interestingly, only one pilocytic tumor in 27 specimens analyzed in two studies was found to harbor a p53 mutation (Lang, 1994; Litofsky, 1994). Though this absence of p53 mutation is in contrast to low-grade adult astrocytomas, p53 protein accumulation is a shared feature (Lang, 1994). The significance of this protein accumulation in the absence of p53 mutation or mdm-2 overexpression is still uncertain.

## Pleomorphic Xanthoastrocytoma

Described as recently as 1979, the pleomorphic xanthoastrocytoma shares several clinicopathologic features with pilocytic astrocytomas (Kepes, 1979). It is a tumor of the second decade of life and possesses histologic atypia that bely its long postoperative survival times. Malignant recurrence as a glioblastoma has been reported only in exceptional cases.

Cytogenetic study of these tumors has been limited. The first reports of cytofluorometry showed an absence of aneuploidy and very low S-phase fractions consistent with a benign clinical behavior (Hosokawa, 1991). More recently, a specific comparison has been made to diffuse fibrillary astrocytomas in regard to allelic gains and losses (Louis, 1995a) and has shown that the genetic events that occur during their genesis and progression differ from the diffuse astrocytoma paradigm. In particular, chromosome 9, 10, and 19q losses are not seen, and EGFR amplification is associated only with the rare recurrence as a glioblastoma.

## Subependymal Giant Cell Astrocytoma

Subependymal giant cell astrocytomas are periventricular, low-grade tumors that appear to be unique to tuberous sclerosis patients. Though these tumors have been reported in patients without obvious features of this syndrome, a detailed search of the Mayo Clinic tissue registry did not reveal patients with these tumors who lacked stigmata of tuberous sclerosis (Shepherd, 1991). These tumors appear to be enlargements of incipient subependymal hamartomatous nodules, and despite possessing atypical histologic features, none shows morphologic or clinical features of malignancy.

Genetic studies have shown linkage of half of families with tuberous sclerosis to the TSC1 gene (located at 9q34), and the other half to the TSC2 gene (located at 16p13.3)(Green, 1994a). Subsequent LOH studies of TSC-associated subependymal giant cell astrocytomas have demonstrated allelic loss of either the TSC1 or TSC2 loci (Green, 1994b), suggesting that these TS genes act as classic tumor suppressors. Detailed mutational analyses of the TSC1 and TSC2 genes in cases of subependymal giant cell astrocytoma, however, have not been reported.

## OTHER GLIAL TUMORS

### Oligodendroglioma/Oligoastrocytoma

Oligodendrogliomas are predominantly slow-growing tumors arising in the supratentorial compartment at sites proportional to white matter content during the fourth or fifth decades of life. Oligodendroglioms are presumably derived from oligodendrocytes, but their occurrence with a significant astrocytic component (oligoastrocytoma) has raised questions as to their origin from bipotential glial precursor cells. This question aside, oligodendrogliomas and oligoastrocytomas exhibit similar clinicopathologic behaviors. In particular, they appear uniquely chemosensitive to the PCV drug regimen, in marked contrast to diffuse astrocytic tumors (Cairncross, 1992).

Some of the first detailed reports of the cytogenetic changes in oligodendrogliomas and mixed oligoastrocytomas noted a karyotypic resemblance to low-grade astrocytomas (Bigner, 1990b). More extensive characterization, however, has revealed consistent chromosomal losses at two particular chromosomal sites, 1p and 19q, which appear to occur in tandem in both low-grade and high-grade tumors (Hashimoto, 1995; Ransom, 1992). Subsequent molecular analysis of these sites has determined that the 19q locus lies between APOC 2 and HRC, markers that span the astrocytoma-associated 19q locus (Rubio, 1994b; Reifenberger, 1994a). Similar mapping of chromosome 1p has implicated the 1p32-36 telomeric region (Bello, 1995). Interestingly, one larger study that included both low- and high-grade oligodendrogliomas as well as oligoastrocytomas also identified allelic losses at 17p, 9p, and 10, and infrequent EGFR amplification (Reifenberger, 1994b). The 9p and 10 losses (and EGFR amplification) occurred exclusively in the anaplastic tumors. And while the 17p deleted region included the p53 gene in all cases, sequencing of the remaining p53 allele did not reveal mutations in any of the tumors. Thus, allelic losses from 19q and 1p, when combined with the rarity of p53 mutations in these tumors, suggest that the early events in their tumorigenesis are distinct from those associated with diffuse astrocytic tumors. However, similarities are indicated by the 9p and 10 allelic losses together with EGFR amplification that occur in the anaplastic tumors. Hence, the later events in oncogenesis for these tumors appear to share common pathways.

Additional work has confirmed the distinctiveness of the early events in these tumors. As opposed to the frequent alterations of the p15, p16, CDK4, and cyclin D1 genes that occur in diffuse astrocytic tumors, alterations of these genes were found to be rare in oligodendrogliomas (Sato, 1996). Similarly, WAF1/CIP1 mutations do not appear to be involved in the formation of these tumors, further arguing against a role for the p53-mediated pathway (Koopmann, 1995). Finally, there are only sporadic reports of oncogene amplification/overexpression in these tumors (Rencic, 1996; Banerjee, 1996).

Considering the range of clinicopathologic features in the spectrum between oligodendrogliomas and diffuse astrocytic tumors, the mixed phenotype oligoastrocytomas merit specific discussion. Once suspected as being two cell lineages comprising one tumor, loss of heterozygosity analysis of chromosome 10, 1p, and 19q loci has now established the clonal origin of oligoastrocytomas (James, 1988; Kraus, 1995). The histologic identity of the cell of origin remains unclear. One potential candidate is the oligodendrocyte-type 2 astrocyte (O-2A) progenitor cells, which are capable of differentiation into either mature oligodendrocytes or astrocytes depending on culture conditions (Raff, 1983). On the other hand, the concept of a mature oligodendrocyte de-differentiating and expressing its O-2A genetic lineage may be supported by the histopathologic identification of transitional forms of oligodendroglial tumors (Herpers, 1984). The mixed oligoastrocytoma may therefore represent the end stage of a transition process that includes the gliofibrillary oligodendroglioma and transitional oligoastrocytoma. Such a conversion to an astrocytic lineage may then predispose these neoplastic cells to the late genetic events more characteristic of diffuse astrocytomas: 9p loss, 10 loss, and EGFR amplification. In addition, bcl-2 and c-myc expression have been detected exclusively in the astrocytic components of mixed tumors (Nakasu, 1994; Banerjee, 1996).

## Ependymoma

Ependymomas are predominantly tumors of childhood and adolescence that may originate from any part of the ventricular cavity, although the fourth ventricle is the most common site. They constitute approximately 5% of intracranial gliomas and show a slight male preponderance. Although atypical histologic features are not usually prominent, some cases show a high mitotic index and cytologic changes that may be associated with poor prognosis (Russell and Rubenstein, 1989). Nevertheless, cases of malignant progression of ependymoma to glioblastoma are rare and have not been subjected to detailed molecular genetic analysis.

Although consistent cytogenetic aberrations have not been identified in these neoplasms, non-random changes include monosomy of 17 and 22, and structural changes of 2, 4, 6, 10, 11, 12, and X (Rogatto, 1993; Stratton, 1989; Weremowicz, 1992). Since these chromosomal segments each include regions known to harbor tumor suppressor genes, more extensive allelic loss studies have been performed using a large number of markers. Although allelic loss at 22q, 17p, 10, 6q, 13q, and 19q coincided in each case with the site of the known tumor suppressors, their relative infrequency raises questions of their biologic importance (Bijlsma, 1995).

Although the most commonly associated tumor in type 2 neurofibromatosis is the acoustic neuroma, meningiomas and spinal cord ependymomas are not rare (Louis, 1995c). Since meningiomas and acoustic neuromas from non-NF2 patients are frequently found to harbor NF2 gene mutations/allelic loss, analysis of NF2 gene mutations in sporadic ependymomas has been performed. In one series, such a search revealed only a single mutation in an ependymoma that had corresponding wild-type allelic loss (Rubio, 1994a), while in contrast, analysis of another series detected NF2 mutations in the majority (70%) of tumors (Birch, 1996). This dichotomy has led to the as yet unproven hypothesis that an as yet unidentified second chromosome 22q tumor suppressor gene may be involved (Louis and Cavenee, 1995). Although the infrequent, but nonrandom, occurrence of 17p loss has suggested a role for p53 in ependymoma genesis, subsequent mutational analysis of a series of 15 tumors showed no mutations (Ohgaki, 1991).

## Choroid Plexus Tumors

Choroid plexus tumors typically arise in the lateral ventricles of children under two years of age, but they may also afflict adults. They more frequently occur as benign choroid plexus papillomas, but on occasion they exhibit the locally invasive features of choroid plexus carcinoma. Because these tumors are rare, it has been difficult to derive useful clinicopathologic data upon which hypotheses of molecular events can be based.

Choroid plexus tumors have now been linked to three familial syndromes, two of which (Li-Fraumeni, Von Hippel Lindau) are associated with the tumor suppressor genes TP53 and VHL, respectively (Blamires, 1992; Garber, 1990; Robinow, 1984). In the single analyzed case of a VHL-associated choroid plexus papilloma, allelic loss of the tip of 3p (which contains the VHL gene) was confirmed (Blamires, 1992). Similar LOH analysis has not yet been described for Li-Fraumeni associated choroid plexus tumors.

As intriguing as these syndromal associations are, alternative genetic analyses of choroid plexus tumors have led to speculation that they may be the sequela of papovavirus infection. One such study identified SV40 T-antigen sequences in 50% of the sampled tumors (Bergsagel, 1992). Although SV40 or related papovaviruses may cause choroid plexus tumorigenesis, the association of viral sequences with a tumor does not prove an etiological role. The association has, however, led to mechanistic insights. Earlier work in transgenic mice had shown that choroid plexus cells are the most sensitive to transformation by the T-antigen (Brinster, 1984), and T-antigen is able to form complexes with pRB and p53 (Dyson, 1990; Symonds, 1991). Therefore, if viral infection is one mechanism for tumorigenesis, LOH/mutation of the suppressor genes that are targeted by the viral T-antigen could be an alternative mechanism. This hypothesis awaits further molecular genetic analysis of these tumors.

## GENETICS OF MIXED GLIAL-NEURONAL TUMORS

Tumors of mixed glial-neuronal composition are generally benign and amenable to cure by surgical resection. Both cellular elements show cytological features of neoplasia, although they vary in their relative proportions and differentiation. Ganglioglioma is one such tumor, and it usually presents itself in young adulthood with a history of seizures. These tumors typically are well-demarcated temporal lobe lesions and only rarely undergo malignant evolution (Kalyan-Raman, 1987). Since they are uncommon, only limited genetic analysis has been reported. In one, neither EGFR gene amplification nor allelic loss on chromosomes 10, 13q, 17p, 19q, and 22q was detected (Louis and Cavenee, 1995).

Desmoplastic supratentorial neuroepithelial tumors of infancy are another category of mixed tumors that have only recently been clinicopathologically described (VandenBerg, 1987). Known also as desmoplastic infantile ganglioglioma (DIG) and desmoplastic cerebral astrocytoma of infancy (DCAI), this rare tumor type has a favorable clinical prognosis despite an aggressive, sarcomatous-like histologic appearance. A preliminary molecular genetic study of two such tumors showed no loss of heterozygosity at multiple loci on chromosomes 10 and 17, including the p53 locus (Louis, 1992). While dysembryoplastic neuroepithelial tumors (DNET) and central neurocytomas are additional members of the mixed glial-neuronal tumor category, no cytogenetic or molecular genetic analysis has yet been reported.

## GENETICS OF PRIMITIVE NEUROECTODERMAL TUMORS

Primitive neuroectodermal tumors (PNET) are central nervous system lesions of childhood that are composed largely of undifferentiated neuroepithelial cells. Because these tumors arise in various sites in the central nervous system and are difficult to separate histologically, they have a nosology that does not emphasize anatomic origin (Kleihues, 1993). Here, the terms medulloblastoma, pineoblastoma, and neuroblastoma are considered to be synonymous with PNETs of the cerebellum, pineal region, and cerebrum, respectively. Regardless of location, these tumors exhibit aggressive behavior characterized by leptomeningeal dissemination, or less commonly direct parenchymal invasion (Nelson, 1993).

The most frequent cytogenetic structural chromosomal changes are deletions and nonreciprocal translocations, with chromosomes 5, 6, 11, 16, 17, and a sex chromosome appearing to be nonrandomly involved (Biegel, 1989). The most consistent of these changes is an isochromosome 17q that occurs in a third to half of reported cases (Griffin, 1988). Deletion mapping of medulloblastoma has shown allelic loss of chromosome 17p sequences in the same region to which the p53 gene is mapped (Cogen, 1990). However, a series of p53 sequencing studies did not reveal p53 mutations in the most highly conserved coding regions of the gene (Adesina, 1994; Biegel, 1992; Saylors, 1991); one notable exception was a series of 20 PNETs from Taiwan in which 35% harbored p53 mutations as determined by PCR-SSCP (Hsieh, 1994). A second locus on distal 17p distinct from p53 has been mapped (Biegel, 1992). The possibility of an alternative mechanism of p53 inactivation that is not revealed by current methods of p53 mutational analysis still exists. Such an alternative mechanism might involve the WAF1/CIP1 gene, which acts as a downstream mediator of p53 via its p21 product. Examination of this locus in 12 PNETs, however, revealed multiple polymorphisms but neither overt mutations nor 6p allelic loss (Koopmann, 1995). Nevertheless, loss of 17p heterozygosity appears to have real biologic importance as indicated by the significantly worse prognosis among patients with medulloblastomas and this pattern of allelic loss (Batra, 1995).

Loss of genetic material from other suspect chromosomal locations does not appear to be as critical to the biology of these tumors. In the limited cytogenetic database of pineoblastomas, an 11q interstitial deletion was uncovered in a cultured cell line, but this may be the result of *in vitro* selection (Sreekantaiah, 1989). One other report characterizing a small sample of primary pineoblastomas did demonstrate a case of 1p loss, but this site was not significantly altered in larger PNET samplings (Bello, 1995; Griffin, 1988). Finally, loss of heterozygosity has been revealed at significant frequencies on chromosomes 6q and 16q (Thomas, 1991). More interesting, though, has been the finding that these two sites, as well as 17p, do not cluster in individual tumors. Hence, PNETs may be a heterogeneous collection of tumors in regard to their underlying genetic mechanisms.

In contrast to these genetic losses, oncogene amplification is not a frequent occurrence in central PNETs. When found in primary tumor specimens, the amplification usually involves either N-myc or c-myc (Bigner, 1990b; Donner, 1991; Fuller, 1992). Additional examination of overexpression of the erbB1, gli, neu, L-myc, H-ras, K-ras, N-ras, sis, and src genes has so far been unrevealing (Wasson, 1990). In derived cell cultures, however, oncogene overexpression appears to be the rule rather than the exception (Kees, 1994; Wasson, 1990), suggesting *in vitro* selection of dominant subpopulations of cells.

Additional genetic loci may be more specifically linked to the aggressive clinicopathologic features of PNETs. The marked propensity for leptomeningeal dissemination may rely upon N-CAM and L1 mediated cell-cell and cell-substrate binding mechanisms, as demonstrated by the uniform expression of these adhesion molecules in medulloblastoma cell lines (Wikstrand, 1991).

The involvement of N-CAM and L1 would correlate with the partial neuronal differentiation of PNETs, and contrasts to patterns of differential adhesion molecule expression in astrocytic tumors. Interestingly, the glioblastoma-associated CD44 molecule that binds leptomeningeal matrix components was only weakly expressed in medulloblastomas (Kuppner, 1992). However, the malignant phenotype of PNETs may yet share a genetic determinant with the astrocytic lineage of tumors. As previously mentioned, in situ studies of human gliomas show overexpression of PDGF-$\alpha$ receptors in glioma cells of high-grade tumors (Westermark, 1995). Similar aberrant expression of PDGF-$\alpha$ receptors is apparent in a study of 14 cases of PNET (Smits, 1996). Finally, there is preliminary data suggesting a link between the recurrence tendency of medulloblastoma and BCL-2 expression (Nakasu, 1994), where the frequency of BCL-2 expression was increased when compared to specimens from initial treatments. There was, however, no corresponding significant relationship between degree of expression and patient survival.

Other molecular genetic analyses of PNETs have shown that the EWS/FLI-1 gene fusion common in Ewing's sarcoma is rare in central PNET tumors (Jay, 1995). They also do not exhibit the deletions of the CDKN2/p16 gene that occur frequently in astrocytoma progression as well as other tumor types (Raffel, 1995).

## GENETICS OF NON-GLIAL TUMORS

### MENINGIOMA

Meningiomas are primary tumors of the leptomeninges derived from arachnoid cap cells, which are generally well-demarcated and slow-growing and represent roughly 15 to 20% of all primary intracranial and 25% of all intraspinal tumors (Russell and Rubinstein, 1989). The incidence of meningiomas is greater for women than men, particularly for intraspinal tumors (Burger and Vogel, 1982), and so the involvement of sex hormones and their transduction mechanisms have been investigated. Recent reports finding active progesterone receptors in meningiomas support this hypothesis (Hsu, 1997; Cahill, 1984), as does a proposed link to breast cancer (Schoenberg, 1975), and the observed growth acceleration during the menstrual cycle and pregnancy (Hsu, 1997). Other noteworthy features of meningiomas include their rarity in children (Chan, 1984), their association with previous trauma (Barnett, 1986), and their occurrence as a sequela to cranial irradiation (Rubinstein, 1984).

Cytogenetic and molecular genetic studies have shown that chromosome 22q allelic loss occurs frequently in the development of meningiomas (Ruttledge, 1994b; Ng, 1995). This cytogenetic localization of a putative tumor suppressor gene together with the non-random representation of meningiomas in NF2 has implicated the NF2 gene. Indeed, subsequent NF2 LOH/mutational analysis of sporadic meningiomas showed this gene to be involved in 60% of the cases (Ruttledge, 1994a), with a strong correlation of NF2 mutations to allelic loss. Furthermore, these mutations were predominantly at the moesin-ezrin-radixin homology region of the coding sequence, resulting in inactivation of the protein product (Wellenreuther, 1995). Despite this strong evidence, the significance of the role of the NF2 gene in meningioma development should not be overstated. First, the higher frequency of NF2 mutations in specific histologic subtypes (transitional and fibroblastic) has unclear meaning given the lack of correlation of these subtypes with prognosis (Wellenreuther, 1995). Second, NF2 mutations do not underlie all sporadic and NF2-associated meningiomas. As an example, multipoint linkage analysis of a pedigree with multiple meningiomas and ependymomas found a familial meningioma locus distinct from the NF2 locus (Pulst, 1993). Additionally, genes for two other 22q loci have been characterized that are also important in meningioma development. One gene was revealed by its association with a 22q12 homozygous deletion site and later found to be a member of the $\beta$-adaptin gene family (named BAM22) (Peyrard, 1994). The second gene, MN1, resides at 22q11 and was disrupted by a balanced translocation t(4;22) in the germline of a patient with multiple meningiomas (Lekanne Deprez, 1995).

Though the majority of meningiomas maintain discrete margins as they enlarge, some have malignant characteristics with cortical invasion and high recurrence rates (Russell and Rubinstein, 1989). How these grade II and III meningiomas originate remains unclear, but some data has emerged describing molecular defects associated with progression. In particular, several genetic studies of malignant meningiomas revealed preferential losses of chromosomes 1p, 10, and 14q, with the frequency of LOH at these sites increasing with grade of tumor (Lindblom, 1994; Rempel, 1993; Simon, 1995). Interestingly, the chromosome 10 loss was observed only in morphologically malignant tumors (i.e., high proliferation index, nuclear pleomorphism, and necrosis) and not in meningiomas classified as malignant by virtue of brain invasion alone (Rempel, 1993). Also, micro-dissection and molecular genetic analysis of a malignant region amidst a benign background in a single tumor revealed a dichotomous karyotype. Whereas cells from both the background and overtly malignant regions of the tumor exhibited 22q and 1p deletions, only the cells from the malignant region possessed 9q and 17p deletions (Lindblom, 1994). These loci associated with progression now await identification of their corresponding candidate suppressor genes.

During the last decade, many studies have concentrated on the hormone and growth factor receptor expression of these tumors with the hope of determining prognostic indices. The absence of progesterone receptor expression in meningioma now appears to be a significant factor for shorter disease-free intervals (Hsu, 1997). In contrast, no such clinical correlates have been established for a variety of other receptors with increased expression in meningiomas. These include receptors for neurotensin (Mailleux, 1990a), type A cholecystokinin (Mailleux, 1990b), somatostatin (Reubi, 1986), prolactin and dopamine D1 (Carroll, 1996), PDGF-B (Wang, 1990), and EGF (Diedrich, 1995). Of these receptors, EGFR warrants an additional comment given its tendency for amplification/mutation in the most malignant diffuse astrocytomas. Of note, previous studies showed that EGF greatly stimulates DNA synthesis and growth of human meningioma cells cultured in serum-free medium when compared to PDGF, FGF, and IGF-1 (Kurihara, 1989; Weisman, 1987). This observation requires caution as only a few meningiomas have been studied for EGFR gene mutations and no amplification of DNA sequences has been detected (Diedrich, 1995).

When a meningioma produces a poor clinical outcome, it is usually the result of either underlying invasiveness, early recurrence, or excessive peritumoral vasogenic cerebral edema. Molecular genetic studies have now begun to find determinants for these phenomena. Regarding invasiveness, the differential expression of CD44 isoforms may contribute to whether meningioma cells infiltrate brain (Ariza, 1995). Second, new clues to the clonal composition of meningiomas may shed light upon the subcategory of tumors that recur early. In a recent report, the finding of polyclonality in 35% of cases using a highly informative X chromosome probe was surprising (Wu, 1996). In that series of tested tumors, clonal status did not correlate with histopathological or radiological features. Nevertheless, this does not exclude a biologic effect of polyclonality on meningioma growth. For instance, regional multicentricity is a growth pattern of meningiomas that has been linked to recurrence (Borovich, 1986), but the factors that cause the acceleration of multicentric cell aggregates after resection of the primary tumor mass remain unclear. Finally, the demonstration of a strong link between VEGF mRNA expression and peritumoral edema may indicate an important role for VEGF in the formation of the edema surrounding meningiomas (Kalkanis, 1996).

## HEMANGIOPERICYTOMA

Previously classified as the "angioblastic" meningioma, the hemangiopericytoma has been the subject of considerable debate regarding its histogenesis. The consensus view is that meningeal hemangiopericytomas derive from the pericyte and bear more resemblance to peripheral soft tissue than to meningiomas. This is particularly true with regard to biologic behavior, since meningeal hemangiopericytomas are among the most aggressive of dural-based tumors. Indeed, metastatic rates of 10 to 30% have been observed (Goellner, 1978; Enziger, 1976).

Consistent with this new classification of hemangiopericytomas has been the results of their molecular genetic analysis. One report demonstrated a 35% mutation frequency for the NF2 gene in meningiomas in contrast to an absence of mutations in both meningeal and peripheral hemangiopericytomas (Joseph, 1995). In addition, there is preliminary reporting of hemangiopericytomas having frequent homozygous deletions of the CDKN2/p16 gene (Louis and Cavenee, 1995), a finding that might implicate this cell-cycle control mechanism in the malignant progression of these tumors. Other candidate genetic loci that have been implicated in peripheral soft tissue hemangiopericytomas await testing in meningeal tumors. These include CHOP/GADD153 and MDM2 — oncogenes that reside in the 12q13 region that commonly undergoes structural rearrangement in the peripheral tumors (Forus, 1994).

## Schwannoma

CNS schwannomas are benign tumors of schwann cells that account for roughly 8% of intracranial tumors and 30% of intraspinal tumors. In patients with NF2, bilateral vestibular schwannomas (acoustic neuromas) are the sine qua non finding, and they occur at a much earlier age than in patients with sporadic unilateral vestibular schwannomas (Louis, 1995a). Therefore, like meningiomas, schwannomas occur frequently in NF2 patients, display frequent 22q chromosomal loss, and harbor frequent NF2 mutations whatever their context of occurrence (Jacoby, 1994; Couturier, 1990; Lekanne-Deprez, 1994). More detailed investigation of NF2 mutations has further revealed that the majority result in truncation of the normal protein product (merlin) and consequent absence of merlin immunostaining (Sainz, 1994). Thus, NF2 inactivation appears to require the loss of merlin function as well. Not excluded, however, is the possibility that some of the NF2 mutations may act as dominant negatives whereby the altered merlin protein interacts with the remaining normal allele product and inactivates it.

## Hemangioblastoma

Hemangioblastomas are benign, slow-growing, vascular tumors that clinically present themselves in the third through fifth decades as cerebellar or spinal cord lesions. Although most cases arise sporadically, a significant proportion are associated with Von Hippel Lindau disease (VHL) (Neumann, 1995), an autosomal dominant hereditary tumor syndrome, characterized by the frequent development of retinal angiomas, pancreatic/renal cysts, renal cell carcinomas, and pheochromocytomas, in addition to hemangioblastomas.

Because of these syndromal associations, studies of the VHL gene have focused on its role in sporadic tumor counterparts. Paradoxically, VHL gene mutations and allelic loss were detected in roughly 60% of sporadic renal cell carcinomas, but not at all in sporadic pheochromocytomas (Neumann, 1995). In sporadic hemangioblastomas, however, somatic mutations of the VHL gene have been found with an estimated frequency of 20% (Kanno, 1994). Though the mutational frequencies in sporadic tumors as a whole differ greatly among tumor types, there may be alternative mechanisms of VHL gene inactivation that are not revealed by the attempted methods of mutational analysis. One such mechanism may involve hypermethylation of normally unmethylated CpG islands in the 5′ region of VHL. Striking evidence for this has been described in sporadic renal cell carcinomas, where 19% of tumors examined failed to express the VHL gene in the absence of detectable mutations. Upon closer inspection, these tumors had either two heavily methylated alleles or one heavily methylated allele combined with corresponding allele loss (Herman, 1994).

## Pituitary Tumors

With an incidence approaching 15% of all intracranial neoplasms, pituitary tumors predominantly arise during the third to sixth decades, though no age group is exempt (Russell and Rubinstein, 1989). Additionally, an association with the inherited syndrome MEN1 (multiple endocrine neoplasia, type I) has been demonstrated (Bahn, 1986). These tumors tend to be well-demarcated adenomatous lesions that can be readily controlled by either pharmacotherapy or surgical resection. However, a deceiving feature of these tumors is their uniformly benign histological appearance (Nelson, 1993). This reality has hindered identification of a subset of tumors that aggressively recur and/or invade locally despite maximal multi-modality management. The genetic characterization of these tumors may hold hope for better prognostication of clinical course.

A central theme in the genetic characterization of pituitary tumors has long been the question of whether an intrinsic pituitary or hypothalamic pathogenesis is dominant in their formation. Evidence for the primacy of excessive hypothalamic stimulation has included a variety of well-known clinical associations. Very early, Nelson recognized the development of ACTH-secreting tumors after bilateral adrenalectomy (Wilson, 1979). Additionally, there have been reports of GH-producing pituitary adenomas arising in patients with hypothalamic gangliocytomas producing GHRH (Asa, 1984). As previously reviewed, ectopic secretion of additional hypothalamic releasing factors has also been implicated (Melmed and Rushakoff, 1987). Conversely, an intrinsic pituitary pathogenesis has been supported by two bodies of evidence. First, histopathologic examination of pituitary tissue surrounding an adenoma has not revealed the presence of a generalized hyperplastic field from which the tumor emerges (Nelson, 1993; Russell and Rubinstein, 1989). Second, the mono-clonality of pituitary tumors has been confirmed using X-inactivation analysis, strongly suggesting that both functional and non-functional tumors arise from a single precursor cell (Herman, 1990). Consequently, a consensus view has emerged in which hypothalamic factors act as promoters by stimulating clonal expansion of a transformed pituicyte.

There now is a wealth of molecular genetic data on pituitary tumors to suggest important mechanisms of pituicyte transformation. Consistent with the paradigm established for diffuse astrocytic tumors, both dominant mutations in growth promoting genes as well as deletions of tumor suppressor genes have been proposed as possible etiolocic factors. Interestingly, several of the dominant structural genetic mutations associated with tumorigenesis are more intimately related to the molecular mechanisms controlling hormone secretion. The gsp mutation, for instance, has been detected in one-third of GH-producing adenomas (Landis, 1989). These tumors characteristically harbor elevated cAMP levels as a consequence of a GHRH receptor-linked G protein defect that leads to constitutive adenyl cyclase activation. Similarly, the association of c-fos was established by the demonstration of c-fos mRNA induction in GHRH-stimulated rat somatotroph cells (Billestrup, 1987). Most compelling, however, has been the characterization of Pit-1, a pituitary-specific homeodomain protein that regulates the expression of the prolactin, growth hormone, and TSH genes (Fox, 1990). The importance of Pit-1 was first demonstrated when mutations that lead to a functional defect of the gene were found to result in dwarfism and pituitary hypoplasia in mice (Castrillo, 1991). Soon afterward, several investigations reported on the expression of Pit-1 mRNA in human pituitary adenomas (Asa, 1993; Friend, 1993; Lloyd, 1993). Their results, however, were contradictory with respect to the differential expression of the gene, warranting further investigation with a more sensitive technique. Using RT-PCR, such an investigation concluded that Pit-1 mRNA was present in most adenomas irrespective of hormonal type or functional status (Hamada, 1996). Additionally, this message appeared overexpressed vs. normal tissue control but without a differential display of splice variants designated Pit-1$\alpha$, Pit-1$\beta$, and Pit-1T. Consequently, these data suggest that the Pit-1 gene has a role in pituitary adenoma formation that extends beyond its involvement in the regulation of hormone production.

Because of the development of pituitary adenomas in MEN1, its susceptibility locus was predicted to be a tumor suppressor gene. This locus was mapped to chromosome 11q13, and subsequent loss of heterozygosity studies demonstrated allelic loss in a minority of sporadic prolactinomas (Herman, 1993). Similar allelic losses, though, were not demonstrated for GH-producing pituitary adenomas, a discrepancy which raises questions concerning the pervasiveness of the MEN1 gene in pituicyte tumorigenesis. Apparently, more complex factors play a role in the expression of MEN1, as highlighted by a report of nonidentical expression of MEN1 in HLA-and blood group-identical twins (Bahn, 1986). In that report, one twin developed corticotropic hyperplasia whereas the other was afflicted with an invasive pituitary adenoma.

As stated before, the minority of pituitary adenomas progress to a malignant phenotype. Genetic analysis of the few invasive adenomas and pituitary carcinomas has nonetheless identified potential markers of aggressive biologic behavior. In one study, H-ras mutations were consistently found in a series of metastatic pituitary secondary tumors, but not in their respective primary pituitary lesions (Pei, 1994). Given the similarity between $p21^{ras}$ and G-protein, it is thus intriguing to speculate that H-ras mutations further derange the signal transduction pathways regulating pituitary cell growth and differentiation. In another series of studies, p53 protein accumulation has been proposed as a marker for malignant progression (Buckley, 1994; Thapar, 1996) despite the caveat that p53 immunopositivity cannot reliably be interpreted as an indication of a critical genetic alteration. Although p53 mutational analysis of pituitary adenomas has so far identified no mutations (Levy, 1994), all of the tested tumors were associated with a noninvasive clinical phenotype.

## PINEAL REGION TUMORS

Pineal region tumors are rare lesions that, as a group, account for approximately 1% of all intracranial tumors (Burger and Vogel, 1982). They typically present in childhood and have been classified as either germinal or pineal parenchymal in origin. Curiously, there is a significantly greater representation of pineal tumors in the Japanese population (4.5% of all intracranial tumors), but there are no familial patterns or syndromal associations to help target a genetic lesion (Araki, 1968).

The limited cytogenetic information on pineal tumors has so far been obtained exclusively for germ cell tumors. Of this subcategory of pineal tumors, the most common is the germinoma, a tumor that, outside the CNS, is histologically identical to the testicular seminoma and ovarian dysgerminoma (Nelson, 1993). Characteristically, germinomas are exquisitely radiosensitive, having ten-year postradiation survival rates greater than 80% (Jennings, 1985). In addition, their presentation at the onset of puberty may be a sign-post for an underlying hormonal promoter mechanism.

The reported intracranial germinoma karyotypes lack a 12p isochromosome, and in roughly 20% of cases they display a homogeneously staining region (HSR) (Albrecht, 1993). Whereas the absence of a 12p isochromosome may merely distinguish an intracranial from an extracranial germinoma, the presence of an HSR may signify a genetic locus for malignant progression. Indeed, the association of this HSR with aggressive behavior in a series of noncerebral extragonadal germinomas complements the single report of an HSR-associated intracranial germinoma that rapidly recurred and disseminated (Samaniego, 1990). Unfortunately, molecular genetic analysis of this HSR did not reveal homology to CMYC, NMYC, INT1, INT2, MYB, MOS, GLI, ERBB2, ERBB, MDR1, or KRAS1. Therefore, the identity of this suspected oncogene amplification must await further scrutiny.

The other type of pineal germ cell tumor that has undergone genetic analysis is the malignant rhabdoid tumor (MRT). These tumors are extremely aggressive lesions of early childhood that are characterized by local brain invasion as well as subarachnoid spread. Cytogenetic studies have been limited but nonetheless have been consistent in implicating a particular chromosome in MRT tumorigenesis. Two groups, in fact, detected monosomy 22 (Biegel, 1990; Muller, 1995), whereas a third group reported a chromosomal translocation involving chromosomes 11 and 22 (Karnes, 1992). This information may be a clue to the identity of a tumor suppressor gene, but it requires further analysis of a larger number of karyotypes.

Though the significance of these chromosome 22 structural changes is unclear, the genetic determinants of MRT invasiveness are gaining resolution. Specifically, a case of MRT was shown to express the highest levels of 72- and 92-kD type-IV collagenase transcripts of any pediatric brain tumor previously examined (Muller, 1995). Furthermore, this tumor did not express any significant amounts of transcripts for the protease inhibitors, TIMP-1 or TIMP-2. Such a pattern of protease up-regulation in combination with protease inhibitor down-regulation has been found to be highly facilitating of matrix degradation and invasion in other tumor systems.

## CNS TUMOR SYNDROMES

As alluded to in previous sections of this chapter, the hereditary tumor syndromes of the nervous system have provided insight into the genetic basis of both hereditary cancer predisposition and sporadic nervous system tumors. This point was particularly highlighted in the discussions of meningioma, schwannoma, and ependymoma: three tumor types whose somatic genetic defects in sporadically occurring tumors correlate well with the phenotypic expression of the NF2 germline mutation. However, an exact correlation between germline and sporadic mutations in individual CNS tumor types has not always been observed. For example, while medulloblastoma was found to segregate with those Turcot syndrome families having APC tumor suppressor gene mutations (Hamilton, 1995), sporadically occurring medulloblastomas have not yielded detectable mutations (Mori, 1994).

As a group, the hereditary tumor syndromes of the nervous system can be divided into those occurring more frequently (neurofibromatosis 1, neurofibromatosis 2, tuberous sclerosis, and Von Hippel-Lindau conditions) and those that occur more rarely (Li-Fraumeni, Turcot, Gorlin, Cowden, familial glioma, multiple endocrine neoplasia (type 1), and the retinoblastoma susceptibility syndromes). Each syndrome has a very characteristic phenotypic expression, and in the majority of cases it has been linked to alterations at a specific genetic locus. For the classic neurocutaneous disorders, these loci include the NF1 gene (17q11) for neurofibromatosis 1; the NF2 gene (22q12) for neurofibromatosis 2; the VHL gene (3p25) for Von Hippel-Lindau; and TSC2 (16p13) for tuberous sclerosis.

Interestingly, the products of two of these genes share a similar function. Both neurofibromin (NF1 gene product) and tuberin (TSC2 gene product) have been shown to act as guanosine triphosphatase-activating proteins, and thereby likely modulate growth factor-mediated signals via an interaction with the p21 product of the ras oncogene (Basu, 1992; Johnson, 1994). In contrast, the merlin NF2 gene product appears to have a morphoregulatory function, as evidenced by its homology to the product of the moesin-, ezrin-, and radixin-like genes (Trofatter, 1993). Finally, the VHL protein appears to be expressed as multiple isoforms, whose function may in part be to regulate the stability of VEGF mRNA in a hypoxia-inducible fashion (Levy, 1996).

Among the less common hereditary tumor syndromes, three have yielded to genetic analysis by revealing an association with specific genes. The retinoblastoma susceptibility syndrome, with its increased incidence of pineoblastoma and malignant glioma (Bader, 1982; Eng, 1993), has been shown to harbor germline mutations of the RB gene (Friend, 1986). Similarly, the predisposition to malignant gliomas that has been described in Li-Fraumeni families derives from germline p53 mutations (Malkin, 1990). In contrast, Turcot's syndrome displays segregation of its associated brain tumors with two separate gene defects. Whereas Turcot families with medulloblastomas harbor germline APC gene mutations, those with glioblastomas have been shown to have defects in the mismatch-repair genes hMLH1 and hPMS2 (Hamilton, 1995).

Although specific genes have not yet been identified for the familial glioma, Gorlin, Cowden, and MEN1 syndromes, there is compelling evidence of the chromosomal position. The Gorlin (associated with medulloblastoma) and MEN1 (associated with pituitary adenoma) syndromes, for instance, have been mapped to the 9q31 and 11q13 loci, respectively (Boggild, 1994; Gailani, 1992). The familial glioma syndrome, interestingly, has not shown a consistent association with p53 mutations and thus does not appear to be a form of Li-Fraumeni. As has been previously discussed, the 19q tumor suppressor gene is an attractive candidate for this syndrome, since 19q alterations are common to multiple glioma subtypes but not with other human tumors (Louis and von Deimling, 1995). Finally, Cowden syndrome is associated with dysplastic gangliocytomas of the cerebellum, but has so far it has eluded any characterization of an involved chromosomal locus (Albrecht, 1992).

## CONCLUSION

In the past, assessment of the clinical behavior of a primary CNS tumor relied primarily on predictions based on histopathologic classification and grading schemes. In contrast, contemporary genetic analysis has clearly shown that human CNS tumors possess molecular alterations reflecting tumor type, associated genetic syndrome, and stage. Moreover, such analysis has shown the power to define new clinically distinct subsets within particular tumor categories. This knowledge of the underlying effector mechanisms associated with these alterations will likely provide new avenues for designing strategies for effective therapeutic intervention. Indeed, such strategies have now emerged with the identification of abnormal target proteins in cancer cells, and the development of agents to selectively control or kill targeted cells.

As for tumors of the central nervous system, the high grade astrocytomas have been the particular focus of such therapeutic strategies. Specifically, the frequent activation of the EGFR gene in these tumors has prompted attempts to interfere with the EGFR signaling pathway. A recent phase I study utilized an anti-EGFR monoclonal antibody (Mab-425), and demonstrated acceptable levels of antibody binding in the tumor with limited toxicity (Faillot, 1996). It should be noted, however, that this antigen is not expressed uniformly on all tumor cells, thus making therapeutic efficacy with a single monoclonal antibody unlikely. Indeed, this expectation is borne out by a pilot phase I/II study of I-125 labelled Mab-425 that showed only slight prolongation of time to recurrence in a small group of recurrent tumors (Brady, 1990). More intriguing, though, has been the investigation of a class of selective tyrosine kinase inhibitors that were the subject of a recent review (Levitzki and Gazit, 1995). Tyrphostin A25, for example, has been shown to inhibit glioblastoma cell invasion when co-cultures of glioblastoma spheroids and fetal rat brain aggregates were examined (Penar, 1997). In addition, tyrphostin AG 1478 exhibited the capacity to preferentially suppress the growth of human glioma cells expressing truncated, tumor-specific EGFRs rather than wild-type EGFRs (Han, 1996). This latter observation is especially important, since it potentially offers a greater therapeutic advantage over those more general inhibitors that unavoidably target the normal receptor as well. Further advances in the diagnosis and treatment of primary CNS tumors will almost certainly be based on similar insights concerning the mechanisms involved in tumor initiation and progression.

## REFERENCES

Adesina, A., Nalbantoglu, J., and Cavenee, W., p53 gene mutation and mdm2 gene amplification are uncommon in medulloblastoma, *Cancer Res.,* 54, 5649, 1994.

Albarosa, R., DiDonato, S., and Finocchiaro, G., Redefinition of the coding sequence of the MXI1 gene and identification of a polymorphic repeat in the 3′ non-coding region that allows the detection of loss of heterozygosity of chromosome 10q25 in glioblastomas, *Hum. Genet.,* 95, 709, 1995.

Albrecht, S., Armstrong, D. L., Mahoney, D. H., Cheek, W. R., and Cooley, L. D., Cytogenetic demonstration of gene amplification in a primary intracranial germ cell tumor, *Genes Chromosom. Cancer,* 6, 61, 1993.

Albrecht, S., Haber, R. M., Goodman, J. C., and Duvic, M., Cowden syndrome and Lhermitte-Duclos disease, *Cancer,* 70, 869, 1992.

Alderson, L. M., Castleberg, R. L., Harsh, G. R. T., Louis, D. N., and Henson, J. W., Human gliomas with wild-type p53 express bcl-2, *Cancer Res.,* 55, 999, 1995.

Araki, C. and Matsumoto, S., Statistical reevaluation of pinealoma and related tumors in Japan, *J. Neurosurg.,* 30, 146, 1968.

Arap, W., Nishikawa, R., Furnari, F. B., Cavenee, W. K., and Huang, H. J., Replacement of the p16/CDKN2 gene suppresses human glioma cell growth, *Cancer Res.,* 55, 1351, 1995.

Ariza, A., Lopez, D., Mate, J. L., Isamat, M., Musulen, E., Pujol, M., Ley, A., and Navas-Palacios, J. J., Role of CD44 in the invasiveness of glioblastoma multiforme and the noninvasiveness of meningioma: an immunohistochemistry study, *Hum. Pathol.,* 26, 1144, 1995.

Asa, S. L., Puy, L. A., Lew, A. M., Sundmark, V. C., and Elsholtz, H. P., Cell type-specific expression of the pituitary transcription activator pit-1 in the human pituitary and pituitary adenomas, *J. Clin. Endocrinol. Metab.*, 77, 1275, 1993.

Asa, S. L., Scheithauer, B. W., Bilbao, J. M., Horvath, E., Ryan, N., Kovacs, K., Randall, R. V., Laws, E. R., Jr., Singer, W., Linfoot, J. A. et al., A case for hypothalamic acromegaly: a clinicopathological study of six patients with hypothalamic gangliocytomas producing growth hormone-releasing factor, *J. Clin. Endocrinol. Metab.*, 58, 796, 1984.

Bader, J. L., Meadows, A. T., Zimmerman, L. E., Rorke, L. B., Voute, P. A., Champion, L. A., and Miller, R. W., Bilateral retinoblastoma with ectopic intracranial retinoblastoma: trilateral retinoblastoma, *Cancer Genet. Cytogenet.*, 5, 203, 1982.

Bahn, R. S., Scheithauer, B. W., van Heerden, J. A., Laws, E. R., Jr., Horvath, E., and Gharib, H., Nonidentical expressions of multiple endocrine neoplasia, type I, in identical twins, *Mayo Clin. Proc.*, 61, 689, 1986.

Banerjee, M., Dinda, A. K., Sinha, S., Sarkar, C., and Mathur, M., c-myc oncogene expression and cell proliferation in mixed oligo-astrocytoma, *Int. J. Cancer*, 65, 730, 1996.

Bargonetti, J., Manfredi, J. J., Chen, X., Marshak, D. R., and Prives, C., A proteolytic fragment from the central region of p53 has marked sequence-specific DNA-binding activity when generated from wild-type but not from oncogenic mutant p53 protein, *Genes Dev.*, 7, 1993.

Barnett, G. H., Chou, S. M., and Bay, J. W., Posttraumatic intracranial meningioma: a case report and review of the literature, *Neurosurgery*, 18, 75, 1986.

Basu, T. N., Gutmann, D. H., Fletcher, J. A., Glover, T. W., Collins, F. S., and Downward, J., Aberrant regulation of ras proteins in malignant tumour cells from type 1 neurofibromatosis patients [see comments], *Nature*, 356, 713, 1992.

Batra, S. K., McLendon, R. E., Koo, J. S., Castelino-Prabhu, S., Fuchs, H. E., Krischer, J. P., Friedman, H. S., Bigner, D. D., and Bigner, S. H., Prognostic implications of chromosome 17p deletions in human medulloblastomas, *J. Neurooncol.*, 24, 39, 1995.

Bello, M. J., Leone, P. E., Nebreda, P., de Campos, J. M., Kusak, M. E., Vaquero, J., Sarasa, J. L., Garcia-Miguel, P., Queizan, A., Hernandez-Moneo, J. L. et al., Allelic status of chromosome 1 in neoplasms of the nervous system, *Cancer Genet. Cytogenet.*, 83, 160, 1995.

Bennett, W. P., Hollstein, M. C., Hsu, I. C., Sidransky, D., Lane, D. P., Vogelstein, B., and Harris, C. C., Mutational spectra and immunohistochemical analyses of p53 in human cancers, *Chest*, 101, 1992.

Bergsagel, D. J., Finegold, M. J., Butel, J. S., Kupsky, W. J., and Garcea, R. L., DNA sequences similar to those of simian virus 40 in ependymomas and choroid plexus tumors of childhood, *N. Engl. J. Med.*, 326, 988, 1992.

Biegel, J., Rorke, L., Packer, R., and Emanuel, B., Monosomy 22 in rhabdoid or atypical tumors of the brain, *J. Neurosurg.*, 73, 710, 1990.

Biegel, J. A., Burk, C. D., Barr, F. G., and Emanuel, B. S., Evidence for a 17p tumor related locus distinct from p53 in pediatric primitive neuroectodermal tumors, *Cancer Res.*, 52, 3391, 1992.

Biegel, J. A., Rorke, L. B., Packer, R. J., Sutton, L. N., Schut, L., Bonner, K., and Emanuel, B. S., Isochromosome 17q in primitive neuroectodermal tumors of the central nervous system, *Genes Chromosom. Cancer*, 1, 139, 1989.

Bigner, S., Humphrey, P., Wong, A. J., Vogelstein, B., Mark, J., Friedman, H., and Bigner, D., Characterization of the epidermal growth factor receptor in human glioma cell lines and xenografts, *Cancer Res.*, 50, 8017, 1990a.

Bigner, S. H., Mark, J., and Bigner, D. D., Cytogenetics of human brain tumors, *Cancer Genet. Cytogenet.*, 47, 141, 1990b.

Bigner, S. H., Mark, J., Burger, P. C., Mahaley, M. S., Jr., Bullard, D. E., Muhlbaier, L. H., and Bigner, D. D., Specific chromosomal abnormalities in malignant human gliomas, *Cancer Res.*, 48, 1988.

Bijlsma, E. K., Voesten, A. M., Bijleveld, E. H., Troost, D., Westerveld, A., Merel, P., Thomas, G., and Hulsebos, T. J., Molecular analysis of genetic changes in ependymomas, *Genes Chromosom. Cancer*, 13, 272, 1995.

Billestrup, N., Mitchell, R. L., Vale, W., and Verma, I. M., Growth hormone-releasing factor induces c-fos expression in cultured primary pituitary cells, *Mol. Endocrinol.*, 1, 300, 1987.

Birch, B. D., Johnson, J. P., Parsa, A., Desai, R. D., Yoon, J. T., Lycette, C. A., Li, Y. M., and Bruce, J. N., Frequent type 2 neurofibromatosis gene transcript mutations in sporadic intramedullary spinal cord ependymomas, *Neurosurgery*, 39, 135, 1996.

Blamires, T. L. and Maher, E. R., Choroid plexus papilloma. A new presentation of von Hippel-Lindau (VHL) disease, *Eye*, 6, 90, 1992.

Boggild, M. D., Jenkinson, S., Pistorello, M., Boscaro, M., Scanarini, M., McTernan, P., Perrett, C. W., Thakker, R. V., and Clayton, R. N., Molecular genetic studies of sporadic pituitary tumors, *J. Clin. Endocrinol. Metab.*, 78, 387, 1994.

Bogler, O., Huang, H. J., and Cavenee, W. K., Loss of wild-type p53 bestows a growth advantage on primary cortical astrocytes and facilitates their in vitro transformation, *Cancer Res.*, 55, 2746, 1995.

Bogler, O., Huang, H. J., Kleihues, P., and Cavenee, W. K., The p53 gene and its role in human brain tumors, *Glia*, 15, 308, 1995.

Borovich, B. and Doron, Y., Recurrence of intracranial meningiomas: the role played by regional multicentricity, *J. Neurosurg.*, 64, 58, 1986.

Brady, L., Markoe, A., Woo, D., Rackover, M., Koprowski, H., and Peyster, R., Iodine-125 labeled anti-epidermal growth factor receptor-425 in the treatment of malignant astrocytomas, *J. Neurosurg. Sci.*, 34, 243, 1990.

Brinster, R. L., Chen, H. Y., Messing, A., van Dyke, T., Levine, A. J., and Palmiter, R. D., Transgenic mice harboring SV40 T-antigen genes develop characteristic brain tumors, *Cell*, 37, 367, 1984.

Buckley, N., Bates, A. S., Broome, J. C., Strange, R. C., Perrett, C. W., Burke, C. W., and Clayton, R. N., p53 Protein accumulates in Cushings adenomas and invasive non-functional adenomas (corrected and republished in *J. Clin. Endocrinol. Metab.*, 80(2), 4, 1995), *J. Clin. Endocrinol. Metab.*, 79, 1513, 1994.

Burger, P. C. and Vogel, F. S., *Surgical Pathology of the Nervous System and Its Coverings*, 2nd ed., John Wiley & Sons Inc., New York, 1982.

Burger, P. C. and Green, S. B., Patient age, histologic features, and length of survival in patients with glioblastoma multiforme, *Cancer*, 59, 1617, 1987.

Bystrom, C., Larsson, C., Blomberg, C., Sandelin, K., Falkmer, U., Skogseid, B., Oberg, K., Werner, S., and Nordenskjold, M., Localization of the MEN1 gene to a small region within chromosome 11q13 by deletion mapping in tumors, *Proc. Natl. Acad. Sci. U.S.A.*, 87, 1968, 1990.

Cahill, D. W., Bashirelahi, N., Solomon, L. W., Dalton, T., Salcman, M., and Ducker, T. B., Estrogen and progesterone receptors in meningiomas, *J. Neurosurg.*, 60, 985, 1984.

Cairncross, J. G., Macdonald, D. R., and Ramsay, D. A., Aggressive oligodendroglioma: a chemosensitive tumor, *Neurosurgery*, 31, 78, 1992.

Carroll, R. S., Schrell, U. M., Zhang, J., Dashner, K., Nomikos, P., Fahlbusch, R., and Black, P. M., Dopamine D1, dopamine D2, and prolactin receptor messenger ribonucleic acid expression by the polymerase chain reaction in human meningiomas, *Neurosurgery*, 38, 367, 1996.

Castrillo, J. L., Theill, L. E., and Karin, M., Function of the homeodomain protein GHF1 in pituitary cell proliferation, *Science*, 253, 197, 1991.

Cavenee, W. K., Dryja, T. P., Phillips, R. A., Benedict, W. F., Godbout, R., Gallie, B. L., Murphree, A. L., Strong, L. C., and White, R. L., Expression of recessive alleles by chromosomal mechanisms in retinoblastoma, *Nature*, 305, 1983.

Cavenee, W. K., Hansen, M. F., Nordenskjold, M., Kock, E., Maumenee, I., Squire, J. A., Phillips, R. A., and Gallie, B. L., Genetic origin of mutations predisposing to retinoblastoma, *Science*, 228, 501, 1985.

Chan, R. C. and Thompson, G. B., Intracranial meningiomas in childhood, *Surg. Neurol.*, 21, 319, 1984.

Cheng, S. Y., Huang, H. J., Nagane, M., Ji, X. D., Wang, D., Shih, C. C., Arap, W., Huang, C. M., and Cavenee, W. K., Suppression of glioblastoma angiogenicity and tumorigenicity by inhibition of endogenous expression of vascular endothelial growth factor, *Proc. Natl. Acad. Sci. U.S.A.*, 93, 8502, 1996.

Chintala, S. K., Sawaya, R., Gokaslan, Z. L., and Rao, J. S., Modulation of matrix metalloprotease-2 and invasion in human glioma cells by alpha 3 beta 1 integrin, *Cancer Lett.*, 103, 201, 1996.

Chou, D., Miyashita, T., Mohrenweiser, H., Ueki, K., Kastury, K., Druck, T., Von Deimling, A., Huebner, K., Reed, J., and Louis, D., The BAX gene maps to glioma candidate region at 19q13.3 but is not altered in human gliomas, *Cancer Genet. Cytogenet.*, 88, 136, 1996.

Cogen, P. H., Daneshvar, L., Metzger, A. K., and Edwards, M. S., Deletion mapping of the medulloblastoma locus on chromosome 17p, *Genomics*, 8, 279, 1990.

Collins, V. P., Amplified genes in human gliomas, *Semin. Cancer Biol.*, 4, 27, 1993.

Collins, V. P., Gene amplification in human gliomas, *Glia*, 15, 289, 1995.

Costello, J. F., Berger, M. S., Huang, H. S., and Cavenee, W. K., Silencing of p16/CDKN2 expression in human gliomas by methylation and chromatin condensation, *Cancer Res.*, 56, 2405, 1996.

Couturier, J., Delattre, O., Kujas, M., Philippon, J., Peter, M., Rouleau, G., Aurias, A., and Thomas, G., Assessment of chromosome 22 anomalies in neurinomas by combined karyotype and RFLP analyses, *Cancer Genet. Cytogenet.,* 45, 55, 1990.

Cunningham, J. M., Kimmel, D. W., Scheithauer, B. W., O'Fallon, J. R., Novotny, P. J., and Jenkins, R. B., Analysis of proliferation markers and p53 expression in gliomas of astrocytic origin: relationships and prognostic value, *J. Neurosurg.,* 86, 121, 1997.

Danks, R. A., Chopra, G., Gonzales, M. F., Orian, J. M., and Kaye, A. H., Aberrant p53 expression does not correlate with the prognosis in anaplastic astrocytoma, *Neurosurgery,* 37, 246, 1995.

Debiec-Rychter, M., Alwasiak, J., Liberski, P. P., Nedoszytko, B., Babinska, M., Mrozek, K., Imielinski, B., Borowska-Lehman, J., and Limon, J., Accumulation of chromosomal changes in human glioma progression. A cytogenetic study of 50 cases, *Cancer Genet. Cytogenet.,* 85, 61, 1995.

Di Fiore, P. P., Pierce, J. H., Fleming, T. P., Hazan, R., Ullrich, A., King, C. R., Schlessinger, J., and Aaronson, S. A., Overexpression of the human EGF receptor confers an EGF-dependent transformed phenotype to NIH 3T3 cells, *Cell,* 51, 1, 1987.

Diedrich, U., Lucius, J., Baron, E., Behnke, J., Pabst, B., and Zoll, B., Distribution of epidermal growth factor receptor gene amplification in brain tumours and correlation to prognosis, *J. Neurol.,* 242, 683, 1995.

Donner, L. R., Cytogenetics and molecular biology of small round-cell tumors and related neoplasms. Current status, *Cancer Genet. Cytogenet.,* 54, 1, 1991.

Dyson, N., Bernards, R., Friend, S. H., Gooding, L. R., Hassell, J. A., Major, E. O., Pipas, J. M., Vandyke, T., and Harlow, E., Large T antigens of many polyomaviruses are able to form complexes with the retinoblastoma protein, *J. Virol.,* 64, 1353, 1990.

Edvardsen, K., Pedersen, P. H., Bjerkvig, R., Hermann, G., Zeuthen, J., Laerum, O. D., Walsh, F., and Bock, E., Transfection of glioma cells with the neural-cell adhesion molecule NCAM: effect on glioma-cell invasion and growth in vivo, *Int. J. Cancer,* 58, 116, 1994.

Eibl, R. H., Pietsch, T., Moll, J., Skroch-Angel, P., Heider, K. H., von Ammon, K., Wiestler, O. D., Ponta, H., Kleihues, P., and Herrlich, P., Expression of variant CD44 epitopes in human astrocytic brain tumors, *J. Neurooncol.,* 26, 165, 1995.

Ekstrand, A., James, C., Cavenee, W., Seliger, B., Pettersson, R., and Collins, V., Genes for epidermal growth factor receptor, transforming growth factor alpha, and epidermal growth factor and their expression in human gliomas in vivo, *Cancer Res.,* 51, 2164, 1991.

Ekstrand, A. J., Sugawa, N., James, C. D., and Collins, V. P., Amplified and rearranged epidermal growth factor receptor genes in human glioblastomas reveal deletions of sequences encoding portions of the N- and/or C-terminal tails, *Proc. Natl. Acad. Sci. U.S.A.,* 89, 4309, 1992.

Eng, C., Spechler, S. J., Ruben, R., and Li, F. P., Familial Barrett esophagus and adenocarcinoma of the gastroesophageal junction, *Cancer Epidemiol. Biomarkers Prev.,* 2, 397, 1993.

Enzinger, F. M. and Smith, B. H., Hemangiopericytoma. An analysis of 106 cases, *Hum. Pathol.,* 7, 61, 1976.

Faillot, T., Magdelenat, H., Mady, E., Stasiecki, P., Fohanno, D., Gropp, P., Poisson, M., and Delattre, J., A phase I study of an anti-epidermal growth factor receptor monoclonal antibody for the treatment of malignant gliomas, *Neurosurgery,* 39, 478, 1996.

Finkelstein, S. D., Black, P., Nowak, T. P., Hand, C. M., Christensen, S., and Finch, P. W., Histological characteristics and expression of acidic and basic fibroblast growth factor genes in intracerebral xenogeneic transplants of human glioma cells, *Neurosurgery,* 34, 136, 1994.

Fischer, U., Muller, H. W., Sattler, H. P., Feiden, K., Zang, K. D., and Meese, E., Amplification of the MET gene in glioma, *Genes Chromosom. Cancer,* 12, 63, 1995.

Fleming, T. P., Saxena, A., Clark, W. C., Robertson, J. T., Oldfield, E. H., Aaronson, S. A., and Ali, I. U., Amplification and/or overexpression of platelet-derived growth factor receptors and epidermal growth factor receptor in human glial tumors, *Cancer Res.,* 52, 4550, 1992.

Forus, A., Florenes, V. A., Maelandsmo, G. M., Fodstad, O., and Myklebost, O., The protooncogene CHOP/GADD153, involved in growth arrest and DNA damage response, is amplified in a subset of human sarcomas, *Cancer Genet. Cytogenet.,* 78, 165, 1994.

Fox, S. R., Jong, M. T., Casanova, J., Ye, Z. S., Stanley, F., and Samuels, H. H., The homeodomain protein, Pit-1/GHF-1, is capable of binding to and activating cell-specific elements of both the growth hormone and prolactin gene promoters, *Mol. Endocrinol.,* 4, 1069, 1990.

Friedlander, D. R., Zagzag, D., Shiff, B., Cohen, H., Allen, J. C., Kelly, P. J., and Grumet, M., Migration of brain tumor cells on extracellular matrix proteins in vitro correlates with tumor type and grade and involves alphaV and beta1 integrins, *Cancer Res.,* 1996, p. 1939.

Friend, K. E., Chiou, Y. K., Laws, E. R., Jr., Lopes, M. B., and Shupnik, M. A., Pit-1 messenger ribonucleic acid is differentially expressed in human pituitary adenomas, *J. Clin. Endocrinol. Metab.,* 77, 1281, 1993.

Friend, S. H., Bernards, R., Rogelj, S., Weinberg, R. A., Rapaport, J. M., Albert, D. M., and Dryja, T. P., A human DNA segment with properties of the gene that predisposes to retinoblastoma and osteosarcoma, *Nature,* 323, 643, 1986.

Fujimoto, M., Weaker, F. J., Herbert, D. C., Sharp, Z. D., Sheridan, P. J., and Story, J. L., Expression of three viral oncogenes (v-sis, v-myc, v-fos) in primary human brain tumors of neuroectodermal origin, *Neurology,* 38, 289, 1988.

Fuller, G. N. and Bigner, S. H., Amplified cellular oncogenes in neoplasms of the human central nervous system, *Mutat. Res.,* 276, 299, 1992.

Fults, D., Pedone, C. A., Thomas, G. A., and White, R., Allelotype of human malignant astrocytoma, *Cancer Res.,* 50, 5784, 1990.

Fults, D., Brockmeyer, D., Tullous, M. W., Pedone, C. A., and Cawthon, R. M., p53 mutation and loss of heterozygosity on chromosomes 17 and 10 during human astrocytoma progression, *Cancer Res.,* 52, 674, 1992.

Fults, D. and Pedone, C., Deletion mapping of the long arm of chromosome 10 in glioblastoma multiforme, *Genes Chromosom. Cancer,* 7, 173, 1993.

Furnari, F. B., Huang, H. J., and Cavenee, W. K., Molecular biology of malignant degeneration of astrocytoma, *Pediatr. Neurosurg.,* 24, 41, 1996.

Gailani, M. R., Bale, S. J., Leffell, D. J., DiGiovanna, J. J., Peck, G. L., Poliak, S., Drum, M. A., Pastakia, B., McBride, O. W., Kase, R. et al., Developmental defects in Gorlin syndrome related to a putative tumor suppressor gene on chromosome 9, *Cell,* 69, 111, 1992.

Garber, J. E., Burke, E. M., Lavally, B. L., Billett, A. L., Sallan, S. E., Scott, R. M., Kupsky, W., and Li, F. P., Choroid plexus tumors in the breast cancer-sarcoma syndrome, *Cancer,* 66, 2658, 1990.

Garcia, D. M. and Fulling, K. H., Juvenile pilocytic astrocytoma of the cerebrum in adults. A distinctive neoplasm with favorable prognosis, *J. Neurosurg.,* 63, 382, 1985.

Giani, C. and Finocchiaro, G., Mutation rate of the CDKN2 gene in malignant gliomas, *Cancer Res.,* 54, 6338, 1994.

Goellner, J. R., Laws, E. R., Jr., Soule, E. H., and Okazaki, H., Hemangiopericytoma of the meninges. Mayo Clinic experience, Am. J. Clin. Pathol., 70, 375, 1978.

Goldberg, M. A. and Schneider, T. J., Similarities between the oxygen-sensing mechanisms regulating the expression of vascular endothelial growth factor and erythropoietin, *J. Biol. Chem.,* 269, 4355, 1994.

Graeber, T. G., Osmanian, C., Jacks, T., Housman, D. E., Koch, C. J., Lowe, S. W., and Giaccia, A. J., Hypoxia-mediated selection of cells with diminished apoptotic potential in solid tumours [see comments], *Nature,* 379, 88, 1996.

Green, A. J., Johnson, P. H., and Yates, J. R., The tuberous sclerosis gene on chromosome 9q34 acts as a growth suppressor, *Hum. Mol. Genet.,* 3, 1833, 1994a.

Green, A. J., Smith, M., and Yates, J. R., Loss of heterozygosity on chromosome 16p13.3 in hamartomas from tuberous sclerosis patients, *Nat. Genet.,* 6, 193, 1994b.

Griffin, C. A., Hawkins, A. L., Packer, R. J., Rorke, L. B., and Emanuel, B. S., Chromosome abnormalities in pediatric brain tumors, *Cancer Res.,* 48, 175, 1988.

Guthrie, B. L. and Laws, E. R., Jr., Supratentorial low grade gliomas, *Neurosurg. Clin. N. Am.,* 1, 37, 1990.

Haines, J. and Short, M., Tuberous sclerosis: hamartomas, subependymal giant cell astrocytomas, and other central nervous system tumors, in *Molecular Genetics of Nervous System Tumors,* Levine, A. and Schmidek, H., Eds., Wiley-Liss Inc., New York, 1993, p. 303.

Hamada, K., Nishi, T., Kuratsu, J., and Ushio, Y., Expression and alternative splicing of Pit-1 messenger ribonucleic acid in pituitary adenomas, *Neurosurgery,* 38, 362, 1996.

Hamilton, S. R., Liu, B., Parsons, R. E., Papadopoulos, N., Jen, J., Powell, S. M., Krush, A. J., Berk, T., Cohen, Z., Tetu, B. et al., The molecular basis of Turcot's syndrome [see comments], *N. Engl. J. Med.,* 332, 839, 1995.

Han, Y., Caday, C. G., Nanda, A., Cavenee, W. K., and Huang, H. J., Tyrphostin AG 1478 preferentially inhibits human glioma cells expressing truncated rather than wild-type epidermal growth factor receptors, *Cancer Res.,* 56, 3859, 1996.

Hanahan, D. and Folkman, J., Patterns and emerging mechanisms of the angiogenic switch during tumorigenesis, *Cell,* 86, 353, 1996.

Hashimoto, N., Ichikawa, D., Arakawa, Y., Date, K., Ueda, S., Nakagawa, Y., Horii, A., Nakamura, Y., Abe, T., and Inazawa, J., Frequent deletions of material from chromosome arm 1p in oligodendroglial tumors revealed by double-target fluorescence in situ hybridization and microsatellite analysis, *Genes Chromosom. Cancer,* 14, 295, 1995.

He, J., Allen, J. R., Collins, V. P., Allalunis-Turner, M. J., Godbout, R., Day, R. S. R., and James, C. D., CDK4 amplification is an alternative mechanism to p16 gene homozygous deletion in glioma cell lines, *Cancer Res.,* 54, 5804, 1994.

He, J., Olson, J. J., Ekstrand, A. J., Serbanescu, A., Yang, J., Offermann, M. K., and James, C. D., Transfection of IFNalpha in human glioblastoma cells and tumorigenicity in association with induction of PKR and OAS gene expression, *J. Neurosurg.,* 85, 1085, 1996.

Hecht, B. K., Turc-Carel, C., Chatel, M., Grellier, P., Gioanni, J., Attias, R., Gaudray, P., and Hecht, F., Cytogenetics of malignant gliomas: I. The autosomes with reference to rearrangements, *Cancer Genet. Cytogenet.,* 84, 1, 1994.

Henson, J. W., Schnitker, B. L., Correa, K. M., von Deimling, A., Fassbender, F., Xu, H. J., Benedict, W. F., Yandell, D. W., and Louis, D. N., The retinoblastoma gene is involved in malignant progression of astrocytomas, *Ann. Neurol.,* 36, 714, 1994.

Herman, J. G., Latif, F., Weng, Y., Lerman, M. I., Zbar, B., Liu, S., Samid, D., Duan, D. S., Gnarra, J. R., Linehan, W. M. et al., Silencing of the VHL tumor-suppressor gene by DNA methylation in renal carcinoma, *Proc. Natl. Acad. Sci. U.S.A.,* 91, 9700, 1994.

Herman, V., Drazin, N. Z., Gonsky, R., and Melmed, S., Molecular screening of pituitary adenomas for gene mutations and rearrangements, *J. Clin. Endocrinol. Metab.,* 77, 50, 1993.

Herman, V., Fagin, J., Gonsky, R., Kovacs, K., and Melmed, S., Clonal origin of pituitary adenomas, *J. Clin. Endocrinol. Metab.,* 71, 1427, 1990.

Hermanson, M., Funa, K., Hartman, M., Claesson-Welsh, L., Heldin, C. H., Westermark, B., and Nister, M., Platelet-derived growth factor and its receptors in human glioma tissue: expression of messenger RNA and protein suggests the presence of autocrine and paracrine loops, *Cancer Res.,* 52, 3213, 1992.

Hermanson, M., Funa, K., Koopmann, J., Maintz, D., Waha, A., Westermark, B., Heldin, C. H., Wiestler, O. D., Louis, D. N., von Deimling, A., and Nister, M., Association of loss of heterozygosity on chromosome 17p with high platelet-derived growth factor alpha receptor expression in human, *Cancer Res.,* 56, 164, 1996.

Herpers, M. J. and Budka, H., Glial fibrillary acidic protein (GFAP) in oligodendroglial tumors: gliofibrillary oligodendroglioma and transitional oligoastrocytoma as subtypes of oligodendroglioma, *Acta. Neuropathol. (Berl.),* 64, 265, 1984.

Hollstein, M., Rice, K., Greenblatt, M. S., Soussi, T., Fuchs, R., Sorlie, T., Hovig, E., Smith-Sorensen, B., Montesano, R., and Harris, C., Database of p53 gene somatic mutations in human tumors and cell lines, *Nucleic Acids Res.,* 22, 3551, 1994.

Hosokawa, Y., Tsuchihashi, Y., Okabe, H., Toyama, M., Namura, K., Kuga, M., Yonezawa, T., Fujita, S., and Ashihara, T., Pleomorphic xanthoastrocytoma. Ultrastructural, immunohistochemical, and DNA cytofluorometric study of a case, *Cancer,* 68, 853, 1991.

Hsieh, L. L., Hsia, C. F., Wang, L. Y., Chen, C. J., and Ho, Y. S., p53 gene mutations in brain tumors in Taiwan, *Cancer Lett.,* 78, 25, 1994.

Hsu, D., Efird, J., and Hedley-Whyte, E., Progesterone and estrogen receptors in meningiomas: prognostic considerations, *J. Neurosurg.,* 86, 113, 1997.

Humphrey, P., Wong, A., Vogelstein, B., Zalutsky, M., Fuller, G. Archer, G., Friedman H., Kwatra, M., Bigner, S., and Bigner, D., Anti-synthetic peptide antibody reacting at the fusion junction of deletion-mutant epidermal growth factor receptors in human glioblastoma, *Proc. Natl. Acad. Sci. U.S.A.,* 87, 4207, 1990.

Hurtt, M. R., Moossy, J., Donovan-Peluso, M., and Locker, J., Amplification of epidermal growth factor receptor gene in gliomas: histopathology and prognosis, *J. Neuropathol. Exp. Neurol.,* 51, 84, 1992.

Iwadate, Y., Fujimoto, S., Tagawa, M., Namba, H., Sueyoshi, K., Hirose, M., and Sakiyama, S., Association of p53 gene mutation with decreased chemosensitivity in human malignant gliomas, *Int. J. Cancer,* 69, 236, 1996.

Izumoto, S., Ohnishi, T., Arita, N., Hiraga, S., Taki, T., and Hayakawa, T., Gene expression of neural cell adhesion molecule L1 in malignant gliomas and biological significance of L1 in glioma invasion, *Cancer Res.*, 56, 1440, 1996.

Jacoby, L. B., MacCollin, M., Louis, D. N., Mohney, T., Rubio, M. P., Pulaski, K., Trofatter, J. A., Kley, N., Seizinger, B., Ramesh, V. et al., Exon scanning for mutation of the NF2 gene in schwannomas, *Hum. Mol. Genet.*, 3, 413, 1994.

James, C. D., He, J., Karlbom, E., Nordenskjold, M., Cavenee, W. K., and Collins, V. P., Chromosome 9 deletion mapping reveals interferon-alpha and interferon-beta-1 gene deletions in human glial tumors, *Cancer Res.*, 51, 1684, 1991.

James, C. D., Carlbom, E., Dumanski, J. P., Hansen, M., Nordenskjold, M., Collins, V. P., and Cavenee, W. K., Clonal genomic alterations in glioma malignancy stages, *Cancer Res.*, 48, 5546, 1988.

James, C. D., He, J., Collins, V. P., Allalunis-Turner, M. J., and Day, R. S. D., Localization of chromosome 9p homozygous deletions in glioma cell lines with markers constituting a continuous linkage group, *Cancer Res.*, 53, 3674, 1993.

Jay, V., Pienkowska, M., Becker, L., and Zielenska, M., Primitive neuroectodermal tumors of the cerebrum and cerebellum: absence of t(11;22) translocation by RT-PCR analysis, *Mod. Pathol.*, 8, 488, 1995.

Jen, J., Harper, J. W., Bigner, S. H., Bigner, D. D., Papadopoulos, N., Markowitz, S., Willson, J. K., Kinzler, K. W., and Vogelstein, B., Deletion of p16 and p15 genes in brain tumors, *Cancer Res.*, 54, 6353, 1994.

Jennings, M. T., Gelman, R., and Hochberg, F., Intracranial germ-cell tumors: natural history and pathogenesis, *J. Neurosurg.*, 63, 155, 1985.

Johnson, M. R., DeClue, J. E., Felzmann, S., Vass, W. C., Xu, G., White, R., and Lowy, D. R., Neurofibromin can inhibit Ras-dependent growth by a mechanism independent of its GTPase-accelerating function, *Mol. Cell. Biol.*, 14, 641, 1994.

Joseph, J. T., Lisle, D. K., Jacoby, L. B., Paulus, W., Barone, R., Cohen, M. L., Roggendorf, W. H., Bruner, J. M., Gusella, J. F., and Louis, D. N., NF2 gene analysis distinguishes hemangiopericytoma from meningioma, *Am. J. Pathol.*, 147, 1450, 1995.

Kaaijk, P., Troost, D., Morsink, F., Keehnen, R. M., Leenstra, S., Bosch, D. A., and Pals, S. T., Expression of CD44 splice variants in human primary brain tumors, *J. Neurooncol.*, 26, 185, 1995.

Kalkanis, S. N., Carroll, R. S., Zhang, J., Zamani, A. A., and Black, P. M., Correlation of vascular endothelial growth factor messenger RNA expression with peritumoral vasogenic cerebral edema in meningiomas, *J. Neurosurg.*, 85, 1095, 1996.

Kalyan-Raman, U. P., and Olivero, W. C., Ganglioglioma: a correlative clinicopathological and radiological study of ten surgically treated cases with follow-up, *Neurosurgery*, 20, 428, 1987.

Kamb, A., Gruis, N. A., Weaver-Feldhaus, J., Liu, Q., Harshman, K., Tavtigian, S. V., Stockert, E., Day, R. S. R., Johnson, B. E., and Skolnick, M. H., A cell cycle regulator potentially involved in genesis of many tumor types, *Science*, 264, 436, 1994.

Kandt, R. S., Pericak-Vance, M. A., Hung, W. Y., Gardner, R. J., Crossen, P. E., Nellist, M. D., Speer, M. C., and Roses, A. D., Linkage studies in tuberous sclerosis. Chromosome 9?, 11?, or maybe 14!, *Ann. N. Y. Acad. Sci.*, 615, 284, 1991.

Kanno, H., Kondo, K., Ito, S., Yamamoto, I., Fujii, S., Torigoe, S., Sakai, N., Hosaka, M., Shuin, T., and Yao, M., Somatic mutations of the von Hippel-Lindau tumor suppressor gene in sporadic central nervous system hemangioblastomas, *Cancer Res.*, 54, 4845, 1994.

Karlbom, A. E., James, C. D., Boethius, J., Cavenee, W. K., Collins, V. P., Nordenskjold, M., and Larsson, C., Loss of heterozygosity in malignant gliomas involves at least three distinct regions on chromosome 10, *Hum. Genet.*, 92, 169, 1993.

Karnes, P. S., Tran, T. N., Cui, M. Y., Raffel, C., Gilles, F. H., Barranger, J. A., and Ying, K. L., Cytogenetic analysis of 39 pediatric central nervous system tumors, *Cancer Genet. Cytogenet.*, 59, 12, 1992.

Kees, U. R., Biegel, J. A., Ford, J., Ranford, P. R., Peroni, S. E., Hallam, L. A., Parmiter, A. H., Willoughby, M. L., and Spagnolo, D., Enhanced MYCN expression and isochromosome 17q in pineoblastoma cell lines, *Genes Chromosom. Cancer*, 9, 129, 1994.

Kepes, J. J., Rubinstein, L. J., and Eng, L. F., Pleomorphic xanthoastrocytoma: a distinctive meningocerebral glioma of young subjects with relatively favorable prognosis. A study of 12 cases, *Cancer*, 44, 1839, 1979.

Kieser, A., Weich, H. A., Brandner, G., Marme, D., and Kolch, W., Mutant p53 potentiates protein kinase C induction of vascular endothelial growth factor expression, *Oncogene*, 9, 963, 1994.

Kim, D. H., Mohapatra, G., Bollen, A., Waldman, F. M., and Feuerstein, B. G., Chromosomal abnormalities in glioblastoma multiforme tumors and glioma cell lines detected by comparative genomic hybridization, *Int. J. Cancer,* 60, 812, 1995.

Kleihues, P., Burger, P. C., and Scheithauer, B. W., *Histological Typing of Tumors of the Central Nervous System,* 2nd ed., Springer-Verlag, Berlin, 1992

Kleihues, P., Burger, P. C., and Scheithauer, B. W., The new WHO classification of brain tumours, *Brain Pathol.,* 3, 255, 1993.

Koopmann, J., Maintz, D., Schild, S., Schramm, J., Louis, D. N., Wiestler, O. D., and von Deimling, A., Multiple polymorphisms, but no mutations, in the WAF1/CIP1 gene in human brain tumours, *Br. J. Cancer,* 72, 1230, 1995.

Kraus, J. A., Koopmann, J., Kaskel, P., Maintz, D., Brandner, S., Schramm, J., Louis, D. N., Wiestler, O. D., and von Deimling, A., Shared allelic losses on chromosomes 1p and 19q suggest a common origin of oligodendroglioma and oligoastrocytoma, *J. Neuropathol. Exp. Neurol.,* 54, 91, 1995.

Krishna, M., Smith, T. W., and Recht, L. D., Expression of bcl-2 in reactive and neoplastic astrocytes: lack of correlation with presence or degree of malignancy, *J. Neurosurg.,* 83, 1017, 1995.

Kuppner, M. C., Van Meir, E., Gauthier, T., Hamou, M. F., and de Tribolet, N., Differential expression of the CD44 molecule in human brain tumours, *Int. J. Cancer,* 50, 572, 1992.

Kurihara, M., Tokunaga, Y., Tsutsumi, K., Kawaguchi, T., Shigematsu, K., Niwa, M., and Mori, K., Characterization of insulin-like growth factor I and epidermal growth factor receptors in meningioma, *J. Neurosurg.,* 71, 538, 1989.

Kyritsis, A. P., Bondy, M. L., Xiao, M., Berman, E. L., Cunningham, J. E., Lee, P. S., Levin, V. A., and Saya, H., Germline p53 gene mutations in subsets of glioma patients, *J. Natl. Cancer Inst.,* 86, 344, 1994.

Landis, C. A., Masters, S. B., Spada, A., Pace, A. M., Bourne, H. R., and Vallar, L., GTPase inhibiting mutations activate the alpha chain of Gs and stimulate adenylyl cyclase in human pituitary tumours, *Nature,* 340, 692, 1989.

Lane, D. P., On the expression of the p53 protein in human cancer, *Mol. Biol. Rep.,* 19, 23, 1994.

Lang, F. F., Miller, D. C., Pisharody, S., Koslow, M., and Newcomb, E. W., High frequency of p53 protein accumulation without p53 gene mutation in human juvenile pilocytic, low grade and anaplastic astrocytomas, *Oncogene,* 9, 949, 1994.

Lekanne Deprez, R. H., Bianchi, A. B., Groen, N. A., Seizinger, B. R., Hagemeijer, A., van Drunen, E., Bootsma, D., Koper, J. W., Avezaat, C. J., Kley, N. et al., Frequent NF2 gene transcript mutations in sporadic meningiomas and vestibular schwannomas, *Am. J. Hum. Genet.,* 54, 1022, 1994.

Lekanne Deprez, R. H., Riegman, P. H., Groen, N. A., Warringa, U. L., van Biezen, N. A., Molijn, A. C., Bootsma, D., de Jong, P. J., Menon, A. G., Kley, N. A. et al., Cloning and characterization of MN1, a gene from chromosome 22q11, which is disrupted by a balanced translocation in a meningioma, *Oncogene,* 10, 1521, 1995.

Levitzki, A. and Gazit, A., Tyrosine kinase inhibition: an approach to drug development, *Science,* 267, 1782, 1995.

Levy, A., Hall, L., Yeudall, W. A., and Lightman, S. L., p53 gene mutations in pituitary adenomas: rare events, *Clin. Endocrinol. (Oxf.),* 41, 809, 1994.

Levy, A. P., Levy, N. S., and Goldberg, M. A., Hypoxia-inducible protein binding to vascular endothelial growth factor mRNA and its modulation by the von Hippel-Lindau protein, *J. Biol. Chem.,* 271, 25492, 1996.

Li, Y. S., Ramsay, D. A., Fan, Y. S., Armstrong, R. F., and Del Maestro, R. F., Cytogenetic evidence that a tumor suppressor gene in the long arm of chromosome 1 contributes to glioma growth, *Cancer Genet. Cytogenet.,* 84, 46, 1995.

Liang, B. C., Ross, D. A., and Reed, E., Genomic copy number changes of DNA repair genes ERCC1 and ERCC2 in human gliomas, *J. Neurooncology,* 26, 17, 1995.

Libermann, T. A., Nusbaum, H. R., Razon, N., Kris, R., Lax, I., Soreq, H., Whittle, N., Waterfield, M. D., Ullrich, A., and Schlessinger, J., Amplification, enhanced expression and possible rearrangement of EGF receptor gene in primary human brain tumours of glial origin, *Nature,* 313, 144, 1985.

Lindblom, A., Ruttledge, M., Collins, V. P., Nordenskjold, M., and Dumanski, J. P., Chromosomal deletions in anaplastic meningiomas suggest multiple regions outside chromosome 22 as important in tumor progression, *Int. J. Cancer,* 56, 354, 1994.

Litofsky, N. S., Hinton, D., and Raffel, C., The lack of a role for p53 in astrocytomas in pediatric patients, *Neurosurgery,* 34, 967, 1994.

Lloyd, R. V., Jin, L., Chandler, W. F., Horvath, E., Stefaneanu, L., and Kovacs, K., Pituitary specific transcription factor messenger ribonucleic expression in adenomatous and nontumorous human pituitary tissues, *Lab. Invest.,* 69, 570, 1993.

Louis, D. N. and Cavenee, W. K., Molecular biology of central nervous system tumors, in *Cancer: Principles and Practice of Oncology,* De Vita, V. T., Hellman, S., and Rosenberg, S. A., Eds., 1995

Louis, D. N., The p53 gene and protein in human brain tumors, *J. Neuropathol. Exp. Neurol.,* 53, 11, 1994.

Louis, D. N. and Gusella, J. F., A tiger behind many doors: multiple genetic pathways to malignant glioma, *Trends Genet.,* 11, 412, 1995.

Louis, D. N., Ramesh, V., and Gusella, J. F., Neuropathology and molecular genetics of neurofibromatosis 2 and related tumors, *Brain Pathol.,* 5, 163, 1995.

Louis, D. N. and von Deimling, A., Hereditary tumor syndromes of the nervous system: overview and rare syndromes, *Brain Pathol.,* 5, 145, 1995.

Louis, D. N., von Deimling, A., Dickersin, G. R., Dooling, E. C., and Seizinger, B. R., Desmoplastic cerebral astrocytomas of infancy: a histopathologic, immunohistochemical, ultrastructural, and molecular genetic study, *Hum. Pathol.,* 23, 1402, 1992.

Lowe, S. W., Bodis, S., McClatchey, A., Remington, L., Ruley, H. E., Fisher, D. E., Housman, D. E., and Jacks, T., p53 status and the efficacy of cancer therapy in vivo, *Science,* 266, 807, 1994.

Mailleux, P., Przedborski, S., Beaumont, A., Verslijpe, M., Depierreux, M., Levivier, M., Kitabgi, P., Roques, B. P., and Vanderhaeghen, J. J., Neurotensin high affinity binding sites and endopeptidase 24.11 are present respectively in the meningothelial and in the fibroblastic components of human meningiomas, *Peptides,* 11, 1245, 1990a.

Mailleux, P. and Vanderhaeghen, J. J., Cholecystokinin receptors of A type in the human dorsal medulla oblongata and meningiomas, and of B type in small cell lung carcinomas, *Neurosci. Lett.,* 117, 243, 1990b.

Malkin, D., Li F., Strong, L., Fraumeni J, Nelson, S., Kim, D., Kassel, J., Gryka, M., Bishoff, F., Taisky, M., and Friend, S., Germ line p53 mutations in a familial syndrome of breast cancer, sarcomas, and other neoplasms, *Science,* 250, 1222, 1990.

Malkin, D., p53 and the Li-Fraumeni syndrome, *Biochim. Biophys. Acta,* 1198, 197, 1994.

Matsuzawa, K., Fukuyama, K., Hubbard, S. L., Dirks, P. B., and Rutka, J. T., Transfection of an invasive human astrocytoma cell line with a TIMP-1 cDNA: modulation of astrocytoma invasive potential, *J. Neuropathol. Exp. Neurol.,* 55, 88, 1996.

McKeever, P. E., Dennis, T. R., Burgess, A. C., Meltzer, P. S., Marchuk, D. A., and Trent, J. M., Chromosome breakpoint at 17q11.2 and insertion of DNA from three different chromosomes in a glioblastoma with exceptional glial fibrillary acidic protein expression, *Cancer Genet. Cytogenet.,* 87, 41, 1996.

Melmed, S. and Rushakoff, R. J., Ectopic pituitary and hypothalamic hormone syndromes, *Endocrinol. Metab. Clin. North. Am.,* 16, 805, 1987.

Mercer, W., Shields, M., Amin, M., Sauve, G., Appella, E., Romano, J., and Ullrich, S., Negative growth regulation in a glioblastoma tumor cell line that conditionally expresses human wild-type p53, *Proc. Natl. Acad. Sci.,* 87, 6166, 1990.

Merlo, A., Herman, J. G., Mao, L., Lee, D. J., Gabrielson, E., Burger, P. C., Baylin, S. B., and Sidransky, D., 5′ CpG island methylation is associated with transcriptional silencing of the tumour suppressor p16/CDKN2/MTS1 in human cancers, *Nat. Med.,* 1, 686, 1995.

Millauer, B., Shawver, L. K., Plate, K. H., Risau, W., and Ullrich, A., Glioblastoma growth inhibited in vivo by a dominant-negative Flk-1 mutant, *Nature,* 367, 576, 1994.

Minchenko, A., Bauer, T., Salceda, S., and Caro, J., Hypoxic stimulation of vascular endothelial growth factor expression in vitro and in vivo, *Lab. Invest.,* 71, 374, 1994.

Miyashita, T., Krajewski, S., Krajewska, M., Wang, H. G., Lin, H. K., Liebermann, D. A., Hoffman, B., and Reed, J. C., Tumor suppressor p53 is a regulator of bcl-2 and bax gene expression in vitro and in vivo, *Oncogene,* 9, 1799, 1994.

Montgomery, R. B., Moscatello, D. K., Wong, A. J., Cooper, J. A., and Stahl, W. L., Differential modulation of mitogen-activated protein (MAP) kinase/extracellular signal-related kinase kinase and MAP kinase activities by a mutant epidermal growth factor receptor, *J. Biol. Chem.,* 270, 30562, 1995.

Mori, T., Nagase, H., Horii, A., Miyoshi, Y., Shimano, T., Nakatsuru, S., Aoki, T., Arakawa, H., Yanagisawa, A., Ushio, Y. et al., Germ-line and somatic mutations of the APC gene in patients with Turcot syndrome and analysis of APC mutations in brain tumors, *Genes Chromosom. Cancer,* 9, 168, 1994.

Nagane, M., Coufal, F., Lin, H., Bogler, O., Cavenee, W. K., and Huang, H. J., A common mutant epidermal growth factor receptor confers enhanced tumorigenicity on human glioblastoma cells by increasing proliferation and reducing apoptosis, *Cancer Res.*, 56, 5079, 1996.

Nakasu, S., Nakasu, Y., Nioka, H., Nakajima, M., and Handa, J., bcl-2 protein expression in tumors of the central nervous system, *Acta Neuropathol. (Berl.)*, 88, 520, 1994.

Nelson, G. A., Bastian, F. O., Schlitt, M., and White, R. L., Malignant transformation in craniopharyngioma, *Neurosurgery*, 22, 427, 1988.

Nelson, J. S., Parisi, J. E., and Schechet, S. S., *Principles and Practice of Neuropathology*, 1st ed., Mosby, St. Louis, 1993.

Neumann, H. P., Lips, C. J., Hsia, Y. E., and Zbar, B., Von Hippel-Lindau syndrome, *Brain Pathol.*, 5, 181, 1995.

Ng, H. K., Lau, K. M., Tse, J. Y., Lo, K. W., Wong, J. H., Poon, W. S., and Huang, D. P., Combined molecular genetic studies of chromosome 22q and the neurofibromatosis type 2 gene in central nervous system tumors, *Neurosurgery*, 37, 764, 1995.

Nishikawa, R., Furnari, F. B., Lin, H., Arap, W., Berger, M. S., Cavenee, W. K., and Su Huang, H. J., Loss of P16INK4 expression is frequent in high grade gliomas, *Cancer Res.*, 55, 1941, 1995.

Nishikawa, R., Ji, X. D., Harmon, R. C., Lazar, C. S., Gill, G. N., Cavenee, W. K., and Huang, H. J., A mutant epidermal growth factor receptor common in human glioma confers enhanced tumorigenicity, *Proc. Natl. Acad. Sci. U.S.A.*, 91, 7727, 1994.

Nobori, T., Miura, K., Wu, D. J., Lois, A., Takabayashi, K., and Carson, D. A., Deletions of the cyclin-dependent kinase-4 inhibitor gene in multiple human cancers, *Nature*, 368, 753, 1994.

Ohgaki, H., Eibl, R. H., Wiestler, O. D., Yasargil, M. G., Newcomb, E. W., and Kleihues, P., p53 mutations in nonastrocytic human brain tumors, *Cancer Res.*, 51, 6202, 1991.

Okada, H., Yoshida, J., Sokabe, M., Wakabayashi, T., and Hagiwara, M., Suppression of CD44 expression decreases migration and invasion of human glioma cells, *Int. J. Cancer*, 66, 255, 1996.

Olopade, O. I., Jenkins, R. B., Ransom, D. T., Malik, K., Pomykala, H., Nobori, T., Cowan, J. M., Rowley, J. D., and Diaz, M. O., Molecular analysis of deletions of the short arm of chromosome 9 in human gliomas, *Cancer Res.*, 52, 2523, 1992.

Paulus, W., Baur, I., Dours-Zimmermann, M. T., and Zimmermann, D. R., Differential expression of versican isoforms in brain tumors, *J. Neuropathol. Exp. Neurol.*, 55, 528, 1996.

Paulus, W., Baur, I., Schuppan, D., and Roggendorf, W., Characterization of integrin receptors in normal and neoplastic human brain, *Am. J. Pathol.*, 143, 154, 1993.

Pei, L., Melmed, S., Scheithauer, B., Kovacs, K., and Prager, D., H-ras mutations in human pituitary carcinoma metastases, *J. Clin. Endocrinol. Metab.*, 78, 842, 1994.

Penar, P., Khoshyomn, S., Bhushan, A., Tritton, T., Inhibition of epidermal growth factor receptor-associated tyrosine kinase blocks glioblastoma invasion of the brain, *Neurosurgery*, 40, 141, 1997.

Pershouse, M. A., Stubblefield, E., Hadi, A., Killary, A. M., Yung, W. K., and Steck, P. A., Analysis of the functional role of chromosome 10 loss in human glioblastomas, *Cancer Res.*, 53, 5043, 1993.

Peyrard, M., Fransson, I., Xie, Y. G., Han, F. Y., Ruttledge, M. H., Swahn, S., Collins, J. E., Dunham, I., Collins, V. P., and Dumanski, J. P., Characterization of a new member of the human beta-adaptin gene family from chromosome 22q12, a candidate meningioma gene, *Hum. Mol. Genet.*, 3, 1393, 1994.

Plate, K. H., Breier, G., Millauer, B., Ullrich, A., and Risau, W., Up-regulation of vascular endothelial growth factor and its cognate receptors in a rat glioma model of tumor angiogenesis, *Cancer Res.*, 53, 5822, 1993.

Plate, K. H., Breier, G., and Risau, W., Molecular mechanisms of developmental and tumor angiogenesis, *Brain Pathol.*, 4, 207, 1994.

Plate, K. H., Breier, G., Weich, H. A., and Risau, W., Vascular endothelial growth factor is a potential tumour angiogenesis factor in human gliomas in vivo, *Nature*, 359, 845, 1992.

Platten, M., Giordano, M. J., Dirven, C. M., Gutmann, D. H., and Louis, D. N., Up-regulation of specific NF 1 gene transcripts in sporadic pilocytic astrocytomas, *Am. J. Pathol.*, 149, 621, 1996.

Prigent, S. A., Nagane, M., Lin, H., Huvar, I., Boss, G. R., Feramisco, J. R., Cavenee, W. K., and Huang, H. S., Enhanced tumorigenic behavior of glioblastoma cells expressing a truncated epidermal growth factor receptor is mediated through the Ras-Shc-Grb2 pathway, *J. Biol. Chem.*, 271, 25639, 1996.

Pulst, S. M., Rouleau, G. A., Marineau, C., Fain, P., and Sieb, J. P., Familial meningioma is not allelic to neurofibromatosis 2, *Neurology*, 43, 2096, 1993.

Raff, M. C., Miller, R. H., and Noble, M., A glial progenitor cell that develops in vitro into an astrocyte or an oligodendrocyte depending on culture medium, *Nature*, 303, 390, 1983.

Raffel, C., Ueki, K., Harsh, G. R. T., and Louis, D. N., The multiple tumor suppressor 1/cyclin-dependent kinase inhibitor 2 gene in human central nervous system primitive neuroectodermal tumor, *Neurosurgery,* 36, 971, 1995.

Ransom, D. T., Ritland, S. R., Kimmel, D. W., Moertel, C. A., Dahl, R. J., Scheithauer, B. W., Kelly, P. J., and Jenkins, R. B., Cytogenetic and loss of heterozygosity studies in ependymomas, pilocytic astrocytomas, and oligodendrogliomas, *Genes Chromosom. Cancer,* 5, 348, 1992.

Rao, J. S., Yamamoto, M., Mohaman, S., Gokaslan, Z. L., Fuller, G. N., Stetler-Stevenson, W. G., Rao, V. H., Liotta, L. A., Nicolson, G. L., and Sawaya, R. E., Expression and localization of 92 kDa type IV collagenase/gelatinase B (MMP-9) in human gliomas, *Clin. Exp. Metastasis,* 14, 12, 1996.

Rasheed, B. K., Fuller, G. N., Friedman, A. H., Bigner, D. D., and Bigner, S. H., Loss of heterozygosity for 10q loci in human gliomas, *Genes Chromosom. Cancer,* 5, 75, 1992.

Rasheed, B. K., McLendon, R. E., Herndon, J. E., Friedman, H. S., Friedman, A. H., Bigner, D. D., and Bigner, S. H., Alterations of the TP53 gene in human gliomas, *Cancer Res.,* 54, 1324, 1994.

Reifenberger, G., Reifenberger, J., Ichimura, K., Meltzer, P. S., and Collins, V. P., Amplification of multiple genes from chromosomal region 12q13-14 in human malignant gliomas: preliminary mapping of the amplicons shows preferential involvement of CDK4, SAS, and MDM2, *Cancer Res.,* 54, 4299, 1994a.

Reifenberger, J., Reifenberger, G., Liu, L., James, C. D., Wechsler, W., and Collins, V. P., Molecular genetic analysis of oligodendroglial tumors shows preferential allelic deletions on 19q and 1p, *Am. J. Pathol.,* 145, 1175, 1994b.

Rempel, S. A., Schwechheimer, K., Davis, R. L., Cavenee, W. K., and Rosenblum, M. L., Loss of heterozygosity for loci on chromosome 10 is associated with morphologically malignant meningioma progression, *Cancer Res.,* 53, 2386, 1993.

Rencic, A., Gordon, J., Otte, J., Curtis, M., Kovatich, A., Zoltick, P., Khalili, K., and Andrews, D., Detection of JC virus DNA sequence and expression of the viral oncoprotein, tumor antigen, in brain of immuno-competent patient with oligoastrocytoma, *Proc. Natl. Acad. Sci. U.S.A.,* 93, 7352, 1996.

Reubi, J. C., Maurer, R., Klijn, J. G., Stefanko, S. Z., Foekens, J. A., Blaauw, G., Blankenstein, M. A., and Lamberts, S. W., High incidence of somatostatin receptors in human meningiomas: biochemical characterization, *J. Clin. Endocrinol. Metab.,* 63, 433, 1986.

Riedel, H., Massoglia, S., Schlessinger, J., and Ullrich, A., Ligand activation of overexpressed epidermal growth factor receptors transforms NIH 3T3 mouse fibroblasts, *Proc. Natl. Acad. Sci. U.S.A.,* 85, 1477, 1988.

Robinow, M., Johnson, G. F., and Minella, P. A., Aicardi syndrome, papilloma of the choroid plexus, cleft lip, and cleft of the posterior palate, *J. Pediatr.,* 104, 404, 1984.

Rogatto, S. R., Casartelli, C., Rainho, C. A., and Barbieri-Neto, J., Chromosomes in the genesis and progression of ependymomas, *Cancer Genet. Cytogenet.,* 69, 146, 1993.

Rubinstein, A. B., Shalit, M. N., Cohen, M. L., Zandbank, U., and Reichenthal, E., Radiation-induced cerebral meningioma: a recognizable entity, *J. Neurosurg.,* 61, 966, 1984.

Rubio, M. P., Correa, K. M., Ramesh, V., MacCollin, M. M., Jacoby, L. B., von Deimling, A., Gusella, J. F., and Louis, D. N., Analysis of the neurofibromatosis 2 gene in human ependymomas and astrocytomas, *Cancer Res.,* 54, 45, 1994a.

Rubio, M. P., Correa, K. M., Ueki, K., Mohrenweiser, H. W., Gusella, J. F., von Deimling, A., and Louis, D. N., The putative glioma tumor suppressor gene on chromosome 19q maps between APOC2 and HRC, *Cancer Res.,* 54, 4760, 1994b.

Rubio, M. P., von Deimling, A., Yandell, D. W., Wiestler, O. D., Gusella, J. F., and Louis, D. N., Accumulation of wild type p53 protein in human astrocytomas, *Cancer Res.,* 53, 3465, 1993.

Russell, D. S. and Rubinstein, L. J., *Pathology of Tumours of the Nervous System,* 5th ed., Williams & Wilkins, Baltimore, 1989.

Rutka, J. T., Hubbard, S. L., Fukuyama, K., Matsuzawa, K., Dirks, P. B., and Becker, L. E., Effects of antisense glial fibrillary acidic protein complementary DNA on the growth, invasion, and adhesion of human astrocytoma cells, *Cancer Res.,* 54, 3267, 1994.

Ruttledge, M. H., Sarrazin, J., Rangaratnam, S., Phelan, C. M., Twist, E., Merel, P., Delattre, O., Thomas, G., Nordenskjold, M., Collins, V. P. et al., Evidence for the complete inactivation of the NF2 gene in the majority of sporadic meningiomas, *Nat. Genet.,* 6, 180, 1994.

Ruttledge, M. H., Xie, Y. G., Han, F. Y., Peyrard, M., Collins, V. P., Nordenskjold, M., and Dumanski, J. P., Deletions on chromosome 22 in sporadic meningioma, *Genes Chromosom. Cancer,* 10, 122, 1994.

Sainz, J., Huynh, D. P., Figueroa, K., Ragge, N. K., Baser, M. E., and Pulst, S. M., Mutations of the neurofibromatosis type 2 gene and lack of the gene product in vestibular schwannomas, *Hum. Mol. Genet.*, 3, 885, 1994.

Samaniego, F., Rodriguez, E., Houldsworth, J., Murty, V. V., Ladanyi, M., Lele, K. P., Chen, Q. G., Dmitrovsky, E., Geller, N. L., Reuter, V. et al., Cytogenetic and molecular analysis of human male germ cell tumors: chromosome 12 abnormalities and gene amplification, *Genes Chromosom. Cancer,* 1, 289, 1990.

Sato, K., Schauble, B., Kleihues, P., and Ohgaki, H., Infrequent alterations of the p15, p16, CDK4 and cyclin D1 genes in non-astrocytic human brain tumors, *Int. J. Cancer,* 66, 305, 1996.

Saylors, R. L. d., Sidransky, D., Friedman, H. S., Bigner, S. H., Bigner, D. D., Vogelstein, B., and Brodeur, G. M., Infrequent p53 gene mutations in medulloblastomas, *Cancer Res.,* 51, 4721, 1991.

Scherer, H. J., The forms of growth in gliomas and their practical significance, *Brain*, 63, 1, 1940.

Schlegel, J., Merdes, A., Stumm, G., Albert, F. K., Forsting, M., Hynes, N., and Kiessling, M., Amplification of the epidermal-growth-factor-receptor gene correlates with different growth behaviour in human glioblastoma, *Int. J. Cancer,* 56, 72, 1994.

Schmidt, E. E., Ichimura, K., Reifenberger, G., and Collins, V. P., CDKN2 (p16/MTS1) gene deletion or CDK4 amplification occurs in the majority of glioblastomas, *Cancer Res.,* 54, 6321, 1994.

Schoenberg, B. S., Christine, B. W., and Whisnant, J. P., Nervous system neoplasms and primary malignancies of other sites. The unique association between meningiomas and breast cancer, *Neurology*, 25, 705, 1975.

Seizinger, B., Klinger, H., Junien, C., Nakamura, Y., Lebeau, M., Cavenee, W., Emanuel, B., Ponder, B., Naylor, S., Mitelman, F., Louis, D., Menon, A., Newsham, I., Decker, J., Kaelbing, M., Henry, I., and von Deimling, A., Report of the committee on chromosome and gene loss in human neoplasia, *Cytogenet. Cell Genet.,* 58, 1080, 1991.

Serrano, M., Hannon, G. J., and Beach, D., A new regulatory motif in cell-cycle control causing specific inhibition of cyclin D/CDK4 [see comments], *Nature*, 366, 704, 1993.

Shepherd, C. W., Scheithauer, B. W., Gomez, M. R., Altermatt, H. J., and Katzmann, J. A., Subependymal giant cell astrocytoma: a clinical, pathological, and flow cytometric study, *Neurosurgery*, 28, 864, 1991.

Shin, T. H., Paterson, A. J., and Kudlow, J. E., p53 stimulates transcription from the human transforming growth factor alpha promoter: a potential growth-stimulatory role for p53, *Mol. Cell. Biol.,* 15, 4694, 1995.

Shweiki, D., Itin, A., Soffer, D., and Keshet, E., Vascular endothelial growth factor induced by hypoxia may mediate hypoxia-initiated angiogenesis, *Nature*, 359, 843, 1992.

Sidransky, D., Mikkelsen, T., Schwechheimer, K., Rosenblum, M. L., Cavanee, W., and Vogelstein, B., Clonal expansion of p53 mutant cells is associated with brain tumour progression, *Nature*, 355, 846, 1992.

Simon, M., von Deimling, A., Larson, J. J., Wellenreuther, R., Kaskel, P., Waha, A., Warnick, R. E., Tew, J. M., Jr., and Menon, A. G., Allelic losses on chromosomes 14, 10, and 1 in atypical and malignant meningiomas: a genetic model of meningioma progression, *Cancer Res.,* 55, 4696, 1995.

Sivaparvathi, M., Sawaya, R., Gokaslan, Z. L., Chintala, K. S., and Rao, J. S., Expression and the role of cathepsin H in human glioma progression and invasion, *Cancer Lett.,* 104, 121, 1996.

Smith, M., Smalley, S., Cantor, R., Pandolfo, M., Gomez, M. I., Baumann, R., Flodman, P., Yoshiyama, K., Nakamura, Y., Julier, C. et al., Mapping of a gene determining tuberous sclerosis to human chromosome 11q14-11q23 [see comments], *Genomics*, 6, 105, 1990.

Smits, A., van Grieken, D., Hartman, M., Lendahl, U., Funa, K., and Nister, M., Coexpression of platelet-derived growth factor alpha and beta receptors on medulloblastomas and other primitive neuroectodermal tumors is consistent with an immature stem cell and neuronal derivation, *Lab. Invest.,* 74, 188, 1996.

Spruck, C. H. R., Gonzalez-Zulueta, M., Shibata, A., Simoneau, A. R., Lin, M. F., Gonzales, F., Tsai, Y. C., and Jones, P. A., p16 gene in uncultured tumours [letter], *Nature*, 370, 183, 1994.

Sreekantaiah, C., Jockin, H., Brecher, M. L., and Sandberg, A. A., Interstitial deletion of chromosome 11q in a pineoblastoma, *Cancer Genet. Cytogenet.,* 39, 125, 1989.

Stratton, M. R., Darling, J., Lantos, P. L., Cooper, C. S., and Reeves, B. R., Cytogenetic abnormalities in human ependymomas, *Int. J. Cancer,* 44, 579, 1989.

Strawn, L., McMahon, G., App, H., Schreck, R., Kuchler, W., Longhi, M., Hui, T., Tang, C., Levitzki, A., and Gazit, A., Flk-1 as a target for tumor growth inhibition, *Cancer Res.,* 56, 3540, 1996.

Sugawa, N., Ekstrand, A., James, C., and Collins, V., Identical splicing of aberrant epidermal growth factor receptor transcripts from amplified rearranged genes in human glioblastomas, *Proc. Natl. Acad. Sci. U.S.A.,* 87, 8602, 1990.

Symonds, H., Chen, J. D., and Van Dyke, T., Complex formation between the lymphotropic papovavirus large tumor antigen and the tumor suppressor protein p53, *J. Virol.,* 65, 5417, 1991.

Takahashi, J. A., Fukumoto, M., Igarashi, K., Oda, Y., Kikuchi, H., and Hatanaka, M., Correlation of basic fibroblast growth factor expression levels with the degree of malignancy and vascularity in human gliomas, *J. Neurosurg.,* 76, 792, 1992.

Takenaka, N., Mikoshiba, K., Takamatsu, K., Tsukada, Y., Ohtani, M., and Toya, S., Immunohistochemical detection of the gene product of Rous sarcoma virus in human brain tumors, *Brain Res.,* 337, 201, 1985.

Thapar, K., Scheithauer, B. W., Kovacs, K., Pernicone, P. J., and Laws, E. R., Jr., p53 expression in pituitary adenomas and carcinomas: correlation with invasiveness and tumor growth fractions, *Neurosurgery,* 38, 763, 1996.

Thomas, G. A. and Raffel, C., Loss of heterozygosity on 6q, 16q, and 17p in human central nervous system primitive neuroectodermal tumors, *Cancer Res.,* 51, 639, 1991.

Tomlinson, F., Scheithauer, B., Hayostek, C., Parisi, J., Meyer, F., Shaw, E., Weiland, T., Katzmann, J., and Jack, C., The significance of atypia and histologic malignancy in pilocytic astrocytoma of the cerebellum: a clinicopathologic and flow cytometric study, *J. Child Neurol.,* 9, 301, 1994.

Tomlinson, F. H., Scheithauer, B. W., Hayostek, C. J., Parisi, J. E., Meyer, F. B., Shaw, E. G., Weiland, T. L., Katzmann, J. A., and Jack, C. R., Jr., The significance of atypia and histologic malignancy in pilocytic astrocytoma of the cerebellum: a clinicopathologic and flow cytometric study, *J. Child. Neurol.,* 9, 301, 1994.

Trofatter, J. A., MacCollin, M. M., Rutter, J. L., Murrell, J. R., Duyao, M. P., Parry, D. M., Eldridge, R., Kley, N., Menon, A. G., Pulaski, K. et al., A novel moesin-, ezrin-, radixin-like gene is a candidate for the neurofibromatosis 2 tumor suppressor, *Cell,* 72, 791, 1993.

Ueba, T., Nosaka, T., Takahashi, J. A., Shibata, F., Florkiewicz, R. Z., Vogelstein, B., Oda, Y., Kikuchi, H., and Hatanaka, M., Transcriptional regulation of basic fibroblast growth factor gene by p53 in human glioblastoma and hepatocellular carcinoma cells, *Proc. Natl. Acad. Sci. U.S.A.,* 91, 9009, 1994.

Ueki, K., Ono, Y., Henson, J. W., Efird, J. T., von Deimling, A., and Louis, D. N., CDKN2/p16 or RB alterations occur in the majority of glioblastomas and are inversely correlated, *Cancer Res.,* 56, 150, 1996.

Ueki, K., Rubio, M. P., Ramesh, V., Correa, K. M., Rutter, J. L., von Deimling, A., Buckler, A. J., Gusella, J. F., and Louis, D. N., MTS1/CDKN2 gene mutations are rare in primary human astrocytomas with allelic loss of chromosome 9p, *Hum. Mol. Genet.,* 3, 1841, 1994.

Vagner-Capodano, A., Gentet, J., and Choux, M., Chromosome abnormalities in sixteen pediatric brain tumors, *Pediatr. Neurosci.,* 14, 159, 1989.

Van Meir, E. G., Kikuchi, T., Tada, M., Li, H., Diserens, A. C., Wojcik, B. E., Huang, H. J., Friedmann, T., de Tribolet, N., and Cavenee, W. K., Analysis of the p53 gene and its expression in human glioblastoma cells, *Cancer Res.,* 54, 649, 1994.

Van Meir, E. G., Polverini, P. J., Chazin, V. R., Su Huang, H. J., de Tribolet, N., and Cavenee, W. K., Release of an inhibitor of angiogenesis upon induction of wild type p53 expression in glioblastoma cells, *Nat. Genet.,* 8, 171, 1994.

Van Meir, E. G., Roemer, K., Diserens, A. C., Kikuchi, T., Rempel, S. A., Haas, M., Huang, H. J., Friedmann, T., de Tribolet, N., and Cavenee, W. K., Single cell monitoring of growth arrest and morphological changes induced by transfer of wild-type p53 alleles to glioblastoma cells, *Proc. Natl. Acad. Sci. U.S.A.,* 92, 1008, 1995.

VandenBerg, S. R., May, E. E., Rubinstein, L. J., Herman, M. M., Perentes, E., Vinores, S. A., Collins, V. P., and Park, T. S., Desmoplastic supratentorial neuroepithelial tumors of infancy with divergent differentiation potential ("desmoplastic infantile gangliogliomas"). Report on 11 cases of a distinctive embryonal tumor with favorable prognosis, *J. Neurosurg.,* 66, 58, 1987.

Venter, D. J. and Thomas, D. G., Multiple sequential molecular abnormalities in the evolution of human gliomas, *Br. J. Cancer,* 63, 753, 1991.

Vertosick, F. T., Selker, R. G., and Arena, V. C., Survival of patients with well-differentiated astrocytomas diagnosed in the era of computed tomography, *Neurosurgery,* 28, 496, 1991.

von Deimling, A., Eibl, R. H., Ohgaki, H., Louis, D. N., von Ammon, K., Petersen, I., Kleihues, P., Chung, R. Y., Wiestler, O. D., and Seizinger, B. R., p53 mutations are associated with 17p allelic loss in grade II and grade III astrocytoma, *Cancer Res.,* 52, 2987, 1992a.

von Deimling, A., Louis, D. N., Menon, A. G., von Ammon, K., Petersen, I., Ellison, D., Wiestler, O. D., and Seizinger, B. R., Deletions on the long arm of chromosome 17 in pilocytic astrocytoma, *Acta. Neuropathol. (Berl.),* 86, 81, 1993a.

von Deimling, A., Louis, D. N., von Ammon, K., Petersen, I., Hoell, T., Chung, R. Y., Martuza, R. L., Schoenfeld, D. A., Ya sargil, M. G., Wiestler, O. D. et al., Association of epidermal growth factor receptor gene amplification with loss of chromosome 10 in human glioblastoma multiforme, *J. Neurosurg.*, 77, 295, 1992b.

von Deimling, A., Louis, D. N., and Wiestler, O. D., Molecular pathways in the formation of gliomas, *Glia*, 15, 328, 1995.

von Deimling, A., Nagel, J., Bender, B., Lenartz, D., Schramm, J., Louis, D. N., and Wiestler, O. D., Deletion mapping of chromosome 19 in human gliomas, *Int. J. Cancer*, 57, 676, 1994.

von Deimling, A., von Ammon, K., Schoenfeld, D., Wiestler, O. D., Seizinger, B. R., and Louis, D. N., Subsets of glioblastoma multiforme defined by molecular genetic analysis, *Brain Pathol.*, 3, 19, 1993.

Wang, J. L., Nister, M., Hermansson, M., Westermark, B., and Ponten, J., Expression of PDGF beta-receptors in human meningioma cells, *Int. J. Cancer*, 46, 772, 1990.

Wasson, J. C., Saylors, R. L. D., Zeltzer, P., Friedman, H. S., Bigner, S. H., Burger, P. C., Bigner, D. D., Look, A. T., Douglass, E. C., and Brodeur, G. M., Oncogene amplification in pediatric brain tumors, *Cancer Res.*, 50, 2987, 1990.

Weisman, A. S., Raguet, S. S., and Kelly, P. A., Characterization of the epidermal growth factor receptor in human meningioma, *Cancer Res.*, 47, 2172, 1987.

Wellenreuther, R., Kraus, J. A., Lenartz, D., Menon, A. G., Schramm, J., Louis, D. N., Ramesh, V., Gusella, J. F., Wiestler, O. D., and von Deimling, A., Analysis of the neurofibromatosis 2 gene reveals molecular variants of meningioma, *Am. J. Pathol.*, 146, 827, 1995.

Weller, M., Malipiero, U., Aguzzi, A., Reed, J. C., and Fontana, A., Protooncogene bcl-2 gene transfer abrogates Fas/APO-1 antibody-mediated apoptosis of human malignant glioma cells and confers resistance to chemotherapeutic drugs and therapeutic irradiation, *J. Clin. Invest.*, 95, 2633, 1995.

Weremowicz, S., Kupsky, W. J., Morton, C. C., and Fletcher, J. A., Cytogenetic evidence for a chromosome 22 tumor suppressor gene in ependymoma, *Cancer Genet. Cytogenet.*, 61, 193, 1992.

Westermark, B., Heldin, C. H., and Nister, M., Platelet-derived growth factor in human glioma, *Glia*, 15, 257, 1995a.

Westermark, B. and Nistér, M., Molecular genetics of human glioma, *Curr. Opin. Oncol.*, 7, 220, 1995b.

Wikstrand, C. J., Friedman, H. S., and Bigner, D. D., Medulloblastoma cell-substrate interaction in vitro, *Invasion Metastasis*, 11, 310, 1991.

Wilson, C. B., Neurosurgical aspects of Cushing's disease and Nelson's syndrome, in *Clinical Management of Pituitary Disorders*, Tindall, G. and Collins, W., Eds., Raven Press, New York, 1979, 229.

Winger, M. J., Macdonald, D. R., and Cairncross, J. G., Supratentorial anaplastic gliomas in adults. The prognostic importance of extent of resection and prior low-grade glioma, *J. Neurosurg.*, 71, 487, 1989.

Wong, A. J., Bigner, S. H., Bigner, D. D., Kinzler, K. W., Hamilton, S. R., and Vogelstein, B., Increased expression of the epidermal growth factor receptor gene in malignant gliomas is invariably associated with gene amplification, *Proc. Natl. Acad. Sci. U.S.A.*, 84, 6899, 1987.

Wong, A. J., Ruppert, J. M., Bigner, S. H., Grzeschik, C. H., Humphrey, P. A., Bigner, D. S., and Vogelstein, B., Structural alterations of the epidermal growth factor receptor gene in human gliomas, *Proc. Natl. Acad. Sci. U.S.A.*, 89, 2965, 1992

Wu, J. K., MacGillavry, M., Kessaris, C., Verheul, B., Adelman, L. S., and Darras, B. T., Clonal analysis of meningiomas, *Neurosurgery*, 38, 1196, 1996.

Yamamoto, M., Mohanam, S., Sawaya, R., Fuller, G. N., Seiki, M., Sato, H., Gokaslan, Z. L., Liotta, L. A., Nicolson, G. L., and Rao, J. S., Differential expression of membrane-type matrix metalloproteinase and its correlation with gelatinase A activation in human malignant brain tumors, *Cancer Res.*, 56, 384, 1996.

Yamamoto, M., Sawaya, R., Mohanam, S., Rao, V. H., Bruner, J. M., Nicolson, G. L., Ohshima, K., and Rao, J. S., Activities, localizations, and roles of serine proteases and their inhibitors in human brain tumor progression, *J. Neurooncol.*, 22, 139, 1994.

Yong, W. H., Chou, D., Ueki, K., Harsh, G. R. T., von Deimling, A., Gusella, J. F., Mohrenweiser, H. W., and Louis, D. N., Chromosome 19q deletions in human gliomas overlap telomeric to D19S219 and may target a 425 kb region centromeric to D19S112, *J. Neuropathol. Exp. Neurol.*, 54, 622, 1995.

Yong, W. H., Ueki, K., Chou, D., Reeves, S. A., von Deimling, A., Gusella, J. F., Mohrenweiser, H. W., Buckler, A. J., and Louis, D. N., Cloning of a highly conserved human protein serine-threonine phosphatase gene from the glioma candidate region on chromosome 19q13.3, *Genomics*, 29, 533, 1995.

# 11 Paraneoplastic Neurologic Disorders and Onconeural Antigens

*Myrna R. Rosenfeld and Josep Dalmau*

## CONTENTS

## INTRODUCTION

Paraneoplastic neurological syndromes are disorders of nervous system function that occur in association with cancer but are not a result of tumor mass or metastasis to the nervous system (Posner, 1995; Black and Loeffler, 1997). These disorders may affect any portion of the nervous system. In some instances, the dysfunction is focal and restricted to a single cell type. Such disorders include paraneoplastic cerebellar degeneration (PCD), where the pathology is mainly restricted to Purkinje cells, and cancer-associated retinopathy (CAR), where there is degeneration of the retinal photoreceptors. In other disorders, such as paraneoplastic encephalomyelitis (PEM), there is widespread neuraxis dysfunction (Posner, 1995; Black and Loeffler, 1997).

Although these disorders are the rarest neurological complications in cancer patients, they are important for several reasons. First, in more than half of the patients, the neurological symptoms precede the diagnosis of the tumor. Second, most of the paraneoplastic syndromes are associated with specific histologic tumor types so their clinical recognition leads to and directs a search for an occult and potentially curable cancer (Henson and Urich, 1989). Additionally, since there is increasing evidence that most paraneoplastic neurologic syndromes may be immune mediated, these disorders provide the neuroscientist and oncologist a unique model to study the relationship between cancer, the nervous system, and the immunological system (Darnell, 1996).

In the past several years, investigators have found that patients with some of these syndromes develop high titers of antibodies in their serum and cerebrospinal fluid (CSF) that specifically react with neuronal antigens that are also expressed by the tumor (Table 11.1) (Graus et al., 1985; Dalmau

**TABLE 11.1**
**Onconeural Antibodies and Antigens**

| Antibody | Associated | Cancer Syndrome | Onconeuronal Antigen Localization | Onconeuronal Antigen |
|---|---|---|---|---|
| anti-Hu | SCLC, neuroblastoma | encephalomyelitis, sensory neuronopathy | all neuronal nuclei, mild in cytoplasm | HuD, HuC, Hel-N1, HuR |
| anti-Yo | gynecologic, breast | cerebellar degeneration | Purkinje cell cytoplasm | cdr-34, cdr-62 |
| anti-Ri | breast, gynecologic, SCLC | cerebellar ataxia, opsoclonus | neuronal nuclei CNS and milder in cytoplasm | Nova-1, 2 |
| anti-amphiphysin | breast, SCLC | stiff-person syndrome | synaptic vesicles | amphiphysin |
| anti-AChR | thymoma | myasthenia gravis | postsynaptic neuromuscular junction | acetylcholine receptor |
| anti-VGCC | SCLC | Lambert-Eaton myasthenic syndrome | presynaptic neuromuscular junction | α1-subunit VGCC |
| anti-MysB | SCLC | Lambert-Eaton myasthenic syndrome | presynaptic neuromuscular junction | β-subunit VGCC |
| anti-Ma | multiple | cerebellar degeneration, brainstem dysfunction | neuronal nuclei and cytoplasm | Ma1, Ma2 (not further characterized) |
| anti-Ta | testicular | limbic encephalitis | neuronal cytoplasm and nuclei | Ma2 (not further characterized) |
| anti-CV2 | SCLC, thymoma | encephalomyelitis, cerebellar degeneration | oligodendrocyte cytoplasm | unc-33 like |
| anti-Tr | Hodgkin's lymphoma | cerebellar degeneration | Purkinje cell cytoplasm and dendrites | (not cloned) |

et al., 1990; Luque et al., 1991; Peterson et al., 1992). In some syndromes, it has been shown that the presence of these antibodies is associated with an effective anticancer immune response. For example, in patients with small cell lung cancer (SCLC), detection of anti-Hu antibodies at the time of cancer diagnosis is a strong and independent predictor of complete response to treatment and prolonged survival (Dalmau et al., 1990; Graus et al., 1997a).

Several different antineuronal antibodies associated with different types of cancers have been described in patients with similar paraneoplastic symptoms. This suggests that several distinct immunological pathways may lead to the same target. For example, patients with SCLC may develop PCD with or without the presence of anti-Hu antibodies (Mason et al., 1997). Similarly, PCD may occur in association with the Lambert-Eaton myasthenic syndrome (LEMS), which results from antibodies against the voltage-gated calcium channel (VGCC) (O'Neill et al., 1988). Other immune responses have been characterized for PCD associated with tumors other than SCLC (O'Neill et al., 1988; Graus et al., 1997b).

The hypothesis that this and other data has generated is that proteins normally expressed by the nervous system are perceived as foreign by the immunological system due to abnormal or dysregulated expression by the tumor. The "irregular" expression of the proteins by the tumor, perhaps in association with other factors such as up-regulation of major histocompatibility complex (MHC) proteins (see below) would then result in an immune response against the tumor and nervous system.

The immune hypothesis for paraneoplastic neurologic disorders has been established for myasthenia gravis and LEMS (Lang et al., 1981; Drachman, 1994). In these disorders, patients develop antibodies that cause the neuromuscular synaptic dysfunction. Removal of the antibodies by plasma exchange improves symptoms, and transfer of antibodies to animals reproduces the disease. Although no pathogenic function has been demonstrated for the antibodies associated with other paraneoplastic disorders, they have proven to be useful tools in identifying tumor and neuronal specific antigens (onconeural). Elucidating the normal cellular roles of these proteins will expand our understanding of tumor immunology, paraneoplastic neurologic disease, and normal neuronal function. This chapter will review the current data regarding several of these onconeural proteins.

## ONCONEURAL ANTIGENS AND THE CENTRAL NERVOUS SYSTEM

### THE HU ANTIGENS

In 1965 Wilkinson and Zeromski first reported the presence of antineuronal antibodies in four patients with SCLC and paraneoplastic sensory neuropathy (Wilkinson and Zeromski, 1965). Similar antibodies were subsequently described in patients with PEM. Graus in 1986 demonstrated that these antibodies, called anti-Hu, reacted with 35- to 40-kDa antigens expressed in the neurons of the central and peripheral nervous system and in SCLC (Graus et al., 1985). Immunohistochemical studies revealed that the majority of the antibody immunoreactivity was with the nuclei of the neurons and to a lesser degree with the cytoplasm (Dalmau et al., 1992a). The human gene *HuD* was cloned in 1991 (Szabo et al., 1991), followed by the cloning of other related genes including *HuC/ple21* (Sakai et al., 1994), *HelN-1* (Levine et al., 1993), *HelN-2* (King, 1994), *HuR* (Ma et al., 1996), and similar genes from *Xenopus laevis*, zebrafish, and mouse (Sakai et al., 1994; Good, 1995; Ma et al., 1996; King, 1996). The Hu proteins are homologous to the *Drosophila* proteins Elav (Robinow et al., 1988), Rbp9 (Kim and Baker, 1993), and Sex-lethal (Bell et al., 1988). These proteins contain three copies of the RNA recognition motif (RRM), which forms the core of functional RNA binding domains (Query et al., 1989). Proteins containing the RRM are involved in many aspects of RNA processing including splicing, transport, and translation (Kenan et al., 1991; Birney et al., 1993). In the Hu antigens, the first two RRMs are tandemly located and connected to the third RRM by a highly basic region (Szabo et al., 1991).

In *Drosophila*, Elav is essential for the development and maintenance of the nervous system (Robinow and White, 1988). Mutations of the *elav* gene result in a lethal phenotype characterized by the abnormal proliferation of immature neuroblasts, a failure of neuronal differentiation, and inappropriate neuroblast migration (Campos et al., 1985). Transcription of *elav* starts early in fly embryogenesis and is restricted to cells destined to become neurons (Robinow and White, 1988, 1991). Similarly, the Hu proteins are expressed early in neurogenesis and may first be detected in neurogenic precursor cells that have exited the ventricular zone of the spinal cord (Marusich and Weston, 1992; Marusich et al., 1994). In chicken embryos the Hu proteins are expressed by cells in the sensory and sympathetic ganglia and in the neural tube, supporting a role in the regulation of early neuronal development and differentiation (Marusich et al., 1994).

The Hu proteins have been shown to bind to AU rich elements in the 3′ untranslated regions (UTRs) of several growth-related mRNAs including *c-myc*, *c-fos*, and GM-CSF (Liu et al., 1995). This sequence is involved in regulating mRNA stability, localization, and translation. The Hu proteins also bind to the transcriptional repressor protein, Id, an inhibitor of DNA binding (King et al., 1994) that has been implicated in muscle cell development in mice.

The strong homology between the Hu antigens and the *Drosophila* proteins rests within the RRM sequences (Kim and Baker, 1993). The Hu proteins are, however, diversified by alternative splicing that involves both the protein coding and noncoding regions. Within the basic region that connects the second RRM to the third, there is a highly conserved submotif that is involved in alternative splicing of the coding region. Recently, a study of the human and mouse Hu homologs revealed numerous splice variants involving both the 5′ and 3′ UTRs (Okano and Darnell, 1997). This analysis suggests the existence of at least 18 different Hu proteins, encoded by four separate genes. This diversity would allow for stringent regulatory control at either the transcriptional or posttranscriptional level.

The homology to Elav, the very early expression during mammalian neuronal development, and the RNA binding properties, support a role of the Hu proteins in neural growth and differentiation. In the neuronal nucleus, based upon homology to the *Drosophila* gene *sex-lethal,* the Hu antigens could function to regulate alternative splicing of neuronal pre-RNAs (Valcarcel et al., 1993). In the cytoplasm, they may modulate RNA processing by binding to mRNA 3′ UTRs (Gao et al., 1994; King et al., 1994; Liu et al., 1995).

Anti-Hu antibodies from patients and those developed in animals immunized with recombinant proteins recognize all members of the Hu family. Epitope mapping reveals two immunodominant regions, contained within the first two RNA binding regions (Manley et al., 1995). It is not known if the binding of antibody affects the RNA binding function of the protein. This could be a mechanism by which the antibodies in the patients' sera cause abnormal neuronal function or even death.

As noted above, the Hu antigens were cloned using serum antibodies from patients with cancer and paraneoplastic neurologic diseases. In the vast majority of these patients, the associated carcinoma is SCLC. There have been isolated cases of other associated tumors and, like SCLC, most of these are of neuroendocrine origin. When examined these tumors express Hu proteins (Dalmau et al., 1992a). Although all SCLC express Hu antigens (but not normal lung or other lung-derived tumors), less than 1% of SCLC patients harbor high titers of anti-Hu antibodies and have neurologic dysfunction. Another 16% of SCLC patients have low titers of anti-Hu antibodies; however, these patients do not develop neurologic dysfunction (Graus et al., 1997a). Patients with high titers of anti-Hu antibodies often die as a result of their neurologic deficits, and the cancer appears to behave more indolently than the same cancer in patients without antibodies or paraneoplastic disorders (de la Monte et al., 1984; Darnell and DeAngelis, 1993). Furthermore, in patients with low titers of anti-Hu antibodies, the presence of the antibodies at the time of cancer diagnosis is a strong and independent predictor of complete response to treatment and prolonged survival (Graus et al., 1997a).

The mechanisms responsible for the association between anti-Hu antibodies and complete response to therapy are unknown as are the reasons why antibodies are produced by only a small percentage of SCLC patients. It is felt that tumors and SCLC cell lines that express neuroendocrine markers appear to be more responsive to treatment (Carney et al., 1983; Berendsen et al., 1987), but what role the Hu antigens play in this remains to be determined.

## THE NOVA ANTIGENS

The Nova antigens are another family of onconeural RNA binding proteins that have been described in association with ataxia and abnormal eye movements such as opsoclonus (Budde-Steffen et al., 1988). This syndrome most often develops in association with breast, gynecologic, or lung carcinomas. Patients harbor in their serum and CSF antibodies, called anti-Ri, that recognize a 55-kDa antigen called Nova-1 (Luque et al., 1991; Buckanovich et al., 1993). These patients also have antibodies that recognize a second antigen of 80 kDa that has not been further characterized. The Nova-1 antigen is a nuclear protein that is highly conserved between humans and rodents. The predicted protein sequence has three putative RNA binding motifs that are homologous to the KH motifs found in hnRNP-K protein (Buckanovich et al., 1993). Other RNA binding proteins that have KH motifs include the *Drosophila* PSI protein (Siebel et al., 1995) and the yeast MER-1 protein (Engebrecht et al., 1991). These and other RNA binding proteins which contain KH domains have been found to be involved with regulation of alternative splicing events and possibly RNA targeting (Gibson et al., 1993; Nandabalan and Roeder, 1995).

The expression pattern of the antigens recognized by anti-Ri antibodies is restricted to neuronal nuclei of the central but not the peripheral nervous system (Graus et al., 1993). Expression of Nova-1 is further restricted to structures in the ventral diencephalon, midbrain, cerebellum, and hindbrain (Buckanovich et al. 1993).

Anti-Ri antibodies from patients with paraneoplastic ataxia and opsoclonus have been shown to recognize the third KH domain and to block Nova-1 RNA binding (Buckanovich et al., 1996). This, in conjunction with a potentially restricted expression pattern by structures likely to be involved in the disorder, suggest an antibody-mediated disruption of Nova-1 function in the pathogenesis of the paraneoplastic disorder.

## THE YO ANTIGENS

In PCD, patients develop an acute to subacute onset of cerebellar dysfunction. Neuropathologic studies show extensive loss of Purkinje cells with gliosis and a variable degree of perivascular and meningeal infiltration by lymphocytes and other mononuclear cells. In some patients, despite extensive autopsy studies, absolutely no Purkinje cells could be found (Schmid and Riede, 1974; Verschuuren et al., 1996). When the associated tumor is an ovarian or breast carcinoma, the sera from these patients contain antibodies, called anti-Yo, that react with Purkinje cells and the tumor (Greenlee and Brashear, 1983; Jaeckle et al., 1985; Rodriguez et al., 1988). Similar to the other disorders previously mentioned, the presence of anti-Purkinje cell antibodies suggests an autoimmune etiology of PCD.

On Western blot analysis of Purkinje cells and tumor tissue, anti-Yo sera react with at least two antigens; a major species of 62 kDa, called cdr-62, and a minor species of 34 kDa, called cdr-34 (Cunningham et al., 1986; Anderson et al., 1988). The genes encoding cdr-62 and cdr-34 have been cloned (Dropcho et al., 1987; Furneaux et al., 1989; Sakai et al., 1990; Fathallah-Shaykh et al., 1991).

The cdr-62 protein contains a zinc finger domain and a leucine zipper motif that is typical of proteins that bind to DNA as hetero- or homodimers (Fathallah-Shaykh et al., 1991). The cdr-34 antigen is unusual in that it consists almost entirely of tandem hexapeptide repeats of the amino acid sequence L/FLEDVE, which gives rise to a number of single Leu-Leu zipper elements. It has been demonstrated that anti-Yo antibodies recognize an epitope within the leucine zipper domain of the cdr-62 protein (Sakai et al., 1990). There has been no functional analysis of the Yo antigens, although their structure would suggest a role in regulation of gene expression, perhaps by acting as transcription factors.

A less studied subgroup of patients with PCD develop an immune response (anti- CV2 antibodies) against a protein called c-22, which is homologous to the unc-33 gene of *Caenorhabditis elegans* (Quach et al., 1997). Mutation of unc-33 leads to defects in neuritic outgrowth and axonal guidance resulting in uncoordinated movements of the nematode (Hedgecock et al., 1985).

## RECOVERIN

Cancer-associated retinopathy (CAR) is a rare paraneoplastic syndrome described in patients with SCLC, melanoma, or gynecologic tumors (Jacobson et al., 1990; Weinstein et al., 1994). Patients develop various visual symptoms that usually progress to complete visual loss. Pathologically, there is degeneration and loss of photoreceptor and ganglion cells with infiltrates of lymphocytes and macrophages. Serum antibodies that react with proteins of the retinal photoreceptor layer or with ganglion cells have been described in some but not all patients (Kornguth et al., 1982; Grunwald et al., 1985; Thirkill et al., 1989). Western blot analysis of human retinal proteins demonstrated that the serum antibodies reacted with recoverin, a calcium binding protein of the calmodulin family (Polans et al., 1991; Adamus et al., 1993).

Recoverin is present in photoreceptors, cone bipolar cells, and a few cells in the retinal ganglion layer. Recoverin regulates visual signal transduction by inhibiting rhodopsin kinase in response to high calcium concentrations (Gray-Keller et al., 1993; Gorodovikova et al., 1994). Epitope mapping studies have shown that the majority of anti-recoverin antibodies are directed against a domain that is in proximity to the calcium binding region of the protein suggesting that antibody binding could disrupt this function (Adamus and Amundson, 1996). As with most other onconeuronal proteins, in addition to expression in the nervous system, recoverin has been found to be expressed by tumors from patients with CAR (Polans et al., 1995; Matsubara et al., 1996).

When anti-recoverin antibodies are coincubated with rat retinal cells, the cells undergo apoptotic cell death (Adamus et al., 1997). The addition of complement does not increase the amount or rate of cell destruction and cells that do not express recoverin are unaffected. These findings are in keeping with the hypothesis that antibodies against onconeural antigens are important, at least in part, for the development of paraneoplastic neurologic dysfunction.

## AMPHIPHYSIN

The stiff-person syndrome is characterized by the development of progressive muscle rigidity and painful spasms. The tumors more commonly involved include SCLC and breast cancer (Bateman et al., 1990; Folli et al., 1993). An autoimmune basis of this disorder was initially suggested by the identification of antibodies against glutamic acid decarboxylase in patients who had nonparaneoplastic stiff-person syndrome (Solimena et al., 1990). More recently, a subset of patients with paraneoplastic stiff-person syndrome and breast cancer were found to harbor antibodies that react with amphiphysin, a 128-kDa synaptic protein (De Camilli et al., 1993; Folli et al., 1993; David et al., 1994). Amphiphysin is localized in presynaptic terminals in brain and adrenal gland and is not found in synaptic terminals in other electrically excitable tissues such as skeletal muscle or heart. It is believed to play a role in synaptic vesicle endocytosis through binding to the vesicle coat protein adapter AP2 and dynamin (David et al., 1996).

## ONCONEURAL ANTIGENS AND THE PERIPHERAL NERVOUS SYSTEM

The autoimmune etiology of the paraneoplastic neurologic disorders has been proven in only two syndromes that affect neuromuscular transmission: myasthenia gravis (MG) and the Lambert-Eaton myasthenic syndrome (LEMS).

Myasthenia gravis is a disorder in which the immune response is directed against the acetylcholine receptor in the neuromuscular junction (Levin and Richman, 1989). The antibody response is T-cell dependent (DeBaets et al., 1982) and results in multiple changes in the postsynaptic membrane, including destruction of the membrane by a complement-mediated process and increased turnover and functional blockade of acetylcholine receptors (Levin and Richman, 1989).

The neoplasm usually associated with MG is thymoma. Thymic abnormalities are reported to occur in approximately 75% of patients with MG (Castleman, 1966). Of these, 15% have either microscopic or gross evidence of thymoma and 85% have evidence of thymic hyperplasia.

At least 90% of patients with MG with an associated thymoma have antiskeletal muscle serum antibodies or anti-titin antibodies (Oosterhuis et al., 1976; Voltz et al., 1997) and approximately 90% have anti-acetylcholine receptor (anti-AChR) antibodies (Drachman et al. 1982). Furthermore, recent studies using a SCID mouse model of MG suggested that seronegative myasthenic patients do in fact have anti-AChR antibodies that are not measured by routine testing (Grimaldi et al., 1992).

The Lambert-Eaton myasthenic syndrome is a disorder of neuromuscular transmission characterized by a defect in the presynaptic release of acetylcholine in response to nerve stimulus (Elmqvist and Lambert, 1968). Approximately 60% of patients with LEMS have an associated SCLC (O'Neill et al., 1988); other malignancies have occasionally been reported.

Patients with LEMS develop serum autoantibodies that interfere with the function of presynaptic VGCCs (Lennon et al., 1995; Motomura et al., 1995). A role of anti-VGCC antibodies in the pathogenesis of LEMS has been confirmed by the ability to passively transfer the electrophysiological and morphological features of LEMS to mice after injection of LEMS patients with IgG (Lang et al., 1981; Fukunaga et al., 1982). In immunoprecipitation assays these antibodies react with L-type (Peers et al., 1990; el Far et al., 1995), N-type (Sher et al., 1989; Leys et al., 1991), and P/Q- type VGCCs (Motomura et al., 1997). Antibodies against P/Q-type VGCCs are likely to be the most relevant in the pathogenesis of LEMS since the P/Q-type channel is responsible for neurotransmitter release at the mammalian neuromuscular junction (Uchitel et al., 1992).

The sera from LEMS patients also contain antibodies directed against other components of the VGCC/synaptic vesicle release complex, including antibodies against synaptotagmin (Leveque et al., 1992) and the $\beta$-subunit of the neuronal VGCC (MysB antibodies) (Rosenfeld et al., 1993). The role of these antibodies in the pathogenesis of LEMS is unknown.

Although the acetylcholine receptor and the VGCC are not classic "onconeural" antigens, the acetylcholine receptor is expressed by thymomas in patients with MG, and VGCC are expressed by SCLC (McCann et al., 1981; Roberts et al., 1985). What is intriguing is that both MG and LEMS respond to treatments that deplete the serum of the autoantibodies, whereas in general, paraneoplastic disorders of the central nervous system (CNS) do not. Furthermore, antibody-associated paraneoplastic disorders of the CNS have not been modeled in animals after passive transfer of paraneoplastic human serum or after animal immunization with recombinant onconeuronal antigens. Animals immunized with these proteins develop high titers of antineuronal antibodies, but do not develop neurological dysfunction (Tanaka et al., 1994; Sillevis Smitt et al., 1995). These findings and the demonstration that the CNS of patients with these disorders usually contains infiltrates of T cells (CD4 and CD8) (Jean et al., 1994) have suggested that cell-mediated mechanisms are also involved in the pathogenesis of some paraneoplastic syndromes. In these disorders the presence of antineuronal antibodies can be used as a highly sensitive marker of the type of disease and specific tumor, but the antibodies may not be pathogenic.

## THE ONCONEURAL ANTIGENS, ANTI-TUMOR IMMUNITY, AND NEUROLOGIC DISEASE

The onconeural antigens as described above are likely to play a role in diverse and possibly critical cellular functions. The presence of specific anti-onconeuronal antibodies and T-cell immune reactions in patients with paraneoplastic neurologic disorders points to more than a coincidental association with cancer. Excluding MG and LEMS, the roles of the autoimmune response in the pathogenesis of the paraneoplastic disease process remains unknown. Does disruption of the normal function of the onconeural proteins by the autoimmune response result in cellular dysfunction or cell death? What breaks normal self-tolerance to these neural antigens, initiating the immune reaction?

In some cancers, autoimmune reactions are associated with disruption of antigen function. For example, melanomas are highly immunogenic and patients with melanoma often have antibodies against antigens expressed by melanoma cells and normal melanocytes. Tyrosinase, an enzyme important in the synthesis of melanin, is one such antigen (Cox et al., 1994; Kawakami et al., 1994). Anti-tyrosinase antibodies are capable of disrupting tyrosinase function, resulting in vitiligo, a depigmenting condition (Naughton et al., 1983; Song et al., 1994). Patients with melanoma who develop vitiligo associated with high titers of anti-tyrosinase antibodies are felt to have a better cancer prognosis suggesting that the antibodies may participate in an immune-mediated destruction or growth inhibition of the melanoma (Nordlund et al., 1983; Duhra and Ilchyshyn, 1991). A similar process may occur with the paraneoplastic neuronal antigens. This is supported by the fact that some onconeuronal antibodies bind to the functional domains of their target antigens (anti-Hu and anti-Ri antibodies to RNA binding domains, anti-cdr-62 to the leucine zipper domain) (Sakai et al., 1990; Manley et al., 1995; Buckanovich et al., 1996).

For antigens that are not surface proteins, it has been difficult to envision how antibodies could affect antigen function. Until recently, it was thought that antibodies could not enter cells (Benacerraf and Unanue, 1979). Studies have now demonstrated that some autoantibodies can penetrate live cells (Alarcon-Segovia et al., 1978; Fabian, 1988; Golan et al., 1993). Most of these react with nuclear antigens and all are of the IgG subtype. Anti-Hu antibodies have been shown to enter cerebellar granule cells (Greenlee et al., 1993), Purkinje cells (Greenlee et al., 1995), and SCLC cells (Hormigo and Lieberman, 1994). For some antibodies, cell penetration results in cell death (Okudaira et al., 1982; Alarcon-Segovia et al., 1995), although this is disputed in the case of anti-Hu antibodies (Greenlee et al. 1993; Hormigo and Lieberman, 1994).

In the paraneoplastic diseases, the immune system recognizes tumor-specific antigens that are normally neuronally restricted. Is the "ectopic" expression of the neural antigen by the tumor sufficient to break self-tolerance? If all SCLCs express Hu antigens why do only a small number of SCLC patients develop anti-Hu antibodies?

It has been speculated that the Hu proteins expressed by the tumor may be altered or mutated, resulting in the development of the anti-Hu immune response. However, in a study of a large series of SCLC cell lines, including three established from tumors of Hu-antibody-positive patients, no mutations of the *HuD* gene were identified (Sekido et al., 1994). Similarly, we studied the SCLC tumors from three patients with Hu antibodies and neurologic dysfunction and found no *HuD* mutations (Carpentier et al., 1998).

Studies suggest that the development of the anti-Hu immune response may depend in part on the ability of the tumor to present the antigen to the immune system. SCLCs have low or no expression of MHC class I proteins (Doyle et al., 1985; Dammrich et al., 1990). However, in one study, 14 of 15 SCLC tumors from patients with anti-Hu antibodies and paraneoplastic disease expressed MHC class I antigens (Dalmau et al., 1995). In contrast, 4 of 11 SCLC tumors from patients without antibodies weakly expressed MHC class I antigens. Additionally, focal expression of MHC class II was seen in 6 of 15 antibody-associated tumors but not in any of the non-antibody-associated tumors. The increased expression of MHC proteins suggests that the tumors of these patients are more immunogenic than those of patients without anti-Hu antibodies, and that cytotoxic cell-mediated mechanisms are also involved in the disease. Additionally, the pathological study of paraneoplastic tumors usually demonstrates conspicuous infiltrates of mononuclear cells, rarely seen in the nonparaneoplastic tumors (Dalmau et al., 1992b).

## SUMMARY

Autoimmune paraneoplastic neurologic disorders are rare but fascinating syndromes. The study of these disorders has lead to the identification of novel proteins that are expressed only by the nervous system and tumor. These disorders provide a model of neuronal degeneration and further studies on the function of the onconeuronal antigens should provide insights into normal neuronal function. Understanding the mechanisms which regulate onconeural protein expression as tumor-specific antigens, and defining events that initiate the anti-onconeuronal immune response may lead to the development of specific antitumor immune therapies.

## REFERENCES

Adamus, G., Guy, J., Schmied, J.L., Arendt, A., and Hargrave, P.A., Role of anti-recoverin autoantibodies in cancer-associated retinopathy, *Invest. Ophthalmol. Vis. Sci.,* 34, 2626–2633, 1993.

Adamus, G., Machnicki, M., and Seigel, G.M., Apoptotic retinal cell death induced by antirecoverin autoantibodies of cancer-associated retinopathy, *Invest. Ophthalmol. Vis. Sci.,* 38, 283–291, 1997.

Adamus, G. and Amundson, D., Epitope recognition of recoverin in cancer associated retinopathy: evidence for calcium-dependent conformational epitopes, *J. Neurosci. Res.,* 45, 863–872, 1996.

Alarcon-Segovia, D., Ruiz-Arguelles, A., and Fishbein, E., Antibody to nuclear ribonucleoprotein penetrates live human mononuclear cells through Fc receptors. *Nature,* 271, 67–69, 1978.

Alarcon-Segovia, D., Llorente, D., Ruiz-Arguelles, A., Richaud-Patin, Y., and Perez-Romano, B., Penetration of anti-DNA antibodies into mononuclear cells (MNC) causes apoptosis, *Arthritis Rheum.,* 38 (Suppl.), 182, 1995.

Anderson, N.E., Rosenblum, M.K., and Posner, J.B., Paraneoplastic cerebellar degeneration: clinical-immunological correlations, *Ann. Neurol.,* 24, 559–567, 1988.

Bateman, D.E., Weller, R.O., and Kennedy, P., Stiffman syndrome: a rare paraneoplastic disorder?, *J. Neurol. Neurosurg. Psychiatr.,* 53, 695–696, 1990.

Bell, L.R., Maine, E.M., Schedl, P., and Cline, T.W., Sex-lethal, a Drosophila sex determination switch gene, exhibits sex-specific RNA splicing and sequence similarity to RNA binding proteins, *Cell,* 55, 1037–1046, 1988.

Benacerraf, B. and Unanue, E.R., *Immunopathology,* Williams & Wilkins, Baltimore, MD, 1979, 250.

Berendsen, H.H., deLeij, L., and Postmus, P.E., Small cell lung cancer. Tumor cell phenotype detected by monoclonal antibodies and response to chemotherapy, *Chest,* 91, 11S–12S, 1987.

Birney, E., Kumar, S., and Krainer, A.R., Analysis of the RNA-recognition motif and RS and RGG domains: conservation in metazoan pre-mRNA splicing factors, *Nucleic Acids Res.,* 21, 5803–5816, 1993.

Buckanovich, R.J., Posner, J.B., and Darnell, R.B., Nova, the paraneoplastic Ri antigen, is homologous to an RNA-binding protein and is specifically expressed in the developing motor system, *Neuron,* 11, 657–672, 1993.

Buckanovich, R.J., Yang, Y.Y., and Darnell, R.B., The onconeural antigen Nova-1 is a neuron-specific RNA-binding protein, the activity of which is inhibited by paraneoplastic antibodies, *J. Neurosci.,* 16, 1114–1122, 1996.

Budde-Steffen, C., Anderson, N.E., Rosenblum, M.K. et al., An antineuronal autoantibody in paraneoplastic opsoclonus, *Ann. Neurol.,* 23, 528–531, 1988.

Campos, A.R., Grossman, D., and White, K., Mutant alleles at the locus elav in Drosophila melanogaster lead to nervous system defects. A developmental-genetic analysis, *J. Neurogenet.,* 2, 197–218, 1985.

Carney, D.N., Mitchell, J.B., and Kinsella, T.J., In vitro radiation and chemotherapy sensitivity of established cell lines of human small cell lung cancer and its large cell morphological variants, *Cancer Res.,* 43, 2806–2811, 1983.

Carpentier, A.F., Rosenfeld, M.R., Delattre, J-Y., Whaten, R.G., Posner, J.B., and Dalmau, J., DNA vaccination with HuD inhibits growth of a neuroblastoma in mice, *Clin. Cancer Res.,* 4, 2819–2824, 1998.

Castleman, B., The pathology of the thymus gland in myasthenia gravis, *Ann. N.Y. Acad. Sci.,* 135, 496–505, 1966.

Cox, A.L., Skipper, J., Chen, Y. et al., Identification of a peptide recognized by five melanoma-specific human cytotoxic T cell lines, *Science,* 264, 716–719, 1994.

Cunningham, J., Graus, F., Anderson, N., and Posner, J.B., Partial characterization of the Purkinje cell antigens in paraneoplastic cerebellar degeneration, *Neurology*, 36, 1163–1168, 1986.

Dalmau, J. and Graus, F., Paraneoplastic syndromes of the nervous system, in *Cancer of the Nervous System*, Black, P.Mc.L. and Loeffler, J.S., Eds., Blackwell Scientific Publications, Boston, 1997, 674–700.

Dalmau, J., Furneaux, H.M., Gralla, R.J., Kris, M.G., and Posner, J.B., Detection of the anti-Hu antibody in the serum of patients with small cell lung cancer — a quantitative western blot analysis, *Ann. Neurol.*, 27, 544–552, 1990.

Dalmau, J., Furneaux, H.M., Cordon-Cardo, C., and Posner, J.B., The expression of the Hu (paraneoplastic encephalomyelitis/sensory neuronopathy) antigen in human normal and tumor tissues, *Am. J. Pathol.*, 141, 881–886, 1992a.

Dalmau, J., Graus, F., Rosenblum, M.K., and Posner, J.B., Anti-Hu-associated paraneoplastic encephalomyelitis/sensory neuronopathy. A clinical study of 71 patients, *Medicine*, 71, 59–72, 1992b.

Dalmau, J., Graus, F., Cheung, N.K. et al., Major histocompatibility proteins, anti-Hu antibodies, and paraneoplastic encephalomyelitis in neuroblastoma and small cell lung cancer, *Cancer*, 75, 99–109, 1995.

Dammrich, J., Muller-Hermelink, H.K., Mattner, A., Buchwald, J., and Ziffer, S., Histocompatibility antigen expression in pulmonary carcinomas as indication of differentiation and of special subtypes, *Cancer*, 65, 1942–1954, 1990.

Darnell, R.B., Onconeural antigens and the paraneoplastic neurologic disorders: at the intersection of cancer, immunity, and the brain, *Proc. Natl. Acad. Sci. U.S.A.*, 93, 4529–4536, 1996.

Darnell, R.B. and DeAngelis, L.M., Regression of small-cell lung carcinoma in patients with paraneoplastic neuronal antibodies, *Lancet*, 341, 21–22, 1993.

David, C., Solimena, M., and De Camilli, P., Autoimmunity in Stiff-Man syndrome with breast cancer is targeted to the C-terminal region of human amphiphysin, a protein similar to the yeast proteins, Rvs167 and Rvs161, *FEBS Lett.*, 351, 73–79, 1994.

David, C., McPherson, P.S., Mundigl, O., and De Camilli, P., A role of amphiphysin in synaptic vesicle endocytosis suggested by its binding to dynamin in nerve terminals, *Proc. Natl. Acad. Sci. U.S.A.*, 93, 331–335, 1996.

De Camilli, P., Thomas, A., Cofiell, R. et al., The synaptic vesicle-associated protein amphiphysin is the 128-kD autoantigen of Stiff-Man syndrome with breast cancer, *J. Exp. Med.*, 178, 2219–2223, 1993.

de la Monte, S.M., Hutchins, G.M., and Moore, G.W., Paraneoplastic syndromes and constitutional symptoms in prediction of metastatic behavior of small cell carcinoma of the lung, *Am. J. Med.*, 77, 851–857, 1984.

DeBaets, M.H., Einarson, B., Lindstrom, J.M., and Weigle, W.O., Lymphocyte activation in experimental autoimmune myasthenia gravis, *J. Immunol.*, 128, 2228–2235, 1982.

Doyle, A., Martin, W.J., Funa, K. et al., Markedly decreased expression of class I histocompatibility antigens, protein, and mRNA in human small-cell lung cancer, *J. Exp. Med.*, 161, 1135–1151, 1985.

Drachman, D.B., Adams, R.N., Josifek, L.F., and Self, S.G., Functional activities of autoantibodies to acetylcholine receptors and the clinical severity of myasthenia gravis, *N. Engl. J. Med.*, 307, 769–775, 1982.

Drachman, D.B., Myasthenia gravis, *N. Engl. J. Med.*, 330, 1797–1810, 1994.

Dropcho, E.J., Chen, Y.T., Posner, J.B., and Old, L.J., Cloning of a brain protein identified by autoantibodies from a patient with paraneoplastic cerebellar degeneration, *Proc. Natl. Acad. Sci. U.S.A.*, 84, 4552–4556, 1987.

Duhra, P. and Ilchyshyn, A., Prolonged survival in metastatic malignant melanoma associated with vitiligo, *Clin. Exp. Dermatol.*, 16, 303–305, 1991.

el Far, O., Marqueze, B., Leveque, C. et al., Antigens associated with N- and L-type calcium channels in Lambert-Eaton myasthenic syndrome, *J. Neurochem.*, 64, 1696–1702, 1995.

Elmqvist, D. and Lambert, E.H., Detailed analysis of neuromuscular transmission in a patient with the myasthenic syndrome sometimes associated with bronchogenic carcinoma, *Mayo Clin. Proc.*, 43, 689–713, 1968.

Engebrecht, J.A., Voelkel-Meiman, K., and Roeder, G.S., Meiosis-specific RNA splicing in yeast, *Cell*, 66, 1257–1268, 1991.

Fabian, R.H., Uptake of plasma IgG by CNS motoneurons: comparison of antineuronal and normal IgG, *Neurology*, 38, 1775–1780, 1988.

Fathallah-Shaykh, H., Wolf, S., Wong, E., Posner, J.B., and Furneaux, H.M., Cloning of a leucine-zipper protein recognized by the sera of patients with antibody-associated paraneoplastic cerebellar degeneration, *Proc. Natl. Acad. Sci. U.S.A.*, 88, 3451–3454, 1991.

Folli, F., Solimena, M., Cofiell, R. et al., Autoantibodies to a 128-kd synaptic protein in three women with the stiff-man syndrome and breast cancer, *N. Engl. J. Med.,* 328, 546–551, 1993.

Fukunaga, H., Engel, A.G., Osame, M., and Lambert, E.H., Paucity and disorganization of presynaptic membrane active zones in the Lambert-Eaton myasthenic syndrome, *Muscle Nerve,* 5, 686–697, 1982.

Furneaux, H.M., Dropcho, E.J., Barbut et al., Characterization of a cDNA encoding a 34-kDa Purkinje neuron protein recognized by sera from patients with paraneoplastic cerebellar degeneration, *Proc. Natl. Acad. Sci.,* 86, 2873–2877, 1989.

Gao, F.B., Carson, C.C., Levine, T., and Keene, J.D., Selection of a subset of mRNAs from combinatorial 3′ untranslated region libraries using neuronal RNA-binding protein Hel-N1, *Proc. Natl. Acad. Sci. U.S.A.,* 91, 11207–11211, 1994.

Gibson, T.J., Thompson, J.D., and Heringa, J., The KH domain occurs in a diverse set of RNA-binding proteins that include the antiterminator NusA and is probably involved in binding to nucleic acid, *FEBS Lett.,* 324, 361–366, 1993.

Golan, T.D., Gharavi, A.E., and Elkon, K.B., Penetration of autoantibodies into living epithelial cells, *J. Invest. Dermatol.,* 100, 316–322, 1993.

Good, P.J., A conserved family of elav-like genes in vertebrates, *Proc. Natl. Acad. Sci. U.S.A.,* 92, 4557–4561, 1995.

Gorodovikova, E.N., Gimelbrant, A.A., Senin, I.I., and Philippov, P.P., Recoverin mediates the calcium effect upon rhodopsin phosphorylation and cGMP hydrolysis in bovine retina rod cells, *FEBS Lett.,* 349, 187–190, 1994.

Graus, F., Cordon-Cardo, C., and Posner, J.B., Neuronal antinuclear antibody in sensory neuronopathy from lung cancer, *Neurology,* 35, 538–543, 1985.

Graus, F., Rowe, G., Fueyo, J., Darnell, R.B., and Dalmau, J., The neuronal nuclear antigen recognized by the human anti-Ri autoantibody is expressed in central but not peripheral nervous system neurons, *Neurosci. Lett.,* 150, 212–214, 1993.

Graus, F., Dalmau, J., Rene, R. et al., Anti-Hu antibodies in patients with small-cell lung cancer: association with complete response to therapy and improved survival, *J. Clin. Oncol.,* 15, 2866–2872, 1997a.

Graus, F., Dalmau, J., Valldeoriola, F. et al., Immunological characterization of a neuronal antibody (anti-Tr) associated with paraneoplastic cerebellar degeneration and Hodgkin's disease, *J. Neuroimmunol.,* 74, 55–61, 1997b.

Gray-Keller, M.P., Polans, A.S., Palczewski, K., and Detwiler, P.B., The effect of recoverin-like calcium-binding proteins on the photoresponse of retinal rods, *Neuron,* 10, 523–531, 1993.

Greenlee, J.E., Parks, T.N., and Jaeckle, K.A., Type IIa ("anti-Hu") antineuronal antibodies produce destruction of rat cerebellar granule neurons in vitro, *Neurology,* 43, 2049–2054, 1993.

Greenlee, J.E., Burns, J.B., Rose, J.W., Jaeckle, K.A., and Clawson, S., Uptake of systemically administered human anticerebellar antibody by rat Purkinje cells following blood-brain barrier disruption, *Acta Neuropathol.,* 89, 341–345, 1995.

Greenlee, J.E. and Brashear, H.R., Antibodies to cerebellar Purkinje cells in patients with paraneoplastic cerebellar degeneration and ovarian carcinoma, *Ann. Neurol.,* 14, 609–613, 1983.

Grimaldi, L.M.E., DuPont, B.L., Martino, G.V. et al., A SCID mouse model of myasthenia gravis: pathological and serological findings, *Neurology,* 42 (Suppl. 3), 233, 1992.

Grunwald, G.B., Klein, R., Simmonds, M.A., and Kornguth, S.E., Autoimmune basis for visual paraneoplastic syndrome in patients with small-cell lung carcinoma, *Lancet,* 1, 658–661, 1985.

Hedgecock, E.M., Culotti, J.G., Thomson, J.N., and Perkins, L.A., Axonal guidance mutants of Caenorhabditis elegans identified by filling sensory neurons with fluorescein dyes, *Devel. Biol.,* 111, 158–170, 1985.

Henson, R.A. and Urich, H., Encephalomyelitis with carcinoma, in *Cancer and the Nervous System,* Henson, R.A. and Urich, H., Eds., Blackwell Scientific Publications, Oxford, 1989, 314–345.

Hormigo, A. and Lieberman, F., Nuclear localization of anti-Hu antibody is not associated with in vitro cytotoxicity, *J. Neuroimmunol.,* 55, 205–212, 1994.

Jacobson, D.M., Thirkill, C.E., and Tipping, S.J., A clinical triad to diagnose paraneoplastic retinopathy, *Ann. Neurol.,* 28, 162–167, 1990.

Jaeckle, K.A., Graus, F., Houghton, A. et al., Autoimmune response of patients with paraneoplastic cerebellar degeneration to a Purkinje cell cytoplasmic protein antigen, *Ann. Neurol.,* 18, 592–600, 1985.

Jean, W.C., Dalmau, J., Ho, A., and Posner, J.B., Analysis of the IgG subclass distribution and inflammatory infiltrates in patients with anti-Hu-associated paraneoplastic encephalomyelitis, *Neurology,* 44, 140–147, 1994.

Kawakami, Y., Eliyahu, S., Delgado, C.H. et al., Identification of a human melanoma antigen recognized by tumor-infiltrating lymphocytes associated with in vivo tumor rejection, *Proc. Natl. Acad. Sci. U.S.A.*, 91, 6458–6462, 1994.

Kenan, D.J., Query, C.C., and Keene, J.D., RNA recognition: towards identifying determinants of specificity, *Trends Biochem. Sci.*, 16, 214–220, 1991.

Kim, Y.J. and Baker, B.S., The Drosophila gene rbp9 encodes a protein that is a member of a conserved group of putative RNA binding proteins that are nervous system-specific in both flies and humans, *J. Neurosci.*, 13, 1045–1056, 1993.

King, P.H., Hel-N2: a novel isoform of Hel-N1 which is conserved in rat neural tissue and produced in early embryogenesis, *Gene*, 151, 261–265, 1994.

King, P.H., Levine, T.D., Fremeau, R.T., Jr., and Keene, J.D., Mammalian homologs of Drosophila ELAV localized to a neuronal subset can bind in vitro to the 3′ UTR of mRNA encoding the Id transcriptional repressor, *J. Neurosci.*, 14, 1943–1952, 1994.

King, P.H., Cloning the 5′ flanking region of neuron-specific Hel-N1: evidence for positive regulatory elements governing cell-specific transcription, *Brain Res.*, 723, 141–147, 1996.

Kornguth, S.E., Klein, R., Appen, R., and Choate, J., Occurrence of anti-retinal ganglion cell antibodies in patients with small cell carcinoma of the lung, *Cancer*, 50, 1289–1293, 1982.

Lang, B., Newsom-Davis, J., Wray, D., Vincent, A., and Murray, N., Autoimmune aetiology for myasthenic (Eaton-Lambert) syndrome, *Lancet*, 2, 224–226, 1981.

Lennon, V.A., Kryzer, T.J., Griesmann, G.E. et al., Calcium-channel antibodies in the Lambert-Eaton syndrome and other paraneoplastic syndromes, *N. Engl. J. Med.*, 332, 1467–1474, 1995.

Leveque, C., Hoshino, T., David, P. et al., The synaptic vesicle protein synaptotagmin associates with calcium channels and is a putative Lambert-Eaton myasthenic syndrome antigen, *Proc. Natl. Acad. Sci. U.S.A.*, 89, 3625–3629, 1992.

Levin, K.H. and Richman, D.P., Myasthenia gravis, *Clin. Aspects Autoimmun.*, 4, 23–31, 1989.

Levine, T.D., Gao, F., King, P.H., Andrews, L.G., and Keene, J.D., Hel-N1: an autoimmune RNA-binding protein with specificity for 3′ uridylate-rich untranslated regions of growth factor mRNAs, *Mol. Cell. Biol.*, 13, 3494–3504, 1993.

Leys, K., Lang, B., Johnston, I., and Newsom-Davis, J., Calcium channel autoantibodies in the Lambert-Eaton myasthenic syndrome, *Ann. Neurol.*, 29, 307–314, 1991.

Liu, J., Dalmau, J., Szabo, A., Rosenfeld, M., Huber, J., and Furneaux, H., Paraneoplastic encephalomyelitis antigens bind to the AU-rich elements of mRNA, *Neurology*, 45, 544–550, 1995.

Luque, F.A., Furneaux, H.M., Ferziger, R. et al., Anti-Ri: an antibody associated with paraneoplastic opsoclonus and breast cancer, *Ann. Neurol.*, 29, 241–251, 1991.

Ma, W.J., Cheng, S., Campbell, C., Wright, A., and Furneaux, H., Cloning and characterization of HuR, a ubiquitously expressed Elav-like protein, *J. Biol. Chem.*, 271, 8144–8151, 1996.

Manley, G.T., Smitt, P.S., Dalmau, J., and Posner, J.B., Hu antigens: reactivity with Hu antibodies, tumor expression, and major immunogenic sites, *Ann. Neurol.*, 38, 102–110, 1995.

Marusich, M.F., Furneaux, H.M., Henion, P.D., and Weston, J.A., Hu neuronal proteins are expressed in proliferating neurogenic cells, *J. Neurobiol.*, 25, 143–155, 1994.

Marusich, M.F. and Weston, J.A., Identification of early neurogenic cells in the neural crest lineage, *Dev. Biol.*, 149, 295–306, 1992.

Mason, W.P., Graus, F., Lang, B. et al., Small-cell lung cancer, paraneoplastic cerebellar degeneration and the Lambert-Eaton myasthenic syndrome, *Brain*, 120, 1279–1300, 1997.

Matsubara, S., Yamaji, Y., Sato, M., Fujita, J., and Takahara, J. Expression of a photoreceptor protein, recoverin, as a cancer-associated retinopathy autoantigen in human lung cancer cell lines, *Br. J. Cancer*, 74, 1419–1422, 1996.

McCann, F.V., Pettengill, O.S., Cole, J.J., Russell, J.A., and Sorenson, G.D., Calcium spike electrogenesis and other electrical activity in continuously cultured small cell carcinoma of the lung, *Science*, 212, 1155–1157, 1981.

Motomura, M., Johnston, I., Lang, B., Vincent, A., and Newsom-Davis, J., An improved diagnostic assay for Lambert-Eaton myasthenic syndrome, *J. Neurol. Neurosurg. Psychiatr.*, 58, 85–87, 1995.

Motomura, M., Lang, B., Johnston, I., Palace, J., Vincent, A., and Newsom-Davis, J., Incidence of serum anti-P/O-type and anti-N-type calcium channel autoantibodies in the Lambert-Eaton myasthenic syndrome, *J. Neurol. Sci.*, 147, 35–42, 1997.

Nandabalan, K. and Roeder, G.S., Binding of a cell-type-specific RNA splicing factor to its target regulatory sequence, *Mol. Cell. Biol.,* 15, 1953–1960, 1995.

Naughton, G.K., Eisinger, M., and Bystryn, J.C., Antibodies to normal human melanocytes in vitiligo, *J. Exp. Med.,* 158, 246–251, 1983.

Nordlund, J.J., Kirkwood, J.M., Forget, B.M., Milton, G., Albert, D.M., and Lerner, A.B., Vitiligo in patients with metastatic melanoma: a good prognostic sign, *J. Am. Acad. Dermatol.,* 9, 689–696, 1983.

O'Neill, J.H., Murray, N.M., and Newsom-Davis, J., The Lambert-Eaton myasthenic syndrome. A review of 50 cases, *Brain,* 111, 577–596, 1988.

Okano, H.J. and Darnell, R.B., A hierarchy of Hu RNA binding proteins in developing and adult neurons, *J. Neurosci.,* 17, 3024–3037, 1997.

Okudaira, K., Searles, R.P., Tanimoto, K., Horiuchi, Y., and Williams, R.C., Jr., T lymphocyte interaction with immunoglobulin G antibody in systemic lupus erythematosus, *J. Clin. Invest.,* 69, 1026–1038, 1982.

Oosterhuis, H.J., Feltkamp, T.E., van Rossum, A.L., van den Berg-Loonen, P.M., and Nijenhuis, L.E., HL-A antigens, autoantibody production, and associated diseases in thymoma patients, with and without myasthenia gravis, *Ann. N.Y. Acad. Sci.,* 274, 468–474, 1976.

Peers, C., Lang, B., Newsom-Davis, J., and Wray, D.W., Selective action of myasthenic syndrome antibodies on calcium channels in a rodent neuroblastoma × lioma cell line, *J. Physiol.,* 421, 293–308, 1990.

Peterson, K., Rosenblum, M.K., Kotanides, H., and Posner, J.B., Paraneoplastic cerebellar degeneration. I. A clinical analysis of 55 anti-Yo antibody-positive patients, *Neurology,* 42, 1931–1937, 1992.

Polans, A.S., Buczylko, J., Crabb, J., and Palczewski, K., A photoreceptor calcium binding protein is recognized by autoantibodies obtained from patients with cancer-associated retinopathy, *J. Cell Biol.,* 112, 981–989, 1991.

Polans, A.S., Witkowska, D., Haley, T.L., Amundson, D., Baizer, L., and Adamus, G., Recoverin, a photoreceptor-specific calcium-binding protein, is expressed by the tumor of a patient with cancer-associated retinopathy, *Proc. Natl. Acad. Sci. U.S.A.,* 92, 9176–9180, 1995.

Posner, J.B., *Neurologic Complications of Cancer,* F.A. Davies, Philadelphia, 1995.

Quach, T.T., Rong, Y., Belin, M.F. et al., Molecular cloning of a new unc-33-like cDNA from rat brain and its relation to paraneoplastic neurological syndromes, *Brain Res. Mol. Brain Res.,* 1–2, 329–332, 1997.

Query, C.C., Bentley, R.C., and Keene, J.D., A common RNA recognition motif identified within a defined U1 RNA binding domain of the 70K U1 snRNP protein, *Cell,* 57, 89–101, 1989.

Roberts, A., Perera, S., Lang, B., Vincent, A., and Newsom-Davis, J., Paraneoplastic myasthenic syndrome IgG inhibits $^{45}Ca^{2+}$ flux in a human small cell carcinoma line, *Nature,* 317, 737–739, 1985.

Robinow, S., Campos, A.R., Yao, K.M., and White, K., The elav gene product of Drosophila, required in neurons, has three RNP consensus motifs, *Science,* 42, 1570–1572, 1988.

Robinow, S. and White, K., The locus elav of Drosophila melanogaster is expressed in neurons at all developmental stages, *Dev. Biol.,* 126, 294–303, 1988.

Robinow, S. and White, K., Characterization and spatial distribution of the ELAV protein during Drosophila melanogaster development, *J. Neurobiol.,* 22, 443–461, 1991.

Rodriguez, M., Truh, L.I., O'Neill, B.P., and Lennon, V.A., Autoimmune paraneoplastic cerebellar degeneration: ultrastructural localization of antibody-binding sites in Purkinje cells, *Neurology,* 38, 1380–1386, 1988.

Rosenfeld, M.R., Wong, E., Dalmau, J. et al., Cloning and characterization of a Lambert-Eaton myasthenic syndrome antigen, *Ann. Neurol.,* 33, 113–120, 1993.

Sakai, K., Mitchell, D.J., Tsukamoto, T., and Steinman, L., Isolation of a complementary DNA clone encoding an autoantigen recognized by an anti-neuronal cell antibody from a patient with paraneoplastic cerebellar degeneration, *Ann. Neurol.,* 28, 692–698, 1990.

Sakai, K., Gofuku, M., Kitagawa, Y. et al., A hippocampal protein associated with paraneoplastic neurologic syndrome and small cell lung carcinoma, *Biochem. Biophys. Res. Commun.,* 199, 1200–1208, 1994.

Schmid, A.H. and Riede, U.N., A morphometric study of the cerebellar cortex from patients with carcinoma. A contribution on quantitative aspects in carcinotoxic cerebellar atrophy, *Acta Neuropathol.,* 28, 343–352, 1974.

Sekido, Y., Bader, S.A., Carbone, D.P., Johnson, B.E., and Minna, J.D., Molecular analysis of the HuD gene encoding a paraneoplastic encephalomyelitis antigen in human lung cancer cell lines, *Cancer Res.,* 54, 4988–4992, 1994.

Sher, E., Gotti, C., Canal, N. et al., Specificity of calcium channel autoantibodies in Lambert-Eaton myasthenic syndrome, *Lancet,* 2, 640–643, 1989.

Siebel, C.W., Admon, A., and Rio, D.C., Soma-specific expression and cloning of PSI, a negative regulator of P element pre-mRNA splicing, *Genes Dev.*, 9, 269–283, 1995.

Sillevis Smitt, P.A., Manley, G.T., and Posner, J.B., Immunization with the paraneoplastic encephalomyelitis antigen HuD does not cause neurologic disease in mice, *Neurology*, 45, 1873–1878, 1995.

Solimena, M., Folli, F., Aparisi, R., Pozza, G., and De Camilli, P., Autoantibodies to GABA-ergic neurons and pancreatic beta cells in stiff-man syndrome, *N. Engl. J. Med.*, 322, 1555–1560, 1990.

Song, Y.H., Connor, E., Li, Y., Zorovich, B., Balducci, P., and Maclaren, N., The role of tyrosinase in autoimmune vitiligo, *Lancet*, 344, 1049–1052, 1994.

Szabo, A., Dalmau, J., Manley, G. et al., HuD, a paraneoplastic encephalomyelitis antigen, contains RNA-binding domains and is homologous to Elav and Sex-lethal, *Cell*, 67, 325–333, 1991.

Tanaka, K., Tanaka, M., Onodera, O., Igarashi, S., Miyatake, T., and Tsuji, S., Passive transfer and active immunization with the recombinant leucine-zipper (Yo) protein as an attempt to establish an animal model of paraneoplastic cerebellar degeneration, *J. Neurol. Sci.*, 127, 153–158, 1994.

Thirkill, C.E., FitzGerald, P., Sergott, R.C., Roth, A.M., Tyler, N.K., and Keltner, J.L., Cancer-associated retinopathy (CAR syndrome) with antibodies reacting with retinal, optic-nerve, and cancer cells, *N. Engl. J. Med.*, 321, 1589–1594, 1989.

Uchitel, O.D., Protti, D.A., Sanchez, V., Cherksey, B.D., Sugimori, M., and Llinas, R., P-type voltage-dependent calcium channel mediates presynaptic calcium influx and transmitter release in mammalian synapses, *Proc. Natl. Acad. Sci. U.S.A.*, 89, 3330–3333, 1992.

Valcarcel, J., Singh, R., Zamore, P.D., and Green, M.R., The protein Sex-lethal antagonizes the splicing factor U2AF to regulate alternative splicing of transformer pre-mRNA, *Nature*, 362, 171–175, 1993.

Verschuuren, J., Chuang, L., Rosenblum, M.K. et al., Inflammatory infiltrates and complete absence of Purkinje cells in anti-Yo-associated paraneoplastic cerebellar degeneration, *Acta Neuropathol.*, 91, 519–525, 1996.

Voltz, R.D., Albrich, W.C., Naegele, A. et al., Paraneoplastic myasthenia gravis: Detection of anti-MGT30 (titin) antibodies predicts thymic epithelial tumor, *Neurology*, 49, 1454–1457, 1997.

Weinstein, J.M., Kelman, S.E., Bresnick, G.H., and Kornguth, S.E., Paraneoplastic retinopathy associated with antiretinal bipolar cell antibodies in cutaneous malignant melanoma, *Ophthalmology*, 101, 1236–1243, 1994.

Wilkinson, P.C. and Zeromski, J., Immunofluorescent detection of antibodies against neurones in sensory carcinomatous neuropathy, *Brain*, 88, 529–538, 1965.

# 12 The Neurofibromatoses: Genetic and Molecular Progress

*Andrea I. McClatchey*

## CONTENTS

## CLINICAL FEATURES OF NF1 AND NF2

The neurofibromatoses (NF1 and NF2) are familial cancer syndromes featuring the development of tumors of the central nervous system. NF1 has been recognized as a medical condition for centuries and as a familial form of cancer since the turn of this century, while NF2 was only recently recognized as a distinct entity in the early 1980s.[89,93] The past decade has yielded tremendous advances in NF research, witnessing the identification and cloning of the genes responsible for both *NF1* and *NF2*, the subsequent revolution in genetic counseling of NF families through the development and application of genetic testing, and progress toward the development of animal models for each. In addition, the NF community is a particularly well-integrated one with well-established national and international networks for communication between patients, clinicians, investigators, pharmaceuticals, and funding agencies. In many ways NF is a paradigm for the coordinated and progressive study and treatment of rare human diseases.

    This chapter summarizes the clinical and genetic features of NF1 and NF2, highlighting both similarities and dissimilarities. The identification of the *NF1* and *NF2* genes is reviewed, with

## TABLE 12.1
## Diagnostic Clinical Criteria for Neurofibromatosis Type 1

The patient should have 2 or more of the following:

1. Six or more café-au-lait spots
   1.5 cm or larger in postpubertal individuals
   0.5 cm or larger in prepubertal individuals
2. Two or more neurofibromas of any type *or* 1 or more plexiform neurofibroma
3. Freckling in the axilla or groin
4. Optic glioma (tumor of the optic pathway)
5. Two or more Lisch nodules (benign iris hamartomas)
6. A distinctive bony lesion
   Dysplasia of the sphenoid bone
   Dysplasia or thinning of the long bone cortex
7. A first-degree relative with NF1

emphasis on the tremendous impact this has had on the clinical evaluation of NF, genetic counseling of NF patients, and redirection of research efforts. The state of current research aimed at understanding and ultimately counteracting the molecular defects underlying NF1 and NF2 is described. In this regard, valuable information has been obtained through the identification and study of *NF1* and *NF2* homologs in other species. Finally, progress toward the development and investigation of animal models of NF1 and NF2 is described. These efforts have yielded important insights about the loss of NF1 and NF2 function in tumorigenesis and the normal function of the *NF1* and *NF2* gene products. These models can be used as vehicles in which to test therapeutic agents directed toward the molecular pathways in which they normally act.

The guidelines for the diagnosis of NF1 and NF2 were delineated at the National Institutes of Health (NIH) Consensus Development Conference in 1988 and were recently modified slightly (Tables 12.1 and 12.2).[41,85] A careful comparison of the two diseases reveals little true clinical overlap but great clinical similarity, which has caused much diagnostic confusion, particularly away from the larger medical facilities which now harbor organized referral centers. The modified criteria move toward further segregation of the two diseases, yet some investigators feel that there remains

## TABLE 12.2
## Diagnostic Clinical Criteria for Neurofibromatosis Type 2

**Individuals with the following clinical features have confirmed (definite) NF2:**
Bilateral vestibular schwannomas
*or*
First-degree relative with NF2
*plus*
1. Unilateral vestibular schwannoma <30 years *or*
2. Any two of the following: meningioma, glioma, schwannoma, juvenile posterior subcapsular lenticular opacities/juvenile cortical cataract

**Individuals with the following clinical features should be evaluated for NF2 (presumptive or probable NF2):**
Unilateral vestibular schwannoma <30 years *plus* at least one of the following: meningioma, glioma, schwannoma, juvenile posterior subcapsular lenticular opacities/juvenile cortical cataract
Multiple meningiomas (two or more) *plus* unilateral vestibular schwannomas <30 years *or* one of the following: glioma, schwannoma, juvenile posterior subcapsular lenticular opacities/juvenile cortical cataracts

**TABLE 12.3**

**Genetic and Molecular Comparison of NF1 vs. NF2**

|  | NF1 | NF2 |
|---|---|---|
| Incidence | 1/4,000 | 1/37,000 |
| Diagnostic features | neurofibromas | vestibular schwannomas |
|  | café-au-lait spots | meningiomas |
|  | Lisch nodules |  |
| Gene location | human 17q | human 22q |
|  | mouse 11q | mouse 11q |
| Identified protein | neurofibromin | merlin |
|  | (rasGAP) | (membrane-cytoskeleton; ezrin/radixin/moesin [ERM]) |

a bona fide overlap. A deeper understanding of the molecular function of the *NF1* and *NF2* gene products will ultimately reveal the nature of any such overlap.

## CLINICAL FEATURES OF NF1

NF1 and NF2 both primarily affect cells of the neural crest lineage, featuring the development of tumors of the nervous system. NF1 (or von Recklinghausen neurofibromatosis) is the more widely recognized form of NF, in part due to a long held public misassociation with Elephant man's disease (this was actually Proteus syndrome).[48,49] Moreover, NF1 is much more common than NF2, affecting approximately 1 in 4000 individuals. The clinical features of NF1 are tremendously variable and include both tumor and nontumor symptoms.[33,48,49] This complexity is heightened by the fact that the symptoms of NF1 are largely age dependent; most are apparent by adolescence but increase in severity with age.

The hallmarks of NF1 are hyperpigmented café-au-lait spots of the skin, freckling in non-sun-exposed regions, Lisch nodules of the iris, and the development of benign neurofibromas of the peripheral nervous system (Tables 12.1 and 12.3). Neurofibromas are mixed tumors composed of Schwann cells, neurons, and perineurial fibroblasts with many infiltrating mast cells. These are most often subcutaneous, arising as nodules around peripheral nerve termini immediately beneath the skin.[144] An individual with NF1 may develop anywhere from a few to hundreds of benign, cutaneous neurofibromas, causing tremendous cosmetic and psychological duress. Neurofibromas also can develop in a diffuse plexiform manner deeper within the soft tissue and extending along the length of the nerve trunk. Plexiform neurofibromas are usually congenital, very painful, and difficult to remove surgically. A small percentage (~15%) of the plexiform type of neurofibroma can apparently progress to become malignant peripheral nerve sheath tumors (MPNSTs), invasive, metastatic tumors which forbode a poor patient prognosis.

NF1 patients are also predisposed to the development of other cancers, notably optic glioma, pheochromocytoma, rhabdomyosarcoma and childhood myeloid leukemia (juvenile monomyelo-cytic; JMML)[6] at frequencies that are low, but nevertheless greatly elevated by comparison with the normal population.[33,41,48] It is interesting to note that while children with NF1 are 200 to 500 times more likely to develop malignant myeloid disorders, adults with NF1 are not at an increased risk, suggesting that NF1 function is growth inhibitory only for an immature myeloid precurser cell.[119] The reduced incidence of malignancies in NF1 patients compared to neurofibromas strongly argues that the mutation of additional loci is also necessary for their formation. In addition to the tumor-associated features of NF1, there is a strong association between NF1 and mild learning impairment, skeletal abnormalities, short stature, and poorly understood "lesions" that appear on T2-weighted magnetic resonance images of the brain.[33,48]

## Clinical Features of NF2

The hallmark of NF2 is the development of benign Schwann cell tumors of the vestibular branch of the acoustic cranial nerve, nearly always leading to loss of hearing. Most NF2 patients also develop multiple schwannomas of the cranial and/or spinal nerves (Tables 12.2 and 12.3). NF2 is quite rare, occuring as infrequently as 1 in 37,000 individuals, and it is associated with higher morbidity than NF1, largely due to the development of many intractable schwannomas.[49] In contrast to the neurofibromas associated with NF1, which contain several cell types including Schwann cells, NF2-associated tumors are composed entirely of Schwann cells (see below). The histological classification of these tumor types is complex and variably based upon pathological and histological examination, the use of histological markers (i.e., antibodies that detect the Schwann-cell-specific antigen S-100 or neuron-specific enolase), and electron microscopy, leading to confusion and frequent misclassification. Whether NF2 patients ever develop true neurofibromas is still under debate. However, in contrast to NF1, malignancy is not a feature of NF2; progression of NF2-associated Schwann cell tumors to malignancy rarely if ever occurs. NF2 patients are also predisposed to the development of meningiomas and ependymomas as well as posterior subcapsular cataracts of the lens.[48] In contrast to the more common crystalline lens opacities, these cataracts are the result of an accumulation of cells in the posterior of the lens and may reflect clonal hyperplasia.[30]

## Related Disorders

A number of variants of NF have been described (i.e., NF3, NF4, and NF5). Most of these patients are likely to have either NF1 or NF2 and a concomitant mutation in either the *NF1* or *NF2* gene. These patients are usually described as exhibiting a combination of NF1- and NF2-like symptoms, a diagnosis complicated by the variability of both disorders and the aforementioned confusion in the pathologic distinction between neurofibromas and schwannomas. In addition, mosaicism for *NF1* or *NF2* mutations (so called segmental NF1 or NF2) is not infrequent and can present additional confusion (i.e., a patient with multiple schwannomas, but without diagnostic vestibular schwannomas). The study of these mosaic patients will be very interesting in terms of understanding the etiology of these diseases.

Despite this confusion, several related disorders do appear to be genetically distinct entities: Noonan syndrome, familial café-au-lait spots, schwannomatosis, and multiple meningiomas. Noonan syndrome exhibits a set of features that overlap with that of NF1, including café-au-lait spots, short stature, developmental delay, congenital heart abnormalities, and facial skeletal abnormalities.[113] Genetic linkage studies indicate that this syndrome is not allelic to NF1; instead a candidate locus has been mapped to chromosome 12.[2,7] However, the related Watson syndrome, featuring café-au-lait spots, short stature, developmental delay, and a similar cardiac defect does appear to be caused by mutation at the *NF1* locus.[127] Conflicting reports have explored linkage of familial café-au-lait spots to chromosome 17, where the *NF1* locus resides.[1,22]

It has been reported that meningiomatosis, or multiple meningiomas, as inherited in an autosomal dominant fashion, is not linked to the *NF2* locus on chromosome 22 and may result from a mutation in an as yet unidentified tumor suppressor gene.[90] Recent evidence suggests that schwannomatosis is likely to involve loss of *NF2* gene function, either through somatic mosaicism or, interestingly, when familial, through an apparently inherited predisposition to develop mutations at the *NF2* locus.[54] The identification of alternative genetic mutations underlying these related disorders will provide valuable information about the molecular pathways that are defective in NF1 and NF2. Moreover, the tremendous symptomatic variability associated with NF1, and to a lesser extent NF2, suggests the existence of strong modifier loci that await identification.

# GENETICS OF NF1 AND NF2

## POSITIONAL CLONING OF THE *NF1* AND *NF2* GENES

Initial NF family studies recognized distinct chromosomal locations for the *NF1* and *NF2* genes by linkage analysis that traced the coinheritance of certain alleles of polymorphic loci with the predisposition to developing symptoms of NF1 or NF2. The *NF1* gene was localized to chromosome 17q[9,108,109] and the *NF2* gene to chromosome 22q.[97] Ultimately, the *NF1* gene was identified in 1990[21,138,140] and the *NF2* gene in 1993.[96,132]

The pattern of genetic inheritance for both NF1 and NF2 is that of an autosomal dominantly inherited predisposition to the development of a number of tumor types, implying that the underlying genetic defects result in the inactivation of tumor suppressor genes.[63] This scenario would then follow Knudson's two-hit model wherein an individual with NF (1 or 2) inherits a constitutional, heterozygous inactivating mutation in the *NF1* or *NF2* tumor suppressor gene. Inactivation of the remaining wild-type copy of the *NF* gene is predicted to occur somatically, resulting in the complete loss of function of the *NF* tumor suppressor gene product and subsequent hyperproliferation of certain cell(s) in which this has occured.[63] Studies based upon this premise aided in the identification of the *NF2* gene, utilizing polymorphisms to identify regions of chromosome 22 that were frequently lost in NF2 patient tumor material.[107-109] So called loss of heterozygosity (LOH) studies identify polymorphisms that are heterozygous in a patient's lymphocyte DNA (presumably constitutional) yet exhibit a monoallelic pattern in the tumor, reflecting the somatic loss of genetic material through interstitial deletion, chromosomal loss, or gene conversion. A large percentage of NF2-associated schwannomas and meningiomas exhibit LOH of regions encompassing the *NF2* locus; the mapping of small regions of allelic loss assisted in the localization of the *NF2* gene.

The use of LOH studies to aid in the localization of the *NF1* gene was not forthcoming because detection of LOH in NF1-associated tumors has been difficult and controversial. The search for somatic inactivation of the *NF1* locus in benign neurofibromas has been hampered by the very large size of the *NF1* gene and by the fact that neurofibromas are tumors composed of admixtures of cell types (see below). This may reflect the clonal outgrowth of one cell type and recruitment of surrounding wild-type cells which would readily contaminate the tumor samples. Alternatively, these tumors may be clonal, reflecting multiple differentiation potentials of an affected cell. While some studies have argued a monoclonal origin for benign neurofibromas,[26,122] initial efforts failed to detect NF1-associated LOH in these tumors.[37,79,118,122] More recent studies have reported evidence for the inactivation of both *NF1* alleles.[23,103,111] In one study of a tumor from a patient carrying a constitutional deletion of the entire *NF1* locus, a region of paraffin-imbedded tissue that was clearly neoplastic was microdissected away from the surrounding normal tissue.[103] The detection of a second, somatic mutation in the tumor tissue but not in surrounding normal tissue provided strong evidence that the inactivation of both *NF1* alleles occur in at least some neurofibromas. It is also possible that alternative molecular mechanisms contribute to the loss of *NF1* expression in tumors. For example, one group has reported that mRNA editing of the *NF1* mRNA can introduce a stop codon into the amino-terminal half of the NF1 protein.[121] Although it is not clear that this altered message is transported to the cytoplasm and is therefore functionally expressed, higher ratios of edited to unedited *NF1* mRNA were detected in tumors compared to lymphocytes from the same NF1 patient.[19] Alternative mechanisms for achieving loss of NF1 function could help to explain the paucity of *NF1* mutations identified in neurofibromas as well as the tremendous phenotypic variability exhibited between NF1 patients and within NF1 families. In contrast to benign neurofibromas, loss of the wild-type NF1 allele can be readily identified in the malignant tumors of NF1 patients, notably neurofibrosarcoma, pheochromocytoma, and some but not all childhood leukemias.[4,67,112,119,148]

An important corollary to the two-hit hypothesis of Knudson is that sporadically occuring examples of NF-associated tumor types are likely to harbor inactivating mutations in both copies of the *NF1* or *NF2* gene, reflecting two individual somatic mutational events. Subsequent research has confirmed that most sporadically occuring schwannomas exhibit inactivation of both copies of the *NF2* gene, either through *NF2* mutation and loss of the wild-type allele or through independent intragenic mutational events.[55] Furthermore, approximately 40% of sporadic meningiomas possess *NF2* mutations, consistent with studies of familial meningiomas that suggest the existence of another meningioma-predisposing locus in addition to *NF2*.[90,143] In contrast, limited information exists regarding the status of the *NF1* locus in sporadic examples of NF1-associated tumors.[119,148] Interestingly, spontaneous neurofibromas rarely, if ever, occur, raising the possibility that *NF1* heterozygosity in surrounding cells may be an integral component of the complex nature of neurofibroma development.

## MUTATIONAL ANALYSIS

The identification of the *NF1* and *NF2* genes had an enormous impact on the counseling and management of NF patients, because of the immediate possibility of prenatal, presymptomatic, and diagnostic genetic testing. However, nearly half of all NF1 and NF2 patients do not exhibit a family history of NF and represent so called new mutations. As a result, predictive genetic testing is only feasible for a small fraction of potential NF patients. The large size of the *NF1* gene and the decreased fitness of NF1 and NF2 patients probably contributes to the bias toward new mutations. In 1992, the NF1 Genetic Analysis Consortium was organized in an effort to facilitate the mutational analysis of NF1 through the establishment of an online database.[64] The majority of germline and somatic NF1and NF2 mutations identified are truncating and presumably inactivating mutations, strengthening the notion that complete loss of function of the encoded protein is critical to the etiology of those tumors.[40,44] Although the technical challenge of detecting mutations in the large *NF1* gene has slowed progress, a procedure for specifically detecting such truncating mutations has recently been developed and successfully utilized to screen the *NF1* transcript.[44,119] The assay couples the (*in vitro*) transcription and translation (IVTT) of segments of the *NF1* coding sequence prepared from patient RNA and amplified by polymerase chain reaction (PCR); truncated protein products are detected by gel electrophoresis. Mutational analysis has also been used to attempt to draw correlations between the type of mutation and disease severity. Genotype-phenotype relationships, if identified, would allow predictions of individual patients' disease course and of the functional regions of the encoded protein. However, the results of these studies have been inconsistent; early studies suggested that large deletions resulting in the removal of the entire *NF1* gene led to a more severe phenotype than mutations that resulted in smaller, intragenic alterations,[145] whereas a more recent study finds that large *NF1* gene deletions are not predictive of a severe phenotype.[131]

Some correlation has similarly been drawn between less severe forms of NF2 and rare missense mutations or splice site alterations predicted to result in the removal of single exons from the *NF2* mRNA.[55,88,99] However, Jacoby et al. recently investigated the nature of the mRNA transcript produced from alleles carrying splice site mutations and discovered the surprising use of cryptic splice sites which resulted in frameshifts and protein truncation instead of the expected excision of a single exon.[55] Even more important, studies of both NF1 and NF2 tumors have rarely included an examination of the nature of the mutant protein product present in the tumor tissue. Indeed, one recent study failed to detect NF2 protein product in any schwannomas, regardless of the type of *NF2* mutation present, arguing that many mutations detrimentally affect NF2 protein stability.[123]

Other tumor suppressors that are associated with familial cancer syndromes have been found to be inactivated in cancers that are not features of the familial disease. For example, mutations in the retinoblastoma (Rb) tumor suppressor gene have also been identified in a large percentage of small cell lung carcinomas, although this tumor does not develop at an elevated frequency in Rb patients.[62] The reasons for this are unclear, but may reflect the need for additional mutational events, or perhaps a specific sequence of mutational events to occur. The search for *NF1* mutations in cancers not associated with NF1 has met with limited success and has relied heavily upon the screening of established cell lines, leaving open the possibility that such lesions arose during their establishment and maintenance. Early studies reported the identification of a specific *NF1* mutation in a single primary colon carcinoma, a myelodysplastic disorder, and an astrocytoma.[69] More substantial evidence exists for *NF1* mutations in neuroblastoma and perhaps melanoma cell lines, both tumors of the neural crest lineage.[57,129] Importantly, the latter studies both evaluated NF1 protein levels and attempted to evaluate the consequences of loss of neurofibromin function biochemically (see below).

A number of studies have similarly attempted to identify *NF2* mutations in cancers not associated with NF2. To date, *NF2* mutations have been detected in a small number of melanomas (6/20) and colon carcinomas (2/64) and in a single breast carcinoma (1/137).[5,14] *NF2* mutations were not identified in lung cancers, pheochromocytomas, hepatocellular carcinomas, astrocytomas, or osteosarcomas, despite the fact that LOH for chromosome 22 is a feature of several of these tumor types.[14,53,58,86,98,110] However, *NF2* mutations have been reproducibly identified in cell lines established from malignant mesotheliomas (~44%), highly invasive tumors of the pleural lung lining that are firmly associated with asbestos exposure.[14,110]

## THE *NF1* AND *NF2* GENE PRODUCTS

### THE *NF1* GENE PRODUCT, NEUROFIBROMIN

The identification of the *NF1* and *NF2* genes has spawned intense investigation into the molecular function of their products. The *NF1* gene is a very large gene (>300 kb, 8.7-kb coding region) and encodes a correspondingly very large (2818 aa; 220–280 kDa) protein product, dubbed neurofibromin.[11,72,147] It was immediately recognized that a central domain of neurofibromin shared striking homology to a family of guanosine triphosphatase (GTPase) activating proteins (RasGAPs) that function molecularly to down-regulate the Ras signal transduction pathway,[18,72,147] providing a testable hypothesis for the molecular mechanism by which NF1 functions as a tumor suppressor.

Hailed as the first known oncogenes, the activated forms of the H-*ras* and K-*ras* genes were originally described as transformation-inducing sequences identified via gene transfer assays and as the transforming sequences carried by the Harvey and Kirsten sarcoma viruses.[24] The discovery of their identity as mutated forms of cellular protooncogenes quickly led to the determination that these genes were commonly mutated in human cancer; it is currently believed that *ras* gene mutations play a role in nearly 30% of all human cancers.[16] RasGAP proteins normally interact specifically with GTP-bound Ras, functioning to accelerate the intrinsic GTPase activity of wild-type Ras proteins, but not mutant (activated/oncogenic) forms, thus promoting the existence of the guanosine diphosphate-(GDP)bound, inactive forms (Figure 12.1). While the central domain of neurofibromin shares significant amino acid similarity with the catalytic domain of mammalian p120RasGAP, it shares similarity with the yeast *IRA1* and *IRA2* genes over its much greater entire length (15%; Figure 12.2). In yeast, the Ras and cAMP pathways are interdependent; IRA1 and IRA2 are critical negative regulators of both. Importantly, the GAP-related domain (GRD) of human NF1 can functionally complement the loss of *IRA1* or *IRA2* function in yeast.[8,146] In addition, like the p120GAP and IRA proteins, the NF1 GRD accelerates GTP hydrolysis of wild-type but not oncogenic Ras proteins.[8,73,146] These data suggest that the growth-suppressive function of neurofibromin lies in its ability to down-regulate Ras signaling. Interestingly, an alternatively spliced exon has been identified which disrupts the GRD and reduces the GAP activity of neurofibromin.[3] Although both forms can be detected in most tissues the function of this alternative NF1 isoform is unknown.

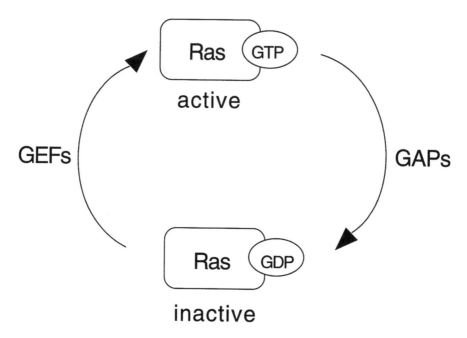

**FIGURE 12.1**   The NF1 gene product neurofibromin is a member of the RasGAP family of molecules that function to down-regulate Ras signaling. Ras proteins cycle between an active, GTP-bound state and an inactive, GDP-bound state. The intrinsic GTPase activity of Ras is accelerated by GTPase activating proteins (GAPs) such as neurofibromin and p120RasGAP, thereby inactivating Ras. In contrast, exchange of GTP for GDP is facilitated by guanine nucleotide exchange proteins (GEFs), which serve to activate Ras.

A corellate of this hypothesis, that elevated levels of GTP-bound Ras should be detected in NF1-deficient tumors has been confirmed in some instances and not in others. For example, neurofibrosarcoma cell lines derived from NF1 patient tumors exhibited elevated levels of RasGTP, while NF1-deficient neuroblastoma or melanoma cell lines did not. Although *NF1* mutations can be identified in each of these neural-crest-derived tumors, NF1 patients are not predisposed to developing neuroblastoma or melanoma.[10,15,27,57,66,129] In general, activating Ras mutations are not commonly found in neural crest tumors. These discrepencies may be explained by the levels of other GAPs functioning in the cell; perhaps only a subset of Ras proteins are being critically acted upon by NF1 (see below). Alternatively, these results may point to a growth-suppressive function for NF1 that is unrelated to its function as a RasGAP.

In contrast, the loss of neurofibromin was found to correlate well with elevated RasGTP levels in NF1-associated myeloid leukemias. Utilizing a specific inhibitor of NF1 GAP activity, dodecyl maltoside (DDM), Bollag et al. determined that neurofibromin function accounts for 20 to 30% of the total GAP activity in bone marrow cells.[15] Although the level of total GAP activity in those cells was unchanged, NF1-GAP activity was found to be specifically reduced in NF1-deficient leukemia samples and total RasGTP levels were found to be consistently elevated by approximately 10%. Activating *ras* mutations are common in childhood leukemias but are conspicuously missing in those of NF1 children, suggesting that in a myeloid precurser the consequences of an activating *ras* mutation and an inactivating *NF1* mutation are the same.[112]

## *NF1* HOMOLOGS IN OTHER SPECIES

It is apparent that the study of signaling molecules in both mammalian systems and lower organisms leads to a synergy in our understanding of the molecular pathways involved. Not only are signaling networks in lower organisms significantly reduced in complexity, but the genetic manipulability of

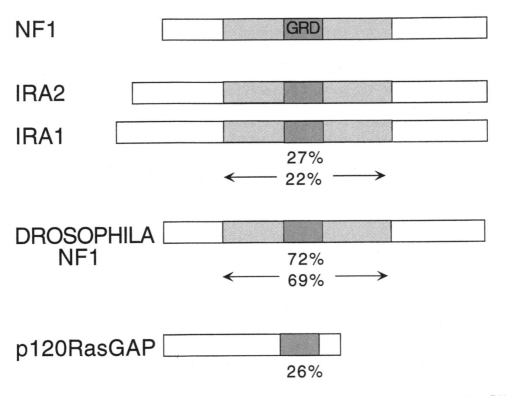

**FIGURE 12.2** In addition to a central GAP-related domain (GRD; dark gray boxes), shared by other GAPs such as p120RasGAP, neurofibromin has extended amino- and carboxy-terminal domains of unknown function that are conserved in the yeast and *Drosophila* homologs (light gray boxes).

lower organisms renders the identification and mutation of individual components of those networks relatively easy. Indeed, important insight has recently emerged through the identification and study of a *NF1* homolog in the fruitfly *Drosophila melanogaster*. The *Drosophila* NF1 homolog shares 60% amino acid identity with its human counterpart across its entire length (Figure 12.2).[128]

Unlike their murine counterparts, flies with a homozygous null mutation in NF1 are viable and fertile yet 20 to 25% smaller at all postembryonic stages than their wild-type or heterozygous counterparts.[128] As determined by a number of criteria, this phenotype does not appear to be due to the perturbation of Ras signaling in *Drosophila*; instead, it is rescued by transgenic overexpression of cAMP-dependent protein kinase (protein kinase A [PKA]). Moreover, *Drosophila* NF1 mutants exhibit a defect in the electrical response of the neuromuscular junction that appears to result from blocked activation of adenylyl cyclase, which catalyzes the production of cAMP.[39] Interestingly, the yeast NF1 homologs, IRA1 and IRA2 are known to participate in Ras-dependent activation of adenylyl cyclase, although adenylyl cyclase signaling is quite different in yeast.[130] In light of these observations, it is interesting to note that murine Nf1-deficient Schwann cells are hypersensitive to forskolin, an activator of adenylyl cyclase.[60] Taken together, these results suggest that in addition to its known biochemical ability to down-regulate Ras signaling, NF1 may also mediate cAMP-PKA signaling, either acting upstream of adenylyl cyclase activation or in a parallel pathway.

The genetic manipulability of *Drosophila* allows the development of genetic screens designed to identify other components of the signaling pathway in which a particular protein functions. So called modifier screens aim to identify mutations in other genes that either enhance or suppress the phenotype(s) of a given mutant allele. It will be of great interest to utilize genetic screens to identify mutations that modify the *NF1* mutant phenotypes. In addition, the identification and study of NF1 homologs in other species will undoubtedly furnish additional information. It is interesting to note that, despite the advanced status of determination of the entire sequence of the genome of the roundworm *Caenorhabditis elegans*, no NF1 homolog has been identified to date.

## THE *NF2* GENE PRODUCT, MERLIN

The identification of the *NF2* gene and determination of the sequence of its encoded protein also provided some clues as to the molecular mechanism by which it functions as a tumor suppressor.[96,132] The *NF2* gene product belongs to the ezrin/radixin/moesin (ERM) family of cytoskeleton-membrane linking proteins, a subfamily of the band 4.1 superfamily. Originally identified through disparate avenues as components of the cortical actin cytoskeleton, ERM proteins participate in cytoskeletal remodeling, such as during cell motility and cell-cell and cell-extracellular matrix adhesion.[134-136] Thus merlin occupies an intriguing physical niche for a tumor suppressor; a role for the cytoskeleton-membrane interface has long been postulated to be important in cell transformation and malignancy, but little direct evidence for this has been achieved.

Evidence for a relationship between merlin and the ERM proteins is suggested by their amino acid similarity and colocalization within the cell. The ERM proteins localize to areas of cortical actin remodeling such as membrane ruffles, microvilli, and the cleavage furrow.[134-136] Merlin has been shown to similarly localize to dynamic cortical actin structures, particularly membrane ruffles.[28,38,101,115] The ERM proteins associate directly with actin via a carboxy-terminal (C-terminal) domain and to membrane partners such as CD44, hNHE-RF/EBP50, ICAM-2, and PKA via an amino-terminal (N-terminal) domain.[29,45,92,133] CD44 is a transmembrane receptor for hyaluronic acid whose function has been linked to metastatic progression of human cancers,[61,84] hNHE-RF/EBP50 is a cofactor for the $Na^+/H^+$ exchanger, and ICAM-2 is a member of the integrin family of cell adhesion receptors.[47,141] The function of an interaction with these proteins is presently unclear. While both CD44 and hNHE-RF/EBP50 can bind to the N-terminal portion of merlin *in vitro*, the C-terminal actin binding domain is not conserved in merlin.[81,101] Interestingly, tissue-specific alternative splicing produces an alternative NF2 C-terminus (isoform II) that also lacks an obvious actin binding domain, but whose predicted secondary structure is quite distinct from that of isoform I.[50,91] Recently, βII-spectrin (fodrin) has been shown to interact specifically with the isoform II C-terminus of merlin *in vitro* and *in vivo*.[106] Spectrin family members bind to components of the cytoskeleton, including actin, suggesting that merlin isoform II association with the cytoskeleton may occur indirectly through an interaction with βII-spectrin.

Phosphorylation of the ERM proteins and merlin occurs upon various stimuli. The ERM proteins are phosphorylated on both serine/threonine and tyrosine residues, while merlin is phosphorylated only on serine/threonine residues.[116,134] It has been suggested that merlin, like the ERM proteins, can exist in an intramolecular/monomeric or intermolecular/homo- or hetero-oligomeric head-to-tail association.[13,34,35,117] This is thought to be the inactive state of the ERMs; phosphorylation and possibly phosphatidylinositol (4,5)-phosphate (PIP2) binding may relieve this inactive conformation, allowing interaction with cytoskeletal and membrane binding partners (Figure 12.3).[134-136] In contrast, the unphosphorylated form of merlin is associated with various forms of growth arrest, and therefore with its active, growth-suppressive function.[116] The levels of total NF2 protein, particularly the unphosphorylated form, are elevated upon increasing confluence, loss of adhesion, and serum deprivation; serum restimulation causes a rapid (~5 min) decrease in merlin levels, including a nearly complete loss of the unphosphorylated form. It has been suggested that merlin isoform I, but not isoform II, forms an intramolecular head-to-tail conformation, and that isoform I and not II can suppress the growth of a rat schwannoma cell line, consistent with the idea that head-to-tail association is required for this function. It will be interesting to further examine the distinct functions of isoforms I and II and to determine whether either isoform can oligomerize with other ERM family members.

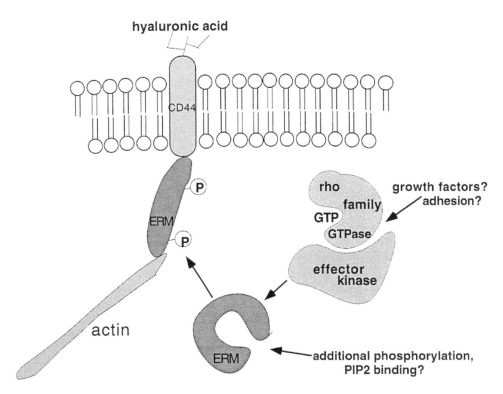

**FIGURE 12.3** ERM proteins can exist in an inactive, intramolecular/monomeric or intermolecular/oligo-meric (not shown) head-to-tail conformation. Phosphorylation, and possibly PIP2 binding, is thought to facilitate a conformational change that unmasks the amino- and carboxy-termini and allows interaction with cytoskeletal proteins such as actin and membrane proteins such as CD44.

Several lines of evidence link ERM function to signaling pathways controlled by the Rho GTPases, a subfamily of the Ras superfamily. The prototypes of this family, RhoA, Rac and Cdc42Hs, effect specific forms of cytoskeletal remodeling in response to various stimuli. The function of these proteins has been linked to such cellular activities as adhesion, migration, and transformation.[43,137] Recently, the ERM proteins have been identified as substrates for Rho kinase, an effector of Rho (see Figure 12.3).[74] Activation of RhoA, but not Rac or Cdc42Hs, affects phosphorylation and relocalization of the ERM proteins.[114] In addition, a requirement for the ERM proteins in Rho- and Rac-dependent actin remodeling in permeabilized cells has been demon-strated.[70] Finally, it has been reported that *in vitro* the ERM proteins can bind directly to both RhoGDI and Dbl, negative and positive regulators of RhoA, respectively.[124,125] Whether or not merlin is similarly a substrate for a Rho family effector kinase is presently unknown.

At the cellular level, increasing evidence points to a role for the ERM proteins and merlin in cellular adhesion and migration. Studies utilizing antisense oligonucleotides to reduce the expres-sion of the ERMs individually or in combination support a role for these proteins in cell-cell and cell-substrate adhesion and in cellular morphology.[126] Similar experimentation designed to eliminate *Nf2* expression in Schwann-like and glioma cells also points to a function for merlin in maintaining cell morphology and adhesion.[51] In addition, ezrin function is required for hepatocyte growth factor (HGF)-induced motility of epithelial cells.[25] Little additional information exists concerning the cell biological function of merlin. However, initial characterization of primary Nf2-deficient mouse embryo fibroblasts supports a role for merlin in cellular adhesion and motility (A. McClatchey, R. Shaw, and T. Jacks unpublished results).

## *NF2* Homologs in Other Species

Both a merlin homolog (*Dmerlin*) and a single ERM homolog, dubbed *Dmoesin*, have been identified in *Drosophila melanogaster* (Figure 12.4).[75] Overlapping, yet distinct expression and localization patterns in *Drosophila* cells and tissues suggest that the functions of Dmerlin and Dmoesin may be different. Notably, Dmerlin is apparently rapidly internalized from its membrane localization into punctate structures that are likely to be endocytic vesicles. Somatic mosaic analysis, which allows the examination of homozygous mutant clones in an otherwise heterozygous mutant tissue, revealed that loss of Dmerlin function led to cell-autonomous hyperproliferation.[65] In addition, the use of *in vitro* mutagenesis, combined with either immunofluorescent assessment of localization or transgenic rescue of Dmerlin mutant phenotypes *in vivo*, led to the generation of both activated and "dominant-interfering" mutant forms of Dmerlin, valuable tools for further examination of the molecular function of merlin *in vitro* and *in vivo*. The simplicity of a single *Drosophila* ERM homolog, together with the utility of genetic screens designed to identify members of the pathway(s) in which Dmerlin and Dmoesin act, render this a valuable system for studying the function of the NF2 tumor suppressor.

Although a merlin homolog has been identified in *C. elegans*, information concerning its expression, localization, or function is lacking (www.sanger.ac.uk/projects/c_elegans/).

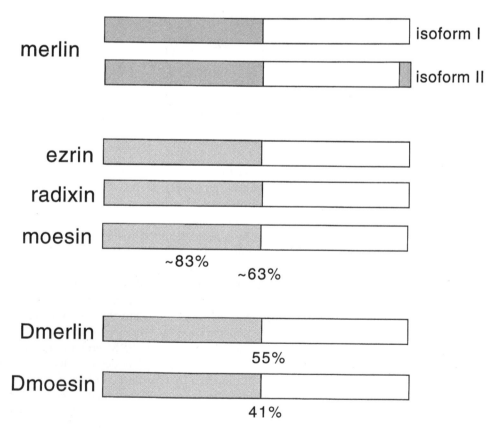

**FIGURE 12.4**   The NF2 protein, merlin, belongs to the ERM subfamily of the band 4.1 family of cytoskeleton-membrane linkers. These proteins share similarity primarily through their amino-terminal halves (gray boxes). The percent amino acid identity between merlin and the amino-terminal half as well as the entire length of the ERMs is depicted. In addition, identity between mammalian merlin and the *Drosophila* merlin and moesin homologs is shown. The mouse and human merlin proteins are 98% identical.

# ANIMAL MODELS OF NF1 AND NF2

## ANIMAL MODELS OF NF1

The occurrence of neurofibromas elsewhere in the animal kingdom is not common, although two naturally occuring animal models of NF1 merit mention and have been studied at some length.[94] First, several Holstein cows that were the offspring of a single sire developed symptoms consistent with NF1, including neurofibromas, one of which progressed to malignancy, and ocular lesions of the iris.[102] While extended genetic analysis was not feasible, the identification of a polymorphism in the bovine *NF1* gene allowed confirmation that the affected cows inherited the same paternal allele at the *NF1* locus. Second, certain bicolored damselfish of southern Florida coral reefs develop multiple neurofibromas, neurofibrosarcomas, and neoplasms of pigmented cells (chromatophoromas) in a syndrome resembling NF1, dubbed DNF.[105] Interestingly, the disease is apparently transmitted by an as yet uncharacterized type C retrovirus.[104] The characterization of this retrovirus and the mechanism by which it causes DNF, coupled with the further characterization of cell lines derived from damselfish tumors, should provide interesting clues toward a molecular understanding of human NF1.

Several models of neurofibroma, neurofibrosarcoma, and/or schwannoma formation have been induced in rodents via chemical carcinogen, retrovirus, or overexpression of tumor virus onco-genes.[94] For example, administration of *N*-nitroso-*N*-ethylurea (NEU) transplacentally to rats or Syrian golden hamsters results in the development of peripheral nerve tumors resembling NF1-associated neurofibromas.[20,82] Interestingly, an early and perhaps initiating event in the formation of these tumors appears to be mutational activation of the *neu* protooncogene (erbB-2), which encodes a receptor for glial growth factor (GGF), a potent Schwann cell mitogen.[83,87] Although *neu* mutations were not identified in human Schwann cell tumors, these observations suggest that activation of the GGF receptor pathway in Schwann cells may be an important mechanism for achieving Schwann cell neoplasia.

Overexpression of the *tax* gene of the human T-lymphotrophic virus type 1 (HTLV-1) under the control of its viral regulatory element (long terminal repeat [LTR]) in transgenic mice induces nerve sheath tumors that may be analogous to NF1-associated neurofibromas and neurofibrosarco-mas.[46] A satisfying explanation for this may reside in the observation that tax functions to repress *Nf1* expression via a cis-acting regulatory element in the *NF1* promoter.[32] Similarly, transgenic mice expressing the SV40 T-antigen in the neural crest lineage develop tumors characterized as schwannomas that may be more like NF2-associated tumors.[56,80] While these models may in fact recapitulate the etiology of NF tumors, histological comparisons of the tumors arising in each of these models have not been sought, and our understanding of species-specific differences is poor, particularly given the complex composition of neurofibromas and the difficulty in distinguishing between neurofibromas, schwannomas, and their malignant counterparts.

The technique of targeted mutagenesis in the mouse has revolutionized the utility of the mouse as a mammalian system to model and study human disease. In particular, mice engineered to carry a heterozygous tumor suppressor mutation genetically mimic humans with familial cancer syn-dromes such as NF1 and NF2. The manipulation of cancer-prone strains of mice in order to better model various forms of human cancer, identify other genetic alterations that participate in cancer development, and as vehicles for the delivery of either therapeutic or environmentally toxic agents is an extremely powerful way to model the study and treatment of human cancer.[76] These models allow the study of the etiology of cancer development from its earliest stages, the study of the normal (often developmental) function of a given tumor suppressor, and the generation of primary cell lines specifically lacking a particular tumor suppressor gene product.

The mouse *Nf1* homolog is highly conserved; the mouse and human NF1 gene products share 98% amino acid identity over their entire 2818 amino acid lengths,[12] predicting similar functional mechanisms. Animals carrying a heterozygous mutation at the *Nf1* locus are cancer prone; however, their phenotype does not closely resemble that of their human counterparts.[17,52] These animals are predisposed to the development of cancer late in life, but they do not develop neurofibromas or Lisch nodules, the hallmarks of human NF1. However, like their human counterparts, *Nf1* heterozygous mice develop pheochromocytomas and myeloid leukemia at rates much higher (~10%) than their wild-type siblings (<<1%). More important, loss of the wild-type *Nf1* allele occurs frequently in the tumors that arise in these animals, confirming the identity of the *Nf1* gene product as a tumor suppressor in the mouse. Moreover, a large percentage of *Nf1+/−* mice have been reported to exhibit learning and memory impairments that are similar to those of NF1 patients, raising the possibility that these mice represent a faithful system for the study of the neurological symptoms of NF1.[120]

Several explanations for the difference in tumor spectra in *NF1+/−* mice and humans are possible, including: (1) significant differences in the rate of loss of the wild-type allele; (2) differences in the ability of other proteins (perhaps family members) to compensate for the loss of Nf1 or Nf2 function; and (3) syntenic differences between the mouse and human genome that affect the rates of mutational events required for tumor formation to occur. The analysis of several strains of tumor suppressor mutant mice has provided evidence for each of these scenarios. In an effort to address the first possibility, chimeric mice composed of both wild-type and Nf1-deficient cells were generated. Many of these chimeric mice develop multiple neurofibroma-like lesions (S. Shih, K. Cichowski, and T. Jacks, unpublished results). Importantly, in this system the *Nf1−/−* cells were marked with a *LacZ* transgene, allowing their ancestors to be traced in the developing embryo and in the adult. It will be of great interest to determine which cell type in these tumors consistently expresses *LacZ* (and is therefore *Nf1−/−*).

The *NF1* heterozygous mutant mouse strain has been the focus of much study, due to the phenotypic overlap with the tumor spectra of human NF1 patients and the likelihood that tumor development in these mice is at least in part due to activation of the Ras pathway, which is activated in a large percentage of human cancers. For example, to develop a better model for NF1-associated leukemia (approximately 10% of *Nf1+/−* mice develop leukemia at 12 to 24 months of age),[52] fetal liver cells from 12.5-day-old (E12.5) *Nf1−/−* embryos were used to reconstitute the immune system of lethally irradiated hosts.[66] All of these animals developed a chronic myeloproliferative disease that shares many features with JMML. Myeloid precursers from donor *Nf1−/−* fetal livers and the reconstituted immune systems of transplanted hosts are hypersensitive to granulocyte-macrophage colony stimulating factor (GM-CSF), a cytokine that supports the proliferation of myeloid precursers *in vivo*. Moreover, serum-starved *Nf1−/−* myeloid cells exhibit a significantly elevated and prolonged activation of Ras following GM-CSF stimulation, perhaps modeling the molecular basis for myeloproliferative disease in young NF1 patients.

Therapy-related myeloid leukemia (t-ML) is an increasingly common and severe complication of chemotherapeutic treatment with alkylating or topoisomerase II-inhibiting agents.[68] Recently, Mahgoub et al. investigated whether treatment with these agents could cooperate with a *Nf1* mutation in predisposing mice to the development of myeloid leukemia.[71] They found that the alkylating agent cyclophosphomide, but not the topoisomerase II inhibitor etoposide, significantly accelerated the development of myeloid leukemia in *Nf1+/−* mice. These results hold obvious ramifications for the consideration of chemotherapeutic choices for NF1 patients. Moreover, the availability of a mouse model of t-ML will allow the investigation of the mechanism whereby these chemotherapeutic agents interact with genetic susceptibility to induce leukemia development.

Genetically engineered mouse models of Nf1-deficient solid tumors and leukemias are logical and powerful vehicles in which to test therapeutic agents that target the inactivation of the Ras pathway, such as the recently developed farnesyl transferase inhibitors (FTIs).[36] Farnesylation of Ras proteins promotes their translocation to the membrane, a requirement for proper interaction with downstream substrates and subsequent activation of Ras-driven signal transduction. FTIs block the farnesylation of Ras (and other cellular substrates) thereby blocking Ras signaling *in vitro* and *in vivo*. Initial experiments designed to test the efficacy of FTIs in NF1-deficient tumors found that an FTI reversibly inhibited the growth of a malignant schwannoma cell line from an NF1 patient.[149] Investigations are undoubtedly underway to examine the effects of FTI treatment of solid tumor and leukemia models derived from *Nf1* mutant mice.

The availability of genetically engineered mice as mouse models of cancer development also allows an investigation of cooperation between cancer predisposing mutations. For example, a mouse model of neuroblastoma has recently been developed; transgenic overexpression of the *MYCN* protooncogene in the neural crest lineage reproducibly leads to the formation of neuro-blastoma.[142] The frequent loss of NF1 function in neuroblastoma cell lines[129] suggests that a heterozygous *Nf1* mutation might cooperate with the *MYCN* transgene in neuroblastoma formation or the development of some other tumor type. Indeed, *MYCN;Nf1+/−* mice developed neuroblastoma with a significantly reduced latency.[142]

Although the identity of many tumor suppressor genes has been revealed through the study of germline heterozygous mutations in humans, their normal growth-suppressive function is often manifest during embryogenesis. The utility of targeted mouse models is underscored by the ability to generate animals that are constitutively homozygous for a mutation in a given gene, something that rarely occurs naturally in the human population. Intercrossing of *Nf1* heterozygous mutant animals revealed that NF1 function is essential for normal embryonic development in the mouse. Embryos carrying a homozygous *Nf1* mutation fail at midgestation due to a failure of the cardiac outflow tracts to septate properly during late stages of cardiac morphogenesis, an abnormality that occurs with some frequency in human infants.[17,52] The defect may be caused by defective neural crest cells which migrate to and populate the outflow tracts. In addition, hyperplasia of neural crest-derived sympathetic ganglion has been described in *Nf1−/−* embryos.[17] Ongoing studies aimed at further understanding the cellular defects in these embryos and analyzing the biochemical properties of the defective cells *in vitro* will provide valuable clues toward understanding NF1 function. Interestingly, targeted deletion of exon 23A, a widely expressed alternatively used exon that reduces the GAP activity of neurofibromin, is compatible with normal mouse embryogenesis (T. Yang and C. Brannan, unpublished results). It will be interesting to examine the tumor spectrum of these mice in comparison to that of the null mutation.

Genetically engineered mice are a valuable source of primary cells that specifically lack the expression of a particular gene of interest. The behavior of these cells can be compared with that of similarly derived wild-type and heterozygous mutant cells. The characterization of a number of Nf1-deficient cell types has been undertaken, demonstrating the powerful use of these tools to help dissect the mechanism of NF1 function. For example, in an effort to identify the defective cell type contributing to neurofibroma formation, the properties of *Nf1−/−* neurons, Schwann cells, and fibro-blasts have been studied in comparison to their +/− and +/+ counterparts. Interestingly, all three cell types were found to behave abnormally in the absence of Nf1 function. Nf1-deficient neurons display neurotrophin-independent survival, suggesting a mechanism for imbalanced homeostasis and increased numbers of neurons during embryonic and perhaps neurofibroma development.[139] In addition, Nf1-deficient Schwann cells exhibit more invasive and angiogenic behavior than their Nf1-expressing counterparts but decreased proliferation in response to GGF, a mitogen for normal Schwann cells.[60] *Nf1−/−* Schwann cells are also dramatically hypersensitive to forskolin treatment, an agent that activates PKA. In cocultures of neurons, Schwann cells, and perineurial fibroblasts, Nf1-deficient perineurial cells fail to form an intact perineurium.[95] Defects in all three cell types may suggest that more than one cell type contributes to tumor formation. Interestingly, intermediate phenotypes were often exhibited by heterozygous mutant cells; perhaps a combination of *Nf1* LOH and haploinsufficiency contributes to the complex nature of neurofibroma formation.

## ANIMAL MODELS OF NF2

There are only eight amino acid differences between the mouse and human NF2 proteins, suggesting that the molecular function of merlin has also been stringently conserved (98%).[42] Targeted disruption of the mouse *Nf2* locus has revealed surprisingly broad consequences of the loss of Nf2 function in tumorigenesis. Heterozygous mice do not develop schwannomas but are predisposed to developing tumors later in life, particularly osteosarcomas, fibrosarcomas, and hepatocellular carcinomas.[77] Importantly, nearly all of the osteo- and fibrosarcomas and a large fraction of the hepatocellular carcinomas exhibit loss of the wild-type *Nf2* allele, confirming the function of Nf2 as a tumor suppressor in the mouse. Interestingly, whereas the tumors that develop in NF2 patients are benign in nature, loss of Nf2 in the mouse leads to the development of malignant and highly metastatic tumors. Given the low rate of spontaneous metastasis in mice in general, this argues that loss of Nf2 function contributes to metastatic potential in these mice. Preliminary experimental evidence does support a role for Nf2 loss in mouse metastasis.[77] These observations suggest that this family of proteins may participate more broadly in human tumorigenesis than previously thought; although some studies have evaluated NF2 expression in cancers not associated with NF2, none of these studies examined the *NF2* locus during tumor progression. Moreover, little information exists concerning the status of the ERM proteins in any type of human cancer. It is interesting to note that overexpression of ezrin has been identified in a mouse sarcoma cell line and in some human tumor cell lines.[31] Moreover, a link between ezrin overexpression and loss of contact inhibition in fibroblasts has been established.[59]

Intercrossing of *Nf2*[+/−] animals established that *Nf2* function is essential for normal mouse development. *Nf2*[−/−] embryos fail immediately prior to gastrulation due to a defect in the extraembryonic lineage.[78] The study of chimeric embryos composed of *LacZ*[+];*Nf2*[−/−] and *LacZ*[−];*Nf2*[+/+] cells identified additional developmental roles for Nf2 including early cardiac development. *Nf2*[−/−] myocardial cells hyperproliferate and exhibit altered morphology, forming lesions in the embryonic wall (A. McClatchey, unpublished results). In fact, *LacZ*[+]; *Nf2*[−/−] myocardial cells invariably form hyperproliferative lesions, indicating that loss of merlin is sufficient for their formation. The nature of these lesions is reminiscent of the benign tumors in NF2 patients; in fact, similar *Nf2*-deficient lesions have been seen in the developing nervous system (A. McClatchey, unpublished results). These data suggest that loss of *Nf2* can result in distinct consequences depending on the cell type, or perhaps the timing of the loss. In some circumstances, *Nf2* loss may initiate hyperproliferation and in others promote tumor progression. Further analysis of the developmental abnormalities in these embryos and characterization of *Nf2*[−/−] cells derived from them will be a powerful way to explore this further.

The embryological defects associated with a homozygous *Nf2* mutation preclude the use of chimeric analysis to determine whether loss of the wild-type allele is rate limiting for schwannoma formation in the mouse. The ability to conditionally target the inactivation of individual genes in a tissue-specific or temporally defined manner will allow the generation of animals in which the *Nf2* gene has been inactivated specifically in Schwann cells.[100] However, an investigation of genetic interaction between an *Nf2* mutation and other tumor suppressor gene mutations has been undertaken (Ref. 77; A. McClatchey, unpublished results). Exploiting the fact that in contrast to humans, the *Nf2*, *Nf1,* and *p53* tumor suppressor loci are syntenic on mouse chromosome 11, mice carrying paired *Nf2;p53* and *Nf2;Nf1* heterozygous mutations both in *trans* (on opposite chromosomes 11) and in *cis* (on the same chromosome 11) were generated. The *cis* configuration had a dramatic effect on the rate to tumorigenesis: both *Nf2*[+/−];*p53*[+/−] and *Nf2*[+/−];*Nf1*[+/−] *cis* animals developed tumors with a strikingly reduced latency compared to either parental strain alone or to mice carrying the same mutations in *trans*. This suggests that linkage between two cancer predisposing mutations (or potentially modifying loci) can dramatically alter the rates to tumorigenesis. Perhaps differences in the configuration of the mouse and human genomes account for much of the discrepancy in tumor spectra of mouse cancer models. Interestingly, while *Nf2*[+/−];*p53*[+/−] *cis* mice rapidly developed multiple osteosarcomas, a tumor that develops in either parental strain with longer latency, *Nf2*[+/−];*Nf1*[+/−] *cis* mice developed a tumor type not seen in either parental genotype. The *Nf2*[+/−];*Nf1*[+/−] *cis* mice consistently developed a tumor that shares many features with malignant nerve sheath tumors — a feature of human NF1 (A. McClatchey and T. Jacks, unpublished results). Although an accurate diagnosis will require a detailed histological and electron microscopic evaluation as well as comparison to other rodent nerve sheath tumor models, these animals may not only represent a model for the study of NF-associated neoplasia, but also provide a point of intersection for the study of the molecular pathways underlying NF1 and NF2.

## FUTURE DIRECTIONS

The cloning of the *NF1* and *NF2* genes provided immediate hope for the development of effective treatments. However, the path toward that goal remains long despite the knowledge of the underlying defect. An understanding of the molecular consequences of mutation of the *NF1* and *NF2* genes and the loss of their encoded products is critical to further progress. Progress in this area has and will continue to come through several avenues. The generation and study of mutant cells *in vitro* will be critical toward this goal. In this context, the existence of animal models as sources of primary mutant cells is invaluable. Further investigation of the cellular defects associated with NF1 and NF2 deficiency during embryogenesis will provide an important window into the functions that these proteins evolved to carry out. Important insight will undoubtedly arise through the facility of genetic dissection of the NF1 and NF2 signaling pathways in other species. The delineation of these biochemical pathways is a critical prerequisite toward developing targeted therapeutic strategies. The fine tuning of mouse models to better recapitulate the etiology of NF1 and NF2 and serve as vehicles in which to test such strategies provides a sound basis for achieving the ultimate goal of NF treatment in humans.

## REFERENCES

1. Abeliovich, D., Gelman-Kohan, Z., Silverstein, S., Lerer, I., Chemke, J., Merin, S., and Zlotogora, J., Familial cafe au lait spots: a variant of neurofibromatosis type 1, *J. Med. Genet.*, 32, 985–6, 1995.
2. Ahlbom, B. E., Dahl, N., Zetterqvist, P., and Anneren, G., Noonan syndrome with cafe-au-lait spots and multiple lentigines syndrome are not linked to the neurofibromatosis type 1 locus, *Clin. Genet.*, 48, 85–9, 1995.
3. Andersen, L. B., Ballester, R., Marchuk, D. A., Chang, E., Gutmann, D. H., Saulino, A. M., Camonis, J., Wigler, M., and Collins, F. S., A conserved alternative splice in the von Recklinghausen neurofibromatosis (NF1) gene produces two neurofibromin isoforms, both of which have GTPase-activating protein activity, *Mol. Cell. Biol.*, 13, 487–95, 1993.
4. Andersen, L. B., Fountain, J. W., Gutmann, D. H., Tarle, S. A., Glover, T. W., Dracopoli, N. C., Housman, D. E., and Collins, F. S., Mutations in the neurofibromatosis 1 gene in sporadic malignant melanoma cell lines, *Nat. Genet.*, 3, 118–21, 1993.
5. Arakawa, H., Hayashi, N., Nagase, H., Ogawa, M., and Nakamura, Y., Alternative splicing of the NF2 gene and its mutation analysis of breast and colorectal cancers, *Hum. Mol. Genet.*, 3, 565–8, 1994.
6. Arico, M., Biondi, A., and Pui, C. H., Juvenile myelomonocytic leukemia [see comments], *Blood*, 90, 479–88, 1997.
7. Bahuau, M., Flintoff, W., Assouline, B., Lyonnet, S., Le Merrer, M., Prieur, M., Guilloud-Bataille, M., Feingold, N., Munnich, A., Vidaud, M., and Vidaud, D., Exclusion of allelism of Noonan syndrome and neurofibromatosis-type 1 in a large family with Noonan syndrome-neurofibromatosis association, *Am. J. Med. Genet.*, 66, 347–55, 1996.
8. Ballester, R., Marchuk, D., Boguski, M., Saulino, A., Letcher, R., Wigler, M., and Collins, F., The NF1 locus encodes a protein functionally related to mammalian GAP and yeast IRA proteins, *Cell*, 63, 851–9, 1990.
9. Barker, D., Wright, E., Nguyen, K., Cannon, L., Fain, P., Goldgar, D., Bishop, D. T., Carey, J., Baty, B., Kivlin, J. et al., Gene for von Recklinghausen neurofibromatosis is in the pericentromeric region of chromosome 17, *Science*, 236, 1100–2, 1987.

10. Basu, T. N., Gutmann, D. H., Fletcher, J. A., Glover, T. W., Collins, F. S., and Downward, J., Aberrant regulation of ras proteins in malignant tumour cells from type 1 neurofibromatosis patients [see comments], *Nature*, 356, 713–5, 1992.

11. Bernards, A., Haase, V. H., Murthy, A. E., Menon, A., Hannigan, G. E., and Gusella, J. F., Complete human NF1 cDNA sequence: two alternatively spliced mRNAs and absence of expression in a neuroblastoma line, *DNA Cell. Biol.*, 11, 727–34, 1992.

12. Bernards, A., Snijders, A. J., Hannigan, G. E., Murthy, A. E., and Gusella, J. F., Mouse neurofibromatosis type 1 cDNA sequence reveals high degree of conservation of both coding and non-coding mRNA segments, *Hum. Mol. Genet.*, 2, 645–50, 1993.

13. Berryman, M., Gary, R., and Bretscher, A., Ezrin oligomers are major cytoskeletal components of placental microvilli: a proposal for their involvement in cortical morphogenesis, *J. Cell. Biol.*, 131, 1231–42, 1995.

14. Bianchi, A. B., Hara, T., Ramesh, V., Gao, J., Klein-Szanto, A. J., Morin, F., Menon, A. G., Trofatter, J. A., Gusella, J. F., Seizinger, B. R. et al., Mutations in transcript isoforms of the neurofibromatosis 2 gene in multiple human tumour types, *Nat. Genet.*, 6, 185–92, 1994.

15. Bollag, G., Clapp, D. W., Shih, S., Adler, F., Zhang, Y. Y., Thompson, P., Lange, B. J., Freedman, M. H., McCormick, F., Jacks, T., and Shannon, K., Loss of NF1 results in activation of the Ras signaling pathway and leads to aberrant growth in haematopoietic cells, *Nat. Genet.*, 12, 144–8, 1996.

16. Bos, J. L., ras oncogenes in human cancer: a review [published erratum appears in *Cancer Res.* 50(4), 1352, 1990], *Cancer Res.*, 49, 4682–9, 1989.

17. Brannan, C. I., Perkins, A. S., Vogel, K. S., Ratner, N., Nordlund, M. L., Reid, S. W., Buchberg, A. M., Jenkins, N. A., Parada, L. F., and Copeland, N. G., Targeted disruption of the neurofibromatosis type-1 gene leads to developmental abnormalities in heart and various neural crest-derived tissues [published erratum appears in *Genes Dev.*, 8(22), 2792, 1994], *Genes Dev.*, 8, 1019–29, 1994.

18. Buchberg, A. M., Cleveland, L. S., Jenkins, N. A., and Copeland, N. G., Sequence homology shared by neurofibromatosis type-1 gene and IRA-1 and IRA-2 negative regulators of the RAS cyclic AMP pathway [see comments], *Nature*, 347, 291–4, 1990.

19. Cappione, A. J., French, B. L., and Skuse, G. R., A potential role for NF1 mRNA editing in the pathogenesis of NF1 tumors [see comments], *Am. J. Hum. Genet.*, 60, 305–12, 1997.

20. Cardesa, A., Ribalta, T., Von Schilling, B., Palacin, A., and Mohr, U., Experimental model of tumors associated with neurofibromatosis, *Cancer*, 63, 1737–49, 1989.

21. Cawthon, R. M., Weiss, R., Xu, G. F., Viskochil, D., Culver, M., Stevens, J., Robertson, M., Dunn, D., Gesteland, R., O'Connell, P. et al., A major segment of the neurofibromatosis type 1 gene: cDNA sequence, genomic structure, and point mutations [published erratum appears in *Cell*, 62(3), 608, 1990], *Cell*, 62, 193–201, 1990.

22. Charrow, J., Listernick, R., and Ward, K., Autosomal dominant multiple cafe-au-lait spots and neurofibromatosis-1: evidence of non-linkage, *Am. J. Med. Genet.*, 45, 606–8, 1993.

23. Colman, S. D., Williams, C. A., and Wallace, M. R., Benign neurofibromas in type 1 neurofibromatosis (NF1) show somatic deletions of the NF1 gene, *Nat. Genet.*, 11, 90–2, 1995.

24. Cooper, G., *Oncogenes*, Jones and Bartlett, London, 1995.

25. Crepaldi, T., Gautreau, A., Comoglio, P. M., Louvard, D., and Arpin, M., Ezrin is an effector of hepatocyte growth factor-mediated migration and morphogenesis in epithelial cells, *J. Cell. Biol.*, 138, 423–34, 1997.

26. Daschner, K., Assum, G., Eisenbarth, I., Krone, W., Hoffmeyer, S., Wortmann, S., Heymer, B., and Kehrer-Sawatzki, H., Clonal origin of tumor cells in a plexiform neurofibroma with LOH in NF1 intron 38 and in dermal neurofibromas without LOH of the NF1 gene, *Biochem. Biophys. Res. Commun.*, 234, 346–50, 1997.

27. DeClue, J. E., Papageorge, A. G., Fletcher, J. A., Diehl, S. R., Ratner, N., Vass, W. C., and Lowy, D. R., Abnormal regulation of mammalian p21ras contributes to malignant tumor growth in von Recklinghausen (type 1) neurofibromatosis, *Cell*, 69, 265–73, 1992.

28. den Bakker, M. A., Riegman, P. H., Hekman, R. A., Boersma, W., Janssen, P. J., van der Kwast, T. H., and Zwarthoff, E. C., The product of the NF2 tumour suppressor gene localizes near the plasma membrane and is highly expressed in muscle cells, *Oncogene*, 10, 757–63, 1995.

29. Dransfield, D. T., Bradford, A. J., Smith, J., Martin, M., Roy, C., Mangeat, P. H., and Goldenring, J. R., Ezrin is a cyclic AMP-dependent protein kinase anchoring protein, *EMBO J.*, 16, 35–43, 1997.

30. Eshagian, J., Human posterior subcapsular cataracts, *Trans. Ophthalmol. Soc. U.K.,* 102(3), 364–8, 1982.

31. Fazioli, F., Wong, W. T., Ullrich, S. J., Sakaguchi, K., Appella, E., and Di Fiore, P. P., The ezrin-like family of tyrosine kinase substrates: receptor-specific pattern of tyrosine phosphorylation and relationship to malignant transformation, *Oncogene,* 8, 1335–45, 1993.

32. Feigenbaum, L., Fujita, K., Collins, F. S., and Jay, G., Repression of the NF1 gene by Tax may expain the development of neurofibromas in human T-lymphotropic virus type 1 transgenic mice, *J. Virol.,* 70, 3280–5, 1996.

33. Ferner, R., Clinical aspects of neurofibromatosis 1, in *Neurofibromatosis Type 1: From Genotype to Phenotype,* Upadhyaya, M. and Cooper, D., Eds., BIOS Scientific Publishers, Oxford, 1998, 21–34.

34. Gary, R. and Bretscher, A., Ezrin self-association involves binding of an N-terminal domain to a normally masked C-terminal domain that includes the F-actin binding site, *Mol. Biol. Cell.,* 6, 1061–75, 1995.

35. Gary, R. and Bretscher, A., Heterotypic and homotypic associations between ezrin and moesin, two putative membrane-cytoskeletal linking proteins, *Proc. Natl. Acad. Sci. U.S.A.,* 90, 10846–50, 1993.

36. Gibbs, J. B., Oliff, A., and Kohl, N. E., Farnesyltransferase inhibitors: Ras research yields a potential cancer therapeutic, *Cell,* 77, 175–8, 1994.

37. Glover, T. W., Stein, C. K., Legius, E., Andersen, L. B., Brereton, A., and Johnson, S., Molecular and cytogenetic analysis of tumors in von Recklinghausen neurofibromatosis, *Genes Chromosomes Cancer,* 3, 62–70, 1991.

38. Gonzalez-Agosti, C., Xu, L., Pinney, D., Beauchamp, R., Hobbs, W., Gusella, J., and Ramesh, V., The merlin tumor suppressor localizes preferentially in membrane ruffles, *Oncogene,* 13, 1239–47, 1996.

39. Guo, H. F., The, I., Hannan, F., Bernards, A., and Zhong, Y., Requirement of Drosophila NF1 for activation of adenylyl cyclase by PACAP38-like neuropeptides, *Science,* 276, 795–8, 1997.

40. Gusella, J. F., Ramesh, V., MacCollin, M., and Jacoby, L. B., Neurofibromatosis 2: loss of merlin's protective spell, *Curr. Opin. Genet. Dev.,* 6, 87–92, 1996.

41. Gutmann, D. H., Aylsworth, A., Carey, J. C., Korf, B., Marks, J., Pyeritz, R. E., Rubenstein, A., and Viskochil, D., The diagnostic evaluation and multidisciplinary management of neurofibromatosis 1 and neurofibromatosis 2 [see comments], *JAMA,* 278, 51–7, 1997.

42. Haase, V. H., Trofatter, J. A., MacCollin, M., Tarttelin, E., Gusella, J. F., and Ramesh, V., The murine NF2 homologue encodes a highly conserved merlin protein with alternative forms, *Hum. Mol. Genet.,* 3, 407–11, 1994.

43. Hall, A., Rho GTPases and the actin cytoskeleton, *Science,* 279, 509–14, 1998.

44. Heim, R. A., Kam-Morgan, L. N., Binnie, C. G., Corns, D. D., Cayouette, M. C., Farber, R. A., Aylsworth, A. S., Silverman, L. M., and Luce, M. C., Distribution of 13 truncating mutations in the neurofibromatosis 1 gene, *Hum. Mol. Genet.,* 4, 975–81, 1995.

45. Helander, T. S., Carpen, O., Turunen, O., Kovanen, P. E., Vaheri, A., and Timonen, T., ICAM-2 redistributed by ezrin as a target for killer cells, *Nature,* 382, 265–8, 1996.

46. Hinrichs, S. H., Nerenberg, M., Reynolds, R. K., Khoury, G., and Jay, G., A transgenic mouse model for human neurofibromatosis, *Science,* 237, 1340–3, 1987.

47. Howe, A., Aplin, A. E., Alahari, S. K., and Uliano, R. L., Integrin signaling and cell growth control, *Curr. Opin. Cell. Biol.,* 10, 220–31, 1998.

48. Huson, S., Neurofibromatosis type 1: historical perspective and clinical overview, in *Neurofibromatosis Type 1: From Genotype to Phenotype,* Upadhyaya, M. and Cooper, D., Eds., BIOS Scientific Publishing, Oxford, 1998, 1–20.

49. Huson, S. M., Neurofibromatosis 2: Clinical features, genetic counseling and management issues, in *The Neurofibromatoses: A Clinical and Practical Overview,* Huson, S. M. and Hughes, R. A. C., Eds., Chapman and Hall Medical Press, London, 1994, 211–233.

50. Huynh, D. P., Nechiporuk, T., and Pulst, S. M., Alternative transcripts in the mouse neurofibromatosis type 2 (NF2) gene are conserved and code for schwannomins with distinct C-terminal domains, *Hum. Mol. Genet.,* 3, 1075–9, 1994.

51. Huynh, D. P. and Pulst, S. M., Neurofibromatosis 2 antisense oligodeoxynucleotides induce reversible inhibition of schwannomin synthesis and cell adhesion in STS26T and T98G cells, *Oncogene,* 13, 73–84, 1996.

52. Jacks, T., Shih, T. S., Schmitt, E. M., Bronson, R. T., Bernards, A., and Weinberg, R. A., Tumour predisposition in mice heterozygous for a targeted mutation in Nf1, *Nat. Genet.*, 7, 353–61, 1994.

53. Jacoby, L., Personal communication.

54. Jacoby, L. B., Jones, D., Davis, K., Kronn, D., Short, M. P., Gusella, J., and MacCollin, M., Molecular analysis of the NF2 tumor-suppressor gene in schwannomatosis, *Am. J. Hum. Genet.*, 61, 1293–302, 1997.

55. Jacoby, L. B., MacCollin, M., Barone, R., Ramesh, V., and Gusella, J. F., Frequency and distribution of NF2 mutations in schwannomas, *Genes Chromosomes Cancer*, 17, 45–55, 1996.

56. Jensen, N. A., Rodriguez, M. L., Garvey, J. S., Miller, C. A., and Hood, L., Transgenic mouse model for neurocristopathy: Schwannomas and facial bone tumors, *Proc. Natl. Acad. Sci. U.S.A.*, 90, 3192–6, 1993.

57. Johnson, M. R., Look, A. T., DeClue, J. E., Valentine, M. B., and Lowy, D. R., Inactivation of the NF1 gene in human melanoma and neuroblastoma cell lines without impaired regulation of GTP.Ras, *Proc. Natl. Acad. Sci. U.S.A.*, 90, 5539–43, 1993.

58. Kanai, Y., Tsuda, H., Oda, T., Sakamoto, M., and Hirohashi, S., Analysis of the neurofibromatosis 2 gene in human breast and hepatocellular carcinomas, *Jpn. J. Clin. Oncol.*, 25, 1–4, 1995.

59. Kaul, S. C., Mitsui, Y., Komatsu, Y., Reddel, R. R., and Wadhwa, R., A highly expressed 81 kDa protein in immortalized mouse fibroblast: its proliferative function and identity with ezrin, *Oncogene*, 13, 1231–7, 1996.

60. Kim, H. A., Ling, B., and Ratner, N., Nf1-deficient mouse Schwann cells are angiogenic and invasive and can be induced to hyperproliferate: reversion of some phenotypes by an inhibitor of farnesyl protein transferase, *Mol. Cell. Biol.*, 17, 862–72, 1997.

61. Kincade, P. W., Zheng, Z., Katoh, S., and Hanson, L., The importance of cellular environment to function of the CD44 matrix receptor, *Curr. Opin. Cell. Biol.*, 9, 635–42, 1997.

62. Knudson, A. G., Antioncogenes and human cancer, *Proc. Natl. Acad. Sci. U.S.A.*, 90, 10914–21, 1993.

63. Knudson, A. G., Jr., Mutation and cancer: statistical study of retinoblastoma, *Proc. Natl. Acad. Sci. U.S.A.*, 68, 820–3, 1971.

64. Korf, B. R., The NF1 analysis consortium, in *Neurofibromatosis Type I: From Genotype to Phenotype*, Upadhyaya, M. and Cooper, D. N., Eds., BIOS Scientific Publishing, Oxford, 1998, 57–63.

65. LaJeunesse, D. R., McCartney, B. M., and Fehon, R. G., Structural analysis of Drosophila merlin reveals functional domains important for growth control and subcellular localization, *J. Cell. Biol.*, 141, 1589–99, 1998.

66. Largaespada, D. A., Brannan, C. I., Jenkins, N. A., and Copeland, N. G., Nf1 deficiency causes Ras-mediated granulocyte/macrophage colony stimulating factor hypersensitivity and chronic myeloid leukaemia, *Nat. Genet.*, 12, 137–43, 1996.

67. Legius, E., Marchuk, D. S., Collins, F. S., and Glover, T. W., Somatic deletion of the neurofibromatosis type 1 gene in a neurofibrosarcoma supports a tumour suppressor gene hypothesis, *Nat. Genet.*, 3, 122–6, 1993.

68. Levine, E. G. and Bloomfield, C. D., Leukemias and myelodysplastic syndromes secondary to drug, radiation, and environmental exposure, *Semin. Oncol.*, 19, 47–84, 1992.

69. Li, Y., Bollag, G., Clark, R., Stevens, J., Conroy, L., Fults, D., Ward, K., Friedman, E., Samowitz, W., Robertson, M. et al., Somatic mutations in the neurofibromatosis 1 gene in human tumors, *Cell*, 69, 275–81, 1992.

70. Mackay, D. J., Esch, F., Furthmayr, H., and Hall, A., Rho- and rac-dependent assembly of focal adhesion complexes and actin filaments in permeabilized fibroblasts: an essential role for ezrin/radixin/moesin proteins, *J. Cell. Biol.*, 138, 927–38, 1997.

71. Mahgoub, N., Taylor, B., Beau, M. L., Gratiot, M., Carlson, K., Jacks, T., and Shannon, K., A mouse model of alkylator-induced leukemia, *Blood*, 90, 385a, 1997.

72. Marchuk, D. A., Saulino, A. M., Tavakkol, R., Swaroop, M., Wallace, M. R., Andersen, L. B., Mitchell, A. L., Gutmann, D. H., Boguski, M., and Collins, F. S., cDNA cloning of the type 1 neurofibromatosis gene: complete sequence of the NF1 gene product, *Genomics*, 11, 931–40, 1991.

73. Martin, G. A., Viskochil, D., Bollag, G., McCabe, P. C., Crosier, W. J., Haubruck, H., Conroy, L., Clark, R., O'Connell, P., Cawthon, R. M. et al., The GAP-related domain of the neurofibromatosis type 1 gene product interacts with ras p21, *Cell*, 63, 843–9, 1990.

74. Matsui, T., Maeda, M., Doi, Y., Yonemura, S., Amano, M., Kaibuchi, K., Tsukita, S., and Tsukita, S., Rho-kinase phosphorylates COOH-terminal threonines of ezrin/radixin/moesin (ERM) proteins and regulates their head-to-tail association, *J. Cell. Biol.,* 140, 647–57, 1998.

75. McCartney, B. M. and Fehon, R. G., Distinct cellular and subcellular patterns of expression imply distinct functions for the Drosophila homologues of moesin and the neurofibromatosis 2 tumor suppressor, merlin, *J. Cell. Biol.,* 133, 843–52, 1996.

76. McClatchey, A. I. and Jacks, T., Tumor suppressor mutations in mice: the next generation, *Curr. Opin. Genet. Dev.,* 8, 304–10, 1998.

77. McClatchey, A. I., Saotome, I., Mercer, K., Crowley, D., Gusella, J. F., Bronson, R. T., and Jacks, T., Mice heterozygous for a mutation at the Nf2 tumor suppressor locus develop a range of highly metastatic tumors, *Genes Dev.,* 12, 1121–33, 1998.

78. McClatchey, A. I., Saotome, I., Ramesh, V., Gusella, J. F., and Jacks., T., The Nf2 tumor suppressor gene product is essential for extraembryonic development immediately prior to gastrulation, *Genes Dev.,* 11, 1253–65, 1997.

79. Menon, A. G., Anderson, K. M., Riccardi, V. M., Chung, R. Y., Whaley, J. M., Yandell, D. W., Farmer, G. E., Freiman, R. N., Lee, J. K., Li, F. P. et al., Chromosome 17p deletions and p53 gene mutations associated with the formation of malignant neurofibrosarcomas in von Recklinghausen neurofibromatosis, *Proc. Natl. Acad. Sci. U.S.A.,* 87, 5435–9, 1990.

80. Messing, A., Behringer, R. R., Wrabetz, L., Hammang, J. P., Lemke, G., Palmiter, R. D., and Brinster, R. L., Hypomyelinating peripheral neuropathies and schwannomas in transgenic mice expressing SV40 T-antigen, *J. Neurosci.,* 14, 3533–9, 1994.

81. Murthy, A., Gonzalez-Agosti, C., Cordero, E., Pinney, D., Candia, C., Solomon, F., Gusella, J., and Ramesh, V., NHE-RF, a regulatory cofactor for Na(+)-H+ exchange, is a common interactor for merlin and ERM (MERM) proteins, *J. Biol. Chem.,* 273, 1273–6, 1998.

82. Nakamura, T., Hara, M., and Kasuga, T., Transplacental induction of peripheral nervous tumor in the Syrian golden hamster by N-nitroso-N-ethylurea. A new animal model for von Recklinghausen's neurofibromatosis, *Am. J. Pathol.,* 135, 251–9, 1989.

83. Nakamura, T., Ushijima, T., Ishizaka, Y., Nagao, M., Nemoto, T., Hara, M., and Ishikawa, T., neu proto-oncogene mutation is specific for the neurofibromas in a N- nitroso-N-ethylurea-induced hamster neurofibromatosis model but not for hamster melanomas and human Schwann cell tumors, *Cancer Res.,* 54, 976–80, 1994.

84. Naot, D., Sionov, R. V., and Ish-Shalom, D., CD44: structure, function, and association with the malignant process, *Adv. Cancer Res.,* 71, 241–319, 1997.

85. Neurofibromatosis, N. C. D. C., Conference statement. *Arch. Neurol.,* 45, 575–578, 1988.

86. Ng, H. K., K. M. Lau, J. Y. Tse, K. W. Lo, J. H. Wong, W. S. Poon, and D. P. Huang, Combined molecular genetic studies of chromosome 22q and the neurofibromatosis type 2 gene in central nervous system tumors, *Neurosurgery,* 37, 764–73, 1995.

87. Nikitin, A., Ballering, L. A., Lyons, J., and Rajewsky, M. F., Early mutation of the neu (erbB-2) gene during ethylnitrosourea-induced oncogenesis in the rat Schwann cell lineage, *Proc. Natl. Acad. Sci. U.S.A.,* 88, 9939–43, 1991.

88. Parry, D. M., MacCollin, M. M., Kaiser-Kupfer, M. I., Pulaski, K., Nicholson, H. S., Bolesta, M., Eldridge, R., and Gusella, J. F., Germ-line mutations in the neurofibromatosis 2 gene: correlations with disease severity and retinal abnormalities, *Am. J. Hum. Genet.,* 59, 529–39, 1996.

89. Preiser, S. and Davenport, C., Multiple neurofibromatosis (von Recklinghausen disease) and its inheritance, *Am. J. Med. Sci.,* 156, 507–41, 1918.

90. Pulst, S. M., Rouleau, G. A., Marineau, C., Fain, P., and Sieb, J. P., Familial meningioma is not allelic to neurofibromatosis 2, *Neurology,* 43, 2096–8, 1993.

91. Pykett, M. J., Murphy, M., Harnish, P. R., and George, D. L., The neurofibromatosis 2 (NF2) tumor suppressor gene encodes multiple alternatively spliced transcripts, *Hum. Mol. Genet.,* 3, 559–64, 1994.

92. Reczek, D., Berryman, M., and Bretscher, A., Identification of EBP50: A PDZ-containing phosphoprotein that associates with members of the ezrin-radixin-moesin family, *J. Cell. Biol.,* 139, 169–79, 1997.

93. Riccardi, V., Neurofibromatosis: clinical heterogeneity, *Curr. Problems Cancer,* 7, 1–34, 1982.

94. Riccardi, V. M., Womack, J. E., and Jacks, T., Neurofibromatosis and related tumors. Natural occurrence and animal models, *Am. J. Pathol.,* 145, 994–1000, 1994.

95. Rosenbaum, T., Boissy, Y. L., Kombrinck, K., Brannan, C. I., Jenkins, N. A., Copeland, N. G., and Ratner, N., Neurofibromin-deficient fibroblasts fail to form perineurium in vitro, *Development*, 121, 3583–92, 1995.

96. Rouleau, G. A., Merel, P., Lutchman, M., Sanson, M., Zucman, J., Marineau, C., Hoang-Xuan, K., Demczuk, S., Desmaze, C., Plougastel, B. et al., Alteration in a new gene encoding a putative membrane-organizing protein causes neuro-fibromatosis type 2 [see comments], *Nature*, 363, 515–21, 1993.

97. Rouleau, G. A., Wertelecki, W., Haines, J. L., Hobbs, W. J., Trofatter, J. A., Seizinger, B. R., Martuza, R. L., Superneau, D. W., Conneally, P. M., and Gusella, J. F., Genetic linkage of bilateral acoustic neurofibromatosis to a DNA marker on chromosome 22, *Nature*, 329, 246–8, 1987.

98. Rubio, M. P., Correa, K. M., Ramesh, V., MacCollin, M., Jacoby, L. B., von Deimling, A., Gusella, J. F., and Louis, D. N., Analysis of the neurofibromatosis 2 gene in human ependymomas and astrocytomas, *Cancer Res.*, 54, 45–7, 1994.

99. Ruttledge, M. H., Andermann, A. A., Phelan, C. M., Claudio, J. O., Han, F. Y., Chretien, N., Ranga-ratnam, S., MacCollin, M., Short, P., Parry, D., Michels, V., Riccardi, V. M., Weksberg, R., Kitamura, K., Bradburn, J. M., Hall, B. D., Propping, P., and Rouleau, G. A., Type of mutation in the neurofi-bromatosis type 2 gene (NF2) frequently determines severity of disease, *Am. J. Hum. Genet.*, 59, 331–42, 1996.

100. Saez, E., No, D., West, A., and Evans, R. M., Inducible gene expression in mammalian cells and transgenic mice, *Curr. Opin. Biotechnol.*, 8, 608–16, 1997.

101. Sainio, M., Zhao, F., Heiska, L., Turunen, O., den Bakker, M., Zwarthoff, E., Lutchman, M., Rouleau, G. A., Jaaskelainen, J., Vaheri, A., and Carpen, O., Neurofibromatosis 2 tumor suppressor protein colocalizes with ezrin and CD44 and associates with actin-containing cytoskeleton, *J. Cell. Sci.*, 110, 2249–60, 1997.

102. Sartin, E. A., Doran, S. E., Riddell, M. G., Herrera, G. A., Tennyson, G. S., D'Andrea, G., Whitley, R. D., and Collins, F. S., Characterization of naturally occurring cutaneous neurofibromatosis in Holstein cattle. A disorder resembling neurofibromatosis type 1 in humans, *Am. J. Pathol.*, 145, 1168–74, 1994.

103. Sawada, S., Florell, S., Purandare, S. M., Ota, M., Stephens, K., and Viskochil, D., Identification of NF1 mutations in both alleles of a dermal neurofibroma, *Nat. Genet.*, 14, 110–2, 1996.

104. Schmale, M. C., Aman, M. R., and Gill, K. A., A retrovirus isolated from cell lines derived from neurofibromas in bicolor damselfish (Pomacentrus partitus), *J. Gen. Virol.*, 77: 1181-7, 1996.

105. Schmale, M. C. and Hensley, G. T., Transmissibility of a neurofibromatosis-like disease in bicolor damselfish, *Cancer Res.*, 48, 3828–33, 1988.

106. Scoles, D. R., Huynh, D. P., Morcos, P. A., Coulsell, E. R., Robinson, N. G., Tamanoi, F., and Pulst, S. M., Neurofibromatosis 2 tumour suppressor schwannomin interacts with betaII-spectrin, *Nat. Genet.*, 18, 354–9, 1998.

107. Seizinger, B. R., Martuza, R. L., and Gusella, J. F., Loss of genes on chromosome 22 in tumorigenesis of human acoustic neuroma, *Nature*, 322, 644–7, 1986.

108. Seizinger, B. R., Rouleau, G. A., Lane, A. H., Farmer, G., Ozelius, L. J., Haines, J. L., Parry, D. M., Korf, B. R., Pericak-Vance, M. A., Faryniarz, A. G. et al., Linkage analysis in von Recklinghausen neurofibromatosis (NF1) with DNA markers for chromosome 17, *Genomics*, 1, 346–8, 1987.

109. Seizinger, B. R., Rouleau, G. A., Ozelius, L. J., Lane, A. H., Faryniarz, A. G., Chao, M. V., Huson, S., Korf, B. R., Parry, D. M., Pericak-Vance, M. A. et al., Genetic linkage of von Recklinghausen neurofibromatosis to the nerve growth factor receptor gene, *Cell*, 49, 589–94, 1987.

110. Sekido, Y., Pass, H. I., Bader, S., Mew, D. J., Christman, M. F., Gazdar, A. F., and Minna, J. D., Neurofibromatosis type 2 (NF2) gene is somatically mutated in mesothelioma but not in lung cancer, *Cancer Res.*, 55, 1227–31, 1995.

111. Serra, E., Puig, S., Otero, D., Gaona, A., Kruyer, H., Ars, E., Estivill, X., and Lazaro, C., Confirmation of a double-hit model for the NF1 gene in benign neurofibromas, *Am. J. Hum. Genet.*, 61, 512–9, 1997.

112. Shannon, K. M., O'Connell, P., Martin, G. A., Paderanga, D., Olson, K., Dinndorf, P., and McCormick, F., Loss of the normal NF1 allele from the bone marrow of children with type 1 neurofibromatosis and malignant myeloid disorders [see comments], *N. Engl. J. Med.*, 330, 597–601, 1994.

113. Sharland, M., Taylor, R., Patton, M. A., and Jeffery, S., Absence of linkage of Noonan syndrome to the neurofibromatosis type 1 locus, *J. Med. Genet.*, 29, 188–90, 1992.

114. Shaw, R. J., Henry, M., Solomon, F., and Jacks, T., RhoA-dependent phosphorylation and relocalization of ERM proteins into apical membrane/actin protrusions in fibroblasts, *Mol. Biol. Cell.*, 9, 403–19, 1998.

115. Shaw, R. J., McClatchey, A. I., and Jacks, T., Localization and functional domains of the neurofibromatosis type II tumor suppressor, merlin, *Cell Growth Differ.*, 9, 287–96, 1998.

116. Shaw, R. J., McClatchey, A. I., and Jacks, T., Regulation of the neurofibromatosis type 2 tumor suppressor protein, merlin, by adhesion and growth arrest stimuli, *J. Biol. Chem.*, 273, 7757–64, 1998.

117. Sherman, L., Xu, H. M., Geist, R. T., Saporito-Irwin, S., Howells, N., Ponta, H., Herrlich, P., and Gutmann, D. H., Interdomain binding mediates tumor growth suppression by the NF2 gene product, *Oncogene*, 15, 2505–9, 1997.

118. Shimizu, E., Shinohara, T., Mori, N., Yokota, J., Tani, K., Izumi, K., Obashi, A., and Ogura, T., Loss of heterozygosity on chromosome arm 17p in small cell lung carcinomas, but not in neurofibromas, in a patient with von Recklinghausen neurofibromatosis, *Cancer*, 71, 725–8, 1993.

119. Side, L., Taylor, B., Cayouette, M., Conner, E., Thompson, P., Luce, M., and Shannon, K., Homozygous inactivation of the NF1 gene in bone marrow cells from children with neurofibromatosis type 1 and malignant myeloid disorders, *N. Engl. J. Med.*, 336, 1713–20, 1997.

120. Silva, A. J., Frankland, P. W., Marowitz, Z., Friedman, E., Lazlo, G., Cioffi, D., Jacks, T., and Bourtchuladze, R., A mouse model for the learning and memory deficits associated with neurofibromatosis type I, *Nat. Genet.*, 15, 281–4, 1997.

121. Skuse, G. R., Cappione, A. J., Sowden, M., Metheny, L. J., and Smith, H. C., The neurofibromatosis type I messenger RNA undergoes base-modification RNA editing, *Nucleic Acids Res.*, 24, 478–85, 1996.

122. Skuse, G. R., Kosciolek, B. A., and Rowley, P. T., The neurofibroma in von Recklinghausen neurofibromatosis has a unicellular origin, *Am. J. Hum. Genet.*, 49, 600–7, 1991.

123. Stemmer-Rachamimov, A. O., Xu, L., Gonzalez-Agosti, C., Burwick, J. A., Pinney, D., Beauchamp, R., Jacoby, L. B., Gusella, J. F., Ramesh, V., and Louis, D. N., Universal absence of merlin, but not other ERM family members, in schwannomas, *Am. J. Pathol.*, 151, 1649–54, 1997.

124. Takahashi, K., Sasaki, T., Mammoto, A., Hotta, I., Takaishi, K., Imamura, H., Nakano, K., Kodama, A., and Takai, Y., Interaction of radixin with Rho small G protein GDP/GTP exchange protein Dbl, *Oncogene*, 16, 3279–84, 1998.

125. Takahashi, K., Sasaki, T., Mammoto, A., Takaishi, K., Kameyama, T., Tsukita, S., and Takai, Y., Direct interaction of the Rho GDP dissociation inhibitor with ezrin/radixin/moesin initiates the activation of the Rho small G protein, *J. Biol. Chem.*, 272, 23371–5, 1997.

126. Takeuchi, K., Sato, N., Kasahara, H., Funayama, N., Nagafuchi, A., Yonemura, S., Tsukita, S., and Tsukita, S., Perturbation of cell adhesion and microvilli formation by antisense oligonucleotides to ERM family members, *J. Cell. Biol.*, 125, 1371–84, 1994.

127. Tassabehji, M., Strachan, T., Sharland, M., Colley, A., Donnai, D., Harris, R., and Thakker, N., Tandem duplication within a neurofibromatosis type 1 (NF1) gene exon in a family with features of Watson syndrome and Noonan syndrome, *Am. J. Hum. Genet.*, 53, 90–5, 1993.

128. The, I., Hannigan, G. E., Cowley, G. S., Reginald, S., Zhong, Y., Gusella, J. F., Hariharan, I. K., and Bernards, A., Rescue of a Drosophila NF1 mutant phenotype by protein kinase A, *Science*, 276, 791–4, 1997.

129. The, I., Murthy, A. E., Hannigan, G. E., Jacoby, L. B., Menon, A. G., Gusella, J. F., and Bernards, A., Neurofibromatosis type 1 gene mutations in neuroblastoma, *Nat. Genet.*, 3, 62–6, 1993.

130. Toda, T., Uno, I., Ishikawa, T., Powers, S., Kataoka, T., Broek, D., Cameron, S., Broach, J., Matsumoto, K., and Wigler, M., In yeast, RAS proteins are controlling elements of adenylate cyclase, *Cell*, 40, 27–36, 1985.

131. Tonsgard, J. H., Yelavarthi, K. K., Cushner, S., Short, M. P., and Lindgren, V., Do NF1 gene deletions result in a characteristic phenotype?, *Am. J. Med. Genet.*, 73, 80–6, 1997.

132. Trofatter, J. A., MacCollin, M. M., Rutter, J. L., Murrell, J. R., Duyao, M. P., Parry, D. M., Eldridge, R., Kley, N., Menon, A. G., Pulaski, K. et al., A novel moesin-, ezrin-, radixin-like gene is a candidate for the neurofibromatosis 2 tumor suppressor, *Cell*, 75, 826, 1993.

133. Tsukita, S., Oishi, K., Sato, N., Sagara, J., Kawai, A., and Tsukita, S., ERM family members as molecular linkers between the cell surface glycoprotein CD44 and actin-based cytoskeletons, *J. Cell. Biol.*, 126, 391–401, 1994.

134. Tsukita, S. and Yonemura, S., ERM (ezrin/radixin/moesin) family: from cytoskeleton to signal transduction, *Curr. Opin. Cell. Biol.,* 9, 70–5, 1997.

135. Tsukita, S., Yonemura, S., and Tsukita, S., ERM proteins: head-to-tail regulation of actin-plasma membrane interaction, *Trends Biochem. Sci.,* 22, 53–8, 1997.

136. Vaheri, A., Carpen, O., Heiska, L., Helander, T. S., Jaaskelainen, J., Majander-Nordenswan, P., Sainio, M., Timonen, T., and Turunen, O., The ezrin protein family: membrane-cytoskeleton interactions and disease associations, *Curr. Opin. Cell. Biol.,* 9, 659–66, 1997.

137. Van Aelst, L. and D'Souza-Schorey, C., Rho GTPases and signaling networks, *Genes Dev.,* 11, 2295–322, 1997.

138. Viskochil, D., Buchberg, A. M., Xu, G., Cawthon, R. M., Stevens, J., Wolff, R. K., Culver, M., Carey, J. C., Copeland, N. G., Jenkins, N. A. et al., Deletions and a translocation interrupt a cloned gene at the neurofibromatosis type 1 locus, *Cell,* 62, 187–92, 1990.

139. Vogel, K. S., Brannan, C. I., Jenkins, N. A., Copeland, N. G., and Parada, L. F., Loss of neurofibromin results in neurotrophin-independent survival of embryonic sensory and sympathetic neurons, *Cell,* 82, 733–42, 1995.

140. Wallace, M. R., Marchuk, D. A., Andersen, L. B., Letcher, R., Odeh, H. M., Saulino, A. M., Fountain, J. W., Brereton, A., Nicholson, J., Mitchell, A. L. et al., Type 1 neurofibromatosis gene: identification of a large transcript disrupted in three NF1 patients [published erratum appears in *Science,* 250(4988), 1749], *Science,* 249, 181–6, 1990.

141. Weinman, E. J., Steplock, D., Wang, Y., and Shenolikar, S., Characterization of a protein cofactor that mediates protein kinase A regulation of the renal brush border membrane Na(+)-H+ exchanger, *J. Clin. Invest.,* 95, 2143–9, 1995.

142. Weiss, W. A., Aldape, K., Mohapatra, G., Feuerstein, B. G., and Bishop, J. M., Targeted expression of MYCN causes neuroblastoma in transgenic mice, *EMBO J.,* 16, 2985–95, 1997.

143. Wellenreuther, R., Kraus, J. A., Lenartz, D., Menon, A. G., Schramm, J., Louis, D. N., Ramesh, V., Gusella, J. F., Wiestler, O. D., and von Deimling, A., Analysis of the neurofibromatosis 2 gene reveals molecular variants of meningioma, *Am. J. Pathol.,* 146, 827–32, 1995.

144. Woodruff, J. M., Pathology of the major peripheral nerve sheath neoplasms, *Monogr. Pathol.,* 38, 129–61, 1996.

145. Wu, B. L., Austin, M. A., Schneider, G. H., Boles, R. G., and Korf, B. R., Deletion of the entire NF1 gene detected by the FISH: four deletion patients associated with severe manifestations, *Am. J. Med. Genet.,* 59, 528–35, 1995.

146. Xu, G. F., Lin, B., Tanaka, K., Dunn, D., Wood, D., Gesteland, R., White, R., Weiss, R., and Tamanoi, F., The catalytic domain of the neurofibromatosis type 1 gene product stimulates ras GTPase and complements ira mutants of S. cerevisiae, *Cell,* 63, 835–41, 1990.

147. Xu, G. F., O'Connell, P., Viskochil, D., Cawthon, R., Robertson, M., Culver, M., Dunn, D., Stevens, J., Gesteland, R., White, R. et al., The neurofibromatosis type 1 gene encodes a protein related to GAP, *Cell,* 62, 599–608, 1990.

148. Xu, W., Mulligan, L. M., Ponder, M. A., Liu, L., Smith, B. A., Mathew, C. G., and Ponder, B. A., Loss of NF1 alleles in phaeochromocytomas from patients with type I neurofibromatosis, *Genes Chromosomes Cancer,* 4, 337–42, 1992.

149. Yan, N., Ricca, C., Fletcher, J., Glover, T., Seizinger, B. R., and Manne, V., Farnesyltransferase inhibitors block the neurofibromatosis type I (NF1) malignant phenotype *Cancer Res.,* 55, 3569–75, 1995.

# 13 Dendritic Localization of mRNA and Local Protein Synthesis

*James Eberwine*

The subcellular transport of substances throughout the neuron facilitates neuronal functioning. A simple example of this is the transport of mRNA from the nucleus to the cytoplasm where translation takes place. These proteins function in various cellular regions depending on their cellular function. At least one class of these proteins, called transcription factors, is transported back into the nucleus where it alters the transcription rates of various genes. A more specialized function for protein transport arises from the functional polarity of a neuron. Neurons receive information about their cellular environment through their dendrites. This information is transduced into a cellular response in the cell body which in turn may be propagated to the next cell via signaling through the axon to the next postsynaptic neuron. This functional polarity results in part from protein transport as demonstrated by the ionotropic glutamate receptors which are translated in the cell body, transported to dendrites, and bind glutamate, which is secreted from the presynaptic neuron. This functional polarity is paralleled in the morphological polarity of neurons where a single neuron usually has multiple dendrites but only one axon.

How and when developing neurons are subdivided into axonal and dendritic domains remains to be determined, but it is hypothesized to be linked to the subcellular localization of several proteins including Map2, Tau, and GAP43 (Kanai and Hirokawa, 1995; Johnson and Jope, 1990; Goslin et al., 1990). Map2 protein is restricted to the somatodendritic regions of mature neurons (Bernhardt and Matus, 1984; Davis et al., 1987; Garner et al., 1988) and interacts with cytoskeletal elements such as tubulin (Johnson and Jope, 1990). Map2 protein appears in late embryogenesis after neurons migrate into the cortical plate, and whereas mRNA levels remain relatively constant, somatodendritic protein expression increases dramatically during dendritic arborization (Figure 13.1*) (Charriere-Bertrand et al., 1991). In contrast, GAP43 is a 46-kDa regulatory protein (Strittmatter et al., 1990) and protein kinase C substrate restricted to the axosomatic regions and is believed to play a central role in axonal growth cone dynamics (Goslin et al., 1990). In cultured chick primary sensory neurons, functional depletion of GAP43 by specific oligonucleotide antisense methodologies impairs axonal growth cone spreading, branching, adhesion, and pathfinding (Aigner and Caroni, 1993, 1995). Hippocampal neurons extend several processes within the first few hours in culture but one process will ultimately become an axon while the remaining arbors will mature into dendrites. This process typically occurs between 24 and 36 h in culture (Dotti et al., 1988; Baas et al., 1989) and is reflected by changes in the orientation of microtubules. These studies were performed using Map2 antibodies to monitor Map2 presence in dendrites and comparison of these immunostained processes with the unstained axon. The study by Dotti et al. showed that the axon grew at a much more rapid rate than the dendrites and was a landmark study in helping to morphologically distinguish axons from dendrites and providing a time frame in which this differentiation occurs. This study did not provide any insight into the trigger for axonal differentiation and its rapid outgrowth from the other neurites.

Dendrites serve not only as the cellular "receivers" of information but also as the initial modulators of postsynaptic responsiveness. Modulation occurs via several mechanisms, including

---

* All color plates appear after p. 272.

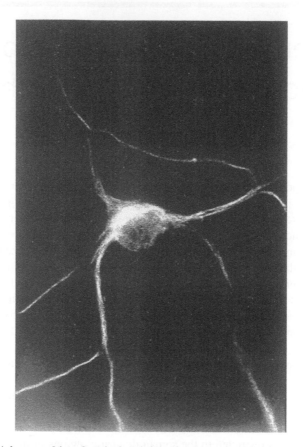

**FIGURE 13.1    Map2A immunohistochemical staining of primary rat hippocampal cells.** E18 embryonic rat hippocampal cells were immunostained with an antibody to microtubule associated protein 2 (Map2A). The secondary antibody was fluorescene labeled, hence the green fluorescence corresponding to Map2A protein. The dendrites are clearly visible as is the perinuclear cytoplasm of the neuronal soma. This staining pattern is known as somatodendritic staining. There is no Map2A in the nucleus of the cell hence the exclusion of nuclear staining. Not visible is the single axon of this neuron which also does not contain Map2A protein. (This photograph was generated by Dr. Stavros Therianos.)

(1) generation of action potentials, (2) influx of $Ca^{2+}$, (3) internalization of receptors, (4) regulation of the phosphorylation state of proteins, and (5) regulation of local protein synthesis.

This last point, regulation of local protein synthesis, suggests that the sorting and transport of mRNAs to the most distal portions of the dendritic arbor may play a key role in the ability of a neuron to respond to the microenvironment. Ideally, the localization of populations of mRNAs in neuronal processes in association with the cell's translational machinery would provide a way for a cell to locally synthesize proteins in response to synaptic signals (Steward and Levy, 1982; Steward and Falk, 1986; Steward and Reeves, 1988). If mRNA localization and local protein synthesis are involved in this process it would be expected that changes in the identity of localized mRNAs and their translational regulation would occur.

The question of which mRNAs are present in dendrites has until recently been hampered by technological limitations. The first two mRNAs that were localized to mammalian neuronal dendrites, Map2A (Kleiman et al., 1990) and CamKII (Burgin et al., 1990), were proposed to be involved in dendritic functioning. Map2A and CamKII mRNAs are relatively prevalent mRNAs that are easily detected by *in situ* hybridization methodologies. The dendritic localization of additional mRNAs was problematic because they are lower in abundance and may be masked by RNA binding proteins or contain a secondary structure that would inhibit hybridization. Using a

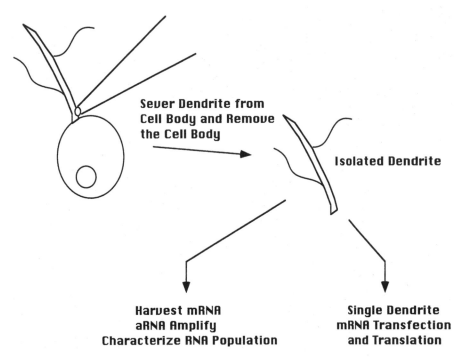

**FIGURE 13.2  Schematic of single-dendrite manipulations.** This schematic depicts the severing of individual dendrites from live cultured neurons by use of a patch pipette. The cell body is removed to eliminate the contribution of any mRNA or protein synthetic machinery to the analysis of dendritic composition. As depicted in the schematic, the single isolated dendrite can be used for either mRNA analysis or mRNA translation studies.

novel set of molecular biological techniques which allow amplification of the mRNA content of single cells, Miyashiro et al. (1994) showed that multiple mRNAs are present in dendrites. These novel techniques included single-dendrite aRNA amplification combined with expression profiling and the polymerase chain reaction (PCR). In these studies a single live dendrite was severed from a cultured primary hippocampal cell and the endogenous mRNA was amplified via the aRNA amplification procedure (a linear nucleic amplification procedure which preserves the relative abundances of mRNAs present in the starting material in the final amplification product) (see Figure 13.2 for schematic). This aRNA product was then characterized by expression profiling, PCR, and differential display. Data from this paper showed that a complex population of mRNAs exists in dendrites including the mRNAs for various ionotropic glutamate receptors.

This study was recently complemented by the demonstration of a complex population of mRNAs in dendritic growth cones (Crino and Eberwine, 1996). Growth cones are lamellipodia-like structures at the tips of immature dendrites that are thought to be involved in dendritic pathfinding and/or synaptogenesis. The localization of mRNAs to dendritic growth cones appears to be developmentally regulated since individual mRNAs were differentially localized as a function of dendritic maturation in culture. These results raise the intriguing possibility that dendritic localization of mRNAs during development, likely followed by local protein synthesis, may play a role in elaborating synaptic pathways in the central nervous system.

The question of regulated mRNA transport must be defined in two ways. The simplest way of examining this is to determine where there is more mRNA in the dendrite after cellular manipulation. This has been shown for some mRNAs; for example, ARC mRNA is increased in dendrites of hippocampal neurons after electroconvulsive shock (ECS) (Lyford et al., 1995). Simultaneously, there is an increase in ARC mRNA in the cell soma, suggesting a linkage between increases in cytoplasmic mRNA levels and those in the dendrite. Importantly, other mRNAs are also increased in the cell soma with ECS and show no concomitant increase in the dendrite, suggesting that ARC mRNA localization is indeed regulated. The coupling of mRNA increases with dendritic localization suggests that nuclear factors may be involved in mediating the dendritic localization of mRNA. A second and more subtle way of thinking about regulated mRNA transport into the dendrite is that perhaps a relative redistribution of mRNA from the cytoplasm to the dendrite (or vice versa) can occur independent of new mRNA synthesis. Redistribution of mRNA in this manner would provide a rapid mechanism for altering the relative abundances of individual mRNAs in dendrites, consequently rapidly changing the composition of the pool of mRNAs which can be locally translated. It is important to note that there are currently no published examples of this phenomenon.

The regulated movement of mRNA within cells has been hypothesized to occur via several mechanisms all of which involve association of the targeted mRNA with RNA binding proteins. While there are clearly mRNAs in mature neuronal processes and neuronal growth cones, the mechanism by which these mRNAs are targeted to these subcellular sites in the mammalian system is unclear. Much more progress has been made in elucidating the molecular mechanisms of RNA transport in simpler systems. In particular, the study of oogenesis in Drosophila and Xenopus have provided clues as to how mRNAs are localized to subcellular sites. For example, the mRNAs for bicoid (Drosophila; MacDonald and Struhl, 1988) and VG1 (Xenopus; Melton, 1987) have been shown to be transported to particular regions of the developing embryo. Bicoid encodes a homeodomain transcription factor which facilitates the transcription of several genes that are responsible for anterior body formation (Dreiver and Nusslein-Volhard, 1988). VG1 is a TGF-b family member that induces mesoderm formation in the oocyte (Weeks and Melton, 1987). The model that is favored for transport of these mRNAs is that a cis-acting regulatory region exists in the mRNAs which interacts with one or more RNA-binding proteins, forming a ribonucleoprotein (RNP) complex. This RNP complex is hypothesized to interact with the cytoskeleton which then traffics the RNP to the appropriate subcellular site.

In yet another system, Ainger et al. (1993) have suggested a unique transport mechanism for mRNA encoding myelin basic protein (MBP). MBP mRNA is transported to the periphery of oligodendrocyte through oligodendrocyte extensions which resemble neuronal dendritic processes. Upon injection of digoxigenin-labeled MBP mRNA into oligodendrocyte cytoplasm the mRNA appears to aggregate into granules, which are transported through the oligodendrocyte process. These granules are visualized by digoxigenin antibody staining of the fixed oligodendrocytes. While it is unclear how these granules form and how they may associate with the cytoskeleton, it is reasonable to assume that a cis-acting element in MBP mRNA must associate with an RNA binding protein for the granules to form.

The clear demonstration of mRNAs in dendrites suggests that they have a functional role in this subcellular domain. The work of Oswald Steward and William Greenough over the last three decades has strongly suggested that protein synthesis can occur in dendrites. Initially, nearly 30 years ago, electron microscopists were able to identify polysomal structures at the base of dendritic spines, suggesting that dendrites had the capacity for protein synthesis. Others have used tritiated amino acids to show that processes that are severed from their cell body can incorporate these amino acids into molecules which are retained in the process, suggesting that proteins containing these labeled amino acids were synthesized. Torres and Steward (1992) have shown that tritiated sugars can be added to molecules in processes that are severed from their cell bodies. These data suggest that mechanisms for posttranslational modification, usually thought to occur in the rough endoplasmic reticulum and golgi apparatus, can occur in dendrites. In a series of experiments which are not as cellularly defined but which give a more robust signal, Greenough and colleagues have isolated functional polysomes from sucrose gradients of synaptodendrosome preparations made from dendritically rich areas of the central nervous system such as the hippocampus (Weiler and Greenough, 1993; Weiler et al., 1994; Weiler et al., 1997). Radiolabeled amino acids were added to these polysome preparations and continued protein synthesis yielded radiolabeled protein. Combined, these data provide strong support for the presumption that translation of dendritically localized mRNAs can occur and consequently may play a role in dendritic functioning. In further support of this hypothesis, Weiler and Greenough have shown that glutamate activation of metabotropic receptors increases the amount of protein synthesis in the synaptodendrosome polysome experiments.

These data have been confirmed and expanded using a single-dendrite transfection assay (Crino and Eberwine, 1996). Briefly, a reporter construct was made fusing a GSK sequence with a myc-epitope tag. This DNA construct was used as a template to synthesize mRNA that was capped *in vitro*. This mRNA was lipid encoated and applied to transected live dendrites through the patch pipette. The reasoning behind this experiment was that the only way in which the myc-epitope would be visualized is if the transfected mRNA was translated (see Figure 13.2). Indeed, in the presence of BDNF or NT3, which had previously been shown by Kang and Schuman (1996) to stimulate protein synthesis, the myc-epitope was visible, thus providing direct proof that proteins can be synthesized in dendrites independent of the cell soma.

Local protein synthesis is a particularly appealing regulatory mechanism for controlling potentiation-induced changes in postsynaptic neuronal responsiveness analogous to the nicotinic acetylcholine receptor expression at the neuromuscular junction. In the postsynaptic muscle cell, individual subunits of the nicotinic acetylcholine receptor mRNAs are concentrated at the neuromuscular junction where they are locally translated whereas other mRNAs are evenly distributed within the muscle syncitia (Merlie and Sanes, 1985; Fontaine et al., 1988). In neurons, functional channel proteins, including sodium channels (Huguenared et al., 1989; Regehr et al., 1992), calcium channels (Westenbroek et al., 1990), and NMDA receptors (Regehr and Tank, 1990), have been shown to be present in dendrites. These channels are known to regulate $Ca^{2+}$ influx and/or action potentials. By analogy with the acetylcholine receptor, it is possible that mRNA for these or subsets of these channels exists in dendritic processes (Miyashiro et al., 1994; Crino and Eberwine, 1996). Furthermore, local protein synthesis and creation of functional channels may be involved in regulating potentiation-induced $Ca^{2+}$ influx, consequently regulating postsynaptic responsiveness.

There are many questions that remain to be answered with respect to the characterization of mRNA localization to dendrites. Over the last few years new methodologies have provided insights into this biological process and it is reasonable to expect that the recognition of the biological importance of dendritic localization of mRNAs will only increase as more information is gained.

## REFERENCES

Aigner, L. and Caroni, P., Deletion of 43-kD growth associated protein in primary sensory neurons leads to diminished formation and spreading of growth cones, *J. Cell. Biol.,* 123, 417–429, 1993.

Aigner, L. and Caroni, P., Absence of spreading, branching, and adhesion in GAP43-depleted growth cones, *J. Cell. Biol.,* 128, 647–660, 1995.

Ainger, K., Avossa, D., Morgan, F., Hill, S., Barry, C., Barbarese, E. and Carson, J., Transport and Localization of Exogenous Myelin Basic Protein mRNA Microinjected into Oligodendrocytes, *J . Cell. Biol.,* 123, 431–441, 1993.

Baas, P.W., Black, M.M., and Banker, G.A., Changes in microtubule polarity orientation during the development of hippocampal neurons in culture, *J. Cell. Biol.,* 109, 3085–3094, 1989.

Bernhardt, R. and Matus, A., Light and electron microscopic studies of the distribution of microtubule associated protein 2 in rat brain, *J. Comp. Neurol.,* 226, 203–231, 1984.

Burgin, K. et al., *In situ* hybridization histochemistry of $Ca^{++}$/calmodulin-dependent protein kinase in developing rat brain, *J. Neurosci.,* 10, 1788–1798, 1990.

Crino, P. and Eberwine, J., Molecular characterization of the dendritic growth cone: regulated mRNA transport and local protein synthesis, *Neuron,* 17, 1173–1187, 1996.

Davis, L., Banker, G., and Steward, O., Selective dendritic transport of RNA in hippocampal neurons in culture, *Nature,* 330, 477–479, 1987.

Dotti, C., Sullivan, C., and Banker, G., The establishment of polarity by hippocampal neurons in culture, *J. Neurosci.*, 8, 1454–1468, 1988.

Garner, C.B., Tucker R.P., and Matus, A., Selective localization of messenger RNA for cytoskeletal protein MAP2 in dendrites, *Nature*, 336, 374–377, 1988.

Goslin, K., Schreyer, D.J., Skene, J.H.P., and Banker, G., Changes in the distribution of GAP-43 during the development of neuronal polarity, *J. Neurosci.*, 10, 588–602, 1990.

Huguenared, J., Hamill, O., and Prince, D., Sodium channels in dendrites of rat cortical pyramidal neurons, *Proc. Natl. Acad. Sci. U.S.A.*, 86, 2473–2477, 1989.

Johnson, G.V.W. and Jope, R.S., The role of microtubule associated protein 2 (MAP-2) in neuronal growth, plasticity and degeneration, *J. Neurosci. Res.*, 33, 505–512, 1989.

Kanai, Y. and Hirokawa, N., Sorting mechanisms of tau and MAP2 in neurons: suppressed axonal transit of MAP2 and locally regulated microtubule binding, *Neuron*, 14, 421–432, 1995.

Kang, H. and Schuman, E., A requirement for local protein synthesis in neurotrophin induced hippocampal synaptic plasticity, *Science*, 273, 1402–1406, 1996.

Kleiman, R., Banker, G., and Steward, O., Differential subcellular localization of particular mRNAs in hippocampal neurons in culture, *Neuron*, 5, 821–830, 1990.

MacDonald, P. and Struhl, G., Cis-acting Sequences responsible for anterior localization of bicoid mRNA in drosophila embryos, *Nature*, 336, 595–598, 1988.

Melton, D., Translocation of a localized maternal mRNA to the vegetal pole of Xenopus oocytes, *Nature*, 328, 80–82, 1987.

Miyashiro, K., Dichter, M., and Eberwine, J., On the nature and distribution of mRNAs in hippocampal neurites: implications for neuronal functioning, *Proc. Natl. Acad. Sci. U.S.A.*, 91, 10800–10804, 1994.

Regehr, W. and Tank, D., Postsynaptic NMDA receptor-mediated calcium accumulation in hippocampal CA1 pyramidal cell dendrites, *Nature*, 345, 807–810, 1990.

Steward, O. and Falk, P., Protein-synthetic machinery at postsynaptic sites during synaptogenesis: a quantitative study of the association between polyribosomes and developing synapses, *J. Neurosci.*, 6, 412–423, 1986.

Steward, O. and Levy, W., Preferential localization of polyribosomes under the base of dendritic spines in granule cells of the dentate gyrus, *J. Neurosci.*, 2, 284–291, 1982.

Steward, O. and Reeves, T., Protein-synthetic machinery beneath postsynaptic sites in CNS neurons: association between polyribosomes and other organelles at the synaptic site, *J. Neurosci.*, 8, 176–184, 1988.

Torres, E. and Steward, O., Demonstration of local protein synthesis within dendrites using a new cell culture system that permits the isolation of living axons and dendrites from their cell bodies, *J. Neurosci.*, 12, 762–772, 1992.

Weeks, D. and Melton, D., A maternal mRNA localized to the vegetal hemisphere in Xenopus eggs codes for a growth factor related to TGF-b, *Cell*, 51, 61–867, 1987.

Weiler, I.J. and Greenough, W.T., Metabotropic glutamate receptors trigger postsynaptic protein synthesis, *Proc. Natl. Acad. Sci. U.S.A.*, 90, 7168–7171, 1993.

Weiler, I.J., Irwin, S., Klintsova, A., Spencer, C., Brazelton, A., Miyashiro, K., Comery, T., Patel, B., Eberwine, J., and Greenough, W., Fragile X mental retardation protein is translated near synapses in response to neurotransmitter activation, *Proc. Natl. Acad. Sci. U.S.A.*, 94, 5395–5400, 1997.

Weiler, I.J., Wang, X., and Greenough, W.T., Synapse-activated protein synthesis as a possible mechanism of plastic neural change, *Prog. Brain Res.*, 100, 189–194, 1994.

Westenbroek, R., Ahlijianian, M., and Catterall, W., Clustering of L-type Ca channels at the base of major dendrites in hippocampal pyamidal neurons, *Nature*, 347, 281–284, 1990.

# 14

# Molecular Biological and Histochemical Analysis on the Functional Significance of Phosphoinositide Metabolism in the Nervous System

*Hisatake Kondo*

## CONTENTS

## ABSTRACT

In view of recent accumulation of evidence for the functional importance of phosphoinositide metabolism in signal transduction, membrane trafficking, and some other cellular responses, the molecular heterogeneity of various enzymes and proteins involved in phosphoinositide metabolism are summarized and the *in situ* hybridization histochemical findings of these molecules in developing and mature brain are described in detail.

## AN OVERVIEW OF PHOSPHOINOSITIDES

Our understanding about the lipids in eukaryotes has recently been changed from being viewed mainly as structural and static components of biological membranes to being recognized as functional and dynamic components that play important roles in signal transduction, membrane traffic, and some, other cellular responses. The discovery of the receptor-controlled splitting of phosphatidylinositol 4,5-bisphosphate in the cell membrane by phospholipase C (PLC) into its hydrophobic and hydrophilic components, 1,2-diacylglycerol (DAG) and inositol 1,4,5-triphosphate (IP3), was a starting event in this change relating to the phospholipids, and the two products are well-established second messengers functioning through activation of protein kinase C (PKC) and through release of intracellular calcium ions. These second messengers are involved in cellular functions such as cell growth and oncogenesis, differentiation, and development, and in the action of neurotransmitters and hormones and sensory perception (Berridge, 1993; Divecha and Irvine, 1995; Liscovitch and Cantley, 1994; Rhee et al., 1989; Rhee and Bae, 1997).

Once phosphatidylinositol 4,5-bisphosphate has been hydrolyzed by PLC, it must be resynthesized. The process of resynthesis begins when DAG is phosphorylated to make phosphatidic acid (PA) by DAG kinase. The next step in the resynthesis is performed by cytidine 5′-diphosphate (CDP-DAG) synthase (CDS) which catalyzes the conversion of PA to CDP-DAG, followed by phosphatidylinositol synthase (PIS), which catalyzes the reaction of CDP-DAG and myo-inositol to form phosphatidylinositol and cytidine 5′-monophosphate (CMP). Phosphatidylinositol is then phosphorylated at the D-4 and D-5 positions of the inositol ring sequentially by phosphatidylinositol 4-kinase (PI4-kinase) and phosphatidylinositol 4-phosphate 5-kinase (PI4P-5-kinase), respectively, to finally re-produce phosphatidylinositol 4,5-bisphosphate. This entire cycle of the metabolic route is conventionally termed the phosphoinositide turnover.

In addition to this classical turnover of phosphoinositide metabolism, it has recently been clarified that another class of second messengers is generated via phosphorylation of phosphoinositides at D-3, but not D-4, of the inositol ring by phosphoinositide 3-kinase (PI3-kinase). The products are phosphatidylinositol 3-phosphate, phosphatidylinositol 3,4-bisphosphate, and phosphatidylinositol 3,4,5-trisphosphate, the latter two of which are acutely regulated in agonist-stimulated cells through several growth factor receptors, while the former is constitutively present (Toker and Cantley, 1997; Vanhaesebroeck et al., 1997). All of the three phospholipids are not substrates for PLC. Although the D-3 phosphoinositides target a variety of downstream signal molecules to be involved in different cellular responses such as cell survival, cell growth, cytoskeletal rearrangement, and membrane trafficking, much progress has recently been made regarding the implication of the D-3 phosphoinositide in cell survival (Datta et al., 1997; Del Peso et al., 1997). The generation of phosphatidylinositol 3,4-bis- and 3,4,5-trisphosphate activates serine/threonine kinases Akt/PKB/rac and PDK1 (phosphoinositide-dependent protein kinase-1) through their pleckstrin homology (PH) domains (Alessi et al., 1997a,b; Stokoe et al., 1997). The full activation of Akt requires its phosphorylation by PDK1. A proapoptotic factor termed Bad is phosphorylated on its serine residue by activated Akt, resulting in its dissociation from Bcl-xL, a cell survival factor, and its subsequent association with the phosphoserine-binding protein 14-3-3. The released Bcl-xL then promotes cell survival by restoration of its gatekeeping function of cytochrome c efflux at the mitochondrial membrane and following blocking the caspase protease cascade. There has also been evidence that phosphoinositol 3,4-bis- and 3,4,5-trisphosphate activate the novel PKC-ε, PKC-η, and atypical PKC-ζ as well as the PKC-related kinase PRK1. It is also likely that activating PKC downstream of PI3-kinase affects the Ras/Raf/MEK/MAP kinase pathway initiated by receptor tyrosine kinases (Nakanishi et al., 1993; Palmer et al., 1995; Toker and Cantley, 1997).

On the other hand, accumulating evidence has indicated that phosphoinositide synthesis is linked to a variety of membrane traffic events, although the exact mechanism remains to be elucidated. This is based on the initial findings that the Vps34 of yeast is required for budding of a specific vesicle type from the trans-Golgi network, and that Vps34 and its mammalian homolog phosphorylate at the D-3 position only phosphatidylinositol but not phosphatidylinositol 4-phosphate and 4,5-bisphosphate (Volinia et al., 1995). PI4-kinase and PI4P-5-kinase have also been implicated in membrane trafficking by the findings that PI4-kinase activity is localized in most intracellular vesicles and granules; phosphatidylinositol 4,5-bisphosphate is involved in the function of ADP-ribosylation factor (ARF), a small G-protein regulating the formation of coated vesicles in the Golgi; and two genes isolated from budding yeast have homology to the catalytic domain of a mammalian PI4P-5-kinase (Gehrmann and Heilmeyer, 1998; Liscovitch et al., 1994; Randazzo and Kahn, 1994; Terui et al., 1994; Wiedemann et al., 1996).

Although no distinct information is so far available about any serious phenotypes originated from deterioration of phosphoinositide metabolism in the mammalian nervous system, such cases are well established in the invertebrate photoreceptor: *Drosophila* mutants termed *norpA, rdgA, rdgB,* and *rdgD* (Ranganathan et al., 1995). norpA mutant flies fail to give any electrical response to light and lack normal PLC activity. The PLC encoded by the gene norpA represents only one of the many isoforms present in *Drosophila* and most closely resembles the mammalian β-subtypes. *rdgA* mutant flies are characterized by light-induced retinal degeneration and are deficient in DAG kinase activity, which suggests that the degeneration of photoreceptor cells may be related to the effect of chronic activation of a PKC or the lack of PAP. *rdgB* mutant flies are characterized by a rapid course of photoreceptor degeneration and the *rdgB* gene encodes a protein with a sequence identity to the rat brain phosphatidylinositol transfer protein (PITP). The light-stimulated electrical response of *Drosophila* photoreceptors is reduced in some mutants of the *CDS* gene whose photoreceptors degenerate in response to light exposure (Wu et al., 1995). This mutant has been identified by some authors as *rdgD*. In addition to these findings in *Drosophila*, there has been evidence that the gene product mutated in ataxia telangiectasia (Savitsky et al., 1995) is a molecule homologous to RAFT1 (FK506-binding protein-rapamycin-associated protein), which contains the PI4-kinase activity in accord with the existence of amino acid sequences homologous to several PI-kinases. Therefore, it is possible that defective expression of any enzymes involved in phosphoinositide metabolism causes some problems in the mammalian central nervous system.

In order to understand more clearly the functional significance of phosphoinositides in the mammalian central nervous system, information on the molecular structure and heterogeneity of the enzymes involved in the phosphoinositide metabolism and on their localization is crucial. Rodent brain tissue is the system of choice for such histological localization analyses. We will therefore review here the published findings concerning the molecular heterogeneity and expression localization of phosphoinositide-related molecules in the rodent brain.

## PHOSPHOLIPASE C (PLC)

This enzyme has been revealed to comprise a family of multiple isozymes by protein purification and molecular cloning. A total of 10 isoforms have so far been cloned from mammalian cells, and they can be classified into three types — PLC β, γ, and δ — based on comparison of amino acid sequences and functional properties. The PLC β family is coupled to seven transmembrane-helix receptors through the α-subunits of the pertussis toxin-insensitive Gq family of G-proteins and/or through G-protein β,γ-subunits. The PLC γ family is activated by growth factor receptors via tyrosine phosphorylation, whereas the activation mechanism for the PLC δ family remains unknown (Rhee and Bae, 1997).

By *in situ* hybridization histochemistry of adult rat brain, expression of mRNA for PLC β-1 is highest in the internal granular cell layer of the olfactory bulb, cerebral cortex, caudate putamen, thalamic reticular nucleus, hippocampus and dentate gyrus, and cerebellar granule, but not Purkinje, cell layer. Expression of mRNA for PLC β-3 is detected weakly only in the adenohypophysis and the cerebellar Purkinje and granule cells, whereas PLC β-4 mRNA is intense in the olfactory mitral cells, thalamic nuclei, medial habenula, adenohypophysis, and cerebellar Purkinje and granule cells (Ross et al., 1989; Tanaka and Kondo, 1994). The β-4 mRNA is also detected discretely in large neurons of presumably cholinergic nature in the caudate putamen and diagonal band although no significant expression is seen in the hippocampal pyramidal cells and dentate granule cells. The expression for PLC γ-1 has a more widespread distribution, with relatively high levels in the internal granular layer of the olfactory bulb, hippocampus and dentate gyrus, and cerebellar Purkinje and granule cells, whereas PLC γ-2 mRNA is localized only in the Purkinje and granule cells in the vermal portions of lobules IX and X of the cerebellum and in the adenohypophysis. The expression for δ-1 is much lower throughout the brain. In embryonic brain, the expression for PLC β-1 and γ-1 is evident in the germinal zone.

As previously described, the defect in *Drosophila* PLC results in a mutant called *norpA* whose photoreceptors fail to give an electrical response to light (Hotta & Benzer, 1970). Molecular cloning studies have suggested that PLC β-4 is the mammalian homolog of PLC-*norpA* (Bloomquist et al., 1988; Ferreira et al., 1993; Lee et al., 1993), and the localization of PLC β-4 has been demonstrated in the photoreceptor cells of rat retina by *in situ* hybridization and immunohistochemistry, although there is still a discrepancy in the photoreceptor cell types, rod vs. cone (Ferreira and Pak, 1994; Peng et al., 1997). In addition, the colocalization of PLC β-4 and G-protein type G α-11 is suggested to occur in the photoreceptor cells by immunohistochemistry, which agrees with the biochemical finding that PLC β-4 is specifically activated by α-subunits of the Gq family (Jhon et al., 1993; Lee et al., 1994). However, it is well-known that vertebrate phototransduction is mediated by the cGMP cascade, in contrast to the fact that the phototransduction cascade of invertebrates is based on the phosphoinositide cycle (Fein, 1986; Stryer, 1986). Because there is no uniform agreement as to whether light actually influences phosphoinositide turnover (Gehm and McConnell, 1990; Panfoli et al., 1990), and because there is no evidence for the presence of an IP3 receptor in the rod outer segment (Peng et al., 1997), the occurrence of PLC β-4 in mammalian photoreceptor cells may represent involvement of phosphoinositide turnover in light adaptation or functions other than phototransduction triggered by a chemical signal rather than light.

In general, phosphoinositide turnover is considered to occur in the plasma membrane, and most PLC isoforms seem to be located in the cytoplasm. However, phosphoinositide turnover has been demonstrated to occur in the nucleus during cell growth and differentiation, and PLC activity has been shown to exist in the internal nuclear matrix and the envelope-depleted nucleus (Divecha et al., 1993; Payrastre et al., 1992; York and Majerus, 1994). PLC β-1 is the first isoform shown to be present in the nuclei of Swiss 3T3 cells. This is activated immediately after stimulation with insulin growth factor-1 (IGF-1) and its activity returns to normal within 30 min, although no significant change during the cell cycle is detected in the level of PLC β-1 (Martelli et al., 1992). PLC δ-4 is another nuclear isoform that has recently been isolated from regenerating rat liver and expressed selectively in cells with high proliferating activity such as hepatoma, intestine, and src-transformed 3Y1 cells (Liu et al., 1996). Different from PLC β-1, the expression of PLC δ-4 is induced by mitogens and its level reaches a maximum at S-phase in the cell cycle, suggesting important roles in cell growth. Because of high homology between PLC δ-4 and δ-2, they may also play similar roles. The localization and function of PLC δ-4 remain to be elucidated.

## DIACYLGLYCEROL KINASE (DAG KINASE)

Upon cell stimulation by a variety of mitogens and growth factors, DAG kinase rapidly causes a marked accumulation of PA. It has also been shown that PA activates *in vitro* a number of enzymes involved in signal transduction, including PKC-ζ (Limatola et al., 1994), unidentified protein kinases (Khan et al., 1994; McPhail et al., 1995), PLC-γ1 (Jones and Carpenter, 1993), and polyphosphoinositide kinases (Moritz et al., 1992). In addition, PA binds to and may regulate the translocation and subsequent activation of the protein kinase Raf-1 (Ghosh et al., 1996) and a protein-tyrosine phosphatase PTP1C (Zhao et al., 1993). Therefore, the conversion of DAG into PA by DAG kinase is regarded not only as attenuation of the protein kinase C activator and the first step in recycling phosphoinositides, but also as generation of the novel second messenger.

A cDNA for DAG kinase was first isolated from the porcine thymus cDNA library and termed α-isoform encoding 734 amino acids (Sakane et al., 1990). The primary structure of this molecule contains two EF hand motifs which confer calcium-dependent activity, two cysteine-rich zinc finger-like sequences which are similar to those found in PKC and Raf kinase, and a putative catalytic domain in the C-terminal half of the protein. The cysteine-rich motifs and the putative catalytic domain sequences are conserved in all DAG kinase isoforms which have been subsequently identified. This identification was followed by a series of studies to clone multiple isoforms from rat tissue, which is the system of choice for localization analysis by *in situ* hybridization histochemistry (Goto and Kondo, 1993, 1996; Goto et al., 1992, 1994). A rat homolog to the porcine DAG kinase molecule, termed the type I isoform, is specifically expressed in the oligodendrocyte and T-lymphocytes. Its expression pattern in postnatal brain is quite similar to that of myelin-specific proteins, suggesting the intimate relation in its function to the genesis and maintenance of myelin. The kinase activity in transfected COS-7 cells with the cDNA for DAG kinase-I is dominant in the soluble fraction. The first neuronal DAG kinase was identified and termed rat DAG kinase-II (92 kDa), whose composition of molecular consensus sequences is similar to type I with 58% amino acid sequence identity. In contrast to type I, kinase activity is recovered entirely from the particulate fraction of cells transfected with cDNA, and gene expression is intense in the caudate putamen, the accumbens nucleus, the olfactory tubercle, and the hippocampal pyramidal cell layer. The former brain regions are the major dopaminoceptive fields, suggesting intimate involvement of rat DAG kinase-II in dopaminergic transmission. However, because no changes in the expression of this isoform are detected in the caudate putamen of adult mice deficient in the gene for D2-dopamine receptor by gene targeting (Yamaguchi et al., 1996), the functional involvement mechanism in the dopaminergic transmission of this DAG kinase isoform remains to be elucidated. A third of the EF hand motif-containing isoforms, rat DAG kinase-III (88 kDa), is expressed intensely in the cerebellar Purkinje cells and moderately in the cerebellar granule cells, hippocampal pyramida, and dentate granule cells. The human homolog to the type III isoform has been identified as DAG kinase γ (Kai et al., 1994). This isoform is associated equally with particulate and supernatant fractions and it remains in the detergent-insoluble cytoskeletal fraction of transfected COS-7 cells and brain. When the expression patterns for the three isoforms are compared with those for multiple isoforms of PLC (as described above) in the brain, some similarity is noticed between PLC β-1 and DAG kinase-II and between PLC γ-1 and DAG kinase-III, suggesting that some PLC and DAG kinase molecules participate in phosphoinositide turnover in a subtype-preferred paired fashion.

Different from the three isoforms, the primary structure of rat DAG kinase-IV (104 kDa) is characterized by the presence of ankyrin repeats at the C-terminal and of a nuclear targeting motif, as well as the absence of EF hand motifs. The human homolog has been identified as DAG kinase ζ (Bunting et al., 1996). The gene expression for this isoform is detected intensely in the olfactory mitral and granule cells, the hippocampus, and the cerebellar Purkinje and granule cells, and moderately in the cerebral cortex and the olfactory tubercle of adult brain. In accord with the structural features, this isoform is localized in the nucleus and its kinase activity is not affected by calcium concentration in cDNA-transfected COS cells. The nuclear localization of DAG kinase-IV as well as PLC β-1 and δ-4 further strengthens the hitherto stated possibility that the nuclear phosphoinositide cascade can be activated in response to occupancy of some specific cell surface receptors, although it remains to be elucidated how the intranuclear enzymes involved in phosphoinositide turnover are targeted to specific intranuclear sites. This may be the case in neurons as well. In this regard, recently data suggest that memory formation involves altered gene expression in neurons: repeated and spaced pulses of serotonin cause an increase in cAMP and give rise to translocation of the catalytic subunit of the cAMP-dependent protein kinase to the neuronal nucleus, which induces protein synthesis-dependent, long lasting structural changes in neurons (Bailey and Kandel, 1993). It has also been shown that PKC is involved in learning and memory, as revealed by a gene knockout study of PKC γ (Abeliovich et al., 1993). Thus, it is plausible that the nuclear DAG kinase and PLC molecules might be related to the molecular mechanism of memory formation by regulating the activity of the nuclear PKC through regulation of intranuclear DAG and PA.

It should be pointed out that the primary structure of rat DAG kinase-IV closely resembles a *Drosophila* DAG kinase encoded by the *rdgA* gene in terms of the presence of ankyrin-like repeats, the absence of EF hand motifs, and a higher identity (40%) in the amino acid sequence than the other isoforms. This suggests that DAG kinase-IV is the mammalian homolog of *Drosophila* DAG kinase-*rdgA*. However, DAG kinase-IV is expressed more abundantly in the brain and thymus than in the eye and its mRNA expression is confined to the inner granular layer of the retina in the rat eye (Goto and Kondo, 1996). This localization contrasts with the specific expression of *Drosophila* DAG kinase-*rdgA* in photoreceptor cells, which implies that the two homologous molecules play discrete roles in the different animal species.

Three more isoforms have recently been identified which are characterized by the presence of a pleckstrin homology (PH) domain found in a number of proteins involved in signal transduction. They are termed human DAG kinase δ (Sakane et al., 1996), hamster DAG kinase η (Klauck et al., 1996), and human/rat DAG kinase θ (Houssa et al., 1997). All of them contain cysteine-rich zinc finger-like structures but no EF hand motifs. The δ-isoform has at the C-terminal tail a domain similar to those of the EPH family of receptor tyrosine kinases. The δ-isoform is undetectable in the brain or retina, but expressed in muscles and testis in Northern blotting, whereas the η-isoform is more widely expressed with the highest in brain and its expression is glucocorticoid inducible. The *in situ* localization of mRNA for the η-isoform in the brain remains to be elucidated. On the other hand, the θ-isoform has a PH domain at a location different from the δ- and η-isoforms and the first half of this PH domain overlaps a Ras-associated domain. This mRNA expression is detected intensely in the cerebellar Purkinje and granule cells and the hippocampal pyramidal and dentate granule cells, and more or less widely throughout the gray matter in the adult rat brain. The expression is discerned to a lesser extent in the small intestine and liver.

There has been evidence suggesting the presence of a DAG kinase isoform which preferentially phosphorylates 1-stearoyl 2-arachidonyl DAG, a molecular species which is primarily derived from the phosphoinositide turnover (MacDonald et al., 1988; Walsh et al., 1994). Although the seven isoforms described above show no marked specificity with regard to the acyl composition of DAG, a DAG kinase isoform of 64 kDa termed ε has been identified from the human and rat cDNA library to have a clear selectivity for arachidonyl-containing species of DAG (Kohyama-Koganeya et al., 1997; Tang et al., 1996). This molecule has two zinc finger-like structures in its N-terminal region and two ATP-binding motifs, but does not contain EF hand motifs. By Northern blotting the expression for the rat ε-isoform is found most intensely in the retina, intensely in the brain, and to a lesser extent in the heart, spleen, lung, kidney, and testis, whereas that for the human isoform is detected in the testis, at a lower level in the ovary, and at a barely detectable level in the skeletal muscle and pancreas. By *in situ* hybridization histochemistry, the mRNA expression is positive in the cerebellar Purkinje and granule cells, the hippocampal pyramidal and dentate granule cells, and retinal cells in the inner and outer nuclear layers. With such a substrate specificity, this DAG kinase isoform might play a role selectively in phosphoinositide metabolism by producing a precursor of phosphatidylinositol, phosphatidic acid, that is enriched in arachidonic acid. It is known that one of eight subtypes of metabotropic glutamate receptors, mGluR1, is linked to the phosphoinositide turnover and subsequent PKC γ activation (Houamed et al., 1991; Masu et al., 1991). It is also known that activation of mGluR1 plays an important role in cerebellar long-term depression and hippocampal long-term potentiation, which are thought to be a molecular mechanism of learning and memory (Zheng and Gallagher, 1992). Since the gene expression for DAG kinase ε is detected in those neurons in which the expression for mGluR1 and PKC γ are also detected (Brandt et al., 1987; Ohishi et al., 1994), it is possible that the arachidonyl-preferred DAG kinase isoform is involved in the same cascade as mGluR1 and PKC γ and the learning memory mechanism.

## CDP-DAG SYNTHASE AND PIS SYNTHASE

The molecular cloning of mammalian CDS has been initially succeeded from the rat brain cDNA library based on the primary sequence of CDS of *Drosophila* and *Escherichia coli* (Saito et al., 1997) and subsequently from human species (Heacock et al., 1996; Weeks et al., 1997). The rat cDNA encodes a 462 amino acid protein whose molecular mass is 53 kDa, and it contains several possible membrane-spanning regions. This rat CDS molecule shows the substrate preference toward 1-stearoyl-2-arachidonyl phosphatidic acid, suggesting that this molecule selectively participates in the phosphoinositide cycle, but not in the synthesis of phosphatidylglycerol and cardiolipin, in which CDS is also involved. By epitope-tag immunocytochemistry, the CDS protein is mainly localized in close association with the membrane of the endoplasmic reticulum in COS-7 cells transfected with cDNA (Saito et al., 1997). In Northern blotting of adult rat tissues, the gene expression is the highest in the testis, followed by the brain and eye ball, and it is weak in the kidney, small intestine, and placenta. By *in situ* hybridization histochemistry of adult rat brain, the most intense expression signal is detected in the cerebellar Purkinje cells and intense expression is in the pineal body. Moderate to low expression is seen in layers II–VI of the cerebral cortex, the hippocampal pyramidal cell layer and subiculum, and the olfactory mitral cells. Low expression is seen in almost all neurons in the fore- and hindbrain regions with much lower expression in the caudate putamen, and also in the choroid plexuses. No significant expression is detected in any embryonic brain regions of rats. In the retina positive expression is confined to the inner segment of the photoreceptor cells. In the testis the intense expression signals are confined to postmitotic spermatocytes and spermatids.

The molecular cloning of a mammalian PIS has recently been made from rat brain by functional complementation of the yeast PIS mutation (Tanaka et al., 1996). The deduced protein comprises 213 amino acids with a calculated molecular mass of 23,613 Da. In Northern blotting the mRNA is expressed in the brain and kidney abundantly and to a much lesser extent in the testis.

When the localization of mRNA for this rat PIS is compared with that of the rat CDS, discrepant as well as parallel expression patterns are noticed in the brain (Saito et al., 1998). The parallel expression is as expected in view of the two consecutive metabolic reactions catalyzed by the two enzyme molecules: no significant expression for CDS and PIS is detected in the germinal zones including the cerebellar external granule cell layer; and the expression for both of the two molecules is seen more or less widely in the gray matters throughout the entire brain during the postnatal stages. On the other hand, the discrepancy is pointed out as follows: substantial expression of PIS mRNA throughout the neuraxis in contrast to the actual absence of the expression for CDS in the prenatal brain; high expression for PIS in the cerebellar granule cells as well as Purkinje cells in contrast to negligible levels of CDS expression in the granule cells; and the substantial expression of PIS mRNA in the ganglion cell layer, inner nuclear layer, and photoreceptor cell layer in contrast to CDS expression confined to the photoreceptor cell layer. Although the reason for the discrepant expression remains to be elucidated, the existence of yet unknown isoforms of CDS or differences in regulation of the region/stage-specific requirement for CDP-DAG and phosphatidylinositol should be considered.

## PI4-KINASE AND PI4P-5 KINASE

Two isoforms of PI4-kinase have been identified by molecular cloning from rat and bovine cDNA libraries: one is composed of 2041 amino acids with a calculated molecular weight of 231,317 Da and the other is composed of 816 amino acids with a calculated molecular weight of 91,654 Da (Gehrmann and Heilmeyer, 1998; Nakagawa et al., 1996a,b). The primary structure of the larger one contains an SH3 domain, proline-rich regions, two nuclear localization signals, leucine zippers, a helix-loop-helix motif, an ankyrin repeat domain, a lipid-kinase-unique domain, and a presumed lipid kinase/protein kinase homology domain. The domain structure in the C-terminal of the larger isoform is similar to that of a yeast PI4-kinase termed STT4 (Yoshida et al., 1994) with the identity of 52.3% to STT4 in the presumed catalytic domain. The smaller molecule contains a lipid-kinase-unique domain and a presumed lipid/protein kinase homology domain. It shares 43.3% identity with the presumed catalytic domain of yeast PI4-kinase termed PIK1 (Garcia-Bustos et al., 1994) and it has a region of similarity that is shared with PIK1 but not conserved in other lipid kinases. Subsequently, bovine counterparts of the larger and smaller rat molecules have been identified and the terms PI4-kinase type III α and β are proposed, respectively (Balla et al., 1997). The terms of type II and III are derived from the conventional classification of PI4-kinase according to the difference in sensitivity of the enzyme to adenosine: type II is inhibited by adenosine whereas type III is resistant to inhibition by adenosine. No information is available on the primary structure of type II. Although a human PI4-kinase molecule of 854 amino acids has already been identified preceeding to the two isoforms of rats and bovine and first termed PI4-kinase α (Wong and Cantley,

1994), it is likely from the comparison in the primary structure that this human molecule is an alternatively spliced or truncated human counterpart of the larger isoform of rats and bovine, PI4-kinase III α (Balla et al., 1997; Gehrmann et al., 1996; Nakagawa et al., 1996a). The human counterpart of the smaller isoform PI4-kinase III β is also identified as a protein of 801 amino acids (Meyers and Cantley, 1997).

By *in situ* hybridization histochemistry the gene expression for PI4-kinase III α and β is detected more or less widely in the gray matters throughout the brain and spinal cord of adult brain, among which the hippocampal pyramidal cells, dentate granule cells, and cerebellar granule cells express the mRNAs more abundantly. In the embryonic brain, the expression of both mRNAs is much higher than in the adult brain and it is evident in the mantle zones including the cortical plate as well as the ventricular germinal zone (Nakagawa et al., 1996a,b). At protein levels, the immuno-cytochemical studies have shown the localization of the two isoforms in Golgi apparatus by Flag-tagged cDNA transfection and anti-Flag antibody (Nakagawa et al., 1996a,b); or the larger isoform in endoplasmic reticulum and the smaller one in the cytoplasmic matrix and Golgi apparatus by epitope-tagged cDNA transfection and individual PI4-kinase antibodies (Wong et al., 1997). No information is available to see whether or not the larger isoform can be localized in the nucleus in accord with the occurrence of the nuclear targeting signals in its primary structure.

With regard to PI4P-5-kinase, two isoforms termed type I and II have been characterized by PA stimulation of type I and by low activity of purified type II toward native membrane (Jenkins et al., 1994). The type I, but not the type II, 5-kinase isoform has been implicated in the secretion of neurotransmitters/hormones and small G-protein regulation. Neurotransmitter/hormone release from permeabilized PC12 cells has been divided into an initial Mg-ATP-dependent step, followed by a Ca-triggered exocytotic event. The type I isoform and the production of PI4,5-bisphosphate are required for the priming step, whereas the type II isoform is ineffective. The type I isoform also appears to associate with, and is putatively stimulated by, the small G-proteins Rac and Rho, whereas the type II isoform does not interact with these G-proteins (Chong et al., 1994; Hay and Martin, 1993; Hay et al., 1995; Ren et al., 1996; Tolias et al., 1995). However, a recent study has indicated that the type II isozyme phosphorylates phosphatidylinositol-5-phosphate at the D-4 position, and thus should be regarded as a 4-OH kinase or PIP(4)kinase (Rameh et al., 1997).

Molecular cloning studies have revealed the primary structure of two type I isoforms (α and β) from human (Loijens and Anderson, 1996), three type I isoforms (α, β, γ) from mouse (Ishihara et al., 1996, 1998), and one type II isoform from human (Boronenkov and Anderson,1995, Divecha et al., 1995). The three mouse type I isoforms are composed of 539, 546, and 635 (alternative 661) amino acids, respectively, whereas the mouse type II consists of 405 amino acids. The central portions of the three type I isoforms are very similar to each other in amino acid sequences with about 80% identity, and the amino-terminal sequence of the type I γ isoform shows partial homology with that of the β isoform with about 40% identity whereas the carboxy-terminal regions differ in length and amino acid sequence among the three. In Northern blotting these isoforms are expressed rather widely with variable intensity: the type I α is highly expressed in the brain and testis; the type I β is highly expressed in the skeletal muscle, testis, brain, and lung; the type I γ in the brain, lung, and kidney; and human type II is expressed intensely in the brain, less intensely in the kidney, placenta, and heart. Because the type I α is 95% identical to the amino acid sequence of STM7, a gene identified in the critical region for Friedreich's ataxia locus which is not responsible for the disease (Campuzano et al., 1996; Carvajal et al., 1995), the *in situ* localization of mRNAs for all these isoforms in developing and adult brains is important for understanding of their functional significance in the brain, which remains to be elucidated.

## PITP

PITP is a cytosolic protein which was originally identified for its ability to transfer phosphatidyli-nositol and, to a lesser extent, phosphatidylcholine between lipid bilayers (Wirtz, 1991). This *in vitro* property suggests that PITP *in vivo* transports these phospholipids from the sites of synthesis in the endoplasmic reticulum and the Golgi apparatus to other plasma membranes (Liscovitch and Cantley, 1995). There have been an increasing number of reports that PITP is required for signaling pathways through phosphoinositide hydrolysis by PLC and that it is also required for ATP-dependent priming of secretory vesicles for fusion with the plasma membrane (Cunningham et al., 1996; Hay et al., 1995; Kauffmann-Zeh et al., 1995; Ohashi et al., 1995; Thomas et al., 1993).

Two isoforms of rat PITP have so far been identified by cDNA cloning: PITP $\alpha$ and $\beta$. They are both composed of 271 amino acids and share 77% identity (Dickeson et al., 1989; Tanaka and Hosaka, 1994). By *in situ* hybridization histochemistry, the gene expression for both isoforms is detected widely throughout the entire neuraxis, though less evident in the ventricular zones, during embryonic stages. In adult brain the expression for the $\alpha$-isoform is positive in almost all neurons throughout the entire brain, whereas that for the $\beta$-isoform markedly decreases in the entire gray matter region except for the cerebellar cortical neurons (Utsunomiya et al., 1997). The rather ubiquitous expression in almost all neurons in embryonic brain and the chronological decrease in the expression in the entire brain during the postnatal stages is similar to the expression patterns for PI4-kinase described above, in favor of the cooperation of PITP with PI-kinase in the regulation of the vesicular traffic which is required for the active membrane biogenesis related to the neuronal differentiation including elongation and synaptogenesis. No good coincidence is found in the expression patterns in the brain between PLC and PITP isoforms, with slight similarities between PITP $\alpha$ and PLC $\beta$-4 and between PITP $\beta$ and PLC $\gamma$-1.

## PI3-KINASE, AKT/PKB, AND PDK-1

Three classes of PI3-kinases are discriminated on the basis of their *in vitro* lipid substrate specificity. Class I molecules are characterized by phosphorylation of all phosphatidylinositol, phosphatidylinositol 4-phosphate, and phosphatidylinositol 4,5-bisphosphate, and they include P110 $\alpha$, $\beta$, and $\gamma$ (Chantry et al., 1997; Hiles et al., 1992; Hu et al., 1993; Stoyanov et al., 1995; Vanhaesebroeck et al., 1997). P110 $\alpha$ and $\beta$ interact with SH2/SH3-domain-containing p85 adaptor proteins to form heterodimers and with GTP-Ras, whereas P110 $\gamma$ does not interact with p85 but associates with a p101 adaptor protein. The p85/p110 heterodimers interact with phosphorylated tyrosine residues in receptors and other proteins, while the activity of P110 $\gamma$ is stimulated by a G-protein subunit. Class II molecules are characterized by phosphorylation of phosphatidylinositol and phosphatidylinositol 4-phosphate, but not phosphatidylinositol 4,5-bisphosphate, and the presence of C2 domain at their C-terminus. They include three molecular species: Cpk-m/p170 from mouse (PI3K-II$\alpha$), HsC2-PI3K from human (PI3K-II$\beta$) and PI3K-II$\gamma$ from rat and mouse (Brown et al., 1997; Misawa et al., 1998; Ono et al., 1998). Class III molecules represent a mammalian homolog of yeast PI3-kinase, Vps34, which is involved in trafficking of proteins from the Golgi to the yeast equivalent of mammalian lysosomes, and is characterized by the substrate specificity confined to phosphatidylinositol (Volinia et al., 1995).

Akt/PKB/rac kinase is a protein-serine/threonine kinase having a catalytic domain closely related to both cAMP-dependent protein kinase (PKA) and PKC. The three different names are derived as follows: "rac" represents "related to A and C kinases"; "PKB" because of simulation to PKA and PKC; and "Akt" because this is a cellular homolog of the viral oncogene, v-akt. Three isoforms of Akt/PKB/rac, termed $\alpha$, $\beta$, and $\gamma$, or $-1$, $-2$, and $-3$, are identified by cDNA cloning and the deduced amino acid sequence of the rat species are composed of 480, 481, and 455, respectively (Konishi et al., 1994, 1995). The three isoforms are characterized by a PH domain and a kinase catalytic domain at the amino- and carboxy-terminal regions, respectively. In Northern blotting the $\alpha$-isoform is expressed intensely in the heart, brain, and lung; less intensely in the spleen, skeletal muscle, and kidney; and very weakly in the testis, whereas the $\beta$-isoform is expressed highly in the heart, skeletal muscle, and testis. The expression for the $\gamma$-isoform is high in the brain and testis, and moderate in the heart, spleen, lung, and skeletal muscle.

Although no information is available yet concerning the *in situ* localization of mRNAs for the class II and III PI3-kinases in the brain, the detailed localization of mRNAs for the class I p110 has been clarified in developing and mature rat brain (Ito et al., 1995). No marked differences in the expression patterns are seen between p110 α and p110 β. The expression for p110 is evident throughout the entire neuraxis in the ventricular and mantle zones during embryonic stages. After birth the expression decreases gradually throughout the brain, although the high expression is seen in the cerebellar external granule cells. In the adult brain expression is eventually weak in almost all neurons, although expression remains evident in the hippocampal pyramidal and dentate granule cells as well as cerebellar Purkinje and granule cells. The expression of mRNAs from Akt/PKB/rac kinase in the brain is also localized by *in situ* hybridization histochemistry without any marked differences among the multiple isoforms (Owada et al., 1997). The expression pattern is similar to that for PI3-kinase p110, although the sustained expression in the hippocampus and the cerebellar cortex is less evident than for p110.

The dominant expression for both PI3-kinase and Akt in the gray matters of immature brain, especially in the neurogenic zones, indicates that these enzymes play roles in the mitogenic signal pathway and DNA synthesis and in early neuronal differentiation including migration, neurite elongation, and synaptogenesis. Although the expression for both PI3-kinase and Akt is low in almost all neurons in the adult brain, dramatic increases in the expression levels for both p110 α and Akt-1 is noticed in the hypoglossal neuron somata after 48 h to 7 days following axotomy of the nerve (Ito et al., 1996; Owada et al., 1997). This synchronous enhancement in expression for both enzyme molecules after axotomy is in accord with the findings that Akt is one of the most likely targets of PI3-kinase and that Akt seems to play a central role in the PI3-kinase-mediated protection against apotosis (Datta et al., 1997; Del Peso et al., 1997; Franke et al., 1997). Since products of PI3-kinase increase in PC12 cells, forming neurites on NGF stimulation, and because the neurite outgrowth of PC12 cells is suppressed by wortmannin, an inhibitor of PI3-kinase (Kimura et al., 1994; Soltoff et al., 1992), it is possible that PI3-kinase and Akt are involved in axonal elongation as a recovery from axonal injury.

As previously stated, full activation of Akt has been accomplished by other serine/threonine protein kinases and one such kinase, PDK1, has recently been purified and cloned (Alessi et al., 1997a,b). This is also termed PKB kinase (PKBK). The *in situ* localization of PDK1 mRNA in the brain remains to be elucidated.

## CONCLUSION

The involvement of phosphoinositide metabolism in signal transduction, membrane trafficking, and some other cellular functions has been a central feature in molecular biology during the past 10 years. It has been clarified that individual enzymes involved in phosphoinositide metabolism exhibit a marked heterogeneity in molecular structure and biochemical characteristics, and the paradigm of the signal transduction cascades has been increasingly complicated in detail year by year. However, most of the present information has come from experiments in cell homogenization and *in vitro* cells, and the *in vivo* significance of the signaling cascades and other cellular responses executed by the phosphoinositide metabolism are not fully understood. The *in situ* spatio-temporal localization of the phosphoinositide-relating molecules in the brain at mRNA and protein levels must first be analyzed, because it is highly possible that the phosphoinositide-related signaling, vesicle trafficking, and so forth are intimately implicated in the development and maturation of the brain, and because the brain is too heterogeneous in terms of cellular composition, morphology, and function to draw any conclusions from analyses of only the tissue homogenization in toto. This chapter has summarized the *in situ* hybridization histochemical findings in the brain and has proposed several new questions. One example is the discrepant expression in a specific brain region of two given enzyme molecules which are considered in molecular biology to function in consecutive turn. Assuming that the expression at the level of a given mRNA represents that at the protein level and further at the activity level, the occurrence of the discrepant expression leads us to search for the possible existence of yet unidentified molecules as exemplified in the case of DAG kinase. Another example is the intracellular localization: although the stimulation-relating translocation of soluble signal-related enzymes between the membrane and the cytoplasmic matrix is a necessary step to understand the involvement mechanisms of membrane-residing phosphoinositides in signal transduction, the idea of such molecular translocation comes largely from biochemical analysis and its confirmation remains to be elucidated by detailed immunohistochemistry at ultrastructural levels. Altogether, more light should now be shed on *in vivo* analysis in the brain of the phosphoinositide-signaling phenomena by various histochemical techniques.

# REFERENCES

Abeliovich, A., Chen, C., Goda, Y., Silva, A.J., Stevens, C.F., and Tonegawa, S., Modified hippocampal long-term potentiation in PKC gamma-mutant mice, *Cell*, 75, 1253–1262, 1993.

Alessi, D.R., Deak, M., Casamayor, A., Caudwell, F.B., Morrice, N., Norman, D.G., Gaffney, P., Reese, C.B., MacDougall, C.N., Harbison, D., Ashworth, A., and Bownes, M., 3-Phosphoinositide-dependent protein kinase-1 (PDK1): structural and functional homology with the Drosophila DSTPK61 kinase, *Curr. Biol.*, 7, 776–789, 1997a.

Alessi, D.R., James, S.R., Downes, C.P., Holmes, A.B., Gaffney, P.R., Reese, C.B., and Cohen, P., Characterization of a 3 phosphoinositide-dependent protein kinase which phosphorylates and activates protein kinase Balpha, *Curr. Biol.*, 7, 261–269, 1997b.

Bailey, C.H. and Kandel, E.R., Structural changes accompanying memory storage, *Annu. Rev. Physiol.*, 55, 397–426, 1993.

Balla, T., Downing, G., Jaffe, H., Kim, S., Zolyomi, A., and Catt, K.J., Isolation and molecular cloning of Wortmannin-sensitive bovine type III phosphatidylinositol 4-kinases, *J. Biol. Chem.*, 272, 18358–18366, 1997.

Berridge, M.J., Inositol trisphosphate and calcium signaling, *Nature*, 361, 315–325, 1993.

Bloomquist, B.T., Shortridge, R.D., Schneuwly, S., Perdew, M., Montel, C., Steller, H., Rubin, G., and Pak, W., Isolation of a putative phospholipase C gene of *Drosophila, norpA*, and its role in phototransduction, *Cell*, 54, 723–733, 1988.

Boronenkov, I.V. and Anderson, R.A., The sequence of phosphatidylinositol-4-phosphate 5-kinase defines a novel family of lipid kinases, *J. Biol. Chem.*, 270, 2881–2884, 1995.

Brandt, S.J., Niedel, J.E., Bell, R.M., and Young, W.S., III, Distinct patterns of expression of different protein kinase C mRNAs in rat tissues, *Cell*, 49, 57–63, 1987.

Brown, R.A., Ho, L.K.F., Wever-Hall, S.J., Shipley, J.M., and Fly, M.J., Identification and cDNA cloning of a novel mammalian C2 domain-containing phosphoinositide 3-kinase HsC2-PI3K, *Biochem. Biophys. Res. Commun.*, 233, 537–544, 1997.

Bunting, M., Tang, W., Zimmerman, G.A., McIntyre, T.M., and Prescott, S.M., Molecular cloning and characterization of a novel human diacylglycerol kinase $\zeta$, *J. Biol. Chem.*, 271, 10230–10236, 1996.

Campuzano, V., Montermini, L., Molto, M.D., Pianese, L., Cossee, M., Cavalcanti, F., Monros, E., Rodius, F., Duclos, F., Monticelli, A., Zara, F., Canizares, J., Koutnikova, H., Nidichandoni, S.L., Gellera, C., Brice, A., Trouillas, P., De Michele, G., Gilla, A., De Frutos, R., Palau, F., Patel, P.I., Di Donato, S., Mandel, J.-L., Cocozza, S., Koenig, M., and Pandolfo, M., Friedreich's ataxia: autosomal recessive disease caused by an intronic GAA triplet repeat expansion, *Science*, 271, 1423–1427, 1996.

Carvajal, J.J., Pook, M.A., Doudney, K., Hillermann, R., Wilkes, D., Almahdawi, S., Williamson, R., and Chamberlain, S., Friedreich's ataxia: a defect in signal transduction?, *Hum. Mol. Genet.*, 4, 1411–1419, 1995.

Chantry, D., Vojtek, A., Kashishian, A., Holtzman, D.A., Wood, C., Gray, P.W., Cooper, J.A., and Hoekstra, M.F., p110δ, a novel phosphatidylinositol 3-kinase catalytic subunit that associates with p85 and is expressed predominantly in leucocytes, *J. Biol. Chem.*, 272, 19236–19241, 1997.

Chong, L.D., Traynor-Kaplan, A., Bokoch, G.M., and Schwartz, M.A., The small GTP-binding protein Rho regulates a phosphatidylinositol 4-phosphate 5-kinase in mammalian cells, *Cell*, 79, 507–513, 1994.

Cunningham, E., Tan, S.K., Swigart, P., Hsuan, J.J. Bankaitis, V., and Cockcroft, S., The yeast and mammalian isoforms of phosphatidylinositol transfer protein can all restore phospholipase C-mediated inositol lipid signaling in cytosol-depleted RBL-2H3 and HL-60 cells, *Proc. Natl. Acad. Sci. U.S.A.*, 93, 6589–6593, 1996.

Datta, S.R., Dudek, H., Tao, X., Masters, S., Fu, H., Gotoh, Y., and Greenberg, M.E., Akt phosphorylation of BAD couples survival signals to the cell-intrinsic death machinery, *Cell*, 91, 231–241, 1997.

Del Peso, L., Gonzalez-Garcia, M., Page, C., Herrera, R., and Nunez G., Interleukin-3-induced phosphorylation of BAD through the protein kinase Akt, *Science*, 278, 687–689, 1997.

Dickeson, S.K., Lim, C.N., Schuyler, G.T., Dalton, T.P., Helmkamp, Jr., G.M., and Yarbrough, L.R., Isolation and sequence of cDNA clones encoding rat phosphatidylinositol transfer protein, *J. Biol. Chem.*, 264, 16557–16564, 1989.

Divecha, N., Banfic, H., and Irvine, R.F., Inosides and the nucleus and inosides in the nucleus, *Cell*, 74, 405–407, 1993.

Divecha, N. and Irvine, R.F., Phospholipid signaling, *Cell*, 80, 269–278, 1995.

Divecha, N., Truong, O., Hsuan, J.J., Hinchliffe, K.A., and Irvine, R.F., The cloning and sequence of the C isoform of PtdIns4P 5-kinase, *Biochem. J.*, 309, 715–719, 1995.

Fein, A., Excitation and adaptation of *Limbus* photoreceptors by light and inositol 1, 4, 5-triphosphate, *TINS*, 93, 110–114, 1986.

Ferreira, P.A. and Pak, W.L., Bovine phospholipase C highly homologous to the *norpA* protein of *Drosophila* is expressed specifically in cones, *J. Biol. Chem.*, 269, 3129–3131, 1994.

Ferreira, P.A., Shortridge, R.D., and Pak, W.L. Distinctive subtypes of bovine phospholipase C that have preferential expression in the retina and high homology to the *norpA* gene product of *Drosophila*, *Proc. Natl. Acad. Sci. U.S.A.*, 90, 6042–6046, 1993.

Franke, T.F., Kaplan, D.R., and Cantley, L.C. PI3K: Downstream AKTion blocks apoptosis, *Cell*, 88, 435–437, 1997.

Garcia-Bustos, J.F., Marini, F., Stevenson, I., Frei, C., and Hall, M.N., PIK1, an essential phosphatidylinositol 4-kinase associated with the yeast nucleus, *EMBO J.*, 13, 2352–2361, 1994.

Gehm, B.D. and McConnell, D.G., Phosphatidylinositol-4, 5-bisphosphate phospholipase C in bovine rod outer segments, *Biochemistry*, 29, 5447–5452, 1990.

Gehrmann, T. and Heilmeyer, Jr., M.G., Phosphatidylinositol 4-kinases, *Eur. J. Biochem.*, 253, 357–370, 1998.

Ghosh, S., Strum, J.C., Sciorra, V.A., Daniel, L., and Bell, R.M., Raf-1 kinase possesses distinct binding domains for phosphatidylserine and phosphatidic acid. Phosphatidic acid regulates the translocation of Raf-1 in12-O-tetradecanoylphorbol-13-acetate-stimulated Madin-Darby canine kidney cells, *J. Biol. Chem.*, 271, 8472–8480, 1996.

Goto, G. and Kondo, H., Molecular cloning and expression of a 90-kDa diacylglycerol kinase that predominantly localizes in neurons, *Proc. Natl. Acad. Sci. U.S.A.*, 90, 7598–7602, 1993.

Goto, K. and Kondo, H., A 104-kDa diacylglycerol kinase containing ankyrin-like repeats localizes in the cell nucleus, *Proc. Natl. Acad. Sci. U.S.A.*, 93, 11196–11201, 1996.

Goto, K., Funayama, M., and Kondo, H., Cloning and expression of a cytoskeleton-associated diacylglycerol kinase that is dominantly expressed in cerebellum, *Proc. Natl. Acad. Sci. U.S.A.*, 91, 13042–13046, 1994.

Goto, K., Watanabe, M., Kondo, H., Yuasa, H., Sakane, F., and Kanoh, H., Gene cloning, sequence, expression and in situ localization of 80 kDa diacylglycerol kinase specific to oligodendrocyte of rat brain, *Mol. Brain Res.*, 16, 75–87, 1992.

Hay, J.C. and Martin, T.F. J., Phosphatidylinositol transfer protein required for ATP-dependent priming of Ca-activated secretion, *Nature*, 366, 572–575, 1993.

Hay, J.C., Fisette, P.L., Jenkins, G.H., Fukami, K., Takenawa, T., Anderson, R.A., and Martin, T.F.J., ATP-dependent inositide phosphorylation required for Ca-activated secretion, *Nature*, 374, 173–177, 1995.

Heacock, A.M., Uhler, M.D., and Agranoff, B.W., Cloning of CDP-diacylglycerol synthase from a human neuronal cell line, *J. Neurochem.*, 67, 2200–2203, 1996.

Hiles, I., Otsu, M., Volinia, S., Fly, M.J., Gout, I., Dhand, R., Panayotou, G., Ruiz-Larrea, F., Thompson, A.S., Totty, N.F., Hsuan, J.J., Courtneidge, S.A., Parker, P.J., and Waterfield, M.D., Phosphatidylinositol 3-kinase: structure and expression of the 110 kd catalytic subunit, *Cell*, 70, 419–429, 1992.

Hotta, Y. and Benzer, S., Genetic dissection of the *Drosophila* nervous system by means of mosaics, *Proc. Natl. Acad. Sci. U.S.A.*, 67, 1156–1163, 1970.

Houamed, K.M., Kuijper, J.L., Gilbert, T.L., Haldeman, B.A., O'Hara, P.J., Mulvihill, E.R., Almers, W., and Hagen, F.S., Cloning, expression, and gene structure of a G protein-coupled glutamate receptor from rat brain, *Science*, 252, 1318–1321, 1991.

Houssa, B., Schaap, D., van der Wal, J., Goto, K., Kondo, H., Yamakawa, A., Shibata, M., Takenawa, T., and van Blitterswijk, W.J., Cloning of a novel human diacylglycerol kinase (DGKθ) containing three cysteine-rich domains, a proline-rich region, and a pleckstrin homology domain with an overlapping Ras-associating domain, *J. Biol. Chem.*, 272, 10422–10428, 1997.

Hu, P., Mondino, A., Skolnik, E.Y., and Schlessinger, J., Cloning of a novel, ubiquitously expressed human phosphatidylinositol 3-kinase and identification of its binding site on p85, *Mol. Cell. Biol.*, 13, 7677–7688, 1993.

Ishihara, H., Shibasaki, Y., Kizuki, N., Katagiri, H., Yazaki, Y., Asano, T., and Oka, Y., Cloning of aDNAs encoding two isoforms of 68-kDa type I phosphatidylinositol-4-phosphate 5-kinase, *J. Biol. Chem.*, 271, 23611–23614, 1996.

Ishihara, H., Shibasaki, Y., Kizuki, N., Wada, T., Yazaki, Y., Asano, T., and Oka, Y., Type I phosphatidylinositol-4-phosphate 5-kinases, *J. Biol. Chem.*, 273, 8741–8748, 1998.

Ito, Y., Sakagami, H., and Kondo, H., Enhanced gene expression for phosphatidylinositol 3-kinase in the hypoglossal motoneurons following axonal crush, *Mol. Brain Res.*, 37, 329–332, 1996.

Ito, Y., Goto, K., and Kondo, H. Localization of mRNA for phosphatidylinositol 3-kinase in brain of developing and mature rats, *Mol. Brain Res.*, 94, 149–153, 1995.

Jenkins, G.H., Fisett, P.L., and Anderson, R.A., Type I phosphatidylinositol 4-phosphate 5-kinase isoforms are specifically stimulated by phosphatidic acid, *J. Biol. Chem.*, 269, 11547–11554, 1994.

Jhon, D.-Y., Lee, H.-H., Park, D., Lee, C.-W., Lee, K.-H., Yoo, O.J., and Rhee, S.G., Cloning, sequencing, purification, and Gq-dependent activation of phospholipase C-beta 3, *J. Biol. Chem.*, 268, 6654–6661, 1993.

Jones, G.A. and Carpenter, G., The regulation of phospholipase C-γ1 by phosphatidic acid, *J. Biol. Chem.*, 268, 20845–20850, 1993.

Kai, M., Sakane, F., Imai, S., Wada, I., and Kanoh, H., Molecular cloning of a diacylglycerol kinase isozyme predominantly expressed in human retina with a truncated and inactive enzyme expression in most other human cells, *J. Biol. Chem.*, 269, 18492–18498, 1994.

Kauffmann-Zeh, A., Thomas, G.M.H., Ball, A., Prosser, S., Cunningham, E., Cockcroft, S., and Hsuan, J.J., Requirement for phosphatidylinositol transfer protein in epidermal growth factor signaling, *Science*, 268, 1188–1190, 1995.

Khan, W.A., Blobe, G.C., Richards, A.L., and Hannun, Y.A., Identification, partial purification, and characterization of a novel phospholipid-dependent and fatty acid-activated protein kinase from human platelets, *J. Biol. Chem.*, 269, 9729–9735, 1994.

Kimura, K., Hattori, S., Kabuyama, Y., Shizawa, Y., Takayanagi, J., Nakamura, S., Toki, S., Matsuda, Y., Onodera, K and Fukui, Y., Neurite outgrowth of PC12 cells is suppressed by Wortmannin, a specific inhibitor of phosphatidylinositol 3-kinase, *J. Biol. Chem.*, 269, 18961–18967, 1994.

Klauck, T.M., Xu, Z., Mousseau, B., and Jaken, S., Cloning and characterization of a glucocorticoid-induced diacylglycerol kinase, *J. Biol. Chem.*, 271, 19781–19788, 1996.

Kohyama-Koganeya, A., Watanabe, M., and Hotta, Y., Molecular cloning of a diacylglycerol kinase isozyme predominantly expressed in rat retina, *FEBS Lett.*, 409, 258–264, 1997.

Konishi, H., Shimomura, T., Kuroda, S., Ono, Y., and Kikkawa, U., Cloning of rat RAC protein kinase alpha and beta and their association with protein kinase C zeta, *BBRC*, 205, 817–825, 1994.

Konishi, H., Kuroda, S., Tanaka, M., Matsuzaki, H., Ono, Y., Kameyama, K., Haga, T. and Kikkawa, U., Molecular cloning and characterization of a new member of the RAC protein kinase family: association of the pleckstrin homology domain of three types of RAC protein kinase with protein kinase C subspecies and bg subunits of G proteins, *BBRC*, 216, 526–534, 1995.

Lee, C.-W., Park, D.J., Lee, K.-H. Kim, C.G., and Rhee, S.G., Purification, molecular cloning, and sequencing of phospholipase C-beta 4, *J. Biol. Chem.*, 268, 21318–21327, 1993.

Lee, C.-W., Lee, K.-H., Lee, S.B., Park, D., and Rhee, S.G., Regulation of phospholipase C-beta 4 by ribonucleotides and the alpha subunit of Gq, *J. Biol. Chem.*, 269, 25335–25338, 1994.

Limatola, C., Schaap, D., Moolenaar, W.H., and van Blitterswijk, W.J., Phosphatidic acid activation of protein kinase C-zeta overexpressed in COS cells: comparison with other protein kinase C isotypes and other acidic lipids, *Biochem. J.*, 304, 1001–1008, 1994.

Liscovitch, M. and Cantley, L.C., Lipid second messengers, *Cell*, 77, 329–334, 1994.

Liscovitch, M. and Cantley, L.C., Signal transduction and membrane traffic: the PITP/phosphoinositide connection, *Cell*, 81, 659–662, 1995.

Liscovitch, M., Chalifa, V., Pertile, P., Chen, C.-S., and Cantley, L.C., Novel function of phosphatidylinositol 4, 5-bisphosphate as a cofactor for brain membrane phospholipase D, *J. Biol. Chem.*, 269, 21403–21406, 1994.

Liu, N., Fukami, K., Yu, H., and Takenawa, T., A new phospholipase C δ4 is induced at S-phase of the cell cycle and appears in the nucleus, *J. Biol. Chem.*, 271, 355–360, 1996.

Loijens, J.C. and Anderson, R.A., Typw I phosphatidylinositol-4-phosphate 5-kinases are distinct members of this novel lipid kinase family, *J. Biol. Chem.*, 271, 32937–32943, 1996.

MacDonald, M.L., Mack, K.D., Williams, B.W., King, W.C., and Glomset, J.A., A membrane-bound diacylglycerol kinase that selectively phosphorylates arachidonyl-diacylglycerol, *J. Biol. Chem.*, 263, 1584–1592, 1988.

Martelli, A.M., Gilmour, R.S., Bertagnoto, V., Neri, L.H., Manzol, L., and Cocco, L., Nuclear localization and signalling activity of phosphoinositidase C beta in Swiss 3T3 cells, *Nature*, 358, 242-245, 1992.

Masu, M., Tanabe, Y., Tsuchida, K., Shigemoto, R., and Nakanishi, S., Sequence and expression of a metabotropic glutamate receptor, *Nature*, 349, 760–765, 1991.

McPhail, L.C., Qualliotine-Mann, D., and Waite, K.A., Cell-free activation of neutrophil NADPH oxidase by a phosphatidic acid-regulated protein kinase, *Proc. Natl. Acad. Sci. U.S.A.*, 92, 7931–7935, 1995.

Meyers, R. and Cantley, L.C., Cloning and characterization of a Wortmannin-sensitive human phosphatidylinositol 4-kinase, *J. Biol. Chem.*, 272, 4384–4390, 1997.

Misawa, H., Ohtsubo, M., Copeland, N.G., Gilbert, D.J., Jenkins, N.A., and Yoshimura, A., Cloning and characterization of a novel class II phosphoinositide 3-kinase containing C2 domain, *Biochem. Biophys. Res. Commun.*, 244, 531–539, 1998.

Moritz, A., DeGraan, P.N.E., Gispen, W.H., and Wirtz, K.W.A., Phosphatidic acid is a specific activator of phosphatidylinositol-4-phosphate kinase, *J. Biol. Chem.*, 267, 7207–7210, 1992.

Nakagawa, T., Goto, K., and Kondo, H., Cloning, expression, and localization of 230-kDa phosphatidylinositol 4-kinase, *J. Biol. Chem.*, 271, 12088–12094, 1996a.

Nakagawa, T., Goto, K., and Kondo, H., Cloning and characterization of a 92 kDa soluble phosphatidylinositol 4-kinase, *Biochem. J.*, 320, 643–649, 1996b.

Nakanishi, H., Brewer, K.A., and Exton, J.H., Activation of the zeta isozyme of protein kinase C by phosphatidylinositol 3, 4, 5-triphosphate, *J. Biol. Chem.*, 268, 13–16, 1993.

Ohashi, M., DeVries, K.J., Frank, R., Snoek, G., Bankaitis, V., Wirtz, K., and Huttner, W.B., A role for phosphatidylinositol transfer protein in secretory vesicle formation, *Nature (London)*, 377, 544–547, 1995.

Ohishi, H., Ogawa-Meguro, R., Shigemoto, R., Kaneko, T., Nakanishi, S., and Mizuno, N., Immunohistochemical localization of metabotropic glutamate receptors, mGluR2 and mGluR3, in rat cerebellar cortex, *Neuron*, 13, 55–66, 1994.

Ono, F. Nakagawa, T., Saito, S., Owada, Y., Sakagami, H., Goto, K., Suzuki, M., Matsuno, S., and Kondo, H., A novel class II phosphoinositide 3-kinase predominantly expressed in the liver and its enhanced expression during liver regeneration, *J. Biol. Chem.*, 273, 7731–7736, 1998.

Owada, Y., Utsunomiya, A., Yoshimoto, T., and Kondo, H., Expression of mRNA for Akt, serine-threonine protein kinase, in the brain during development and its transient enhancement following axotomy of hypoglossal nerve, *J. Mol. Neurosci.*, 9, 27–33, 1997.

Palmer, R.H., Dekker, L.V., Woscholski, R., Le Good, J.A., Gigg, R., and Parker, P.J., Activation of PRK1 by phosphatidylinositol 4, 5-bisphosphate and phosphatidylinositol 3, 4, 5-trisphosphate, *J. Biol. Chem.*, 270, 22412–22416, 1995.

Panfoli, I., Morelli, A., and Pepe, I., Calcium ion-regulated phospholipase C activity in bovine rod outer segments, *BBRC*, 173, 283–288, 1990.

Payrastre, B., Nievers, M., Boonstra, J., Breton, H., Verkleiji, A.J., and Van Begen en Henegouwen, P.M.P., A differential location of phosphoinositide kinases, diacylglycerol kinase, and phospholipase C in the nuclear matrix, *J. Biol. Chem.*, 267, 5078–5084, 1992.

Peng, Y.-W., Rhee, S.G., Yu, W.-P., Ho, Y.-K., Shoen, T., Chader, G.J., and Yau, K.-W., Identification of components of a phosphoinositide signaling pathway in retinal rod outer segments, *Proc. Natl. Acad. Sci. U.S.A.*, 94, 1995–2000, 1997.

Rameh, L.E., Tolias, K.F., Duckworth, B.C., and Cantley, L.C., A new pathway for synthesis of phosphatidylinositol-4, 5-bisphosphate, *Nature*, 390, 192–196, 1997.

Randazzo, P.A. and Kahn, R.A., GTP hydrolysis by ADP-ribosylation factor is dependent on both an ADP-ribosylation factor GTPase-activating protein and acid phospholipids, *J. Biol. Chem.*, 269, 10758–10763, 1994.

Ranganathan, R., Malicki, D.M., and Zuker, C.S., Signal transduction in Drosophila photoreceptors, *Annu. Rev. Neurosci.*, 18, 283–317, 1995.

Ren, X.D., Bokoch, G.M., Traynor-Kaplan, A., Jenkins, G.H., Anderson, R.A., and Schwartz, M.A., Physical association of the small GTPase Rho with a 68-kDa phosphatidylinositol 4-phosphate 5-kinase in Swiss 3T3 cells, *Mol. Biol. Cell.*, 7, 435–442, 1996.

Rhee, S.G. and Bae, Y.S., Regulation of phosphoinositide-specific phospholipase C isozymes, *J. Biol. Chem.*, 272, 15045–15048, 1997.

Rhee, S.G., Suh, P.-G., Ryu, S.-H., and Lee, S.Y., Studies of inositol phospholipid-specific phospholipase C, *Science*, 244, 546–550, 1989.

Ross, C.A., MacCumber, M.W., Glatt, C.E., and Snyder, S.H., Brain phospholipase C isozymes: differential mRNA localizations by in situ hybridization, *Proc. Natl. Acad. Sci. U.S.A.*, 86, 2923–2927, 1989.

Saito, S., Goto, K., Tonosaki, A., and Kondo, H., Gene cloning and characterization of CDP-diacylglycerol synthase from rat brain, *J. Biol. Chem.*, 272, 9503–9509, 1997.

Saito, S., Sakagami, H., Tonosaki, A., and Kondo, H., Localization of mRNAs for CDP-diacylglycerol synthase and phosphatidylinositol synthase in the brain and retina of developing and adult rats, *Dev. Brain Res.*, 21–30, 1998.

Sakane, F., Yamada, K., Kanoh, H., Yokoyama, C., and Tanabe, T., Procine diacylglycerol kinase sequence has zinc finger and E-F hand motifs, *Nature*, 344, 345–348, 1990.

Sakane, S., Imai, S.-I. Kai, M., Wada, I., and Kanoh, H. Molecular cloning of a novel diacylglycerol kinase isozyme with a pleckstrin homology domain and a C-terminal tail similar to those of the EPH family of protein-tyrosine kinases, *J. Biol. Chem.*, 271, 8394–8401, 1996.

Savitsky, S., Bar-Shira, A., Gilad, S., Rotman, G., Ziv, Y., Vanagaite, L., Tagle, D.A., Smith, S., Uziel, T., Sfez, S., Ashkerazi, M., Pecker, I., Frydman, M., Harnik, R., Patanjali, S.R., Simmons, A., Clines, G.A., Sartiel, A., Gatti, R.A., Chessa, L., Sanal, O., Lavin, M.F., Jaspers, N.G.J., Malcolm, A., Taylor, R., Arlett, C.F., Miki, T., Weissman, S.M., Lovett, M., Collins, F.S., and Shiloh, Y., A single ataxia telangiectasia gene with a product similar to PI-3 kinase, *Science*, 268, 1749–1753, 1995.

Soltoff, S.P., Rabin, S.L., Cantley, L.C., and Kaplan, D.R. Nerve growth factor promotes the activation of phosphatidylinositol 3-kinase and its association with the trk tyrosine kinase, *J. Biol. Chem.*, 267, 17472–17477, 1992.

Stokoe, D., Stephens, L.R., Copeland, T., Gaffney, P.R., Reese, C.B., Painter, G.F., Holmes, A.B., McCormick, F., and Hawkins, P.T., Dual role of phosphatidylinositol-3, 4, 5-trisphosphate in the activation of protein kinase B, *Science*, 277, 567–70, 1997.

Stoyanov, B., Volinia, S., Hanck, T., Rubio, I., Loubtchenkov, M., Malek, D., Stoyanov, S., Vanhaesebroeck, B., Dhand, R., Nurnberg, B., Giershik, P., Sedorf, K., Hsuan, J.J., Waterfield, M.D., and Wetzker, R., Cloning and characterization of a G protein-activated human phosphoinositide-3 kinase, *Science*, 269, 690–693, 1995.

Stryer, L., (1986) Cyclic GMP cascade of vision, *Annu. Rev. Neurosci.*, 9, 87–119, 1986.

Tanaka, O. and Kondo, H., Localization of mRNAs for three novel members ($\beta$3, $\beta$4 and $\gamma$2) of phospholipase C family in mature rat brain, *Neurosci. Lett.*, 182, 17–20, 1994.

Tanaka, S. and Hosaka, K., Cloning of a cDNA encoding a second phosphatidylinositol transfer protein of rat brain by complementation of the yeast sec 14 mutation, *J. Biochem.*, 115, 981–984, 1994.

Tanaka, S., Nikawa, J.-I., Yamashita, S., and Hosaka, K., Molecular cloning of rat phosphatidylinositol synthase cDNA by functional complementation of the yeast Saccharomyces cervisiae pis mutation, *FEBS Lett.*, 393, 89–92, 1996.

Tang, W., Bunting, M., Zimmerman, G.A., McIntyre, T.M., and Prescott, S.M., Molecular cloning of a novel human diacylglycerol kinase highly selective for arachidonate-containing substrates, *J. Biol. Chem.*, 271, 10237–10241, 1996.

Terui, T., Kahn, R.A., and Randazzo, P.A., Effects of acid phospholipids on nucleotide exchange properties of ADP-ribosylation factor 1. Evidence for specific interaction with phosphatidylinositol 4, 5-bisphosphate, *J. Biol. Chem.*, 269, 28130–28135, 1994.

Thomas, G.M.H., Cunningham, E., Fensome, A., Ba, A., Totty, N.F., Truong, O., Hsuan, J.J., and Cockcroft, S., An essential role for phosphatidylinositol transfer protein in phospholipase C-mediated inositol lipid signaling, *Cell*, 74, 919–928, 1993.

Toker, A. and Cantley, L.C., Signalling through the lipid products of phosphoinositide-3-OH kinase, *Nature*, 387, 673–676, 1997.

Tolias, K.F., Cantley, L.C., and Carpenter, C.L., Rho family GTPases bind to phosphoinositide kinases, *J. Biol. Chem.*, 270, 17656–17659, 1995.

Utsunomiya, A., Owada, Y., Yoshimoto, T., and Kondo, H., Localization of gene expression for phosphatidylinositol transfer protein in the brain of developing and mature rats, *Mol. Brain Res.*, 45, 349–352, 1997.

Vanhaesebroeck, B., Leevers, S.J., Panayotou, G., and Waterfield, M.D., Phosphoinositide 3-kinases: a conserved family of signal transducers, *TIBS*, 22, 267–272, 1997.

Vanhaesebroeck, B., Welham, M.J., Kotani, K., Stein, R., Downward, J., and Waterfield, M.D., p110δ, a novel phosphoinositide 3-kinase in leukocytes, *Proc. Natl. Acad. Sci. U.S.A.*, 94, 4330–4335, 1997.

Volinia, S., Dhand, R., Vanhaesebroeck, B., MacDougall, L.K., Stein, R., Zvelebil, M.J., Domin, J., Penaretou, C., and Waterfield, M.D., A human phosphatidylinositol 3-kinase complex related to the yeast Vps34p-Vps15p protein sorting system, *EMBO J.*, 14, 3339–3348, 1995.

Walsh, J.P., Suen, R., Lemaitre, R.N., and Glomset, J.A., Arachidonyl-diacylglycerol kinase from bovine testis, *J. Biol. Chem.*, 269, 21155–21164, 1994.

Weeks, R., Dowhan, W., Shen, H., Balantac, N., Meengs, B., Nudelman, B.E., and Leung, D.M., Isolation and expression of an isoform of human CDP-diacylglycerol synthase cDNA, *DNA Cell Biol.*, 16, 281–289, 1997.

Wiedemann, C., Schafer, T., and Burger, M.M.B., Chromaffin granule-associated phosphatidylinositol 4-kinase activity is required for stimulated secretion, *EMBO J.*, 15, 2094–2101, 1996.

Wirtz, K.W.A., Phospholipid transfer proteins, *Annu. Rev. Biochem.*, 60, 73–99, 1991.

Wong, K. and Cantley, L.C., Cloning and characterization of a human phosphatidylinositol 4-kinase, *J. Biol. Chem.*, 269, 28878–28884, 1994.

Wong, K., Meyers, B., and Cantley, L.C., Subcellular locations of phosphatidylinositol 4-kinase isoforms, *J. Biol. Chem.*, 272, 13236–13241, 1997.

Wu, L., Niemeyer, B., Colley, N., Socolich, M., and Zuker, C.S., (1995) Regulation of PLC-mediated signalling in vivo by CDP-diacylglycerol synthase, *Nature*, 373, 216–222, 1995.

Yamaguchi, H., Aiba, A., Nakamura, K., Nakao, K., Sakagami, H., Goto, K., Kondo, H., and Katsuki, M., Dopamine D2 receptor plays a critical role in cell proliferation and proopiomelanocortin expression in the pituitary, *Genes Cells*, 1, 253–268, 1996.

York, J.D. and Majerus, P.W., Nuclear phosphatidylinositols decrease during S-phase of the cell cycle in HeLa cells, *J. Biol. Chem.*, 269, 7847–7850.

Yoshida, S., Ohya, Y., Goebl, M., Nakano, A., and Anraku, Y., A novel gene, STT4, encodes a phosphatidylinositol 4-kinase in the PKC1 protein kinase pathway of Saccharomyces cerevisiae, *J. Biol. Chem.*, 269, 1166–1171, 1994.

Zhao, Z., Shen, S.-H., and Fischer, E.H., Stimulation by phospholipids of a protein-tyrosine-phosphatase containing two src homology 2 domains, *Proc. Natl. Acad. Sci. U.S.A.*, 90, 4251–4252, 1993.

Zheng, F. and Gallagher, J.P., (1992) Metabotropic glutamate receptors are required for the induction of long-term potentiation, *Neuron*, 9, 163–172, 1992.

# 15 Genes in Differentiation, Survival, and Degeneration of Midbrain Dopaminergic Neurons

*Jin H. Son*

## CONTENT

## INTRODUCTION

Midbrain dopaminergic (DA) neurons are divided anatomically into three nuclear groups: substantia nigra (SN, A9), ventral tegmental area (VTA, A10), and retrorubral field (A8). Based on their efferent projections, these neurons are classified into the mesostriatal system and the mesolimbic and mesocortical systems. The former system is involved in the integration of incoming sensory stimuli and control of voluntary movement. Selective degeneration of the mesostriatal DA system results in Parkinson's disease (PD), which lowers dopamine levels in striatum and disrupts the motor control circuit (Jankovic and Tolosa, 1993; Koller and Paulson, 1995; Fallon and Loughlin, 1994). Because severity of the clinical symptom in PD directly correlates with the extent of DA cell loss in SN, a vast majority of current research has focused on both the prevention of further DA cell death and repopulation of DA neurons in PD brain. The mesolimbic and mesocortical systems are believed to be concerned with cognitive, reward, and emotional behavior and functionally important in psychotic disorders as well as in the therapeutic effects of neuroleptic drugs (Leonard, 1997). This chapter will first review much of the current progress in the biology of DA neuronal development in SN and its survival and degeneration in PD model systems, primarily at the molecular level, and will conclude by raising the important issues in delineating the developmental, genetic, and environmental factors regarding the degeneration of DA neurons in PD, in addition to genetic tools to manipulate DA neurons both *in vivo* and *in vitro*.

## DIFFERENTIATION OF MIDBRAIN DOPAMINERGIC NEURONS

### MONITORING DOPAMINERGIC NEUROGENESIS BY IMMUNO- AND X-GAL HISTOCHEMICAL STAINING

The early ontogenic study of midbrain DA neurons was performed by using histofluorescence technique, which was later replaced by immunohistochemical detection of tyrosine hydroxylase (TH) in DA neurons (Specht et al., 1981a). TH catalyzes the first and rate limiting step in the biosynthesis of dopamine, which is an important phenotypic hallmark of DA neurons (Nagatsu et al., 1964). In the adult brain the expression of TH is restricted to a small number of distinct neuronal groups, including the DA neurons of the SN, VTA, hypothalamus, olfactory bulb, and retina, as well as the noradrenergic neurons of the locus ceruleus and subceruleus, and several smaller groups of noradrenergic and adrenergic neurons in the lower brainstem (see Figure 15.1B) (Hökfelt et al., 1984; Moore and Bloom, 1978). This highly specific localization made the expression of TH an important marker gene to monitor DA differentiation in the midbrain. In the developing mouse brain, between embryonic days 9.5 and 10.5 (E9.5–E10.5), the first visible DA neurons were identifiable in the ventral portion of the midbrain by both TH immunohistochemistry and X-gal histochemical staining of lacZ reporter transgene (Figure 15.1*). In our ontogenic study, using transgenic mouse embryos expressing TH-lacZ transgene, abundance of midbrain DA neurons was first visible at E10.5. After the exuberant appearance of lacZ-positive cells, a localized and diminished adult-like pattern of lacZ expression emerged in SN-VTA on E13.5 when DA fibers enter the striatum (Son et al., 1996). In developing rat brain, the first TH-immunopositive and catecholamine fluorescent progenitor cells for the midbrain DA neurons appear in the ventral midbrain at E12.5, and these DA progenitor cells migrate into three cell groups, A8–A10, by E17 (Specht et al., 1981b). In human the sequence of developmental events for the midbrain DA neurons is very similar to that in mice (Freeman et al., 1991; Freeman et al., 1995) except the duration of the developmental period. In the midbrain of human embryos, TH-immunopositive cells are first detected at 5 to 6 weeks gestational age. Distinct DA neuritic processes begin to extend by 8 weeks and DA neurites are first seen in the developing putamen at 9 weeks gestation.

### MOLECULAR SIGNALS FOR LOCALIZATION IN NEURAL PLATE: SHH AND FGF8

Rapid advances in molecular analysis of vertebrate nervous system have uncovered various molecules which appear to be crucial to the phenotypic differentiation of midbrain DA neurons. The embryonic induction of midbrain DA neurons depends on the interaction between induction signals and target progenitor cells competent to respond to specific signaling molecules. The most recent findings demonstrate that midbrain DA neurons are induced in stereotypic locations along the dorsal-ventral (D-V) and anterior-posterior (A-P) axes of the neural tube through a cooperative interaction between two distinct signaling molecules, Sonic hedgehog (Shh) and fibroblast growth factor 8 (FGF8) (Ye et al., 1998). Shh is one of the vertebrate homologs of the Drosophila segment polarity gene hedgehog encoding a family of signaling proteins, which is expressed in the signaling centers, such as notochord, floor plate, and the zone of polarizing activity (Echelard et al., 1993). These signaling centers are thought to mediate central nervous system (CNS) and limb polarity. Thus, Shh is thought to play a key role in the CNS signaling centers and is implicated in the regulation of CNS polarity. In other words, Shh has a ventralizing activity along the D-V axis of the neural tube, which is known to induce multiple ventral neuronal phenotypes, such as DA,

---

* All color plates appear after p. 272.

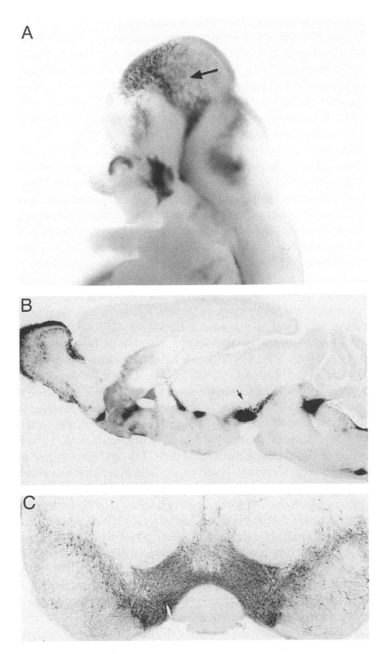

**FIGURE 15.1**   Monitoring the phenotype of DA neurons by X-gal histochemistry and TH immunostaining. (A) X-gal staining of the transgenic embryonic brain carrying the TH 9.0-kb lacZ transgene at E10.5 (Son et al., 1996). The reporter β-galactosidase-positive (or TH-positive) progenitor neurons are visible in the mesencephalon (see arrow). (B) The tissue-specific expression of TH monitored by X-gal staining in the sagittal section of the adult transgenic mouse brain carrying the TH 9.0-kb lacZ (Min et al., 1996). The DA neurons in the SN-VTA region are marked by an arrow. The other DA neurons in the olfactory bulb and the hypothalamus and noradrenergic neurons in locus ceruleus are clearly visible. (C) Coronal section through the SN-VTA region of the adult mouse brain showing DA neurons and neuronal fibers immunostained for TH.

serotonergic, and motor neurons in a concentration-dependent manner (Ericson et al., 1995; Ye et al., 1998). In particular, DA neuronal phenotype develops near the midbrain floor plate by the floor plate-produced inductive signal, Shh, which was recapitulated in *in vitro* explant cultures by Hynes et al. (1995). It was demonstrated that the *in vitro* induction of DA neuronal differentiation could be achieved by the amino-terminal product of Shh (Shh-N). The Shh-N can induce very effectively DA neurons in forebrain/midbrain explant cultures from E9 rat embryos. However, it is not known the true identity of the DA progenitor cells present in the brain explants, which could be the Ptx3-expressing cells (see the next section).

Of interest, agonists of cyclic AMP-dependent protein kinase A (PKA), which is known to antagonize hedgehog signaling pathways, are able to block DA neuronal induction in the explant culture (Hynes et al., 1995). The repression of PKA signaling pathways by Shh signaling appears to be essential for DA phenotype induction during development. Despite the importance of Shh in DA induction, little is known about its receptor, signaling mechanism, and target genes in the early DA progenitor cells. From studies in Drosophila hedgehog and its receptors Smoothened (Smo) and Patched (Ptc), the vertebrate homologs of Smo and Ptc were identified as a receptor complex to which Shh-N binds (Stone et al., 1996). Since Ptc binds Shh-N with high affinity and is coexpressed with Smo in many tissues, Ptc is considered a ligand-binding component for Shh and Smo a signaling component of the Shh-Ptc-Smo receptor complex. In this model, it is still not known whether Smo itself is constitutively active and Smo requires a specific ligand for activation in the absence of Ptc. Although the details of Shh signaling pathways are unknown, Gli1 is identified as a midline target of Shh signaling as a transcriptional regulator during induction of floor plate cells and ventral DA neurons (Lee et al., 1997; Hynes et al., 1997). Gli1 is a zinc finger transcription factor and a vertebrate homolog of the Drosophila gene Cubitus interruptus (Ci). When Gli1 was ectopically expressed in the dorsal midbrain and hindbrain of transgenic mice using the En-2 promoter, the cell pattern in the neural tube was strikingly similar to that of transgenic mice overexpressing Shh-N in the same tissues. Gli1 induced the ventral neural tube markers, such as Shh, Ptc, and the winged-helix transcription factor HNF-3β, and suppressed the dorsal markers, such as AL-1 and Pax-3. In addition, the ectopic expression of Gli1 induced two ventral cells, DA, and serotonergic neurons in ectopic dorsal neural tube. Inconsistent with these observations, Gli1 expression was dramatically enhanced in dorsal midbrain-hindbrain regions in the Shh-N transgenic mice. Since the Shh signaling seems to repress PKA signaling pathways, it is the open possibility that Gli1 can interact with cAMP-responsive transcription factor, CREB, or be modulated by cAMP-dependent PKA.

FGF8 is expressed in the locally recognized organizing (or signaling) centers that regulate the growth and patterning of different regions of the embryo including the limb, the elongating body axis, the face, and the midbrain/hindbrain region (Crossley and Martin, 1995). In particular, FGF8 expression is identified in early brain development in both the forebrain and the isthmus at the junction of midbrain and hindbrain. The isthmus is the region of a constriction in the embryonic midbrain/hindbrain boundary and is known as an organizing center for midbrain development (Martin and Puelles, 1994). For the establishment of the normal mid/hindbrain (isthmic) organizer, Gbx2, a vertebrate homolog of the Drosophila unplugged homeobox gene, is thought to be essential (Wassarman et al., 1997). In the mouse embryos Gbx2 was detected at E7.5 in all three germ layers and at E9.5, expressed in a transverse isthmic ring by coinciding with FGF8 expression. However, it remains to be determined how Gbx2 protein regulates the FGF expression in the isthmus. The expression of FGF8 in the isthmus was first detected between E8.0 and E8.5 and lasted until E12.5 in developing mouse embryo. During this period midbrain DA neurons were born and being positioned at the appropriate location for further maturation.

In the chicken embryos a FGF8-soaked bead was able to induce the transformation of the caudal diencephalon to ectopic midbrain by acting as a new isthmus-like organizing center (Crossley et al., 1996). After a FGF8-bead was implanted three known isthmus marker genes — En2, Fgf8, and Wnt1 — were also induced in the diencephalon in the vicinity of the bead. Thus, it appears that FGF8 is sufficient to induce a complete midbrain in competent neuroepithelium. During midbrain development the isthmus-derived signal molecule, FGF8, appears to control the position of the midbrain DA neurons along the A-P axes of the neural tube (Ye et al., 1998). Because the isthmus graft can induce DA neurons in ectopic ventrocaudal forebrain explants, whereas the floor plate graft can induce DA neurons only in ectopic dorsal locations in the midbrain and rostral forebrain, but not in the caudal forebrain explants. Moreover, Shh can induce ectopic DA neurons in the dorsal midbrain and rostral forebrain, but not in ventro- and dorsocaudal forebrain or hindbrain (Ye et al., 1998). The ectopic expression of FGF8 in midbrain is sufficient to activate En-2 and ELF-1 expression, which are normally expressed in a decreasing caudal-to-rostral gradient in the posterior midbrain. Thus, in a possible molecular hierarchy, FGF8 signaling produces the graded En-2 expression, which in turn specifies the graded expression of ELF-1 for A-P polarity (Lee et al., 1997). Shh can control the position of DA neurons along the D-V axis, but not the A-P axis. Therefore, isthmus-derived signal FGF8 and floor plate-derived signal Shh control the position of DA neurons along the A-P and D-V axes, respectively. It remains to be determined how and what two different signaling pathways by Shh and FGF8 induce the midbrain DA phenotype in DA progenitor cells of competent neuroepithelium.

## COMMITMENT TO FINAL DIFFERENTIATION: NURR1 AND PTX3

Nur-related factor 1 (Nurr1) is an orphan member of the steroid/thyroid hormone receptor super-family of transcription factors and an immediate early gene product (Law et al., 1992). In adult rat brain Nurr1 is highly expressed in the piriform and entorhinal cortex, hippocampus, medial habe-nular, and paraventricular thalamic nuclei. Moderate expression is detected in layers II–V of most of the cerebral cortex, the dorsal lateral geniculate nucleus, SN, interpeduncular nucleus, and the cerebellar internal granular cell layer and Purkinje cell layer (Xiao et al., 1996). In particular DA neurons in SN, olfactory bulb, and hypothalamus persistently expressed Nurr1 in adult mouse and rat brains (Zetterstrom, 1996). However, only midbrain DA neurons failed to develop in the Nurr1-deficient mice throughout all prenatal stages (from E11.5 to newborn), while the other TH-positive neurons, such as diencephalic DA neurons, and olfactory periglomerular DA neurons, locus ceruleus noradrenergic neurons were intact. The Nurr1-deficient mutants were hypoactive and died soon after birth, probably due to inability to suckle, which certainly resulted from the lack of proper development of neuronal circuitry (Zetterstrom et al., 1997; Saucedo-Cardenas et al., 1998). In contrast, the neurotransmitter dopamine-deficient mice maintained the normal midbrain DA neu-ronal projections to striatum and the normal morphological organization, in which the continued administration of L-dihydroxyphenylalanine (L-DOPA), precursor of dopamine, promoted a near normal growth (Zhou and Palmiter, 1995).

Moreover, Saucedo-Cardenas et al. (1998) demonstrated the persistent expression of a mesen-cephalic marker, Ptx3, in the midbrain of Nurr1-deficient mutant comparable to the level of wild-type embryo at E11.5. Without the expression of Nurr1, neuroepithelial DA progenitor cells at E11.5 still positioned a normal ventral localization, while maintaining appropriate expression of Ptx3. However, in Nurr1-deficient neonates, a significant loss of Ptx3-positive cells was observed exclusively in ventral midbrain. Ptx3 is a bicoid-related homeobox transcription factor and a member of the pituitary homeobox (Ptx) subfamily (Smidt et al., 1997). At E11.5 Ptx3 expression was first detected in the small layer at the ventral surface of mouse midbrain. Ptx3 expression persists in midbrain DA neurons in adult mouse and human. Thus, in E11.5 mouse embryonic brain, Ptx3-expressing cells seem to be a population of neuroepithelial DA progenitor cells. These DA pro-genitor cells appear to fail in induction of further typical DA phenotypic differentiation in Nurr1-deficient neonates, which demonstrates an increase of apoptotic cell death in the ventral midbrain region (Saucedo-Cardenas et al., 1998). The heterozygous animals contained reduced striatal dopamine levels, which also implies the possible role of Nurr1 in maintaining the full phenotypic differentiation of DA neurons in midbrain. Thus, Nurr1 appears to be essential for both survival and final differentiation of ventral DA progenitor cells into the complete DA phenotype.

Furthermore, the fact that Ptx3 and Nurr1 expression persists in midbrain DA neurons in adults confirms the notion that both gene products may be essential to maintain DA cell function and phenotype in adult brain. Ptx3 is not expressed in other Nurr1-positive neurons, including DA neurons of the olfactory bulb and hypothalamus and the limbic system (Smidt et al., 1997). Moreover, Ptx3 expression in developing midbrain DA neurons, which coincides with the full DA phenotypic differentiation, made this gene as an ideal marker for late DA progenitor cells in the midbrain. Therefore, future work will need to address the cascadal interaction and molecular mechanism of these gene products in DA neuronal development: How does Nurr1 increase the survival of midbrain DA progenitor cells and enhance the DA phenotype? Can retinoic acid or a putative Nurr1 ligand can influence DA neuronal survival and phenotypic differentiation? What are the interactive roles between Shh, FGF8, Ptx3, and Nurr1 in DA progenitor cells? When these issues are addressed, we may be able to find a better way of handling DA neurons of SN *in vitro* and provide an ample opportunity for medical benefit for PD patients.

## SURVIVAL OF MIDBRAIN DOPAMINERGIC NEURONS: NEUROTROPHINS AND GROWTH FACTORS

Recent studies show that many growth factors are able to promote survival and phenotypic differentiation of midbrain DA neurons in culture and to protect them against 1-methyl-4-phenyl- 1,2,3,6-tetrahydropyridine (MPTP)- or 6-hydroxydopamine (6-OHDA)-induced neurotoxicity via as yet unidentified signaling mechanisms. For instance, a prototype member of DA neurotrophic factors, brain-derived neurotrophic factor (BDNF), exerted both neurotrophic and neuroprotective effects on DA neurons of SN. It increased the survival of DA neurons in fetal mesencephalic cultures, enhanced high-affinity dopamine uptake, TH activity, dopamine content, and neurite outgrowth and protected against the neurotoxic effects of $MPP^+$, the active metabolite of MPTP (Hyman et al., 1991; Hyman et al., 1994; Knusel et al., 1991). Several *in vivo* studies, delivering BDNF by direct infusion or transplantation of genetically modified cells, confirmed some functional roles of BDNF, such as the activation of the DA system, amelioration of Parkinsonian symptoms in the 6-OHDA lesion model, and increased neuronal survival against $MPP^+$ lesions (Altar et al., 1994; Frim et al., 1994). However, the definitive molecular mechanisms for neuroprotection have yet to be firmly established in mesencephalic cultures and animal models of PD. BDNF appears to be expressed in both DA neurons of SN and their innervation target, the striatum (Gall et al., 1992; Seroogy et al., 1994). Thus, BDNF may have autocrine, paracrine, and retrograde transport effects on DA neurons of SN, which express TrkB receptor mRNA (Hyman et al., 1994).

The most recently discovered neurotrophic factor, glial cell line-derived neurotrophic factor (GDNF), is a new member of the transforming growth factor-β (TGF-β) family and the first member of the GDNF family including neurturin (NTN) and persephin (PSP) (Kotzbauer et al., 1996; Milbrandt et al., 1998). GDNF is much more potent than BDNF in enhancing survival, morphological differentiation, high-affinity dopamine uptake, and protection against $MPP^+$ toxicity in embryonic mesencephalic culture (Lin et al., 1993; Poulsen et al., 1994; Krieglstein et al., 1995). Recent *in vivo* studies have demonstrated that intranigral, intrastriatal, or intracerebral GDNF injection protects and enhances the functional recovery of DA neurons against axotomy-induced degeneration and MPTP- or 6-OHDA-induced lesion (Tomac et al., 1995; Beck et al., 1995; Kearns and Gash, 1995). In addition, GDNF promoted DA neurite outgrowth of developing mesencephalic DA neurons (Stromberg et al., 1993) and growth of TH-immunoreactive fibers into encapsulated DA- and GDNF-producing cells (Lindner et al., 1995). GDNF was demonstrated to be widely expressed in the developing peripheral tissues and brain including ventral region of midbrain and its innervation targets (Strömberg et al., 1993; Poulsen et al., 1994; Schaar et al., 1993). But no *in situ* signal of GDNF mRNA was observed in SN of neonatal (P1) rat brain, while it is up-regulated in the innervation targets of the midbrain DA systems. Similarly, NTN is also expressed sequentially in the ventral midbrain and striatum during development and a modest level of NTN expression is maintained only within the striatum in the adult brain (Horger et al., 1998). Thus, both GDNF and NTN appear to be locally acting and/or target-derived neurotrophins for DA neurons in SN. In contrast, PSP is expressed in most embryonic and adult rat tissues at very low levels. It also promotes the survival of ventral DA neurons in culture and protects against 6-OHDA-induced neurotoxicity (Milbrandt et al., 1998). GDNF and NTN exert their biological functions through a receptor complex that contains the Ret receptor tyrosine kinase and a member of the GDNF family receptor (GFR) containing GFRα1 and GFRα2 (Jing et al., 1996; Klein et al., 1997). In the proposed model, GFRα1 and GFRα2 display a high affinity of binding specificity for GDNF and NTN, respectively, as a glycosylphosphatidylinositol (GPI)-linked coreceptor protein. This ligand-coreceptor complex is able to activate the shared Ret receptor tyrosine kinase (RTK) (Klein et al., 1997). Surprisingly, mutant mice homozygous for a loss-of-function allele of BDNF, TrkB, GDNF, ret, or GFRα1 exhibited no apparent abnormalities in DA neurons of the SN area and the striatum (Jones et al., 1994; Klein et al., 1993; Schuchardt et al., 1994; Moore et al., 1996; Cacalano et al., 1998). However, further careful investigation is necessary to evaluate any subtle defect in SN DA neuronal survival and phenotypic differentiation in various pathophysiological conditions as well as occurrence of any developmental compensation and redundancy in the mutant animals.

BDNF and GDNF exert their various functions via activation of intracellular signaling pathways, which are mediated by TrkB and Ret RTK, respectively. The RTK activity of growth factor and neurotrophic factor receptors is crucial for the signal transduction pathways required for differentiation and proliferation. In particular, the MAP (mitogen-activated protein) kinase pathway is known to be activated by a variety of extracellular signals leading to cellular differentiation or proliferation, via activation of gene transcription, depending on cell context. For example, treatment of neuroendocrine PC12 cells with NGF leads to differentiation, such as outgrowth of neurites and cessation of cell division. NGF stimulation in PC12 cells produces prolonged activation of the MAP kinase pathway, which is necessary and sufficient for differentiation (Cowley et al., 1994). However, treatment of PC12 cells with epidermal growth factor (EGF) leads to a proliferation signal by short-lived activation of the MAP kinase pathway (Traverse et al., 1992). The Ras-Raf-MEK-p44/42 (Erk1/2) pathway is one example of the MAP kinase pathway. In this signaling pathway the small G-protein Ras is activated to the GTP form through Shc-Grb2-Sos or Grb2-Sos complex being recruited to TrkA or EGF RTK, respectively (Buday and Downward, 1993; Stephens et al., 1994). In brain a neuron-specific adaptor molecule, N-Shc, becomes tyrosine phosphorylated and forms a complex with Grb2 adaptor, which seems to bind to the activated TrkB and EGF RTKs (Nakamura et al., 1996). In fact, the activation of the MAP kinase pathway appears to be an essential step in suppression of free radical formation and blocking cell death in various cell contexts (Dugan et al., 1997; Villalba et al., 1997; Hiraiwa et al., 1997). In neuronal cell cultures both BDNF and GDNF were able to activate the MAP kinase (p44/42) signaling pathway through TrkB RTK and Ret RTK, respectively (Marsh et al., 1993; Kotzbauer et al., 1996). Other neurotrophins and growth factors also have been shown to promote the survival and differentiation of the DA neurons in mesencephalic cultures, including EGF, FGFs, ciliary neurotrophic factor (CNTF), and insulin-like growth factors (IGFs) (Casper et al., 1991; Knusel et al., 1990; Takayama et al., 1995), but their molecular mechanisms of action are largely unknown. Thus, further elucidation of molecular signaling pathways of DA neuronal survival and differentiation supported by neurotrophins and growth factors will provide new therapeutic targets in PD.

## DEGENERATION OF DOPAMINERGIC NEURONS OF SUBSTANTIA NIGRA: NEUROTOXIN, OXIDATIVE STRESS, AND α-SYNUCLEIN

Although the initial cause of SN DA neuronal degeneration in PD is unknown, a substantial number of studies support the key pathophysiological roles of reactive oxygen free radicals and mitochondrial defects. In human, nonhuman primate, and mouse, the neurotoxin MPTP produces biochemical and neuropathological changes very similar to those observed in idiopathic PD (Burns et al., 1983; Bloem et al., 1990; Kopin, 1994). These include marked reduction in the levels of striatal dopamine and its metabolites dihydroxyphenylacetic acid (DOPAC) and homovanillic acid (HVA). MPTP also causes a significant reduction in the number of DA cell bodies in the SN and induces the formation of intraneuronal eosinophilic inclusions resembling Lewy bodies, which are the neuropathological characteristics of PD and dementia with Lewy bodies. In the *in vitro* model of PD, employing the primary mesencephalic culture system, MPP$^+$ is selectively taken up by DA neurons and results in the selective neurotoxicity, at least in part, by inhibiting enzymes in the mitochondrial electron transport chain, which results in the leakage of the free radical, superoxide anion ($O_2 \cdot^-$) (Kopin, 1994; Nicklas et al., 1985; Krueger et al., 1990). Thus, a close molecular examination of the MPTP model of PD may provide important new insights into DA cell death in PD. The most recently identified neurotransmitter, nitric oxide (NO), is known to mediate glutamate neurotoxicity via the formation of the potent oxidant peroxynitrite (ONOO$^-$), which can initiate cell death by various toxic insults (Przedborski et al., 1996; Dawson et al., 1993). Strikingly, in MPTP models of mice and baboon (Schulz et al., 1995; Hantraye et al., 1996), 7-nitroindazole (7-NI), a selective inhibitor of nNOS, markedly reduced MPTP-induced neurotoxicity, such as striatal DA depletion and loss of SN DA neurons. Moreover, nNOS knockout mutant mice were significantly resistant to MPTP-induced neurotoxicity (Przedborski et al., 1996). In fact, in the SN of PD patients unusually active microglia, known to produce NO, have been found (McGeer et al., 1988). Therefore, NO appears to play a significant pathophysiological role in PD and animal models of PD. In addition to the generation of potent oxidants, NO itself acts as a signaling molecule by nitrosylation of various thiol- and metal-containing proteins, which occurs at either active or allosteric sites (Stamler, 1994; Lander, 1997). For example, by a change in ambient redox milieu, NO becomes neuroprotective by down-regulation of NMDA-receptor activity via nitrosylation of the receptor's redox modulatory site (Lipton et al., 1993). Thus the major neuronal target sites for NO are receptors, G-proteins, transcription factors, enzymes, ion channels, and transporters, which shuttle information from the cell surface to the nucleus (Lander, 1997). In the brain NO influences apoptosis, neural development, synaptic plasticity, and behavioral changes. In addition, the reactive oxygen species (ROS; i.e., $O_2 \cdot^-$ and $\cdot OH$) are emerging as important intracellular messengers targeting transcription factors, G-protein, ion channels, protein kinase, and NOS. Thus, strict regulation of ROS concentrations in SN neuronal tissues appears to be crucial for neuronal survival by both enhancing antioxidative defensive mechanisms (i.e., superoxide dismutase, catalase, glutathione peroxidase, glutathione reductase, Bcl$_2$ family, thioredoxin, vitamin E, etc.) and tightly regulating the source of ROS (i.e., NAD(P)H oxidases, xanthine oxidase, ionizing radiation, NOS) (Lipton et al., 1993; Sampath et al., 1994; Mattson et al., 1995; Reiter et al., 1995; Sen and Packer, 1996). Recent findings, such as protection of the cerebral cortex from ischemia-induced injury and blocking the increase in NO by GDNF (Wang et al., 1997), suppression of free radical formation (Dugan et al., 1997), and the up-regulation of glutathione peroxidase and catalase mRNA expression by NGF (Sampath et al., 1994), strongly suggest that a common mechanism may be involved in neurodegenerative diseases. Thus, neurotrophic factors seem to exert neuroprotective function by altering free radical formation and detoxification via modulation of redox-sensitive proteins and gene transcription via unidentified signal transduction pathways in particular DA neurons of SN.

α-Synuclein is a 140 amino acid protein and a major component of the Lewy bodies and Lewy neurites in SN from idiopathic PD patients (Spillantini et al., 1997). Two independent point mutations in the α-synuclein gene were discovered as a cause of familial PD (Polymeropoulos et al., 1997; Kruger et al., 1998). The major expression site of α-synuclein is the nervous system and it is concentrated in presynaptic nerve terminals (Maroteaux et al., 1998). In the amino-terminal of human α-synuclein seven repeats, with the consensus sequence KTKEGV, are identified, which are followed by a hydrophobic middle region and a negatively charged carboxy-terminal region. The first point mutation (A53T), found in one Italian and three Greek families with autosomal dominant inheritance for PD phenotype, lies in the linker region between repeats four and five of α-synuclein. Surprisingly, rodent and zebra finch α-synuclein carry a threonine (T) residue at position 53, like the human A53T mutant α-synuclein protein (Polymeropoulos et al., 1997). The other point mutation (A30P) lies in the linker region between repeats two and three of α- synuclein, which is either alanine (A) or threonine (T) depending on the species at position 30. Although the normal function of human α-synuclein is not known, these mutations may result in an increased aggregation in Lewy bodies and synaptic malfunction, which plays an important role in neurode-generation in PD.

## GENETIC MANIPULATION OF MIDBRAIN DOPAMINERGIC NEURONS: CELL CULTURE AND TRANSGENIC ANIMALS

Pharmacological effects of various neurotrophins on midbrain DA neurons have been tested primarily using DA neurons in mixed primary cultures of ventral mesencephalon derived from E15–E16 rat embryos (Knusel et al., 1990; Rousselet et al., 1988). Culture studies have often resulted in confounding data primarily due to the heterogeneity of embryonic cell cultures and the paucity of DA neurons in the cultures. For instance, BDNF enhanced the survival of not only DA neurons but also GABAergic neurons and glial cells present in the mesencephalic cultures (Hyman et al., 1994; Engele et al., 1991). As a result it is not clear whether the action of BDNF on DA neurons is dependent on production by, or is secondary to, its effect on glial cells comprising the majority of mesencephalic cultures. In fact, DA neurons comprise less than a few percent of the mesencephalic cultures. Typically the rat contains $2 \times 10^4$ mesencephalic DA neurons on each side. Together with the presence of large numbers of non-DA cells in the mesencephalic cultures, the short survival time and an inability of genetic manipulation have been major obstacles to the molecular mechanistic study of SN DA neuronal survival and cell death.

Thus, it is crucial to acquire an abundant source of homogeneous DA neurons *in vitro* for the molecular dissection of SN DA neurons. To generate a consistent and abundant source of SN DA neurons *in vitro*, several approaches, employing developmental, anatomical, or genetic manipulations, have been investigated. First, Hynes et al. (1995) tried to recapitulate the ontogeny of SN DA neuronal differentiation in explant cultures. Shh can induce effectively DA neurons in rat embryonic forebrain/midbrain explant. But, to practically manipulate the DA neurons *in vitro*, it is essential to elucidate the DA progenitor cells present in the brain explants and establish them as a cell line. The second approach was the purification of fluorescence-labeled DA neurons from embryos by flow cytometry (Kerr et al., 1994). However, both the contamination with non-DA neurons and the low yield requiring a large number of embryos made this approach impractical for a routine pure DA neuronal culture. In the final approach, the midbrain progenitor cells were randomly immortalized by introducing an oncogene via retroviral infection (Anton et al., 1994). Apparently the lack of a cell type specificity in the retroviral vector made it extremely difficult to immortalize specifically DA progenitor cells in the heterogeneous mesencephalic cultures. Therefore, we have developed a unique SN-derived DA progenitor cell line (Figure 15.2) via combined genetic and developmental manipulations by employing a DA neuron-specific promoter (Min et al., 1994), the temperature-sensitive mutant form of an oncogene (Jat and Sharp, 1989), and the anatomical dissection of transgenic embryonic SN (Son et al., 1999). The advantage of this modulatable oncogene is that it provided isolation of immortalized neuronal precursor cells at early developmental stages without alteration of the genetic content in the normal cells, which often occurs in the tumor cell lines obtained from the full-blown tumors in adulthood. It also provided an experimental means for conditional immortalization by shifting the cultivation temperature from the proliferative (33°C) to the nonproliferative (38 to 39°C), similar to mammalian body temperature (Stringer et al., 1994; Whitehead et al., 1993). Thus, the conditional immortalization of specific neuronal progenitor cells will provide a new tool to investigate molecular mechanistic studies of

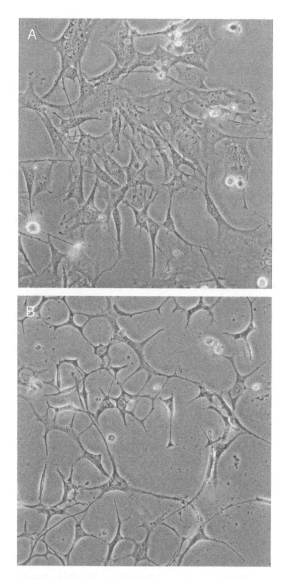

**FIGURE 15.2**  The morphological differentiation of the conditionally immortalized DA progenitor cells derived from SN. (A) Under proliferation condition (33°C) the cells have a fibroblast-like morphology with less prominent neurite growth, whereas (B) the morphological differentiation with extensive neurite outgrowth and multipolar processes is induced at the nonpermissive temperature (39°C) with a minimal serum concentration.

DA differentiation, survival, cell death, and neuroprotective actions of BDNF and GDNF. If necessary, it will be possible to isolate independent DA cell clones at various degrees of DA phenotypic differentiation by preparing clones from appropriate embryonic brains representing different ontogenic stages.

To investigate the role of neurotrophins on midbrain DA neurons *in vivo,* they have been delivered to *in vivo* DA neurons by several different means: intracerebral infusion by minipump (Altar et al., 1994), gene transfer by viral vectors (Kaplitt et al., 1994; During et al., 1994; Choi-Lundberg et al., 1997), and transplantation of genetically modified cells (Lindner et al., 1995), by which one cannot be delivered simultaneously in a neuronal phenotype-specific manner. These *in vivo* neurotrophin applications via surgical intervention are designed to rescue or prevent damage to DA neurons partially. Moreover, it is necessary to confirm the previous *in vitro* cell culture studies *in vivo*. However, the major obstacle to the *in vivo* manipulation of DA survival and cell death is the inaccessibility for genetic manipulation as well as lack of means for the controlled DA-specific gene expression in developing embryos and adult animals. For this purpose, we have employed the TH gene promoter-regulated tetracycline binary systems whose temporal control of a foreign gene expression can be directly manipulated in an SN DA cell type-specific manner as a genetic on-off switch, which is under investigation in our laboratory (Son et al., 1998). The inducible tetracycline binary system is known to have extremely low basal activity and over several thousand-fold inducibility in both cell culture and transgenic mice (Gossen et al., 1995; Furth et al., 1994). This system is based on two regulatory elements derived from the tetracycline-resistant operon, the tet repressor protein (TetR), and the tet operator DNA sequence (tetO) to which TetR binds. The level of induction was tetracycline dose dependent. Initially, this system was developed as a tetracycline repressible transcription system (Gossen and Bujard, 1992). In this system a tetracycline-controlled transactivator (tTA), which is a fusion gene composed of the TetR with the activating domain of VP 16 of herpes simplex virus, acts as a tet-responsive transcriptional activator to express a gene of interest in the absence of tetracycline. In the inductive system, the rtTA (the reverse tetracycline-controlled transactivator) exhibited an inductive mode of reporter gene expression in double transgenic mice in a tetracycline dose-dependent manner (Gossen et al., 1995). Thus, in these models the duration of a specific gene expression in all DA neurons of SN can be regulated simultaneously and temporally by external genetic switches without surgical intervention. In addition, the phenotype of the targeted loss-of-function mutation is often compensated by up- or down-regulation of the related gene family during embryonic development. This is the case for the apparently normal phenotype of DA neurons in several loss-of-function mutations in BDNF, TrkB, GDNF, ret, or GFRα1 gene, as mentioned earlier. Therefore, it is essential to employ an inducible and/or tissue-specific loss-of-function mutation, in which the gene-targeting event can be initiated in DA neurons at a discrete time point via either a Cre/loxP or Flp/frt recombination system (Kuhn et al., 1995; Dymecki, 1996). Thus, the animals will experience no specific mutation during development and may demonstrate any subtle effect of a targeted loss-of-function mutation only in SN DA neurons.

## CONCLUSIONS

Recent studies in differentiation and survival of DA neurons in SN have uncovered key molecules such as Shh, FGF8, Nurr-1, Ptx3, BDNF, GDNF, and NTN, which appear to be implicated in the positioning of DA neurons along the D-V and A-P axis, the final phenotypic differentiation and the maintenance of DA phenotype and appropriate synaptic contacts. Gli1 and ELF-1 transcription factors are identified as a midline target of Shh and FGF8 signaling pathways, respectively. The interaction between Shh and FGF8 signaling in DA progenitor cells of competent neuroepithelium will be a subject of major importance in the ontogeny of DA neurons in SN. Ptx3-positive cells in ventral midbrain appear to be a population of late DA progenitor cells. Nurr-1 has an essential functional role to maintain the survival and final differentiation of the Ptx3-positive cells, which may be achieved in part via regulation of neurotrophin (BDNF, GDNF, and NTN), their receptors (TrkB, Ret, GFRα1, and GFRα2), and DA phenotype (TH, dopamine, and dopamine transporter). In addition, oxidative stress and mitochondrial defect are strongly implied in various neurodegenerative diseases including PD. In particular, mutations in α-synuclein were found as a cause of familial PD, raising the questions of how these mutations play a role in DA neuronal degeneration in PD and whether there is a possible crosstalk between oxidative stress and mutations in α-synuclein. New genetic tools, such as development of DA progenitor cell lines, an inducible tetracycline binary system in transgenic mice, and inducible and DA neuron-specific loss-of-function mutant mice, will permit us to address these issues at the molecular level, provide new insight into the complex molecular mechanisms of DA neuronal cell death in PD, and reveal new molecular targets for PD therapy.

## ACKNOWLEDGMENTS

This work was supported in part by National Institutes of Health Grant AG14093.

## REFERENCES

Altar, C.A., Boylan, C.B., Fritsche, M., Jones, B., Jackson, C., Wiegand, S.J., Lindsay, R.M., and Hyman, C., Efficacy of brain-derived neurotrophic factor and neurotrophin-3 on neurochemical and behavioral deficits associated with partial nigrostriatal dopamine lesions, *J. Neurochem.*, 63, 1021–1032, 1994.

Anton, R., Kordower, J.H., Maidment, N.T., Manaster, J.S., Kane, D.J., Rabizadeh, S., Schueller, S.B., Yang, J., Rabizadeh, S., Edwards, R.H., Markham, C.H., and Bredesen, D.E., Neural-targeted gene therapy for rodent and primate hemiparkinsonism, *Exp. Neurol.*, 127, 207–218, 1994.

Beck, K., Valverde, J., Alexi, T., Poulsen, K., Moffat, B., Vandlen, R., Rosenthal, A., and Hefti, F., Mesen-cephalic dopaminergic neurons protected by GDNF from axotomy-induced degeneration in the adult brain, *Nature*, 373, 339–341, 1995.

Bloem, B.R. et al., The MPTP model: versatile contributions to the treatment of idiopathic Parkinson's disease, *J. Neurol. Sci.*, 97, 273–293, 1990.

Buday, L. and Downward, J., EGF regulates p21ras through the formation of a complex of receptor, Grb2 adaptor protein and Sos nucleotide exchange factor, *Cell*, 73, 611–620, 1993.

Burns, R.S. et al., A primate model of parkinsonism: selective destruction of dopaminergic neurons in the pars compacta of the substantia nigra by MPTP, *Proc. Natl. Acad. Sci. U.S.A.*, 80, 4546–4550, 1983.

Cacalano, G., Farinas, I., Wang, L., Hagler, K., Forgie, A. et al., GFRα1 is an essential receptor component for GDNF in the developing nervous system and kidney, *Neuron*, 21, 53–62, 1998.

Casper, D., Mytilineou, C., and Blum, M., EGF enhances the survival of dopamine neurons in rat embryonic mesencephalon primary cell culture, *J. Neurosci. Res.*, 30, 372–381, 1991.

Choi-Lundberg, D., Lin Q., Chang Y. et al., Dopaminergic neurons protected from degeneration by GDNF gene therapy, *Science*, 275, 838–841, 1997.

Cowley, S., Paterson, H., Kemp, P., and Marshall, C.J., Activation of MAP kinase kinase is necessary and sufficient for PC12 differentiation and for transformation of NIH 3T3 cells, *Cell*, 77, 841–852, 1994.

Crossley, P.H. and Martin G.R., The mouse Fgf8 gene encodes a family of polypeptides and is expressed in regions that direct outgrowth and patterning in the developing embryo, *Development*, 121, 439–451, 1995.

Crossley, P.H., Martinez, S., and Martin, G.R., Midbrain development induced by FGF8 in the chick embryo, *Nature*, 380, 66–68, 1996.

Dawson, V.L., Dawson, T.M., Bartley, D.A., Uhl, G.R., and Snyder, S.H., Mechanisms of nitric oxide-mediated neurotoxicity in primary brain cultures, *J. Neurosci.*, 13, 2651–2661, 1993.

Dugan, L.L., Creedon, D.J., Hohnson, E.M., and Holtzman, D.M., Rapid suppression of free radical formation by NGF involves the MAP kinase pathway, *Proc. Natl. Acad. Sci. U.S.A.*, 94, 4086–4091, 1997.

During, M., Naegele, J., O'Malley, K., and Geller, A., Long-term behavioral recovery in Parkinsonian rats by a HSV vector expressing tyrosine hydroxylase, *Science*, 266, 1399–1403, 1994.

Dymecki, S. M., Flp recombinase promotes site-specific DNA recombination in embryonic stem cells and transgenic mice, *Proc. Natl. Acad. Sci. U.S.A.*, 93, 6191–6196, 1996.

Echelard, Y., Epstein, D.J., St. Jacques, B., Shen, L., Mohler, J., McMahon, J.A., and McMahon, A.P., Sonic hedgehog, a member of a family of putative signaling molecules, is implicated in the regulation of CNS polarity, *Cell*, 75, 1417–1430, 1993.

Engele, J., Schubert, D., and Bohn, M., Conditional media derived from glial cell lines promote survival and differentiation of dopaminergic neurons in vitro: role of mesencephalic glia, *J. Neurosci. Res.*, 30, 359–371, 1991.

Ericson, J., Muhr, J., Placzek, M., Lints, T., Jessell, T.M., and Edlund, T., Sonic hedgehog induces the differentiation of ventral forebrain neurons: a common signal for ventral patterning within the neural tube, *Cell*, 81, 747–756, 1995.

Fallon, J.H. and Loughlin, S.E., Substantia nigra, in *The Rat Nervous System*, Paxinos, G., Ed., Academic Press, San Diego, 1994, 215–237.

Freeman, T.B, Spence M.S., Boss, B.D., Spector, D.H., Stecker, R.E., Olanow, C.W., and Kordower, J.H., Development of dopaminergic neurons in the human substantia nigra, *Exp. Neurol.*, 113, 344–353, 1991.

Freeman, T.B., Sanberg, P.R., Nauert, G.M., Boss, B.D., Spector, D., Olanow, C.W., and Kordower, J.H., The influence of donor age on the survival of solid and suspension intraparenchymal human embryonic nigral grafts, *Cell. Transplant.*, 4, 141–154, 1995.

Frim, D.M., Uhler, T.A., Galpern, W.R., Beal, M.F., Breakefield, X.O., and Isacson, O., Implanted fibroblasts genetically engineered to produce BDNF prevent MPP+ toxicity to dopaminergic neurons in the rat, *Proc. Natl. Acad. Sci. U.S.A.*, 91, 5104–5108, 1994.

Furth, P., Onge, L., Boger, H., Gruss, P., Gossen, M. et al., Temporal control of gene expression in transgenic mice by a tetracycline-responsive promoter, *Proc. Natl. Acad. Sci. U.S.A.*, 91, 9302–9306, 1994.

Gall, C.M., Gold, S.J., Isackson, P.J., and Seroogy, K.B., Brain-derived neurotrophic factor and neurotrophin-3 mRNAs are expressed in ventral midbrain regions containing dopaminergic neurons, *Mol. Cell. Neurosci.*, 3, 56–63, 1992.

Gossen, M. and Bujard, H., Tight control of gene expression in mammalian cells by tetracycline-responsive promoters, *Proc. Natl. Acad. Sci. U.S.A.*, 89, 5547–5551, 1992.

Gossen, M., Freundlieb, S., Bender, G., Müller, Hillen, W., and Bujard, H., Transcriptional activation by tetracycline in mammalian cells, *Science*, 268, 1766–1769, 1995.

Hantraye, P., Brouillet, E., Ferrante, R., Palfi, S., Dolan, R., Matthews, R., and Beal, F., Inhibition of neuronal nitric oxide synthase prevents MPTP-induced parkinsonism in baboons, *Nat. Med.*, 2, 1017–1021, 1996.

Hiraiwa, M., Taylor, E., Campana, W., Darin, S., and O'Brien, J., Cell death prevention, MAP kinase stimulation, and increased sulfide concentrations in Schwann cells and ologodendrocytes by prosaposin and prosaptides, *Proc. Natl. Acad. Sci. U.S.A.*, 94, 4778–4781, 1997.

Hökfelt, T., Martensson, A., Björklund, Kleinau, S., and Goldstein, M., Distribution maps of tyrosine hydroxylase immunoreactive neurons in the rat brain, in *Handbook of Chemical Neuroanatomy, Vol. 2: Classical Transmitters in the CNS, Part I.*, Björklund, A. and Hökfelt, T., Eds., Elsevier Science, New York, 1984, 277–379.

Horger, B., Nishimura, M., Armanini, M., Wang. L., Poulsen, K. et al., Neurturin exerts potent actions on survival and function of midbrain dopaminergic neurons, *J. Neurosci.*, 18, 4929–4937, 1998.

Hyman, C., Hofer, M., Barde, Y.A., Juhasz, M., Yancopoulos, G.D., Squinto, S.P., and Lindsay, R.M., BDNF is a trophic factor for dopaminergic neurons of the substantia nigra, *Nature*, 350, 230–233, 1991.

Hyman, C., Juhasz, M., Jackson, C., Wright, P., Ip, N.Y., and Landsay, R.M., Overlapping and distinct actions of neurotrophins BDNF, NT-3 and NT-4/5 on cultured dopaminergic and GABAergic neurons of the ventral mesencephalon, *J. Neurosci.*, 14, 335–347, 1994.

Hynes, M., Porter, J.A., Chiang, C., Chang, D., Tessier-Lavigne, M., Beachy, P., and Rosenthal, A., Induction of midbrain dopaminergic neurons by sonic hedgehog, *Neuron*, 15, 35–44, 1995.

Hynes, M., Stone, D., Dowd, M., Pitts-Meek, S., Goddard, A., Gurney, A., and Rosenthal, A., Control of cell pattern in the neural tube by the zinc finger transcription factor and oncogene Gli1, *Neuron*, 19, 15–26, 1997.

Jankovic, J. and Tolosa, E., Eds., *Parkinson's Disease and Movement Disorders,* 2nd ed., Williams & Wilkins, Baltimore, MD, 1993.

Jat, P.S. and Sharp, P.A., Cell lines established by temperature-sensitive simian virus 40 large T-antigen gene are growth restricted at the nonpermissive temperature, *Mol. Cell. Biol.*, 9, 1672–1681, 1989.

Jing, S., Wen, D., Yu, Y., Holst, P., Luo, Y. et al., GDNF-induced activation of the ret protein tyrosine kinase is mediated by GDNFR-α, a novel receptor for GDNF, *Cell*, 85, 1113–1124, 1996.

Jones, K.R., Farnas, I., Backus, C., and Reichardt, L.F., Targeted disruption of the brain-derived neurotrophic factor gene perturbs brain and sensory neurons but not motor neuron development, *Cell*, 76, 989–1000, 1994.

Kaplitt, M.G., Leone, P., Samulski, R.J., Xiao, X., Pfaff, D.W., O'Malley, K.L., and During, M.J., Long term gene expression and phenotypic correction using adeno-associated viral vectors in the mammalian brain, *Nat. Genet.*, 8, 148–154, 1994.

Kearns, C.M. and Gash, D.M., GDNF protects nigral dopamine neurons against 6-hydroxydopamine in vivo, *Brain Res.*, 672, 104–111, 1995.

Kerr, C., Lee, L., Romero, A., Stull, N., and Iacovitti, L., (1994) Purification of dopamine neurons by flow cytometry, *Brain Res.*, 665, 300–306, 1994.

Klein, R., Sherman, D., Ho, W., Stone, D., Bennett, G., Moffat B. et al., A GPI-linked protein that interacts with Ret to form a candidate neurturin receptor, *Nature*, 387, 717–721, 1997.

Klein, R., Smeyne, R.J., Wurst, W., Long, L.K., Auerbach, B.A., Joyner, A.L., and Barbacid, M., Targeted disruption of the trkB neurotrophin receptor gene results in nervous system lesions and neonatal death, *Cell*, 75, 113–122, 1993.

Knusel, B., Michel, P., Schwaber, J., and Hefti, F., (1990) Selective and nonselective stimulation of central cholinergic and dopaminergic development in vitro by nerve growth factor, basic fibroblast growth factor, epidermal growth factor, insulin and the insulin-like growth factors I and II, *J. Neurosci.,* 10, 558–570, 1990.

Knusel, B., Winslow, J.W., Rosenthal, A., Burton, L.E., Seid, D.P., Nikolics, K., and Hefti, F., Promotion of central cholinergic and dopaminergic neurons differentiation by brain-derived neurotrophic factor but not neurotrophin-3, *Proc. Natl. Acad. Sci. U.S.A.,* 88, 961–965, 1991.

Koller, W.C. and Paulson, G., Eds., *Therapy of Parkinson's Disease,* 2nd ed., Marcel Dekker, New York, 1995.

Kopin, I.J., (1994) Tips from toxins: the MPTP model of Parkinson's disease, in *Neurodegenerative Diseases,* Jolles, G. and Stutzmann, J.M., Eds., Academic Press, San Diego, 1994, 143–154.

Kotzbauer, P., Lampe, P., Heuckeroth, R., Golden, J., Creedon, D., Johnson, E., and Milbrandt J., Neurturin, a relative of glial-cell-line-derived neurotrophic factor, *Nature,* 384, 467–470, 1996.

Krieglstein, K., Suter-Crazzolara, C., Fischer, W.H. and Unsicker, K. TFG-β superfamily members promote survival of midbrain dopaminergic neurons and protect them against MPP+ toxicity, *EMBO J.,* 14, 736–742, 1995.

Krueger, M.J., Singer, T.P., Casida, J.E., and Ramsey, R.R., Evidence that the blockade of mitochondrial respiration by the neurotoxin MPP+ involves binding at the same site as the respiratory inhibitor, rotenone, *Biochem. Biophys. Res. Commun.,* 169, 123–128, 1990.

Kruger, R., Kuhn, W., Muller, T., Woitalla, D., Graeber, M. et al., Ala30Pro mutation in the gene encoding α-synuclein in Parkinson's disease, *Nat. Genet.,* 18, 106–108, 1998.

Kuhn, R., Schwenk, F., Aguet, M., and Rajewsky, K., Inducible gene targeting in mice, *Science,* 269, 1427–1429, 1995.

Lander, H.M., An essential role for free radicals and derived species in signal transduction, *FASEB J.,* 11, 118–124, 1997.

Law, S.W., Conneely, O.M., DeMayo, F.J., and O'Malley, B.W., Identification of a new brain-specific transcription factor, Nurr1, *Mol. Endocrinol.,* 6, 2129–2135, 1992.

Lee, J., Platt, K.A., Censullo, P., and Altaba, A., Gli1 is a target of Sonic hedgehog that induces ventral neural tube development, *Development,* 124, 2537–2552, 1997.

Leonard, B.E., *Fundamentals of Psychopharmacology,* 2nd ed., John Wiley & Sons, New York, 1997.

Lin, L., Doherty, D.H., Lile, J.D., Bektesh, S., and Collins, F., GDNF: a glial cell line-derived neurotrophic factor for midbrain dopaminergic neurons, *Science,* 260, 1130–1132, 1993.

Lindner, M., Winn, S., Baetge, E., Hammang, J., Gentile, F., Doherty, E., McDermott, P., Frydel, B., Ullman, M., Schallert, T., and Emerich, D., Implantation of encapsulated catecholamine and GDNF-producing cells in rats with unilateral dopamine depletions and Parkinsonian symptoms, *Exp. Neurol.,* 132, 62–76, 1995.

Lipton, S.A., Choi, Y.B., Pan, Z.H., Leis, Z. et al., A redox-based mechanism for the neuroprotective and neurodestructive effects of nitric oxide and related nitroso-compounds, *Nature,* 364, 626–632, 1993.

Maroteaux, L., Campanelli, J., and Scheller, R., Synuclein: a neuron-specific protein localized to the nucleus and presynaptic nerve terminal, *J. Neurosci.,* 8, 2804–2815, 1998.

Marsh, H., Scholz, W., Lamballe, F., Klein, R. et al., Signal transduction events mediated by the BDNF receptor gp145trkB in primary hippocampal pyramidal cell culture, *J. Neurosci.,* 13, 4281–4292, 1993.

Martin, F. and Puelles, L., Patterning of the embryonic avian midbrain after experimental inversions: a polarizing activity from the isthmus, *Dev. Biol.,* 163, 19–37, 1994.

Mattson, M.P., Lovell, M.A., Furukawa, K., and Markesbery, W.R., Neurotrophic factors attenuate glutamate-induced accumulation of peroxides, elevation of intracellular $Ca^{2+}$ concentration, and neurotoxicity and increase antioxidant enzyme activities in hippocampal neurons, *J. Neurochem.,* 65, 1740–1751, 1995.

McGeer, P.L., Itagaki, S., Boyes, B.E., and McGeer, E.G., Reactive microglia are positive for HLA-DR in the substantia nigra of Parkinson's and Alzheimer's disease brains, *Neurology,* 38, 1285–1291, 1988.

Milbrandt, J., de Sauvage, F., Fahrner, T., Baloh, R., Leitner, M. et al., Persephin, a novel neurotrophic factor related to GDNF and neurturin, *Neuron,* 20, 245–253, 1998.

Min, N., Joh, T., Corp, E., Baker H., Cubells, J., and Son, J.H., A transgenic mouse model to study transsynaptic regulation of tyrosine hydroxylase gene expression, *J. Neurochem.,* 67, 11–18, 1996.

Min, N., Joh, T.H., Kim, K.S., Peng, C., and Son, J.H., 5′ Upstream DNA sequence of the rat tyrosine hydroxylase directs high-level and tissue-specific expression to catecholaminergic neurons in the central nervous system of transgenic mice, *Mol. Brain Res.,* 27, 281–289, 1994.

Moore, R.Y. and Bloom, F.E., Central catecholamine neuron systems: anatomy and physiology of the dopamine system, *Annu. Rev. Neurosci.*, 1, 129–169, 1978.

Moore, M.W., Klein, R.D., Farinas, I., Sauer, H., Armanini, M. et al., Renal and neural abnormalities in mice lacking GDNF, *Nature*, 382, 76–79, 1996.

Nagatsu, T., Levitt, M., and Udenfriend, S., Tyrosine hydroxylase: the initial step in norepinephrine biosynthesis, *J. Biol. Chem.*, 239, 2910–2917, 1964.

Nakamura, T., Sanokawa, R., Sasaki, Y., Ayusawa, D., Oishi, M., and Mori, N., N-Shc: a neural-specific adaptor molecule that mediates signaling from neurotrophin/Trk to ras? MAPK pathway, *Oncogene*, 13, 1111–1121, 1996.

Nicklas, W.J., Vyas, I., and Heikkila, R.E., Inhibition of NADH-linked oxidation in brain mitochondria by MPP+, a metabolite of the neurotoxin, MPTP, *Life Sci.*, 36, 2503–2508, 1985.

Polymeropoulos, M., Lavedan, C., Leory, E., Ide, S., Dehejia, A. et al., Mutation in the α-Synuclein gene identified in families with Parkinson's disease, *Science*, 276, 2045–2047, 1997.

Poulsen, K., Armanini, M., Klein, R., Hynes, M., Phillips, H., and Rosenthal, A., TGFβ2 and TGFβ3 are potent survival factors for midbrain dopaminergic neurons, *Neuron*, 13, 1245– 1252, 1994.

Przedborski, S., Jackson-Lewis, V., Yokoyama, R., Shibata, T., Dawson, V., and Dawson, T.M., Role of neuronal nitric oxide in MPTP-induced dopaminergic neurotoxicity, *Proc. Natl. Acad. Sci. U.S.A.*, 93, 4565–4571, 1996.

Reiter, R., Oxidative processes and antioxidative defense mechanisms in the aging brain, *FASEB J.*, 9, 526–533, 1995.

Rousselet, A., Fetler, L., Chamak, B., and Prochiantz, A., Rat mesencephalic neurons in culture exhibit different morphological traits in the presence of media conditioned on mesencephalic or striatal astroglia, *Dev. Biol.*, 129, 495–504, 1988.

Sampath, D., Jackson, G.R., Werrbach-Perez, K., and Perez-Polo, J.R., Effects of nerve growth factor on glutathione peroxidase and catalase in PC12 cells, *J. Neurochem.*, 62, 2476–2479, 1994.

Saucedo-Cardenas, O., Quintana-Hau, J.D., Le, W., Smidt, M.P., Cox, J.J., De Mayo, F., Burbach, J., and Conneely, O.M., Nurr1 is essential for the induction of the dopaminergic phenotype and the survival of ventral mesencephalic late dopaminergic precursor neurons, *Proc. Natl. Acad. Sci. U.S.A.*, 95, 4013–4018, 1998.

Schaar, D., Sieber, B., Dreyfus, C., and Black I., Regional and cell-specific expression of GDNF in rat brain, *Exp. Neurol.*, 124, 368–371, 1993.

Schuchardt, A., D'Agati, V., Larsson-Blomberg, L., Costantini, F., and Pachnis, V., Defects in the kidney and enteric nervous system of mice lacking the tyrosine kinase receptor Ret, *Nature*, 367, 380–383, 1994.

Schulz, J., Matthews, R., Muqit, M., Browne, S., and Beal, M., Inhibition of neuronal nitric oxide synthase by 7-nitroindazole protects against MPTP-induced neurotoxicity in mice, *J. Neurochem.*, 64, 936–939, 1995.

Sen, C.K. and Packer, L., Antioxidant and redox regulation of gene transcription, *FASEB J.*, 10, 709–720, 1996.

Seroogy, K.M., Lundgren, K.H., Tran, T., Guthrie, K.M., Isackson, P.J., and Gall, C.M., Dopaminergic neurons in rat ventral midbrain express brain-derived neurotrophic factor and neurotrophin-3 mRNAs, *J. Comp. Neurol.*, 342, 321–334, 1994.

Smidt, M.P., van Schaick, H., Lanctot, C., Tremblay, J., Cox, J., van der Kleij, A., Wolterink, G., Drouin, J., and Burbach, J., A homeodomain gene Ptx3 has highly restricted brain expression in mesencephalic dopaminergic neurons, *Proc. Natl. Acad. Sci. U.S.A.*, 94, 13305–13310, 1997.

Son, J.H., Chun, H.S., Joh, T.H., and Peng, C.H. Temporal control study of tetracycline-regulated BDNF expression in substantia nigra dopaminergic and locus coeruleus noradrenergic neurons in transgenic mice, *Soc. Neurosci. Abstr.*, 24, 36, 1998.

Son, J.H., Chun, H.S., Joh, T.H., Cho, S., Conti, B., and Lee, J.W., Neuroprotection and neuronal differentiation studies using substantia nigra dopaminergic cells derived from transgenic mouse embryos, *J. Neurosci.*, 19, 10–20, 1999.

Son, J.H., Min, N., and Joh, T.H., Early ontogeny of catecholaminergic cell lineage in brain and peripheral neurons monitored by tyrosine hydroxylase-lacZ transgene, *Mol. Brain Res.*, 36, 300–308, 1996.

Specht, L.A., Pickel, V.M., Joh, T.H., and Reis, D.J., Light-microscopic immunocytochemical localization of tyrosine hydroxylase in prenatal rat brain. I. Early ontogeny, *J. Comp. Neurol.*, 199, 233–253, 1981a.

Specht, L.A., Pickel, V.M., Joh, T.H., and Reis, D.J., Light-microscopic immunocytochemical localization of tyrosine hydroxylase in prenatal rat brain. II. Late ontogeny, *J. Comp. Neurol.*, 199, 255–276, 1981b.

Spillantini, M., Schmidt, M., Lee, V., Trojanowski, J., and Goedert, R., α-Synuclein in Lewy bodies, *Nature*, 388, 839–840, 1997.

Stamler, J.S., Redox signaling: nitrosylation and related target interactions of nitric oxide, *Cell*, 78, 931–936, 1994.

Stephens, R.M., Loeb, D.M., Copeland, T.D., Pawson, T., Greene, L.A., and Kaplan, D.R., Trk receptors use redundant signal transduction pathways involving SHC and PLC-γ1 to mediate NGF responses, *Neuron*, 12, 691–705, 1994.

Stone, D.M., Hynes, M., Armanini, M., Swanson, T., Gu, Q., Johnson, R., Scott, M., Pennica, D., Goddard, A., Phillips, H., Noll, M., Hooper J.N., de Sauvage, F., and Rosenthal, A., The tumor-suppressor gene patched encodes a candidate receptor for Sonic hedgehog, *Nature*, 384, 129–134, 1996.

Stringer, B., Verhofstad, A. and Foster, G. (1994) Raphe neuronal cells immortalized with temperature-sensitive oncogene: differentiation under basal conditions down an APUD cell lineage. Dev. Brain Res. 79, 267–274.

Strömberg, I., Björklund, L., Johansson, M., Tomac, A., Collins, F., Olson, L., Hoffer, B., and Humpel, C., Glial cell line-derived neurotrophic factor is expressed in the developing but not adult striatum and stimulates developing dopamine neurons in vivo, *Exp. Neurol.*, 124, 401–412, 1993.

Takayama, H., Ray, J., Raymon, H., Baird, A., Hogg, J., Fisher, L., and Gage, F., Basic fibroblast growth factor increases dopaminergic graft survival and function in a rat model of Parkinson's disease, *Nat. Med.*, 1, 53–58, 1995.

Tomac, A., Lindqvist, E., Lin, L.-F.H., Ögren, S.O., Young, D., Hoffer, B.J., and Olson, L., Protection and repair of the nigrostriatal dopaminergic system by GDNF in vivo, *Nature*, 373, 335–339, 1995.

Traverse, S., Gomez, N., Paterson, H., Marshall, C., and Cohen, P., Sustained activation of the MAP kinase cascade may be required for differentiation of PC12 cells. Comparision of the effects of NGF and EGF, *Biochem. J.*, 288, 351–355, 1992.

Villalba, M., Bockaert, J., and Journot, L., Pituitary adenylate cyclase-activating polypeptide (PACAP-38) protects cerebellar granule neurons from apoptosis by activating the MAP kinase pathway, *J. Neurosci.*, 17, 83–90, 1997.

Wang, Y., Lin, S., Chiou, A., Williams, L., and Hoffer, B.L., GDNF protects against ischemia-induced injury in the cerebral cortex, *J. Neurosci.*, 17, 4341–4348, 1997.

Wassarman, K.M., Lewandoski, M., Campbell, K., Joyner, A.L., Rubenstein, J., Martinez, S., and Martin, G.R., Specification of the anterior hindbrain and establishment of a normal mid/hindbrain organizer is dependent on Gbx2 gene function, *Development*, 124, 2923–2934, 1997.

Whitehead, R., VanEeden, P., Noble, M., Ataliotis, P., and Jat, P., Establishment of conditionally immortalized epithelial cell lines from both colon and small intestine of adult H-2K^b- tsA58 transgenic mice, *Proc. Natl. Acad. Sci. U.S.A.*, 90, 587–591, 1993.

Xiao, Q., Castillo, S.O., and Nikodem, V.M., Distribution of messenger RNAs for the orphan nuclear receptors Nurr1 and Nur77 (NGFI-B) in adult rat brain using in situ hybridization, *Neuroscience*, 75, 221–230, 1996.

Ye, W., Shimamura, K., Rubenstein, J., Hynes, M., and Rosenthal, A., FGF and Shh signals control dopaminergic and serotonergic cell fate in the anterior neural plate, *Cell*, 93, 755–766, 1998.

Zetterstrom, R.H., Solomin, L., Jansson L., Hoffer, B.J., Olson, L., and Perlmann, T., Dopamine neuron agenesis in Nurr1-deficient mice, *Science*, 276, 248–250, 1997.

Zetterstrom, R.H., Williams, R., Perlmann, T., and Olson, L., Cellular expression of the immediate early transcription factors, Nurr1 and NGFI-B suggests a gene regulatory role in several brain regions including the nigrostriatal dopamine system, *Mol. Brain Res.*, 41, 111–120, 1996.

Zhou, Q. and Palmiter, R.D., (1995) Dopamine-deficient mice are severely hypoactive, adipsic and aphagic, *Cell*, 83, 1197–1209, 1995.

# 16  Genetic Influences on Circadian Rhythms in Mammals

*Sharon S. Low-Zeddies and Joseph S. Takahashi*

## CONTENTS

## INTRODUCTION

All living organisms on Earth exist in an environment that varies regularly with a period of 24 hours. The individual organism experiences a daily cycle not only of abiotic parameters such as light and temperature, but also in all aspects of the biotic sphere that it inhabits. There has been strong evolutionary pressure for animals to develop and maintain genetic programs which temporally optimize their physiology and behavior on a daily basis, such as the cycle of rest and activity. Just as strictly as a species is organized in space, so also is its temporal organization precisely regulated.[1] Functional control over overt circadian (about 24 h) rhythms and their phasing relative to the environment delineates a critical temporal niche for each organism within its own ecosystem. The endogenous source of this potent control mechanism is what we refer to as its circadian "clock."

This chapter focuses on what is known about the genetic basis of the central circadian pacemaker in mammals, specifically mouse and hamster. At present, we report in the midst of a remarkable phase of growth in mammalian circadian gene discovery. Current advances build upon decades worth of insight into the physiological underpinnings of mammalian circadian rhythmicity.[2,3] Although comparative study of clock systems has prompted less emphasis on neural specialization as a basis for circadian pacemakers, as a neurobiological phenomenon, circadian rhythmicity in animals represents an uncommonly tractable link between brain and behavior. The conceptual frameworks which structure the hypotheses governing our explorations of mammalian clocks at the molecular level are largely derived from the advanced genetic dissection of circadian clock components in the fungus

*Neurospora crassa* and in the fruitfly *Drosophila melanogaster*, which we will cover only briefly.[4] This seminal work has brought evolutionarily distant organisms closer together in our understanding of the fundamental form in which daily temporal information is biologically generated.

## CORE CLOCK CHARACTERISTICS

Several experimental observations are key to our further consideration of genetic influences on circadian rhythmicity. The following reflect properties that are thought of as being intrinsic to the core clock mechanism. The first is that animals continue to express precise physiological and behavioral rhythms in an environment devoid of external temporal cues. The period of the rhythm under these conditions, also referred to as the "free-running" period, is close to but usually not exactly 24 hours; period length is normally consistent over an animal's lifetime and is a species-specific characteristic, that is, it is largely genetically determined. Second, a necessary feature of a pacemaker is the ability to integrate pertinent environmental information, principally light, to appropriately align its period and phase with the external oscillation.[5] These phase-response characteristics of an organism's circadian pacemaker, in conjunction with period length, determine how it entrains to a light/dark cycle.[6] Light alters the phase of the rhythm differentially depending on the phase at which it interacts with the ongoing circadian cycle. The phase is delayed by light during the early part of the clock's subjective night, and advanced by light during the late night. Light exposure at points during the day phase of the clock does not change its phase. A final universal feature of circadian clocks, which perhaps only rarely has relevance to homeotherms such as mammals, is the capacity to compensate for temperature effects on the rate of the biochemical processes underlying timekeeping.[7] Evidence continues to mount suggesting that the core circadian oscillator mechanism is fundamentally similar in all organisms at a molecular, and, to some extent, even a genetic level.

In addition to the central pacemaker, or clock, the circadian timing system includes input and output signaling pathways. Input to the central oscillator transduces environmental information, such as light, as well as feedback about the state of the organism itself, such as whether the animal is awake or asleep. Output signals are hooked up to temporally regulate a wide range of physiological functions, many of which sustain their own entrainable oscillations. The details of these pathways diverge considerably across the animal kingdom, a consequence of species-specific evolutionary tailoring.[8]

## HOW WILL WE KNOW A CLOCK GENE WHEN WE SEE ONE?

The behavior of clock molecules in *Neurospora* and *Drosophila* led circadian biologists to propose that the genes essential for the progression of the core driving rhythm will fulfill certain criteria; these conditions were designed to exclude those genes affecting behavior peripherally as elements of input or output coupling pathways:

1. Mutations in clock genes should alter significantly one or more of the following:
   a. the free-running period length,
   b. phase of the entrained rhythm relative to external cycles, and/or
   c. phase response characteristics to light or other relevant stimuli.
2. Deletion of the gene product, or holding its level or activity constant, should abolish or halt the circadian rhythm (unless there are redundancies or overlapping functions).
3. Perturbing the level or activity of the gene product transiently should alter the phase of the rhythm, whereas constitutive changes should affect the period or persistence of the rhythm.
4. The level or activity of the gene product *in vivo* should be acutely responsive to light stimuli.

Although this set of criteria provides a useful starting point for classifying genes as clock components, we feel that these expectations may prove to be too restrictive once specific examples are analyzed.

## MUTATIONAL ANALYSIS

The tangle of genes underlying circadian rhythmic behavior began to unravel in the 1970s with the advent of Seymour Benzer's chemical mutagenesis approach in fruitflies.[9] This method, whereby single genes were disrupted then mutants identified based upon aberrant behavior, represented a significant advance in overcoming the traditional obstacles of quantitative-genetic behavioral analysis. Success in this kind of endeavor depends upon efficient and sensitive behavioral screening: circadian clock mutants reveal themselves most reliably through alterations in the overt free-running period or phase angle of entrainment relative to a light/dark cycle. In this manner, new mutations with strong phenotypic effects could be experimentally generated that would never be propagated in nature's laboratory. In these induced mutants, the signal-to-noise ratio of individual genes was increased against a background of polygenic influences. The random nature in which mutations were created made it theoretically possible to "hit" all genes affecting any given trait. This strategy resulted in the isolation of the *period* mutant, the most tangible evidence at that time that a single gene could have a substantial influence over circadian period length.[10] From that time to the present, *Drosophila* mutants have proved an invaluable genetic tool for studying neural oscillators at both the cellular and the molecular level. As we will see, the mutagenesis approach also remains an important legacy for behavior-genetic analysis.

## HOW FRUITFLIES AND FUNGI TELL TIME

Current investigation of circadian timing in *Drosophila* centers around two genes, *period (per)* and *timeless (tim),* whose protein products (PER and TIM) together appear to constitute a functional unit of the central clock mechanism. Within pacemaker cells in the central brain of the fly, circadian oscillations occur in both mRNA and proteins of these genes, which have direct consequences for behavioral rhythmicity in these animals.[11-14] The two proteins interact before entering the nucleus,[15-17] where the proteins ultimately down-regulate their own transcription.[12,14,18] A 4- to 6-hour time lag between transcription and translation completes a feedback loop that repeats about every 24 hours. An additional feature of the TIM protein is its rapid degradation in response to light,[19-22] which is consistent with a role in integrating light stimuli into the ongoing molecular oscillation.

The molecular characterization of the circadian system in another genetic model system, *N. crassa,* though itself lacking a nervous system, can also inform us about properties of neural clocks. In this species, the primary gene of circadian significance is the *frequency (frq)* gene.[23-25] Multiple alleles of *frq* exist and have different effects on rhythmic phenotype (cycles of spore production). As with the *per* and *tim* genes, the product of *frq* also negatively regulates its own transcription,[26] and peak protein level is delayed about 4 hours relative to peak *frq* mRNA.[27] In addition, *frq* transcription itself appears to be directly inducible by light.[28]

Empirical observations of the ways in which biological clocks function in these species point to some similarities that may generalize to clocks in vertebrates. A conspicuous commonality is that in each of these systems, negatively autoregulatory feedback loops of gene transcription and translation occurring within single cells function as the basic unit of the circadian pacemaker. There is a strong suspicion that the PER-TIM complex and FRQ are capable of transcriptional regulation, although DNA-binding domains in these proteins have not yet been identified. The ongoing expression of these clock genes is integral to continuous rhythm generation, and any pleiotropic effects on traits other than circadian rhythms appear to be minimal. It has also been observed that the majority of mutant alleles of *per, tim,* and *frq* behave in a semidominant fashion[29] and tend to affect multiple clock properties (e.g., period, phase, and temperature compensation), where there is no necessity that these phenotypic characteristics be interdependent at a functional level. Various alleles of these genes can alter period length in both directions — shortening or lengthening — in addition to abolishing circadian cycles. Finally, one interesting difference: whereas light effects the break-down of a gene product (TIM) in Drosophila, in Neurospora, the level of *frq* mRNA is enhanced by photic input. Clock genes, then, can be either "day-phased" or "night-phased," as defined by the time of their peak in transcription relative to the environmental cycle.

## CIRCADIAN RHYTHMIC BEHAVIOR IN MAMMALS

Clearly, one reason to study circadian rhythms in mammals is that genetic parallels exist among all mammalian species, including humans. Circadian clock genes are candidates for clinical intervention in human patients who suffer from maladaptive circadian function. To the genetic neuro-biologist, the circadian system represents, in addition, a means by which to study how the mammalian nervous system controls and organizes a complex behavioral function, all the way from the gene product to the cell to the integrated, active animal. Particularly in higher vertebrates, where behavior often seems unnavigable at a genetic level, we consider this opportunity valuable. One of the main reasons why this is so is that in mammals, the dominant circadian pacemaker responsible for rhythmicity at the organismal level is discretely anatomically localized amongst some 16,000 cells, comprising the suprachiasmatic nucleus (SCN) in the hypothalamus.[2,30-33] Circadian oscillators in vertebrates also reside in other neural tissues: the pineal gland (in birds, reptiles and fish),[34] and in the retinae of amphibians, birds, and mammals.[35-38] Whereas circadian pacemaking is functionally distributed to various degrees between these multiple rhythmic structures in nonmammalian verte-brates, in mammals, the SCN has evolved to govern a hierarchy of circadian oscillations. Through anatomical tracing of pathways, pharmacological manipulation of the clock, and *in vitro* analyses, much has been revealed about the routes and messengers trafficking information into and out of the SCN (for reviews see Refs. 2 and 39). The rhythm-generating processes intrinsic to the SCN itself, however, are still not fully understood.

## GENETIC INFLUENCES ON THE CIRCADIAN SYSTEM IN RODENTS

Endogenous circadian rhythmic characteristics in rodents have been known to be species specific, and therefore, heritable, for decades.[7,40,41] However, variability among strains of mice, for example, seemed to indicate that no one gene would ever be found to significantly affect circadian pheno-type.[42-45] To some extent this was true: alleles with extreme effects on circadian behavior did not appear to be maintained in existing natural populations, presumably a testament to the ecological import of temporal coordination. Amazingly, about a decade ago, the first single-gene mammalian circadian mutant spontaneously appeared in a laboratory stock of hamsters.[46] As in fruitfly mutants, the so-called *tau* mutant hamster exhibited a cluster of clock phenotypic abnormalities. The period of the circadian rhythm in activity is shortened by this autosomal, semidominant mutation from a reliable 24 hours in wild-type hamsters to about 22 hours in heterozygous mutants, and 20 hours in homozygotes, with hardly any overlap between genotypes.[46] Furthermore, their phase shifts in response to both light and also to nonphotic stimulation (activity-induced phase shifts) are exag-gerated in amplitude and variability.[47-49] *tau* mutants do not entrain normally to 24-hour light/dark cycles in either the heterozygous or homozygous state.[47,50]

The dramatically altered phenotype of the *tau* mutant clock was subsequently found to affect rhythmic aspects of physiology besides activity. Male reproductive responses to photoperiod length as well as the estrous cycle in females, both known to be intimately coupled with the circadian cycle,[51,52] are clearly altered by the *tau* mutation.[50,53] *In vitro* isolated retinae were also found to exhibit an autonomous circadian oscillation of melatonin synthesis, whose period in constant conditions is shortened in the mutant;[38] similarly shortened is the period of a rhythm in shedding of photoreceptor outer segment disks in the intact eye.[54] The apparent slowing of rhythmic luteinizing hormone release in *tau* mutant females may demonstrate an even more general effect of the mutation on higher frequency neural oscillations.[55] It seems significant that the *period* mutation also affects a range of high-frequency oscillatory parameters in *Drosophila* which are mediated by tissues separate from the primary circadian oscillators in the brain[56] (see also Ref. 5).

Finally, and perhaps most importantly, the *tau* mutant's circadian phenotypic alteration served as a unique marker of pacemaker tissue function. At the time of the isolation of the mutant, the SCN was well established as a key circadian oscillator (for review, see Ref. 2), but endogenous clock properties, such as period length, expressed in behavioral rhythms could not be decisively attributed to this tissue alone. It had been demonstrated previously that complete lesions of the SCN abolished behavioral rhythmicity in rodents, which could then be restored by transplanting perinatal hypothalamic tissue containing the SCN, close to the lesion.[59,60] In 1990, Ralph et al. were able to use donor SCN tissue of contrasting *tau* genotype to demonstrate categorically that the SCN functions as the master circadian pacemaker in mammals, driving behavioral output; concomitantly, the main effect of the *tau* mutation on circadian rhythmicity was localized to this brain region. Indeed, experiments recording from SCN neurons in dissociated cell culture have since indicated that the *tau* mutation alters circadian oscillations expressed by single cells.[61]

## A GENE IN HAND — A MUTAGENESIS SUCCESS IN THE MOUSE

The discovery of the *tau* mutant hamster inspired a quest for an equivalent in a more genetically tractable system. Ample genomic resources in the mouse, combined with a robust and quantifiable rhythm in wheel-running activity, made this species an obvious choice.[62] As simple a behavior as it may appear, embedded in records of wheel-running activity, or actograms, are multiple measures of clock output, including period, phase of entrainment, phase shift responses to light, and profiles of activity distribution across the day (Figure 16.1). It was the genes behind these traits that were sought in implementing a mutagenesis and screening strategy capable of detecting genes with a dominant or semidominant effect on behavioral phenotype.

Offspring of mice treated with the chemical mutagen *N*-ethyl-*N*-nitrosourea (ENU) were systematically behaviorally screened. ENU tends to induce point mutations, the restricted disruptive effects of which are likely to still permit meaningful performance of the behavioral phenotype. A single mouse was recovered, surprisingly quickly, which showed a clearly aberrant (>6 standard deviations from the wild-type mean) circadian period. This was the first carrier of a mutation in what was subsequently named the *Clock* gene.[63] The predominant phenotypic effect of the *Clock* mutation is to alter properties of the endogenous circadian clock: its period, phase response and entrainment characteristics, and the ability to sustain circadian rhythmicity in an environment without temporal cues (see Figure 16.1B). The absence of apparent anatomical or developmental defects associated with *Clock* suggests that it is foremost a circadian "behavioral gene"; in spite of the gene's bodywide expression, any pleiotropic effects of the *Clock* mutation appear limited.[63,64]

The homozygous *Clock* mutant mouse, in most cases, expresses a long circadian period (about 28 hours) when transferred from a light/dark cycle into constant darkness, after which rhythmicity in the circadian range is gradually lost. The *Clock* mutation is semidominant, such that a heterozygote in constant conditions has a sustained circadian rhythm, but one whose period and stability are altered to a degree intermediate between the homozygous and wild-type (WT) phenotypes. Additional copies of the normal *Clock* gene, inserted via transgenesis, appear to dose-dependently shorten the free-running period of activity;[65] this effect of gene dosage is reminiscent of a similar correlation between the level of *period* gene expression and period length in *Drosophila*.[66,67] *Clock* mutants can also exhibit much larger phase shifts than WT mice in response to light pulses during the subjective night.[68] This kind of altered response to perturbation, associated with both *tau* and *Clock* mutations, is consistent with either a change at the level of the input pathway or a destabilization at the level of the underlying oscillator. Homozygous *Clock* mutant mice entrain only marginally to a 24-hour light/dark cycle and often activity is abnormally phased relative to the

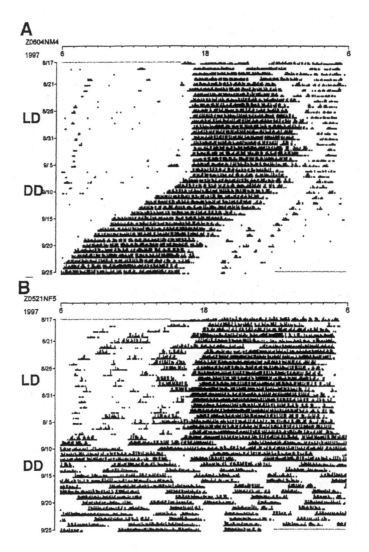

**FIGURE 16.1**  Examples of records of wheel-running activity in mice. Both animals were initially exposed to a 12:12 light:dark cycle (LD) (lights on at 0500, lights off at 1700) for three weeks, followed by constant darkness (DD). A. In LD, mice normally entrain to 24-h light/dark cycle, and running activity occurs almost exclusively during the dark phase of the cycle. In DD, the animal's activity rhythm free-runs, revealing its endogenous period (in mice, < 24 h). B. In *Clock* homozygous mutant mice, entrainment in LD is often unstable. In DD, a 28-h free-running period is sometimes exhibited initially, but behavioral rhythmicity can also be immediately lost, as in the record shown here.

light/dark transition. Both homozygous and heterozygous mutants, however, entrain to long circadian cycles which are closer to their own endogenous periods, but beyond the range to which normal mice are able to entrain (28 hour).[68] As will be discussed further, the *Clock* mutant phenotype can be modified by genetic background.

## MOLECULAR THEMES AMONG RHYTHM GENES

The *Clock* gene was cloned[64] and functionally confirmed to be responsible for circadian phenotype by the transgenic insertion of copies of normal *Clock* into mutant mouse embryos, which was shown to rescue circadian rhythmic behavior.[65] The gene, located in the midportion of mouse chromosome 5 (syntenic to a region of human chromosome 4), encodes a new member of a family of basic-helix-loop-helix (bHLH)-PAS domain-containing transcription factors. The mutant *Clock* protein has a 51 amino acid deletion occurring in what is presumed to be a transactivation domain. The standing hypothesis which would account for its semidominant character is that the mutant form of *Clock* competes with the WT form but is then functionally subeffective. *Clock* mRNA abundance does not itself show a circadian rhythm,[69,70] although its protein activity has not yet been assayed at the time of this writing.

The presence of a PAS domain in the *Clock* gene product has raised some interesting functional and theoretical possibilities. Named for the original three members which defined the family of genes, most notably in this case *Period* (the *Drosophila* clock gene), PAS domains can mediate protein-protein interactions.[71] It is the PAS domain in the PER protein that binds the TIM molecule in flies.[13,14,16] Probably not coincidentally, components of the PAS domain have been found in two *Neurospora* genes: the *white collar* genes (*wc-1; wc-2*), which are involved in transducing photic information into the clock, and *wc-2*, which also comprises part of the clock itself.[72] It has been further noted that sequences in photosensory molecules in both plants and bacteria show homology to PAS sequences (reviewed in Ref. 73). The PAS domain, then, may be a theme that has ramified evolutionarily from a common progenitor light-dependent molecular oscillation, and its phylogenetic conservation could signify that it plays an important functional role in light responsiveness and timing mechanisms. The generally close relationship between light reception and circadian rhythmicity in modern organisms embodies Pittendrigh's theory that the original impetus for circadian rhythmicity in single cells was to restrict enzymatic reactions sensitive to light energy to the dark hours of the night.[1]

## A CLOCK GENE FAMILY REUNION

Since the time that the *Drosophila period* gene was cloned[74,75] comparative studies have intensified the belief that if basic mechanisms of circadian clocks are similar, perhaps their molecular elements may also be homologous at the sequence level, and even orthologous at the functional level. It was not until recently, however, that mammalian *period* homologs finally emerged, and in an embarrassment of riches, emerged in triplicate (so far) as members of what appears to be a family of mammalian *per* homologs.[69,70,76,77] *mPer1, mPer2,* and *mPer3* in the mouse and their human counterparts have now been cloned and are the focus of much energy aimed at determining what their products are doing, and when they are doing it. It is now known that both *mPer1* and *mPer2* transcripts oscillate in both the SCN and the eye, sites of autonomous circadian oscillators.[69,70,76,77] Both *mPer1* and *mPer2* are densely expressed in the SCN, and at least *mPer1*, if not both genes, are induced in this tissue by acute light exposure, at phases during which light shifts behavioral rhythms.[76-78] That the *period* genes encode PAS-domain-containing proteins presents the obvious possibility of potential interactions with the *Clock* gene product(s); *Clock* too is highly conserved among vertebrate species, including humans.[64,79]

## ... TWO IN THE BUSH

Although single genes can have major effects even on complex behavioral regulation, such overt organismal behavior will always emerge from interactions between multiple genetic and epigenetic factors. The challenge we now face is to define more of the elements in the pathway to circadian rhythmicity, even in cases where their phenotypic effects may be somewhat subtle.

Having one mutation in hand represents a tremendous advantage for subsequent mutagenesis screens in the mouse. By conducting a genetic screen over the *Clock* mutation, one can detect both additional alleles of *Clock* and mutations in modifier genes — genes which interact with the *Clock* mutation, suppressing or enhancing its phenotypic effects. Such a screen is further sensitized to detect any *Clock* modifiers that may be segregating in the genetic backgrounds of other strains of mice. The positions of these genes could then be resolved through quantitative trait loci (QTL) analysis.

QTL analysis, implemented in our laboratory, has proven capable of identifying multiple loci involved in circadian behavioral differences between strains of mice. Genome-wide QTL analysis comparing the inbred strains C57BL/6J and BALB/cJ has revealed three loci correlated with a strain difference in free-running period length (LOD score range: 3.3–5.6), one locus involved in determining the phase angle of entrainment to a 24-hour light/dark cycle (LOD score: 3.5), and two loci associated with differences in levels of wheel-running activity (LOD score range: 2.9–3.3).[80]

An additional application of QTL analysis centers around the observation that details of the *Clock* mutant phenotype are quantitatively different against contrasting inbred strain backgrounds. In *Clock* heterozygote mice, the free-running period is initially short upon release from a light/dark cycle into constant darkness, then gradually lengthens, but with a latency that is influenced by strain;[81] the stabilized free-running circadian period length is also contingent upon genetic background.[82] Furthermore, the degree to which the mutation destabilizes circadian rhythms may also vary by strain. QTL analysis of selected behavioral traits such as these could uncover genetic factors responsible for modulating the effects of the *Clock* mutation.

## USING MUTATIONS TO DISSECT CLOCK PHYSIOLOGY IN RODENTS

One reason to clone genes in the mouse is the feasibility of transgenic and gene targeting techniques in this species. Knocking out the *Clock* gene, inducibly or otherwise, is clearly a goal for the future, in order to gauge effects on circadian physiology. With so widely expressed a gene as *Clock*, however, the possibility exists that completely removing its gene product would be incompatible with viability.

We have elaborated upon another approach that, like mutagenesis, is a legacy of Benzer's work in the fruitfly. Mosaic analysis was used by Hotta and Benzer to characterize the neural basis of various behaviors.[83,84] The particular merit of the mosaic approach is that by drawing correlations between behavior and the pattern of WT versus mutant tissue across a range of individual animals, the location of the critical tissue focus responsible for determining a given behavioral phenotype can theoretically be inferred.[85] Konopka et al.[86] analyzed fruitfly mosaics for the *period* mutation, demonstrating that *per* expression in the head exclusively was responsible for circadian behavioral rhythmicity. Since then, Ewer et al.,[87] analyzing *per* mosaics, defined at a finer, cellular level the location of candidate pacemaker cells. This type of experiment provides powerful functional evidence of the role of specific tissues or cells in behavioral generation. In contrast, merely assessing patterns of gene expression cannot be conclusive, especially with genes as ubiquitously expressed as *per* and *Clock*.[64,88,89]

Until now, this idea of creating genetically mosaic animals could be approached in higher vertebrates only in hamsters through SCN transplantation using a contrasting *tau* genotype to functionally "tag" the donor tissue (reviewed in Ref. 90). Other studies have since demonstrated that introducing SCN tissue of different *tau* genotype into a hamster with its own partially intact SCN can organize behavior according to two concurrent but distinct rhythmic components. The dual discrete free-running periods that result are characteristic of the donor genotypes and do not appear to interact with one another functionally.[91] The drawbacks of transplantation, however, are the requirement for multiple surgical procedures and disruptive effects on the SCN tissue.

In light of evidence that period is a property of single SCN cells,[61,92] we sought to attach pacemaker attributes to specific cell types within the SCN while exploring the behavioral consequences of noninvasively combining *Clock* mutant with normal cells within the SCN of individual, intact, behaving animals. Hence, we have undertaken to produce chimeric mice by aggregating WT embryos, carrying a cell marker, together with *Clock* mutant embryos, at the eight-cell stage. This procedure generates animals composed of unique proportions and distributions of *Clock* mutant and WT cells in every tissue, including the SCN. Wheel-running activity has been recorded in a number of chimeric mice and we find that intermediate phenotypes can be achieved by intercalating mutant and WT cells within the master pacemaker.[93] Further analyses will focus on the relationship between behavioral phenotype and distribution of the contrasting cellular genotypes in the SCN. By this means we aim to use the *Clock* mutant mouse to genetically dissect SCN function at a cellular level *in vivo*.

Chimeric mouse analysis potentially can be adopted to hone in on the tissue substrates responsible for other behavioral and physiological phenomena, capitalizing on the vast number of mouse mutants, which is increasing rapidly with developing gene targeting capabilities in this species. Indeed, in some cases, viable chimeras can be produced by combining otherwise prenatally lethal mutant embryos (e.g., certain homozygous gene knockout embryos) with WT embryos.

## SUMMARY

Circadian clock function and its behavioral manifestation in mice has proven to be an ideal system in which to foray into the formidable complexity of behavior-genetic analysis in mammals. That a discrete tissue focus, the SCN, has been functionally identified as the control center for this behavior is a considerable advantage. Phenotypically, rhythms in activity are also readily recorded and quantified. Furthermore, the ubiquity of circadian rhythmicity as a biological phenomenon suggests a reasonable likelihood for its having been phylogenetically, and hence mechanistically, conserved. Finally, the accessibility of molecular genetic resources in mice facilitates gene identification and manipulation.

In this chapter we have articulated a number of approaches by which specific questions about nervous system function and its behavioral consequences are being answered at a genetic level in the mouse:

1. Genes intimately involved in particular behaviors can be identified through deliberate mutagenesis and screening in mice. The ability to map, clone, and transgenically express genes in this species permits further functional characterization.
2. Comparative studies and cloning genes by homology to known genes in other organisms can be fruitful where the behavior in question is likely to have been evolutionarily conserved.
3. Quantitative trait loci analysis is an effective means of resolving the influences of multiple loci on complex behavioral traits.
4. Chimeric or mosaic analysis recruits existing mutants as tools to genetically discriminate the anatomic substrate(s) critical for generating normal behavior.

These strategies adopted in analyzing the genes behind circadian clock function are relevant and applicable to the genetic analysis of mammalian brain-behavior functions in general. Questions about which, how, and where genes affect nervous system physiology to influence behavior are not only of interest to science, but are also of considerable clinical and societal import. Determining how those genes then continue to interact with the environment and with one another throughout life to effect adaptive organismal behavior is a challenge for the future. Overall, it seems that behavior in mammals is like a language for which we are just beginning to form a genetic alphabet.

## ACKNOWLEDGMENTS

Thanks to David G. Zeddies for encouragement. Research supported by grants from the NSF Center for Biological Timing, National Institutes of Health, Bristol Myers-Squibb Unrestricted Grant in

Neuroscience, and Air Force Office of Scientific Research. J.S. Takahashi is an Investigator in the Howard Hughes Medical Institute.

## REFERENCES

1. Pittendrigh, C. S., Temporal organization: reflections of a Darwinian clock-watcher, *Annu. Rev. Physiol.*, 55, 16, 1993.
2. Klein, D. C., Moore, R. Y., and Reppert, S. M., Eds., *Suprachiasmatic Nucleus: The Mind's Clock*, Oxford University Press, New York, 1991.
3. Aschoff, J., Ed., *Handbook of Behavioral Neurobiology, Biological Rhythms*, Vol. 4, Plenum Press, New York, 1981.
4. Takahashi, J. S., Molecular neurobiology and genetics of circadian rhythms in mammals, *Annu. Rev. Neurosci.*, 18, 531, 1995.
5. Pittendrigh, C. S., Circadian organization and the photoperiodic phenomena, in *Biological Clocks in Seasonal Reproductive Cycles,* Follett, B. K., Ed., Wright & Sons, Bristol, England, 1981, 1.
6. Pittendrigh, C. S. and Daan, S., A functional analysis of circadian pacemakers in nocturnal rodents. IV. Entrainment: pacemaker as clock, *J. Comp. Physiol.*, 106, 291, 1976.
7. Barrett, R. K. and Takahashi, J. S., Temperature compensation and temperature entrainment of the chick pineal cell circadian clock, *J. Neurosci.*, 15, 5681, 1995.
8. Takahashi, J. S., Circadian rhythms: from gene expression to behavior, *Curr. Opin. Neurobiol.*, 1, 556, 1991.
9. Benzer, S., From the gene to behavior, *J. Am. Med. Assoc.*, 218, 1015, 1971.
10. Konopka, R. J. and Benzer, S., Clock mutants of *Drosophila melanogaster, Proc. Natl. Acad. Sci. U.S.A.*, 68, 2112, 1971.
11. Edery, I., Zwiebel, L. J., Dembinska, M. E., and Rosbash, M., Temporal phosphorylation of the *Drosophila* Period protein, *Proc. Natl. Acad. Sci. U.S.A.,* 91, 2260, 1994.
12. Hardin, P. E., Hall, J. C., and Rosbash, M., Feedback of the *Drosophila period* gene product on circadian cycling of its messenger RNA levels, *Nature*, 343, 536, 1990.
13. Myers, M. P., Wager-Smith, K., Wesley, C. S., Young, M. W., and Sehgal, A., Positional cloning and sequence analysis of the *Drosophila* clock gene, *timeless, Science*, 270, 805, 1995.
14. Sehgal, A., Rothenfluh-Hilfiker, A., Hunter-Ensor, M., Chen, Y., Myers, M. P., and Young, M. W., Rhythmic expression of *timeless*: a basis for promoting circadian cycles in *period* gene autoregulation, *Science*, 270, 808, 1995.
15. Sehgal, A., Price, J. L., Man, B., and Young, M. W., Loss of circadian behavioral rhythms and *per* RNA oscillations in the *Drosophila* mutant *timeless, Science*, 263, 1603, 1994.
16. Gekakis, N., Saez, L., Delahaye-Brown, A.-M., Myers, M., Sehgal, A., Young, M. W., and Weitz C. J., Isolation of *timeless by* PER protein interaction: defective interaction between *timeless* protein and long-period mutant PER[L], *Science*, 270, 811, 1995.
17. Saez, L. and Young, M. W., Regulation of nuclear entry of the *Drosophila* clock proteins *period* and *timeless, Neuron*, 17, 911, 1996.
18. Zeng, H., Hardin, P. E., and Rosbash, M., Constitutive overexpression of the *Drosophila period* protein inhibits *period* mRNA cycling, *EMBO J.*, 13, 3590, 1994.
19. Zeng, H., Qian, Z., Myers, M. P., and Rosbash, M., A light-entrainment mechanism for the *Drosophila* circadian clock, *Nature*, 380, 129, 1996.
20. Myers, M. P., Wager-Smith, K., Rothenfluh-Hilfiker, A., and Young, M. W., Light-induced degradation of TIMELESS and entrainment of the *Drosophila* circadian clock, *Science*, 271, 1736, 1996.
21. Hunter-Ensor, M., Ousley, A., and Sehgal, A., Regulation of the *Drosophila* protein Timeless suggests a mechanism for resetting the circadian clock by light, *Cell*, 84, 677, 1996.
22. Lee, C., Parikh, V., Itsukaichi, T., Bae, K., and Edery, I., Resetting the *Drosophila* clock by photic regulation of PER and a PER-TIM complex, *Science*, 271, 1740, 1996.
23. Feldman, J. F. and Hoyle, M. N., Isolation of circadian clock mutants of *Neurospora crassa, Genetics*, 75, 605, 1973.
24. Gardner, G. F. and Feldman, J. F., The *frq* locus in *Neurospora crassa*: a key element in circadian clock organization, *Genetics*, 96, 877, 1980.

25. Dunlap, J. C., Loros, J. J., Aronson, B. D., Merrow, M., Crosthwaite, S., Bell-Pedersen, D., Johnson, K., Lindgren, K., and Garceau, N. Y., The genetic basis of the circadian clock: identification of *frq* and FRQ as clock components in *Neurospora*, in *Circadian Clocks and Their Adjustment, Ciba Foundation Symposium,* Vol. 183, Chadwick D. J. and Ackrill K., Wiley, Chichester, 1995, 3.

26. Aronson, B., Johnson, K., Loros, J. J., and Dunlap, J. C., Negative feedback defining a circadian clock: autoregulation in the clock gene *frequency, Science,* 263, 1578, 1994.

27. Garceau, N. Y., Liu, Y., Loros, J. J., and Dunlap, J. C., Alternative initiation of translation and time-specific phosphorylation yield multiple forms of the essential clock protein *frequency, Cell,* 89, 469, 1997.

28. Crosthwaite, S. K., Loros, J. J., and Dunlap, J. C., Light-induced resetting of a circadian clock is mediated by a rapid increase in *frequency* transcript, *Cell,* 81, 1003, 1995.

29. Dunlap, J. C., Genetic analysis of circadian clocks, *Annu. Rev. Physiol.,* 55, 683, 1993

30. Moore, R. Y. and Eichler, V. B., Loss of a circadian adrenal corticosterone rhythm following supra-chiasmatic lesions in the rat, *Brain Res,* 42, 201, 1972.

31. Moore, R. Y., Organization and function of a central nervous system circadian oscillator: the supra-chiasmatic hypothalamic nucleus, *Fed. Proc.,* 42, 2783, 1983.

32. Meijer, J. H. and Reitveld, W. J., Neurophysiology of the suprachiasmatic nucleus circadian pacemaker in rodents, *Physiol. Rev.,* 69, 671, 1989.

33. Ralph, M. R., Foster, R. G., Davis, F. C., and Menaker M., Transplanted suprachiasmatic nucleus determines circadian period, *Science,* 247, 975, 1990.

34. Takahashi, J. S., Murakami, M., Nikaido, S. S., Pratt, B. L., and Robertson, L. M., The avian pineal, a vertebrate model system of the circadian oscillator: cellular regulation of circadian rhythms by light, second messengers, and macromolecular synthesis, *Recent Prog. Horm. Res.,* 45, 279, 1989.

35. Underwood, H., Barrett, R. K., and Siopes, T., The quail's eye: a biological clock, *J. Biol. Rhythms,* 5, 257, 1990.

36. Cahill, G. M., Grace, M. S., and Besharse, J. C., Rhythmic regulation of retinal melatonin: metabolic pathways, neurochemical mechanisms, and the ocular circadian clock, *Cell. Mol. Neurobiol.,* 11, 529, 1991.

37. Pierce, M. E., Sheshberadaran, H., Zhang, Z., Fox, L. E., Applebury, M. L., and Takahashi, J. S., Circadian regulation of iodopsin gene expression in embryonic photoreceptors in retinal cell culture, *Neuron,* 10, 579, 1993.

38. Tosini, G. and Menaker, M., Circadian rhythms in cultured mammalian retina, *Science,* 272, 419, 1996.

39. Moore-Ede, M. C., Sulzman, F. M., and Fuller, C. A., *The Clocks That Time Us,* Harvard University Press, Cambridge, MA, 1982.

40. Aschoff, J., Exogenous and endogenous components in circadian rhythms, *Cold Spring Harbor Symp. Quant. Biol.,* 25, 1, 1960.

41. Pittendrigh, C. S. and Daan, S., A functional analysis of circadian pacemakers in nocturnal rodents. I. Stability and lability of spontaneous frequency, *J. Comp. Physiol.,* 106, 233, 1976.

42. Possidente, B. and Hegmann, J. P., Circadian complexes: circadian rhythms under common gene control, *J. Comp. Physiol.,* 139, 121, 1980.

43. Possidente, B. and Stephan, F. K., Circadian period in mice: analysis of genetic and maternal contributions to inbred strain differences, *Behav. Genet.,* 18, 109, 1988.

44. Beau, J., Activity rhythms in inbred mice. I. Genetic analysis with recombinant inbred strains, *Behav. Genet.,* 21, 117, 1991.

45. Schwartz, W. J. and Zimmerman, P., Circadian timekeeping in BALB/c and C57BL/6 inbred mouse strains, *J. Neurosci.,* 10, 3685, 1990.

46. Ralph, M. R. and Menaker, M., A mutation of the circadian system in golden hamsters, *Science,* 241, 1225, 1988.

47. Ralph, M. R., Suprachiasmatic nucleus transplant studies using the *tau* mutation in golden hamsters, in *Suprachiasmatic Nucleus: The Mind's Clock,* Klein, D. C., Moore, R. Y., and Reppert, S. M., Eds., Oxford University Press, New York, 1991, 341.

48. Mrosovsky, N., Salmon, P., Menaker, M., and Ralph, M. R., Nonphotic phase shifting in hamster clock mutants, *J. Biol. Rhythms,* 7, 41, 1992.

49. Shimomura, K. and Menaker, M., Light-induced phase shifts in *tau* mutant hamsters, *J. Biol. Rhythms,* 9, 97, 1994.

50. Shimomura, K., Nelson, D. E., Ihara, N. L., and Menaker, M., Photoperiodic time measurement in *tau* mutant hamsters, *J. Biol. Rhythms*, 12, 423, 1997.

51. Alleva, J. J., Waleski, M. V., and Alleva, F. R., A biological clock controlling the estrous cycle of the hamster, *Endocrinology*, 88, 1368, 1971.

52. Fitzgerald, K. M. and Zucker, I., Circadian organization of the estrous cycle of the golden hamster, *Proc. Natl. Acad. Sci. U.S.A.*, 73, 2923, 1976.

53. Refinetti, R. and Menaker, M., Evidence for separate control of estrous and circadian periodicity in the golden hamster, *Behav. Neural Biol.*, 58, 27, 1992.

54. Grace, M. S., Wang, L. M., Pickard, G. E., Besharse, J. C., and Menaker, M., The *tau* mutation shortens the period of rhythmic photoreceptor outer segment disk shedding in the hamster, *Brain Res.*, 735, 93, 1996.

55. Loudon, A. S. I., Wayne, N. L., Krieg, R., Iranmanesh, A., Veldhuis, J. D., and Menaker, M., Ultradian endocrine rhythms are altered by a circadian mutation in the Syrian Hamster, *Endocrinology*, 135, 712, 1994.

56. Hall, J. C., Complex brain and behavior functions disrupted by mutations in *Drosophila*, *Dev. Genet.*, 4, 355, 1984.

57. Stephan, F. K. and Zucker, I., Circadian rhythms in drinking behavior and locomotor activity of rats are eliminated by hypothalamic lesions, *Proc. Natl. Acad. Sci. U.S.A.*, 69, 1583, 1972.

58. Stephan, F. K. and Nunez, A. A., Elimination of circadian rhythms in drinking, activity, sleep, and temperature by isolation of the suprachiasmatic nuclei, *Behav. Biol.*, 20, 1, 1977.

59. Drucker-Colin, R., Aguilar-Roblero, R., Garcia-Hernandez, F., Fernandez-Cancino, F., and Bermudez-Rattoni, F., Fetal suprachiasmatic nucleus transplants: diurnal rhythm recovery of lesioned rats, *Brain Res.*, 311, 353, 1984.

60. Sawaki, Y., Nihonmatsu, I., and Kawamura, H., Transplantation of the neonatal suprachiasmatic nucleus into rats with complete bilateral SCN lesion, *Neurosci. Res.*, 1, 67, 1984.

61. Liu, C., Weaver, D. R., Strogatz, S. H., and Reppert, S. M., Cellular construction of a circadian clock: period determination in the suprachiasmatic nuclei, *Cell*, 91, 855, 1997.

62. Takahashi, J. S., Pinto, L. H., and Vitaterna, M. H., Forward and reverse genetic approaches to behavior in the mouse, *Science*, 264, 1724, 1994.

63. Vitaterna, M. H., King, D. P., Chang, A.-M., Kornhauser, J. M., Lowrey, P. L., McDonald, J. D., Dove W. F., Pinto, L. H., Turek, F. W., and Takahashi, J. S., Mutagenesis and mapping of a mouse gene, *Clock*, essential for circadian behavior, *Science*, 264, 719, 1994.

64. King, D. P., Zhao, Y., Sangoram, A. M., Wilsbacher, L. D., Tanaka, M., Antoch, M. P., Steeves, T. D. L., Vitaterna, M. H., Kornhauser, J. M., Lowrey, P. L., Turek, F. T., and Takahashi, J. S., Positional cloning of the mouse circadian *Clock* gene, *Cell*, 89, 641, 1997.

65. Antoch, M. P., Song, E.-J., Chang, A.-M., Vitaterna, M. H., Zhao, Y., Sangoram, A. M., Wilsbacher, L. D., King, D. P., Pinto, L. H., and Takahashi, J. S., Functional identification of the mouse circadian *Clock* gene by transgenic BAC rescue, *Cell*, 89, 655, 1997.

66. Cote, G. G. and Brody, S., Circadian rhythms in *D. melanogaster*: analysis of *period* as a function of gene dosage at the *per* (period) locus, *J. Theor. Biol.*, 121, 487, 1986.

67. Baylies, M. K., Bargiello, T. A., Jackson, F. R., and Young, M. W., Changes in abundance or structure of the *per* gene products can affect periodicity of the *Drosophila* clock, *Nature*, 326, 390, 1987.

68. Vitaterna, M. H., unpublished data, 1998.

69. Sun, Z. S., Albrecht, U., Zhuchenko, O., Bailey, J., Eichele, G., and Lee, C. C., *RIGUI*, a putative mammalian ortholog of the *Drosophila period* gene, *Cell*, 90, 1003, 1997.

70. Tei, H., Okamura, H., Shineyoshi, Y., Fukuhara, C., Ozawa, R., Hirose, M., and Sakaki, Y., Circadian oscillation of a mammalian homologue of the *Drosophila period* gene, *Nature*, 389, 512, 1997.

71. Huang, Z. J., Edery, I., and Rosbash, M., PAS is a dimerization domain common to *Drosophila* period and several transcription factors, *Nature*, 364, 259, 1993.

72. Crosthwaite, S. K., Dunlap, J. C., and Loros, J. J., *Neurospora wc-1* and *wc-2*: transcription, photo-responses, and the origins of circadian rhythmicity, *Science*, 276, 763, 1997.

73. Kay, S. A., PAS, present, and future: clues to the origins of circadian clocks, *Science*, 276, 753, 1997.

74. Bargiello, T. A., Jackson, F. R., and Young, M. W., Restoration of circadian behavioral rhythms by gene transfer in *Drosophila*, *Nature*, 312, 752, 1984.

75. Zehring, W. A., Wheeler, D. A., Reddy, P., Konopka, R. J., Kyriacou, C. P., Rosbash, M., and Hall, J. C., P-element transformation with *period* locus DNA restores rhythmicity to mutant, arrhythmic *Drosophila melanogaster*, *Cell*, 39, 369, 1984.

76. Albrecht, U., Sun, Z. S., Eichele, G., and Lee, C. C., A differential response of two putative mammalian circadian regulators, *mper1* and *mper2* to light, *Cell*, 91, 1055, 1997.

77. Shearman, L. P., Zylka, M. J., Weaver, D. R., Kolakowski, L. F., Jr., and Reppert, S. M., Two *period* homologs: circadian expression and photic regulation in the suprachiasmatic nuclei, *Neuron*, 19, 1261, 1997.

78. Shigeyoshi, Y., Taguchi, K., Yamamoto, S., Takekida, S., Yan, L., Tei, H., Moriya, T., Shibata, S., Loros, J. J., Dunlap, J. C., and Okamura, H., Light-induced resetting of a mammalian circadian clock is associated with rapid induction of the *mPer1* transcript, *Cell*, 91, 1043, 1997

79. Steeves, T. D. L., King, D. P., Zhao, Y., Sangoram, A., Moore, R. Y., and Takahashi, J. S., Molecular cloning, chromosomal mapping, mRNA expression and suprachiasmatic nucleus localization of the human *CLOCK* gene, *Genomics,* in press.

80. Shimomura, K., Vitaterna, M. H., Low-Zeddies, S. S., Whiteley, A. R., and Takahashi, J. S., Genetic dissection of strain differences in mouse circadian rhythmicity, *Soc. Res. Biol. Rhythms Abstr.*, 1998.

81. Shimomura, K., unpublished observation, 1998.

82. King, D. P., Vitaterna, M. H., Chang, A.-M., Dove, W. F., Pinto, L. H., Turek, F. W., and Takahashi, J. S., The mouse *Clock* mutation behaves as an antimorph and maps within the $W^{19H}$ deletion, distal of *Kit*, *Genetics*, 146, 1049, 1997.

83. Hotta, Y. and Benzer, S., Mapping of behavior in *Drosophila* mosaics, *Nature*, 240, 527, 1972.

84. Hotta, Y. and Benzer, S., Courtship in *Drosophila* mosaics: sex-specific foci for sequential action patterns, *Proc. Natl. Acad. Sci. U.S.A.*, 73, 4154, 1976.

85. Benzer, S., Genetic dissection of behavior, *Sci. Am.*, 229, 24, 1973.

86. Konopka, R. J., Wells, S., and Lee, T., Mosaic analysis of a *Drosophila* clock mutant, *Mol. Gen. Genet.*, 190, 284, 1983.

87. Ewer, J., Frisch, B., Hamblen-Coyle, M. J., Rosbash, M., and Hall, J. C., Expression of the *period* clock gene within different cell types in the brain of *Drosophila* adults and mosaic analysis of these cells' influence on circadian behavioral rhythms, *J. Neurosci.*, 12, 3321, 1992.

88. Liu, X., Lorenz, L., Yu, Q., Hall, J. C., and Rosbash, M., Spatial and temporal expression of the *period* gene in *Drosophila melanogaster*, *Genes Dev.*, 2, 228, 1988.

89. Siwicki, K. K., Eastman, C., Petersen, G., Rosbash, M., and Hall, J. C., Antibodies to the *period* gene product of *Drosophila* reveal diverse tissue distribution and rhythmic changes in the visual system, *Neuron*, 1, 141, 1988.

90. Ralph, M. R. and Hurd, M. W., Circadian pacemakers in vertebrates, in *Circadian Clocks and Their Adjustment, Ciba Foundation Symposium,* Vol. 183, Chadwick, D. J. and Ackrill, K., Eds., Wiley, Chichester, 1995, 67.

91. Vogelbaum, M. A. and Menaker, M., Temporal chimeras produced by hypothalamic transplants, *J. Neurosci.*, 12, 3619, 1992.

92. Welsh, D. K., Logothetis, D. E., Meister, M., and Reppert, S. M., Individual neurons dissociated from rat suprachiasmatic nucleus express independently phased circadian firing rhythms, *Neuron*, 14, 697, 1995.

93. Low-Zeddies, S. S. and Takahashi, J. S., Circadian behavior of *Clock* mutant wild-type chimeric mice, *Soc. Res. Biol. Rhythms Abstr.,* 1998.

# 17 Genetics of Sleep and Its Disorders

*Bruce F. O'Hara and Emmanuel Mignot*

## CONTENTS

## INTRODUCTION

Sleep is a fundamental biological need in all birds and mammals; however, its basic functions are still unclear. In humans, sleep typically consumes more than half of the first year of life and roughly a third of adult life. In rats, a complete lack of sleep leads to death in a matter of weeks.[1,2] Clinically, the most frequent sleep disorders are insomnia (whether or not of chronobiological origin), obstructive sleep apnea, restless leg syndrome/periodic leg movements, and narcolepsy. In this review, we will discuss a variety of evidence supporting the existence of genetic factors for both normal and abnormal variations in sleep. In addition to explaining such variation, the homeostatic and circadian regulation of sleep in all birds and mammals almost certainly involves regulated changes in gene expression. An understanding of this genetic variation, and of changes in gene expression relevant to sleep, may suggest novel approaches in the treatment of sleep disorders.

Electrophysiological studies have long shown that sleep is a heterogeneous state, most classi-cally separated into rapid eye movement (REM) and non-REM sleep. Non-REM sleep can also be subdivided into light NREM sleep (stages I and II) and slow wave sleep (SWS, stages III and IV). Independent of this organization by sleep stage, the propensity to sleep or to stay awake is regulated by homeostatic (sleep debt dependent) and circadian (clock dependent) processes. The importance of circadian factors can be best illustrated in the absence of time cues. In these conditions, sleep, wakefulness, core temperature, and various other behaviors still fluctuate with a periodicity close to 24 h called the free-running circadian period ($\tau$). Circadian and homeostatic factors have distinct anatomical substrates, circadian factors regulating sleep and wakefulness being mostly if not exclusively localized in the hypothalamus within the suprachiasmatic nuclei (SCN). Finally, sleep is associated with a host of physiological changes that have a pathological impact. These include well-established sleep-state-specific or circadian controlled endocrine release including changes in convulsive thresholds, regulation of breathing, cardiovascular control, gastrointestinal physiology, and muscle tone control.

## IMPORTANCE OF GENES AND GENE REGULATION

The involvement of genetic factors in sleep and sleep disorders can be viewed from a number of different perspectives. At the very least we know that genes are necessary to make all of the proteins required for the proper development and functioning of the brain. It is conceivable, however, that the actual regulation of sleep at the molecular level might occur via posttranslational modifications of proteins and changes in the abundance of other nonprotein molecules. Genes might just provide the building blocks. Variations in sleep between individuals might be primarily due to environmental factors. One argument in support of this view is that transcription and translation are relatively slow processes requiring many minutes to hours to substantially alter protein levels. In contrast, state transitions in sleep such as NREM to REM to wake, can occur in seconds to minutes, especially in some rodents. Therefore, it does not seem feasible that gene expression plays a central role in state-to-state transitions. However, the longer term homeostatic regulation of these sleep states and the circadian influences both occur over many hours and almost certainly require genetic regulation. The strongest case for such regulation is in the circadian system (see Chapter 16).

## CIRCADIAN RHYTHMS AND GENES

Circadian rhythmicity is an almost universal property observed in most organisms, including some unicellular organisms.[3] A variety of mutations have been reported to alter circadian rhythmicity in *Drosophila, Neurospora,* and *Arabidosis*.[4,5] Research in this field has already led to the isolation of two genes ("period" or *per* and "timeless" or *tim*) in *Drosophila* and of one gene in *Neurospora* ("frequency" or *frq*) whose mutations can suppress, decrease, or increase the circadian free-running period $\tau$.[3,6-8] How these genes contribute to the generation of 24-h rythmicity is still uncertain[9] but transcription-translation autoregulatory feedback loops are probably involved.[3,8] In *Drosophila*, for example, PER protein and per mRNA levels fluctuate with a 3- to 4-h difference in phase, and TIM is necessary for these fluctuations to occur. It is thus hypothesized that TIM interacts with PER to enter into the nucleus and directly or indirectly regulates the transcription of the per locus with a delay to produce the 24-h rhythmicity.[7-9]

In mammals, research in the area is facilitated by the fact that most circadian rhythms are generated in a discrete region of the hypothalamus, the SCN.[11] Lesions of these nuclei suppress all behavioral rhythms in the absence of time cues, while transplanting fetal hypothalamic tissue into lesioned animals restores circadian fluctuations. Mouse strains and mutant hamsters with abnormal free-running periods (too long or too short $\tau$) have been identified.[12-13] Transplanting mutant SCN tissue into the third ventricle of a normal animal whose SCN were lesioned results in restoration of circadian rhythmicity, but with the mutant period.[11,14] This supports the view that the SCN is necessary and sufficient for the generation of behavioral activity rhythms. The neuronal, metabolic, and neurochemical activity of SCN tissue also varies with a circadian periodicity *in vitro*.[11] The mechanism by which the SCN generates rhythmicity is still debated. The transplant probably generates rhythmicity independent of any synaptic connection with the host tissue. Rhythmicity within the nuclei also seems independent of the existence of synaptic connections, because individual neurons and glial cells can generate circadian fluctuations independently. These results appear to be parallel to what has been found in avian pineal glands or retina and in the marine mollusc bulla.[3] Based on the knowledge gathered from lower organisms, it is likely that transcription-translation mechanisms within individual cells are a core phenomenon initiating overall behavioral circadian rhythmicity across the animal kingdom. Synaptic organization in the SCN may thus coordinate and relay rather than generate rhythmicity.

It is generally assumed that the genetic control of circadian rhythmicity is polygenic in mammals as it is in lower organisms, and several mutations with a strong effect on free-running period have been reported.[3,12,13,15] One of these phenotypes is a spontaneous semidominant mutation in the golden Hamster, *Tau*,[12] that induces a shorter free-running period and no other apparent abnormalities. The two others, *Clock* and *Wheel*, are dominant mutations in mice that were produced through *N*-ethyl-*N*-nitrosourea (ENU) germline mutagenesis. *Wheel* is a complex neurological mutation that associates a complex array of abnormal behaviors such as circling, hyperactivity, and abnormal circadian rhythmicity and was mapped to mouse chromosome 4.[15] *Clock* is a pure circadian mutation associated with a long free-running period that was mapped to the midportion of mouse chromosome 5, in a region of conserved synteny with human chromosome 4 (see Chapter 16).

Recently, the identity of the *Clock* gene was established by a combination of positional cloning[16] and rescue of the mutant phenotype by transgenic approaches.[17] Of particular interest is the fact that *Clock* is a member of the bHLH-PAS family of proteins. The PAS domain is named for three proteins: PER, SIM, and ARNT. The *per* gene in *Drosophila*, as discussed above, is probably the best-studied clock gene. However, *Clock* does not share significant sequence identity with *per* outside of the PAS domain and therefore is not a clear homolog of this gene. Subsequent to the cloning of *Clock*, a true *per* homolog in both humans and mice was discovered independently by two groups.[18,19] The human gene designated h*PER* and the mouse gene designated m*Per* have several regions of high sequence identity with the *Drosophila* per gene outside of the PAS domain, and like the *Drosophila per*, have rhythmic expression of their mRNAs across the day. *Clock* mRNA, in contrast, has only minor variations across the day. Interestingly, the mammalian PERs have a bHLH domain lacking in *Drosophila* PER. The CLOCK protein also has this domain, which can be utilized for DNA binding and transcriptional activation. Two other PAS domain genes, White-collar-1 and 2 (*wc-1* and *wc-2*), were recently shown to be important for circadian rhythmicity in *Neurospora*, further extending the evolutionary significance of this family in clock designs.[20] WC-1 and WC-2 appear to act as positive regulators of gene transcription as opposed to the negative feedback loops established for *per* in *Drosophila* and *frq* in *Neurospora*.[7,9,10] In mammals, it is not yet clear whether *Clock* and *Per* form feedback loops. They may even act together as heterodimers or combine with other PAS domain proteins (with or without the bHLH motif). Two possible candidates are the recently cloned *NPAS1* and *NPAS2* genes expressed in neurons.[21] In summary, protein-protein interactions and regulation of transcription are likely to be involved in generating circadian rythmicity across all organisms.

## SLEEP AND GENES

### CHANGES IN GENE EXPRESSION ACROSS AROUSAL STATES

The genes that have been examined most thoroughly in relation to sleep are the immediate early genes (IEGs), most commonly *c-fos*. Changes in IEG expression are of interest for at least two reasons. Since *c-fos* and other IEG expression generally increases with neuronal activity,[22] *in situ* hybridization to the mRNAs and/or immunocytochemistry to the proteins can be used to examine which brain regions are most responsive to changes in arousal state. Second, since IEG products are themselves transcription factors, they are presumably altering the expression of many other genes, although to date very few examples of target genes have been documented.

Several groups have utilized IEG expression to examine different questions relevant to arousal state control. Across the day, *c-fos* expression is clearly correlated with the rest/activity cycle, with higher expression at night in nocturnal animals and lower levels during the day.[23] This effect was reversed by preventing sleep during the day with resultant increased sleep at night. In addition, diurnal rodents also have the expected pattern of *c-fos* expression reversed, with high levels in the day and lower levels at night.[24] Sleep deprivation increases *c-fos* and other IEGs whereas recovery sleep reduces such expression.[23,25,26] Given the potential stress confounds of sleep deprivation, Pompeiano et al.[27] examined *c-fos* and NGFI-A mRNA and protein following periods of spontaneous wakefulness and sleep and found similar results, greatly strengthening the argument that IEG expression is correlated with wakefulness. Whether or not IEG expression plays a functional role in sleep regulation is still unclear; however, one study has suggested such a role in the preoptic hypothalamus.[28] Antisense oligonucleotides to *c-fos* mRNA were injected in this region and shown to reduce C-FOS protein. These animals subsequently had greater wakefulness on the day following these injections.

IEG expression has also been used to look at the neuroanatomy of sleep. For example, carbachol-induced REM sleep activates *c-fos* in several nuclei implicated in the control of REM sleep.[29] The locus ceruleus (LC), in particular, seems to be important not only because *c-fos* expression changes across arousal states in this nucleus, but also because the LC appears to control a large amount of *c-fos* expression throughout the forebrain. Unilateral lesions of the LC reduce C-FOS and NGFI-A levels during wake on the ipsalateral, but not contralateral side of the brain.[30] The EEG of these rats appeared normal. The reduced level of C-FOS and NGFI-A during wake on the lesioned side was comparable to the levels seen during typical periods of extensive sleep. Recently, an exception to the normal correlation of wake and C-FOS levels was found in certain cells of the ventrolateral preoptic area.[31] These neurons express high levels of *C-FOS* during sleep and may play a key role in arousal state control.

In summary, the IEG data are suggestive of an important role for gene expression in sleep; however, to date no genes or their protein products have been convincingly established as sleep factors. Given that the mammalian brain expresses approximately half of the estimated 100,000 genes, more global approaches to identify the differential expression of genes important in arousal state control will probably be needed. Such approaches could include substractive hybridization, differential display polymerase chain reaction (PCR), and the use of microarrays of many cDNAs. One attempt at using subtractive hybridization has been made which utilized forced locomotion to deprive rats of sleep for 24 h.[32] Four clones were isolated whose mRNA appeared to be reduced following this treatment and six clones were found to have increased mRNA.

## EVIDENCE FOR GENETIC FACTORS IN NORMAL SLEEP: TWIN STUDIES

Another approach to investigate whether genetic factors play a role in sleep is to study the extent to which variability in some aspect of sleep is heritable. Twin studies provide one such model by comparing similarities in monozygotic (MZ) vs. dizygotic (DZ) twins. Research in this area is primarily based on questionnaire studies comparing sleep habits in MZ and DZ twins (duration of sleep, schedules and quality of night sleep, frequency of napping).[33-36] For most of the variables analyzed, correlations are higher for MZ vs. DZ twins. This effect remains significant even when twins do not share the same environment[34] and does not correlate strongly with depression or anxiety.[37] Environmental factors, however, do contribute significantly to the variance.[34,35] Measures of the residual variance between MZ twins (1-rmz) quantify the influence of environmental factors specific to each twin pair.[38] In all studies, correlations barely reach 0.60, thus suggesting that half of the variance is associated with environmental factors. Since twins live in similar environments, this difference probably corresponds to short-term environmental variance.

Several authors have studied sleep in MZ and DZ twins using polygraphic techniques.[39-42] These studies generally confirm the results obtained with questionnaires but sample sizes are always small. Linkowski et al.,[41,42] studying 26 pairs during three consecutive nights, demonstrated significant differences between MZ and DZ twin pair correlations for all stages of sleep except REM sleep. More recently, Van Beijsterveldt et al.,[43] studying awake resting EEG frequencies in a large number of MZ and DZ twins, also demonstrated high average heritabilities (0.76 to 0.89) for all analyzed EEG frequency bands. Other evidence that basic EEG characteristics are genetically determined comes from Vogel,[44] who studied EEG variants measured during waking periods, which are present in a few percentage of the population. Several of these variants appear to be inherited in a simple mendelian pattern, suggesting a single gene. One such variant is referred to as the low-voltage EEG, characterized by an almost complete lack of alpha waves upon eye closure. This trait appears to have an autosomal dominant mode of transmission and has recently been mapped to human chromosome 20q.[45,46] These studies provide evidence that EEG genetic variations are not only quantitative but also qualitative.[44,45]

Most of the early studies did not take into account the fact that sleep is independently regulated by circadian and homeostatic factors. Linkowski et al.[47,48] tried to address this issue by measuring cortisol and prolactin levels in twins. Results suggest that genetic factors play a major role in the regulation of cortisol secretion but not of prolactin. Drennan et al.[36] used the Horne-Ostberg questionnaire to examine morningness/eveningness in 238 twin pairs and found higher correlations in MZ pairs, thus suggesting the existence of human circadian genetic factors. Such studies could certainly be extended. As of today, there has been no twin study measuring SWS/REM sleep homeostasis or circadian rhythm properties under optimal experimental conditions.

## EVIDENCE FOR GENETIC FACTORS IN NORMAL SLEEP: ANIMAL STUDIES

Animal studies also support the concept of genetic influences on sleep. Major differences in SWS versus REM sleep amount and distribution can be observed within the same species, these differences being resistant to prolonged manipulations such as forced immobilization or sleep deprivation,[49-52] Significant variations in sleep/wake architecture and EEG profiles are also observed between rodent inbred strains.[50-58] C57BL or C57BR strains are characterized by long REM sleep episodes, short SWS episodes, and significant circadian variation under light/dark conditions.[51,55] At the opposite end of the spectrum, BALB/c is characterized by REM sleep episodes of very short duration and poor diurnal rhythm, and DBA mice are intermediary for these characteristics.[51,55] The characteristic free-running period is also 50 min longer in C57BL/6J than in BALB/cByJ.[59] Qualitative differences in EEG signals are also observed. DBA and BALB/c but not C57BR display high amplitude spindles while REM-associated theta frequency varies significantly between strains.[51,55] Franken et al.[60] have recently extended this work with an eye toward the quantitative trait locus (QTL) approach (see below), and also have shown differences between strains in the homeostatic delta rebound following sleep deprivation.

These phenotypic differences are genetically transmitted. Diallelic methods,[53] simple segregation analysis in a backcross setting,[55] and recombinant inbred strain studies[59,61] suggest that many genes are involved in the expression of each trait.[52-54,59,61] The interactions observed are complex and not strictly additive, with hybrids of inbred strains occasionally presenting important deviations when compared to the average of parental strains.[53]

## PHARMACOGENETIC APPROACHES IN RODENTS

The basis of this technique is the selection of animal strains relatively sensitive or resistant to pharmacologic agents, for example, ethanol,[62-64] benzodiazepines,[65] barbiturates,[66] or cholinergic compounds.[67,68] These models can then be studied pharmacologically, physiologically, and genetically. Mice that have been selected for their hypersensitivity to cholinergic compounds display, for example, an increase in paradoxical sleep,[68] which confirms the role of acetylcholine in REM sleep regulation. Another group of actively studied mice are the long-sleep (LS) and short-sleep (SS) mouse strains, which differ in sleep time following ethanol administration.[64,69]

The utility of these model strains for studying the genetic control of sleep remains uncertain. Indeed, as of today, LS and SS animals have not been studied for sleep and circadian rhythms using either polygraphic recordings or wheel running activity. Moreover, the effect of benzodiazepines and alcohol on sleep seems to be indirect and very dependent on previously accumulated sleep debt.[70-73] A better analysis of the physiology of these model animals, specifically circadian rhythms and sleep during baseline condition and after deprivation, is needed.

## Cloning Sleep Related Genes in Mammals

Even for a relatively simple, anatomically localized function such as the regulation of circadian rhythmicity in the SCN, multiple genes appear to be involved. The situation is thus likely to be even more complex for normal and abnormal sleep regulation. Mutations and variations in some of these genes will cause pathological phenotypes in animals and humans, while others may contribute to interindividual differences or variations in sleep patterns between species.

Mouse models are likely to be one of the best tools for discovering sleep-related genes.[74] Not only are mice easy to breed and study, but high-density marker maps, such as the Whitehead Institute/MIT map, are now available. One possible approach, already exemplified for circadian rhythmicity, is to use mutagenesis to produce mutants with sleep or circadian abnormalities and to isolate the mutant genes through positional cloning. This is clearly one of the most promising avenues but the feasibility of this approach is limited by the relatively large number of animals (200 to 1,000) that need to be screened to find a mutation of interest.[3] Another strategy is to use QTL analysis and inbred mouse strains. This usually involves studying recombinant inbred (RI) strains to first identify possible genetic effects and phenotypes of interest and then verify the QTLs by breeding experiments, genetic typing, and building congenic lines. These protocols have been used successfully for numerous other multifactorial traits from autoimmune diabetes in the nonobese diabetic (nod) mouse,[75] to drug response for addiction research.[76,77] Candidate QTLs for circadian rhythmicity in mice have recently been reported using available RI strains.[61,78] One limitation in these studies is the small number of RI strains and inbred strains used to conduct them. Therefore, this approach may miss many important genes because no allelic differences exist between the inbred strains. This limitation can be reduced by testing more divergent strains of different mouse species or subspecies which can still interbred. However, this involves constructing large intercrosses or backcrosses and then generating complete genetic maps, clearly a large undertaking. Further, in many cases the allelic differences may be weak or dependent upon genetic background. This can make the next step, gene isolation, extremely difficult if not impossible. An advantage of the QTL technique is that it may lead to the isolation of naturally polymorphic factors that are involved in phenotypic variations; it is thus a complementary technique to mutagenesis. The QTL approach is also possible in humans whereas mutagenesis, of course, is not.

Another approach consists of direct examination of "candidate" genes in phenotypically distinct animal strains. Korpi et al.,[65] for example, recently demonstrated that a strain of rats that is more resistant to benzodiazepines and alcohol carried a specific mutation of the $\alpha_6$-subunit of the GABA$_A$ receptor. This result agrees with the idea that sensitivity to alcohol and to benzodiazepines proceeds from a common GABAergic mechanism. Nevertheless, other genes and factors also seem to be involved because mutations at this locus are not found in other strains of rodents sensitive to sedatives or to alcohol.[65] The candidate gene approach will become more feasible as more and more genes are isolated and sequenced, and it is likely to be most powerful in humans. In this species, gene isolation and sequencing is moving forward at a faster pace than in mice, and human disorders offer a wide-open field of investigation.

Other research strategies that look promising use genetically manipulated animal strains (transgenic or "knockout" mice).[79,80] If the animal is viable, the analysis of the obtained phenotype provides information on the normal function of the gene which has been eliminated or modified. A recent example of this research strategy was provided by the study of the prion knockout mouse.[81,82] In these studies, mice with a null mutation in the prion protein gene (PrP0/0), a gene associated with fatal familial insomnia and Creutzfeldt-Jakob, were reported to display alterations in both circadian activity and sleep patterns, thus suggesting a role for the prion protein in sleep regulation. New mouse strains manipulated for one or multiple candidate genes (neuroreceptors and enzymes, candidate disease genes) are being developed at an increasing pace. Ultimately, sleep and circadian rhythms will also be studied in these mutants and conclusions will be drawn regarding the involvement of a given system in the control of sleep.

The study of genetically altered strains will soon lead to the identification of numerous genetic factors involved in the physiological control of sleep in rodents. The potential clinical implications of these research avenues are still difficult to measure. Only a fraction of the genes identified in rodents will play a role in human disease. The rodent models will, however, remain attractive for the design of better controlled behavioral and genetic studies.

## GENETIC ASPECTS OF PATHOLOGICAL SLEEP

Numerous sleep pathologies such as narcolepsy, fatal familial insomnia, delayed and advanced sleep phase syndrome, sleep paralysis, hypnagogic hallucinations, sleep apnea, and restless leg syndrome are well-known for recurring in certain families with a high frequency.[35,83-90] All these results confirm the existence of a group of genes whose function is more specifically related to sleep. Genome screening is therefore another possible research strategy for identification of pathological factors in sleep disorders.

### MOLECULAR GENETICS AND NARCOLEPSY-CATAPLEXY

Narcolepsy is a disorder characterized by excessive daytime sleepiness and abnormal REM sleep. Disease onset usually occurs at 15 to 25 years of age and only exceptionally before puberty. Daytime somnolence is usually the most disabling symptom; it frequently requires life-long treatment with amphetamine-like stimulants. Sleepiness can be objectively demonstrated in a sleep laboratory using the Multiple Sleep Latency Test (MSLT). In this simple test, latencies to falling asleep are measured during four or five short naps taken at 2-h intervals during the daytime, and the presence or absence of a REM sleep transition is recorded. In narcolepsy, sleep latencies are decreased (mean less than 8 min) and multiple REM sleep transitions (more than two) are observed. Cataplexy, sleep paralysis, and hypnagogic hallucinations are frequently associated symptoms. In cataplexy, brief episodes of muscle weakness resulting in knees buckling, jaw sagging, head dropping, or, less frequently, full body paralysis are observed when the patient is laughing or elated. In sleep paralysis, the patient finds himself unable to move for a few seconds to several minutes when waking up or while falling asleep. Sleep paralysis and cataplexy are both pathological manifestations of REM sleep atonia but only cataplexy is specific for the narcolepsy syndrome. Hypnagogic hallucinations are dream-like experiences occurring at sleep onset or during sleep attacks.

Narcolepsy with cataplexy affects 0.02 to 0.06% of the general population in the United states and Western European countries.[91-94] It may be more frequent (0.16 to 0.18%) in Japan,[94,95] and rarer in Israel.[96] Since its description in 1880 by Gélineau, familial cases have been reported by numerous authors,[88,97-101] thus suggesting a genetic basis to narcolepsy. This pathology thus offers a unique opportunity to discover genes involved in the control of sleep.

More recent studies, however, suggest that narcolepsy is not a simple genetic disorder.[94] The development of human narcolepsy involves environmental factors on a specific genetic background, and only 25 to 31% of MZ twins reported in the literature are concordant for narcolepsy.[94] One of the predisposing genetic factors is located in the major histocompatibility complex (MHC) DQ region. Between 90 and 100% of all narcoleptic patients with definite cataplexy share a specific human leukocyte antigen (HLA) class II allele, HLA DQB1*0602 (most often in combination with HLA DR2), versus 12 to 38% of the general population in various ethnic groups.[102-106] The finding of an HLA association in narcolepsy, together with the fact that HLA DQB1*0602 is likely to be the actual HLA narcolepsy susceptibility gene,[107] suggests that narcolepsy might be an autoimmune disorder. However, all attempts to demonstrate an immunopathology in narcolepsy have failed to date and the mode of action of HLA DQB1*0602 is still uncertain.[108-113]

Between 12 and 38% of the general population carry HLA DQB1*0602 and only a small fraction have narcolepsy; DQB1*0602 is thus a weakly penetrant genetic factor (lHLA = 2), even if genetic association with the disorder is high. Other genetic factors, possibly more penetrant than HLA, are likely to be involved. One to two percent of the first-degree relatives of a patient with narcolepsy-cataplexy are affected by the disorder versus 0.02 to 0.06% in the general population in various ethnic groups, suggesting a 20- to 40-fold increased risk for first-degree relatives.[88,94,101] However, familial aggregation cannot be explained by the sharing of HLA haplotypes alone[94] because some families are non-HLA DQB1*0602 positive,[88] thus suggesting the importance of non-HLA susceptibility genes that could be positionally cloned using genome screening approaches in human multiplex families or isolated populations.

Studies using a canine model of narcolepsy also illustrate the importance of non-MHC genes. In this model, narcolepsy-cataplexy is transmitted as a single autosomal recessive trait with full penetrance, canarc-1.[73,114,115] This high-penetrance narcolepsy gene is unlinked to MHC class II but cosegregates with a DNA segment with high homology to the human immunoglobulin μ-switch sequence.[116] This linkage marker is located very close to the actual narcolepsy gene (current LOD score 15.3 at 0% recombination), and gene isolation is ongoing both in canines and in the corresponding human region of conserved synteny.

## Genetics and Dissociated REM Sleep Events

Sleep paralysis and hypnagogic hallucinations, two symptoms of dissociated REM sleep, occur frequently in the general population independent of narcolepsy.[117,118] Sleep paralysis is highly familial, and autosomal dominant transmission has been observed in some cases.[82,96,119,120] Twin studies suggest a much higher concordance in MZ vs. DZ twins for this symptom,[121] which may be more frequent in the black population.[120] There is no association with HLA DQB1*0602.[117]

In REM sleep behavior disorder (RBD), motor behaviors arise during REM sleep and disturb sleep continuity.[122] RBD is frequently associated with other pathologies such as narcolepsy but may occur in isolation.[122] The familiality of isolated RBD is not established but the disorder may be weakly associated with HLA DQ1.[123]

Cataplexy without sleepiness is exceptional[93] but some rare familial cases have been described with or without associated sleep paralysis.[124-126] In many of these cases, however, clinical presentation seems to differ quite significantly from narcolepsy-cataplexy and cataplexy presented in the first months of life.[125,126] HLA typing has not been done for these families.

## Molecular Genetics and Fatal Familial Insomnia

Fatal familial insomnia is a rare neurological condition characterized by severe insomnia, neurovegetative symptoms, intellectual deterioration, and death.[87,127-129] Insomnia is an early sign and sleep disruption is associated with a disappearance of stage II light sleep and SWS while brief episodes of REM sleep are usually maintained. Neuropathologic lesions are mostly limited to a spongiform degeneration of the anterior ventral and mediodorsal thalamic nuclei and of the inferior olive.[129] This pathology is typically associated with a mutation of the codon 178 in the prion protein gene but one recent report detected a codon 200 mutation.[130] These same mutations are also found in some forms of dementia, like Creutzfeldt-Jakob, but a polymorphism at codon 129 seems to determine the phenotypic expression of fatal familial insomnia versus Creutzfeldt-Jakob disease.[128,130]

The prion protein is encoded by a gene located on human chromosome 20. The normal function of the protein is unknown but the gene is expressed in neurons. Mice homozygous for mutations disrupting the prion protein gene are behaviorally normal but may display sleep abnormalities, as mentioned above.[81,82] Prions are involved in a group of human and animal disorders with more or less anatomically confined spongious degeneration and neuronal atrophy (spongiform encephalopathies). A proteinase-resistant form of the prion protein is probably involved in the pathology.[131] These diseases can appear either in a familial context or in an infectious context, the prion protein (or an agent that cannot be distinguished from the protein element) acting as the transmitting agent. The mechanism by which certain isoforms of the protein are infectious remains a widely discussed topic.[132-134]

How a simple additional polymorphism on codon 129 alters the symptomatology from Creutzfeldt-Jakob to fatal familial insomnia is not understood, but molecular studies are underway to evaluate the effect of these mutations on the metabolism of the protein.[135] The differences in symptomatology are probably due to a differential anatomic localization of the lesions. In fatal familial insomnia, degeneration primarily is localized in the anterior ventral and mediodorsal thalamic nuclei whereas lesions are much more diffuse in Creutzfeldt-Jakob disease.[132,136] The well-established role of the thalamus (albeit mostly of the intralaminar thalamus) and of its cortical projections in the generation of the cortical synchronization of SWS and sleep spindles[137] suggests that thalamic lesions may cause the insomnia in this disorder.[136] To date, however, no study has convincingly demonstrated that the destruction of these nuclei can produce a fatal insomnia in animal models. Bilateral lesions of these nuclei produce a persistent insomnia which is not fatal.[138] Therefore, other more discrete anatomical lesions or a distinct pathophysiological mechanism could also play a role. Transgenic mice carrying the human prion allele specific for fatal familial insomnia have now been generated and are under study to answer these questions.

The implication of the thalamus in the pathophysiology of fatal familial insomnia suggests that this brain structure may be involved in the genesis of other, more frequent insomnias. Insomnia is a very frequent symptom that affects at least 10% of the general population.[139,140] Many insomnias appear to be constitutional[141] and genetic factors influencing the thalamus and homeostatic abnormalities in the regulation of sleep may be involved in some cases. Other genetic factors, such as those regulating circadian rhythmicity at the level of the SCN, could be involved in other cases.

## GENETIC ASPECTS OF RESTLESS LEG SYNDROME AND PERIODIC LIMB MOVEMENTS

Restless leg syndrome (RLS) is a frequent (2 to 5% of general population)[142-145] syndrome that worsens with age and affects both sexes. RLS is almost always associated with periodic leg movement (PLM) during sleep. RLS is best defined as uncomfortable or painful sensations in the legs which force the patient to get up several times each night.[145,146] PLMs are brief and repetitive muscular jerks of the lower limbs occurring mostly during stage II sleep.[145] When these movements increase in strength and frequency, sleep is altered. RLS is highly familial and up to one third of the reported cases may transmit the condition as an autosomal dominant trait[83,147-152] with possible genetic anticipation.[152] Unfortunately, no twin study is available and both the prevalence and the proportion of familial cases seem to vary widely according to the geographical origin of the population studied. These differences may reflect founder effects, such as in Quebec, where one finds a high proportion of familial cases[145] and a higher prevalence[153] or the influence of local environmental factors.

Population-based risk estimations in first- and second-degree relatives are not yet available for this interesting pathology. In a recent study published only as an abstract, risks to first- and second-degree relatives were 19.9 and 4.1%, respectively.[154] This compared to 3.5 and 0.5% for first- and second-degree relatives of control subjects, and suggested a $\lambda_{siblings}$ of approximately 5.[154] Linkage studies using either microsatellite markers or candidate genes in multiplex families are ongoing in order to identify the gene(s) involved[155] (Montplaisir, personal communication). Possible candidate genes are enzymes and receptors of the dopaminergic and enkephalinergic metabolisms, two neurotransmitters involved in the pharmacological treatment of the syndrome. As of today, however, no result suggestive of linkage has been published.

## Genetic Aspects of Sleepwalking, Sleeptalking, and Sleep Terrors

These parasomnias generally occur during SWS (stages III and IV).[156] They are usually grouped together and considered to share a common or related pathophysiologic mechanism,[157] although this notion is sometimes disputed.[156] The prevalence of these symptoms is several percent among children and only scarcely requires a medical consultation. Symptoms generally disappear in adulthood.[156,158]

The familial nature of these symptoms has been recognized by most authors,[159-162] but the exact mode of transmission is uncertain. Twin studies have shown a high degree of concordance for sleepwalking and sleep terrors (50% in MZ, 10 to 15% in DZ).[121,163,164] The genetic predisposition to sleepwalking, sleeptalking, and, to a lesser degree, sleep terrors and enuresis may overlap. Indeed, the frequency of sleep terrors and enuresis might be more frequent in families with somnambulism.[159,161,162] This suggests a related pathophysiologic mechanism and similar genetic control. To date, however, there has not been any molecular study initiated on these pathologies.

## Obstructive Sleep Apnea Syndrome and Related Breathing Abnormalities during Sleep

Obstructive sleep apnea syndrome (OSAS) is a complex syndrome in which the upper airway collapses repetitively during sleep, thus blocking breathing.[165] Snoring is one of the cardinal symptoms. Repeated apneas prevent the patient from sleeping soundly, and the patient is frequently excessively sleepy the following day. Between 4 and 5% of the general population suffers from OSAS,[166,167] which may be a risk factor for high blood pressure and for cardiovascular accidents.[168-170] Recent studies suggest increased vulnerability in African Americans.[171]

Twin studies are lacking in OSAS but two recent studies have shown higher concordance in MZ versus DZ twins for habitual snoring.[121,172] Multiplex families of patients suffering from OSAS have also been reported in the literature[89,173-180] and one study found a substantial increase in HLA A2 and B39 in Japanese patients with OSAS.[181] Familial aggregation is generally explained by the fact that most risk factors involved in the physiopathology of sleep apnea are, in large part, genetically determined. These include obesity, alcoholism, facial soft tissue, and bone anatomy, which all predispose to upper airway obstruction.[89,178,182,183] In some cases, the genetic factor primarily involves abnormal ventilatory control by the central nervous system.[89,174] A possible genetic overlap between OSAS and sudden infant death[174,175,184] and the high degree of concordance in chemoreceptor responses observed in MZ twins[185,186] suggests the importance of genetic factors regulating the central control of ventilation in OSAS.

A multiplicity of genetic factors is likely to correspond to the multifactorial aspect of OSAS. A genetic linkage approach in OSAS would thus be facilitated by a careful phenotypic analysis; for example, studying sleepy or nonsleepy subjects, nonobsese vs. obese OSAS patients,[182] or subjects with selected morphological features.[187]

## CHROMOSOMAL AND GENETIC ABNORMALITIES AND SLEEP DISTURBANCES

The coincidental association of specific chromosomal breakpoints with specific pathologies can be very helpful to localize susceptibility gene(s). In practice, however, karyotypes are rarely requested when a sleep disorder is the primary abnormality and very few sleep studies have been performed in patients with chromosomal or genetic abnormalities. These disorders frequently produce behavioral and medical problems that have secondary effects on sleep, particularly disturbed nocturnal sleep, so it may be difficult to identify a disease-specific sleep phenotype.[188] In spite of these limitations, fragile X subjects have been reported to experience sleep disturbances and low melatonin levels,[189,190] whereas subjects with Norrie disease (genetic alterations in a region encompassing the monoamine oxidase genes at Xp11.3) or Nieman Pick Type C (18q11–q12) may experience cataplexy and sleep disturbances.[191,192] An interesting family with autosomal dominant cerebral ataxia, deafness, normal karyotype, and clinically defined narcolepsy with cataplexy was also described and shown to be non-HLA DR2 associated.[193] OSAS are also frequently observed as a result of anatomic malformations, adenotonsilar enlargement, or morbid obesity.[188,194,195] In a few instances, however, polygraphic studies suggest that central factors are also involved in addition to or independently of abnormal breathing during sleep. This may be the case for the Prader-Willi and Angelman syndromes [del(15q)][195-197] or the Smith Magenis syndrome [del(17)(p11.2)].[198,199]

In spite of their relatively high population frequency, the effects of sex chromosomal anapleudies on sleep have been only marginally studied, but XXY subjects may display increased 24-h sleep time.[200] The general topic of abnormal hormonal control and sleep in these patients would be worth investigating more thoroughly because puberty is associated with established changes in sleep needs.[200] Moreover, narcolepsy often starts in adolescence, and in two cases narcolepsy started at the unusual age of 6 in a Turner syndrome patient [XO][202] and in coincidence with a precocious puberty.[203]

## OTHER SLEEP PATHOLOGIES

Insomnia, obstructive sleep apnea, narcolepsy, PLM and RLS, parasomnias, and circadian disorders are the most frequent sleep pathologies. There are few or no studies on other forms of hypersomnias or parasomnias. One twin study suggested increased frequency of bruxism (teeth grinding during sleep) in MZ vs. DZ twins,[121] and bruxism has been reported in a multiplex context.[204] Familial forms of essential hypersomnia,[98] of hypersomnias associated with dystrophia myotonica[205] or sleep-responsive extrapyramidal dystonias[206,207] and jactatio capitis nocturna[208] have also been reported. A possible association of idiopathic hypersomnias with HLA Cw2 and of hypersomnia in dystrophia myotonica with DR6 has also been found[205,209] but would need independent confirmation.

## PERSPECTIVES IN HUMAN GENETIC RESEARCH IN SLEEP DISORDERS

The complexity of sleep as a physiological phenomenon is matched by a vast number of pathologies. Most of these pathologies are multifactorial and to a large extent genetically determined. The recent progress of molecular genetics has enabled researchers to undertake a purely genetic approach to understanding the pathophysiology of these disorders. This approach will most likely first lead to the identification of genes involved in etiologically homogeneous sleep disorders such as narcolepsy. Genome screening studies in more frequent and complex sleep disorders, such as OSAS or RLS, will require the inclusion of a large number of multiplex families but are now feasible. These disorders may also benefit from studies in isolated populations or even of association studies using very large numbers of single case families; this last design has the advantage of being easily used for secondary candidate gene studies. Those are likely to become a more viable research strategy as more and more genes are cloned and positioned on the human map and possible candidate genes identified in mouse models.

## CONCLUSIONS

Considerable evidence supports a role for genes in many aspects of sleep and its disorders. Changes in gene expression are important in the circadian clock and, by extension, the circadian regulation of sleep propensity. The role of gene expression in the homeostatic regulation of sleep propensity is less clear, but almost certainly important. Evidence also has been presented that variations in sleep and variation in EEG patterns among individuals are highly genetic. Similarly, when such variation extends to pathological conditions, differences in individual genes are implicated. The identification of these genes can only help our understanding of sleep and the treatment of sleep disorders.

## ACKNOWLEDGMENTS

Part of the work described in this chapter has been funded by a National Institutes of Health grant from the National Institute of Neurological Disease and Stroke NS23724 to E. Mignot and a scientist development award DA00187 to B. O'Hara.

## REFERENCES

1. Rechtschaffen, A., Gilliland, M. A., Bergmann, B. M., et al., Physiological correlates of prolonged sleep deprivation in rats, *Science*, 221, 182, 1983.
2. Kushida, C. A., Bergman, B. M., and Rechtschaffen, A., Sleep deprivation in the rat. IV. Paradoxical sleep deprivation, *Sleep*, 12, 22, 1989.
3. Takahashi, J. S., Molecular neurobiology and genetics of circadian rhythms in mammals, *Annu. Rev. Neurosci.*, 18, 531, 1995.
4. Dunlap, J. C., Genetic analysis of circadian clocks, *Annu. Rev. Physiol.*, 55, 683, 1993.
5. Millar, A. J., Carré, I. A., Strayer, C. A., Chau, N. H., and Kay, S. A., Circadian clock mutants in Arabidopsis identified by luciferase imaging, *Science*, 267, 1161, 1995.
6. Hall, J. C., Tripping along the trail to the molecular mechanism of biological clocks, *Trends. Neurosci.*, 18, 230, 1995.
7. Dunlap, J. C., The genetic and molecular dissection of a prototypic circadian system, *Annu. Rev. Genet.*, 30, 579, 1996.
8. Sehgal, A., Ousley, A., and Hunter-Enso,r M., Control of circadian rhythms by a two component clock, *Mol. Cell. Neurosci.*, 7, 165, 1996.
9. Hall, J. C., Are cycling gene products as internal zeitgebers no longer the zeitgeist of chronobiology?, *Neuron*, 17, 799, 1996.
10. Hardin, P. E., Hall, J. C., and Rosbash, M., Feedback of the Drosophilia period gene product on circadian cycling of its messenger RNA levels, *Nature*, 343(6258), 536, 1990.
11. Klein, D., Moore, R. Y., and Reppert, S. M., Suprachiasmatic nucleus, in *The Mind's Clock*, Oxford University Press, New York, 1991, 467.
12. Ralph, M. R. and Menaker, M., A mutation of the circadian system in golden hamsters, *Science*, 241, 1225, 1988.
13. Vitaterna, M. H., King, D. P., Chang, A. M., Kornhauser, J. M., Lowrey, P. L., McDonald, J. D., Dove, W. F., Pinto, L.H., Turek, F. W., and Takahashi, J. S., Mutagenesis and mapping of a mouse gene clock, essential for circadian behavior, *Science*, 264, 719, 1994.
14. Ralph, M. R., Foster, R. G., Davis, F. C., and Menaker, M., Transplanted suprachiasmatic nucleus determines circadian period, *Science*, 247(4945), 975, 1990.

15. Nolan, P., Sollars, P. J., Bohne, B. A., Ewens, W. J., Pickard, G. E., and Bucan, M., Heterozygosity mapping of partially congenic lines: mapping of a semi dominant neurological mutation, wheels, on mouse chromosome 4, *Genetics*, 140, L245, 1995.

16. King, D. P., Zhao, Y., Sangoram, A. M., Wilsbacher, L. D., Tanaka, M., Antoch, M. P., Steeves, T. D. L., Vitaterna, M. H., Kornhauser, J. M., Lowery, P. L., Turek, F. W., and Takahashi, J. S., Positional cloning of the mouse circadian clock gene, *Cell*, 89, 641, 1997.

17. Antoch, M. P., Song, E. J., Chang, A. M., Vitaterna, M. H., Zhao, Y., Wilsbacher, K. S., Sangoram, A. M., King, D. P., Pinto, L. H., and Takahashi, J. S., Functional identification of the mouse circadian clock gene by transgenic BAC rescue, *Cell*, 89, 655, 1997.

18. Sun, ZS, Albrecht, U., Zhuchenko, O., Bailey, J., Eichele, G., and Lee, C.C., *RIGUI*, a putative mammalian ortholog of the Drosophila *period* gene, *Cell*, 90, 1003, 1997.

19. Okamura, H.T.H., Shigeyoshi, Y., Fukuhara, C., Ozawa, R., and Hirose, M., Circadian oscillation of a mammalian homologue of the *Drosophila period* gene, *Nature*, 389, 512, 1997.

20. Crosthwaite, S. K., Dunlap, J. C., and Loros, J. J., Neurospora wc-1 and wc-2 transcription, photoresponses, and the origin of circadian rhythmicity, *Science*, 276, 763, 1997.

21. Zhou, Y. D., Barnard, M., Tian, H., Li, X., Ring, H. Z., Francke, U., Shelton, J., Richardson, J., Russell, D. W., and McKnight, S. L., Molecular characterization of two mammalian bHLH-PAS domain proteins selectively expressed in the central nervous system, *Proc. Natl. Acad. Sci. U.S.A.*, 94, 713, 1997.

22. Morgan, J.I. and Curran, T., Stimulus-transcription coupling in the nervous system: involvement of the inducible proto-oncogenes *fos* and *jun*, *Annu. Rev. Neurosci.*, 14, 421, 1991.

23. Grassi-Zucconi, G., Menegazzi, M., Carcereri De Prati, A., Bassetti, A., Montagnese, P., Mandile, P., Cosi, C., and Bentivoglio, M., c-fos mRNA is spontaneously induced in the rat brain during the activity period of the circadian cycle, *Eur. J. Neurosci*, 5, 1071, 1993.

24. O'Hara, B. F., Watson, F. L., Andretic, R., Wiler, S. W., Young, K. A., Bitting, L., Heller, H. C., and Kilduff, T. S., Daily variation of CNS gene expression in nocturnal vs. diurnal rodents and in the developing rat brain, *Mol. Brain Res.*, 48, 73, 1997.

25. Grassi-Zucconi, G., Giuditta, A., Mandile, P., Chen, S., Vescia, S., and Bentivoglio, M., c-fos spontaneous expression during wakefulness is reversed during sleep in neuronal subsets of the rat cortex, *J. Physiol.*, 88, 91, 1994.

26. O'Hara, B. F., Young, K. A., Watson, F. L., Heller, H. C., and Kilduff, T. S., Immediate early gene expression in brain during sleep deprivation: preliminary observations, *Sleep*, 16, 1, 1993.

27. Pompeiano, M., Cirelli, C., and Tononi, G., Immediate-early genes in spontaneous wakefulness and sleep: expression of c-fos and NGFI-A mRNA and protein, *J. Sleep Res.*, 3, 80, 1994.

28. Cirelli, C., Pompeiano, M., Arrighi, P., and Tononi, G., Sleep-waking changes after c-fos antisense injections in the medial preoptic area, *Neuroreport*, 6(5), 801, 1995.

29. Shiromani, P. J., Kilduff, T. S., Bloom, F. E., and McCarley, R. W., Cholinergic induced REM sleep triggers Fos-like immunoreactivity in dorsolateral pontine regions associated with REM sleep, *Brain Res.*, 580, 351, 1992.

30. Cirelli, C., Pompeiano, M., and Tononi, G., Neuronal gene expression in the waking state: a role for the locus coeruleus, *Science*, 274, 1211, 1996.

31. Sherin, J. E., Shiromani, P. J., McCarley, R. W., and Saper, C. B., Activation of ventrolateral preoptic neurons during sleep, *Science*, 271, 216, 1996.

32. Rhyner, T. A., Borbely, A. A., and Mallet, J., Molecular cloning of forebrain mRNAs which are modulated by sleep deprivation, *Eur. J. Neurosci.*, 2, 1063, 1990.

33. Gedda, L. and Brenci, G., Twins living apart test: progress report, *Acta. Genet. Med. Gemellol.*, 32, 17, 1983.

34. Partinen, M., Kaprio, J., Koskenvuo, M., Koskenvuo, M., Putkonen, P., and Langinvainio, H., Genetic and environmental determination of human sleep, *Sleep*, 6, 179, 1983.

35. Heath, A., Kendler, K. S., Eaves, L. J., and Martin, N. G., Evidence for genetic influences on sleep disturbance and sleep patterns in twins, *Sleep*, 13, 318, 1990.

36. Drennan, M. D., Shelby, J., Kripke, D. F., et al., Morningness/eveningness is heritable, *Soc. Neurosci. (Abstr.)*, 18, 196, 1992.

37. Kendler, K. S, Heath, A. C., Martin, N. G., and Eaves, L. J., Symptoms of anxiety and symptoms of depression: same genes, different environments?, *Arch. Gen. Psychiatr.*, 122, 451, 1987.

38. Hrubec, Z. and Robinette, C. D., The study of human twins in medical research, *N. Engl. J. Med.*, 310(7), 435, 1984.
39. Webb, W. B. and Campbell, S. S., Relationship in sleep characteristics of identical and fraternal twins, *Arch. Gen. Psychiatr.*, 40, 1093, 1983.
40. Hori, A., Sleep characteristics in twins, *Jpn. J. Psychiatr. Neurol.*, 40(1), 35, 1986.
41. Linkowski, P., Kerkhofs, M., Hauspie, R., Susanne, C., and Mendlewicz, J., EEG sleep patterns in man: a twin study, *Electroencephalogr. Clin. Neurophysiol.*, 73, 279, 1989.
42. Linkowski, P., Kerkhofs, M., Hauspie, R., and Mendlewicz, J., Genetic determinants of EEG sleep: a study in twins living apart, *Electroencephalogr. Clin. Neurophysiol.*, 79, 114, 1991.
43. Van Beijsterveldt, C. E. M., Molenaar, P. C. M., and De Geus, E. J. C., Boomsma, D. I., Heritability of human brain functioning as assessed by electroencephalography, *Am. J. Hum. Genet.*, 58, 562, 1996.
44. Vogel, F., Brain physiology: genetics of the EEG, in *Human Genetics*, Vogel, F. and Motulsky, A.G., Eds., Springer-Verlag, New York, 1986, 590.
45. Anokhin, A., Steinlein, O., Fischer, C., Mao, Y., Vogt, P., Schalt, E., and Vogel, F., A genetic study of the human low-voltage electroencephalogram, *Hum. Genet.*, 90, 99, 1992.
46. Steinlein, O., Anokhin, A., Yping, M., Schalt, E., and Vogel, F., Localization of a gene for low-voltage EEG on 20q and genetic heterogeneity, *Genomics*, 12(1), 69, 1992.
47. Linkowski, P., Kerhofs, M., V and an Cauter, E., Sleep and biological rhythms in man: a twin study, *Clin. Neuropharmacol.*, 15(Suppl.), 42A, 1992.
48. Linkowski, P., Van Onderbergen, A., Kerkhofs, M., Bosson, D., Mendlewicz, J., and Van Cauter, E., Twin study of the 24-h cortisol profile: evidence for genetic control of the human circadian clock, *Am. J. Physiol.*, 264, E173, 1993.
49. Webb, W. B. and Friedmann, J. K., Attempts to modify the sleep patterns of the rat, *Physiol. Behav.*, 6, 459, 1971.
50. Kitahama, K. and Valatx, J. L., Instrumental and pharmacological paradoxical sleep deprivation in mice: strain differences, *Neuropharmacology*, 19, 529, 1980.
51. Valatx, J. L., Genetics as a model for studying the sleep-waking cycle, *Exp. Brain Res.*, (Suppl. 8), 135, 1984.
52. Rosenberg, R. S., Bergmann, B. M., Son, H. J., Arnason, BG., and Rechtschaffen, A., Strain differences in the sleep of rats, *Sleep*, 10(6), 537, 1987.
53. Friedmann, J. K., A diallel analysis of the genetic underpinnings of mouse sleep, *Physiol. Behav.*, 12, 169, 1974.
54. Valatx, J. L., Buget, R., and Jouvet, M., Genetic studies of sleep in mice, *Nature*, 238, 226, 1972.
55. Valatx, J. L. and Buget, R., Facteurs génétiques dans le déterminisme du cycle veille-sommeil chez la souris, *Brain Res.*, 69, 315, 1974.
56. Van Twyver, H., Webb, W. B., Dube, M., et al., Effects of environment and strain differences on EEG and behavioral measurement of sleep, *Behav. Biol.*, 9, 105, 1973.
57. Benca, R. M., Bergmann, B. M., Leung, C., Nummy, D., and Rechtschaffen, A., Rat strain differences in response to dark pulse triggering of paradoxical sleep, *Physiol. Behav.*, 49, 83, 1991.
58. Leung, C., Bergmann, B. M., Rechtschaffen, A., and Benca, R. M., Heritalility of dark pulse triggering of paradoxical sleep in rats, *Physiol. Behav.*, 52, 127, 1994.
59. Schwartz, W. J. and Zimmerman, P., Circadian time keeping in BALB/c and C57BL/6 inbred mouse strains, *J. Neurosci.*, 10(1), 3685, 1990.
60. Franken, P., Malafosse, A., and Tafti, M., Genetic variation in EEG activity during sleep in inbred mice, *Am. J. Physiol.*, 275, R1127, 1998.
61. Hofstetter, J. R., Mayeda, A. R., Possidented, B., and Nurnberger, J. I., Jr., Quantitative trait loci for circadian rhythms of locomotor activity in mice, *Behav. Genet.*, 25(6), 545, 1995.
62. McClearn, G. E. and Kakihana, R., Selective breeding for ethanol sensitivity in mice, *Behav Genet*, 3, 409, 1973.
63. Morzorati, S., Lamishaw, B., Lumeng, L., Li, T. K., Bemis, K., and Clemens, J., Effects of low dose ethanol on the EEG of alcohol-preferring and nonpreferring rats, *Brain Res. Bull.*, 21, 101, 1988.
64. Philips, T. J., Feller, D. M., and Crabbe, J. C., Selected mouse lines, alcohol and behavior, *Experientia*, 45, 805, 1989.
65. Korpi, E. R., Kleingoor, C., Kettenmann, H., and Seeburg, P. H., Benzodiazepine-induced motor impairment linked to point mutation in cerebellar GABA-A receptor, *Nature*, 361, 356, 1993.

66. Stino, F. K. R., Divergent selection for pentobarbital-induced sleeping times in mice, *Pharmacology*, 44, 257, 1992.

67. Overstreet, D. H, Rezvani, A. H., and Janowsky, D. S., Increased hypothermic responses to ethanol in rats selectively bred for cholinergic supersensitivity, *Alcohol Alcohol*, 25(1), 59, 1990.

68. Shiromani, P. J., Velazquez-Moctezuma, J., Overstreet, D., Shalauta, M., Lucero, S., and Floyd, C., Effects of sleep deprivation on sleepiness and increased REM sleep in rats selectively bred for cholinergic hyperactivity, *Sleep*, 14(2), 116, 1991.

69. Markel, P. D., Fulker, D. W., Bennet, B., Corley, R. P., DeFries, J. C., Erwin, V. G., and Johnson, T. E, Quantitative trait loci for ethanol sensitivity in the LSXSS recombinant inbred stains: interval mapping, *Behav. Genet.*, 26(4), 447, 1996.

70. Roehrs, T., Zwyghuizen-Doorenbos, A., Timms, V., Zorick, F., and Roth, T., Sleep extension, enhanced alertness and the sedating effects of ethanol, *Pharmacol. Biochem. Behav.*, 34, 321, 1989.

71. Zwyghuizen-Doorenbos, A., Roehrs, T., Timms, V., and Roth, T., Individual differences in the sedating effects of ethanol, *Alcohol. Clin. Exp. Res.*, 14(3), 400, 1990.

72. Edgar, D. M., Seidel, W. F., Martin, C. E., Sayeski, P. P., and Dement, W. C., Triazolam fails to induce sleep in suprachiasmatic nucleus-lesioned rats, *Neurosci. Lett.*, 125, 125, 1991.

73. Mignot, E., Guilleminault, C., Dement, W. C., and Grumet, F. C., Genetically determined animal models of narcolepsy, a disorder of REM sleep, in *Genetically Defined Animal Models of Neurobehavioral Dysfunction*, Driscoll, P., Ed., Birkhäuser Boston, Cambridge, 1992, 90.

74. Takahashi, J. S., Pinto, L. H., and Vitaterna, M. H., Forward and reverse genetic approaches to behavior in the mouse, *Science*, 264, 1724, 1994.

75. Todd, J. A., Aitman, T. J., Cornall, R. J., Ghosh, S., Hall, J. R., Hearne, C. M., Knight, A. M., Love, J. M., McAleer, M. A., and Prins, J. B., Genetic analysis of autoimmune type 1 diabetes mellitus in mice, *Nature*, 351(6327), 542, 1991.

76. Crabbe, J. C., Belknap, J., and Buck, K. J., Genetic animal models of alcohol and drug abuse, *Science*, 264, 1715, 1994.

77. Dudek, B. C. and Tritto, T., Classical and nonclassical approaches to the genetic analysis of alcohol-related phenotypes, *Alcohol. Clin. Exp. Res.*, 19(4), 802, 1995.

78. Mayeda, A. R., Hofstetter, J. R., Belknap, J. R., and Nurnberger, Jr., J. I., Hypothetical quantitative trait loci (QTL) for circadain period of locomotor activity in CxB recombinat inbred strains of mice, *Behav. Genet.*, 26(5), 505, 1996.

79. Roemer, K., Johnson, P. A., and Friedmann, T., Knock-in and knock-out. Transgenes, development and disease, *N. Biologist*, 3(4), 331, 1991.

80. Travis, J., Scoring a technical knockout in mice, *Science*, 256, 1392, 1992.

81. Tobler, I., Gaus, S. E., Deboer, T., Achermann, P., Fischer, M., Aulicke, T., Moser, M., Oesch, B., McBride, P. A., and Manson, J. C., Altered circadian activity rhythms and sleep in mice devoid of prion protein, *Nature*, 380, 6439, 1996.

82. Tobler, I., Deboer, T., and Fischer, M., Sleep and sleep regulation in normal and prion protein-deficient mice, *J. Neurosci.*, 17, 1869, 1997.

83. Bornstein, B., Restless leg syndrome, *Psychiatr. Neurol.*, 141, 165, 1961.

84. Roth, B., Bruhova, S., and Berkova, L., Familial sleep Paralysis, *Arch. Suisses Neurol. Neurochir. Psychiatr.*, 102, 321, 1968.

85. Kales, A., Soldatos, C. R., Bixler, E. O., Ladda, R. L., Charney, D. S., Weber, G., and Schweitzer, P. K., Hereditary factors in sleepwalking and night terrors, *Br. J. Psychiatr.*, 137, 111, 1980.

86. Montplaisir, J., Godbout, R., Boghden, D., DeChamplain, J., Young, S. N., and Lapierre, G., Familial restless legs with periodic movements in sleep, electrophysiological, biochemical, and pharmacological study, *Neurology*, 35, 130, 1985.

87. Lugaresi, E. X., Medori, R., Montagna, P., Baruzzi, A., Cortelli, P., Lugaresi, A., Tinuper, P., Zucconi, M., and Gambetti, P., Fatal familial insomnia and dysautonomia with selective degeneration of thalamic nuclei, *N. Engl. J. Med.*, 315, 997, 1986.

88. Guilleminault, C., Mignot, E., and Grumet, F. C., Familial patterns of narcolepsy, *Lancet*, 2(8676),1376, 1989.

89. El Bayadi, S., Millman, R. P., Tishler, P. V., Rosenberg, C., Saliski, W., Boucher, M. A., and Redline, S., A family study of sleep apnea. Anatomic and physiologic interactions, *Chest*, 98(3), 554, 1990.

90. Fink, R. and Ancoli-Israel, S., Pedigree of one family with delayed sleep phase syndrome, *Sleep Res.*, 26, 713, 1997.

91. Dement, W. C., Carskadon, M., and Ley, R., The prevalence of narcolepsy II, *Sleep Res.*, 2, 147, 1973.

92. Solomon, P., Narcolepsy in negroes, *Dis. Nerv. Syst.*, 6, 179, 1990.

93. Aldrich, M., Narcolepsy, *Neurology*, 42(Suppl. 6), 34, 1992.

94. Mignot, E., Genetic and familial aspects of narcolepsy, *Neurology*, 50, 2(Suppl. 1), 516, 1998.

95. Honda, Y., Consensus of narcolepsy, cataplexy and sleep life among teenagers in Fujisawa city, *Sleep Res.*, 8, 191, 1979.

96. Lavie, P. and Peled, R., Narcolepsy is a rare disease in Israel, *Sleep (Lett.)*, 10(6), 608, 1987.

97. Daly, D. and Yoss, R., A family with narcolepsy, *Proc. Staff Meet. Mayo Clinic*, 34, 313, 1959.

98. Nevsimalova-Bruhova, S., On the problem of heredity in hypersomnia, narcolepsy and related disturbance, *Acta. Univ. Carol. Med.*, 18, 109, 1973.

99. Kessler, S., Guilleminault, C., and Dement, W. C., A family study of 50 REM narcoleptics, *Acta. Neurol. Scand.*, 50, 503, 1979.

100. Singh, S., George, C. F. P., Kryger, M. H., et al., Genetic heterogeneity in narcolepsy. *Lancet*, 335, 726, 1990.

101. Billiard, M., Pasquie-Magentto, V., Heckman, M., Carlander, B., Besset, A., Zachariev, Z., Eliaou, J. F., and Malafosse, A., Family studies in narcolepsy, *Sleep*, 17, S54, 1994.

102. Honda, Y., Asaka, A., Tanaka, Y. et al., Discrimination of narcolepsy by using genetic markers and HLA, *Sleep Res.*, 12, 254, 1983.

103. Honda, Y. and Matsuki, K., Genetic aspects of narcolepsy, in *Handbook of Sleep Disorders*, Thorpy, M.J., Ed., Marcel Dekker, New York, 1990, 217.

104. Matsuki, K., Grumet, F. C., Lin, X., Gelb, M., Guilleminault, C., Dement, W. C., and Mignot, E., DQ (rather than DR) gene marks susceptibility to narcolepsy, *Lancet*, 339, 1052, 1992.

105. Rogers, A. E., Meehan, J., Guilleminault, C., Grumet, F. C., and Mignot, E., HLA DR15(DR2) and DQB1*0602 typing studies in 188 narcoleptic patients with cataplexy, *Neurology*, 48, 1550, 1997.

106. Mignot, E., Hayduk, R., Black, J., Grumet, F. C., and Guilleminault, C., HLA DQBI* 0602 is associated with cataplexy in 509 narcoleptic patients, *Sleep,* 20, 1012, 1997.

107. Mignot, E., Kimura, A., Lattermann, A., Lin, X., Yasunaga, S., Mueller-Eckhard, G., Rattazzi, C., Lin, L., Guilleminault, C., Grumet, F. C., Mayer, G., Dement, W. C., and Underhill, P., Extensive HLA Class II studies in 58 non DRB1*15 (DR2) narcoleptic patients with cataplexy, *Tissue Antigen*, 49, 329, 1997.

108. Matsuki, K., Juji, T., and Honda, Y., HLA and Narcolepsy, in *Immunological Features in Japan*, Honda, Y. and Juji, T., Eds., Springer Verlag, New York, 1988, 58.

109. Rubin, R. L., Hajdukovic, R. M., and Mitler, M. M., HLA DR2 association with excessive somnolence in narcolepsy does not generalize to sleep apnea and is not accompanied by systemic autoimmune abnormalities, *Clin. Immunity Immunopathol.*, 49, 149, 1988.

110. Fredrikson, S., Carlander, B., Billiard, M., and Link, H., CSF Immune variable in patients with narcolepsy, *Acta. Neurol. Scand.*, 81, 253, 1990.

111. Carlander, B., Eliaou, J. F., and Billiard, M., Autoimmune hypothesis in narcolepsy. *Neurophysiol. Clin.*, 23, 15, 1993.

112. Mignot, E., Tafti, M., Dement, W. C., and Grumet, F. C., Narcolepsy and immunity, *Adv. Neuroimmunol.*, 5, 23, 1995.

113. Tafti, M., Nishino, S., Aldrich, M. S., Liao, W., Dement, W. C., and Mignot, E., Major histocompatibility class II molecules in the CNS: increased microglial expression at the onset of narcolepsy in a canine model, *J. Neurosci.*, 16(15), 4588, 1996.

114. Baker, T. L. and Dement, W. C., Canine narcolepsy-cataplexy syndrome: evidence for an inherited monoaminergic-cholinergic imbalance, in *Brain Mechanisms of Sleep*, McGinty, D.J., Drucker-Colin, R., and Morrisson, A., Eds., Raven Press, New York, 1985, 199.

115. Mignot, E., Nishino, S., Hunt-Sharp, L., Arrigoni, J., Siegel, J. M., Reid, M. S., Edgar, D. M., Ciaranello, R. D., and Dememt, W. C., Heterozygocity at the canarc-1 locus can confer susceptibility for narcolepsy: induction of cataplexy in heterozygous asymtomatic dogs after administration of a combination of drugs acting on monoaminergic and cholinergic systems, *J. Neurosci.*, 13(3), 1057, 1993.

116. Mignot, E., Wang, C., Rattazzi, C., Gaiser, C., Lovett, M., Guilleminault, C., Dement, W. C., and Grumet, F. C., Genetic linkage of autosomal recessive canine narcolepsy with an immunoglobulin μ chain switch like segment, *Proc. Natl. Acad. Sci. U.S.A.*, 88, 3475, 1991.

117. Dahlitz, M., Parles, J. D., Vaughan, R., et al., The sleep paralysis — excessive daytime sleepiness syndrome, *J. Sleep Res.*, 1(Suppl. 1), 52, 1992.

118. Ohayon, M. M., Priest, R., Caulet, M., and Guilleminault, C., Hypnagogic and hypnopompic hallucinations: pathological phenomena?, *Br. J. Psychiatr.*, 169, 459, 1996.

119. Goode, G. B., Sleep paralysis, *Arch. Neurol.*, 6(3), 228, 1962.

120. Bell, G. C., Dixie-Bell, D. D., and Thompson, B., Further studies on the prevalence of isolated sleep paralysis in black subjects, *J. Natl. Med. Assoc.*, 78(7), 649, 1986.

121. Hori, A. and Hirose, G., Twin studies on parasomnias, *Sleep Res.*, 24A, 324, 1995.

122. Mahowald, M. W. and Schenck, C. H., REM sleep behavior disorder, in *Principles and Practice of Sleep Medicine*, 2nd ed., Kryeger, M., Roth, T., and Dement, W.C., Eds., W.B. Saunders, Philadelphia, 1994, 574.

123. Schenck, C. H., Garcia-Rill, E., Segall, M., Noreen, H., and Mahowald, M. W., HLA Class II genes associated with REM sleep behavior disorder, *Annu. Neurol.*, 39, 261, 1996.

124. Gelardi, J. A. M. and Brown, J. W., Hereditary cataplexy, *J. Neurol. Neurosurg. Psychiatr.*, 30, 455, 1967.

125. Vela Bueno, A., Campos Castello, J. C., et al., Hereditary cataplexy — is it primary cataplexy?, *Waking Sleeping*, 2, 125, 1978.

126. Hartse, K. M., Zorick, F. J., Sicklesteel, J. M., and Roth, T., Isolated cataplexy: afamily study, *Henry Ford Hosp. Med. J.*, 36(1), 24, 1988.

127. Julien, J., Vital, C., Deleplanque, B., et al., Athrophie thalamique subaigue familiale. Troubles mnésiques et insomnie totale, *Rev. Neurol.* (Paris), 146(3), 173, 1990.

128. Goldfarb, L. G., Petersen, R. B., and Tabaton, M., Fatal familial insomnia and familial Creutzfeldt-Jakob disease: a disease phenotype determined by a DNA polymorphism, *Science*, 258, 806, 1980.

129. Manetto, V., Medori, R., Cortelli, P., Montagna, P., Tinuper, P., Baruzzi, A., Rancurel, G., Hauw, J. J., Vanderhaeghen, J. J., and Mailleux, P., Familial fatal insomnia: clinical and pathological study of five new cases, *Neurology* 42, 312, 1992.

130. Chapman, J., Arlazoroff, A., Goldfarb, L. G. Cervenakova, L., Neufeld, M. Y., Weber, E., Herbert, M., Brown, P., Gajdusek, D. C., and Korczyn, A. D., Fatal Insomnia in a case of familial Creuztfeldt-Jakob disease with codon 200(lys) mutation, *Neurology*, 46(3), 758, 1996.

131. Prussiner, S. B., Molecular biology of prion diseases, *Science*, 252, 1515, 1991.

132. Weissman, C., A "unified theory" of prion propagation, *Nature*, 352, 679, 1991.

133. Mestel, R., Putting prions to the test, *Science* 273, 184, 1996.

134. Horwich, A. L. and Weissman, J. S., Deadly conformations — protein misfolding in prion disease, *Cell*, 89(4), 499, 1997.

135. Petersen, R. B., Parchi, P., Richardson, S. L., Urig, C., B., and Gambetti, P., Effect of the D178N mutation and the codon 129 polymorphism on the prion protein, *J. Biol. Chem.*, 271(21), 12661, 1996.

136. Lugaresi, E., The thalamus and insomnia, *Neurology*, 42(Suppl. 6), 28, 1992.

137. Stériade, M., Basic mechanisms of sleep generation, *Neurology*, 42(Suppl. 6), 9, 1992.

138. Marini, G., Imeri, L., and Mancia, M., Changes in sleep waking cycle induced by lesions of medialis dorsalis thalamic nuclei in the cat, *Neurosci. Lett.*, 85, 223, 1988.

139. National Institute of Mental Health, Consensus Development Conference Drugs and Insomnia The use of medications to promote sleep, *JAMA*, 251, 2410, 1984.

140. Angst, J., Vollrath, M., Koch, R., and Dobler-Mikola, A., The Zurick study. VII. Insomnia: symptoms, classification and prevalence, *Eur. Arch. Psychiatr. Neurol. Sci.*, 238, 285, 1989.

141. Hauri, P., Primary insomnia, in *Principles and Practice of Sleep Medicine*, Kryeger, M., Roth, T., Dement, W.C., Eds., W.B. Saunders, Philadelphia, 1989, 442.

142. Ekbom, K., Restless legs syndrome, *Neurology*, 10, 868, 1960.

143. Strang, P. G., The symptom of restless legs, *Med. J. Aust.*, 1, 1211, 1967.

144. Cirignotta, F., Zucconi, M., Mondini, D., et al., Epidemiological data on sleep disorder, *Sleep Res*, 11, 211, 1982.

145. Montplaisir, J., Godbout R., Prllettier G., et al., Restless legs syndrome and periodic movements during sleep, in *Principles and Practice of Sleep Medicine*, 2nd ed., Kryeger, M., Roth, T., and Dement, W.C., Eds., W.B. Saunders, Philadelphia, 1994, 589.

146. Walters, A. S., Toward a better definition of the restless legs syndrome. The international restless legs syndrome study group, *Mov. Disord.*, 10, 634, 1995.

147. Ambrosetto, C., Lugaresi, E., Coccagna, G., et al., Clinical and polygraphic remarks on the restless leg syndrome, *Rev. Pathol. Nerv. Ment.*, 86, 244, 1965.

148. Montagna, P., Coccagna, G., Cirignotta, F., et al., Familial restless leg syndrome, in *Sleep/Wake Disorders: Natural History, Epidemiology and Long Term Evolution*, Guilleminault, C. and Lugaresi, E., Eds., Raven Press, New York, 1983, 231.

149. Jacobsen, J. H., Rosenberg, R. S., Huttenlocher, P. R., et al., Familial nocturnal cramping, *Sleep*, 9(1), 54, 1986.

150. Walters, A., Picchietti, D., Hening, W., and Lazzarini, A., Variable expressivity in familial restless legs syndrome, *Arch. Neurol.*, 47, 1219, 1990.

151. Walters, A. S., Picchietti, D. I., Ehrenberg, B. L., and Wagner, M.L., Restless legs syndrome in childhood and adolescence, *Pediatr. Neurol.*, 1193, 241, 1994.

152. Trenkwalder, C., Colladoso-Seidel, V., Gasser, T., and Oertel, W. H., Clinical symptoms and possible anticipation in a large kindred of familial restless legs syndrome, *Mov. Disord.*, 11, 389, 1996.

153. Lavigne, G. and Montplaisir, J., Restless legs syndrome and sleep bruxism: prevalence and association among Canadians, *Sleep*, 17, 739, 1994.

154. Montplaisir, J., Boucher, S., Poirier, G., Lavigne, G., Lapierre, O., and Lesperance, P., Clinical, polysomnographic, and genetic characteristics of restless legs syndrome: a study of 133 patients diagnosed with new standard criteria, *Mov. Disord.*, 12, 61, 1997.

155. Johnson, W., Walter, A., Lehner, T., et al., Affected only linkage analysis of autosomal dominant restless legs syndrome, *Sleep Res.*, 21, 214, 1992.

156. Keefauver, S. P. and Guilleminault, C., Sleep Terrors and sleep walking, in *Principles and Practice of Sleep Medicine*, 2nd ed., Kryeger, M., Roth, T., and Dement, W.C., Eds., W.B. Saunders, Philadelphia, 1994, 567.

157. Broughton, R. J., Sleep disorders: disorders of arousal?, *Science*, 159, 1070, 1968.

158. Abe, K. and Shimakawa, M., Genetic and developmental aspects of sleeptalking and teeth-grinding, *Acta Paedopsychiatr.*, 33(11), 339, 1966.

159. Debray, P. and Huon, H., A propos de trois cas de somnambulisme familial, *Annu. Méd. Intern.*, 124(1), 27, 1973.

160. Hälstrom, T., Night terror in adults through three generations, *Acta. Psychiatr. Scand.*, 48, 350, 1972.

161. Kales, A., Soldatos, C. R., Bixler, E. O., Ladda, R. L., Charney, D. S., Weber, G., and Schweitzer, P. K., Hereditary factors in sleepwalking and night terrors, *Br. J. Psychiatr.*, 137, 111, 1980.

162. Abe, K., Amatomi, M., and Oda, N., Sleepwalking and recurrent sleeptalking in children of childhood sleepwalkers, *Am. J. Psychiatr.*, 14, 800, 1984.

163. Bakwin, H., Sleepwalking in twins, *Lancet* 2, 466, 1970.

164. Hublin, C., Kaprio, J., Partinen, M., Heikki, K., and Koskenvuo, M., Prevalence and genetics of sleepwalking: a population based twin study, *Neurology*, 48(1), 177, 1997.

165. Gastaut, H., Tassinari, C., and Duron, B., Etude polygraphique des manifestations épisodiques (hypniques et respiratoires) diurnes et nocturnes du syndrome de Pickwick, *Rev. Neurol.* (Paris), 112, 573, 1965.

166. Lugaresi E., Medori R., Montagna P., Baruzzi A., Cortelli P., Lugaresi A., Tinuper P., Fatal familial insomnia and dysautonomia with selective degeneration of thalamic nuclei, *N. Engl. J. Med.*, 315, 997, 1986.

167. Young, T., Palta, M., Dempsey, J., Skatrud, J., Weber, S., and Badr, S., The occurrence of sleep-disordered breathing among middle-aged adults, *N. Engl. J. Med.*, 328, 1230, 1993.

168. Guilleminault, C., Eldridge, F. L., and Simmons, F. B., Sleep apnea syndrome: can it induce hemodynamic changes?, *West J. Med.*, 123, 7, 1975.

169. Koskenvuo, M., Kaprio, J. et al., Snoring as a risk for hypertension and angina pectoris, *Lancet*, 1, 893, 1985.

170. Hall, M. J. and Bradley, T. D., Cardiovascular disease and sleep apnea, *Curr. Opin. Pulm. Med.*, 1(6), 512, 1995.

171. Redline, S., Tishler, P. V., Hans, M. G., Tosteson, T. D., Strohl, K. P., and Spry, K., Racial differences in sleep-disordered breathing in African-Americans and Caucasians, *Am. J. Respir. Crit. Care Med.*, 155, 186, 1997.

172. Ferini-Strambi, L., Calori, G., Oldani, A., Della Marca, G., Zucconi, M., Castronovo, V., Gallus, G., and Smirne, S., Snoring in twins, *Respir Med*, 89, 337, 1995.
173. Strohl, K. P., Saunders, N. A., Feldman, N. T., and Hallett, M., Obstructive sleep apnea in family members, *N. Engl. J. Med.*, 229(18), 969, 1978.
174. Adickes, E. D., Buehler, B. A., and Sanger, W. G., Familial sleep apnea, *Hum. Genet.*, 73, 39, 1986.
175. Oren, J., Kelly, D. H., and Shannon, D. C., Familial occurrence of sudden infant death syndrome and apnea of infancy, *Pediatrics*, 80, 355, 1987.
176. Manon-Espaillat, R., Gothe, B., Adams, N., Newman, C., and Ruff, R., Familial "sleep apnea plus" syndrome: report of a family, *Neurology*, 38, 190, 1988.
177. Wittig, R. M., Zorick, F. J., Roehrs, T. A., and Roth, T., Familial childhood sleep apnea, *Henry Ford Hosp. Med. J.*, 36(1), 13, 1988.
178. Mathur, R. and Douglas, N. J., Family study in patients with the sleep apnea-hypopnea syndrome, *Annu. Intern. Med.*, 122, 174, 1995.
179. Pillar, G. and Lavie, P., Assessment of the role of inheritence in sleep apnea syndrome, *Am, J. Respir. Crit. Care Med.*, 151, 688, 1995.
180. Redline, S., Tishler, P. V., Tosteson, T. D., et al., The familial aggregation of obstructive sleep apnea, *Am. J. Respir. Crit. Care Med.*, 151, 682, 1995.
181. Yoshizawa, T., Akashiba, T., and Kurashina, K., Genetics and obstructive sleep apnea syndrome, *Int. Med.*, 32(2), 94, 1993.
182. Guilleminault, C., Partinen, M., Hollman, K., Powell, N., and Stoohs, R., Familial aggregates in obstructive sleep apnea syndrome, *Chest*, 107, 1545, 1995.
183. Kronholm, E., Aunola, S., Hyyppa, M. T., Kaitsaari, M., Koskenvuo, M., Mattlar, C. E., and Rannemaa, T., Sleep in monozygotic twin pairs discordant for obesity, *J. Appl. Physiol.*, 80(1), 14, 1996.
184. Tishler, P. V., Redline, S., Ferrette, V., Hans, M. G., and Altose, M. D., The association of sudden unexpected infant death with obstructive sleep apnea, *Am. J. Respir. Crit. Care Med.*, 153(6), 1857, 1996.
185. Kawakami, Y., Yamamoto, H., Yoshikawa, T., and Shida, A., Chemical and behavioral control of breathing in twins, *Am. Rev. Respir. Dis.*, 129, 703, 1982.
186. Thomas, D. A., Swaminathan, S., Beardsmore, C. S., McArdle, E. K., MacFadyen, U. M., Goodenough, P. C., Carpenter, R., and Simpson, H., Comparison of peripheral chemoreceptor responses in monozygotic and dizygotic twin infants, *Am. Rev. Respir. Dis.*, 148(6), 1605, 1993.
187. Kushida, C. A., Guilleminault, C., Mignot, E., Ahmed, O., Won, C., O'Hara, B., and Clerk ,A. A., Genetics and craniofacial dysmorphism in family studies of obstructive sleep apnea, *Sleep Res.*, 25, 275, 1996.
188. Carskadon, M. A., Pueschel, S. M., and Millman, R. P., Sleep-disordered breathing and behavior in three risk groups: preliminary findings from parental reports, *Childs Nerv. Syst.*, 9(8), 452, 1993.
189. O'Hare, J.P., O'Brien, I.A., Arendt, J. et al., Does melatonin deficiency cause the enlarged genitalia of the fragile-X syndrome?, *Clin. Endocrinol.*, 24(3), 327, 1986.
190. Staley-Gane, M. S., Hollway, R. J., and Hagerman, M. D., Temporal sleep characteristics of young fragile X boys, *Am. J. Hum. Genet.*, 59(4), A105, 1996.
191. Challamel, M. J., Mazzola, M. E., Nevsimalova, S. et al., Narcolespy in children, *Sleep*, 17, S17, 1994.
192. Vossler, D. G., Wyler, A. R., Wilkus, R. J., Gardner-Walker, G., and Vlcek, B. W., Cataplexy and monoamine oxidase deficiency in Norrie disease, *Neurology*, 46, 1258, 1996.
193. Melberg, A., Hetta, J., Dahl, N., Stalberg, E., Raininko, R., Oldfors, A., Bakall, B., Lundberg, P. O., and Holme, E., Autosomal dominant cerebellar ataxia deafness and narcolepsy, *J. Neurol. Sci.*, 124, 119, 1995.
194. Goldberg, R., Fish, B., Ship, A., and Shprintzen, R. J., Deletion of a portion of the long arm of chromosome 6, *Am. J. Med. Genet.*, 5, 73, 1980.
195. Kaplan, J., Frederickson, P. A., and Richardson, J. W., Sleep and breathing in patients with the Prader-Willi syndrome, *Mayo Clin. Proc.*, 66, 1124, 1991.
196. Summers, J. A., Lynch, P. S., Harris, J. C., Burke, J. C., Allison, D. B., and Sandler, L., A combined behavioral/pharmacological treatment in sleep-wake schedule disorder in Angelman syndrome, *Dev. Behav. Pediatr.*, 13(4), 284, 1992.
197. Vgontzas, A. N., Kales, A., Seip, J., Mascari, M. J., Bixler, E. O., Myers, D. C., Vela-Bueno, A. V., and Rogan, P. K., Relationship of sleep abnormalities in Prader-Willi syndrome, *Am. J. Hum. Genet.*, 67, 478, 1996.

198. Greenberg, F., Guzzetta, V., Montes de Oca-Luna, R. et al., Molecular analysis of the Smith-Magenis syndrome: a possible contiguous syndrome associated with del 17 (p11.2), *Am. J. Hum. Genet.*, 49, 1207, 1991.

199. Fischer, H., Oswald, H. P., Duba, H. C., Doczy, L., Simma, B., Utermann, G., and Haas, O. A., Constitutional interstitial deletion of 17 (p11.2) (Smith Magenis syndrome): a clinically recognizable microdeletion syndrome. Report of two cases and review of the literature, *Klin. Paediatr.*, 205(3), 162, 1993.

200. Higurashi, M., Kawai, H., Segawa, M., Iijima, K., Ikeda, Y., Tanaka, F., Egi, S., and Kamashita, S., Growth, psychologic characteristics, and sleep-wakefulness cycle of children with sex chromosomal abnormalities, *Birth Defects,* 22(3), 251, 1986.

201. Carskadon, M. A., Patterns of sleep and sleepiness in adolescents, *Pediatrician*, 17(1), 5, 1990.

202. George, C. F. and Singh, S. M., Juvenile onset narcolepsy in an individual with Turner syndrome. A case report, *Sleep*, 14(3), 267, 1991.

203. Chisolm, R. C., Brooks, C. J., Harrison, G. F., et al., Prepubescent narcolepsy in a six year old girl, *Sleep Res.*, 15, 113, 1985.

204. Hartman, E., Bruxism, in *Principles and Practice of Sleep Medicine*, Kryeger, M., Roth, T., and Dement, W.C., Eds., W.B. Saunders, Philadelphia, 1989, 385.

205. Manni, R., Zucca, C., Martinetti, M., Ottolini, A., Lanzi, G., and Tartara, A., Hypersomnia in dystrophia myotonica: a neurophysiological and immunogenetic study, *Acta. Neurol. Scand.*, 84, 498, 1991.

206. Byrne, E., White, O., and Cook, M., Familial dystonic choreoathetosis with myokymia: a sleep responsive disorder, *J. Neurol. Neurosurg. Psychiatr.*, 54, 1090, 1991.

207. Ishikawa, A. and Miyatake, T., A family with hereditary juvenile dystonia-parkinsonism, *Mov. Disord.*, 10(4), 482, 1995.

208. Thorpy, M. J. and Glovinsky, P. B., Headbanging (Jactatio Capitis Nocturna), in *Principles and Practice of Sleep Medicine*, Kryeger, M., Roth, T., and Dement, W. C., Eds., W.B. Saunders, Philadelphia, 1989, 648.

209. Poirier, G., Montplaisir, J., Decary, F., Momege, D., and Lebrun, A., HLA antigens in narcolepsy and idiopathic central nervous system hypersomnolence, *Sleep*, 9, 153, 1986.

# 18 Hearing

*Adam J. W. Paige and Stephen D. M. Brown*

## CONTENTS

## INTRODUCTION

Helen Keller claimed that "To be deaf is a greater affliction than to be blind." Indeed it is difficult to imagine the sense of isolation and the resulting social withdrawal experienced by those with a hearing impairment. The inner ear is a marvel of evolution, capable of amplifying and analyzing complex sound patterns and also providing information about spatial orientation and motion, and yet it is no larger than a marble. The evolutionary success of the inner ear is reflected in the structural similarity of the auditory system in all vertebrates. Indications of the semicircular canals and otolithic organs of the ear can be seen in fossil remains of jawless fishes, which date from 500 million years ago.[1] Another remarkable feature of the ear is the speed of its response to stimuli, which can be measured in milliseconds compared to hundreds of milliseconds for the visual and olfactory systems. A human cochlea is thus able to respond to sound frequencies as great as 20 kHz, whilst bats and whales can detect frequencies almost 10-fold greater.[2,3] Despite its great sensitivity, each ear contains only 16,000 auditory hair cells and 67,000 vestibular hair cells compared to the hundred million photoreceptors and olfactory receptor neurons present in each individual.[4] This paucity of sensory cells makes the inner ear very sensitive to hair cell damage and loss. In a British survey conducted in 1989, 16% of adults were found to have a significant hearing loss (>25 dB)[5] and severe hearing impairment was prevalent among the aged. The available treatments consist of hearing aids, which boost stimulation of surviving hair cells, and cochlear implants, which replace

0-8493-2688-5/00/$0.00+$.50
© 2000 by CRC Press LLC

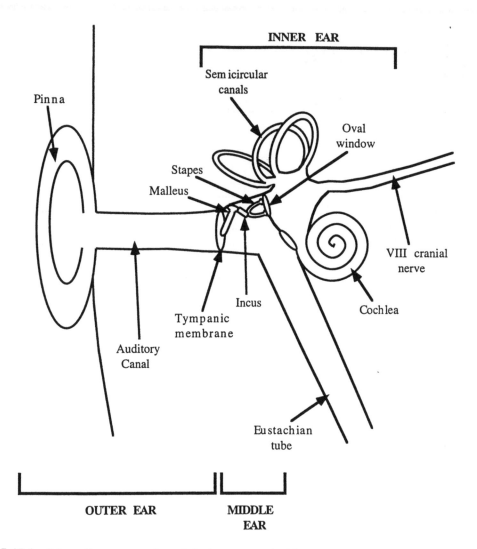

**FIGURE 18.1**   Schematic representation of the human ear detailing the major components of the outer, middle, and inner ear.

the 32,000 hair cells with 20 electrical sensors. A greater understanding of auditory function should lead to specific treatments for both genetic and environment-related hearing impairment.

## STRUCTURE AND FUNCTION OF THE EAR

The mammalian ear is subdivided into three parts: the outer, middle, and inner ear (Figure 18.1). The outer ear consists of the pinna and the auditory canal, which direct sound waves to the tympanic membrane (ear drum) causing it to vibrate.[6,7] The middle ear contains three small bones or ossicles: the malleus, incus, and stapes. Movement of the tympanic membrane causes the ossicles to pivot, and transfers the vibration, via the stapes, to the membranous oval window of the inner ear. The middle ear also includes the eustachian tube, which runs from the ear to the pharynx and acts to balance the air pressures on either side of the ear drum.

The inner ear consists of the auditory sensory system, the cochlea, and the vestibular apparatus, which responds to gravity and acceleration. Both the cochlea and the vestibular system are bony cavities filled with perilymphatic and endolymphatic fluid. A cross-section of the cochlea shows it

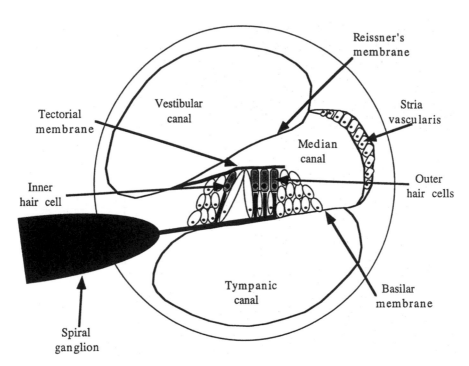

**FIGURE 18.2** Schematic representation of a cross-section through the cochlear duct. The vestibular and tympanic canals are filled with perilymph and the median canal contains endolymph. Three rows of outer hair cells and one row of inner hair cells can be seen in the organ of Corti.

to contain three canals (Figure 18.2). The perilymphatic fluid-filled vestibular and tympanic canals are separated from the endolymph-filled median canal by Reissner's membrane and the basilar membrane, respectively. Resting on the elastic basilar membrane is the organ of Corti, which contains the sensory neuroepithelia, one row of inner hair cells and three or four rows of outer hair cells, which run all the way from the base to the apex of the cochlea.[1,8] The hair cells at the base of the cochlea respond to high frequencies whilst those toward the apex respond to progressively lower frequencies. On the apical surface of each hair cell is a bundle of actin-filled stereocilia which brush against the hard tectorial membrane and are immersed in the potassium ion-rich endolymph. The large concentration gradient of potassium ions across the hair cell membrane creates a potential gradient between the endolymph and the hair cell cytoplasm.

The tapping of the stapes against the oval window creates a traveling wave through the perilymph-filled canals of the cochlea, causing the basilar membrane to vibrate and deflecting the hair bundle. The bending of the stereocilia opens cation channels in the cell membrane and allows the influx of potassium and calcium ions, which stimulate the release of neurotransmitters activating the afferent neurons surrounding the inner hair cells. Similarly, changes in gravity and acceleration cause motion in the endolymph of the vestibular apparatus, deflecting vestibular hair cells and stimulating the afferent vestibular neurons. The neurons from the auditory and vestibular systems meet and become the eighth cranial nerve, which carries signals to the brainstem's medulla and pons.[8]

Despite extensive study, hair cell transduction is still poorly understood but the available evidence suggests a mechanism known as the gating- spring model.[3,9-12] This hypothesis posits that an elastic element connected to the transduction channels is responsible for transmitting the mechanical stimulus of the hair bundle deflection. This model is supported by the speed of the hair cell response, mere milliseconds, which is far too fast to involve second messengers, and the observation that hair cell deflection along only a single axis results in depolarization. Deflection in the other direction along that axis results in hyperpolarization, whilst deflection in a transverse direction

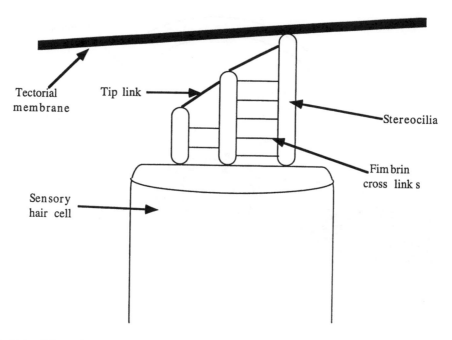

**FIGURE 18.3**  Schematic representation of a sensory hair cell showing the tiered arrangement of stereocilia on the apical surface. The stereocilia are joined together by fimbrin cross-linking and tip links.

produces no effect. A likely candidate for this gating-spring has already been identified, the tip links[13] (Figure 18.3). Stereocilia atop the hair cells show a variation in size with a progressive increase in length across the apical surface of each hair cell. Each stereocilium is attached to adjacent stereocilia by fimbrin cross-linking and also possesses a tip link, a fine extracellular filament, which runs from the tip of the stereocilium to the side of its tallest neighbor. The tip links only join adjacent stereocilia along the axis of transduction, as would be predicted for the gating-spring, and loss of tip links has been shown to be associated with loss of transduction.[14] The gating-spring model suggests that transduction channels are located at one or both ends of these tip links[15,16] and that subsequent deflection of the hair bundle increases the tension in the tip links and opens the ion channels, allowing the influx of $K^+$ and $Ca^{2+}$ ions and resulting in the activation of the afferent cochlear neurons.

There are still several gaps in our understanding of auditory mechanotransduction: For example, are tip links the gating-springs? What type of channels are involved in transduction?[1,8] Many of these questions may be answered by isolating genes involved in auditory signaling using human pedigrees and animal models displaying hearing impairment.

## OVERVIEW OF HEREDITARY DEAFNESS

It was in the mid-19th century that otology became an accepted clinical discipline due to a debate between two eminent scientists, Wilheim Kramer and William Wilde. Whereas Kramer did not accept that deafness was heritable, Wilde identified pedigrees displaying dominant, recessive, and sex-linked inheritance of deafness.[17] Deafness affects 1 in 1000 live births and at least one half of these are attributed to genetic causes.[4,18] A further 1 in 1000 children become deaf before adulthood and the proportion of the population displaying hearing loss greater than 65 dB increases from 0.3 to 2.3% between the ages of 30 and 70 years.[4] Sixteen percent of British adults suffer a hearing loss of >25 dB in the better ear according to a survey by Davis et al.[5] Deafness is also seen in several animals including mice, cats, dogs (particularly Dalmatians), and mink.[19-21]

Seventy percent of congenital genetic deafness is nonsyndromic (not associated with other clinical features) and 60 to 75% of cases are autosomal recessive forms. Of the remaining cases, 20 to 30% show autosomal dominant inheritance and about 2% are either X linked or of mitochondrial origin. Despite the prevalence of nonsyndromic deafness, the identification of causative genes in humans has proved a complicated task (for a review see Ref. 22). One problem encountered by geneticists is the high heterogeneity of the disorder. There are presently 45 loci for nonsyndromic deafness mapped in humans,[23] including 15 autosomal dominant and 20 autosomal recessive loci (Table 18.1), and yet many deaf pedigrees still fail to show linkage to any of these loci, indicating that further genes may be involved. This is further complicated by small pedigree size and frequent intermarrying between affected pedigrees (90% of deaf adults in the U.S. marry another deaf person[17]). As a result of these difficulties, only two deafness loci had been mapped prior to 1994.[24,25] The past few years have seen rapid progress in gene mapping through the study of large, consanguineous families from geographically isolated regions[26] and from the application of homozygosity mapping analysis[27] of families isolated from immigration.[28] In spite of this progress, only three genes for nonsyndromic deafness in humans have been identified,[29-32] along with a gene causing combined conductive and sensorineural hearing loss[33] and two mitochondrial genes[25,34] Table 18.1). One of the three genes for nonsyndromic hearing loss, *MYO7A*, was originally identified from a mouse model of deafness.[35] The mouse is a well-established genetic model and possesses an inner ear which is very similar to that of humans.[36] In addition, more than 100 mouse mutants are known or are believed to lead to hearing deficits, and approximately 25 of these exhibit nonsyndromic deafness.[37]

## CENTRAL AUDITORY DEAFNESS

Hearing loss caused by a primary central auditory system defect is rare, but several examples have been identified in the mouse. The *reeler* mutant exhibits impaired motor coordination with a circling and head-tossing behavior, tremors, ataxia, and hyperactivity.[37,38] Studies of the neocortex in *reeler* homozygotes reveal widespread morphological abnormalities.[39] Normal cortical development occurs in two stages: (1) formation of a preplate containing early generated neurons such as Cazal-Retzius (CR) neurons, and (2) division of this preplate into the superficial marginal zone, containing the CR neurons, and the deep subplate.[40] The preplate division is due to the insertion of late-generated neurons which migrate along glial fibers across the subplate and accumulate before the superficial marginal zone to form the cortical plate. Cortical plate maturation occurs "inside out" with the younger neurons migrating into the more superficial positions. However, in the *reeler* mutant, the late-generated neurons do not split the preplate, but pile up beneath it with the younger neurons positioned at deeper, rather than more superficial, levels.[40] Almost all of the central nervous structures are believed to be affected,[41] including the dorsal cochlear nucleus, which is disorganized and contains reduced numbers of granule cells.[42] The *reeler* gene has been cloned and encodes a ~3500 amino acid residue protein (reelin) containing several epidermal growth factor (EGF)-like domains.[38,39,43] These EGF motifs closely resemble those found in extracellular matrix proteins such as tenascin-C, tenascin-X, and the β-integrins. Reelin is expressed by CR neurons and is then believed to be secreted into the local extracellular matrix.[43] This is supported by the findings of Ogawa et al., who identified an antibody, CR-50, which recognizes an epitope on the surface of CR neurons in wild-type mice but not in the *reeler* homozygotes.[40] The authors performed reaggregation experiments using cultured cortical cells in the presence and absence of CR-50. Aggregates of *reeler* cortical cells consistently showed patches of clustered neurons, making them easily distinguishable from wild-type aggregates. However, when wild-type cortical cells were incubated with CR-50 prior to reaggregation the resulting aggregates resembled those of the *reeler* cells. These findings support a model whereby secreted reelin protein interacts with the migrating, later-generated neurons, possibly through the EGF-like domains, thereby providing the neurons with the

**TABLE 18.1**
**Nonsyndromic Deafness Loci in Humans**

| Locus Name | Location of Human Gene | Gene Name | Mouse Chromosome | Possible Mouse Homolog |
|---|---|---|---|---|
| **Autosomal** | **Dominant** | | | |
| DFNA1 | 5q31 | *HDIA1* | 11 or 18 | *sh2, sy* |
| DFNA2 | 1p34 | unknown | | |
| DFNA3 | 13q12 | *Cx26* | | |
| DFNA4 | 19q13 | unknown | 7, 9, or 10 | *qv, nv* |
| DFNA5 | 7p15 | unknown | 6 or 13 | *Hoxa1* |
| DFNA6 | 4p16.3 | unknown | 5 or 11 | *tlt, sh2, bv* |
| DFNA7 | 1q21–q23 | unknown | | *dr, Lp* |
| DFNA8 | 11q22–24 | unknown | 7, 9, or 19 | *Fgf3, Dc, tub* |
| DFNA9 | 14q12–q13 | unknown | 12 or 14 | |
| DFNA10 | 6q22–q23 | unknown | | |
| DFNA11 | 11q13.5 | *MYO7A* | | *sh1* |
| DFNA12 | 11q22–q24 | unknown | 7, 9, or 19 | *Fgf3, Dc, tub* |
| DFNA13 | 6p21 | unknown | | *Fu* |
| DFNA14 | 4p16 | unknown | 5 or 11 | *tlt, sh2, bv* |
| DFNA15 | 5q31 | unknown | 11 or 18 | *sh2, sy* |
| | | | | |
| **X-Linked** | | | | |
| DFN1[a] | Xq22 | *DDP* | X | |
| DFN2 | Xq22 | unknown | X | |
| DFN3 | Xq21.1 | *POU3F4* | X | |
| DFN4 | Xp21.2 | unknown | X | |
| DFN5 | | unknown | X | |
| DFN6 | Xp22 | unknown | X | *Gy* |
| DFN7 | | unknown | X | |
| DFN8 | | unknown | X | |
| | | | | |
| **Autosomal** | **Recessive** | | | |
| DFNB1 | 13q12 | *Cx26* | 5 | |
| DNFB2 | 11q13.5 | *MYO7A* | 7 | *sh1* |
| DNFB3 | 17p11.2 | unknown | 11 | *sh2* |
| DFNB4 | 7q31 | unknown | 6 | *Sig* |
| DFNB5 | 14q12 | unknown | 14 | |
| DFNB6 | 3p14–p21 | unknown | 6, 9, or 14 | *sr, mi* |
| DFNB7 | 9q13–q21 | unknown | 13 or 19 | *dn* |
| DFNB8 | 21q22 | unknown | 16 | *mh* |
| DFNB9 | 2p22–p23 | unknown | 12 or 17 | |
| DFNB10 | 21q22.3 | unknown | 10, 16, or 17 | *mh* |
| DFNB11 | 9q13–q21 | unknown | 13 or 19 | *dn* |
| DFNB12 | 10q21–q22 | unknown | 10 | *v, jc, av, mh* |
| DFNB13 | 7q34–36 | unknown | 5 or 6 | |
| DFNB14 | | unknown | | |
| DFNB15 | 3q21–q25 & | unknown | 3, 6, or 9 | *dfw* |
| | 19p13 | unknown | 8 or 10 | *mh* |
| DFNB16 | 15q21–q22 | unknown | 2 or 9 | |
| DFNB17 | 7q31 | unknown | 6 | *Sig* |
| DFNB18 | 11p14–15.1 | unknown | 7 | *tub* |
| DFNB19 | 18p11 | unknown | 18 | *sy* |
| DFNB20 | | unknown | | |

**TABLE 18.1** *(continued)*
**Nonsyndromic Deafness Loci in Humans**

| Locus Name | Location of Human Gene | Gene Name | Mouse Chromosome | Possible Mouse Homolog |
|---|---|---|---|---|
| **Mitochondrial** | | | | |
| mitochondria1 | | *12S rRNA* | | |
| mitochondria2 | | *tRNA-Ser* | | |

ᵃ DFN1 has been reassessed as a syndromic deafness locus since patients also exhibit blindness and dystonia.[69] For more detailed tables see Ref. 23.

positional information required for aligning within the preplate. The mechanism of this cell-cell interaction has yet to be determined, however.

Defects in myelin, the multilaminar sheath surrounding axons, can also result in impaired hearing.[37] The myelin sheath is composed of several different proteins including myelin basic protein (MBP), myelin-associated glycoprotein (MAG), peripheral myelin protein (PMP22), and protein zero (P0) and serves to increase the axonal condutance velocity. Loss of myelin therefore results in reduced nerve conductance and neuropathy. The consequences of myelin deficiency are illustrated by the *quaking* mouse mutant, which displays severe demyelination in the central nervous system (CNS) and a more moderate myelin loss in the peripheral nervous system (PNS).[44,45] The *quaking* mice show a rapid tremor when moving and are prone to seizures, during which they remain motionless for many seconds. In addition, the auditory brainstem response (ABR) of *quaking* mice, which is a measure of both peripheral and more central auditory processing, exhibits prolonged latencies reflecting impaired auditory function.[46] The *quaking* gene has been cloned and encodes three different sized transcripts (5kb, 6 kb, and 7 kb) which differ from one another at the C-terminus.[47,48] The *quaking viable* mutant mouse exhibits a 1-Mb deletion that results in the loss of the 6-kb and 7-kb transcripts in oligodendrocytes and Schwann cells, which synthesize myelin, although their expression in other glial cells remains unaffected. All three transcripts contain a STAR domain closely related to the *quaking* gene in *Xenopus* (*Xqua*), the *Gld-1* gene in *Caenorhabditis elegans*, and the human gene *SAM68*. A variety of STAR proteins are known from several diverse species and appear to be involved in both signal transduction and the activation of RNA.[49] However, the function of the *quaking* protein in myelination has yet to be determined.

The *shiverer* mouse mutant also displays tremors and seizures due to severe demyelination of the CNS, and the ABR shows prolonged latencies and raised thresholds.[50] The *shiverer* locus was mapped to chromosome 18, close to the *Mbp* gene.[51] Analysis of the *Mbp* locus identified a partial deletion of the gene, suggesting that loss of MBP results in neuropathy. Transgenic introduction of a functional *Mbp* gene into *shiverer* homozygotes restored MBP levels to 25% that of wild-type mice, and this was sufficient to prevent shivering and premature death, thus confirming the role of MBP in the *shiverer* phenotype.[52]

Demyelination of the PNS has also been linked to hearing loss. The *Trembler* mutant exhibits tremor, convulsions in early life, and spasticity in the lower back and limbs. The myelin sheath of the PNS is greatly reduced or absent, and many onion bulbs (axons surrounded by layers of Schwann cell basement membrane and cytoplasm) can be found.[53] Again the lack of myelin along the cochlear neuron results in prolonged latencies of the auditory response waveforms and impaired hearing.[54] *Trembler* has been shown to be caused by point mutations in the *Pmp22* gene on mouse chromosome 11,[55] a gene which has been implicated in two peripheral neuropathies in humans. Hereditary neuropathy with liability to pressure palsies (HNPP) is caused by deletion of the *PMP22* gene on human chromosome 17[56] and results in impaired myelination of the PNS and the formation of sausage-like swellings protruding from the myelin sheaths (tormacula). Charcot-Marie-Tooth disease type 1A (CMT1A), however, results from a duplication of the *PMP22* gene, leading to hypomyelination and the presence of onion bulbs.[57] Both disorders have been associated with deafness in some families,[58] revealing that demyelination and hearing impairment are often linked in humans as well as mice.

## HEREDITARY INNER EAR ABNORMALITIES

Abnormalities of the inner ear were originally classified by Ormerod based on descriptions of specimens of human cochleae from the 18th, 19th, and early 20th centuries. He defined four classes of anomaly, three of which, the Michel, Mondini-Alexander, and Bing-Siebenmann types, described structural malformations of the inner ear. The Michel type describes cases where there is a total lack of development of the inner ear, whilst the Mondini-Alexander type involves an underdevelopment of the ear resulting in a curved tube in place of the normally spiral cochlea. The Bing-Siebenmann type describes cases in which the bony labyrinth is formed but the membranous labyrinth is poorly developed. The fourth class, or Scheibe type, is defined by degenerative abnormalities limited to the cochlea and saccule and not affecting the bony labyrinth. This classification proved limited, however, and many abnormalities could not be easily categorized. Later studies on mice prompted a revision of the classification system by Gruneberg, who defined two types of abnormality, morphogenetic and degenerative, and this was further refined by Steel and Bock into the three categories still used today: morphogenetic, cochleosaccular, and neuroepithelial.[59]

### MORPHOGENETIC DEAFNESS

Morphogenetic deafness defines all cases of structural deformity of the membranous or bony labyrinths and includes the Michel, Mondini-Alexander, and Bing-Siebenmann types from Ormerod's original classification. The severity of the malformation is variable, ranging from the absence of a recognizable cochlea in *kreisler* mutants to merely a constricted or absent lateral semicircular canal in *Nijmegen waltzer* mice.[59,60] Indeed, a wide variability in the severity of hearing impairment and vestibular dysfunction can be seen between different animals carrying the same mutation and even between the two cochleae of a single affected animal.[60]

Morphogenetic abnormalities result from mutations affecting inner ear development, which has long been recognized to be influenced by the neural tube. The mammalian inner ear derives from part of the neural and surface ectoderm called the otic placode, which forms early in embryonic development and then invaginates to form a closed sack called the otic vesicle. This vesicle then divides into ventral and dorsal components which differentiate into the cochlea and vestibule, respectively. Many of the known causes of morphogenetic deafness are transcription factors which control this developmental process. The *kreisler* gene on mouse chromosome 2 encodes a basic-domain-leucine zipper transcription factor which is expressed in the caudal hindbrain prior to the formation of both rhombomere boundaries and the otic vesicle.[61] As the rhombomeres form, kreisler protein is expressed throughout rhombomere 5 (r5) and midway through r6. In the mutant, the boundaries dividing r4–r6 do not form, creating a broad rhombomere band between r3 and r7 and altering the expression domains of other embryonic transcription factors including several *Hoxb* genes, *Krox20,* and *Fgf3.* The otic vesicle, which should lie adjacent to r5, is laterally displaced and develops abnormally forming a membranous labyrinth which is barely recognizable. No expression of *kreisler* can be detected in the otic vesicle and therefore the abnormalities of the inner ear are believed to be secondary effects caused by the malformation of rhombomere segmentation.

The *Pax3* gene maps to mouse chromosome 1 and human chromosome 2q35–37 and is responsible for both the mouse *splotch* mutant and the human disorder Waardenburg's syndrome (WS) types 1 and 3.[62,63] WS is a dominant disorder characterized by deafness and pigmentary disturbances. Heterozygous *splotch* mice exhibit the pigmentary disturbances but are not deaf. Homozygous *splotch* mice display abnormalities in the neural plate and the neural tube fails to close.[64] This neural defect is so severe that mice homozygous for the mutation are embryonic lethal. Although the combination of ear defects and pigmentary disturbances suggests that *Pax3* mutations lead to cochleosaccular abnormalities (see later), the inner ear anomalies in the *splotch* homozygotes are morphogenetic. The otic vesicle forms in the correct position but develops abnormally so that not a single structure in the labyrinth appears normal.

Two genes have been the subject of targeted gene disruption in the mouse which resulted in morphogenetic defects: the homeobox gene *Hox-1.6* on mouse chromosome 6 and the fibroblast growth factor receptor 3 (*Fgfr3*) gene.[65,66] The vertebrate *Hox* genes are transcription factors believed to specify cell identity along the anteroposterior axis of the embryo. The *Hox-1.6⁻* homozygotes display profound deafness due to defects in the outer, middle, and inner ears and also show abnormalities in the cranial nerves and the hindbrain. In the outer ear the auricle and the external acoustic meatus are distorted, whilst in the middle ear the ossicles are absent. In the inner ear, both the cochlea and the vestibular system are grossly malformed and there is a lack of cochlear and vestibular nerves. The VII and VIII ganglia are displaced rostrally and in the hindbrain the glossopharyngeal and vagus nerves are poorly formed and possess no preganglionic connections to the brainstem. The involvement of Hox-1.6 with the development of the outer, middle, and inner ears is difficult to explain since each was formed by very different embryonic pathways. Although the inner ear forms from the surface ectoderm of the otic placode, the ossicles are formed from the cartilage of the first and second pharyngeal arches, the acoustic meatus from the first pharyngeal cleft, and the auricle from the mesenchymal proliferations at the ends of the first and second pharyngeal arches. The remaining affected structures are all derived from the region bounded by r4–7. It is likely that this region represents the limit of expression of the *Hox-1.6* gene. The suspected influences of r4–6 on the development of the otic vesicle have already been described (see above), and again the disruption of this region results in the rostral and ventral displacement of the otic vesicle. Whether Hox-1.6 has a direct effect on the primordial outer and middle ears or whether the defects seen are secondary characteristics is still undetermined.

*Fgfr3* is a tyrosinase kinase transmembrane receptor expressed in developing bone, cochlea, brain, and spinal cord. In the cochlea, expression of *Fgfr3* is limited to differentiating hair cells and their underlying support cells, and no expression can be detected in the vestibule.[67] Mutations in the gene have been implicated in achondroplasia through abnormal activation of the signaling pathway. In *Fgfr3⁻* homozygotes skeletal overgrowth, kyphosis, and deafness are seen. The deafness appears to be caused by the absence of pillar cells supporting the cochlear hair cells. An excess of Dieters' cells can be seen in place of the pillar cells and the tunnel space, which suggest either that Dieters' cells and pillar cells share a common progenitor or that some Dieters' cells can differentiate into pillar cells. The *Fgfr3⁻* mice also show a reduction in innervation of outer hair cells, which may be caused by disrupted Fgf signaling or may be the result of the lack of pillar cell differentiation. The *Fgfr3⁻/⁻* mutant is unusual in that the organ of Corti does not undergo degeneration, therefore indicating that Fgf signaling is required for cochlear differentiation but not maintainance.

A fifth example of a mutation causing morphogenetic deafness is *his*, which has been implicated in both mouse and human hearing impairment.[68] *his* encodes the enzyme histidase which is required to reduce levels of histidine. In *his* mutants the levels of histidine are raised causing a teratogenic effect on embryonic development such that the offspring of *his⁻* homozygote mothers are deaf. Some transient neural tube anomalies have been seen during embryogenesis but the role of histidine levels on inner ear development is still unclear.

Further investigation of these and other morphogenetic mutations should provide a clear understanding of the process of inner ear formation.

## COCHLEOSACCULAR DEAFNESS

Cochleosaccular abnormalities are named for the primary sites of lesions in the ear, the cochlea, and the sacculus, and include the Scheibe-type anomalies from Ormerod's original classification.

The major defect is in the stria vascularis of the cochlea (see Figure 18.2), which appears thinner than normal and never reaches histological maturity. The stria is positioned on the lateral wall of the cochlear duct and is involved in generating the endocochlear potential in the endolymphatic fluid which surrounds the hair cells. Following the incomplete development of the stria, the collapse of Reissner's membrane can be seen and the organ of Corti begins to degenerate. In the vestibular system the saccular wall collapses, damaging the macula, but no other abnormalities can be detected. Like the morphogenetic defects, cochleosaccular abnormalities tend to be highly variable in their severity, even within a single pedigree, and the hearing loss can affect one or both ears to different degrees.

Cochleosaccular deafness is often seen to be associated with pigmentary disturbances such as in the human Waardenburg's syndrome and a variety of spotting mouse mutants, for instance, the *dominant spotting*, *Steel,* and *piebald* loci. In addition, deafness is often seen in blue-eyed, white cats and in Dalmatian dogs that do not display the color patch, a dark region of fur present at birth. The connection between deafness and color abnormalities can be explained by the presence of melanocytes, the pigment-producing cells, in the intermediate layer of the stria vascularis. Melanoblasts, the precursors of the inner ear melanocytes, are produced in the neural crest during embryogenesis and migrate to the area around the developing cochlear duct before entering the stria vascularis. In spotting mouse mutants such as *dominant spotting* and *Steel*, no recognizable melanocytes can be found in the stria and there is an associated lack of endocochlear potential (EP).[69] In certain alleles of the *dominant spotting* mutation, low levels of pigment can be detected in the stria and these are associated with a positive resting potential in the scala media, although these potentials are always lower than in wild-type mice. Thus, there seems to be a direct correlation between the proportion of pigmentation in the stria and the size of the EP.[70] The EP is not dependent on the presence of pigment but rather on the presence of functional melanocytes. This is supported by the fact that albino mutations, which affect the biochemical pathway producing melanin but otherwise result in healthy melanocytes, display no hearing impairment. It is therefore likely that any gene affecting neural crest-derived melanocyte development or migration will result in cochleosaccular deafness.

One such gene is *Pax3*, implicated in WS types 1 and 3 and the mouse *splotch* mutant. *Sp* homozygotes exhibit no migration of melanocytes or melanoblasts from the neural crest, suggesting an involvement for Pax3 in the differentiation of melanoblasts or their migration pathway.[71] Curiously, though, the inner ear abnormalities displayed by *Sp/Sp* mice are morphogenetic in origin (see above).

The *microphthalmia* mouse, however, shows typical cochleosaccular defects, a lack of melanocytes in the skin and ear causing lack of pigment and deafness. They also exhibit small, unpigmented eyes for which they were named. The *mi* gene was mapped to mouse chromosome 6 and found to encode a basic-helix-loop-helix zipper protein.[72] It is a member of a group of transcription factors which include TFEB and TFE3 and are defined by the presence of three domains: a basic domain involved in DNA binding and an HLH and ZIP domain through which they dimerize. Subsequently, the human homolog of *mi* was found to be responsible for 20% of WS2 cases, a form of Waardenburg's syndrome without the dystopia canthorum which is characteristic of type 1. Analysis of the Mi protein showed that it was capable of forming either homodimers or heterodimers with other members of the bHLH-ZIP family and that these dimers were able to bind to a DNA motif called the M-box, found in the promotor region of three pigmentation genes: tyrosinase and tyrosinase-related proteins 1 and 2.[73] Since *mi* also affects the melanocyte population of the eye, which is not derived from the neural crest but rather the eye anlage, it seems unlikely that Mi is involved in melanoblast differentiation or migration like the Pax3 protein. This is supported by the expression pattern of *mi* mRNA, which is limited to the developing eye, inner ear and skin.[72] Ectopic expression of *MITF*, the human homolog of *mi*, was able to transform fibroblasts into cells with characteristics of melanocytes.[74] Refractile cells with dendritic processes and containing membrane-bound vesicles with homogeneous matrices were seen and are characteristic of immature melanosomes. In addition, expression of the melanogenic markers tyrosinase and tyrosinase-related protein 1 was detected, although no melanin was produced. This data support the hypothesis that *mi* is expressed in melanoblast-containing tissues subsequent to their migration from the neural crest and is important for their differentiation into mature melanocytes.

Other causes of cochleosaccular abnormalities include mutations in the genes *ednrb* and *edn3*, which encode a G-protein coupled receptor and its ligand endothelin-3. Mutations in *edn3* have been shown to be present in *lethal spotting* mice and in WS4, a disorder combining Waardenburg's syndrome and aganglionic megacolon (Hirschsprung disease).[75,76] *Edn3-* homozygote mice exhibit white spotting of the skin across 70 to 80% of the coat and aganglionic megacolon consistent with the human phenotype in WS4.[77] Mutations in *ednrb* are found in *piebald lethal* mice and some cases of Hirschsprung disease (with or without WS) and present an identical phenotype to *edn3-* mutants, although the white spotting is more pronounced, covering greater than 90% of the coat.[77,78] By studying the melanogenic marker tyrosinase-related protein 2 (TRP2), Pavan et al. found that ednrb is involved in the development of melanocytes before or coincident with their migration from the neural crest.[79] The combined data from these studies suggest that both edn3 and ednrb are essential for the normal development of enteric ganglion neurons and epidermal and choroidal melanocytes in the neural crest.

## Variability of the Phenotype

One common feature between morphogenetic and cochleosaccular defects is that they both display extreme variability in phenotype even within a single pedigree. This is clearly illustrated by the genes causing Waardenburg's syndrome and their murine homologs. WS is subdivided into four types by clinical observations, and different genes are known to be causative, even within a single subtype. WS1 is defined by the presence of dystopia canthorum, as well as the deafness and pigmentary disturbances, and this is the most consistent feature of this subtype with 98% penetrance. WS3 includes all the features of WS1 as well as skeletal anomalies of the upper arms. Mutations in *Pax3* have been shown to cause WS1 and WS3 in 78% of cases according to Tassabehji et al., although their analysis was not 100% sensitive and did not screen the first exon or 5' untranslated region (UTR).[80] It is believed that *Pax3* mutations will represent 100% of WS1 cases; however, deafness is found in only 20% of patients, although it can be as high as 78% in some pedigrees.[63] *Pax3* is also implicated in craniofacial deafness hand syndrome (CDHS), which is similar but distinct from WS3. This variable phenotype is presumed to be due to modifier loci, genes which interact with the mutated gene and alter the resulting phenotype. Studies on the *splotch* mutant, the murine homolog of WS1, suggest at least two other loci interacting with *Pax3* to affect skull shape, one of which is sex linked and affects posterior shape while the other is autosomal and affects anterior shape.[81] Similarly, 68% of heterozygotes displayed white belly spots on a $BC_1$ background compared to 100% on a C57BL/6J background, implicating at least two modifiers: one sex linked and one mapping close to the *agouti* allele on mouse chromosome 2.

Modifier genes are also implicated in WS4, which combines the features of WS1, except dystopia canthorum, with aganglionic megacolon. This disorder has been linked to mutations in both the *edn3* and *ednrb* genes. Analysis of one human pedigree segregating for a mutation in *ednrb* revealed some patients displaying WS4 whereas others exhibited only hearing loss or pigmentary disturbances.[76] Another pedigree segregating an *ednrb* missense mutation showed only 74% of homozygotes and 21% of heterozygotes exhibiting megacolon.[78] The heterozygotes also showed a sex bias, with males being more susceptible to megacolon than females.

These studies reveal highly complicated genic interactions in these syndromes. Identification of these modifier genes will be required to fully understand the disease phenotype and to provide an accurate diagnosis and risk assessment for sufferers.

## NEUROEPITHELIAL DEAFNESS

Neuroepithelial defects involve primary abnormalities of the sensory neuroepithelia which can be found in the organ of Corti, the saccular and utricular macula, and the cristae of the vestibular system, although not all of these need be affected. The developing cochlea appears grossly normal until about the time the organ of Corti reaches its histological maturation and can then be seen to degenerate,[82] although there are often defects noticeable prior to this. The degeneration often involves the organ of Corti and the spiral ganglion and thus may appear to be similar to that seen in cochleosaccular mutants; however, the stria vascularis appears normal at first, the EP is present, and Reissner's membrane is in place. The degenerative process resembles that described following experimental trauma with the outer hair cells degenerating before the inner hair cells and the process beginning in the upper basal turn of the cochlea and proceeding toward both the apex and base. Finally, the supporting cells collapse and leave a mass of dedifferentiated cells on the basal membrane.[59]

A characteristic feature of neuroepithelial defects is the involvement of the vestibular system, and many animal models of this type display a vestibular dysfunction. In fact, most mouse neuroepithelial mutants were originally identified because of behavioral anomalies such as hyperactivity, head-tossing, and circling, which all result from a nonfunctional vestibular system. Similar hearing defects have been described in other animals such as dogs and guinea pigs.

In contrast to other classes of hearing impairment, neuroepithelial deafness is typically uniformly penetrant in all affected members of a single pedigree and is symmetrical across both ears. The severity of the defects are quite variable between different mutants, however, with 15% of hair cells remaining in *deafness* mice 6 weeks after birth, compared to complete loss of hair cells in 6-week-old *Snell's waltzer* mice, and cochlear responses which decay from 12 days, as in *shaker1* mice, compared to a total lack of cochlear response from birth in *deafness* mice.[35,83,84]

Identifying genes involved in neuroepithelial deafness is of particular interest, first, because this class of mutation is most likely to affect proteins involved in auditory transduction, and second, because it is predicted to represent a large proportion of human deafness. One gene for neuroepithelial deafness was identified from studies of a spontaneously occurring deaf mouse mutant, *shaker1*.

The *shaker1* mouse exhibits deafness and vestibular dysfunction. The mutation was mapped to chromosome 7 and linked to the olfactory marker protein. A positional cloning strategy identified a mutation in a myosin type VII gene and screening of the human population identified mutations in the human *MYO7A* gene in Usher's syndrome type 1b patients.[35,85] Usher's syndrome type 1 associates severe congenital deafness and vestibular dysfunction with progressive retinitis pigmentosa (RP) leading to blindness. Curiously, in the mouse no blindness is detected, which may result from compensation for the loss of Myo7a in the eye. An alternative explanation is suggested by a recent study of *MYO7A* expression in humans and mice. Similar patterns of expression were seen in the sensory hair cells of the cochlea and the pigmented epithelium of the retina in the two species; however, *MYO7A* was expressed in the photoreceptor cells of the retina in humans but not mice.[86] The lack of blindness in mice may therefore reflect the lack of Myo7a protein in the photoreceptor cells. Alternatively, since mutations in the *MYO7A* gene have also been implicated in nonsyndromic deafness in humans[29,30] (see later), the lack of RP in mice may represent genetic background effects modifying the *shaker1* phenotype (Figure 18.4).

Although the inner ear of *shaker1* mice appears grossly normal until about 12 days when the hair cells begin to degenerate, electron microscopy reveals hair cell defects at earlier stages, with the hair cell stereocilia failing to organize into the hair bundle from 18 days into gestation.[87] Myosin genes are known to be involved in vesicle transport and it has been proposed that Myo7a is involved in vesicle transport and membrane turnover in the cochlear hair cells.[88]

A further two genes were the subject of gene targeting strategies and were subsequently found to result in hearing impairment. Thyroid hormone has long been known to affect inner ear devel-

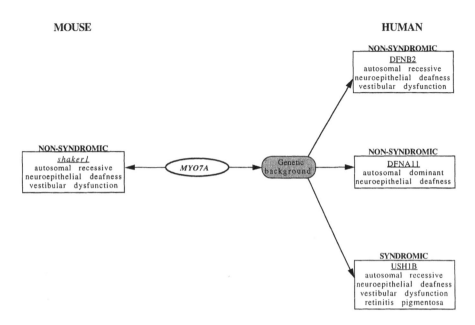

**FIGURE 18.4** Diagrammatic representation showing the phenotypic modifying effect of genetic background on the *MYO7A* gene. Different backgrounds can result in either nonsyndromic deafness or syndromic deafness with retinitis pigmentosa, and also in both dominant and recessive forms of deafness.

opment and humans with hypothyroidism present hearing loss in 20 to 80% of cases. Pendred syndrome in humans exhibits deafness and is associated with impaired thyroid hormone synthesis, and rats made hypothyroid in the laboratory show abnormal development of the organ of Corti with anomalies in the sensory epithelium and the tectorial membrane.[89] Thyroid hormone acts through two types of receptors, *Trα* and *Trβ*, both of which are expressed in the cochlea, but only *Trα* is expressed in the vestibular system. Mice homozygous for a *Trβ⁻* mutation displayed severe to profound hearing loss (70 to 100 dB).[90] Histological analysis of the inner ear revealed no gross morphological changes but showed some hair cell loss, mainly in the basal turn of the cochleae. This indicates that Trb is required for development of auditory function but not for the morphogenesis of the cochlea, and that this role cannot be compensated for by Tra.

A second gene targeting experiment in mice studied the role of several class IV POU domain transcription factors in auditory and visual development. This family includes the genes *Brn3.0*, *Brn3.1*, and *Brn3.2*, which have been shown to be important for the development of specific populations of neurons. *Brn3.1* shows a pattern of expression that overlaps that of the genes *Brn3.0* and *Brn3.2*; however, *Brn3.1* is also uniquely expressed in the cochlear and vestibular hair cells of the inner ear. *Brn3.1⁻* homozygotes show the characteristic hyperactivity and vestibular dysfunction and also exhibit severe hearing loss (>80 dB).[91] Morphological examination of these mice at birth shows a lack of differentiation of inner and outer hair cells in both the cochlea and the vestibular system. Hair cell stereocilia have not grown and the nuclei have not segregated to a level above the plane of the supporting cells. By 2 weeks after birth neither hair cells nor the supporting cells of the cochlea can be identified and the vestibular labyrinth and spiral and vestibular ganglions have degenerated. These findings support a role for Brn3.1 in the terminal differentiation of sensorineural hair cells in the inner ear.

Neuroepithelial defects are typically confined to the inner ear and therefore give rise to nonsyndromic hearing impairment. In addition, they also tend to be recessive traits. Since the majority of human deafness is nonsyndromic, autosomal recessive, study of neuroepithelial mutants should therefore identify genes for human nonsyndromic deafness.

## GENES FOR NONSYNDROMIC DEAFNESS

Seventy percent of human deafness is nonsyndromic and 45 loci for nonsyndromic hearing impairment have been mapped in humans (see Table 18.1). Only five genes for nonsyndromic deafness have been identified, however. The X-linked deafness, DFN1, has been shown to be caused by mutations in the *DDP* gene, although the role of this gene is still unknown. The clinical phenotype for DFN1 has since been reexamined, however, and is now classed as syndromic, with patients exhibiting dystonia, blindness, fractures, and mental retardation as well as deafness.[92,93] The remaining genes known to be involved in nonsyndromic deafness are *POU3F4*, which is responsible for another X-linked deafness; connexin26, implicated in both DFNA3 and DFNB1; myosin VIIa, mutated in both DFNA11 and DFNB2; and *diaphanous*, which is disrupted in DFNA1. A further gene, *Myo6*, has been identified in the mouse but no deafness has been associated with the human homolog.

### POU3F4

Several studies of pedigrees segregating nonsyndromic X-linked deafness have identified a locus, DFN3, mapping to Xq21.1 and characterized by profound sensorineural deafness with or without a conductive component and associated with stapes fixation.[58] A mouse transcription factor belonging to the POU domain family was found to map to the syntenic region in mouse and was known to be expressed in the brain, neural tube, and otic vesicle in the developing rat embryo. Subsequent screening of this *POU3F4* gene in DFN3 pedigrees identified three deletions and two missense mutations in the coding sequence and several microdeletions and duplications upstream of the gene which are thought to affect transcription. The role of the gene on cochlear development is still unknown.

### Cx26

Connexins are the monomeric protein subunits of gap junctions, intracellular channels allowing molecules smaller than 1000 kDa to pass between neighboring cells. *Cx26* maps to human chromosome 13q12, close to two nonsyndromic deafness loci, the dominant locus DFNA3 and the recessive locus DFNB1. Analysis of the gene identified mutations in a pedigree cosegregating palmoplantar keratoderma and dominant deafness (presumably DFNA3) and also in three families segregating the DFNB1 locus.[31] Connexin 26 is expressed in the stria vascularis and the basement membrane of the cochlea and is predicted to be involved in the recycling of potassium ions from the hair cells to the endolymph in order to maintain the high EP.[94]

### MYO7A

The myosin VIIa gene was originally identified as being responsible for the *shaker1* mouse mutant and was subsequently shown to be involved in Usher's syndrome type 1b (see above). The locus for USH1b maps to human chromosome 11q13.5 and lies close to the nonsyndromic deafness locus DFNB2.[95] It has long been suggested that the two loci may be allelic. DFNB2 is a recessive form of deafness with vestibular dysfunction but without the progressive blindness that characterizes Usher's syndrome. Mutation screening of the *MYO7A* gene in pedigrees segregating nonsyndromic recessive deafness has identified three mutations: one family homozygous for a missense mutation, and a second family carrying a splice site mutation in one allele and a single base insertion leading to a frameshift in the other allele.[29] A gene for dominant nonsyndromic deafness (DFNA11) has also been mapped to 11q13[96] and mutation screening of the *MYO7A* gene in this pedigree identified a 9 bp in frame deletion.[30] Thus *MYO7A* mutations have been shown to result in both dominant and recessive nonsyndromic, as well as syndromic, deafness. It has yet to be determined whether the phenotypic differences resulting from different *MYO7A* alleles are related to the type of mutation or the involvement of modifier genes (see Figure 18.4).

## *MYO6*

Like *shaker1*, the *Snell's waltzer* mouse displays the phenotype of typical neuroepithelial abnormalities: deafness, hyperactivity, and vestibular dysfunction. A myosin gene, myosin VI, was found to contain an intragenic deletion in *sv* mice and lie close to the breakpoint of a chromosomal inversion in the allelic *se^sv* mice.[84] The human homolog maps to the pericentromeric region of chromosome 6,[97] but no known genes for human deafness are linked to this region.

The cochlea of *sv* mice showed complete loss of both inner and outer hair cells by 6 weeks of age and these tissues were the only parts of the inner ear to show expression of myosin VI. Although the organ of Corti appeared normal up to 12 days, by 3 weeks the hair cells appeared pycnotic and no stereocilia were observed. Myosin VI is therefore likely to be required for maintainance of the structural integrity of the hair cells, perhaps through vesicle transport and membrane recycling as proposed for Myo7a.

## *HDIA1*

The latest human deafness locus to be cloned, at the time of writing, is DFNA1 on chromosome 5q31.[32] Sequencing 800 kb of genomic DNA containing the locus identified a gene showing extensive homology to the *Drosophila diaphanous* and mouse *p140mDia* genes, both of which are members of the formin gene family. Mutation screening of the human *diaphanous* gene revealed a guanine-to-thymine substitution which disrupts a splice donor site in the penultimate exon, resulting in a 4-bp insertion. This leads to a frameshift producing 21 aberrant amino acids followed by a premature termination signal which truncates a further 32 residues. Diaphanous proteins in yeast and mouse are involved in the recruitment to the membrane of profilin, a regulator of actin polymerization. Actin filaments provide the rigidity of hair cell stereocilia and the interconnecting framework between the inner and outer hair cells, and also anchor the stereocilia within the cuticular plate of the hair cells.[32] Three of the five genes for nonsyndromic deafness cloned to date appear to affect either actin polymerization or actin-binding motor proteins (e.g., myosins), thus emphasizing the importance of the actin cytoskeleton in inner ear structure and function.

## LATE-ONSET DEAFNESS

In addition to nonsyndromic (neuroepithelial) deafness, progressive deafness is another important form of hearing loss in humans. Much of human progressive deafness appears to be age related, a condition known as presbycusis, with the proportion of the population exhibiting severe hearing loss rising from 0.3 to 2.3% between the ages of 30 and 70 years.[4] Although much of this increase can be ascribed to environmental factors, genetic influences are known to be involved. Human presbycusis is characterized as one of four pathological types — (1) sensory, resulting in an abrupt high-tone signal loss; (2) neural, causing loss of word discrimination; (3) strial, producing a flat threshold pattern; and (4) cochlear conductive, defined by a gradual decrease of all tones with no pathological correlate — although many cases display a combination of these types.[98] Mice have also been shown to exhibit presbycusis. The mouse strains SAMP1 and SAMR1 are, respectively, accelerated senescence-prone and -resistant mice and are used to compare the effects of accelerated, overnormal, aging. ABR measurements in the two mouse strains showed that the auditory function of SAMP1 mice at 12 months was comparable to that of SAMR1 mice at 20 months.[99] SAMP1 also displayed a more pronounced decrease in cell density and size of the spiral ganglion neurons, an accelerated loss of inner and outer hair cells, and earlier atrophy of the stria vascularis than the SAMR1 mice.[100] These observations suggest that hearing loss in SAM may be a combination of sensory, strial, and neural presbycusis, reflecting that seen in humans, and suggesting that the mouse represents a good model of late-onset human deafness.

Two further mouse strains, CBA/H-T6J and C57BL/6J, display a gradual, progressive, high-frequency sensorineural hearing loss resembling that seen in humans.[101] However, whereas CBA mice maintain good hearing until late in life, C57BL/6J mice show hearing loss as early as 2 months of age and adult mice exhibit severe deafness. The difference in these two strains is due to a single gene named age-related hearing loss 1 (*Ahl1*), which was found to map to mouse chromosome 10.[102] Another two age-related hearing loss loci (*Ahl2* and *Ahl3*) were identified in the mouse strains WB/ReJ and BALB/cBY, and the strain DBA/2J was found to be homozygous for all three loci.[103] In each strain, the severity of the hearing loss was found to correspond to the degree of cochlear pathology, with mice homozygous for any of the three loci showing severe cochlear pathology.[103] Although these genes have not yet been cloned, the similarity in electrophysiology and histology of the inner ear between the mouse strains and humans suggests that these loci will provide an insight into presbycusis in humans.

## CONCLUSIONS

The auditory and vestibular systems of mammals are highly complicated and organized structures capable of reacting to displacements of less than 1 nm with response times of several milliseconds. Many of the genes involved in the auditory transduction process are still unknown. However, recent advances in genetic mapping in humans and positional cloning in mice combined with the wealth of information produced by gene targeting experiments is beginning to unravel this small but intricate organ.

## REFERENCES

1. Corwin, J. and Warchol, M., Auditory hair cells: structure, function, development and regeneration, *Annu. Rev. Neuroscii*, 14, 301, 1991.
2. Hudspeth, A.J. and Gillespie, P., Pulling springs to tune transduction: adaptation by hair cells, *Neuron*, 12, 1, 1994.
3. Gillespie, P., Molecular machinery of auditory and vestibular transduction, *Curr. Opinion Neurobiol.*, 5, 449, 1995.
4. Petit, C., Genes responsible for hereditary deafness: symphony of a thousand, *Nat. Genet.*, 14, 385, 1997.
5. Davis, A., The prevalence of hearing impairment and reported hearing disability among adults in Great Britain, *Int. Jo. Epidemiol.*, 18, 911, 1989.
6. 3D-based textbook for *Human Body Education*, Chiyokura Laboratory, Keio University and the Information-technology Promotion Agency, Japan, 1995; World Wide Web URL: http://www.edu.-ipa.go.jp/chiyo/HuBEd/HTML1/en/3D/intro.html, 1995.
7. Ballantyne, J. and Martin, J., *Deafness*, 4th ed. Churchill Livingstone, Edinburgh, 1984.
8. Hudspeth, A.J., How the ear's works work, *Nature*, 341, 397, 1989.
9. Corey, D. and Hudspeth, A., Kinetics of the receptor current in bullfrog saccular hair cells, *J. Neurosci.*, 3, 962, 1983.
10. Howard, J. and Hudspeth, A., Compliance of the hair bundle associated with gating of mechanoelectrical transduction channels in the bullfrog's saccular hair cell, *Neuron*, 1, 189, 1988.
11. Pickles, J. and Corey, D., Mechanoelectrical transduction by hair cells, *Trends Neurosci.*, 15, 254, 1992.
12. Markin, V. and Hudspeth, A., Gating-spring models of mechanotransduction by hair cells of the internal ear, *Annu. Rev. Biophys. Biomol. Struct.*, 24, 59, 1995.
13. Pickles, J., Comis, S., and Osborne, M., Cross-links between stereocilia in the guinea pig organ of Corti, and their possible relation to sensory transduction, *Hearing Res.*, 15, 103, 1984.
14. Assad, J., Shepherd, G., and Corey, D., Tip-link integrity and mechanical transduction in vertebrate hair cells, *Neuron*, 7, 985, 1991.

15. Denk, W. et al., Calcium imaging of single stereocilia in hair cells: localization of transduction channels at both ends of tip links, *Neuron*, 15, 1311, 1995.

16. Jaramillo, F. and Hudspeth, A., Localization of the hair cell's transduction channels at the hair bundle's top by iontophoretic application of a channel blocker, *Neuron*, 7, 409, 1991.

17. Reardon, W., Genetic deafness, *J. Med. Genet.*, 29, 521, 1992.

18. Brown, S.D.M. and Steel, K.P., Genetic deafness — progress with mouse models, *Hum. Mol. Genet.*, 3, 1453, 1994.

19. Lyon, M., Rastan, S., and Brown, S., *Genetic Variants and Strains of the Laboratory Mouse*, 3rd ed., Oxford University Press, Oxford, 1996.

20. Searle, A., *Comparative Genetics of Coat Colour in Mammals*, Logos Press, London, 1986.

21. Famula, T.R., Oberbauer, A.M., and Sousa, C.A., A threshold model analysis of deafness in Dalmatians, *Mammalian Genome*, 7, 650, 1996.

22. Cremers, F. et al., Mapping and cloning hereditary deafness genes, *Curr. Opinion Genet. Dev.*, 5, 371, 1995.

23. Camp, G.V. and Smith, R., Hereditary hearing loss homepage, World Wide Web, http://dnalab-www.uia.ac.be/dnalab/hhh/index.html, 1996.

24. Leon, P. et al., The gene for an inherited form of deafness maps to chromosome 5q31, *Proc. Natl. Acad. Sci. U.S.A.*, 89, 5181, 1992.

25. Prezant, T. et al., Mitochondrial ribosomal RNA mutation associated with both antibiotic-induced and non-syndromic deafness, *Nat. Genet.*, 4, 289, 1993.

26. Guilford, P. et al., A non-syndromic form of neurosensory, recessive deafness maps to the pericentromeric region of chromosome 13q, *Nat. Genet.*, 6, 24, 1994.

27. Lander, E. and Botstein, D., Homozygosity mapping: a way to map human recessive traits with the DNA of inbred children, *Science*, 236, 1567, 1987.

28. Friedman, T. et al., A gene for congenital, recessive deafness DFNB3 maps to the pericentric region of chromosome 17, *Nat. Genet.*, 9, 86, 1995.

29. Liu, X. et al., Mutations in the myosin VIIA gene cause non-syndromic recessive deafness, *Nat. Genet.*, 16, 188, 1997.

30. Liu, X. et al., Autosomal dominant non-syndromic deafness caused by a mutation in the myosin VIIA gene, *Nat. Genet.*, 17, 268, 1997.

31. Kelsell, D. et al., Connexin 26 mutations in hereditary non-syndromic sensorineural deafness, *Nature*, 387, 80, 1997.

32. Lynch, E. et al., Nonsyndromic deafness DFNA1 associated with mutation of a human homolog of the *Drosophila* gene *diaphanous*, *Science*, 278, 1315, 1997.

33. Kok, Y.D. et al., Associations between X-linked mixed deafness and mutations in the POU domain gene *POU3F4*, *Science*, 267, 685, 1995.

34. Reid, F., Vernham, G., and Jacobs, H., A novel mitochondrial point mutation in a maternal pedigree with sensorineural deafness, *Hum. Mutation*, 3, 243, 1994.

35. Gibson, F. et al., A type VII myosin encoded by the mouse deafness gene *Shaker1*, *Nature*, 374, 62, 1995.

36. Steel, K.P., Similarities between mice and humans with hereditary deafness, *Ann. N. Y. Acad. Sci.*, 630, 68, 1991.

37. Steel, K.P., Inherited hearing defects in mice, *Annu. Rev. Genet.*, 29, 675, 1995.

38. Hirotsune, S. et al., The *reeler* gene encodes a protein with an EGF-like motif expressed by pioneer neurons, *Nat. Genet.*, 10, 77, 1995.

39. Bar, I. et al., A YAC contig of the *reeler* locus with preliminary characterization of candidate gene fragments, *Genomics*, 26, 543, 1995.

40. Ogawa, M. et al., The *reeler* gene-associated antigen on Cajal-Retzius neurons is a crucial molecule for laminar organization of cortical neurons, *Neuron*, 14, 899, 1995.

41. Goffinet, A., A real gene for *reeler*, *Nature*, 374, 675, 1995.

42. Martin, M., Morphology of the cochlear nucleus of the normal and *reeler* mutant mouse, *J. Comp. Neurol.*, 197, 141, 1981.

43. D'Arcangelo, G. et al., A protein related to extracellular matrix proteins deleted in the mouse mutant *reeler*, *Nature*, 374, 719, 1995.

44. Sidman, R., Dickie, M., and Appel, S., Mutant mice (*quaking* and *jimpy*) with deficient myelination in the central nervous system, *Science*, 144, 309, 1964.

45. Samorajski, T., Friede, R., and Reimer, P., Hypomyelination in the *quaking* mouse. A model for the analysis of disturbed myelin formation, *J. Neuropathol. Exp. Neurol.*, 29, 507, 1970.

46. Shah, S. and Salamy, A., Auditory-evoked far-field potentials in myelin deficient mutant *quaking* mice, *Neuroscience*, 5, 2321, 1980.

47. Ebersole, T. et al., The *quaking* gene product necessary in embryogenesis and myelination combines features of RNA binding and signal transduction proteins, *Nat. Genet.*, 12, 260, 1996.

48. Hardy, R. et al., Neural cell type-specific expression of QK1 proteins is altered in *quakingviable* mutant mice, *J. Neurosci.*, 16, 7941, 1996.

49. Vernet, C. and Artzt, K., STAR, a gene family involved in signal transduction and activation of RNA, *Trends Genet.*, 13, 479, 1997.

50. Fujiyoshi, T., Hood, L., and Yoo, T., Restoration of brain stem auditory-evoked potentials by gene transfer in *shiverer* mice, *Ann. Otol., Rhonol. Laryngol.*, 103, 449, 1994.

51. Sidman, R., Conover, C., and Carson, J., *Shiverer* gene maps near the distal end of chromosome 18 in the house mouse, *Cytogenet. Cell Genet.*, 39, 241, 1985.

52. Readhead, C. and Hood, L., The dysmyelinating mouse mutations *shiverer* (*shi*) and *myelin deficient* (*shimld*), *Beh. Genet.*, 20, 213, 1990.

53. Low, P., The evolution of "onion bulbs" in the hereditary hypertrophic neuropathy of the *trembler* mouse, *Neuropathol. Appl.Neurobiol.*, 3, 81, 1977.

54. Zhou, R. et al., Using a mutant model of peripheral myelin deficiency to evaluate electrophysiological alterations in the auditory system, in 17th Meeting of Associated Research in Otolaryngology, Florida, 1994.

55. Suter, U. et al., *Trembler* mouse carries a point mutation in a myelin gene, *Nature*, 356, 241, 1992.

56. Chance, P. et al., DNA deletion associated with hereditary neuropathy with liability to pressure palsies, *Cell*, 72, 143, 1993.

57. Pentao, L. et al., Charcot-Marie-Tooth type 1A duplication appears to rise from recombination at repeat sequences flanking the 1.5Mb monomer unit, *Nat. Genet.*, 2, 292, 1992.

58. *Online Mendelian Inheritance in Man,* OMIN(TM), Center for Medical Genetics, Johns Hopkins University (Baltimore, MD) and National Center for Biotechnology Information, National Library of Medicine (Bethesda, MD), 1999; World Wide Web URL: http://www3.ncbi.nlm.nih.gov/omim/.

59. Steel, K. and Bock, G., Hereditary inner-ear abnormalities in animals, *Arch. Otolaryngol.*, 109, 22, 1983.

60. Deol, M., The abnormalities of the inner ear in *kreisler* mice, *J. Embryol. Exp. Morphol.*, 12, 475, 1964.

61. Cordes, S. and Barsh, G., The mouse segmentation gene *kr* encodes a novel basic domain-leucine zipper transcription factor, *Cell*, 79, 1025, 1994.

62. Tassabehji, M. et al., Waardenburg's syndrome patients have mutations in the human homologue of the *Pax-3* paired box gene, *Nature*, 355, 635, 1992.

63. Baldwin, C. et al., An exonic mutation in the *HuP2* paired domain gene causes Waardenburg's syndrome, *Nature*, 355, 637, 1992.

64. Deol, M., Influence of the neural tube on the differentiation of the inner ear in the mammalian embryo, *Nature*, 209, 219, 1966.

65. Chisaka, O., Musci, T., and Capecchi, M., Developmental defects of the ear, cranial nerves and hindbrain resulting from targeted disruption of the mouse homeobox gene *Hox-1.6*, *Nature*, 355, 516, 1992.

66. Colvin, J. et al., Skeletal overgrowth and deafness in mice lacking fibroblast growth factor receptor 3, *Nat. Genet.*, 12, 390, 1996.

67. Peters, K. et al., Unique expression pattern of the FGF receptor 3 gene during mouse organogenesis, *Dev. Biol.*, 155, 423, 1993.

68. Kacser, H. et al., Maternal histidine metabolism and its effect on foetal development in the mouse, *Nature*, 265, 262, 1977.

69. Steel, K., Barkway, C., and Bock, G., Strial dysfunction in mice with cochleosaccular abnormalities, *Hearing Res.*, 27, 11, 1987.

70. Cable, J., Jackson, I., and Steel, K., Mutations at the *W* locus affect survival of neural crest-derived melanocytes in the mouse, *Mech. Dev.*, 50, 139, 1995.

71. Auerbach, R., Analysis of the developmental effects of a lethal mutation in the house mouse, *J. Exp. Zool.*, 127, 305, 1954.

72. Hodgkinson, C. et al., Mutations at the mouse microphthalmia locus are associated with defects in a gene encoding a novel basic-helix-loop-helix-zipper protein, *Cell*, 74, 395, 1993.

73. Hemesath, T. et al., *Microphthalmia*, a critical factor in melanocyte development, defines a discrete transcription factor family, *Genes Dev.*, 8, 2770, 1994.

74. Tachibana, M. et al., Ectopic expression of *MITF*, a gene for Waardenburg syndrome type 2, converts fibroblasts to cells with melanocyte characteristics, *Nat. Genet.*, 14, 50, 1996.

75. Edery, P. et al., Mutation of the endothelin-3 gene in the Waardenburg-Hirschsprung disease (Shah-Waardenburg syndrome), *Nat. Genet.*, 12, 442, 1996.

76. Hofstra, R. et al., A homozygous mutation in the endothelin-3 gene associated with a combined Waardenburg type 2 and Hirschsprung phenotype (Shah-Waardenburg syndrome), *Nat. Genet.*, 12, 445, 1996.

77. Baynash, A. et al., Interaction of endothelin-3 with endothelin-B receptor is essential for development of epidermal melanocytes and enteric neurons, *Cell*, 79, 1277, 1994.

78. Puffenberger, E. et al., A missense mutation of the endothelin-B receptor gene in multigenic Hirschsprung's disease, *Cell*, 79, 1257, 1994.

79. Pavan, W. and Tilghman, S., Piebald lethal ($s^l$) acts early to disrupt the development of neural crest-derived melanocytes, *Proc. Natl. Acad. Sci. U.S.A.*, 91, 7159, 1994.

80. Tassabehji, M. et al., The mutational spectrum of Waardenburg syndrome, *Hum. Mol. Genet.*, 4, 2131, 1995.

81. Asher, J. et al., Effects of *Pax3* modifier genes on craniofacial morphology, pigmentation and viability: a murine model of Waardenburg syndrome variation, *Genomics*, 34, 285, 1996.

82. Deol, M., The anatomy and development of the mutants pirouette, shaker-1 and waltzer in the mouse, *Pro. R. Soc. London, Ser. B*, 145, 206, 1956.

83. Steel, K.P. and Bock, G., The nature of inherited deafness in *deafness* mice, *Nature*, 288, 159, 1980.

84. Avraham, K. et al., The mouse *Snell's waltzer* deafness gene encodes an unconventional myosin required for structural integrity of inner ear hair cells, *Nat. Genet.*, 11, 369, 1995.

85. Weil, D. et al., Defective myosin VIIA gene responsible for Usher syndrome type 1B, *Nature*, 374, 60, 1995.

86. El-Amraoui, A. et al., Human Usher 1B/mouse *shaker-1*: the retinal phenotype discrepancy explained by the presence/absence of myosin VIIA in the photoreceptor cells, *Hum. Mol. Genet.*, 5, 1171, 1996.

87. Self, T. et al., *Shaker1* mutations reveal roles for myosin VIIA in both development and function of cochlear hair cells, *Development*, 125, 557, 1998.

88. Mburu, P. et al., Mutation analysis of the mouse myosin VIIA deafness gene — a putative myosin motor-kinesin tail hybrid, *Genes Function*, 1, 191, 1997.

89. Corey, D. and Breakefield, X., Transcription factors in inner ear development, *Proc. Natl. Acad. Sci. U.S.A.*, 91, 433, 1994.

90. Forrest, D. et al., Thyroid hormone receptor β is essential for development of auditory function, *Nat. Genet.*, 13, 354, 1996.

91. Erkman, L. et al., Role of transcription factors Brn-3.1 and Brn-3.2 in auditory and visual system development, *Nature*, 381, 603, 1996.

92. Mohr, J. and Mageroy, K., Sex-linked deafness of a possibly new type, *Acta Genet. Stat. Med.*, 10, 54, 1960.

93. Tranebjaerg, L. et al., A new X-linked recessive deafness syndrome with blindness, dystonia, fractures and mental deficiency is linked to Xq22, *J. Med. Genet.*, 32, 257, 1995.

94. Kikuchu, T. et al., Gap junctions in the rat cochlea: immunohistochemical and ultrastructural analysis, *Anat. Embryol.*, 191, p.101, 1995.

95. Guilford, P. et al., A human gene responsible for neurosensory, non-syndromic recessive deafness is a candidate homologue of the mouse *sh-1* gene, *Hum. Mol. Genet.*, 3, 989, 1994.

96. Tamagawa, Y. et al., A gene for a dominant form of non-syndromic sensorineural deafness (DFNA11) maps within the region containing the DFNB2 recessive deafness gene, *Hum. Mol. Genet.*, 5, 849, 1996.

97. Avraham, K. et al., Characterization of unconventional *MYO6*, the human homologue of the gene responsible for deafness in *Snell's waltzer* mice, *Hum. Mol. Genet.*, 6, 1225, 1997.

98. Schucknecht, H. and Gacek, M., Cochlear pathology in presbycusis, *Ann. Otol. Rhinol. Laryngol.*, 102, 1, 1993.

99. Saitoh, Y. et al., Age-related hearing impairment in senescence-accelerated mouse (SAM), *Hearing Res.*, 75, 27, 1994.

100. Saitoh, Y. et al., Age-related cochlear degeneration in senescence-accelerated mouse, *Neurobiol. Aging*, 16, 129, 1995.

101. Hultcrantz, M. and Li, H., Inner ear morphology in CBA/Ca and C57BL/6J mice in relationship to noise, age and phenotype, *Eur. Arch. Otorhinolaryngol.*, 250, 257, 1993.

102. Johnson, K., Cook, S., and Erway, L., A C57BL/6J congenic strain of mice without presbycusis, *Hereditary Deafness Newsl.*, 13, 22, 1997.

103. Willott, J. et al., Genetics of age-related hearing loss in mice. II. Strain differences and effects of caloric restriction on cochlear pathology and evoked response thresholds, *Hearing Res.*, 88, 143, 1995.

# 19 Peripheral Mechanisms of Somatovisceral Sensation

*Armen N. Akopian*

## CONTENTS

The major function of primary sensory neurons is to transmit information to the central nervous system (CNS) about the nature, intensity, duration, and location of sensory stimuli. The aims of this chapter are

- To present current opinions of molecular genetic aspects of the peripheral mechanisms of hyperalgesia and chronic pain
- To discuss the molecular mechanisms of the activation and sensitization of the primary sensory neurons of the somatovisceral sensory system (SVSS), with particular emphasis on nociceptors
- To discuss molecular genetic approaches to mechanoreceptor, thermoreceptor, and nociceptor development and function

## GENERAL INTRODUCTION

It has become apparent that functional differences among the sensory neurons, at least within the SVSS, can authentically be described if they are considered and studied as afferent units, which consist of receptive terminals innervating peripheral tissues, peripheral nerve fibers, the associated neuronal cell bodies, and interspinal terminals (or terminals to brainstem) (Table 19.1).

The receptive terminals of afferent units either bear a free nerve end or are associated with specialized nonneuronal capsules/cells surrounding the axon terminals. These specialized end capsules/cells can be found in muscle and in a great number and variety in skin (Table 19.1).

Five subsets of different classes of fibers, which are distinguished by conduction velocities (CVs), have been established (Table 19.1).[1,2] Lloyd and Chang proposed a different classification nomenclature to label muscle afferent fibers as I, II, III, and IV.[3] Electrophysiological parameters

**TABLE 19.1**
**Properties of Afferent Units**

| Fiber Type | Velocity (m/s) | AP[a] Duration (ms) | Laminae | API (%)[b] | Soma[c] | Receptive Terminal |
|---|---|---|---|---|---|---|
| Aα | 30–55 | 0.5–1.35 | III IV V VI VII IX | ? | L | Meissner corpuscle Ruffini corpuscle Hair follicle Merkel cells Pacinian corpuscle Annulospiral Golgi organ Free nerve end |
| Aβ | 12–30 | 0.6–2.9 | II$_i$ III,IV V,VI VII IX | 22–53 | L | Ruffini corpuscle Hair follicle Flower spray Pacinian corpuscle Free nerve end |
| Aδ | 2–12 | 0.5–1.7 | I II$_i$ III,IV V,X | 61–70 | L | Paciniform corpuscle Hair follicle Tunicaplanar complex Free nerve end |
| Aδ/C | 1.3–2 | ? | I,II | ? | L/SD | Free nerve end |
| C | <1.3 | 0.6–7.4 | I,II (III,IV,V,X)[d] | 100 | SD | Free nerve end |

[a] Action potential.
[b] Percentage of cells showing an inflection on the falling phase of APs.
[c] L, large light neurons; SD, small dark neurons.
[d] For visceral C-fibers and C-cold thermoreceptors.

for peripheral fibers and related neuronal cell bodies also differ in membrane potential, action potential (AP) amplitude and duration, after-potential height and duration, and input resistance (Table 19.1).

The morphology of neuronal cell bodies in cranial and dorsal root ganglia has been extensively studied.[4,5] Two types of neuronal cell bodies — large light (L) and small dark (SD) neurons — have been identified on the basis of the distribution of their organelles. Intracellular recording and dye-injection techniques were combined with immunocytochemistry to establish a correlation between the morphology of neuronal cell bodies (L and SD) and the CV of fiber groups (Aα, Aβ, Aδ, Aδ/C, and C)[6] (Table 19.1). Anatomical and electrophysiological studies of the primary sensory neurons of the SVSS have revealed that their neuronal cell bodies are located in trigeminal, facial (geniculate), inferior glossopharyngeal (petrosal), inferior (nodose) and superior (jugular) vagus, and dorsal root ganglia.

There are four somatovisceral sensory modalities: mechanosensitivity, thermal sensitivity, proprioception, and pain. Each modality is associated with a functionally separate class of primary sensory neurons, which can be classified into three basic types — (1) mechanoreceptors, including proprioceptors, (2) thermoreceptors, and (3) nociceptors — according to the capability to respond more and less selectively to mechanical, thermal, and noxious (the Latin "noxa" means "damage") stimuli, respectively. Within each individual modality, it is possible to distinguish submodalities with regard to the type of activating stimuli (Table 19.2). Furthermore, some sensory neurons are

**TABLE 19.2**
**Functional Heterogeneity of Primary Sensory Neurons**

| Receptor Type | Submodalities | Fiber | Receptive Terminal |
|---|---|---|---|
| Mechanoreceptor | Pressure(hairless) | Aα | Merkel cells |
| | Pressure(hairy type I) | Aα | Tactile disks |
| | Pressure(glabrous SA) | Aα/Aβ | Ruffini corpuscle |
| | Pressure(hairy type II) | Aβ | Ruffini corpuscle |
| | Touch (glabrous RA) | Aα | Meissner corpuscle |
| | Touch (G1-hair) | Aα | Hair follicle |
| | Touch (G2-hair) | Aα/Aβ | Hair follicle |
| | Touch (D-hair) | Aδ | Hair follicle |
| | Touch (field 1&2) | Aα/Aβ | FE[a] |
| | Vibration | Aα | Pacinian corpuscle |
| | C-mechanoreceptors | C | FE |
| Mechanoreceptor/ proprioreceptor | Joint (movement/position) | Aβ/Aδ | FE |
| | Muscle spindle static | Aβ | Flower spray |
| | Tendon organ | Aα | Golgi organ |
| | Joint (phasic) | Aα/Aβ | FE |
| | Muscle spindle phasic | Aα | Annulospiral |
| Visceral mechanoreceptor | Baroreceptors | Aδ | FE |
| | Tension/stretch (colon) | Aδ/C | FE |
| | Pressure (uterine) | C | FE |
| Thermoreceptor | Warm | C | FE |
| | Cold | Aδ/C | FE |
| Nociceptor | Mechanical (HTMN) | Aβ/Aδ | FE |
| | Mechanical (joint) | Aδ | FE |
| | Mechanical (muscle) | Aδ | FE |
| | Mechanical (muscle) | C | FE |
| | C-mechanical (skin) | C | FE |
| | Cold | C | FE |
| | Heat | Aδ/C | FE |
| | Heat-mechanical | Aδ | FE |
| | Heat-mechanical | C | FE |
| | Cold-mechanical | C | FE |
| | C-polymodal | C | FE |
| | C-polymodal (visceral) | C | FE |
| | Chemical (visceral) | C | FE |

[a] Free nerve end.

polymodal and can respond to two or more stimuli which are distinct. A functional correlation has also been established between sensory modality and lamina-specific termination in the spinal cord[7-13] (Table 19.1).

The common features of functionally distinctive sensory neurons can be summarized as follows:

1. Virtually all cutaneous mechanoreceptors have specialized receptive terminals of various sorts. Some vibration mechanoreceptors in joint and visceral organs contain Pacinian corpuscles as receptive terminals.
2. All thermoreceptors and nociceptors bear free nerve ends.

3. The C-fiber population is connected to SD neurons, whereas all types (sizes) of L neurons have A-fibers.

4. All C axons are unmyelinated, whereas all A axons are myelinated.

5. Sixty to eighty-five percent (depending on species and innervated tissues) of C-fiber units in dorsal root ganglia (DRG) are polymodal nociceptors, including cutaneous and visceral C-polymodal, C-heat-mechanical, and C-cold-mechanical nociceptors.

6. Spikes of AP in sensory neurons with C-fibers, regardless of sensory neuron type, are broad, and a subset of C-fibers has an inflection on the falling phase of APs.

7. Aβ- and Aδ-fibers are more variable in the duration of APs. This variability likely is related to sensory neuron type. Thus, A-fiber mechanoreceptors have narrow spikes and A-fiber nociceptors have broad spikes with inflection.

8. Mechanical nociceptors excited by high-intensity stimuli are predominantly Aδ-fiber units, whereas mechanical nociceptors activated by tissue-damaging stimuli are C-fiber units.

9. Nociceptors and thermoreceptors mainly terminate in laminae I and II of the spinal cord, whereas mechanoreceptors extend to laminae III and IV.

## MOLECULAR MARKERS OF PRIMARY SENSORY NEURONS

Cell-type-specific molecular markers are broadly applied to visualize the development and plasticity of particular defined types of cells among heterogeneous cell populations. Moreover, specific transcripts/proteins may play an important functional role in the subset of cells where they are selectively expressed. The use of molecular markers for different functional classes of primary sensory neurons is especially necessary, because different subsets of sensory neurons are mixed, and the size and morphology of sensory neurons cannot be used as an indicator of their function. Markers detecting exclusively particular functional groups of sensory neurons are not yet available. In contrast, many available markers for sensory neurons define a subset of sensory neurons that consist of: (a) both L and SD neurons, (b) either only part of L neurons or only part of SD neurons, or (c) either all L neurons or all SD neurons. In many cases, the precise functional meaning of these subsets is unknown.

Lawson and colleagues used RT97, a monoclonal antibody against the phosphorylated form of the 200-kDa neurofilament, to define the L neuronal cell bodies in the trigeminal, vagal, and dorsal root ganglia.[5] It has been revealed that all A-fiber somata were RT97 positive and all C-fiber somata were RT97 negative.[6]

Peripherin (type III neuron-specific intermediate filament), which is broadly expressed in peripheral nervous system (PNS), including sensory neurons of trageminal and dorsal root ganglia, enteric nervous system, and sympathetic ganglia, stains predominantly SD neurons.[14,15] In rat L2 DRG a minor population, 8.1%, of mainly intermediate-size neurons were double immunostained by peripherin and RT97.[15]

The calcitonin gene-related peptide (CGRP) is encoded by two genes: αCGRP and the homologous βCGRP. Small- and medium-sized neurons expressed both types of mRNA, whereas large neurons express a predominance of αCGRP mRNA.[16] Forty percent of the C-, 33% of the Aδ-, and 17% of the Aβ-fiber neurons have CGRP-like immunoreactivity (CGRP-LI).[17] What is the possible correlation of CGRP-positive neurons with their functions? It may be premature to draw conclusions when direct functional evidence is not available. However, the great bulk of the Aδ and Aβ neurons with CGRP-LI, 28 to 53% of all Aα/Aβ- and 70% of all Aδ-fiber neurons, may be high-threshold mechanical nociceptors (HTMRs), although some of the Aδ and Aβ neurons with CGRP-LI may represent muscle and visceral afferent units and cutaneous mechanoreceptors.[17] The correlation between CGRP-positive C-fiber neurons and their function is not clear.

Many other neuropeptides — substance P (SP), neurokinin A, neuropeptide K, somatostatin (SOM), vasoactive intestinal polypeptide (VIP), bombesin, galanin, among others — are expressed in sensory neurons of rat, pig, and cat DRG. These neuropeptides, apart from SP and SOM, cannot be exploited as molecular markers to visualize different groups of sensory neurons, because (a) their expression depends on the spinal level of DRG where transcripts/proteins have been analyzed; (b) the pattern of their expression in DRG sensory neurons varies from species to species; and (c) subsets of sensory neurons expressing these neuropeptides are not well characterized by double labeling and electrophysiological techniques. Readers interested in the vast literature on the quantification of sensory neurons expressing neuropeptides, the percentage of sensory neurons labeled by two different neuropeptides, and peripheral projections of neuropeptide-labeled subsets of sensory neurons are referred to an excellent review of this subject.[18]

It was expected that antibodies directed against different sequences of glycolipids and glycoproteins would label multiple overlapping subsets of sensory neurons.[19] Thus, LA4-like immunoreactivity (LI) neurons constitute 30% of trigeminal and dorsal root ganglion cells, have unmyelinated fibers, and do not exhibit RT97 and CGRP immunoreactivity.[20] However, LA4 stains predominantly the inner layer of laminae II (IIi).[21]

Lectin *Griffonia simplicifolia* antibody, named IB4, behaves similarly to LA4. IB4 expression has been negatively correlated with CGRP expression.[22]

The trk family of protooncogenes encodes receptors that mediate the biological effects of neurotrophins. The trkA receptor binds nerve growth factor (NGF) with high affinity and exhibits tyrosine kinase activity after NGF binding; trkB binds and is activated by brain-derived neurotrophic factor (BDNF), and also by neurotrophin-3 (NT-3) and neurotrophin-4 (NT-4). trkC is a high-affinity tyrosine kinase receptor for neurotrophin-3 (NT-3).[23] trk receptors can be considered molecular markers of defined distinct populations of sensory neurons in cranial and dorsal root ganglia.[24,25]

trkA is expressed in virtually all CGRP-positive neurons (92%), and trkA, as well as CGRP, stain laminae I and IIo in the dorsal horn; this suggests that the trkA-positive population does not overlap with the LA4/IB4 subset.[20,26] It may be speculated that trkA-positive neurons represent, at least, a subset of nociceptors, because NGF modulates and sensitizes nociceptors, and also supports the development of nociceptors but not mechanoreceptors.[27-29]

Half of SD neurons in mice and rats subsequently down-regulate trkA during the first 3 weeks after birth.[30] The down-regulation of trkA occurs selectively in a subset that is labeled by IB4 and the receptor tyrosine kinase Ret.[31] NGF dependence of this group of putative nociceptors switches to GDNF dependence in early postnatal life.[31] Functional meaning of ret-positive nociceptors is still unknown.

trkC presents in 16 to 25% of all L neurons, which are largest.[24,25] NT-3 supports DRG neurons retrogradely labeled from muscle.[32] Mutant trkC (–/–) mice display behavioral phenotypes consistent with proprioceptive abnormalities.[33,34] On the basis of the aforesaid information, it may be concluded that, at day of birth, trkC is expressed predominantly by muscle proprioceptors. Unlike trkA and trkC, the function of sensory neurons expressing trkB remains unknown.

Carbonic anhydrase (CA) stains 20 to 38% of DRG sensory neurons, which are nearly all RT97 positive and comprise a subpopulation (40 to 70%) of L neurons, although some of the SD neurons in the rat DRG (10%) are CA positive.[35] CA-positive neuronal somata are connected to groups I and II proprioceptive afferents.[36] However, CA does not exclusively mark proprioceptors because some neurons, with afferents that terminate in skin, also express CA.[37] Further, a great proportion of Aα/Aβ neurons contain CA, but some Aδ neurons bear CA as well.[35]

Parvalbumin, a calcium-binding protein, can be considered as another marker for a subpopulation (28%) of L neurons. Parvalbumin coexists extensively with CA.[38] Parvalbumin may be considered as a marker for sensory neurons that innervate muscle spindles.[39] L neurons expressing parvalbumin are not labeled by Ret antibodies.[31]

In summary, virtually all studies devoted to determining modalities and/or submodalities for subsets of sensory neurons expressing particular markers were carried out indirectly; that is, researchers strove to determine the electrophysiological and morphological properties of sensory neurons expressing a marker of interest, but not their modalities. This indirect correlation is problematic. Therefore, in spite of vast efforts, a number of questions are still unresolved.

## ACTIVATION OF PRIMARY SENSORY NEURONS

How do a variety of subsets of sensory neurons code different modalities and submodalities? The differential activation of functionally distinct subsets of sensory neurons through ligand-gated ion channels and poorly characterized temperature or pressure receptors may be the key to understanding this problem. So far, the available information does not fit into any unified theory of the differential activation of sensory neurons. Thus, there is little correlation between the types of sensation (touch, pressure, stretch, cooling, heating, and so on) and release of particular ligands in the periphery (ATP, proton, 5-HT, bradykinin, and so on), or between the types of stimuli and the channel/receptor subtypes differentially activated on subsets of sensory neurons (GluR5, $P2X_3$, $P2Y_1$, ASIC, $5-HT_3$, and so on). For example, it is not clear if chemical mediators transduce mechanical and thermal stimuli.

Peripheral tissues can be exposed to three types of stimuli: (1) mechanical stimuli, which can activate mechanoreceptors, proprioreceptors, or nociceptors; (2) thermal stimuli, which can activate thermoreceptors or nociceptors; and (3) chemical stimuli, which activate only nociceptors. Stimulated peripheral tissue may either release ligands, which activate the sensory neurons, or directly react with channels/receptors with no release of ligands. The action of different extracellular ligands on subsets of sensory neurons is beginning to be understood, and receptors mediating these actions are well characterized. The mechanisms of the activation of sensory neurons by extracellular ligands, the possible stimuli giving rise to these extracellular ligands, and the functional consequences of neuronal activation will all be discussed below.

ATP and, to a lesser extent, adenosine have been claimed to cause a sensation of pain on human blister bases.[40] The actions of ATP on cells are mediated by purinoreceptors classified into P2Y purinoreceptors, G-protein-coupled seven-transmembrane receptors, and P2X purinoreceptors, ATP-gated ion channels.[41] At present, seven P2X and eight P2Y receptors have been isolated and expressed either in *Xenopus* oocytes or in transfected cell lines (G. Burnstock, personal communication).[42] What channels account for the properties of ATP-induced current described in sensory neurons? Electophysiological recordings from cultured sensory neurons of nodose, trigeminal, and dorsal root ganglia have shown that almost all neurons respond to extracellular ATP. Both slow and fast desensitizing ATP-gated currents have been recorded in sensory neurons.[43,44] Six P2X channels are expressed in sensory neurons of the DRG, nodose, and trigeminal ganglia.[45,46] However, only one subtype, $P2X_3$, is exclusively expressed in a subset of SD: peripherin-positive sensory neurons.[45] Nociceptors (tooth-pulp afferents) have currents that are similar to those of heterologously expressed channels containing $P2X_3$ (fast desensitizing ATP current). Muscle-stretch receptors have slow desensitizing current and have no $P2X_3$ immunoreactivity.[47] However, $P2X_3$ desensitizes faster than some nodose ganglion sensory neurons in response to α-β methylene ATP.[43] The kinetics and pharmacology of ATP-gated currents in nociceptors suggest that the channels are comprised of either homomeric $P2X_3$ receptors or heteromeric combinations of $P2X_2$ and $P2X_3$ receptors.[48,49] To prove this observation, specific antibodies to $P2X_3$ and $P2X_2$ have been used.[50]

Burnstock and Wood have recently proposed an elegant hypothesis for nociception mediated by ATP.[51] ATP can be released from many types of cells damaged either by different noxious stimuli or in pathological conditions, and it activates nociceptors through $P2X_3$ (or the $P2X_3/P2X_2$ complex) acting on receptive terminals. Thus, in pathological conditions, such as causalgia, reflex sympathetic dystrophy, and "sympathetically maintained pain," ATP may be released, as a cotransmitter with noradrenaline and neuropeptide Y, from sympathetic nerves.[52] A surgical sympathectomy and guanethidine, which prevents a release of the sympathetic cotransmitters, are equally effective in blocking pain in these conditions and are more effective than adrenoreceptor antagonists and reserpine, which depletes noradrenaline, but not ATP.[53] Vascular pain, including angina, migraine, ischemic muscle pain, and lumbar pain, occurs during the reactive hyperemic phase, which is accompanied by the release of large amounts of ATP from damaged endothelial cells. Thus, it was proposed that ATP could be a pain mediator in migraine.[54] Tumor cells contain very high levels of ATP. The release of ATP from broken tumor cells, which can happen during abrasive movements of cells in a mature tumor, may cause a sensation of pain. ATP released from damaged muscle and endothelial cells, which contain ATP at high concentrations, following surgery or intense trauma may also be involved in local pain. Recently, to test whether ATP contributes to nociception, a tissue culture system, which allows comparison of nociceptive (tooth-pulp afferent) and nonnociceptive (muscle stretch sensory neuron) rat sensory neurons, was developed.[47] Obtained results support the theory that $P2X_3$ receptors mediate nociception.[47]

Bradykinin (BK), an oligopeptide generated by proteolysis of a kininogen precursor, is a potent activator of sensory neurons and is able to evoke a sensation of pain.[55] The subset of sensory neurons that respond to BK is predominantly unmyelinated C-fibers.[56] BK can potentially acivate sensory neurons through G-protein-coupled B1 and B2 receptors.[57,58] Studies using selective B1 and B2 receptor antagonists and agonists suggest that the B2 receptor, but not the B1 receptor, is involved in activation of nociceptors.[55] The generation of B2 (–/–) mutant mice provided the opportunity to determine the contribution of B2 receptors in nociception.[56,59] B2 receptor (–/–) mice show no acute nociceptive response to BK, but do respond to other stimuli, such as formalin and noxious heat.[59]

Serotonin (5-HT) can modulate nociceptive pathways via central and peripheral mechanisms. The activation of DRG neurons by 5-HT was investigated by intracellular recording from cells classified by conduction velocity.[60] 5-HT at relatively high concentrations (3 to 100 $\mu M$) depolarized 82% of the A-fiber neurons and 41% of the C-fiber neurons. The depolarizing responses were mediated through either $5\text{-HT}_2$ or $5\text{-HT}_3$ receptors, which were observed in both A- and C-fiber neurons.[60,61] $5\text{-HT}_{1C}$ may also be involved in activation of sensory neurons. These results suggest that 5-HT may activate not only nociceptors but also thermoreceptors and mechanoreceptors, although C-fiber neurons respond particularly strongly to 5-HT through $5\text{-HT}_3$ receptors.[61]

Capsaicin is a hot chili pepper-derived compound which is able to induce a strong sensation of pain. Capsaicin selectively activates a subset of nodose, trigeminal, and dorsal root ganglion neurons (C-polymodal, C-mechanoheat, visceral C-polymodal, C-chemical, and A$\delta$-mechano-heat nociceptors, and also C-warm thermoreceptors) and has no effect on other neurons except at high concentrations.[62,63] Sensory neurons of nonmammals either are weakly capsaicin sensitive or are not sensitive at all.[64] Capsaicin activates sensory neurons through a putative capsaicin-gated ion channel.[65] Recently, an expression cloning strategy based on calcium influx has been used to isolate cDNA encoding a capsaicin receptor.[66] The receptor (vanilloid receptor 1, VR1) is a very slow nonselective cation channel that is structurally related to the six-transmembrane TRP family of ion channels. VR1 is competitevly antagonized by capsazepine, whereas resiniferatoxin (RTX) is a 20-fold more potent agonist for VR1 than capsaicin.[65,66] VR1 expression is exclusively restricted to a subset of sensory neurons of trigeminal and dorsal root ganglia.[66] How many capsaisin receptor subtypes are present in sensory neurons? Liu and Simon have demonstrated an amazing variability of capsaicin-induced responses in patch-clamped sensory neurons of dorsal root and trigeminal ganglia.[63,67] In situ hybridization of VR1 with trigeminal and dorsal root ganglia clearly shows that expression of the receptor predominates in 15 to 20% SD neurons,[66] whereas capsaicin-induced currents have been observed in 40 to 60% of sensory neurons of small to medium size.[64] Further, VR1 has not been detected in nodose ganglia the sensory neurons of which respond to capsaicin.[64,66] These results taken together support the viewpoint that many subtypes of capsaicin receptors are expressed by sensory neurons. Where is the capsaicin-binding site? One view holds that the capsaicin-binding site is located on the intracellular part of receptors,[69] whereas Vlachova and Vyklicky have suggested that the binding site is not intracellular.[68] It has been found that capsaicin is able to produce identical responses when added to either side of a patch excised from a cell expressing VR1.[66] What are the endogenous activators of capsaicin receptors? Bevan and Geppetti have proposed a hypothesis that protons might act as endogenous activators of capsaicin-gated ion channels.[70] Whereas, Oh and colleagues were not able to activate capsaicin receptors in outside-out patches nor activate capsaicin-sensitive neurons with acid solution (pH 5.9 to 6.0).[69] It was also found that protons potentiate sevenfold the capsaicin responsiveness in trigeminal and dorsal root ganglion neurons.[72] The effects of hydrogen ions on the cloned capsaicin receptor were examined using the oocyte and HEU293 cell expression system.[66,71] Results suggest that protons decrease the temperature threshold for VR1 activation.[71] Moreover, protons activate VR1 at around pH 6.4, at normal physiological temperature.[71] Up to now, the precise mechanisms of activation of capsaicin channels have been the subject of much debate. VR1 is structurally related to the TRP family of ion channels; therefore, capsaicin receptors could be gated either by diffusible small molecules released from intracellular calcium stores or by direct allosteric interactions with store-associated proteins.

Protons have also been implicated as activators of nociceptive neurons. Tissue acidosis, which can result from either an accumulation of inflammatory cells or a reduced oxygenated blood supply during ischemia, produces sustained graded and spatially restricted pain.[73-75] $A\beta$- and $A\delta$-fiber LTMRs (low-threshold mechanical nociceptors) are not excited by acid pH, whereas $A\delta$-fiber nociceptors are activated with poor response characteristics.[75] However, C-mechanoheat and C-polymodal nociceptors show stimulus-related responses increasing with proton concentration; the threshold level ranges from pH 6.9 to 6.1 with maximum discharge at pH 5.2.[75] The action of acid solution on sensory neurons is mediated through proton-gated receptors. Two types of the proton-induced currents can be distinguished in sensory neurons: a rapidly inactivating proton-gated ion current[76] and a slowly inactivating current.[77] The slow proton-induced current was detected in the majority of capsaicin-sensitive neurons.[77] Similarly, the fast proton-gated current was recorded predominantly in small neurons (74%) with diameters less than 26 $\mu$m.[76] Fast proton-gated currents are activated at pH 6.9 and show $Na^+$ selectivity,[76] while slow proton-gated currents are activated at pH 6.2 and are nonselective to monovalent cation and $Ca^{2+}$.[77] Recently, a proton-gated ion channel, ASIC (BNaCI2), that shares more than 67% identity with mammalian degenerin MDEG (BNaCI1), has been cloned from a rat brain cDNA library.[78-80] ASIC expressed in *Xenopus* oocytes shows biophysical and pharmacological properties of the fast proton-induced ion current described in sensory neurons.[76,79] ASIC mRNA is expressed in the brain and in 20 to 25% of sensory neurons, which represents a subset of adult rat DRG SD neurons.[81] However, fast proton-gated channels were recorded in 50 to 70% of sensory neurons.[76] Furthermore, proton-gated channels recorded from different sensory neurons revealed a variety of kinetic properties, although they could be considered fast channels.[76] This implies that many subtypes of proton-gated channels are expressed by sensory neurons. Thus, Waldmann and colleagues have recently cloned a proton-gated channel subunit, named DRASIC, that is permeable only to $Na^+$ ions, and its kinetic closely resembles the properties of the slow inactivating proton current described in sensory neurons.[77,82] DRASIC is predominantly expressed in DRG sensory neurons.[82] However, a DRASIC transcript can be found in sympathetic cervical ganglia (SCG), spinal cord, and brain stem.[81] Clearly, DRASIC and ASIC cannot explain all proton-gated currents in sensory neurons. For example, the slow proton-gated currents in sensory neurons are permeable $Ca^{2+}$. Therefore, neither ASIC nor DRASIC nor their heteromultimerization can account for the native slow proton-gated current.[77,79,82] Numerous sub-units of proton-gated channels would thus be expected to be expressed in sensory neurons, and heteromultimerization of these channels is probably used for creating a diversity of proton-gated currents described in sensory neurons.[83]

The receptors and ion channels mediating the action of the pain-inducing chemical ligands, such as capsaicin, BK, ATP, protons, and 5-HT, are likely to play an important functional role in nociception. By contrast, the functional significance of the channels gated by excitatory amino acids or acetylcholine (ACh), which have been discovered in DRG sensory neurons, remains uncertain. Moreover, the peripheral sources of glutamate and ACh are also uncertain. ACh or nicotine activated inward currents in 51% of rat sensory neurons, whereas approximately 90% of the sensory neurons isolated from E18 chicken DRG were sensitive to ACh.[84,85] Using *in situ* and Northern hybridizations, researchers found that both α-3 and α-4 nAChR subtypes were expressed in sensory neurons.[86] A novel putative AChR, which binds α-bungarotoxin and reveals nicotinic pharmacology, has recently been identified in ciliary, sympathetic, and dorsal root ganglia, but not in brain, spinal cord, or retina.[87]

At least two types of amino acid-gated channels, non-NMDA ("kainate") and NMDA receptors, are present in sensory neurons.[88] Kainate receptors were revealed only in a subset of primary afferent C-fibers.[88] The AMPA/kainate-gated channel expressed in DRG is comparable to some non-NMDA channels described in the CNS. However, the DRG non-NMDA channel has several distinct properties.[89] What channels account for the properties of DRG non-NMDA current described in sensory neurons? Almost all known subunits of non-NMDA receptors (GluR1-7, KA-1, and KA-2) are expressed in sensory neurons. However, KA-2, GluR6, and especially GluR5 are the most abundant subtypes.[90,91] A glutamate receptor subunit, GluR5, displaying 40 to 41% amino acid identity with the kainate/AMPA receptor subunits GluR1–4, forms homomeric ion channels in a mammalian cell line or *Xenopus* oocytes and displays the pharmacological profile of a receptor described in a subset of rat DRG neurons.[90] Heterologous expression of KA-2 did not yield a functional ligand-gated ion channel, but KA-2 coexpressed with GluR5 or GluR6 produced heteromeric channels, which are activated and desensitized by kainate.[92] Therefore, KA-2 is also a strong candidate as a subunit in DRG heteromeric non-NMDA ("kainate") receptors. Because the pharmacological profile and biophysical properties of GluR5/KA-2 and GluR6/KA-2 have not been investigated fully, it is not clear which heteromeric receptors account for DRG-expressed "kainate" receptors.

There is much evidence for an important role for non-NMDA as well as NMDA receptors in DRG-evoked synaptic potentials and amino acid-evoked responses in dorsal horn neurons.[93] Are there NMDA channels activated on primary sensory neurons? Lovinger and Weight demonstrated that glutamate-induced depolarization of DRG neurons (30%) is mediated predominantly by NMDA receptors.[94] It was shown that the C-fiber neurons do not express NMDA receptors;[88] possibly, NMDA currents were recorded mainly in large neurons, although the authors did not determine the cell sizes from which records were made. NMDAR-1, but not NMDAR-2A, 2B, 2C, and 2D, mRNA is present in sensory neurons of mouse DRG from embryonic stages through P 21.[95] These findings suggest that the NMDA receptor channel in sensory neurons is homomeric, although Petralia and colleagues showed that sensory neurons can express NMDAR-2A and 2B as well.[96] Whereas the role of NMDA receptors in central sensitization and transmission of slow ventral root potentials is beyond question, the activation, at least, of nociceptors through NMDA receptors is not readily apparent.

Three different mechanisms may underlie the activation of sensory neurons by light mechanical stimuli. First, peripheral tissues are able to release chemical ligands with light mechanical stimulation.[97] These ligands, acting through specific receptors, can directly excite sensory neurons. This mechanism may constitute the basis for a mechanical activation of nociceptors. Indeed, strong mechanical stimuli may damage cells of peripheral tissues that release a range of mediators (5-HT, ATP, BK) which are known activators of nociceptors. However, the nature of ligands released by light mechanical stimulation is difficult to imagine, although ATP may be released from intact Merkel cells, which contain exceptionally high levels of ATP, at light mechanical stimulation. Second, sensory neurons may be excited by light mechanical stimuli through receptors responding to direct touch, stretch, or tension. Such a type of channel has been cloned from *Escherichia coli*,[98] and reconstructed by expression of the α-subunit of the epithelial sodium channel cloned from osteoblasts.[99] At present, no compound has been found that displays high specificity and affinity for mechano-gated (MG) channels on sensory neurons. Nevertheless, amiloride, gentamicin, and gadolinium clearly block MG channels.[100] Third, receptive terminals of mechanosensitive sensory neurons are attached to either the extracellular matrix or intracellular matrix of specialized cells such as hair follicles, Meisner and Pacinian corpuscles, Merkel cells, and so on. Light mechanical stimulus may change the configuration of the matrix, which may in turn open or activate receptors on the nerve ends of sensory neurons. This means that these putative MG receptors have to be connected somehow with the matrix. Thus, the unc-105 gene product represents a putative channel, which is a homolog of degenerins, interacting with type IV collagen in the extracellular matrix underlying the muscle cell.[101] This interaction may serve as a mechanism for stretch-activated muscle contraction.

During the last few years, strategies for functional cloning of the MG ion channels have been proposed.[100] Nakamura and Strittmatter, using the strategy of functional cloning, isolated a single cRNA derived from sensory neurons that renders *Xenopus* oocytes mechanosensitive.[102] Surprisingly, this cRNA encodes a $P2Y_1$ purinoreceptor that had originally been isolated from brain. $P2Y_1$ mRNA is concentrated in L DRG neurons, unlike $P2X_3$ mRNA, which is localized to SD DRG neurons and produces 100-fold less mechanosensitivity in oocytes. The authors have shown that, in this experiment, ATP was released from oocytes by mechanical stimulation. Analogous mechanisms may work *in vivo;* ATP could be released from tissues adjacent to peripheral nerves by light or strong mechanical stimulation. The discovery by Nakamura and Strittmatter has important implications for understanding the molecular basis of the differential activation of subsets of sensory neurons.

What are the cellular mechanisms involved in activation of mechanoreceptors? Mechanical stimulation reversibly increased intracellular calcium concentration ($[Ca^{2+}]_i$) by sixfold in only 62% of neurons.[103] The mechanically induced rise in $[Ca^{2+}]_i$ was essentially abolished by gadolinium, a blocker of stretch-activated ion channels. Recently, Svichar and colleagues have shown that the application of ATP in $Ca^{2+}$-free solution, like mechanical stimulus, triggers a rise of $[Ca^{2+}]_i$ in 93% of DRG L neurons, but in no SD neurons.[104] Adenosine did not elevate $[Ca^{2+}]_i$, and ATP action was blocked by suramin. This suggests an involvement of P2Y purinoreceptors in the rise of $[Ca^{2+}]_i$.

Heat-evoked outwardly rectifying nonselective cation channels have been revealed on 60 to 70% of sensory neurons that are predominantly small-size neurons.[105] The "burning" sensation of capsaicin-induced pain suggests that VR1 could represent heat-activated receptors. Indeed, VR1 is activated by thermal stimuli within the noxious temperature range (>45°C).[66] In cultured sensory neurons, some heat-activated currents are reported to be insensitive to antagonists of capsaicin receptors.[105] Therefore, responses to noxious thermal stimuli could be transduced through multiple molecular pathways. What are the mechanisms activating heat receptors? We partly discussed this problem above. It might be well to point out that the receptor may be activated by conformational changes of the receptor structure which are induced by either high or low temperature.

## THE MECHANISMS OF HYPERALGESIA AND CHRONIC PAIN: SENSITIZATION OF PRIMARY SENSORY NEURONS

Sensitization of sensory neurons, which is measured by electophysiological methods, is the reduction of an activation threshold for sensory neurons. Hyperalgesia, which is measured at the behavioral level, is a decrease in pain threshold on exposure of peripheral tissues to mechanical (mechanical hyperalgesia), heat, cold (heat or thermal hyperalgesia), and chemical noxious stimuli, and during inflammation (inflammatory hyperalgesia), and peripheral nerve injury as well. What are the mechanisms of peripheral nerve injury- and inflammation-induced hyperalgesia, and what is the relationship between hyperalgesia and sensitization of nociceptors?

Neurotrophic factors support survival of the sensory neurons during ontogeny and play an important role in functional maintenance of mature sensory neurons, for example, following axotomy and neurotoxic damage. In addition, they also regulate function of sensory neurons. Strong evidence suggests that NGF is linked to hyperalgesia.[106] A single systemic or local injection of NGF in neonatal and adult rats and mice leads to a rapid, profound, and long-lasting heat and mechanical hyperalgesia.[27,107] Analogously, a single intravenous or subcutaneous administration of recombinant human NGF in healthy human volunteers results in muscle pain, and the duration and severity of pain varied in a dose-dependent manner.[108] A synthetic protein, trkA-IgG, has been used to sequester endogenous NGF. Acute administration of the trkA-IgG blocks the inflammatory hyperalgesia developed by the administration of carrageenan.[109] It was noted that NGF levels are rapidly increased in all known models of inflammatory hyperalgesia produced by the intraplantar injection of Freund's adjuvant, carrageenan, cytokines, and so on. Therefore, there is a view that NGF is the linchpin that links inflammation to hyperalgesia.[27] In neonatal rats (0 to 14 days) treated with NGF, a mechanical hyperalgesia and sensitization of $A\delta$ nociceptive afferents to mechanical stimuli was revealed. Treatment of animals older than 2 weeks of age led to a similar behavioral hyperalgesia but not to a corresponding sensitization of $A\delta$ nociceptors. Therefore, the NGF-induced mechanical hyperalgesia may have different mechanisms in neonatal and adult rats. Moreover, the mechanical and heat hyperalgesia in neonatal animals (0 to 14 days) may also involve two different mechanisms, because heat hyperalgesia appears faster (15 min) than mechanical hyperalgesia (5 to 7 h) following NGF injection. Probably, a hyperalgesia following peripheral nerve injury is also directed by different mechanisms. It has been suggested that increased NGF levels, peripheral nerve injury, and inflammation may potentially act via peripheral and central mechanisms to produce hyperalgesia.[110] These mechanisms include: (1) sensitization of the peripheral afferent fibers, (2) alteration of peripheral modalities, (3) changes of primary afferent connectivity to the spinal cord, and (4) alteration of central sensitization states.

Sensitization of peripheral afferents can proceed in two different ways. An indirect way of sensory neuron sensitization is likely to work via inflammatory cells, whose activity can be changed by NGF or inflammation. Freund's adjuvant and carrageenan activate many types of inflammatory cells, including macrophages and neutrophils. NGF is a very potent degranulator of mast cells. Degranulated and/or activated inflammatory cells may release different inflammatory mediators,[27,107,110] which in turn modulate the function of nociceptors by alteration of their activation threshold. Thus, hyperalgesia and sensitization of the nociceptive afferent fibers can be induced by 5-HT,[111] oxygen radicals,[112] prostacyclin,[113] $PGE_2$,[113] BK,[114] some types of cytokines,[114] histamine,[115] and leukotrine B4 (LTB4).[116] What is the molecular basis of the sensitization of sensory neurons by these ligands?

5-HT may sensitize sensory neurons through a number of serotonin receptor subtypes. Since 5-HT can be a sensitizer, as well as an activator of sensory neurons, different receptors and pathways would presumably be involved in activation and sensitization by 5-HT. Selective agonists for the G-protein-coupled $5-HT_{1A}$ receptor produce dose-dependent hyperalgesia and sensitization of sensory neurons, whereas no sensitization of sensory neurons was seen after application of $5-HT_{1B}$ agonists, $5-HT_2/5-HT_{1C}$ agonists, or $5-HT_3$ agonists.[111,117] It was suggested that the $5-HT_{1A}$ receptor in primary afferent neurons is coupled to the cAMP second messenger system to produce sensitization.[117] Interestingly, within rat lumbar DRG, the presence of the $5-HT_{1B}$, $5-HT_{1D}$, $5-HT_{2A}$, $5-HT_{2C}$, 5-HT3 and $5-HT_7$ receptor subtype mRNAs was detected by PCR (polymerase chain reaction), whereas $5-HT_{1A}$ was detected in SCG, but not in DRG.[118] There is, however, a different view point. Rueff and Dray[119] studied the relative involvement of 5-HT receptors in sensitization of peripheral nociceptive fibers in preparations of the neonatal rat spinal cord with attached tail (Otsuka, in preparation); nociceptors in the rat tail were activated by chemical (capsaicin) and thermal (heat) stimuli. They claimed that $5-HT_2$ receptors play a role in sensitization,[119] whereas Taiwo and Levine did not detect 5-HT-mediated sensitization through $5-HT_2$ receptors.[111,117] To define specific roles for 5-HT receptor subtypes, Tecott et al. generated mutant mice lacking functional $5-HT_{2C}$ receptors.[120] However, no analysis of their pain behavior was carried out. An analysis of sensory neuron-specific $5\text{-HT}_{1A}$-, $5\text{-HT}_{1C}$-, $5\text{-HT}_{1D}$-, $5\text{-HT}_{2A-C}$-, and $5\text{-HT}_3$-deficient mice would be extremly useful to study the *in vivo* role of 5-HT in sensory neuron sensitization and activation.

BK sensitizes cutaneous C-polymodal and A$\delta$-heat-mechanical nociceptors in carrageenan inflamed tissue.[114,121] It also may potentially sensitize sensory neurons through B1 and B2 receptors. The selective antagonists for the B1 and B2 receptors have been used to study the role of these receptors in sensitization of nociceptors.[122] Data suggest that following inflammation, the B2 receptor is involved in the sensitization of nociceptors, but when inflammation is prolonged, the B1 receptor, which is not expressed in healthy tissues to a significant level, may play an important role in the sensitization. Studies using B2-receptor-deficient mice confirmed the role of the B2 receptor in hyperalgesia.[57] Unlike in wild-type controls, the injection of carrageenan did not induce thermal hyperalgesia in B2 receptor (–/–) mice. In contrast, chronic thermal hyperalgesia induced by Freund's complete adjuvant was indistinguishable in wild-type and B2 receptor (–/–) mice.

$PGE_2$ decreases the mechanical threshold for 94% of C-polymodal, 60% of A$\delta$/C-mechano-heat, 42% of C-mechano-cold, 60% of C-mechanical, and 70% of A$\delta$-high-threshold (HTMRs) nociceptors in the saphenous nerve of rats.[123] Responses evoked by BK, capsaicin, or thermal stimulation are enhanced by $PGE_1$, $PGE_2$, $\alpha PGF_2$, $PGI_2$ (prostacyclin), and cicaprost (analog of $PGI_2$), but not by $PGD_2$. What are the mechanisms of sensitization by prostaglandins? Kumazawa and colleagues suggested that different prostaglandin receptors for the E series (EP receptors) are involved in sensitization mediated by $PGE_2$ depending on whether nociceptors are activated by heating, capsaicin, or BK.[124] It was shown that the EP3 and EP2 receptor subtypes are implicated in the sensitization by $PGE_2$ of the BK- and heat-activated polymodal nociceptors, respectively. To determine the role of prostacyclin in inflammatory processes, researchers have disrupted prostacyclin receptors in mice using homologous recombination.[125] Inflammatory and pain responses in mice without the receptors are reduced to the levels observed in indomethacin-treated wild-type mice.[125] These results are in agreement with the theory that aspirin-like drugs, including indomethacin, suppress inflammatory swelling and pain via inhibition of prostanoid biosynthesis.

Leukotriene B4 (LTB4) is a product of the 5-lipoxygenase pathway of arachidonic acid metabolism. It was noted that LTB4 sensitizes nociceptors to chemical (glacial acetic acid, BK, capsaicin), mechanical, and heat stimuli, and the degrees of sensitization of cutaneous nociceptive fibers by LTB4 and $PGE_2$ were highly correlated.[126] They even sensitize the same sensory neurons. The mechanism of LTB4 action on sensitization of nociceptors is unknown. Recently, using a subtractive cloning strategy, researchers successfully cloned a G-protein-coupled receptor for LTB4.[127] This should contribute to understanding of the mechanism of LTB4 action on sensory neurons.

Histamine (His) sensitizes about 30% of C-polymodal visceral sensory neurons to heat stimuli in a dose-dependent manner.[115] To study the histamine receptor subtype involved in the sensitization, researchers have used antagonists for H1, H2, and H3 receptors.[128] The sensitization of the heat response induced by His is mediated through the H1 receptor.[128]

Cytokines, such as $\beta$IL-1, IL-2, and IL-8, as well as IL-6 and $\alpha$TNF, induce a mechanical hyperalgesia and sensitization of nociceptive neurons.[114] What are the mechanisms of hyperalgesia and sensitization by cytokines? There are two different viewpoints, both involving BK: either B2 receptors play a primary role in cytokine-induced mechanical hyperalgesia,[114] or BK can initiate a cascade of cytokine release that mediates hyperalgesia through cytokine receptors.[129]

Acidosis can contribute to sensitization of nociceptors in ischemic and in inflamed tissues, where pH levels can fall to as low as 5.4.[75] Repeated or prolonged treatment with low pH induced a significant and long-lasting decrease of mechanical and thermal thresholds in almost all tested C-fibers when they were activated by capsaicin.[75] The inflammatory mediators (BK, 5-HT, His, $PGE_2$) potentiate the algogenic effect of low pH rather than vice versa. Therefore, tissue acidosis may be one of the dominant factors in inflammatory pain. Mechanisms of sensitization by low pH are unknown. However, many different receptors may be potentiated by low pH. The study of this phenomenon may shed light on this problem.

Direct modification of membrane properties of afferent fibers by NGF acting through trkA, which may decrease the activation threshold of nociceptors, can be considered a direct way of sensitization.

The key feature of sensitized sensory neurons is a change in their membrane properties that is characterized by both spike shape and frequency of APs. How can the inflammatory mediators and NGF alter the membrane properties of sensory neurons?

## NGF

Primary culture of DRG sensory neurons and DRG neurons from NGF-treated animals were used to study the alteration of expression of different voltage-gated ion channels, as well as the modification of different currents by NGF. The threshold for spike generation was lower in untreated cells than in NGF-treated cells, whereas no significant differences in resting membrane potential and input resistance were noted.[130] NGF can up-regulate (40 to 60%) tetrodotoxin-resistant (TTXr) voltage-gated Na$^+$ current in adult rat DRG neurons.[130] Chronic treatment with NGF has been reported to increase the duration of the falling phase of somal APs in HTMRs, but not in LTMRs (mechanoreceptors),[131] although abnormal activity of C- and Aδ-fibers was not recorded 1 day following NGF injection.[107] Two molecular mechanisms may be involved in regulation of Na$^+$ currents by NGF. First, NGF may induce the transcription of different voltage-gated sodium channel (VGSC) genes in sensory neurons that finally alter membrane properties of nociceptors. Second, NGF may increase the density, magnify Na$^+$ current, or shift the current-voltage relationship in a hyperpolarizing direction for VGSCs by posttranslational and/or posttranscriptional modification of the channels. Two α-subunits (SNS and PN1), β1- and β2-subunits of VGSCs, are predominantly expressed in DRG neurons (A. N. Akopian, S. England, and J. N. Wood, unpublished data).[132,133] It seems likely that SNS, which exclusively is expressed in SD neurons, accounts for TTXr Na$^+$ current described in all C-fiber neurons and some A-fiber neurons.[133,134] TTX-sensitive current is probably represented mainly by PN1, which is expressed only in PNS (SCG and DRG).[132] NaCh6 and brain type I subunits of VGSC can be detected in adult rat DRG only by reverse transcription polymerise chain reaction amplification (RT-PCR) but not Northern hybridization (A. N. Akopian, S. England, and J. N. Wood, unpublished data). However, Black and colleagues, using *in situ* hybridization and RT-PCR, have detected eight subunits of VGSC in rat adult DRG.[135] Studies using cultured PC12 cells indicated that both expression and targeting of PN1 is induced by treatment of the cells with NGF. NGF regulation of PN1 expressed in DRG neurons has not been reported. The transcriptional regulation of SNS in cultured DRG sensory neurons treated with NGF was examined by using RNase protection. Total SNS mRNA levels are unaffected by this treatment.[136] It is pertinent to note that up-regulation of SNS mRNA in cultured DRG neurons by NGF was detected using RT-PCR.[137] Posttranscriptional or posttranslational forms of regulation by NGF, when the density of protein is increased at the same level of transcription, for both TTXr and TTXs VGSCs are unknown. However, such regulation was detected for voltage-gated potassium channels (VGPCs) in PC12 cells. Thus, NGF treatment of PC12 cells led to a fourfold elevation of VGPC (Kv2.1) density, but no rise in transcription levels was detected.[138] The mechanisms of this regulation are also unknown. Regulation of K$^+$ and Ca$^{2+}$ currents in DRG sensory neurons by NGF have not yet been studied.

## Inflammatory Mediators

The functional modulation of TTXr Na$^+$ current, but not TTXs Na$^+$ current, in sensory neurons by inflammatory mediators has recently been demonstrated.[139,140] PGE$_2$, 5-HT, and adenosine increase the amplitude as well as the rates of activation and inactivation of TTXr Na$^+$ current, and shift the peak of conductance-voltage relation in a hyperpolarizing direction in SD neurons in culture. It was reported that prostaglandins and prostacyclin sensitize the rat sensory neurons, in part, through the inhibition of an outward K$^+$ current that may modulate the firing threshold for generation of APs.[140,141] APs of some putative nociceptors, especially those of the visceral afferents, contain a Ca$^{2+}$-dependent component that is responsive for the afterhyperpolarization (AHP) on the falling phase of APs. Gold and colleagues suggested that inhibition of this $Ca^{2+}$-dependent AHP contributes to the $PGE_2$-induced sensitization of nociceptors.[142] However, inhibition of the AHP is not sufficient to explain $PGE_2$-induced sensitization in the majority of DRG neurons, because the population of DRG neurons expressing prolonged AHPs is less than half of the population of DRG neurons sensitized by $PGE_2$. It is anticipated that the inflammatory mediators, which act predominantly through G-protein-coupled receptors, would turn on different signal transduction cascades that may utilize second messengers playing an important role in the sensitization. Both forskolin, as activator of adenyl cyclase, and dibutyryl cAMP can mimic the action of $PGE_2$ on TTXr $Na^+$ current, and a peptide inhibitor of protein kinase A (PKA) abolished the modification of TTXr $Na^+$ current. Therefore, it was suggested that phosphorylation of TTXr VGSC can diminish the threshold of activation of SD neurons.[140] It is worthy of note that phosphorylation of TTXs VGSCs usually leads to a reduction in $Na^+$ current amplitude. Thus, brain VGSCs are phosphorylated by protein kinase C (PKC) in interdomain 1 (ID1), which results in reduced current amplitude.[143] Phosphorylation of brain type II by PKC at $Ser^{1506}$ decreases $Na^+$ current magnitude.[144] SNS is therefore of particular interest in this regard. Evidence for the modulation of SNS by phosphorylation has recently been reported. SNS contains several potential PKA and/or PKC phosphorylation sites in ID1.[132]

The contribution of PKA to inflammation has been sudied in mice that carry a null mutation in the gene that encodes the neuronal-specific isoform of the type I regulatory subunit (RIbeta) of PKA.[146] Inflammation was significantly reduced in the mutant mice, whereas PKA was not involved in a model of neurophatic pain.[146] Authors suggested that RIbeta PKA is specifically required for nociceptive processing in the terminals of primary afferent fibers, because RIbeta is expressed only in the nervous system.[146] The deletion of RIbeta also leads to impairments in tests of learning and memory. Furthermore, RIbeta is required for many processes in the brain, including normal development of long-term potentiation. Therefore, a sensory neuron-specific deletion of PKA may provide clearer insight into the involvement of PKA in nociceptive processing.

Despite the fact that there has been substantial progress in understanding the mechanisms of sensory neuron sensitization, a number of issues remain unresolved. What signal transduction pathways are turned on by different inflammatory mediators? What channels are responsible for the $Ca^{2+}$ component of APs? What are the molecular mechanisms of the modification of TTXr VGSC(s)? It seems likely that considerable insight into these problems may be provided by: (a) extensive molecular biological characterization of voltage-gated ion channels expressed by subsets of nociceptors that code different submodalities; (b) generation and analysis of inducible and sensory neuron-specific null mutants for various components involved in the sensitization process; and (c) investigation of the molecular basis of the modification of voltage-gated channels under different conditions.

The hyperalgesia, which appears several hours after NGF application, may result from sensitization of dorsal horn neurons to peripheral inputs.[110] This central sensitization has been attributed to the enhanced activity of Aδ- and C-fibers, which increase the excitability of their postsynaptic targets by releasing glutamate and substance P. It was suggested that activation of NMDA and substance P receptors (NK1) play an important role in central sensitization.[110,147] In an attempt to define the role of SP in sensitization, researchers disrupted the gene encoding the SP receptor (NK1) using homologous recombination in ES cells.[148] Acute nociception was unaffected in NK1 (–/–) mice. However, NK1 contributes to descrete components of the hyperalgesic state.[148] Thus, NK1 (–/–) mice showed a 30% attenuation of the behavioral responses in the second phase of the formalin paw test. Cao and colleagues created recombinant mice that express neither SP nor neurokinin A (NKA).[149] The mice (+/+ and –/–) did not differ in tests of the behavioral response to mildly painful stimuli.[149] Whereas the response to moderate and intense pain is significantly reduced.[149]

Another central mechanism of hyperalgesia has recently been proposed.[150] An inflammation can result in a phenotypic switch in a subpopulation of Aβ-fibers. They, like C-fibers, start to express SP, which leads to Aβ-fibers also acquiring the capacity to increase the excitability of spinal cord neurons and to contribute to inflammatory hypersensitivity.

What roles do sympathetic neurons play in hyperalgesia? Sympathetic postganglionic neurons may be involved in hyperalgesia and inflammation under pathophysiological conditions. Sympathetic nerve terminals in peripheral tissues may serve as mediator elements in hyperalgesia and inflammation following tissue trauma without nerve lesion. Thus, in animals subjected to surgical or chemical sympathectomy NGF-induced hyperalgesia was markedly reduced.[151] It was suggested that the prostaglandin (probably $PGE_2$) is synthesized and released from the sympathetic terminal to afferent units.[151] The postganglionic sympathetic neuron mediator, norepinephrine, may also lead to sensitization of nociceptive afferents under inflammatory conditions.[152] This sensitization is presumably mediated by alpha 2-adrenoceptors and $PGI_2$, which is synthesized and released from the sympathetic terminals.[152]

Alteration in peripheral and central connectivity may also play a role in hyperalgesia. Chronic pain syndromes that occur after nerve injury may induce plastic changes of the afferent and sympathetic postganglionic neurons. Both afferent and postganglionic neurons exhibit degenerative and regenerative changes and unlesioned neurons may show collateral sprouting in the periphery as well as in DRG.[151] This reorganization of the peripheral neurons may lead to coupling between sympathetic and afferent neurons and activation of primary afferent neurons by substances, such as norepinephrine, released from the sympathetic neurons. It was suggested that neurotrophic factors trigger these changes. Indeed, novel projections from adrenergic sympathetic neurons to sensory ganglia are induced in transgenic mice that overexpress NGF in the skin.[153] Electron microscopic analysis of NGF transgenic mice revealed that trigeminal neurons were surrounded by numerous tyrosine hydroxylase (TH)-positive axons. These results suggest a model in which increased NGF expression plays a role in the development of chronic pain after nerve injury.

Woolf and colleagues have shown that axotomy of the posterior cutaneous nerve of the thigh or the saphenous nerve induces sprouting of the central terminals of axotomized myelinated afferents, which normally terminate in laminae I, III, and IV, into lamina II, an area that is normally innervated by unmyelinated C-fibers.[154] Interestingly, the NMDA antagonist MK801 is able to block the reorganization of dorsal horn input after peripheral axotomy.[155] These changes may contribute to chronic pain and sensory hypersensitivity in denervated skin. However, the molecular basis of this phenomenon is unknown.

# MOLECULAR GENETIC APPROACHES TO NOCICEPTORS, MECHANORECEPTORS, AND THERMORECEPTORS: DEVELOPMENT AND FUNCTION

The great functional heterogeneity of primary sensory neurons implies the existence of a variety of molecular mechanisms in their activation and sensitization. Six major strategies could be adopted to study the molecular basis of function and development of sensory neurons.

## MOLECULAR ANALYSIS OF FUNCTIONALLY DEFINED INDIVIDUAL NEURONS

The molecular mechanisms involved in nociception, mechanoreception, and thermoreception have ideally must be studied separately for each "functionally homogenous" subset of sensory neurons (e.g., for touch cutaneous mechanoreceptors innervating G2 hair, and extended to lamina III; or for cutaneous C-polymodal nociceptors extended to lamina II; etc.). However, with present-day molecular biology, this strategy is difficult to implement. Moreover, it is impractical, although in theory it can be adopted, when we either look for unknown genes/transcripts or study known genes/transcripts. First, functionally different subsets of sensory neurons are mixed in cranial and dorsal root ganglia. Therefore, "functionally homogeneous" subsets of sensory neurons have to be identified by direct stimulus-response experiments, during which electrophysiological recordings are made in somata of sensory neurons. Second, cytoplasmic extracts, containing RNA, DNA, proteins, and so on, have to be isolated by micropipettes from identified soma(ta). If the type of sensory neuron is identified by recording from a single nerve fiber then it will be impossible to isolate cytoplasmic extracts from the sensory neuron soma that has not been identified. Third, it is apparent that the amount of isolated material from single identified cells will be insufficient to look for the novel transcripts. However, this approach might be applicable if: (a) A number (>100) of identical cells are isolated. Thus, novel transcripts have been cloned from snail giant interneurons, the sizes of which are comparable with the sizes of 100 sensory neurons.[156] (b) We would like to isolate the novel genes that belong to a characterized gene family. Thus, the genes encoding putative pheromone receptors, which are members of G-protein-coupled seven-transmembrane segment receptors, were cloned from a single cell.[157] In summary, the above approach is not generally applicable and is extremely difficult; so far it has never been used to establish the molecular basis of function and development of sensory neurons. Nevertheless, manipulation with tiny amounts of material, using molecular biological problems, is a rapidly advancing field.

## ISOLATION OF SPECIFIC TRANSCRIPTS FOR FUNCTIONALLY IDENTIFIED GROUP OF NEURONS

It could be assumed that several different functional groups of sensory neurons that share some physiological, molecular, and/or biochemical features may be involved in nociception, mechanoreception, and thermoreception and may respond to inflammatory mediators, axotomy, or cronic pain syndrome, and so on. One way to understand these processes is to look for transcripts/proteins that are selectively expressed by subsets of cells that are involved in these events. Such specific molecules seem likely to play a functional role for this particular subset, which may consist of several distinct functional groups of sensory neurons. For example, CGRP, which is specifically expressed by the subset including SD and L neurons, may play an important functional role in mechanical (HTMR) and heat-mechanical nociception. Another way to understand these specific processes is to search for transcripts/proteins, expression of which is dramatically changed during these events. There are three approaches to look for specific transcripts, and to search for transcripts the expression of which is dramatically changed.

### Differential Display

Differential display, a PCR-based method using partially degenerate primers, allows the identification of differentially expressed transcripts.[158] This method is relatively simple but does have several pitfalls. First, differential display is not a very sensitive method. Second, the method is susceptible to the generation of artefacts. Third, isolated cDNAs often represent 3'- untranslated regions (3'-UTR) of the transcript. Therefore, additional screening of a full-length cDNA library is required to understand the nature of isolated transcripts.

### Differential Screening

Differential screening of full-length cDNA libraries has been used to isolate specific abundant transcripts.[159] To make possible the interpretation of the results of differential screening, a full-length cDNA library should be plated at low-density. Screening of low-density cDNA libraries automatically impairs the sensitivity and efficiency of the method, because these cDNA libraries do not contain medium and rare represented transcripts. Alternatively, if cDNA libraries are plated on many dishes, it will require a large volume of hybridization solution, which will also decrease the sensitivity of screening.

## Generation of a cDNA Subtractive Library and Its Differential Screening

This powerful method was developed by Davis and colleagues.[160] Subtractive libraries avoid the problems noted above. Removal of common transcripts from the tissue of interest allows one not only to construct a representative and low-density cDNA library, but also to enrich cDNA libraries for specific transcripts. Original methods of construction of subtractive cDNA libraries required many grams of both the tissue of interest and "irrelevant" tissues, the RNA of which is used to remove common transcripts from the tissue of interest. Since the PCR method appeared, it has been successfully extended to identify transcripts specifically expressed in very small numbers of cells.[161,162] The choice of "irrelevant" tissues is the key factor to the production of a representative library. Both the spectrum of transcripts as well as their abundance in the subtractive library are influenced by the RNA from "irrelevant" tissues, because the relative proportion of various tissue (cell)-specific genes in the library may be altered by different levels of subtraction. It is impossible to predict the level of specificity for unknown transcripts/proteins. They can play an important functional role in the tissue of interest, and at the same time, may be expressed in "irrelevant" tissues. Therefore, obviously, there are no standard, commonly accepted recommendations for choosing the "irrelevant" tissues.

This strategy has been used to identify novel proteins that may play an important role in mechanoreceptor, thermoreceptor, and nociceptor function and development.[162] The identified clones are divided into three classes: (1) known DRG cell markers (myelin $P_o$, CGRP, peripherin, parvalbumin, carbonic anhydrase, etc.), (2) DRG cell-specific proteins that are related to known proteins (synuclein-like, vanillin-like, etc.), and (3) unknown DRG cell-specific proteins, including P2X3, SNS, and so on.[45,132,162] Another example of the application of this strategy has been reported by Saito and colleagues.[163] A method that combines subtractive hybridization with degenerate RT-PCR, named differential RT-PCR, has been used to rapidly screen gene families for members exhibiting differential expression in DRG sensory neurons. Thus, a novel protein containing a paired homeodomain, named DRG-11, was identified that is expressed in the trkA-positive subset of DRG sensory neurons and in the dorsal horn neurons, but not in glia and sympathetic neurons.[266]

## Homology Screening

It is now known that the regulatory mechanisms involved in specifying different cell fates show some similarities across evolution. The search for vertebrate homologs of neurogenic factors that specify sensory neurogenesis in flies and worms may result in the discovery of regulators of mechanoreceptor, thermoreceptor, and nociceptor fate.

The nematode *Caenorhabditis elegans* and the fruitfly *Drosophila* exhibit a variety of responses to touch. It has been demonstrated that *C. elegans* has at least six morphologically distinct classes of mechanosensory neurons.[164] Interestingly, the ASH neuron also acts as a chemosensory neuron; it mediates the avoidance of noxious chemicals. Since ASH possesses both chemosensory and mechanosensory modalities, this neuron might be functionally analogous to polymodal nociceptors.[164] Generation and screening of mutations in lower organisms is relatively simple. Many defects associated with chemotransduction and mechanotransduction in *C. elegans* have been identified, when specific sets of mechanosensory neurons are killed with a laser. These mutations have been classified into developmental and nondevelopmental. Developmental defects define genes that (a) generate precursor cells of sensory neurons; (b) specify those cells to differentiate as chemoreceptors or mechanoreceptors; (c) maintain the differentiated state; (d) guide axon outgrowth for sensory neurons; and (e) control the formation of the peripheral and central innervation pattern. Nondevelopmental defects define genes that control the expression of products needed for mechanoreceptor and chemoreceptor function, but not development.

Until now, many identified mutations that caused nondevelopmental and developmental defects in *Drosophila* and *C. elegans,* have not been characterized by molecular biological methods; in addition, vertebrate homologs of many defined fly and worm genes remain to be isolated. mec-3, which contains a cysteine-rich domain characteristic of this class of proteins (LIM domain), is essential for proper differentiation of the set of six touch receptor neurons in *C. elegans.* In mutants lacking mec-3 activity, the touch receptors express none of their unique differentiated features and appear to be transformed into other types of neurons.[165] The mammalian homolog of mec-3 is unknown. pox neuro, a paired domain gene of *Drosophila,* is expressed in four neuronal precursors: two neuronal stem cells (neuroblasts) in the CNS and two sensory mother cells (SMCs) in the PNS.[166] The pox-neuro-expressing SMCs produce the poly-innervated (chemosensory) external sense organs of the larva, whereas in pox-neuro mutated embryos, poly-innervated sense organs are transformed into mono-innervated (mechanosensory) sense organs.[166] The vertebrate homolog of pox-neuro is also unknown.

However, a number of vertebrate homologs of strikingly interesting transcription factors of *C. elegans* and *Drosophila*, which play an important role in sensory neurogenesis, have been isolated. Thus, the sensory neuron-expressed genes Brn 3.0, Brn 3.1, and Brn 3.2 are mammalian homologs of the *C. elegans* unc-86 gene, which is responsible for some sensory neuron specification in worms.[167-169] The reader interested in detailed references to this vast literature, which describes known factors controlling peripheral neurogenesis (including migration and differentiation of neuronal crest cells and development enteric, autonomic, and sympathetic neurons) in *Drosophila* and *C. elegans*, is referred to a review of this subject.[170]

Of special interest are the nondevelopmental defects of mechanoreceptors and chemoreceptors in lower organisms. In *C. elegans* necrosis-like acute excitotoxic neuronal death is induced by mutations in two genes, mec-4 and deg-1. The dominant mutation deg-1 results in a toxic gene product that leads to the late degeneration of a small number of neurons, including mechanoreceptors.[171] The mutations of the mec-4 gene, which is similar in sequence to deg-1 and expressed only in mechanoreceptors, causes degeneration of the touch receptor neurons.[172] Genetic screening has identified another mechanosensory transcript, mec-10, which potentially encodes a component of the mechanosensory apparatus. mec-10 and mec-4 are expressed in the same cells and, together with deg-1, comprise the degenerin gene family that encodes proteins containing two transmembrane hydrophobic segments.[173] As things now stand, three mammalian neurospecific homologs of *C. elegans* degenerins have been isolated: *MDEG (BNC-1), ASIC (BNC-2),* and *DRASIC* (see above).[79,80,82]

Molecular attachments of ion channels, both to extracellular and intracellular matrix components, such as specific integral membrane and cytoskeletal proteins, may be required for mechanical gating. Moreover, cytoskeletal proteins may be important in defining the overall cellular structure of both mechanoreceptors and chemoreceptors. Thus, mutations in the gene mec-7, which encodes a β-tubulin with 90 to 93% identity to vertebrate β-tubulins, result in touch-insensitive animals whose touch cells lack the 15-protofilament microtubules.[174] The mec-2 gene, which encodes a stomatin-like integral protein, is required for the function of a set of six touch receptor neurons in C. elegans, although touch sensory neurons appear morphologically normal in mec-2 mutants.[175] Gene interaction studies suggest that mec-2 positively regulates the activity of the putative mechanosensory transduction channel.[175]

The above strategy is of considerable current interest. However, organization of the SVSS of higher organisms is much more complex than organization of the analogous system (mechanoreception and chemoreception) in flies and nematodes. In addition, as a rule, a substantial proportion of the members of some gene families are the mammalian transcripts; that is, a particular gene family is often more ramified in mammals than the same gene family in lower organisms. Therefore, in many instances, mammalian homologs and members of a gene family have an unexpected pattern of expression and/or function.[170]

## Expression Cloning

Proteins as diverse as ligand-gated and voltage-gated ion channels, and G-protein-coupled receptors, which may play an important role in activation and sensitization of subsets of sensory neurons, have been isolated by expression cloning. To choose a method for expression cloning, investigators need the following information: (a) properties of protein(s) of interest, which were obtained by pharmacological and electrophysiological methods; (b) heterologous expression systems (*Xenopus* oocytes, cell lines, and so on) where protein(s) of interest might be expressed; and (c) subset of cells expressing protein(s) of interest.

## Mammalian Genetics

The next strategy is based on classical genetic approaches used in combination with modern molecular biology techniques. A number of mouse and rat mutants as well as human neurological disorders show somatovisceral sensation-related phenotypes. It is obvious that chromosomal mapping of these mutations and subsequent isolation of genes associated with the mutations would allow huge progress in the understanding of function and development of sensory neurons of the SVSS. These mouse and rat lines were created by several methods: (a) spontaneous mutation; (b) repeated sibling mating for at least 20 generations; (c) selective breeding of individuals displaying abnormal somatovisceral sensation; and (d) recombinant inbreeding between two different strains of rats or mice. Genetics of pain-related phenomena in mice and rats have recently been reviewed.[176]

Several neurological disorders in man that have been classified as congenital indifference to pain (or hereditary sensory neuropathy, or congenital analgesia) are known. The hereditary sensory neuropathies can be subdivided into five categories according to the loss of peripheral sensory modalities and/or the phenotype of DRG sensory neurons in individuals with these neurological disorders.[177]

Hereditary sensory neuropathy type I (HSN-I or HSAN-I), also known as hereditary sensory radicular neuropathy, is an autosomal dominant disorder that is the most common of a group of degenerative disorders of sensory neurons. The disease involves a progressive degeneration of DRG, autonomic, and motor neurons, that leads to distal sensory loss and later distal muscle wasting and weakness and variable neural deafness. Sensory deficits include loss of all modalities, particularly loss of sensation to pain and temperature. A genome screen has been undertaken to map the HSN1 gene.[178] Multipoint linkage analysis suggests the HSN1 gene is likely to be located within an 8-centiMorgan (cM) region flanked by D9S318 and D9S176 on chromosome 9q22.1–q22.3. D9S287, within a 4.9-cM confidence interval.

Hereditary sensory neuropathy type II (HSN-II) is an autosomal recessive disorder characterized by the loss of some peripheral sensory modalities, myelinated A-fibers, which leads to deficits in low- and high-threshold mechanosensitivity in individuals with otherwise normal development. Recent characterization of a similar phenotype in mice carrying a targeted mutation in the low-affinity NGF receptor, p75NGFR, suggested the possibility that mutations in this gene or other members of the NGF family of genes and their receptors might be responsible for HSN-II.[179] However, the segregation of polymorphic alleles at and around loci for p75NGFR, trkA, trkB, BDNF, and familial dysautonomia virtually excluded these genes as the candidate for HSN-II.[179]

Hereditary sensory neuropathy type III (HSN-III), also known as familial dysautonomia (DYS) or the Riley-Day syndrome, is an autosomal recessive disorder characterized by developmental loss of unmyelinated C-fiber sensory neurons and some neurons from the autonomic nervous system. The DYS gene was mapped to chromosome 9q31–q33 by linkage analysis. D9S53 and D9S105 represent the closest flanking markers for the disease gene.[180] The neurotrophins and their receptors have also been considered candidate genes for the DYS gene. Therefore, the chromosomal localization of the candidate gene NTRK2, the human homolog of trkB, was determined. NTRK2 was assigned to chromosome 9, but the NTRK2 gene is located approximately 22 cM proximal to DYS gene. Therefore, the NTRK2 gene can now be excluded as a candidate gene for HSN-III.[181]

Hereditary sensory neuropathy type IV (HSN-IV), also known as congenital insensitivity to pain with anhidrosis (CIPA), is an autosomal recessive disorder characterized by absence of both sweating and reaction to noxious stimuli, self-mutilating behavior, and mental retardation. A distinguishing feature of DRG in individuals with CIPA is the loss of Aδ- and C-fiber sensory neurons. Mice lacking NGF or its high-affinity receptor, trkA, show the phenotypic features of CIPA. Therefore, it has been considered that the human NGF and trkA homologs are candidates for CIPA gene.[182] The trkA gene and transcript was analyzed in three unrelated CIPA patients who had consanguineous parents. A deletion, a splicing defect, and a missense mutation in the tyrosine kinase domain in these patients were detected.[182] However, no effects were found on p75 and NGF transcripts in these patients.[182]

Hereditary sensory neuropathy type V (HSN-V) is also known as congenital insensitivity to pain without anhidrosis (CIP). Perception of the other sensory modalities in individuals with CIP is normal. Nerve biopsy has rarely been reported in this condition. However, in some cases, nerve biopsies have been demonstrated. In four patients a severe loss of myelinated Aδ-fibers was detected. The CIP gene is not yet mapped.

## Transgenic Analysis

The isolation of sensory neuron-specific transcripts by subtractive library construction and differential screening or expression cloning, or the search for vertebrate homologs of functionally valuable transcripts cloned using genetic models of flies and worms does not guarantee defining a physiologically significant protein. To prove functional and physiological significance of proteins in subsets of sensory neurons, transgenic analysis of mice carrying misexpression or deletion mutations of genes of interest is particularly valuable.

Gene targeting in embryonic stem cells allows the generation of mice containing a deletion in a defined gene of interest.[183] Analysis of mouse null mutants is a direct and powerful approach for the functional identification of a gene product. In the field of sensory neurophysiology, the gene knockout approach has been particularly useful. Thus, a plethora of sensory neuron-expressing genes that could play a role in sensory neuron specification, differentiation, and function were targeted, and resultant null mutants have been analyzed. The phenotypes of these null mutant mice are listed in Table 19.3.

Conventional gene knockout techniques provide mice that inherit deletions in all cell types. Therefore, unrestricted gene deletion may lead to (a) severe developmental defects, and prenatal death of mutated mice, which excludes the possibility to analyze protein function in adult animals; (b) abnormal phenotype in adult animals that appeared indirectly from developmental defects; and (c) abnormal phenotypes of many cell types, which makes it difficult to attribute phenotypes to a particular type of cell. One method to overcome these difficulties is to generate cell type- or tissue type-restricted gene knockouts using the Cre/loxP system.[184,185] Site-specific recombination between molecules of bacteriophage P1 DNA occurs at sites called loxP (34 bp) and requires the action of a protein that is the product of the P1 Cre gene. The loxP sites can be inserted into the genome by homologous recombination, such that they flank one or more exons of a gene of interest, named a "floxed gene." It is essential that this insertion should not alter normal expression of the gene. Mice homozygous for the floxed gene are crossed with transgenic mice bearing Cre-recombinase that is controlled by a cell type- or tissue-specific promoter. In progeny that are both homozygous for the floxed gene and possess the Cre transgene, a gene of interest will be deleted by the Cre/lox recombination system only in the subset of cells where the Cre-recombinase is expressed. Promoters

**TABLE 19.3**

**Sensory Neurons Abnormalities in Null Mutant Mice**

| Mutant Mice | Phenotype(s) of Sensory Neurons | | Ref. |
|---|---|---|---|
| | **Behavioral** | **Developmental (% Neuron Loss)** | |
| trkA/NGF | loss of responses to painful stimuli | SCG, 99%; DRG 70-80%; | 28 |
| | | TG, 70%; NG, no loss; | 29 |
| | | proper phenotypic development of HTMRs | |
| trkB/BDNF | in BDNF: head turning, spinning | SCG, no loss; NG, 45-65% | 188 |
| | | DRG, 30-50%; TG, 25-55%; | 189 |
| | | VG, >80% | |
| trkC/NT-3 | abnormal movement, loss of proprioception | SCG, 55% (NT-3); DRG, 20% (trkC), | 33 |
| | | 55% (NT-3) | 34 |
| | | NG, 50%; TG, 65% (NT-3); | 191 |
| | | postnatal loss of D hair and SA1 | |
| | | mechanoreceptors | |
| NT-4/5 | ND | SCG and TG, no loss; | 192 |
| | | NG, PG, and GG, 50-70% | |
| p75 | loss of heat sensitivity | decreased innervation of the skin by CGRP | 190 |
| | | and SP-positive sensory neurons | |
| Brn3a | motor and ingestive behavioral phenotype, similar | neuronal death of subsets of | 194 |
| | to BDNF (–/–) and NT-3 (–/–) mice | proprioreceptors, mechanoreceptors, and | |
| | | nociceptors | |
| Brn3b | blindness | loss of most retinal ganglion cells; sensory | 193 |
| | | neurons of DRG are not characterized | |
| Brn3c | deafness | loss of hair cells in the innear ear; sensory | 193 |
| | | neurons of DRG are not characterized | |
| B2 | loss of response to BK; normal pain responses to | NR | 56 |
| | mechanical and heat stimuli; no changed | | |
| | hyperalgesia | | |
| MOR | loss of morphine-induced analgesia; normal pain | NR | 195 |
| | responses to mechanical heat stimuli | | |
| Enk | modulate responses to painful stimuli | NR | 196 |
| β-end | loss of analgesia in responses to the opioid and | ND | 197 |
| | non-opioid | | |
| PGI$_2$ | inflammatory and pain responses are reduced to | NR | 125 |
| | the levels observed in indomethacin-treated | | |
| | wild-type mice | | |
| A$_{2a}$ | slow response of mice to acute pain stimuli | NR | 198 |
| PKCγ | normal responses to acute pain stimuli; but failed | NR | 145 |
| | to develop a neuropathic pain syndrome after | | |
| | sciatic nerve section | | |
| RIbeta | diminished inflammation and is not required for | NR | 146 |
| | neuropathic pain | | |
| NK1 | normal response to acute pain stimuli and | ND | 148 |
| | hyperalgesia; but essential for the full | | |
| | development of stress-induced analgesia | | |
| PPT-A | required to produce moderate to intense pain; | ND | 149 |
| | mildly pain intact in KO mice | | |

*Abbreviations:* ND — no obvious defects; NR — no reported; TG — trigeminal ganglia; DRG — dorsal root ganglia; NG — nodose ganglia: VG — vestibular ganglia; PG — petrosal ganglia; GG — geniculate ganglia; SCG — superior cervical ganglia; B2 — bradykinin receptor 2; p75 — low-affinity NGF receptor; MOR — μ-opioid-receptor; Enk — pre-proenkephalin; β-end — β-endorphin; PGI$_2$ — prostacyclin; PKCγ — γ subunit of protein kinase C; RIbeta — the neuronal-specific isoform of the type I regulatory subunit of PKA; NK1 — SP receptor; PPT-A — preprotachykinin-A (SP and neurokinin A).

that are specific for subsets of sensory neurons coding different modalities and submodalities are not known. One of the major limitations in the use of the Cre/lox recombination system is the lack of temporal control of gene function. To overcome this problem, tissue- and cell type-specific promoter could be combined with inducible promoters, such as the interferon-inducible promoter Mx1,[186] or the tetracycline-inducible promoter,[187] to achieve both regional and temporal control of gene expression. All the more attractive is the idea of constructing mutant animals where levels of transcription of a gene of interest can be reversibly regulated in living animals.

In summary, a number of key proteins playing a role in development and function have already been identified by exploiting the approaches and strategies discussed above, and more will follow.

## CONCLUSION

The preponderance of evidence concerning the mechanisms of differential activation and sensitization of sensory neurons, the peripheral and central mechanisms of hyperalgesia, neuropathic, and chronic pain, have so far been obtained by pharmacological and electrophysiological methods. These methods have three obvious limitations: (1) many drugs are not completely specific, (2) the effects of drugs depend on their concentrations, and (3) some properties of drugs, which are in use, have not yet been fully characterized in their actions. However, the combination of molecular biological, pharmacological, and electrophysiological approaches, as well as the possibility to measure functional consequences at the behavioral level, will become an attractive strategy to study the function and development of sensory neurons. Moreover, with the increasing subtlety of manipulation with tiny amounts of RNA, possibilities to investigate sensory neuron-specific promoters and inducible and tissue/cell-specific knockout technology, (a) many proteins that are important for sensory neuron function and development will be isolated and (b) the roles of known receptors, channels, transcription factors, cytoskeletal proteins, and proteins involved in signal transduction should soon become clear.

## AKNOWLEDGMENTS

I thank J. N. Wood for advice and support, N. Abson, S. England, and A. Dickenson for critical advice, and C.-C. Chen, V. Souslova and K. Okuse for helpful discussion. I am very grateful to the Wellcome Trust for generous support.

## REFERENCES

1. Waddell, P. J. and Lawson, S. N., Electrophysiological properties of subpopulations of rat dorsal root ganglion neurons in vitro, *Neuroscience,* 36, 811, 1990

2. Harper, A. A. and Lawson, S. N., Electrical properties of rat dorsal root ganglion neurons with different peripheral nerve conduction velocities, *J. Physiol.,* 359, 47, 1985.

3. Lloyd, D. P. C. and Chang, H.-T., Afferent fibers in muscle nerves, *J. Neurophysiol.,* 11, 199, 1948.

4. Lawson, S. N., The postnatal development of large light and small dark neurons in mouse dorsal root ganglia: a statistical analysis of cell numbers and size, *J. Neurocytol.,* 8, 275, 1979.

5. Lawson, S. N., Harper, A. A., Harper, E. I., Garson, J. A., and Anderton, B. H., A monoclonal antibody against neurofilament protein specifically labels a subpopulation of rat sensory neurones, *J. Comp. Neurol.,* 228, 263, 1984.

6. Lawson, S. N. and Waddell, P. J., Soma neurofilament immunoreactivity is related to cell size and fibre conduction velocity in rat primary sensory neurons, *J. Physiol.,* 435, 41, 1991.

7. Light, A. R. and Perl, E. R. Reexamination of the dorsal root projection to the spinal dorsal horn including observations on the differential termination of coarse and fine fibers, *J. Comp. Neurol.,* 186, 117, 1979.

8.  Light, A. R. and Perl, E. R., Spinal termination of functionally identified primary afferent neurons with slowly conducting myelinated fibers, *J. Comp. Neurol.,* 186, 133, 1979b.
9.  Brown, A. G. and Fyffe, R. E. W., The morphology of Group Ia afferent fiber collaterals in the spinal cord of the cat, *J. Physiol.,* 274, 111, 1978.
10. Fyffe, R. E. W., The morphology of Group II muscle afferent fiber collaterals, *J. Physiol.,* 296, 39P, 1979.
11. Brown, A. G. and Fyffe, R. E. W., The morphology of Group Ib afferent fiber collaterals in the spinal cord of the cat, *J. Physiol.,* 296, 215, 1979.
12. Rethelyi, M., Light, A. R., and Perl, E. R., Synaptic complexes formed by functionally defined primary afferent units with fine myelinated fibers, *J. Comp. Neurol.,* 207, 381, 1982.
13. Sugiura, Y., Lee, C. L., and Perl, E. R., Central projections of identified, unmyelinated (C) afferent fibers innervating mammalian skin, *Science,* 234, 358, 1986.
14. Troy, C. M., Brown, K., Greene, L. A., and Shelanski, M. L., Ontogeny of the neuronal intermediate filament protein, peripherin, in the mouse embryo, *Neuroscience,* 36, 217, 1990.
15. Ferri, G. L., Sabani, A., Abelli, L., Polak, J. M., Dahl, D., and Portier, M. M., Neuronal intermediate filaments in rat dorsal root ganglia: differential distribution of peripherin and neurofilament protein immunoreactivity and effect of capsaicin, *Brain Res.,* 515, 331, 1990.
16. Mulderry, P. K., Ghatei, M. A., Spokes, R. A., Jones, P. M., Pierson, A. M., Hamid, Q. A., Kanse, S., Amara, S. G., Burrin, J. M., Legon, S., et al., Differential expression of alpha-CGRP and beta-CGRP by primary sensory neurons and enteric autonomic neurons of the rat, *Neuroscience,* 25, 195, 1988.
17. Lawson, S. N., McCarthy, P. W., and Prabhakar, E., Electrophysiological properties of neurones with CGRP-like immunoreactivity in rat dorsal root ganglia, *J. Physiol.,* 365, 355, 1996.
18. Lawson, S. N., Morphological and biochemical cell types of sensory neurons, in *Sensory Neurons: Diversity, Development, and Plasticity,* Scott, S. A., Ed., Oxford University Press, New York, 1992, 27.
19. Jessell, T. M. and Dodd, J., Functional chemistry of primary afferent neurons, in *Textbook of Pain,* Wall, P. D. and Melzack, R., Eds., Churchill Livingstone, Edinburgh, 1989, 82.
20. Alvarez, F. J., Morris, H. R., and Priestly, J. V., Sub-populations of small diameter trigeminal primary afferent neurons defined by expression of calcitonin gene-related peptide and the cell surface oligosaccharide recognized by monoclonal antibody LA4, *J. Neurocytol.,* 20, 716, 1991.
21. Dodd, J. and Jessell, T. M. Lactoseries carbohydrates specify subsets of dorsal root ganglion neurons projecting to the superficial dorsal horn of rat spinal cord, *J. Neurosci.,* 5, 3275, 1985.
22. Bennett, D. L., Dmietrieva, N., Priestley, J. V., Clary, D., and McMahon, S. B., trkA, CGRP and IB4 expression in retrogradely labelled cutaneous and visceral primary sensory neurones in the rat, *Neurosci. Lett.,* 206, 33, 1996.
23. Snider, W. D., Functions of the neurotrophins during nervous system development: what the knockouts are teaching us, *Cell,* 77, 627, 1994.
24. Mu, X., Silos-Santiago, I., Carroll, S. L., and Snider, W. D., Neurotrophin receptor genes are expressed in distinct patterns in developing dorsal root ganglia, *J. Neurosci.,* 13, 4029, 1993.
25. Wright, D. E. and Snider, W. D., Neurotrophin receptor mRNA expression defines distinct populations of neurons in rat dorsal root ganglia, *J. Comp. Neurol.,* 351, 329, 1995.
26. Averill, S., McMahon, S. B., Clary, D. O., Reichardt, L. F., and Priestley, J. V., Immunocytochemical localization of trkA receptors in chemically identified subgroups of adult rat sensory neurons, *Eur. J. Neurosci.,* 7, 1484, 1995.
27. Lewin, G. R. and Mendell, L. M., Nerve growth factor and nociception, *Trends Neurosci.,* 16, 353, 1993.
28. Crowley, C., Spencer, S. D., Nishimura, M. C., Chen, K. S., Pitts-Meek, S., Armanini, M. P., Ling, L. H., MacMahon, S. B., Shelton, D. L., Levinson, A. D., et al., Mice lacking nerve growth factor display perinatal loss of sensory and sympathetic neurons yet develop basal forebrain cholinergic neurons, *Cell,* 76, 1001, 1994.
29. Smeyne, R. J., Klein, R., Schnapp, A., Long, L. K., Bryant, S., Lewin, A., Lira, S. A., and Barbacid, M., Severe sensory and sympathetic neuropathies in mice carrying a disrupted Trk/NGF receptor gene, *Nature,* 368, 246, 1994.
30. Molliver, D. C. and Snider, W. D., The NGF receptor TrkA is downregulated during postnatal development by a subset of dorsal root ganglion neurons, *J. Comp. Neurol.,* 381, 428, 1997.

31. Molliver, D. C., Wright, D. E., Leitner, M. L., Parsadanian, A. Sh., Doster, K., Wen, D., Yan, Q., and Snider, W. D., IB4-binding DRG neurons switch from NGF to GDNF dependence in early postnatal life, *Neuron,* 19, 849, 1997.

32. McMahon, S. B., Armanini, M. P., Ling, L. H., and Phillips, H. S., Expression and coexpression trk receptors in subpopulation of adult primary sensory neurons projecting to identified peripheral targets, *Neuron,* 12, 1161, 1994.

33. Ernfors, P., Lee, K.-F., Kucera, J., and Jaenisch, R., Lack of neurotrophin-3 leads to deficiencies in the peripheral nervous system and loss of limb proprioceptive afferents, *Cell,* 77, 503, 1994.

34. Klein, R., Silos-Santiago, I., Smeyne, R. J., Lira, S., Brambrilla, R., Bryant, S., Zhang, L., Snider, W. D., and Barbacid, M., Targeted disruption of trkC, the neurotrophin-3 receptor gene, eliminates Ia muscle afferents and results in loss of proprioception, *Nature,* 368, 249, 1994.

35. Prabhakar, E. and Lawson, S. N., The electrophysiological properties of rat primary afferent neurones with carbonic anhydrase activity, *J. Physiol.,* 482, 609, 1995.

36. Szabolcs, M. J., Kopp, M., and Schaden, G. E., Carbonic anhydrase activity in the peripheral nervous system of rat: the enzyme as a marker for muscle afferents, *Brain Res.,* 492, 129, 1989.

37. Peyronnard, J. M., Charron, L. F., Messier, J. P., and Lavoie, J., Differential effects of distal and proximal nerve lesions on carbonic anhydrase activity in rat primary sensory neurons, ventral and dorsal root axons, *Exp. Brain Res.,* 70, 550, 1988.

38. Carr, P. A., Yamamoto, T., Staines, W. A., Whittaker, M. E., and Nagy, J. I., Parvalbumin is highly co-localized with calbindin D28k and rarely with calcitonin gene-related peptidein dorsal root ganglia of rat, *Brain Res.,* 497, 163, 1989.

39. Copray, J. C., Mantingh-Otter, I. J., and Brouwer, N., Expression of calcium-binding proteins in the neurotrophin-3-dependent subpopulation of rat embryonic dorsal root ganglion cells in culture, *Dev. Brain Res.,* 81, 57, 1994.

40. Bleehen, T. and Keele, C. A., Observations on the algogenic actions of adenosine compounds on human blister base preparation, *Pain,* 3, 367, 1977.

41. Burnstock, G. and Kennedy, C., Is there a basis for distinguishing two types of P2 purinoceptors?, *Genet. Pharmacol.,* 16, 433, 1985.

42. North, A. R., Families of ion channels with two hydrophobic segments, *Curr. Opin. Cell Biol.,* 8, 474, 1996.

43. Khakh, B. S., Humphrey, P. A., and Surprenant, A., Electrophysiological properties of P2X purinoceptors in rat superior cervical, nodose and guinea-pig coeliac neurones, *J. Physiol.,* 484, 385, 1995.

44. Robertson, S. J., Rae, M. G., Rowan, E. G., and Kennedy, C., Characterization of P2X purinoceptors in cultured neurones of the rat dorsal root ganglia, *Br. J. Pharmacol.,* 118, 951, 1996.

45. Chen, C.-C., Akopian, A. N., Sivilotti, L., Colquhoun, D., Burnstock, G., and Wood, J. N., A P2X purinoceptor expressed by a subset of sensory neurons, *Nature,* 377, 428, 1995.

46. Collo, G., North, R. A., Kawshima, E., Merlo-Pich, E., Neidhardt, S., Surprenant, A., and Buell, G., Cloning of P2X5 and P2X6 receptors and the distribution and properties of an extended family of ATP-gated ion channels, *J. Neurosci.,* 16, 2495, 1996.

47. Cook, S. P., Vulchanova, L., Hargreaves, K. M., Elde, R., and McCleskey, E. W., Distinct ATP receptors on pain-sensing and stretch-sensing neurons, *Nature,* 387, 505, 1997.

48. Lewis, C., Neldhart, S., Holy, C., North, R. A., Buell, G., and Surprenant, A., Coexpression of P2X2 and P2X3 receptor subunits can account for ATP-gated currents in sensory neurons, *Nature,* 377, 432, 1995.

49. Cook, S. P. and McCleskey, E. W., Desensitization, recovery and $Ca^{+2}$-dependent modulation of ATP-gated P2X receptors in nociceptors, *Neuropharmacology,* 36, 1303, 1997.

50. Vulchanova, L., Riedl, M. S., Shuster, S. J., Buell, G., Surprenant, A., North, R. A., and Elde, R., Immunohistochemical study of the P2X2 and P2X3 receptor subunits in rat and monkey sensory neurons and their central terminals, *Neuropharmacology,* 36, 1229, 1997.

51. Burnstock, G. and Wood, J. N., Purinergic receptors: their role in nociception and primary afferent neurotransmission, *Curr. Opin. Neurobiol.,* 6, 526, 1996.

52. Burnstock, G., Noradrenaline and ATP as cotransmitters in sympathetic nerves, *Neurochem. Int.,* 17, 357, 1990.

53. Yasuda, J. M., Guanethidine for reflex sympathetic dystrophy, *Ann. Pharmacother,* 28, 338, 1994.

54. Burnstock, G., Pathophysiology of migraine: a new hypothesis, *Lancet,* 1, 1397, 1981.

55. Dray, A. and Perkins, M., Bradykinin and inflammatory pain, *Trends Neurosci.,* 16, 99, 1993.

56. Borkowski, J. A., Ransom, R. W., Seebrook, G. R., Trimbauer, M., Chen, H., Hill, R. G., Strader, C. D., and Hess, J. F., Targeted disruption of a B2 bradykinin receptor gene in mice eliminates bradykinin action in smooth muscle and neurons, *J. Biol. Chem.*, 270, 13706, 1995.

57. Menke, J. G., Borkowski, J. A., Bierilo, K. K., MacNeil, T., Derrick, A. W., Schneck, K. A., Ransom, R. W., Strader, C. D., Linemeyer, D. L., and Hess, J. F., Expression cloning of a human B1 bradykinin receptor, *J. Biol. Chem.*, 269, 21583, 1994.

58. McEachern, A. E., Shelton, E. R., Bhakta, S., Obernolte, R., Bach, C., Zuppan, P., Fujisaki, J., Aldrich, R. W., and Jarnagin, K., Expression cloning of a rat B2 bradykinin receptor, *Proc. Natl. Acad. Sci. U.S.A.*, 88, 7724, 1991.

59. Boyce, S., Rupniak, N. M. J., Carlson, E. J., Webb, J., Borkowski, J. A., Hess, J. F., Strader, C. D., and Hill, R. G., Nociception and inflammatory hyperalgesia in B2 bradykinin receptor knockout mice, *Immunopharmacology*, 33, 333, 1996.

60. Todorovic, S. and Anderson, E. G., Serotonin preferentially hyperpolarizes capsaicin-sensitive C type sensory neurons by activating 5-HT1A receptors, *Brain Res.*, 585, 212, 1992.

61. Robertson, B. and Bevan, S., Properties of 5HT3 receptor gated currents in adult rat DRG neurones, *Br. J. Pharmacol.*, 102, 272, 1991.

62. Bevan, S. and Forbes, C. A., Membrane effects of capsaicin on dorsal root ganglion neurones in cell culture, *J. Physiol.*, 398, 28P, 1988.

63. Liu, L. and Simon, S. A., A rapid capsaicin-activated current in rat trigeminal ganglion neurons, *Proc. Natl. Acad. Sci. U.S.A.*, 91, 738, 1994.

64. Holzer, P., Capsaicin: cellular targets, mechanisms of action, and selectivity for thin sensory neurones, *Pharmacol. Rev.*, 43, 143, 1991.

65. Wood, J. N., Winter, J., James, I. F., Rang, H. P., Yeats, J., and Bevan, S., Capsaicin-induced ion fluxes in dorsal root ganglion cells in culture, *J. Neurosci.*, 8, 3208, 1988.

66. Caterina, M. J., Schumacher, M. A., Tominaga, M., Rosen, T. A., Levine, J. D., and Julius, D., The capsaicin receptor: a heat-activated ion channel in the pain pathway, *Nature*, 389, 816, 1997.

67. Liu, L. and Simon, S. A., Capsaicin-induced currents with distinct desensitization and Ca2+ dependence in rat trigeminal ganglion cells, *J. Neurophysiol.*, 75, 1503, 1996.

68. Vlachova, V. and Vyklicky, L., Capsaicin-induced membrane currents in cultured sensory neurons of the rat, *Physiol. Res.*, 42, 301, 1993.

69. Oh, U., Hwang, S. W. and Kim, D., Capsaicin activates a nonselective cation channel in cultured neonatal rat dorsal root ganglion neurons, *J. Neurosci.*, 16, 1659, 1996.

70. Bevan, S. and Geppetti, P., Protons: small stimulants of capsaicin-sensitive sensory nerves, *Trends Neurosci.*, 17, 509, 1994.

71. Tominaga, M., Caterina, M. J., Malmberg, A. B., Rosen, T. A., Gilbert, M., Skinner, K., Raumann, B. E., Basbaum, A. I., and Julius, D., The cloned capsaicin receptor integrates multiple pain-producing stimuli, *Neuron*, 21, 531, 1998.

72. Martenson, M. E., Ingram, S. L., and Baumann, T. K., Potentiation of rabbit trigeminal responses to capsaicin in a low pH environment, *Brain Res.*, 651, 143, 1994.

73. Steen, K. H. and Reeh, P. W., Sustained graded pain and hyperalgesia from harmless experimental tissue acidosis in human skin, *Neurosci. Lett.*, 154, 113, 1993.

74. Steen, K. H., Steen, A. E., and Reeh, P. W., A dominant role of acid pH in inflammatory excitation and sensitization of nociceptors in rat skin, in vitro, *J. Neurosci.*, 15, 3982, 1995.

75. Steen, K. H., Reeh, P. W., Anton, F., and Handwerker, H. O., Protons selectively induce lasting excitation and sensitization to mechanical stimulation of nociceptors in rat skin, in vitro, *J. Neurosci.*, 12, 86, 1992.

76. Krishtal, O. A. and Pidoplichko, V. I., A receptor for protons in the membrane of sensory neurons may participate in nociception, *Neuroscience*, 6, 2599, 1981.

77. Bevan, S. and Yeats, J., Protons activate a cation conductance in a sub-population of rat dorsal root ganglion neurones, *J. Physiol.*, 433, 145, 1991.

78. Waldmann, R., Champigny, G., Voilley, N., Lauritzen, I., and Lazdunski, M., The mammalian degenerin MDEG, an amiloride-sensitive cation channel activated by mutations causing neurodegeneration in Caenorhabditis elegans, *J. Biol. Chem.*, 271, 10433, 1996.

79. Waldmann, R., Champigny, G., Bassilana, F., Heurteaux, C., and Lazdunski, M., A proton-gated channel involved in acid-sensing, *Nature*, 386, 173, 1997.

80. Garcia-Anoveros, J., Derfler, B., Neville-Golden, J., Hyman, B. T., and Corey, D. P., BNaC1 and BNaC2 constitute a new family of human neuronal sodium channels related to degenerins and epithelial sodium channels, *Proc. Natl. Acad. Sci. U.S.A.*, 94, 1459, 1997.

81. Chen, C. C., England, S., Akopian, A. N., and Wood, J. N., A sensory-neuron-specific, proton-gated ion channel, *Proc. Natl. Acad. Sci. U.S.A.*, 95, 10240, 1998.

82. Waldmann, R., Bassilana, F., de Weille, J., Champigny, G., Heurteaux, C., and Lazdunski, M., Molecular cloning of a non-inactivating proton-gated Na+ channel specific for sensory neurons, *J. Biol. Chem.*, 272, 20975, 1997.

83. Bassilana, F., Champigny, G., Waldmann, R., de Weille, J. R., Heurteaux, C., and Lazdunski, M., The acid-sensitive ionic channel subunit ASIC and the mammalian degenerin MDEG form a heteromultimeric H+-gated Na+ channel with novel properties, *J. Biol. Chem.*, 272, 28819, 1997.

84. Sucher, N. J., Cheng, T. P., and Lipton, S. A., Neural nicotinic acetylcholine responses in sensory neurons from postnatal rat, *Brain Res.*, 533, 248, 1990.

85. Margiotta, J. F. and Howard, M. J., Eye-extract factors promote the expression of acetylcholine sensitivity in chick dorsal root ganglion neurons, *Dev. Biol.*, 163, 188, 1994.

86. Boyd, R. T., Jacob, M. H., McEachern, A. E., Caron, S., and Berg, D. K., Nicotinic acetylcholine receptor mRNA in dorsal root ganglion neurons, *J. Neurobiol.*, 22, 1, 1991.

87. Pugh, P. C., Corriveau, R. A., Conroy, W. G., and Berg, D. K., Novel subpopulation of neuronal acetylcholine receptors among those binding α-bungarotoxin, *Mol. Pharmacol.*, 47, 717, 1995.

88. Agrawal, S. G. and Evans, R. H., The primary afferent depolarizing action of kainate in the rat, *Br. J. Pharmacol.*, 87, 345, 1986.

89. Huettner, J. E. Glutamate receptor channels in rat DRG neurons: activation by kainate and quisqualate and blockade of desensitization by Con A, *Neuron*, 5, 255, 1990.

90. Bettler, B., Boulter, J., Hermans-Borgmeyer, I., O'Shea-Greenfield, A., Deneris, E. S., Moll, C., Borgmeyer, U., Hollmann, M., and Heinemann, S., Cloning of a novel glutamate receptor subunit, GluR5: expression in the nervous system during development, *Neuron*, 5, 583, 1990.

91. Petralia, R. S., Wang, Y. X., and Wenthold, R. J., Histological and ultrastructural localization of the kainate receptor subunits KA2 and GluR6/7, in the rat nervous system using selective antipeptide antibodies, *J. Comp. Neurol.*, 349, 85, 1994.

92. Herb, A., Burnashev, N., Werner, P., Sakmann, B., Wisden, W., and Seeburg, P., The KA-2 subunit of excitatory amino acid receptors shows widespread expression in brain and forms ion channels with distantly related subunits, *Neuron*, 8, 775, 1992.

93. Jahr, C. E. and Jessell, T. M., Synaptic transmission between dorsal root ganglion and dorsal horn neurons in culture: antagonism of monosynaptic excitatory postsynaptic potentials and glutamate excitation by kynurenate, *J. Neurosci.*, 5, 2281, 1985.

94. Lovinger, D. M. and Weight, F. F., Glutamate induces a depolarization of adult rat dorsal root ganglion neurons that is mediated predominantly by NMDA receptors, *Neurosci. Lett.*, 94, 314, 1988.

95. Watanabe, M., Mishina, M., and Inoue, Y., Distinct gene expression of the N-methyl-D-aspartate receptor channel subunit in peripheral neurons of the mouse sensory ganglia and adrenal gland, *Neurosci. Lett.*, 165, 183, 1994.

96. Petralia, R. S., Wang, Y, X., and Wenthold, R. J., The NMDA receptor subunits NR2A and NR2B show histological and ultrastructural localization patterns similar to those of NR1, *J. Neurosci.*, 14, 6102, 1994.

97. Juusola, M., Seyfarth, E. A., and French, A. S., Sodium-dependent receptor current in a new mechanoreceptor preparation, *J. Neurophysiol.*, 72, 3026, 1994.

98. Sukharev, S. I., Blount, P., Martinac, B., Blattner, F. R., and Kung, C., A large conductance mechanosensitive channel in E. coli encoded by mscL gene, *Nature*, 368, 265, 1994.

99. Kizer. N., Guo, X.-L., and Hruska, K., Reconstitution of stretch-activated cation channels by expression of the α-subunit of the epithelial sodium channel cloned from osteoblasts, *Proc. Natl. Acad. Sci. U.S.A.*, 94, 1013, 1997.

100. Hamill, O. P. and McBride, D. W., Jr., The pharmacology of mechanogated membrane ion channels, *Pharmacol. Rev.*, 48, 231, 1996.

101. Liu, J., Schrank, B. and Waterston, R. H., Interaction between a putative mechanosensory membrane channel and a collagen, *Science*, 273, 361, 1996.

102. Nakamura, F. and Strittmatter, S. M., P2Y1 purinergic receptors in sensory neurons: contribution to touch-induced impulse generation, *Proc. Natl. Acad. Sci. U.S.A.,* 93, 10465, 1996.

103. Sharma, R. V., Chapleau, M. W., Hajduczok, G., Wachtel, R. E., Waite, L. J., Bhalla, R. C., and Abboud, F. M., Mechanical stimulation increases intracellular calcium concentration in nodose sensory neurons, *Neuroscience,* 66, 433, 1995.

104. Svichar, N., Shmigol. A., Verkhratsky, A., and Kostyuk, P., ATP induces $Ca^{2+}$ release from $InsP_3$-sensitive $Ca^{2+}$ stores exclusively in large DRG neurones, *Neuroreport,* 8, 7, 1997.

105. Cesare, P. and McNaughton, P., A novel heat-activated current in nociceptive neurons and its sensitization by bradykinin, *Proc. Natl. Acad. Sci. U.S.A.,* 93, 15435, 1996.

106. McMahon, S. B., NGF as a mediator of inflammatory pain, *Philos. Trans. R. Soc. Lond., B. Biol. Sci.,* 351, 431, 1996.

107. Lewin, G. R., Ritter, A. M., and Mendell, L. M., Nerve growth factor-induced hyperalgesia in the neonatal and adult rat, *J. Neurosci.,* 13, 2136, 1993.

108. Petty, B. G., Cornblath, D. R., Adornato, B. T., Chaudhry, V., Flexner, C., Wachsman, M., Sinicropi, D., Burton, L. E., and Peroutka, S. J., The effect of systemically administered recombinant human nerve growth factor in healthy human subjects, *Ann. Neurol.,* 36, 244, 1994.

109. McMahon, S. B., Bennett, D. L., Priestley, J. V., and Shelton, D. L., The biological effects of endogenous nerve growth factor on adult sensory neurons revealed by a trkA-IgG fusion molecule, *Nat. Med.,* 1, 774, 1995.

110. McMahon, S. B., Lewin, G. R., and Wall, P. D., Central hyperexcitability triggered by noxious inputs, *Curr. Opin. Neurobiol.,* 3, 602, 1993.

111. Taiwo, Y. O. and Levine, J. D., Serotonin is a directly-acting hyperalgesic agent in the rat, *Neuroscience,* 48, 485, 1992.

112. Kress, M., Riedl, B., and Reeh, P. W., Effects of oxygen radicals on nociceptive afferents in the rat skin in vitro, *Pain,* 62, 87, 1995.

113. Ferreira, S. H., Nakamura, M., and de Abreu-Castro, M. S., The hyperalgesic effects of prostacyclin and prostaglandin E2, *Prostaglandins,* 16, 31, 1978.

114. Davis, A. J. and Perkins, M. N., The involvement of bradykinin B1 and B2 receptor mechanisms in cytokine-induced mechanical hyperalgesia in the rat, *Br. J. Pharmacol.,* 113, 63, 1994.

115. Mizumura, K., Minagawa, M., Koda, H., and Kumazawa, T., Histamine-induced sensitization of the heat response of canine visceral polymodal receptors, *Neurosci. Lett.,* 168, 93, 1994.

116. Martin, H. A., Leukotriene B4 induced decrease in mechanical and thermal thresholds of C-fiber mechanonociceptors in rat hairy skin, *Brain Res.,* 509, 273, 1990.

117. Taiwo, Y. O., Heller, P. H., and Levine, J. D., Mediation of serotonin hyperalgesia by the cAMP second messenger system, *Neuroscience,* 48, 479, 1992.

118. Pierce, P. A., Xie, G. X., Levine, J. D., and Peroutka, S. J., 5-Hydroxytryptamine receptor subtype messenger RNAs in rat peripheral sensory and sympathetic ganglia: a polymerase chain reaction study, *Neuroscience,* 70, 553, 1996.

119. Rueff, A. and Dray, A., 5-Hydroxytryptamine-induced sensitization and activation of peripheral fibres in the neonatal rat are mediated via different 5-hydroxytryptamine-receptors, *Neuroscience,* 50, 899, 1992.

120. Tecott, L. H., Sun, L. M., Akana, S. F., Strack, A. M., Lowenstein, D. H., Dallman, M. F., and Julius, D., Eating disorder and epilepsy in mice lacking 5-HT2c serotonin receptors, *Nature,* 374, 542, 1995.

121. Kirchhoff, C., Jung, S., Reeh, P. W., and Handwerker, H. O., Carrageenan inflammation increases bradykinin sensitivity of rat cutaneous nociceptors, *Neurosci. Lett.,* 111, 206, 1990.

122. Perkins, M. N., Campbell, E. A., and Dray, A., Antinociceptive activity of the bradykinin $B_1$ and $B_2$ receptor antagonists, des-$Arg^9$[$Leu^8$]BK and Hoe-140, in two models of persistent hyperalgesia in the rat, *Pain,* 53, 191, 1993.

123. Martin, H. A., Basbaum, A. I., Kwiat, G. C., Goetzl, E. J., and Levine, J. D., Leukotriene and prostaglandin sensitization of cutaneous high-threshold C- and A-delta mechanonociceptors in the hairy skin of rat hindlimbs, *Neuroscience,* 22, 651, 1987.

124. Kumazawa, T., Mizumura, K., Koda, H., and Fukusako, H., EP receptor subtypes implicated in the PGE2-induced sensitization of polymodal receptors in response to bradykinin and heat, *J. Neurophysiol.,* 75, 2361, 1996.

125. Murata, T., Ushikubi, F., Matsuoka, T., Hirata, M., Yamasaki, A., Sugimoto, Y., Ichikawa, A., Aze, Y., Tanaka, T., Yoshida, N., Ueno, A., Oh-ishi, S., and Narumiya, S., Altered pain perception and inflammatory response in mice lacking prostacyclin receptor, *Nature,* 388, 678, 1997.

126. Martin, H. A., Basbaum, A. I., Goetzl, E. J., and Levine, J. D., Leukotriene B4 decreases the mechanical and thermal thresholds of C-fiber nociceptors in the hairy skin of the rat, *J. Neurophysiol.,* 60, 438, 1988.

127. Yokomizo, T., Izumi, T., Chang, K., Takuwa, Y., and Shimuzi, T., A G-protein-coupled receptor for leukotriene $B_4$ that mediates chemotaxis, *Nature,* 387, 620, 1997.

128. Koda, H., Minagawa, M., Si-Hong, L., Mizumura, K., and Kumazawa, T., H1-receptor-mediated excitation and facilitation of the heat response by histamine in canine visceral polymodal receptors studied in vitro, *J. Neurophysiol.,* 76, 1396, 1996.

129. Ferreira, S. H., Lorenzetti, B. B., and Poole, S., Bradykinin initiates cytokine-mediated inflammatory hyperalgesia, *Br. J. Pharmacol.,* 110, 1227, 1993.

130. Aguayo, L. G. and White, G., Effects of nerve growth factor on TTX- and capsaicin-sensitivity in adult rat sensory neurons, *Brain Res.,* 570, 61, 1992.

131. Ritter, A. M. and Mendell, L. M., Somal membrane properties of physiologically identified sensory neurons in the rat: effects of nerve growth factor, *J. Neurophysiol.,* 68, 2033, 1992.

132. Akopian, A. N., Sivilotti, L., and Wood, J. N., A tetrodotoxin-resistant voltage-gated sodium channel expressed by sensory neurons, *Nature,* 379, 257, 1996.

133. Toledo-Aral, J. J., Moss, B. L., He, Z. J., Koszowski, A. G., Whisenand, T., Levinson, S. R., Wolf, J. J., Silos-Santiago, I., Halegoua, S., and Mandel, G., Identification of PN1, a predominant voltage-dependent sodium channel expressed principally in peripheral neurons, *Proc. Natl. Acad. Sci. U.S.A.,* 94, 1527, 1997.

134. Elliott, A. A. and Elliott, J. R., Characterization of TTX-sensitive and TTX-resistant sodium currents in small cells from adult rat dorsal root ganglia, *J. Physiol.,* 463, 39, 1993.

135. Black, J. A., Dib-Hajj, S., McNabola, K., Jeste, S., Rizzo, M. A., Kocsis, J. D., and Waxman, S. G., Spinal sensory neurons express multiple sodium channel alpha-subunit mRNAs, *Brain Res. Mol. Brain Res.,* 43, 117, 1996.

136. Okuse, K., Chaplan, R. S., McMahon, S. B., Liuo, D. Z., Calcutt, N. A., Scott, B. P., Akopian, A. N., and Wood, J. N., Regulation of expression of the sensory neuron-specific sodium channel SNS in inflammatory and neuropathic pain, *Mol. Cell. Neurosci.,* 10, 196, 1997.

137. Black, J. A., Langworthy, K., Hinson, A. W., Dib-Hajj, S. D., and Waxman, S. G., NGF has opposing effects on Na+ channel III and SNS gene expression in spinal sensory neurons, *Neuroreport,* 8, 2331, 1997.

138. Sharma, N., D'Arcangelo, G., Kleinlaus, A., Halegoua, S., and Trimmer, J. S., Nerve growth factor regulates the abundance and distribution of K+ channels in PC12 cells, *J. Cell Biol.,* 123, 1835, 1993.

139. Gold, M. S., Reichling, D. B., Shuster, M. J., and Levine, J. D., Hyperalgesic agents increase a tetrodotoxin-resistant Na current in nociceptors, *Proc. Natl. Acad. Sci. U.S.A.,* 93, 1108, 1996.

140. England, S., Bevan, S., and Docherty, R. J., PGE2 modulates the tetrodotoxin-resistant sodium current in neonatal rat DRG neurons via the cAMP-protein kinase A cascade, *J. Physiol.,* 495.2, 429, 1996.

141. Nicol, G. D., Vasko, M. R., and Evans, A. R., Prostaglandins suppress an outward potassium current in embryonic rat sensory neurons, *J. Neurophysiol.,* 77, 167, 1997.

142. Gold, M. S., Shuster, M. J., and Levine, J. D., Role of a Ca(2+)-dependent slow afterhyperpolarization in prostaglandin E2-induced sensitization of cultured rat sensory neurons, *Neurosci. Lett.,* 205, 161, 1996.

143. Smith, R. D. and Goldin, A. L., Phosphorylation of brain sodium channels in the I-II linker modulates channel function in Xenopus oocytes, *J. Neurosci.,* 16, 1965, 1996.

144. Dascal, N. and Lotan, I., Activation of protein kinase C alters voltage-dependence of Na channel, *Neuron,* 6, 165, 1991.

145. Malmberg, A. B., Chen, C., Tonegawa, S., and Basbaum, A. I., Preserved acute pain and reduced neuropathic pain in mice lacking PKCγ, *Science,* 278, 279, 1997.

146. Malmberg, A. B., Brandon, E. P., Idzerda, R. L., Liu, H., McKnight, G. S., and Basbaum, A. I., Diminished inflammation and nociceptive pain with preservation of neuropathic pain in mice with a targeted mutation of the type I regulatory subunit of cAMP-dependent protein kinase, *J. Neurosci.,* 17, 7462, 1997.

147. Lewin, G. R., Rueff, A., and Mendell, L. M., Peripheral and central mechanisms of NGF-induced hyperalgesia, *Eur. J. Neurosci.,* 6, 1903, 1994.

148. De Felipe, C., Herrero, J. F., O'Brien, J. A., Palmer, J. A., Doyle, C. A., Smith, A. J., Laird, J. M., Belmonte, C., Cervero, F., and Hunt, S. P., Altered nociception, analgesia, and aggression in mice lacking the receptor for substance P, *Nature,* 392, 394, 1998.

149. Cao, Y. Q., Manthy, P. W., Carlson, E. J., Gillespie, A. M., Epstein, C. J., and Basbaum, A. I., Primary afferent tachykinins are required to experience moderate to intense pain, *Nature,* 392, 390, 1998.

150. Neumann, S., Doubell, T. P., Leslie, T., and Woolf, C. J., Inflammatory pain hypersensitivity mediated by phenotypic switch in myelinated primary sensory neurons, *Nature,* 384, 360, 1996.

151. Janig, W., Levine, J. D., and Michaelis, M., Interactions of sympathetic and primary afferent neurons following nerve injury and tissue trauma, *Prog. Brain Res.,* 113, 161, 1996.

152. Gonzales, R., Sherbourne, C. D., Goldyne, M. E., and Levine, J. D., Noradrenaline-induced prostaglandin production by sympathetic postganglionic neurons is mediated by alpha 2-adrenergic receptors, *J. Neurochem.,* 57, 1145, 1991.

153. Davis, B. M., Albers, K. M., Seroogy, K. B., and Katz, D. M., Overexpression of nerve growth factor in transgenic mice induces novel sympathetic projections to primary sensory neurons, *J. Comp. Neurol.,* 349, 464, 1994.

154. Woolf, C. J., Shortland, P., and Coggeshall, R. E., Peripheral nerve injury triggers central sprouting of myelinated afferents, *Nature,* 355, 75, 1992.

155. Lewin, G. R., McKintosh, E., and McMahon, S. B., NMDA receptors and activity dependent tuning of the receptive fields of spinal cord neurons, *Nature,* 369, 482, 1994.

156. Bogdanov, Y. D., Ovchinnikov, D. A., Balaban, P. M., and Belyavsky, A., V. Novel gene HCS1 is specifically expressed in the giant interneurones of the terrestrial snail, *Neuroreport,* 5, 589, 1994.

157. Dulac, C. and Axel, R., A novel family of genes encoding putative pheromone receptors in mammals, *Cell,* 83, 195, 1995.

158. Liang, P. and Pardee, A. B., Differential display of eukaryotic messenger RNA by means of the polymerase chain reaction, *Science,* 14, 967, 1992.

159. Dworkin, M. B. and Dawid, I. B., Construction of a cloned library of expressed embryonic gene sequences from Xenopus laevis, *Dev. Biol.,* 76, 435, 1980.

160. Hedrick, S. M., Cohen, D. I., Nielsen, E. A., and Davis, M. M., Isolation of cDNA clones encoding T cell-specific membrane-associated proteins, *Nature,* 308, 149, 1984.

161. Zaraisky, A. G., Lukyanov, S. A., Vasiliev, O. L., Smirnov, Y. V., Belyavsky, A. V., and Kazanskaya, O. V., A novel homeobox gene expressed in the anterior neural plate of the Xenopus embryo, *Dev. Biol.,* 152, 373, 1992.

162. Akopian, A. N. and Wood, J. N., Peripheral nervous system-specific genes identified by subtractive cDNA cloning, *J. Biol. Chem.,* 270, 21264, 1995.

163. Saito, T., Greenwood, A., Sun, Q., and Anderson, D. J., Identification by differential RT-PCR of a novel paired homeodomain protein specifically expressed in sensory neurons and a subset of their CNS targets, *Mol. Cell. Neurosci.,* 6, 280, 1995.

164. Kaplan, J. M. and Horvitz, H. R., A dual mechanosensory and chemosensory neuron in Caenorhabditis elegans, *Proc. Natl. Acad. Sci. U.S.A.,* 90, 2227, 1993.

165. Way, J. C. and Chalfie, M., mec-3, a homeobox-containing gene that specifies differentiation of the touch receptor neurons in C. elegans, *Cell,* 54, 5, 1988.

166. Dambly-Chaudiere, C., Jamet, E., Burri, M., Bopp, D., Basler, K., Hafen, E., Dumont, N., Spielmann, P., Ghysen, A., and Noll, M., The paired box gene pox neuro: a determinant of poly-innervated sense organs in Drosophila, *Cell,* 69, 159, 1992.

167. Finney, M., Ruvkin, G., and Horvitz, H. R., The C. elegans cell lineage and differentiation gene unc-86 encodes a protein with a homeodomain and extended similarity to transcription factors, *Cell,* 55, 757, 1988.

168. Gerrero, M. R., McEvilly, R. J., Turner, E., Lin, C. R., O'Connell, S., Jenne, K. J., Hobbs, M. V., and Rosenfeld, M. G., Brn-3.0: a POU-domain protein expressed in the sensory, immune, and endocrine systems that functions on elements distinct from known octamer motifs, *Proc. Natl. Acad. Sci. U.S.A.,* 90, 10841, 1993.

169. Turner, E. E., Jenne, K. J., and Rosenfeld, M. G., Brn-3.2: a Brn-3-related transcription factor with distinctive central nervous system expression and regulation by retinoic acid, *Neuron,* 12, 205, 1994.

170. Akopian, A. N., Abson, N. C., and Wood, J. N., Molecular genetic approaches to nociceptor development and function, *Trends Neurosci.*, 19, 240, 1996.

171. Chalfie, M. and Wolinsky, E., The identification and suppression of inherited neurodegeneration in Caenorhabditis elegans, *Nature*, 345, 410, 1990.

172. Driscoll, M. and Chalfie, M., The mec-4 gene is a member of a family of Caenorhabditis elegans genes that can mutate to induce neuronal degeneration, *Nature*, 349, 588, 1991.

173. Huang, M. and Chalfie, M., Gene interactions affecting mechanosensory transduction in Caenorhabditis elegans, *Nature*, 367, 467, 1994.

174. Savage, C., Hamelin, M., Culotti, J. G., Coulson, A., Albertson, D. G., and Chalfie, M., mec-7 is a beta-tubulin gene required for the production of 15-protofilament microtubules in Caenorhabditis elegans, *Genes Dev.*, 3, 870, 1989.

175. Huang, M., Gu, G., Ferguson, E. L., and Chalfie, M., A stomatin-like protein necessary for mechanosensation in C. elegans, *Nature*, 378, 292, 1995.

176. Mogil, J. S., Sterenberg, W. F., Marek, P., Sadowski, B., Belknap, J. K., and Liebeskind, J. C., The genetics of pain and pain inhibition, *Proc. Natl. Acad. Sci. U.S.A.*, 93, 3048, 1996.

177. Dyck, P. J., Mellinger, J. F., Reagan, T. J., Horowitz, S. J., McDonald, J. W., Litchy, W. J., Daube, J. R., Fealey, R. D., Go, V. L., Kao, P. C., Brimijoin, W. S., and Lambert, E. H., Not 'indifference to pain' but varieties of hereditary sensory and autonomic neuropathy, *Brain*, 106, 373, 1983.

178. Nicholson, G. A., Dawkins, J. L., Blair, I. P., Kennerson, M. L., Gordon, M. J., Cherryson, A. K., Nash, J., and Bananis, T., The gene for hereditary sensory neuropathy type I (HSN-I) maps to chromosome 9q22.1-q22.3, *Nat. Genet.*, 13, 101, 1996.

179. Davar, G., Shalish, C., Blumenfeld, A., and Breakfield, X. O., Exclusion of p75NGFR and other candidate genes in a family with hereditary sensory neuropathy type II, *Pain*, 67, 135, 1996.

180. Blumenfeld, A., Slaugenhaupt, S. A., Axelrod, F. B., Lucente, D. E., Maayan, C., Liebert, C. B., Ozelius, L. J., Trofatter, J. A., Haines, J. L., Breakefield, X. O., et al., Localization of the gene for familial dysautonomia on chromosome 9 and definition of DNA markers for genetic diagnosis, *Nat Genet.*, 4, 160, 1993.

181. Slaugenhaupt, S. A., Blumenfeld, A., Liebert, C. B., Mull, J., Lucente, D. E., Monahan, M., Breakefield, X. O., Maayan, C., Parada, L., Axelrod, F. B., et al., The human gene for neurotrophic tyrosine kinase receptor type 2 (NTRK2) is located on chromosome 9 but is not the familial dysautonomia gene, *Genomics*, 25, 730, 1995.

182. Indo, Y., Tsuruta, M., Hayashida, Y., Karim, M. A., Ohta, K., Kawano, T., Mitsubuchi, H., Tonoki, H., Awaya, Y., and Matsuda, I., Mutation in the TRKA/NGF receptor gene in patients with congenital insensitivity to pain with anhidrosis, *Nat. Genet.*, 13, 485, 1996.

183. Cappecchi, M. R., Altering the genome by homologous recombination, *Science*, 244, 1288, 1989.

184. Gu, M., Marth, J. D., Orban, P. C., Mossmann, H., and Ragewsky, K., Deletion of a DNA polymerase b gene segment in T cells using cell type-specific gene targeting, *Science*, 265, 103, 1994.

185. Tsein, J. Z., Chen, D. F., Gerber, D., Tom, C., Mercer, E. H., Anderson, D. J., Mayford, M., Kandel, E. R., and Tonegawa, S., Subregion- and cell type-restricted gene knockout in mouse brain, *Cell*, 87, 1317, 1996.

186. Kuhn, R., Schwenk, F., Aguet, M., and Rajewsky, K., Inducible gene targeting in mice, *Science*, 269, 1427, 1995.

187. Mayford, M., Bach, M. E., Huang, Y. Y., Wang, L., Hawkins, R. D., and Kandel, E. R., Control of memory formation through regulated expression of a CaMKII transgene, *Science*, 274, 1678, 1996.

188. Klein, R., Smeyne, R. J., Wurst, W., Long, L. K., Auerbach, B. A., Joyner, A. L., and Barbacid, M., Targeted disruption of the trkB neurotrophin receptor gene results in nervous system lesions and neonatal death, *Cell*, 75, 113, 1993.

189. Jones, K. R., Farinas, I., Backus, C., and Reichardt, L. F., Targeted disruption of the BDNF gene perturbs brain and sensory neuron development but not motor neuron development, *Cell*, 76, 989, 1994.

190. Lee, K. F., Li, E., Huber, L. J., Landis, S. C., Sharpe, A. H., Chao, M. V., and Jaenisch, R., Targeted mutation of the gene encoding the low affinity NGF receptor p75 leads to deficits in the peripheral sensory nervous system, *Cell*, 69, 737, 1992.

191. Airaksinen, M. S., Koltzenburg, M., Lewin, G. R., Masu, Y., Helbig, C., Wolf, E., Brem, G., Toyka, K. V., Thoenen, H., and Meyer, M., Specific subtypes of cutaneous mechanoreceptors require neurotrophin-3 following peripheral target innervation, *Neuron*, 16, 287, 1996.

192. Liu, X., Ernfors, P., Wu, H., and Jaenisch, R., Sensory but not motor neuron deficits in mice lacking NT4 and BDNF, *Nature,* 375, 238, 1995.

193. Erkman, L., McEvilly, R. J., Luo, L., Ryan, A. K., Hooshmand, F., O'Connell, S. M., Keithley, E. M., Rapaport, D. H., Ryan, A. F., and Rosenfeld, M. G., Role of transcription factors Brn-3.1 and Brn-3.2 in auditory and visual system development, *Nature,* 381, 603, 1996.

194. McEvilly, R. J., Erkman, L., Luo, L., Sawchenko, P. E., Ryan, A. F., and Rosenfeld, M. G., Requirement for Brn-3.0 in differentiation and survival of sensory and motor neurons, *Nature,* 384, 574, 1996.

195. Matthes, H. W. D., Maldonado, R., Simonin, F., Valverde, O., Slowe, S., Kitchen, I., Befort, K., Dierich, A., LeMeur, M., Dolle, P., Tzavara, E., Hanoune, J., Roques, B. P., and Kieffer, B. L., Loss of morphine-induced analgesia, reward effect and withdrawal symptoms in mice lacking the $\mu$-opioid-receptor gene, *Nature,* 383, 819, 1996.

196. Konig, M., Zimmer, A. M., Steiner, H., Holmes, P. V., Crawley, J. N., Brownstein, M. J., and Zimmer, A., Pain responses, anxiety and aggression in mice deficient in pre-proenkephalin, *Nature,* 383, 535, 1996.

197. Rubinstein, M., Mogil, J. S., Japon, M., Chan, E. C., Allen, R. G., and Low, M. J., Absence of opioid stress-induced analgesia in mice lacking beta-endorphin by site-directed mutagenesis, *Proc. Natl. Acad. Sci. U.S.A.,* 93, 3995, 1996.

198. Ledent, C., Vaugeois, J.-M., Schiffmann, S. N., Pedrazzini, T., El Yacoubi, M., Vanderhaeghen, J.-J., Costentin, J., Heath, J. K., Vassart, G., and Parmentier, M., Aggressiveness, hypoalgesia and high blood pressure in mice lacking the adenosine $A_{2a}$ receptor, *Nature,* 388, 674, 1997.

# 20 Feeding and Obesity Genes

*Frank H. Koegler and Lori K. Singer*

## CONTENT

## INTRODUCTION

Because of its prevalence and the costs associated with treating its primary and secondary complications, obesity is a major health concern. Recent analysis from the National Health and Nutrition Examination Survey (NHANES) indicates that almost 35% of adults in the United States are overweight. In addition, approximately 14% of children and 12% of adolescents are overweight.[53]

These figures show increases in the proportion of overweight individuals in all age groups compared to the NHANES conducted between 1976 and 1980.[44] Common medical hazards associated with obesity include diabetes mellitus, hypertension, cardiovascular disease, dyslipidemia, and several forms of cancer. A better understanding of the mechanisms controlling the development and maintenance of obesity will lead to the discovery of novel anti-obesity treatments and the improvement of existing anti-obesity treatments, thus reducing the occurrence of secondary complications associated with obesity.

The basic mechanism responsible for obesity is an imbalance between energy intake and energy expenditure although the precise relationship between these two is not fully understood. It is clear that, as in most forms of human obesity, genetic and environmental factors play a role in determining both behavioral and physiologic controls of energy balance. Although certain syndromes which cause obesity, such as Prader-Willi and Bardet-Biedl (BBS), have been traced to specific chromosomes, these syndromes and the majority of human obesities are probably highly multigenic in nature. The number of genetic loci that contribute to human obesity is unknown and because of the complex relationship between genetic and environmental factors, the use of animal models for the study of obesity is warranted. Because obesities are frequently multigenic in nature and can be the result of both metabolic and neural/behavioral causes, this chapter will discuss both neural and metabolic components implicated in the genetics of obesity. The following will focus on information gathered from new and established, genetically based models of obesity in animals, notably rodents, as well as several of the common genes for individual neurochemicals and hormones known to be involved in nonpathologic control of food intake. The characteristics and genetic components of the animal obesities will be described and potential relationships will be suggested in regard to the genetic basis of human obesity. A table juxtaposing quantitative trait loci for obesity and human obesity syndromes with the relative positions of genes involved in food intake behavior, metabolism, and energy partition is included.

## GENETIC MODELS

### SINGLE GENES

#### *ob/ob* Mouse and *db/db* Mouse

Perhaps the most well-known gene in obesity research using animal models is the *ob* gene. The popularity of the *ob/ob* (obese) and *db/db* (diabetic) mice was instrumental to the current activity and growth in the field of obesity research using animal models. Parabiotic experiments with these mice performed by Coleman[10] led to the notion that there are circulating factors which must be adequately detected in order to prevent excessive weight gain. In 1994, the *ob* gene was cloned and identified as a 16-kDa protein that is released from adipocytes.[91] It was shown that administering the *ob* gene product, leptin, to the *ob/ob* mouse causes a reduction in body weight,[61] suggesting that the lack of this signal from adipose tissue was a determinator of obesity in the *ob/ob* mouse. It was shown that the *ob/ob* mouse has a nonsense mutation at codon 105 which prevents production of normal leptin as determined by plasma measurements. Administration of leptin to the phenotypically indistinguishable obese *db/db* mouse, however, was without effect on body weight.[30] This ineffectiveness of leptin in the *db* mouse has been attributed to a defect in the leptin receptor gene,[23] and it has been shown that plasma leptin protein levels are indeed elevated in the *db/db* mouse. The leptin receptor is known to have several splice variants and in the case of the *db/db* mouse, the intracellular domain of the receptor, thought to be critical for signaling, is not translated. Of interest is that high levels of leptin receptor message are expressed in lung and kidney with lower levels present in the hypothalamus. The highest concentration of leptin receptor mRNA in the brain is in the choroid plexus, suggesting that leptin may influence transport mechanisms within the brain. The relationship between leptin and neuropeptides and transmitters involved in the control of food intake has been firmly established. Notably, the administration of leptin causes a marked reduction in neuropeptide Y, the most potent orexigenic peptide in the rat.[70,82] A discussion of leptin's role in human obesity follows in this chapter.

## *fa/fa* Rat

Similar to the *db/db* mouse, the fatty rat has a leptin receptor defect. The fatty (*fa/fa*) rat, popularly known as "Zucker" rat,[92] develops hyperinsulinemia and obesity early and maintains an obese phenotype despite the uncharacteristic absence of hyperglycemia. In contrast, when the *fa* gene is carried on the Wistar diabetic rat, hyperglycemia results. The obesity generated in the Zucker fatty rat is generalized throughout all fat depots and cannot be normalized with food restriction. The *fa* allele demonstrates heterozygote effects; heterozygotes display an intermediate form of obesity and show a greater susceptibility to diet-induced adiposity than do lean control animals.[50,81] The finding that *ob* gene expression was significantly augmented in the adipose tissue of Zucker fatty rats strengthened the notion that the *fa* mutation was a leptin receptor alteration.[60] This mutation is mapped to rat chromosome 5 and corresponds to the mouse *db* gene on mouse chromosome 4. Indeed, as in the *db* mouse, the *fa* gene mutation is one of the leptin receptor. However, in contrast to the *db* mutation, the *fa* mutation affects the extracellular, rather than intracellular, domain of the leptin receptor isoforms[78] and causes reduced binding affinity and reduced intracellular signaling.[6,88]

## Yellow Agouti Mouse

Several major advances have been made in the understanding of the basis of the agouti obesity syndromes and this area of research is extremely active at the time of this writing. In the wild-type mouse, the 131 amino acid agouti protein is transiently expressed and antagonizes melanocyte stimulating hormone at the melanocortin 1 receptor (MC1-R). This causes a decrease in eumelanin synthesis and an increase in phaeomelanin synthesis in the hair follicle, resulting in a yellow banding on black coat hairs. The Yellow mouse (Ay$^y$) and its variations (Av$^y$, A$^{sy}$, A$^{iy}$, Ah$^{vy}$, A$^{iapy}$) have pronounced ectopic expression of the agouti protein due to noncoding, promoter region mutations. Ectopic agouti protein overexpression results in a yellow coat color and an obese phenotype, as well as hyperinsulinemia and hyperglycemia. In addition, we have shown that when maintained on a diet that allows the selection of pure fat, carbohydrate, and protein, yellow agouti mice select a greater proportion of fat for daily intake relative to wild-type mice. The different agouti locus mutations result in obese phenotypes not as severe as those for the *ob/ob* and *db/db* mice. Several of the nonlethal, allele series mutations (i.e., viable yellow; A$^{vy}$) lead to development of an abnormal yellow coat due to agouti antagonism of melanocortin receptors, which regulate eumelanin synthesis. The distinct yellow coloration of the mutant mice is proportional to the severity of the obesity. With regard to obesity, the agouti protein antagonizes the melanocortin 4 receptor (MC4-R) found in the hypothalamus[49] and it has recently been demonstrated that agonists of the MC4-R reduce food intake in the agouti mouse and in the *ob/ob* mouse.[18] It is proposed that normal inhibition of feeding behavior via the action of hypothalamic MC4-Rs is interrupted by the overexpression of agouti protein. This notion that the effects of agouti are via the MC4-R is corroborated by the results of transgenic studies in which MC4-R expression is eliminated; MC4-R knockout mice become hyperphagic and spontaneously develop obesity, yet do not share the coat color abnormalities with the yellow mouse.[35] The agouti gene is located on mouse chromosome 2, and the location in humans of the 85% homologous sequence, ASIP, is 20q11.2. This location is overlapping with the region containing the PPCD gene (corneal dystrophy, hereditary polymorphous), which has been associated with human obesity.[32]

## *tub/tub* Mouse

The tubby (*tub*) mutation, first described by Coleman and Eicher in 1990[9] and cloned in 1996,[40] results in a gradually developing, yet pronounced obesity. Several other members of this gene family have been identified to date (e.g., TULP1 and TULP2, which map to 6p21.3 and 19q13.1, respectively.)[59] The *tub* gene resides on chromosome 7[7] near the mouse hemoglobin gene and near the chromosome 7 obesity quantitative trait locus (QTL). The product of *tub* may be a phosphodiesterase because sequence analysis shows ~60% similarity to known phosphodiesterases. Although the physiology of obesity due to *tub* is not known, a potential candidate gene thought to be responsible for the tubby phenotype is expressed in hypothalamic brain nuclei known to be important to energy regulation; the ventromedial, paraventricular, and arcuate nuclei of the hypothalamus.[40] Of interest is the mapping of the cholecystokinin (CCK) B receptor gene in the same region as the tubby mutation. CCK is a potent stimulator of short-term, meal-related satiety and may be another candidate gene contributing to the development of obesity in the tubby mouse.[67] The human homolog of tubby, TUB, resides at 11p15 and therefore the human HBB locus (11p15.5) may be used as a linkage marker for obesity studies in humans.[38] In addition, the BBS1 locus for the Bardet-Biedl human obesity syndrome, which shares retinopathy as well as obesity with the tubby phenotype, is also found at 11q.[47]

### *Cpe^fat* Mouse

The *fat* mouse is another genetic model exhibiting late-onset obesity and what was originally thought to be pronounced hyperinsulinemia. Interestingly, the *fat* mouse remained sensitive to exogenous insulin suggesting an abnormality with the endogenous insulin. This mutation maps to chromosome 8 in the vicinity of the carboxypeptidase E (*Cpe*) gene and results in the reduction of activity in this prohormone processing enzyme.[56] Indeed, the *fat* mouse has an excessive amount of proinsulin because of the *Cpe* mutation. Not only is insulin aberrantly processed, but CCK, a putative satiety-signaling peptide, has also been shown to be markedly reduced (75%) in the brain of the *fat* mouse.[4,22] Improper prohormone and proneuropeptide processing may affect the proopiomelanocortin derivatives adrenocorticotropic hormone (ACTH) and melanocyte-stimulating hormone (α-MSH), as well as neuropeptide Y, leading to a multiple neuroendocrine deficiency and resultant obesity.

## Multiple Gene Mouse Models

### AKR/J and SWR/J

The AKR/J mouse model of obesity has produced several quantitative trait loci (QTL) for obesity and is a useful model for dissecting multigenic effects on phenotype. The AKR/J, or "diet-induced obese," mouse has approximately sixfold greater adipose mass and is more susceptible to dietary obesity from a high-fat diet than is the lean SWR/J counterpart mouse. West et al.[85] have shown that the obesity-related traits are determined by several genetic loci, designated "Do" (dietary obese), and that there is significant dominance of the obese AKR/J genotype. Analysis has not revealed stringent linkage with markers associated to the known mouse obesity genes *ob*, *db*, *tub*, and *fat*. However, a QTL, Do1, was identified distal to the *db* gene on chromosome 4. This QTL corresponds to human chromosome 1, distal to the leptin receptor (LEPR) gene. Further investigation of this strain revealed QTL for susceptibility to dietary obesity on chromosomes 9 and 15 (corresponding lod-scores of (LODs) 4.85 and 3.93, respectively).[86] Furthermore, using an intercross between the lean CAST/Ei and the obese C57BL/6J strains, researchers found a QTL for body fat on chromosome 15 at the Do3 locus previously identified in the AKR mouse.[89] Do3 maps to human chromosome 8, which contains the gene for the $\beta_3$-adrenergic receptor which is known to stimulate lipolysis.

### BSB Mouse

The BSB mouse displays spontaneous polygenic obesity and has been the subject of several studies of the genetic basis of obesity. This mouse is the result of a *mus spretus* X C57BL6J backcross and is known to be deficient in the response to ACTH. QTL on chromosome 7 in this cross have been established for adiposity, hepatic lipase, and total plasma cholesterol, suggesting a link between plasma lipoproteins and obesity.[83] The multigenic obesity 4 (Mob-4) QTL in the BSB mouse is at the same location as the previously mentioned Do3 locus derived from the AKR/J SWR/J cross and overlaps the human lipoprotein lipase gene.

## NZO (New Zealand Obese) Mouse

The less-well-studied New Zealand Obese (NZO) mouse displays a polygenic syndrome of obesity, hyperphagia, hyperglycemia, and hyperinsulinemia. This syndrome, as in many other obesities, is accompanied by a marked elevation of leptin in adipose tissue and serum. The promoter region and the complementary DNA of the *ob* gene in the NZO mouse are identical with the wild-type sequence (C57BL, BALB/c), except that the transcription start is located 5 bp upstream of the reported site. NZO mice fail to reduce food intake in response to peripheral leptin. However, NZO mice are responsive to i.c.v. leptin.[29] A substitution in the leptin receptor gene coding for the intracellular domain of the receptor is found in the NZO mouse; however, an identical substitution is found in the relatively lean New Zealand Black strain, suggesting that this receptor difference is not a mediator of the obesity seen in the NZO. Because of the insensitivity to peripheral leptin, in contrast to central leptin, it is proposed that an abnormality in leptin transport into the brain may be a more important determinant of the NZO phenotype.[36]

### TRANSGENIC ANIMALS

### Brown Adipose Tissue and Uncoupling Protein

Brown adipose tissue (BAT) has the capacity for diet-induced thermogenesis and thus may play a role in the prevention of obesity in rodents. Transgenic mice that are deficient in brown adipose tissue were created by Lowell et al.[48] by targeted expression of a diphtheria toxin-A chain inserted into the uncoupling protein (UCP1) gene. Animals with a deficiency in BAT become obese, but not when they live in a thermoneutral environment.[52] A similar approach targeting the diphtheria toxin to the adipocyte aP2 gene also produced obesity-resistant animals.[66] A null mutation in the aP2 promoter for fatty acid binding protein (FABP) produced nondiabetic obesity and reduced levels of tumor necrosis factor alpha, which reduces food intake.[34] BAT-deficient mice are normophagic during the early development of obesity, indicating that these animals have increased metabolic efficiency. Hyperphagia becomes evident, however, as the obesity progresses. More recently, transgenic mice were produced by targeted inactivation of the UCP1 gene. Normally, uncoupling protein allows thermogenesis in BAT via uncoupling of the proton-electrochemical gradient from ATP synthesis. Interestingly, these knockout mice were neither hyperphagic nor obese.[15] It was proposed that there may have been a compensation for the lack of UCP1 by the induced expression of the recently described UCP2.[19] Transgenic mice overexpressing UCP1 via the aP2 fat-specific promoter demonstrated a reduction in subcutaneous fat.[41] The role of UCP1 in human energy regulation and obesity is likely to be minor because of the small amounts of brown fat in adults and the relative thermoneutrality of human environments. In contrast, UCP2 may be an important mediator of energy expenditure in humans as it is expressed in many human tissues and its expression is responsive to dietary fat. Furthermore, an association between UCP2 and resting metabolic rate has been demonstrated. UCP2 maps to the same region as the tubby mutation and several mouse obesity QTL. (For a review, see Ref. 20.)

### $\beta_3$-Adrenergic Receptors and Dopamine $\beta$-Hyroxylase

The $\beta_3$ receptor stimulates lipolysis when sympathetically activated and synthetic agonists of this receptor are of interest with respect to the treatment of obesity. In the mouse, both white and brown adipose tissue express all three $\beta$-adrenergic receptor ($\beta$-AR) subtypes; however, the ratio of $\beta_3$-AR is much higher than $\beta_1$- or $\beta_2$-AR. Transgenic mice completely lacking $\beta_3$-AR do not display cAMP stimulation by $\beta_3$-agonists and have been examined with respect to body weight and adiposity.[77] Although the $\beta_3$-AR knockout mice are not significantly heavier than the wild-type mice, they do exhibit moderately increased fat stores. Interestingly, $\beta_1$- but not $\beta_2$-ARs are up-regulated in both white and brown adipose tissue of the knockout mice. Because the $\beta_3$ knockout mice had only a slight tendency towards obesity compared to the wild-type mice and the fact that in other species, including humans, $\beta_1$- and $\beta_2$-AR levels are higher than $\beta_3$-AR levels in adipose tissue, researchers recently expressed the human $\beta_1$-AR in white and brown adipose tissue of mice.[74] Overexpression of the $\beta_1$-AR in mice results in a partial resistance to obesity. These mice gain weight more slowly and have smaller adipose tissue stores than wild-type mice, especially when maintained on a high-fat diet. The human $\beta_3$-AR (ADRB3) maps to 8p12–p11.2. A mutation in the ADRB3 gene has been associated with morbid obesity in humans in one study; however, several different studies failed to find a significant relationship with obesity.[8,24,33,39,55]

Transgenic animals designed to not express dopamine $\beta$-hydroxylase, the synthetic enzyme for adrenaline and noradrenaline, become hyperphagic yet resistant to obesity.[80] The unexplained obesity resistance is not due to shivering thermogenesis, or up-regulation of UCP2, but rather to an increase in metabolic rate, despite the lack of the major sympathetic neurotransmitters.

## GLUT4

There are six glucose transporters found in mammalian tissues, of which the major one is the insulin-sensitive GLUT4. Since GLUT4 is found mainly in muscle and fat tissue, and is elevated in adipocytes in several models of animal obesity, it is possible that this transporter is a mediator in the development of obesity. To investigate this, Shepherd et al.[71] overexpressed GLUT4 using the fat-specific fatty acid binding protein (FABP) promoter, aP2. Transgenic mice show increased GLUT4 in white and brown adipose tissue but not in other tissues. Compared to nontransgenic littermates, these mice had increased body fat stores, primarily due to an increase in number, but not size, of adipocytes. These transgenic animals can become obese when maintained on a low-fat diet; however, the increased adiposity in nontransgenics due to high-fat feeding is prevented with GLUT4 overexpression.[26]

## Glycerol-3-Phosphate Dehydrogenase

The cytoplasmic enzyme glycerol-3-phosphate dehydrogenase (GPDH) forms glycerol-3-phosphate from dihydroxyacetone phosphate in the glycolytic cycle. This enzyme is present in almost all tissues but has high levels of activity in adipocytes. Mice which overexpress the GPDH gene (*Gdc*-1) have a significant reduction of white fat and hypertrophy of brown fat that progresses throughout life.[43] The human GPD2 (mitochondrial) gene maps to 2q24.1 which is encompassed by a mouse QTL for obesity (Obq2).

## Type II Glucocorticoid Receptor

It is well known that glucocorticoids are essential for all forms of experimental obesity. This is probably not surprising since glucocorticoids are known to promote neuropeptide Y synthesis and secretion, inhibit corticotropin-releasing hormone (CRH) production, affect transcription of the proopiomelanocortin gene, and suppress sympathetically mediated thermogenesis. The effect is mediated through actions on the type II receptor, although glucocorticoid stimulation of type I receptors in the central nervous system also promotes food intake and growth in lean animals.[14] Transgenic mice with decreased glucocorticoid receptor activity were produced by incorporating type II glucocorticoid receptor antisense RNA into mice.[62,63] The construct was linked to a human neurofilament gene promoter so that expression was concentrated in neural tissue. Although the transgenic mice had reduced expression of glucocorticoid type II receptors, they had lower energy intake and energy expenditure and had increased body fat content. The increase in body fat was not expected in light of the overwhelming evidence that removal of adrenal steroids prevents obesity, and this increased adiposity may have been due to the imbalance between central and peripheral glucocorticoid receptor activity. In contrast, transgenic mice overexpressing CRH develop a cushingoid increase of fat distribution in the trunk.[76] Of particular recent significance is the demonstration that glucocorticoids stimulate the expression of the leptin gene in adipose tissue and attenuate the inhibition of food intake in response to leptin administered intracerebroventricularly.[90]

## 5-HT$_{2C}$ Receptor

There is considerable evidence implicating the biogenic amine neurotransmitter, serotonin, in the control of food intake. Drugs which increase serotoninergic activity have been used as a treatment for human obesity, and serotoninergic agonists have been shown to selectively reduce the intake of fat in rodents.[31] A targeted disruption of 5-HT$_{2C}$ receptors in mice results in mice that are overweight.[79] This increased weight of mutant mice is due to hyperphagia and not a metabolic effect because transgenic mice that were pair-fed with the wild-type mice are not overweight. 5-HT$_{2C}$ knockout mice are also susceptible to audiogenic seizures and premature death. The 5-HT$_{2C}$ locus (formerly named HTR1C) maps to human Xq24 and to the mouse X chromosome, region D-F4. A mutation in the human 5-HT receptor was shown not to be linked to obesity; despite that 2C receptor knockout mice become obese, and that many human obesities respond to serotoninergic treatment, in at least one study, linkage between obesity and a 5-HT$_{2C}$ mutation was not established.[46] However, other serotonin receptor subtypes have yet to be examined in the context of human obesity.

## Neuropeptide Y

Neuropeptide Y (NPY) is one of the most potent central stimulators of feeding in the rat and has been associated with nearly all hyperphagias and many obesities in rodents and primates. Indeed, NPY is elevated in the brain of the *ob/ob* mouse. Chronic central infusion of NPY leads to obesity,[75] and food deprivation leads to increases of mRNA for NPY in the hypothalamus.[70] Interestingly, transgenic mice lacking NPY do not exhibit hypophagia and have normal refeeding responses following deprivation,[16] suggestive of compensation by other neural elements during development. When leptin-deficient *ob/ob* mice are made not to express NPY, food intake and the characteristic obesity of the obese mouse is reduced.[17] There are at least six NPY receptor subtypes expressed centrally and peripherally, the most likely of which to be involved with food intake are the Y1 and Y5 subtypes. The Y1 subtype is expressed in peripheral tissues as well as in the brain; however, its role in food intake is not as well established as the Y5 subtype. The Y5 receptor is not expressed outside of the mouse brain and is thought to be the subtype involved in feeding responses.[25] Consistent with this idea is the presence of the Y5 receptor in the paraventricular nucleus of the hypothalamus, where NPY is thought to be most potent at stimulating food intake. Intracerebroventricular injections of antisense oligonucleotides targeted to the preproNPY gene prevent the increase in hypothalamic NPY levels during food deprivation and inhibit deprivation-induced food intake.[1,68] Likewise, antisense treatment with oligos directed at the Y5 receptor gene reduces both basal food intake and NPY-induced feeding.[68] Both the Y1 and Y5 receptor genes map to human chromosome 4, syntenic with the mouse Do6 locus. However, a human study using a French population which attempted to associate NPY and the NPY1 and Y5 receptor genes to obesity failed to establish a link using either polymorphic marker or linkage analysis techniques.[65]

## ART

Agouti-related transcript (ART, agrt, agrp), first isolated from mouse, is implicated in feeding behavior and obesity. It is overexpressed in the hypothalamus of the *ob/ob* and *db/db* mice[72] and, like agouti, is an antagonist of melanocyte-stimulating hormone at the MC3-R and MC4-R, albeit with much stronger binding affinity.[21] Transgenic mice overexpressing ART via a beta-actin promoter recapitulate the obesity seen in the agouti syndrome[28] and the MC4-R deficient mouse.[35] However, the characteristic A$^{vy}$ coat colorations are not exhibited, suggesting that ART does not antagonize the MC1-R in this model. The human ART gene maps to 16q22 and is near the BBS2 locus.

## Lipoprotein Lipase

Fatty acid entry into adipose and muscle tissue is regulated by lipoprotein lipase (LPL) and the relative activity in either tissue is a mediator of net utilization versus storage of fat. Transgenic animals deficient in total LPL are extremely hypertriglyceridemic and die within days of birth, whereas overexpression in muscle leads to lower plasma triglyceride levels and a decrease of total body fat as a percentage of body weight. Muscle overexpression of LPL also reduces the susceptibility to high-fat-diet-induced obesity.[37] *ob/ob* mice rendered deficient in adipose tissue LPL have increased endogenous fatty acid synthesis but diminished weight and fat mass.[84] The human LPL gene resides on chromosome 8 syntenic with the mouse Do6 obesity QTL.

## HUMAN OBESITY GENES AND SYNDROMES

Body fat, relative to height, is correlated within families and it is generally accepted that obesity has a familial nature. The degree to which genetics influences obesity in humans is not undisputed; however, several studies with monozygotic twins raised in separate environments suggest that the heritability of body mass index is between 40 and 70%.[49a,76a] The pursuit of specific genes responsible for human obesities has provided several, albeit few, cases of single gene mutations leading to severe obesity. However, as the prevalence of cases suggests, single gene mutations are likely to be the exception in the determination of body mass. Rather, it is likely that development of less dramatic obesity is due to the regulation of many genes involved in the control of food intake, metabolism, and energy partition. The following section will address some of the human obesity populations/syndromes with suggestive or established genetic components, as well as reports of the leptin/leptin receptor mutation in humans.

### LEPTIN/LEPTIN RECEPTOR

As in rodents, serum leptin concentrations in humans are typically related to total adiposity. In at least several cases of extreme obesity, elevated adiposity was not associated with increased leptin levels, suggesting a leptin defect. Analysis of the leptin gene in these cases revealed a mutation resulting in congenitally leptin-deficient patients who are extremely obese,[54] analogous to the *ob/ob* mouse obesity. While this clearly demonstrates the importance of leptin in adiposity regulation, to date, these cases are rare, supporting multigenic explanations for obesity.

To examine the multigenic component of leptin's role in adiposity regulation, a genome-wide linkage analysis in a Mexican American population was performed to identify genes which could explain variations in leptin levels.[11] A QTL with a high LOD score was found on chromosome 2, representing nearly 50% of the variation in leptin levels in this population. A microsatellite polymorphism from this QTL mapped to chromosome 2p21 in the vicinity of the glucokinase regulatory protein and proopiomelanocortin, suggesting that these genes may influence leptin levels and adiposity.

Several attempts at directly implicating the leptin receptor in human obesity have failed. A study of obese and lean Japanese subjects detected several differences in the LEPR gene; however, receptor defects analogous to those seen in the Zucker rat were not detected, nor was there a difference between the obese and lean subjects.[51] Furthermore, another study failed to find a difference in leptin receptor mRNA in hypothalamic tissue obtained from lean and obese subjects.[12] As stated, however, combination defects between leptin production and receptor activation have potential for contributing to aberrant adiposity regulation.

## OBESITY QTL

One QTL for percentage body fat and total adiposity discovered on mouse chromosome 2 has recently been demonstrated to be a QTL for human obesity (OQTL) and maps syntenically to human 20q13.11–q13.2.[45] The OQTL has linkage to body mass index and percentage body fat. As shown in Table 20.1, this chromosome 20 locus encompasses the adenosine deaminase and melanocortin 3 receptor genes, suggesting that they may be involved in the regulation of body weight.

## PRADER-WILLI AND WILSON TURNER SYNDROMES

Obesity is one of many symptoms of both Prader-Willi (PWS) and Wilson Turner syndromes (WTS). Both syndromes exhibit characteristic mental retardation and brachydactyly/acromicria. PWS, which is an autosomal dominant disease, includes complications such as hypogonadism, and hypotonic musculature. This syndrome is caused by deletion of a segment of paternal chromosome 15 or replacement of the entire paternal chromosome resulting in two maternal homologs. An alteration of PWS in which the maternal, rather than paternal, deletion takes place leads to a different phenotype known as the Angelman syndrome. In PWS, diminished fetal growth and activity is followed by hyperphagia which typically begins in early childhood. The hyperphagia can be difficult to control and food restriction in young PWS patients is behaviorally problematic.[13] Fat metabolism in PWS is altered; stimulated lipolysis is reduced and fat synthesis is enhanced. This also occurs in the fasted state, which provides a possible explanation for the hyperphagia and may provide a basis for the increased adiposity. This disorder maps to 15q11 in the same region as ubiquitin-protein ligase E3A and the SNRPN (small nuclear ribonucleoprotein N) gene and is one of the most common human obesity syndromes to date.[3] WTS, which is less common than PWS, is an X-linked recessive syndrome described with primary mental retardation, acromicria, gynecomastia, obesity, and difficulty with speech.[87] WTS is described in successive generations from a single family and has been mapped from Xp21.1–q22.

## BARDET-BIEDL AND COHEN SYNDROMES

Bardet-Biedl syndrome (BBS) is another disorder which includes obesity as a manifestation. Typical of BBS are mental retardation, polydactyly, retinopathy, and hypogenitalism/hypogonadism. This autosomal recessive disorder maps to four different loci (BBS1–4). The BBS4 locus is associated with the most severe obesity and is in the same region as the neuromedin B gene. Interestingly, neuromedin B is a member of the bombesin-like family of peptides, some of which potently inhibit food intake when given centrally in the rat. This locus is syntenic to mouse Do2 QTL, whereas the BBS2 locus, located near the ART gene, is associated with the least severe obesity.[5] In addition to the retinopathy, mental retardation, and obesity seen in BBS, Cohen syndrome exhibits facial macrocephaly. This syndrome maps to 8q22–q23 and is syntenic with several of the mouse obesity QTL as well as the thyrotropin-releasing hormone receptor and proenkephalin.

## PIMA POPULATION

The Pima Native American population has an increased prevalence of obesity and has been the subject of recent studies attempting to determine the relevant/responsible genetic components of the obesity. Initial attempts to establish obesity-related linkages between regions syntenic with known mutant mouse obesity loci failed.[57] Since then, however, Norman et al.[58] have reported suggestive (LOD < 3) linkages of loci to obesity based on genomic searches of this population. Chromosomal regions 11q, 6p, and 3p were shown to be associated with body fat levels. Chromosome 3pter-p21 contains the gene for cholecystokinin, which can potently inhibit food intake in rodents and monkeys. The chromosome 11 locus encodes the genes for the progesterone and melatonin 1B receptors, as well as an inotropic glutamate receptor.

## TABLE 20.1
## Relative Positions in Human Genome of Obesity Quantitative Trait Loci (QTL), Obesity Syndromes and Proteins Related to Food Intake and Obesity

| Human | QTL/Syndrome | Gene | Mouse | Comments |
|---|---|---|---|---|
| 1p36-p32 | DO1 | | 4 | |
| 1p33-p31 | | fatty acid binding protein 3 | 4 | muscle/heart |
| 1p31-p22 | | leptin receptor | 4 | db/db mouse |
| 2p23.3 | adipose tissue | proopiomelanocortin | 12 | |
| 2p23.3-p23.2 | | glucokinase regulatory protein | | |
| 2p21 | microsatellite polymorphism | | | Pima serum leptin |
| 2p11 | | fatty acid binding protein 1 | 6 | liver |
| 2q12-q32 | Obq2 | | 1 | |
| 2q24.1 | | glycerol phosphate dehydrogenase 2 | | mitochondrial |
| 3pter-p21 | | cholecystokinin | 9 | |
| 3p24.2-p22 | body fat QTL | | | Pima population |
| 3p22-q22 | DO2 | | 9 | |
| 3p13-p12 | BBS3 | | | |
| 4 (4p16.2- | | CCK-A receptor | 5 | |
| 4q28-q31 | DO6 | | 8 | |
| 4q28-q31 | | fatty acid binding protein 2 | 3 | intestinal |
| 4q31 | | uncoupling protein 1 | 8 | |
| 4q31-q32 | | NPY1,5 receptors | 8 | |
| 4q32 | | Carboxypeptidase E (Cpe) | 8 | |
| 5p13 | Mob4 | | 15 | |
| 5q31 | | glucocorticoid receptor | 18 | |
| 6p12-q12 | DO2 | | 9 | |
| 6p12-q14 | Obq2 | | 1 | |
| 7p15.1 | | neuropeptide Y | 6 | |
| 7q21-q34 | Mob2 | | 6 | |
| 7q31.3 | | leptin | 6 | leptin deficiency |
| 8p22 | DO6 | | 8 | |
| 8p22 | | lipoprotein lipase | 8 | |
| 8p12-p11.2 | | $\beta_3$-adrenergic receptor | 8 | |
| 8q21-q23 | DO3 | | 15 | |
| 8q22-q23 | Mob4 | | 15 | |
| 8q23 | | thyrotropin-releasing hormone receptor | 15 | |
| 8q22-q23 | Cohen syndrome | | | |
| 8q23-q24 | | proenkephalin | | |
| 9q34 | | dopamine $\beta$-hydroxylase | 2 | |
| 10q25-q26 | Mob1 | | 7 | |
| 11p15.5- | | CCK-B receptor | 7 | |
| 11p15.5 | | insulin (proinsulin) | 6 (Ins1), 7 (Ins2) | |
| 11p15 | | Tubby (tub) | 7 | |
| 11p15 | Mob1 | | 7 | |
| 11p15 | Obq1 | | 7 | |
| 11q13 | | uncoupling protein 2 | 7 | human white adipose tissue |
| 11q13 | BBS1 | | | |
| 11q13.3- | | galanin | 17 | |
| 11q21-q22 | body fat QTL | | | Pima population |
| 12 | | glycerol phosphate dehydrogenase 1 | 15 | cytosolic |
| 14q24-q32 | Mob3 | | 12 | |
| 15q11 | Prader-Willi | | | |
| 15q11-q26 | Obq1 | | 7 | |
| 15q22.3-q23 | BBS4 | | | "most obese" BBS |
| 15q22-q24 | DO2 | | 9 | |
| 16p13-p11 | Mob1 | | 7 | |
| 16q21 | BBS2 | | | "least obese" BBS |
| 16q22 | | agouti-related transcript | | |
| 16q24.3 | | melanocortin receptor 1 | 8 | |
| 17p13 | | GLUT4 | | "SLC2A4" |
| 18q21.3-q22 | | melanocortin receptor 4 | | |
| 18q23 | | galanin receptor 1 | 18 | |
| 19p13.2 | | insulin receptor | 8 | |
| 19p13 | DO6 | | 8 | |
| 19q13 | Obq1 | | 7 | |
| 20p11-q13 | Mob5 | | 2 | |
| 20p11.2- | PPCD | | | |
| 20q11.2 | | agouti signaling protein | 2 | |
| 20q13.2 | | melanocortin 3 receptor | 2 | |
| 20q13.11 | | adenosine deaminase | 2 | |
| 20q13.11- | OQTL | | | "obesity" |
| Xp21.1-q22 | Wilson-Turner | | | |
| Xq24 | | 5-HT$_{2C}$ receptor | X | |

Mouse chromosomal location is given when known. Shading signifies overlap of map location for QTL/syndromes or genes listed.

## CONCLUSIONS

The genetic influence on a fundamental, necessary behavior, such as eating, does not completely explain variation in that behavior or its resultant effects in a complex environment. However, understanding the genetic component of a behavior will provide insight for devising relevant treatment strategies for disorders related to that particular behavior. It seems likely that the majority of human obesity is not due to major mutations in single genes resulting in nonfunctional proteins. Rather, it is likely that the *regulation* of the genes controlling food intake, metabolism, and energy partition is responsible for the majority of obesities. Future research will continue to reveal the nature of this regulation and its heritable component. Current trends in the pharmacological treatment of obesity include interfering with candidate neurotransmitters involved in food intake (serotonin/Sibutramine); undoubtedly, future trends will target the genetic regulatory components of these candidates. Increasing interest in obesity genetics is leading to the discovery of more receptor proteins involved in body weight regulation and food intake; separating the major effectors from those which are not as influential will become a more challenging and critical pursuit. Experimental animal models are becoming more valuable as tools for studying energy balance, and as they are further characterized, they will provide increased impetus and insight to the development of treatments for human obesity. Furthermore, case, cohort, and subpopulation studies with humans will continue to reveal specific, important mediators of obesity, and comprehensive analysis will lead to a greater understanding of multigenic "common" obesities. With the prevalence of obesity and associated health care expenses increasing, it is imperative that rapid, concise obesity research continues, and that safe, effective treatments are developed and made readily available.

## ACKNOWLEDGMENTS

Special thanks to Dr. George Bray and to Dr. David York for contributions and suggestions.

## REFERENCES

1. Akabayashi, A., Wahlestedt, C., Alexander, J.T., and Leibowitz, S.F., Specific inhibition of endogenous neuropeptide Y synthesis in arcuate nucleus by antisense oligonucleotides suppresses feeding behavior and insulin secretion, *Brain Res. Mol. Brain Res.,* 21(1–2), 55–61, 1994.
2. Bouchard, C., Perusse, L., Chagnon, Y. C., Warden, C., and Ricquier, D., Linkage between markers in the vicinity of the uncoupling protein 2 gene and resting metabolic rate in humans, *Hum. Mol. Genet.,* 6, 1887–1889, 1997.
3. Butler, M. G., Prader-Willi syndrome: current understanding of cause and diagnosis, *Am. J. Med. Genet.,* 35, 319–332, 1990.
4. Cain. B.M., Wang, W., and Beinfeld, M.C., Cholecystokinin (CCK) levels are greatly reduced in the brains but not the duodenums of Cpe(*fat*)/Cpe(*fat*) mice: a regional difference in the involvement of carboxypeptidase E (Cpe) in pro-CCK processing, *Endocrinology,* 138(9), 4034–4037, 1997.
5. Carmi, R., Elbedour, K., Stone, E.M., and Sheffield, V.C., Phenotypic differences among patients with Bardet-Biedl syndrome linked to three different chromosome loci, *Am. J. Med. Genet.,* 59(2), 199–203, 1995.
6. Chua, S.C., Jr., White, D.W., Wu-Peng, X.S., Liu, S.M., Okada, N., Kershaw, E.E., Chung, W.K., Power-Kehoe, L., Chua, M., Tartaglia, L.A., and Leibel, R., Phenotype of fatty due to Gln269Pro mutation in the leptin receptor (Lepr), *Diabetes,* 45(8), 1141–1143, 1996.
7. Chung, W.K., Goldberg-Berman, J., Power-Kehoe, L., and Leibel, R.L., Molecular mapping of the tubby (*tub*) mutation on mouse chromosome 7, *Genomics,* 32(2), 210–217, 1996.
8. Clement, K., Vaisse, C., Manning, B. S. J., Basdevant, A., Guy-Grand, B., Ruiz, J., Silver, K. D., Shuldiner, A. R., Froguel, P., and Strosberg, A. D., Genetic variation in the beta-3-adrenergic receptor and an increased capacity to gain weight in patients with morbid obesity, *N. Eng. J. Med.,* 333, 352–354, 1995.

9. Coleman, D.L., Eicher, E.M., Fat (*fat*) and tubby (*tub*): two autosomal recessive mutations causing obesity syndromes in the mouse, *J. Hered.,* 81(6), 424–427, 1990.

10. Coleman, D.L., Obese and diabetes: two mutant genes causing diabetes-obesity syndromes in mice, *Diabetologia,* 14(3), 141–148, 1978.

11. Comuzzie, A.G., Hixson, J.E., Almasy, L., Mitchell, B.D., Mahaney, M.C., Dyer, T.D., Stern, M.P., MacCluer, J.W., and Blangero, J., A major quantitative trait locus determining serum leptin levels and fat mass is located on human chromosome 2, *Nat. Genet.,* 15(3), 273–276, 1997.

12. Considine, R.V., Considine, E.L., Williams, C.J., Hyde, T.M., and Caro, J.F., The hypothalamic leptin receptor in humans: identification of incidental sequence polymorphisms and absence of the *db/db* mouse and *fa/fa* rat mutations, *Diabetes,* 45(7), 992–994, 1996.

13. Curfs, L. M. G., Verhulst, F. C., and Fryns, J. P., Behavioral and emotional problems in youngsters with Prader-Willi syndrome, *Genet. Counseling,* 2, 33–41, 1991.

14. Devenport, L., Knehans, L., Sundstrom, A., and Thomas, T., Corticosterone's dual metabolic actions, *Life Sci.,* 45, 1389–1396, 1989.

15. Enerbäck, S., Jacobsson, A., Simpson, E.M., Guerra, C., Yamashita, H., Harper, M.E., and Kozak, L.P., Mice lacking mitochondrial uncoupling protein are cold-sensitive but not obese, *Nature,* 387, 90–94, 1997.

16. Erickson, J.C., Clegg, K.E., and Palmiter, R.D., Sensitivity to leptin and susceptibility to seizures of mice lacking neuropeptide Y, *Nature,* 381(6581), 415–421, 1996.

17. Erickson, J.C., Hollopeter, G., and Palmiter, R.D., Attenuation of the obesity syndrome of *ob/ob* mice by the loss of neuropeptide Y, *Science,* 274(5293), 1704–1707, 1996.

18. Fan, W., Boston, B.A., Kesterson, R.A., Hruby, V.J., and Cone, R.D., Role of melanocortinergic neurons in feeding and the agouti obesity syndrome, *Nature,* 385(6612), 165–168, 1997.

19. Fleury, C., Neverova, M., Collins, S., Raimbault, S., Champigny, O., Levi-Meyrueis, C., Bouillaud, F., Seldin, M.F., Surwit, R.S., Ricquier, D., and Warden, C.H., Uncoupling protein-2: a novel gene linked to obesity and hyperinsulinemia, *Nat. Genet.,* 15(3), 269–272, 1997.

20. Flier, J. S. and Lowell, B. B., Obesity research springs a proton leak, *Nat. Genet.,* 15, 223–224, 1997.

21. Fong, T.M., Mao, C., MacNeil, T., Kalyani, R., Smith, T., Weinberg, D., Tota, M.R., and Van der Ploeg, L.H., ART (protein product of agouti-related transcript) as an antagonist of MC-3 and MC-4 receptors, *Biochem. Biophys. Res. Commun.,* 237(3), 629–631, 1997.

22. Fricker, L.D., Berman, Y.L., Leiter, E.H., and Devi, L.A., Carboxypeptidase E activity is deficient in mice with the *fat* mutation. Effect on peptide processing, *J. Biol. Chem.,* 271(48), 30619–30624, 1996.

23. Friedman, J.M., Leibel, R.L., and Bahary, N., Molecular mapping of obesity genes, *Mammalian Genome,* 1(3), 130–144, 1991.

24. Gagnon, J., Mauriege, P., Roy, S., Sjostrom, D., Chagnon, Y. C., Dionne, F. T., Oppert, J.-M., Perusse, L., Sjostrom, L., and Bouchard, C., The trp64arg mutation of the beta-3 adrenergic receptor gene has no effect on obesity phenotypes in the Quebec Family Study and Swedish Obese Subjects cohorts, *J. Clin. Invest.,* 98, 2086–2093, 1996.

25. Gerald, C., Walker, M.W., Criscione, L., Gustafson, E.L., Batzl-Hartmann, C., Smith, K.E., Vaysse, P., Durkin, M.M., Laz, T.M., Linemeyer, D.L., Schaffhauser, A.O., Whitebread, S., Hofbauer, K.G., Taber, R.I., Branchek, T.A., and Weinshank, R.L., A receptor subtype involved in neuropeptide-Y-induced food intake, *Nature,* 382(6587), 168–171, 1996.

26. Gnudi, L., Shepherd, P.R., and Kahn, B.B., Over-expression of GLUT4 selectively in adipose tissue in transgenic mice: implications for nutrient partitioning, *Proc. Nutr. Soc.,* 55(1B), 191–199, 1996.

27. Gnudi, L., Tozzo, E., Shepherd, P.R., Bliss, J.L., and Kahn, B.B., High level overexpression of glucose transporter-4 driven by an adipose-specific promoter is maintained in transgenic mice on a high fat diet, but does not prevent impaired glucose tolerance, *Endocrinology,* 136(3), 995–1002, 1995.

28. Graham, G., Shutter, J.R., Sarmiento, U., Sarosi, I., and Stark, K., Overexpression of Agrt leads to obesity in transgenic mice, *Nat. Genet.,* 17, 273–274, 1997.

29. Halaas, J.L., Boozer, C., Blair-West, J., Fidahusein, N., Denton, D.A., and Friedman, J.M., Physiological response to long-term peripheral and central leptin infusion in lean and obese mice, *Proc. Natl. Acad. Sci. U.S.A.,* 94(16), 8878–8883, 1997.

30. Halaas, J.L., Gajiwala, K.S., Maffei, M., Cohen, S.L., Chait, B.T., Rabinowitz, D., Lallone, R.L., Burley, S.K., and Friedman, J.M., Weight-reducing effects of the plasma protein encoded by the obese gene, *Science,* 269(5223), 543–546, 1995.

31. Heisler, L.K., Kanarek, R.B., and Gerstein, A., Fluoxetine decreases fat and protein intakes but not carbohydrate intake in male rats, *Pharmacol. Biochem. Behav.,* 58(3), 767–773, 1997.

32. Heon, E., Linkage of posterior polymorphous corneal dystrophy to 20q11, *Hum. Mol. Genet.,* 4(3), 485–488, 1995.

33. Hinney, A., Lentes, K.U., Rosenkranz, K., Barth, N., Roth, H., Ziegler, A., Hennighausen, K., Coners, H., Wurmser, H., Jacob, K., Romer, G., Winnikes, U., Mayer, H., Herzog, W., Lehmkuhl, G., Poustka, F., Schmidt, M.H., Blum, W.F., Pirke, K.M., Schafer, H., Grzeschik, K.H., Remschmidt, H., and Hebebrand, J., Beta 3-adrenergic-receptor allele distributions in children, adolescents and young adults with obesity, underweight or anorexia nervosa, *Int. J. Obesity Relat. Metab. Disorders,* 21(3), 224–230, 1997.

34. Hotamisligil, G.S., Johnson, R.S., Distel, R.J., Ellis, R., Papaioannou, V.E., and Spiegelman, B.M., Uncoupling of obesity from insulin resistance through a targeted mutation in aP2, the adipocyte fatty acid binding protein, *Science,* 274(5291), 1377–1379, 1996.

35. Huszar, D., Lynch, C.A., Fairchild-Huntress, V., Dunmore, J.H., Fang, Q., Berkemeier, L.R., Gu, W., Kesterson, R.A., Boston, B.A., Cone, R.D., Smith, F.J., Campfield, L.A., Burn, P., and Lee, F., Targeted disruption of the melanocortin-4 receptor results in obesity in mice, *Cell,* 88(1), 131–141, 1997.

36. Igel, M., Becker, W., Herberg, L., and Joost, H.G., Hyperleptinemia, leptin resistance, and polymorphic leptin receptor in the New Zealand obese mouse, *Endocrinology,* 138(10), 4234–4239, 1997.

37. Jensen, D.R., Schlaepfer, I.R., Morin, C.L., Pennington, D.S., Marcell, T., Ammon, S.M., Gutierrez-Hartmann, A., and Eckel, R.H., Prevention of diet-induced obesity in transgenic mice overexpressing skeletal muscle lipoprotein lipase, *Am. J. Physiol.,* 273(2 Pt. 2), R683–689, 1997.

38. Jones, J. M., Meisler, M. H., Seldin, M. F., Lee, B. K., and Eicher, E. M., Localization of insulin-2 (Ins-2) and the obesity mutant tubby (*tub*) to distinct regions of mouse chromosome 7, *Genomics,* 14, 197–199, 1992.

39. Kim-Motoyama, H., Yasuda, K., Yamaguchi, T., Yamada, N., Katakura, T., Shuldiner, A. R., Akanuma, Y., Ohashi, Y., Yazaki, Y., and Kadowaki, T., A mutation of the beta-3-adrenergic receptor is associated with visceral obesity but decreased serum triglyceride, *Diabetologia,* 40, 469–472, 1997.

40. Kleyn, P.W., Fan, W., Kovats, S.G., Lee, J.J., Pulido, J.C., Wu, Y., Berkemeier, L.R., Misumi, D.J., Holmgren, L., Charlat, O., Woolf, E.A., Tayber, O., Brody, T., Shu, P., Hawkins, F., Kennedy, B., Baldini, L., Ebeling, C., Alperin, G.D., Deeds, J., Lakey, N.D., Culpepper, J., Chen, H., Glucksmann-Kuis, M.A., Moore, K.J., et al., Identification and characterization of the mouse obesity gene tubby: a member of a novel gene family, *Cell,* 85(2), 281–290, 1996.

41. Kopecky, J., Clarke, G., Enerbäck, S., Spiegelman, B., and Kozak, L.P., Expression of the mitochondrial uncoupling protein gene from the aP2 gene promoter prevents genetic obesity, *J. Clin. Invest.,* 96(6), 291–2923, 1995.

42. Kozak, L.P. and Birkenmeier, E.H., Mouse glycerol-3-phosphate dehydrogenase: molecular cloning and genetic mapping of a cDNA sequence, *Proc. Natl. Acad. Sci. U.S.A.,* 80(10), 3020–3024, 1983.

43. Kozak, L.P., Kozak, U.C., and Clarke, G.T., Abnormal brown and white fat development in transgenic mice overexpressing glycerol 3-phosphate dehydrogenase, *Genes Dev.,* 5(12A), 2256–2264, 1991.

44. Kuczmarski, M.F., Moshfegh, A., and Briefel, R., Update on nutrition monitoring activities in the United States, *J. Am. Diet. Assoc.,* 94(7), 753–760, 1994.

45. Lembertas, A. V., Perusse, L., Chagnon, Y. C., Fisler, J. S., Warden, C. H., Purcell-Huynh, D. A., Dionne, F. T., Gagnon, J., Nadeau, A., Lusis, A. J., and Bouchard, C., Identification of an obesity quantitative trait locus on mouse chromosome 2 and evidence of linkage to body fat and insulin on the human homologous region 20q, *J. Clin. Invest.,* 100, 1240–1247, 1997.

46. Lentes, K.U., Hinney, A., Ziegler, A., Rosenkranz, K., Wurmser, H., Barth, N., Jacob, K., Coners, H., Mayer, H., Grzeschik, K.H., Schafer, H., Remschmidt, H., Pirke, K.M., and Hebebrand, J., Evaluation of a Cys23Ser mutation within the human 5-HT2C receptor gene: no evidence for an association of the mutant allele with obesity or underweight in children, adolescents and young adults, *Life Sci.,* 61(1), PL9–16, 1997.

47. Leppert, M., Baird, L., Anderson, K. L., Otterud, B., Lupski, J. R., and Lewis, R. A., Bardet-Biedl syndrome is linked to DNA markers on chromosome 11q and is genetically heterogeneous, *Nat. Genet.,* 7, 108–112, 1994.

48. Lowell, B.B., Susulic, V., Hamann, A., Lawitts, J.A., Himms-Hagen, J., Boyer, B.B., Kozak, L.P., and Flier, J.S., Development of obesity in transgenic mice after genetic ablation of brown adipose tissue, *Nature*, 366(6457), 740–742, 1993.

49. Lu, D., Willard, D., Patel, I.R., Kadwell, S., Overton, L., Kost, T., Luther, M., Chen, W., Woychik, R.P., Wilkison, W.O., and Cone, R.D., Agouti protein as an antagonist of the melanocyte-stimulating-hormone receptor, *Nature (Lond.)*, 371, 799–802, 1994.

49a. MacDonald, A. and Stunkard, A., Body mass index of British separated twins, *N. Engl. J. Med.*, 322(21), 1530, 1990.

50. Maher, M.A., Banz, W.J., Truett, G.E., and Zemel, M.B., Dietary fat and sex modify heterozygote effects of the rat fatty (*fa*) allele, *J. Nutr.*, 126(10), 2487–2493, 1996.

51. Matsuoka, N., Ogawa, Y., Hosoda, K., Matsuda, J., Masuzaki, H., Miyawaki, T., Azuma, N., Natsui, K., Nishimura, H., Yoshimasa, Y., Nishi, S., Thompson, D.B., and Nakao, K., Human leptin receptor gene in obese Japanese subjects: evidence against either obesity-causing mutations or association of sequence variants with obesity, *Diabetologia*, 40(10), 1204–1210, 1997.

52. Melnyk, A., Harper, M.E., and Himms-Hagen, J., Raising at thermoneutrality prevents obesity and hyperphagia in BAT-ablated transgenic mice, *Am. J. Physiol.*, 272(4 Pt. 2), R1088–R1093, 1997.

53. MMWR Update: prevalence of overweight among children, adolescents, and adults — United States, 1988–1994, *Morb. Mortal. Wkly. Rep.*, 46(9), 198–202, 1997.

54. Montague, C.T., Farooqi, I.S., Whitehead, J.P., Soos, M.A., Rau, H., Wareham, N.J., Sewter, C.P., Digby, J.E., Mohammed, S.N., Hurst, J.A., Cheetham, C.H., Earley, A.R., Barnett, A.H., Prins, J.B., and O'Rahilly, S., Congenital leptin deficiency is associated with severe early-onset obesity in humans, *Nature*, 387(6636), 903–908, 1997.

55. Nagase, T., Aoki, A., Yamamoto, M., Yasuda, H., Kado, S., Nishikawa, M., Kugai, N., Akatsu, T., and Nagata, N., Lack of association between the trp64arg mutation in the beta-3-adrenergic receptor gene and obesity in Japanese men: a longitudinal analysis, *J. Clin. Endocrol. Metab.*, 82, 1284–1287, 1997.

56. Naggert, J.K., Fricker, L.D., Varlamov, O., Nishina, P.M., Rouille, Y., Steiner, D.F., Carroll, R.J., Paigen, B.J., and Leiter, E.H., Hyperproinsulinaemia in obese *fat/fat* mice associated with a carboxypeptidase E mutation which reduces enzyme activity, *Nat. Genet.*, 10(2), 135–142, 1995.

57. Norman, R.A., Leibel, R.L., Chung, W.K., Power-Kehoe, L., Chua, S.C., Jr., Knowler, W.C., Thompson, D.B., Bogardus, C., and Ravussin, E., Absence of linkage of obesity and energy metabolism to markers flanking homologues of rodent obesity genes in Pima Indians, *Diabetes*, 45(9), 1229–1232, 1996.

58. Norman, R. A., Thompson, D. B., Foroud, T., Garvey, W. T., Bennett, P. H., Bogardus, C., Ravussin, E., Allan, C., Baier, L., Bowden, D., Hanson, R., Knowler, W., Kobes, S., Pettitt, D., and Prochazka, M., Genomewide search for genes influencing percent body fat in Pima Indians: suggestive linkage at chromosome 11q21-q22, *Am. J. Hum. Genet.*, 60, 166–173, 1997.

59. North, M.A., Naggert, J.K., Yan, Y., Noben-Trauth, K., and Nishina, P.M., Molecular characterization of TUB, TULP1, and TULP2, members of the novel tubby gene family and their possible relation to ocular diseases, *Proc. Natl. Acad. Sci. U.S.A.*, 94(7), 3128–3133, 1997.

60. Ogawa, Y., Masuzaki, H., Isse, N., Okazaki, T., Mori, K., Shigemoto, M., Satoh, N., Tamura, N., Hosoda, K., Yoshimasa, Y., et al., Molecular cloning of rat obese cDNA and augmented gene expression in genetically obese Zucker fatty (*fa/fa*) rats, *J. Clin. Invest.*, 96(3), 1647–1652, 1995.

61. Pelleymounter, M.A., Cullen, M.J., Baker, M.B., Hecht, R., Winters, D., Boone, T., and Collins, F., Effects of the obese gene product on body weight regulation in ob/ob mice, *Science*, 269(5223), 540–543, 1995.

62. Pepin, M.C. and Barden, N., Decreased glucocorticoid receptor activity following glucocorticoid receptor antisense RNA gene fragment transfection, *Mol. Cell. Biol.*, 11(3), 1647–1653, 1991.

63. Pepin, M.C., Pothier, F., and Barden, N., Impaired type II glucocorticoid-receptor function in mice bearing antisense RNA transgene, *Nature*, 355(6362), 725–728, 1992.

64. Richard, D., Chapdelaine, S., Deshaies, Y., Pepin, M.C., and Barden, N., Energy balance and lipid metabolism in transgenic mice bearing an antisense GCR gene construct, *Am. J. Physiol.*, 265(1 Pt. 2), R146–R150, 1993.

65. Roche, C., Boutin, P., Dina, C., Gyapay, G., Basdevant, A., Hager, J., Guy-Grand, B., Clement, K., and Froguel, P., Genetic studies of neuropeptide Y and neuropeptide Y receptors Y1 and Y5 regions in morbid obesity, *Diabetologia*, 40(6), 671–675, 1997.

66. Ross, S.R., Graves, R.A., and Spiegelman, B.M., Targeted expression of a toxin gene to adipose tissue: transgenic mice resistant to obesity, *Genes Dev.*, 7(7B), 1318–1324, 1993.

67. Samuelson, L.C., Isakoff, M.S., and Lacourse, K.A., Localization of the murine cholecystokinin A and B receptor genes, *Mammalian Genome*, 6(4), 242–246, 1995.

68. Schaffhauser, A.O., Stricker-Krongrad, A., Brunner, L., Cumin, F., Gerald, C., Whitebread, S., Criscione, L., and Hofbauer, K.G., Inhibition of food intake by neuropeptide Y Y5 receptor antisense oligodeoxynucleotides, *Diabetes*, 46(11), 1792–1798, 1997.

69. Schwartz, M.W., Seeley, R.J., Campfield, L.A., Burn, P., and Baskin, D.G., Identification of targets of leptin action in rat hypothalamus, *J. Clin. Invest.*, 98(5), 1101–1106, 1996.

70. Schwartz, M.W., Sipols, A.J., Grubin, C.E., and Baskin, D.G., Differential effect of fasting on hypothalamic expression of genes encoding neuropeptide Y, galanin, and glutamic acid decarboxylase, *Brain Res. Bull.*, 31(3–4), 361–367, 1993.

71. Shepherd, P.R., Gnudi, L., Tozzo, E., Yang, H., Leach, F., and Kahn, B.B., Adipose cell hyperplasia and enhanced glucose disposal in transgenic mice overexpressing GLUT4 selectively in adipose tissue, *J. Biol. Chem.*, 268(30), 22243–22246, 1993.

72. Shutter, J.R., Graham, M., Kinsey, A.C., Scully, S., Lüthy, R., and Stark, K.L., Hypothalamic expression of ART, a novel gene related to agouti, is up-regulated in obese and diabetic mutant mice, *Genes Dev.*, 11(5), 593–602, 1997.

73. Smith, B.K., West, D.B., and York, D.A., Carbohydrate versus fat intake: differing patterns of macronutrient selection in two inbred mouse strains, *Am. J. Physiol.*, 272(1 Pt. 2), R357–R362, 1997.

74. Soloveva, V., Graves, R.A., Rasenick, M.M., Spiegelman, B.M., and Ross, S.R., Transgenic mice overexpressing the beta 1-adrenergic receptor in adipose tissue are resistant to obesity, *Mol. Endocrinol.*, 11(1), 27–38, 1997.

75. Stanley, B.G., Kyrkouli, S.E., Lampert, S., and Leibowitz, S.F., Neuropeptide Y chronically injected into the hypothalamus: a powerful neurochemical inducer of hyperphagia and obesity, *Peptides*, 7(6), 1189–1192, 1986.

76. Stenzel-Poore, M.P., Cameron, V.A., Vaughan, J., Sawchenko, J.V., and Vale, W., Development of Cushing's syndrome in corticotropin-releasing factor transgenic mice, *Endocrinology*, 130, 3378–3386, 1992.

76a. Stunkard, A.J., Harris, J.R., Pedersen, N.L., and McClearn, G.E., The body mass index of twins who have been reared apart, *N. Engl. J. Med.*, 322(21), 1483–1487, 1990.

77. Susulic, V.S., Frederich, R.C., Lawitts, J., Tozzo, E., Kahn, B.B., Harper, M.E., Himms-Hagen, J., Flier, J.S., and Lowell, B.B., Targeted disruption of the beta 3-adrenergic receptor gene, *J. Biol. Chem.*, 270(49), 29483–29492, 1995.

78. Takaya, K., Ogawa, Y., Isse, N., Okazaki, T., Satoh, N., Masuzaki, H., Mori, K., Tamura, N., Hosoda, K., and Nakao, K., Molecular cloning of rat leptin receptor isoform complementary DNAs — identification of a missense mutation in Zucker fatty (*fa/fa*) rats, *Biochem. Biophys. Res. Commun.*, 225(1), 75–83, 1996.

79. Tecott, L.H., Sun, L.M., Akana, S.F., Strack, A.M., Lowenstein, D.H., Dallman, M.F., and Julius, D., Eating disorder and epilepsy in mice lacking 5-HT2c serotonin receptors, *Nature*, 374(6522), 542–546, 1995.

80. Thomas, S. A. and Palmiter, R. D., Thermoregulatory and metabolic phenotypes of mice lacking noradrenaline and adrenaline, *Nature*, 387, 94–97, 1997.

81. Truett, G.E., Tempelman, R.J., and Walker, J.A., Codominant effects of the fatty (*fa*) gene during early development of obesity, *Am. J. Physiol.*, 268(1 Pt. 1), E15–E20, 1995.

82. Wang, Q., Bing, C., Al-Barazanji, K., Mossakowska, D.E., Wang, X.M., McBay, D.L., Neville, W.A., Taddayon, M., Pickavance, L., Dryden, S., Thomas, M.E., McHale, M.T., Gloyer, I.S., Wilson, S., Buckingham, R., Arch, J.R., Trayhurn, P., and Williams, G., Interactions between leptin and hypothalamic neuropeptide Y neurons in the control of food intake and energy homeostasis in the rat, *Diabetes*, 46(3), 335–341, 1997.

83. Warden, C.H., Fisler, J.S., Shoemaker, S.M., Wen, P.Z., Svenson, K.L., Pace, M.J., and Lusis, A.J., Identification of four chromosomal loci determining obesity in a multifactorial mouse model, *J. Clin. Invest.*, 95(4), 1545–1552, 1995.

84. Weinstock, P.H., Levak-Frank, S., Hudgins, L.C., Radner, H., Friedman, J.M., Zechner, R., and Breslow, J.L., Lipoprotein lipase controls fatty acid entry into adipose tissue, but fat mass is preserved by endogenous synthesis in mice deficient in adipose tissue lipoprotein lipase, *Proc. Natl. Acad. Sci. U.S.A.,* 94(19), 10261–10266, 1997.

85. West, D.B., Goudey-Lefevre, J., York, B., and Truett, G.E., Dietary obesity linked to genetic loci on chromosomes 9 and 15 in a polygenic mouse model, *J. Clin. Invest.,* 94(4), 1410–1416, 1994.

86. West, D.B., Waguespack, J., York, B., Goudey-Lefevre, J., and Price, R.A., Genetics of dietary obesity in AKR/J x SWR/J mice: segregation of the trait and identification of a linked locus on chromosome 4, *Mammalian Genome,* 5(9), 546–552, 1994.

87. Wilson, M., Mulley, J., Gedeon, A., Robinson, H., and Turner, G., New X-linked syndrome of mental retardation, gynecomastia, and obesity is linked to DXS255, *Am. J. Med. Genet.,* 40, 406–413, 1991.

88. Yamashita, T., Murakami, T., Iida, M., Kuwajima, M., and Shima, K., Leptin receptor of Zucker fatty rat performs reduced signal transduction, *Diabetes,* 46(6), 1077–1080, 1997.

89. York, B., Lei, K., and West, D.B., Sensitivity to dietary obesity linked to a locus on chromosome 15 in a CAST/Ei x C57BL/6J F2 intercross, *Mammalian Genome,* 7(9), 677–681, 1996.

90. Zakrzewska, K.E., Cusin, I., Sainsbury, A., Rohner-Jeanrenaud, F., and Jeanrenaud, B., Glucocorticoids as counterregulatory hormones of leptin: toward an understanding of leptin resistance, *Diabetes,* 46, 717–719, 1997.

91. Zhang, Y., Proenca, R., Maffei, M., Barone, M., Leopold, L., and Friedman, J.M., Positional cloning of the mouse *ob* gene and its human homologue, *Nature,* 372, 425–432, 1994.

92. Zucker, L.M. and Zucker, F.T., Fatty, a new mutation in the rat, *J. Hered.,* 52, 275–278, 1961.

# 21 The Genetics of Eating Disorders

*Dorothy E. Grice*

## CONTENTS

## INTRODUCTION

This chapter will provide an overview of the genetic aspects of eating disorders, namely anorexia nervosa and bulimia nervosa. They will be introduced as clinical conditions and diagnostic criteria presented. A review of basic genetic concepts will follow. These concepts then will be applied to a review and summary of genetic studies in eating disorders.

## ANOREXIA NERVOSA

Anorexia nervosa (AN) is a serious eating disorder that typically has its onset during adolescence. There are four essential features of AN according to the *Diagnostic and Statistical Manual of Mental Disorders, fourth edition.*[1] The first criterion is the refusal to maintain a minimally normal body weight for age and height. This weight loss (or, if in a growth period, failure to make appropriate gains) results in a body weight of less than 85% of the expected normal weight. An alternative guideline given by the ICD-10 Diagnostic Criteria for Research is a body mass index (BMI) equal to or less than $17.5 \text{ kg/m}^2$, where BMI equals weight in kilograms divided by height in meters[2]. In AN there also is an intense fear of gaining weight even in the face of significant weight loss. The third criterion is a significant disturbance in perception of one's body weight and shape or denial of the seriousness of one's weight loss. The last criterion is the absence of three or more consecutive menstrual cycles in postpubertal females. The lifetime prevalence of AN is estimated to be 0.5 to 1.6%.[2,3]

## BULIMIA NERVOSA

A related eating disorder is bulimia nervosa (BN). BN is characterized by binge eating and inappropriate compensatory measures to prevent weight gain.[1] Recurrent binges are episodes in which during a discrete period of time an amount of food is consumed that is definitely larger than most people would consume under similar circumstances. During the binge, individuals with BN feel a lack of control over their eating. As in AN, there is a preoccupation and significant disturbance in perception of one's body weight and shape. Recurrent inappropriate behavior such as self-induced vomiting, fasting, misuse of medications, or excessive exercise is undertaken to prevent weight gain. To meet full criteria for BN diagnosis, these symptoms must occur at least twice a week for 3 months. Lifetime prevalence estimates for BN are approximately 2 to 3%.[2,4]

## METHODS OF GENETIC ANALYSIS

Genetic study of disease yields evidence about the transmission and underlying biology of the illness, two important factors related to treatment and prevention. Many psychiatric disorders aggregate in families and most psychiatric disorders are thought to be complex traits.[5,6] Although genetic factors have been implicated in several medical and neurological disorders that have complex presentations (e.g., insulin-dependent diabetes mellitus,[7] cystic fibrosis[8] and Huntington's disease[9]), genes for psychiatric disorders have been difficult to isolate. This is due in part to the complex nature of psychiatric disorders, nonmendelian inheritance patterns, genetic heterogeneity, variability in clinical expression and phenotypes, and difficulty in discerning genetic factors from environmental factors.[10]

Traditional methods to determine the role genetic factors play in the etiology of a given disorder are twin, family, and adoption studies. Heritability, the fraction of the population variance of a trait that can be explained by genetic transmission, can be estimated from these types of studies. With the advent of molecular genetic techniques, genetic linkage analysis, case control studies, and candidate gene surveys become tools in the search for genes related to a particular disorder. Regardless of study method, unbiased ascertainment of probands and accurate assessment of probands and family members are extremely important for the reliable interpretation of the findings.

### TWIN STUDIES

Twin studies help to establish that genetic factors are important in the manifestation of a disorder.[11] Monozygotic (MZ) twins are genetically identical and are presumed to share virtually the same environment whereas dizygotic (DZ) twins have, on average, half of their genes in common and also are presumed to share virtually the same environment. According to mendelian theory, if a trait is due solely to genetic factors MZ twins should have 100% concordance rates. DZ twins should be concordant for the trait 50% of the time. Since twin studies assume that differential environmental influences are minimized in this unique population, differences in concordance rates between MZ and DZ twins are attributed to genetic factors. In the case of complex, nonmendelian psychiatric disorders, a significantly higher concordance rate between MZ twins than DZ twins is taken as evidence that genetic factors play an important role in the etiology of the disorder.

Unbiased ascertainment of subjects is important in twin studies. However, due to the low population prevalence of AN and BN it is extremely difficult to ascertain an epidemiological sample of affected twins large enough for robust statistical analyses. Nonepidemiological samples comprise most of the published literature on AN and BN twin studies. In these studies, accurate clinical assessment of both MZ and DZ pairs becomes even more important. One criticism of this type of sample is that concordant twins are more likely than discordant twins to be brought to the attention of researchers (because they are striking clinically) and may result in biased concordance rates. In the absence of an epidemiological sample, systematic recruitment of a sufficiently representative sample becomes an important study component. Zygosity also must be accurately determined. Finally, assessments of co-twins should continue through the age of risk for the disorder. A summary of twin studies in eating disorders follows.

In a set of AN twin pairs drawn from published case reports dating from 1940 and from the authors' own research, MZ twins had a 52% (14/27) concordance rate, significantly higher than the 11% (1/9) concordance rate of DZ twins.[12] Since the majority of pairs in this analysis are case reports, the possible effect of ascertainment bias should be taken into account. Additionally, because the authors included data from previously published case reports, diagnostic criteria may not have been equivalent across all twin pairs. However, the rates seen in this study are similar to later, more methodologically sound reports.

Holland et al.[13] describe a study of 34 AN twin pairs and a set of triplets drawn from two regional eating disorder units and through volunteerism. Of the 30 female twin pairs, 9 of 16 (55%) MZ twins were concordant for AN compared to 1 of 14 (7%) DZ twins, yielding a statistically significant difference. Zygosity was determined through blood analysis in the majority of the cases while a physical resemblance questionnaire was used for 6 twin pairs. In this study clinical assessment was used rather than structured diagnostic interviews.[13,14]

In a combined twin and family study of AN, Holland et al.[15] reported on 57 AN twin pairs ascertained from various clinical sites, through support groups or advertisement. This study contained some subjects who overlapped with a previous twin report.[13] Structured interviews were administered to subjects who had passed the age of risk for AN. Zygosity was determined through blood analysis on all but two pairs. Significant differences in concordance rates were found in this study. Of the 25 MZ twin pairs 14 (56%) were concordant for AN. In contrast to the MZ rate, only one of the 20 DZ twin pairs (5%) was concordant for AN. Heritability scores were calculated to be over 80% for all relatives.

A study of 11 AN twin pairs (5 MZ and 6 DZ) found no concordant AN twin pairs.[16] However, the sample size was small, data were based on secondary interviews of subjects' mothers and it is not clear if all subjects had passed through the age of risk for onset of AN making it difficult to generalize these results.

In a study designed to examine genetic influences on eating attitudes, a volunteer twin population was surveyed using self-report scales on eating attitudes and behaviors ($n = 580$ twin pairs).[17] High scorers on the survey were identified ($n = 66$) and a subset was interviewed directly, but not in a blinded fashion ($n = 42$). Zygosity was determined using a questionnaire reported by authors to have a reliability of 96%. Of the three concordant AN twin pairs identified, two were MZ pairs and one was a DZ pair. Prevalence rates are likely under-estimated in this study since there was a relatively low (54%) response rate of subjects and not all high scorers agreed to be interviewed.

Walters and Kendler[3] undertook a large population-based study of twins to examine risk factors for AN and AN-like syndromes. Structured interviews were used to study an epidemiological sample of 2163 female twins. Significant comorbidity was found between AN and BN, supporting the premise that shared etiologic factors exist for these disorders. Co-twins of probands were found to be at significantly higher risk for lifetime AN, BN, major depression, and current low BMI. Due to the very low prevalence of AN in the general population, the data obtained in this study could not be reliably examined for MZ and DZ concordance rates and heritability estimates were not undertaken.

Given that family studies support the hypothesis that AN and BN share a common genetic etiology (see discussion below), it is appropriate also to examine twin studies of BN for concordance rates. In a study of 21 BN twins, significant differences in concordance rates of MZ and DZ twins were found.[18] Study subjects drawn from advertisements and clinic records were assessed using structured self-report instruments. Zygosity determinations were done using a questionnaire reported by the authors to have a reliability of 93%. Of 6 MZ twin pairs, 5 (83%) were concordant for BN. In contrast, of 15 DZ twin pairs, only 4 (27%) were concordant for BN. MZ rates remained elevated (71.4%) if the one male MZ pair was included.

Hsu et al.[19] described a sample of 11 BN twin pairs ascertained over a 3-year period at an eating disorder clinic. Clinical diagnoses (without structured interviews) were made in probands. A difficulty in this study is that concordant co-twins were not assessed directly nor were 3 of the 9 discordant co-twins. Zygosity was determined through the use of a questionnaire, except in the case of one adopted twin pair in which blood analysis was done. Of 6 MZ twin pairs, 2 (33%) were concordant for BN. In contrast, for the 5 DZ twins, none (0%) were concordant (neither the 2 female DZ twins nor the 3 male/female DZ twins).

Using the same population-based sample of twins discussed above,[3] Kendler et al.[4] found statistically significant differences in twin concordance rates for BN. In MZ twins, 22.9% were concordant compared to only 8.7% of DZ twins. The best-fitting model indicated that familial aggregation was due solely to genetic factors with a heritability of liability of 55%.

If one examines AN and BN twin studies with the most robust number of subjects,[12,15,18] average concordance rates for AN and BN combined are 64% for MZ twins and 14% for DZ twins. Using Holzinger's index[20] over 50% of the variance in risk for eating disorders is attributable to genetic factors.

## FAMILY STUDIES

In family studies, clinical expression patterns and genetic mechanisms of transmission can be examined. Family studies have established that most psychiatric disorders are inherited.[5,6] Controlled family studies yield important data about clinical expression of a disorder and morbid risk factors. However, controlled family studies that minimize ascertainment bias (e.g., through consecutive selection procedures) and use strong research methodology (such as blind direct interviews) require substantial resources and, as such, are very difficult to carry out and are in limited number in the literature.

Vertical transmission of a disorder through a family pedigree gives indirect evidence that the disorder may be genetic. In general, an increased risk (over population prevalence) to first- and second-degree relatives of affected probands is evidence for a genetic susceptibility to a disorder. In the subset of disorders suitable for segregation analysis, family studies can yield information about the modes of vertical transmission within affected families. However, in both family and segregation studies the presence of strong environmental influences can compromise the analyses.

Through clinical research in eating disorders there has been an increasing appreciation for high rates of AN and BN comorbidity. During the course of AN almost 40% of patients will satisfy diagnostic criteria for BN.[21-23] To date, a variety of family studies have been done in eating disorders (i.e., AN, BN, and eating disorders not other specified [EDNOS]). Controlled family studies comprise a small group of the family studies in AN and BN.[24-28] Several family studies have shown that increased rates of eating disorders are found in families of AN and BN probands compared to control families.[24,25,28,29] Taken together with twin data, AN and BN can be considered heritable eating disorders with overlapping genetic vulnerabilities. A brief review and summary of selected family studies of eating disorders follows.

A subset of older family studies found an association between eating disorders and affective illness. The link between these disorders came mainly in the form of increased risk for affective illness in families of AN probands and higher rates of affective illness in AN individuals compared to other psychiatric controls. Interpretations of these data included that AN and affective illness may share common etiologic factors or that the subgroup of AN patients with affective illness (or with affective illness in their families) have a genetic variant of an affective illness trait. Although not unequivocally resolved, recent studies with more robust methodology indicate that there is aggregation of eating disorders (AN, BN, and EDNOS) in families. However, although affective illness often is found in AN and BN families, these disorders appear to segregate independently from one another and most likely result from distinct genetic etiologies.[4,24,27,28,30]

In one of the earlier studies, Winokur et al.[31] found significantly higher rates of affective disorder in families of AN probands (22%) compared to families of control probands (10%). However, rates of eating disorders in family members were not presented. Hudson et al.[32] studied the first-degree relatives of 89 eating disorder probands ascertained through consecutive admissions (both inpatient and outpatient) as well as advertisements. The family history method, which may underestimate the presence of psychiatric illness compared to the family interview method, was used. Risk for eating disorders in relatives was not calculated. However, the risk for affective illness in families of eating disorder probands was elevated and similar to that found in families of bipolar probands.

Similar results were obtained in a controlled family study of AN derived from a referred clinic population.[24] In this study the lifetime prevalence of affective disorder was significantly higher in relatives of AN probands ($n = 24$) compared to relatives of control probands (medical patients with no psychiatric diagnoses, $n = 44$). Eating disorder rates were determined and also were found to be more common in relatives of AN probands compared to relatives of control probands. The combined rate of AN and BN was six times higher in first-degree relatives of AN probands compared to relatives of control probands. Heritability was estimated to be approximately 60%.

In an attempt to identify clinical subgroups of AN related to affective illness, a related study examined eating disorders and affective disorders in first-degree relatives of AN probands compared first-degree relatives of control probands.[33] Of 364 first-degree relatives, 54% were diagnostically interviewed and indirect diagnoses were made on the remainder. Morbid risk comparisons between AN and clinical AN subgroups (AN with affective illness, AN with BN, and AN with self-induced vomiting) suggested that these AN groups did not increase risk of affective disorders. The authors concluded that although familial and clinical features common to AN and affective illness may reflect shared genetic vulnerability, this study could not define pertinent clinical groupings.

In a controlled family history study, Rivinus et al.[34] found significantly higher rates of psychiatric illness (namely depression and substance use disorders) in first- and second-degree relatives of AN probands ($n = 40$) compared to the families of control probands ($n = 23$). AN subjects were a prospective sample of hospital and eating disorder clinic patients. Controls were volunteers. Of interest, two patients with AN had a MZ twin with AN compared to none in the control families. An uncontrolled family history study that looked specifically at risk for eating disorders in AN families found that both first- and second-degree relatives of AN probands ($n = 45$) had significantly higher rates of AN than one would expect based on population prevalence rates.[15]

Logue et al.[26] also studied risks for eating disorders (and affective disorders) in a controlled family study of women with AN and BN. Although not a consecutive study design, 79% of all clinic patients who had been treated over a previous 18-month period agreed to participate when they were contacted. Three family types were examined: probands with AN and/or BN ($n = 30$), probands with affective illness ($n = 16$), and randomly selected undergraduates who served as controls ($n = 20$). For the 20% of first-degree relatives who were not interviewed directly, two informants were used to gather diagnostic information. None of the first-degree relatives met criteria for AN although there was one case of BN in the family of an affectively ill proband. Significantly higher rates of major depression, but not bipolar disorder, were found in families of eating disorder probands compared to families of affectively ill probands and of control probands. In support of the conclusion that eating disorders and affective illnesses segregate independently, higher rates of depression were found in families of eating disorder probands who themselves had histories of depression. Interestingly, familial aggregation of affective illness was not found in relatives of the affectively ill probands although this would be expected based on current data from the affective disorder field.

The theory that eating disorders (but not affective disorders) segregate though families via a common underlying genetic etiology is supported by data from a population-based sample[3,4] and controlled family studies of AN and BN.[25,27,28] A large epidemiological sample of twins ($n = 2163$) found evidence for significant comorbidity between AN and BN.[3] The authors concluded that these findings support the hypothesis that there is a set of common predisposing factors, including a genetic predisposition as well as familial environment, for both AN and BN. Analysis of the sample for BN twin concordance also gave evidence for significant genetic familial effects.[4]

In Strober et al.,[25] prevalence rates of eating disorders were determined in families of 60 AN and AN+BN probands and 95 non-eating disorder psychiatrically ill probands. Probands were ascertained from inpatient admissions and structured direct interviews were used. Significantly higher rates of eating disorders (AN, BN, and EDNOS) were found in AN and AN/BN families compared to control families. The odds ratio for any eating disorder in relatives of AN probands was 5.5 times greater than in relatives of control probands.

In a study of consecutive inpatient admissions Strober et al.[27] determined the lifetime prevalence rates of eating disorders and affective disorders in families of AN ($n = 97$), affectively ill ($n = 66$), and other psychiatric control ($n = 117$) probands. Direct interviews of probands and families were carried out. AN was found at significantly higher rates in the first-degree relatives of the AN proband group compared to the other proband groups. AN was eight times more common in female first-degree relatives of AN probands than in the general population. Significantly higher rates of affective disorder were documented only in relatives of AN probands with coexisting depression, not in relatives of probands with AN alone. These data support the hypotheses that there is a genetic susceptibility to AN and that the liability for AN is different from that operating in affective illness. Assuming additive gene effects, heritability was estimated to be over 60% — a moderately strong familial effect.

In a more recent controlled study of AN and BN families, first-degree relatives of AN and BN probands had significantly higher rates of eating disorders compared to relatives of control probands.[27] Direct interviews blind to the proband diagnosis were carried out on 26 AN families, 47 BN families, and 44 control families. Higher rates of affective illness, obsessive compulsive disorder (OCD), and generalized anxiety disorder (GAD) also were documented in the eating disorder families compared to control families. Data were stratified to examine aggregation patterns of these non-eating disorder psychiatric disorders in the eating disorder families. Elevated rates of affective illness and OCD in relatives were found only in families containing a proband with comorbid diagnoses (AN/BN + affective illness or AN/BN + OCD, respectively), supporting the conclusion that although eating disorders, affective disorders, and OCD are familial, these disorders are transmitted independently.

## ADOPTION/SEPARATION STUDIES

Adoption/separation studies can be used to examine the role environmental influences play on the expression of a disorder. If a disorder is determined by environment, adoptees will resemble their adoptive parents more so than their biological parents. Conversely, if a disorder is determined primarily by genetic factors, adoptees will more closely resemble their biological parents. However, a significant limitation to adoption/separation studies is the difficulty in obtaining reliable clinical information about biological parents. Often these data are available only from hospital chart notations and confidentiality issues complicate collection of extensive family history data. An extensive review of the literature identified no published adoption/separation studies of AN or BN.

## HERITABILITY

Examination of twin and family studies can lead to estimates of heritability for eating disorders. Heritability is the fraction of the population variance of a trait that can be explained by genetic transmission. Heritability estimates for eating disorders range from 55 to 80%.[4,14,15,27] The most likely model to account for susceptibility to eating disorders will involve gene-environment interaction such that familial, social, and cultural factors interact with preexisting genetic vulnerabilities in complex ways to produce the final expression of a clinical eating disorder.

## MOLECULAR GENETIC STUDIES

Despite the evidence for genetic factors at work in increasing vulnerability to eating disorders, to date, no genetic linkage studies of eating disorders have been published, although at least one large affected sibling pair study currently is underway (W. Berrettini, personal communication). Genetic case control studies and candidate gene analyses can identify gene variants that are statistically associated with illness. Study populations must be of sufficient size to have the power to detect differences between the groups of cases and controls. In these types of studies, as in twin and family studies, proband and family assessments must be rigorous and controlled. When the genotypes of affected individuals are compared to nonfamily controls, great care must be taken in the selection of control populations. This is because the natural variation of gene frequencies can be different among ethnic or racial populations. This phenomenon, known as population stratification, can introduce errors into gene frequency analyses. As a result, differences in gene frequencies between an affected population and a poorly selected control group can be misinterpreted as indication of a genetic association, or lack thereof, when in fact, the results are due to population stratification. A summary of molecular genetic studies follows.

Although largely abandoned now, the possible association of human leukocyte antigen (HLA) markers in AN and BN has been examined in a series of papers. This approach was based on early work that suggested an association between antigens in the HLA system and affective disorders[35] and the subset of AN family studies in which higher rates of affective illness were found (reviewed above). Although an initial study reported a higher frequency of HLA-Bw16 in AN patients compared to controls,[36] subsequent studies were unable to confirm this finding in AN, BN, or affectively ill patients.[37-39] Issues of population stratification cannot be ruled out in any of these studies, as relatively small numbers of subjects ($n = 37–55$) were used and control populations were not family-based.

Polymorphic variants of genes thought to be relevant to the pathophysiology of AN also have been studied. Published works examine genes for the dopamine D3 receptor (DRD3), tryptophan hydroxylase (TPH), the $\beta_3$-adrenergic receptor, the serotonin transporter, and the serotonin receptor 5-HT$_{2A}$. Studies such as these often have limited power to detect differences due to methodological issues (e.g., case control design, insufficient sample size).

Bruins-Slot et al.[40] used a restriction fragment length polymorphism (RFLP) found in the DRD3 gene to genotype 39 AN patients and 42 controls. There were no differences in allele frequencies nor genotype count between the two groups. A case control study of 109 AN subjects and 49 controls found no role for a TPH promoter polymorphism in the predisposition to AN.[41]

Based on findings that the $\beta_3$-adrenergic receptor gene may be associated with obesity and other serious weight conditions, Hinney et al.[42] surveyed obese, normal weight, underweight, and AN patients to compare frequencies of a missense mutation (64 Trp to 64 Arg) in the $\beta_3$-adrenergic receptor gene. Using the transmission disequilibrium test (TDT),[45] the authors found no evidence of linkage or linkage disequilibrium. The TDT uses family-based controls to circumvent the issues of population stratification. A similarly designed study found no evidence for the role of the serotonin transporter gene in AN.[44]

A series of papers has examined the serotonin receptor 5-HT$_{2A}$ gene promoter polymorphism (–1438G/A) in eating disorders. The ultimate conclusion of these studies remains unclear. In the first published paper to examine the –1438A/G polymorphism, Collier et al.[45] employed a case control design. The frequency of the –1438A allele was significantly higher in AN subjects ($n = 81$) compared to controls ($n = 226$, females only = 88). Using the TDT, Hinney et al.[46] were unable to replicate this finding in 57 AN patients. No evidence for transmission disequilibrium or association between the −1438A allele and AN was found. Similar results came from a case control study of Campbell et al.[47]

Two subsequent case control studies of the −1438A allele in AN and BN subjects yielded further interesting results.[48,49] In both studies the −1438A allele was found at a significantly higher frequency in AN subjects compared to BN subjects[48] and controls.[48,49] It is not clear that the sample sizes in any of the case control studies are sufficiently powerful to detect association with confidence. The field needs extension and replication of these findings in rigorous controlled studies to enlighten us as to the ultimate relevance of the serotonin receptor 5-HT$_{2A}$ gene on the genetics of AN and BN.

## CONCLUSIONS

Family and twin studies have generated significant data regarding the familial and genetic components of eating disorders. AN and BN aggregate in families and this familial aggregation appears to be independent from other psychiatric disorders although some findings suggest that eating disorders may share susceptibility factors with affective disorders. Risk for an eating disorder among first-degree relatives of AN or BN probands is significantly increased, perhaps as much as five- to eightfold, compared to control probands. MZ and DZ twin concordance rates also indicate significant genetic factors at work in the expression of AN and BN. Over 50% of the variance in risk for illness can be attributed to genetic factors. Molecular genetic studies are now being employed for candidate gene and genetic linkage analyses. Additional methodologically robust and controlled analyses should help to clarify the complex relationships observed thus far. In conclusion, current family data taken together with twin studies suggest that AN and BN are heritable eating disorders which result from complex interactions of common genetic etiologies and environmental factors. Molecular genetic techniques should identify specific susceptibility areas in the human genome in the near future.

## REFERENCES

1. American Psychiatric Association, *Diagnostic and Statistical Manual of Mental Disorders, 4th ed.,* Washington, D.C., 1994.
2. Berrettini, W., Genetic aspects of anorexia nervosa and bulimia nervosa, *Dir. Psychiatr.,* 17, 53–57, 1998.
3. Walters, E. E. and Kendler, K. S., Anorexia nervosa and anorexic-like syndromes in a population-based female twin sample, *Am. J. Psychiatr.,* 152, 64–71, 1995.
4. Kendler, K. S., MacLean, C., Neale, M. et al., The genetic epidemiology of bulimia nervosa, *Am. J. Psychiatr.,* 148, 1627–1637, 1991.
5. Nurnberger, J. I. and Berrettini, W., *Psychiatric Genetics,* Chapman & Hall Medical, London, 1998.
6. Blum, K. and Noble, E. P., *Handbook of Psychiatric Genetics,* CRC Press, Boca Raton, FL, 1997.
7. She, J. X. and Marron, M. P., Genetic susceptibility factors in type 1 diabetes: linkage, disequilibrium and functional analyses, *Curr. Opin. Immunol.,* 10(6), 682–689, 1998.
8. Rosenstein. B. J. and Zeitlen, P. L., Cystic fibrosis, *Lancet,* 351(9098), 277–282, 1998.
9. Albin, R. L. and Tagle, D. A., Genetics and molecular biology of Huntington's disease, *Trends Neurosci.,* 18, 11–14, 1995.
10. Gershon, E. S. and Cloninger, C. R., *Genetic Approaches to Mental Disorders,* American Psychiatric Press, Washington, D.C., 1994.
11. LaBuda, M. C., Gottesman, I. I., and Pauls, D. L., Usefulness of twin studies for exploring the etiology of childhood and adolescent psychiatric disorders, *Am. J. Med. Genet.,* 48, 47–59, 1993.
12. Garfinkle, P. E. and Garner, D. M., *Anorexia Nervosa: A Multidimensional Perspective,* Brunner/Mazel, New York, 1982.
13. Holland, A. J., Hall, A., Murray, R. et al., Anorexia nervosa: a study of 34 twin pairs and one set of triplets, *Br. J. Psychiatry,* 145, 414–419, 1984.

14. Treasure, J. and Holland, A., Genetic factors in eating disorders, in *Handbook of Eating Disorders: Theory, Treatment, and Research,* Szmukler, G., Dare, C., and Treasure, J., Eds., John Wiley & Sons, New York, 1995, 65.

15. Holland, A. J., Sicotte, N., and Treasure, J., Anorexia nervosa: evidence for a genetic basis, *J. Psychosom. Res.,* 32, 561–571, 1988.

16. Waters, B. G. H., Beumont, P. J. V., Touyz, S. et al., Behavioral differences between twin and non-twin female sibling pairs discordant for anorexia nervosa, *Intl. J. Eating Disorders,* 9, 265–273, 1990.

17. Rutherford, J., McGuffin, P., Katz, R. J, et al., Genetic influences on eating attitudes in a normal female twin population, *Psychol. Med.,* 23, 425–436, 1993.

18. Fichter, M. M. and Noegel, R., Concordance for bulimia nervosa in twins, *Intl. J. Eating Disorders,* 9, 255–263, 1990.

19. Hsu, L. K. G., Chesler, B. E., and Santhouse, R., Bulimia nervosa in eleven sets of twins: a clinical report, *Intl. J. Eating Disorders,* 9, 275–282, 1990.

20. Holzinger, K. J., The relative effect of nature and nurture influences on twin differences, *J. Educ. Psychol.,* 20, 241–252, 1929.

21. Caspar, R. C., Eckert, E. D., Halmi, K. A. et al., Bulimia: its incidence in patients with anorexia nervosa, *Arch. Gen. Psychiatry,* 37, 1030–1035, 1980.

22. Strober, M., Salkin, B., Burroughs, J. et al., Validity of the bulimia-restricter distinction in anorexia nervosa, *J. Nerv. Ment. Dis.,* 170, 345–351, 1982.

23. Hsu, L. K. G., Crisp, A. H., and Harding, B., Outcome of anorexia nervosa, *Lancet,* 1(8107), 61–65, 1979.

24. Gershon, E. S., Hamovit, J. R., Schreiber, J. L. et al., Anorexia nervosa and major affective disorders associated in families: a preliminary report, in *Childhood Psychopathology and Development,* Guze, S. B., Earls, F. J., and Barrett, J. E., Eds., Raven Press, New York, 1983, 279.

25. Strober, M., Morrell, W., Burroughs, J. et al., A controlled family study of anorexia nervosa, *J. Psychiatr. Res.,* 19, 239–246, 1985.

26. Logue, C. M., Crowe, R. R., and Bean, J. A., A family study of anorexia nervosa and bulimia, *Compr. Psychiatr.,* 30, 179–188, 1989.

27. Strober, M., Lampert, C., Morrell, W. et al., A controlled family study of anorexia nervosa: evidence of familial aggregation and lack of shared transmission with affective disorders, *Intl. J. Eating Disorders,* 9, 239–253, 1990.

28. Lilenfeld, L. R., Kaye, W. H., Greeno, C. et al., A controlled family study of anorexia nervosa and bulimia nervosa, *Arch. Gen. Psychiatr.,* 55, 603–610, 1998.

29. Hudson, J. I., Pope, H. G., Jonas, J. M. et al., A controlled family history study of bulimia, *Psychol. Med.,* 17, 883–890, 1987.

30. Stern, S. L., Dixon, K. N., Sansone, R. A. et al., Affective disorder in the families of women with normal weight bulimia, *Am. J. Psychiatr.,* 141, 1224–1227, 1984.

31. Winokur, A., March, V., and Mendels, J., Primary affective disorder in relatives of patients with anorexia nervosa, *Am. J. Psychiatr.,* 137, 695–698, 1980.

32. Hudson, J. I., Pope, H. G., Jonas, J. M. et al., Family history study of anorexia nervosa and bulimia, *Br. J. Psychiatr.,* 142, 133–138, 1983.

33. Gershon, E. S., Schreiber, J. L., Hamovit, J. R. et al., Clinical findings in patients with anorexia nervosa and affective illness in their relatives, *Am. J. Psychiatr.,* 141, 1419–1422, 1984.

34. Rivinus, T. M., Biederman, J., Herzog, D. B. et al., Anorexia nervosa and affective disorders: a controlled family history study, *Am. J. Psychiatr.,* 141, 1414–1418, 1984.

35. Weitkamp, L. R., Stancer, H. C., Persad, E. et al., Depressive disorders and HLA: a gene on chromosome 6 that can affect behavior, *N. Engl. J. Med.,* 305, 1301–1306, 1981.

36. Biederman, J., Rivinus, T. M., Herzob, D. B. et al., High frequency of HLA-Bw16 in patients with anorexia nervosa, *Am. J. Psychiatr.,* 141, 1109–1110, 1984.

37. Biederman, J., Keller, M., Lavori, P. et al., HLA haplotype A26-B38 in affective disorders: lack of association, *Biol. Psychiatr.,* 22(2), 221–224, 1987.

38. Kiss, A., Hajek-Rosenmayr, A., Wiesnagrotzki, S. et al., Lack of association between HLA antigens and bulimia, *Biol. Psychiatr.,* 25, 803–806, 1989.

39. Kiss, A., Hajek-Rosenmayr, A., Haubenstock, A. et al., Lack of association between HLA antigens and anorexia nervosa, *Am. J. Psychiatr.,* 145, 876–877, 1988.

40. Bruins-Slot, L., Gorwood, P., Bouvard, M. et al., Lack of association between anorexia nervosa and D3 dopamine receptor gene, *Biol. Psychiatr.,* 43, 76–78, 1998.

41. Rotondo, A., Schuebel, K. E., Nielsen, D. A. et al., Tryptophan hydroxylase promoter polymorphisms and anorexia nervosa, *Biol. Psychiatr.,* 42, 99S, 1997.

42. Hinney, A., Lentes, K.-U., Rosenkranz, K. et al., Beta$_3$ adrenergic receptor allele distributions in children, adolescents, and young adults with obesity, underweight, and anorexia nervosa, *Intl. J. Obesity,* 21, 224–230, 1997.

43. Spielman, R. S., McGinnis, R. E., and Ewens, W. J., Transmission test for linkage disequilibrium: the insulin gene region and insulin-dependent diabetes mellitus (IDDM), *Am. J. Hum. Genet.,* 52, 506–516, 1993.

44. Hinney, A., Barth, N., Ziegler, A. et al., Serotonin transporter gene-linked polymorphic region: allele distributions in relationship to body weight and in anorexia nervosa, *Life Sci.,* 61, PL 295–303, 1997.

45. Collier, D. A., Arranz, M. J., Li, T. et al., Association between 5-HT$_{2A}$ gene promoter polymorphism and anorexia nervosa, *Lancet,* 350, 412, 1997.

46. Hinney, A., Ziegler, A., Nother, M. M. et al., 5-HT$_{2A}$ receptor gene polymorphisms, anorexia, and obesity, *Lancet,* 350, 1324–5, 1997.

47. Campbell, D. A., Sundaramurthy, D., Markham, A. F. et al., Lack of association between 5-HT$_{2A}$ gene promoter polymorphism and susceptibility to anorexia nervosa, *Lancet,* 351, 499, 1998.

48. Enoch, M. A., Kaye, W. H., Rotondo, A. et al., 5-HT$_{2A}$ promoter polymorphism –1438G/A, anorexia nervosa, and obsessive compulsive disorder, *Lancet,* 351, 1785–1786, 1998.

49. Sorbi, S., Nacmias, B., Tedde, A. et al., 5-HT$_{2A}$ promoter polymorphism in anorexia nervosa, *Lancet,* 351, 1785, 1998.

# 22 Genetic Influences on Aggressive Behavior

*Stephen C. Maxson*

## CONTENTS

## INTRODUCTION

Aggressive behavior has been defined in terms of the form and nature of movement used or antecedent conditions or accompanying internal state or immediate consequences or some combination of these.[1] For example, it has been defined as "overt behavior involving intent to inflict noxious stimulations to or behave destructively toward another organism"[2] or as "any form of behavior directed toward the goal of harming or injuring another living being who is motivated to avoid such treatment."[3]

It has long been recognized that, at least in animals, there are many types of aggressive behavior. Widely used classification systems have been proposed by Moyer,[4] Brain,[5] and Archer.[6] In male and female mice, four types of aggressive behavior are offense, defense, infanticide, and predation.[7] There are also female-specific forms of aggression which occur during pregnancy or lactation.[8] There are differences between inbred and/or selected strains for each type of aggression,[9-11] which is consistent with there being genetic variants for each type of aggressive behavior. This chapter focuses on chromosome regions (two) and genes (fifteen) reported to affect offense in male mice. Offense in mice is characterized by motor patterns and bite targets.[12] The motor patterns are chase, sideways offense posture, upright offense posture, and attacks. The bite targets of attacks are back, rump, and tail. A gene with effects on male and female dominance is also described.

## CHROMOSOME REGIONS

Several groups have indicated that variants of one or more genes on the male-specific part of the Y chromosome affect offense, as measured by latency to attack, proportion of animals attacking, or frequency of offense motor patterns in neutral or home cage with the same or different strain of opponent. This has been proposed for the Y chromosomes of the DBA1 and C57BL10,[13,14] DBA/2 and DBA/1,[15,16] CBA/Fa and C57BL6,[17] NZB and CBA/H,[18,19] SAL and LAL,[20,21] and PHL and PHH,[22] pairs of inbred strains. Also, the effect of the male-specific part of the Y on offense depends on the genetic background.[13,15,17,21] The use in these studies of reciprocal F1s, segregating populations, and congenic strains as evidence for effects of the male-specific part of the Y on offense has been critically reviewed.[12] There are many genes in this region of the Y. However, Maxson[23] has recently proposed that Sry (sex-determining region on the Y) is a candidate gene for effects of the male-specific part of the Y on offense.

There is also evidence that one or more genes in the t region of chromosome 17 have effects on offense in males.[24] There are many genes in this region. The array of genes is a haplotype; + and t are two of these haplotypes. When male mice of the +/+ and +/t genotypes were paired in laboratory cages, in a large arena, and in an outdoor enclosure, more +/t males initiated fights than did +/+ males.[24,25] Also, in the laboratory setting but not in the semi-natural situation, more of the +/t males won the fights. There is also an effect of genetic background. The difference between +/+ and +/t males was seen on the background of wild but not laboratory mice. The t region of chromosome 17 is involved in normal testis development, and it is very closely linked to the major histocompatibility complex (MHC) region, which has an effect on cortisone levels, testosterone levels, and sensitivity to testosterone.

## GENES

Spontaneous mutations, knockout mutations, overexpressed transgenes, and linkage analysis have been used to identify individual genes with effects on aggressive behaviors of males and females. For male offense, these genes affect aspects of steroid hormone, neurotransmitter, second-message, growth factor, and developmental systems. These genes are listed in Table 22.1. Some gene symbols are nonstandard; these are indicated by parentheses.

### STEROID HORMONES

The Ar gene on the X chromosome codes for the intracellular androgen receptor. The gene was formerly designated Tfm (testicular feminization). In rat brain, the receptor is localized at high to medium density in neocortex, thalamus, septum, amygdala, hippocampus, bed nucleus of the stria terminalis, hypothalamus, median eminence, pituitary, brainstem, and spinal cord.[26] When this receptor binds to its ligand, it acts as a transcription factor. There is a spontaneous mutant of this gene,[27] which is on the C57BL6 background. In brain, the mutant mice lack the larger size of the receptor protein and have less than 15% of the smaller size of the receptor protein.[28] In a resident-intruder test of 5 min duration, wild-type males attacked wild-type intruders, whereas mutant mice never attacked wild-type intruders. Both residents and intruders were isolated. This finding is consistent with effects of gonadectomy and testosterone replacement on male offense.[29]

The Estr gene on chromosome 10 codes for the intracellular estrogen α receptor. In rat brain, this receptor is found in high to medium densities in hypothalamus, amygdala, septum, hippocampus, median eminence, and pituitary.[26] When this receptor binds its ligand, it acts as a transcription factor. There is a knockout mutant of this receptor which reduces by at least 95% the binding of this estrogen receptor to 17-β estradiol, and which is on a mixed 129/Sv × C57BL6 background.[30] In resident-intruder tests of 15 min duration, the cumulative duration for aggression without attacks and for attacks was less for mutant than for wild-type males. Intruders were olfactory-bulbectomized Swiss-Webster males who rarely show aggression. Similar results were reported for a homogeneous

**TABLE 22.1**
**Genes and Male Offense**

| System | Name (Symbol) | Chromosome |
|---|---|---|
| Steroids | Androgen receptor (Ar) | X |
| | Estrogen receptor (Estr) | 10 |
| | Steroid sulfatase (Sts) | X/Y PAR |
| Neurotransmitters | Adenosine 2a receptor (Adra2a) | 19 |
| | Enkephalin (Penk) | 4 |
| | Histamine 1 receptor (H1r) | — |
| | 5-HT$_{1B}$ receptor (Htr1b) | 9 |
| | Monoamine oxidase A (Maoa) | X |
| | Neurokinin-1 receptor ("Nkr-1") | — |
| | Nitric oxide synthase (Nos1) | 5 |
| | Oxytocin (Oxt) | 2 |
| | Tachykinin-1 Receptor (Tac1r) | — |
| Second messengers | α-Calcium/calmodulin kinase II (CamK2a) | 18 |
| Growth factors | Transforming growth factor α (Tgfa) | 6 |
| Development | Neural cell adhesion molecule (Ncam) | 9 |
| | Tailless (Tlx) | — |

set test in a neutral cage; in a homogeneous set test, the genotype is the same for the two mice of an encounter. These findings are consistent with the hypothesis that some effects of testosterone on offense by males are due to its aromatization to estradiol in the brain.[29] This is based on studies of gonadectomy and replacement with estrogens and nonaromatizable androgens. Also, in a resident-intruder test of 10 min duration, the same mutant increases the frequency of attacks of resident females toward diestrous intact or ovariectomized and hormone-treated intruder females of the Swiss-Webster strain.[31] Thus, variants of the same gene can have opposite effects on male and female offensive behavior.

The Sts gene on the pseudoautosomal region (PAR) of the heterosomes codes for the enzyme steroid sulfatase. This enzyme regulates the concentration of steroid sulfates in liver and brain. In crosses derived from CBA/H and NZB inbred strains, there was cosegregation of alleles for Sts with proportion of males attacking male opponents of the A/J inbred strain.[32] The test may last as long as 6 min. In crosses derived from SAL and LAL selected strains, there is also cosegregation of the PAR with latency to attack in resident-intruder tests of 10 min duration.[21] The intruders were males of the MAS strain. Since the Sts gene is the only one in the PAR of mice, it is reasonable to assume that this was also a cosegregation of offense with alleles for Sts. Roubertoux (personal communication) has suggested that steroid sulfatase regulates the ratio of sulfated to nonsulfated neurosteroids, and that these neurosteroids affect offensive attack behaviors by acting on ligand-gated ion channels. For example, pregnenolone facilitates GABA$_A$ channels. Conversely, dehydroepiandrosterone sulfate (DHEA-S) and pregnenolone sulfate inhibit GABA$_A$ channels. Consistent with Roubertoux's hypothesis is the finding that selected strains differing in intensity of offense by males and females also differ in GABA-dependent chloride uptake by cortical synaptoneurons and in effects of benzodiazepines on male offense.[33] This Sts gene may also be involved in one type of female aggression. In resident-intruder tests, treatment of gonadectomized females with DHEA reduced their aggression toward lactating female intruders.[34]

### NEUROTRANSMITTERS

The Htr1b gene on chromosome 9 codes for the 5-HT$_{1B}$ receptor. There is a knockout mutant of this gene which results, throughout the brain, in the failure of the 5-HT$_{1B}$ receptor to bind 5-HT, and which is in the 129/Sv-ter strain.[35] $5HT1_B$ is located in many brain structures, including central gray, raphe nuclei, basal ganglia, and hippocampus, and it appears to be primarily an axonal autoreceptor.[36] In resident-intruder tests of 5 min duration, the mutants have shorter latency to attack and a higher frequency of attacks than the wild-type males. This effect occurs both with and without 4 weeks of isolation of the residents prior to the test. The intruders were group-housed males of the 129/Sv-ter strain. These findings are consistent with reduction of offense behaviors by action of mixed $5\text{-}HT1_A$ and $5\text{-}HT1_B$ agonists.[37] Also, in resident-intruder tests, this mutant decreases the latency to attack and increases the number of attacks by lactating females against intruder males of the +/+ genotype;[38] these attacks are primarily directed at the head, back, and flanks. Thus, variants of this gene may have the same effects on some types of male and female aggression.

The Maoa gene on the X chromosome codes for the enzyme monoamine oxidase A (MAOA), which degrades monoamines in presynaptic terminals. The MAOA enzyme is widely distributed in the brain. There is an insertional mutant of this gene which abolishes its activity in brain and liver from as early as 8 days of age.[39] This mutant occurred in the C3H/HeJ strain. In resident-intruder test of 10 min duration, the mutant had shorter latencies and inflicted more wounds than wild-type males. The intruders were group-housed males of the C3H strain. Also, mutant males had higher brain levels of 5-HT, 5-HIAA, dopamine, and norepinephrine from 1 to 90 days of age. Pharmacological studies on offense suggest that lower rather than higher levels of 5-HT should be associated with an increase in offense behaviors.[40] Regardless, postnatal but not prenatal MAOA inhibitors decrease the latency and increase the frequency of attack by CD-1 mice in a resident-intruder test.[41] Also, prenatal and postnatal MAOB inhibitors reduce the latency to attack and increase the frequency of attack in the same tests.[41] However, there may be no effect of a knockout mutant for Maob on attack behaviors.[42] MAOA preferentially metabolizes 5-HT and norpeinephrine, whereas MAOB preferentially metabolizes B-phenylethylamine. Both degrade dopamine.

The H1r gene codes for a histamine receptor. In the rat, histamine neurons project to broad areas of the telencephalon. There is a knockout mutant of this gene which results in failure of the receptor to specially bind its ligands in hypothalamus, thalamus, amygdala, and neocortex. In a resident-intruder test, the mutant males had longer latency to attack and a lower frequency of attacks than did wild-type males.[43] The residents were isolated 4 weeks prior to the test. The intruders were group-housed males of the 129/Sv-ter strain.

The Penk gene codes for pre-proenkephalin. The enkephalins are widely distributed from telencephalon to spinal cord. There is a knockout mutant of this gene which results in reduced enkephalin mRNA levels in brain and testis and in reduced levels of the neuropeptide met-enkephalin but not dynorphin or endorphin in the striatum.[44] The mutant was backcrossed onto the CD-1 strain. In a resident-intruder paradigm, the latency to attack was lower and the intensity of attacks was higher for mutants than for wild types on the first but not the second test. Residents were individually housed for about 4 weeks prior to the tests. The intruders were group-housed, wild-type males. There may also be an effect of the male-specific part of the Y chromosome on brain levels of met-enkephalin which is associated with a Y effect on proportion of males attacking an opponent male.[45]

The Oxt gene on chromosome 2 codes for oxytocin. Oxytocin is also widely distributed from telencephalon to spinal cord. There is a knockout mutant of this gene which results in the complete absence of oxytocin in the nervous system and the pituitary, and which is on a mixed 120/Sv and C57BL background.[46] In resident-intruder tests of 5 min duration in the home cage, wild-type and mutant mice did not differ in latency to attack, the frequency of attacks, or the total time spent in aggressive encounters, but the average duration of attacks was longer for wild-type than mutant males. The intruders were CF-1 males. However, in neutral arena tests of 5 min duration, the total time spent in aggressive encounters was also longer for wild-type than for mutant males. The opponents in the neutral cage test were also CF-1 males. Central administration of oxytocin increases aggression in other species, which is consistent with the findings for this knockout mutant of the oxytocin gene. However, it has been reported that another knockout mutant for the Oxt gene increases aggression of male mice.[47]

The Tac1r gene codes for the tachykinin-1 receptor. This is the postsynaptic receptor for substance P which is found throughout the brain, notably in the limbic system and hypothalamus. There is a knockout mutant for this gene which eliminates substance P binding sites from the brain and spinal cord, and which is on the C57BL6 background.[48] In a resident-intruder test, the mutant males had longer latency to attack and lower total fighting scores than wild-type males. The residents were isolated males, and the intruders were group-housed wild-type males. These findings need to be reconciled with those showing that intravenous injection of substance P increased latency to attack and reduced frequency of attacks.[49]

The gene Adra2a codes for an adenosine receptor. In the rat brain, the A2a receptor is located in striatum, nucleus acumbens, and olfactory tubercle. There is a knockout mutant of the Adra2a gene which eliminates specific binding of adenosine to this receptor, and which is on a mixed 129/Sv and CD-1 background.[50] In resident-intruder tests of 5 min duration, the latency to attack was lower for mutant than for wild-type males, and the frequency of tail rattles and attacks was higher for mutant than for wild-type males. The residents were isolated for 4 weeks prior to the tests, and the intruders were group-housed males of the CD-1 strain. The findings of this study are consistent with a report that adenosine analogs inhibit fighting in male mice.[51] Also, there was no effect of this mutant on aggression in females.

Nos1 on chromosome 5 codes for nitric oxide synthase of the nervous system. This enzyme controls the synthesis of nitric oxide, a diffusable neurotransmitter. In the nervous system, this enzyme is localized in discrete populations of neurons in the cerebellum, brainstem, hypothalamus, striatum, basal forebrain, hippocampus, olfactory bulb, and neocortex. It is also found in posterior pituitary and adrenal medulla. There is a knockout mutant of this gene which eliminates NOS1 enzymatic activity and which is on a mixed 129/Sv and C57BL6 background.[52] In resident-intruder tests of 5 min duration, the proportion of total attacks and the total number of agonistic encounters were higher for mutant than wild-type males, but they did not differ in attack latency. Intruders were wild-type males. In homogeneous set tests with groups of four males and of 15 min duration in a neutral arena, the attack latency was shorter for the mutant than for the wild type, and the total number of attacks and duration of agonistic encounters was higher for mutant than wild-type males. In both tests, the bite target of the attacks appeared to be the neck region rather than the back, rump, or base of tail. This shift in bite target may explain the lethality of attacks by the mutant mice. Although there is no effect of this mutant on plasma levels of testosterone,[52] the effect of the mutant on attack behaviors is testosterone dependent, as shown by gonadectomy and testosterone replacement.[53] Recently, it has been shown that pharmacological inhibition of neuronal nitric oxide synthase had the same effects on attack behaviors in both resident-intruder and neutral arena tests as had the Nos1 mutant.[54] Also, there appears to be no effect on female aggression in these tests.

## SECOND MESSENGERS

The CamK2a gene on chromosome 18 codes for $\alpha$-calcium/calmodulin kinase II, which is widely distributed in the nervous system. This enzyme is involved in the second messenger for calcium. It phosphorylates proteins which are involved presynaptically in neurotransmitter release and postsynaptically in ionic and genomic effects of neurotransmitters. There is a knockout mutant which is on a mixed 129/Sv and BALB/c background.[55] The mutant lacks mRNA for this gene. In a resident-intruder test of 5 min duration, mutant males made fewer (essentially none) attack bites than did wild-type males. The residents were individually housed for 4 weeks before the test. The intruders were group-housed, wild-type males. In other tests, wild-type males were the residents, and the intruders were either heterozygote, mutant, or wild-type males. The heterozygote intruders made more defensive attacks on the residents than did wild-type intruders, whereas the mutants made fewer attacks on the resident males than did wild-type intruders. There was also an effect of this mutant on 5-HT release from raphe neurons, but it had no effect on 5-HT receptor response or on 5-HT reuptake.

## GROWTH FACTORS

The Tgfa gene on chromosome 6 codes for transforming growth factor α, which interacts with cells through the epidermal growth factor receptor. Tgfa is expressed in embryonic and adult tissues, and Tgfa mRNA and protein as well as EGF/TGFα receptors have been demonstrated in rodent brain. There is a strain of mouse with overexpression of TGFα;[56] the transgene is human in origin. This transgene is on the CD-1 background. In a resident-intruder test of 5 or 8 min duration, transgenic males spent more time being aggressive (defined as tail rattle, sideways offense posture, biting, and fighting) than nontransgenic males. The residents were isolated for at least 7 days; the intruders were group-housed males of the CD-1 strain. Conversely, in a resident-intruder test of 8 min duration, transgenic females spent less time being aggressive than did nontransgenic females.[57] The residents were individually housed for 7 days; the intruders were group-housed CD-1 females.

Transgenic and nontransgenic males differed in plasma levels of estradiol but not of testosterone or corticosterone.[58] Also, gonadectomy eliminated the difference in offense between transgenic and nontransgenic males, and the difference between transgenic and nontransgenic males was restored by treatment with estradiol.[58]

The ratio of 5-HIAA/5-HT was lower in the brainstem of transgenic males than in nontransgenic males.[59] This difference was not reported for females. Rather, females had elevated levels of norepinephrine in the hypothalamus and of 5-HT in the cortex and brainstem. Also, the 5-HT uptake inhibitors, zimelidine and clomipramine, decreased time spent fighting for transgenic males to below the level of that for nontransgenics.[60] In males, there was also a difference in GABA$_A$ receptor binding, and in males and females, there were differences in dopamine transporter binding.

## DEVELOPMENT

The Tlx (tailless) gene codes for a transcription factor of the ligand-activated nuclear receptor type, which is normally expressed during development in telencephalic and diencephalic ventricular zone, eye, and nasal placode. Its ligand has not yet been identified. There is a knockout mutant of this gene which lacks the functional protein in brain and presumably eye and nasal placode.[61] In adult mutants, there is a reduction in size of olfactory cortex, infrarinal cortex, entorhinal cortex, anterior commissure, amygdala, and dentate gyrus. All Tlx mutant males mutilated or killed their littermates; similarly, 45% of females were highly aggressive. Also, Tlx mutant males attacked nonestrous females eight times more frequently than did wild-type males. Further details on behavior tests and results have not yet been published.

Ncam on chromosome 9 codes for the neural cell adhesion molecule, which is a glycoprotein of the immunoglobulin superfamily and is involved in cell-cell and cell-matrix interactions. There is a knockout mutant of this gene that lacks a functional NCAM protein and that has a reduction in size of the olfactory bulb and some subtle changes elsewhere in the brain.[62] The knockout mutant is on a mixed 129/Ola/Hsd and C57BL6J background. In a resident-intruder test of 5 min duration, males homozygous and heterozygous for the mutant had shorter attack latencies and made more attacks than wild-type males. The residents were isolated for 6 weeks prior to the test; the intruders were group-housed C57BL6J males. There were no differences among mutant homozygote, heterozygote, or wild-type homozygote males in plasma testosterone at 1 week before testing and in plasma corticosterone immediately after testing. However, after testing, mutant homozygotes and heterozygotes had higher mRNA levels for c-fos than wild-type males in the septum, amygdala, preoptic area, lateral hypothalamus, dorsal raphe nucleus, and gigantocellular reticular nucleus but not in other brain areas.

Dvl1 (dishevelled 1) on chromosome 4 is a homolog of the segment polarity gene, dishevelled, of *Drosophila*; in mice, it is expressed in embryos and adults, including cerebellum, hippocampus, and olfactory bulb. There is a knockout mutant for this gene which is on the 129/SvEv background.[63] There is no Dvl1 protein in mutant mice. The brain of mutants is structurally normal. Social dominance rather than aggression was assessed for these mice. Dominance behavior was assessed in a tube test. Mice of the same sex were used on a test. A test pair consisted of a mutant and a wild-type mouse. A subject was declared a winner when its opponent backed out. By this criterion, the mutant mice were subordinate to the wild-type mice. There was no sex difference.

## CONCEPTUAL AND METHODOLOGICAL ISSUES

The above describes the reported findings for each chromosome region or gene. Now, I turn to conceptual and methodological issues which are requisite to a critical evaluation of these reports and to new ones on the genetics of aggressive behavior.

### GENETIC

There may be prenatal or postnatal maternal effects on aggressive behaviors,[12] and genes may act on the biology or behavior of the mother and thereby on the aggressive behavior of her progeny. Sometimes, this pathway may be confounded with effects of genes acting directly in an individual. For example, the mothers of males mutant for Maoa were themselves also mutant for Maoa; the wild-type males also had wild-type mothers.[39] For transgenics and null mutants, this maternal pathway may be controlled by breeding wild types and mutants or transgenics from crosses of heterozygotes.

There is also a genetic confound for some knockout mutants.[64] Sometimes, a knockout is made in one strain, and then the mutant is backcrossed to another strain. These strains are now congenic rather than coisogenic. This potentially confounds effects of the mutant with effects of genes from the donor strain that are linked to the mutant. For example, the Nos1 mutant was made in a 129/Sv strain, and the mutant was then backcrossed into the C57BL6 strain.[52] Here, genes linked to the Nos1 rather than Nos1 itself may be those with effects on aggressive behavior. There are controls for this possibility, and these were used in the study of the effects of Nos1. They are (1) derive wild types and mutants from crosses of heterozygotes and (2) test the background strains. For Nos1, the background strains were 129/Sv and C57BL6; both of these were less aggressive than the Nos1 mutants.

There may also be epistatic interactions between a single gene and genetic background. For example, it may be that a null mutant may have an effect on aggressive behavior in one genetic background but not another. This may be revealed when the mutant is backcrossed from one strain to another. The possibility of such epistatic interactions has been reported for effects of the Estr mutant on aggressive behavior.[30] Furthermore, the absence of the effect of a null mutant on one genetic background need not mean that across backgrounds, it never has an effect on aggressive behavior.

Thirteen of the 15 genes reported to have effects on aggressive behavior were identified with null mutants. Two (Sts and Tgfa) of the 15 genes are identified by allelic variants with effects on amount of functional protein. It has been suggested that more might be learned about the functional effects of allele substitution by relating gene dose and thereby amount of functional product to phenotype, including aggressive behavior.[65] This could be done by studying effects on aggressive behavior of null mutants and transgenics for the same gene.

### BEHAVIORAL

Effects of the mutant or transgene on aggressive behavior may be due to sensorimotor or other behavioral deficits. For example, if a mutant or transgene eliminated olfaction for pheromonal or other odors, it would reduce or eliminate aggressive behavior. Alternatively, if a mutant or transgene increased motor activity, it might increase aggressive behavior by facilitating contact between resident and intruder. These and other possibilities to account for effects of a mutant or transgene on aggressive behavior should be explored by using other behavioral tests. Some of these are described in Refs. 46, 52, and 66.

Animal husbandry, test conditions, and test measures are not the same in each of the above described studies on chromosome regions or genes and aggressive behavior. For example, the strain or genotype of intruder or opponent is not the same across studies. The strain and type of intruder or opponent can influence whether or not a gene affects aggressive behavior.[12] Also, some studies use attack latency and frequency as dependent measures, and others use attack duration or composite scores of aggression. Again, the dependent variable can influence whether or not a gene affects aggressive behavior.[12] Thus, in order to relate the findings with one gene to those with another, animal husbandry, test conditions, and test measures should be standardized.[66] This might include latency, frequency, and duration of attack in a resident-intruder test of isolated males paired with an intruder of a standard strain. Also, effects of the gene on all aspects of offense should be characterized.[12]

## MECHANISMS

Some of the knockout mutants have effects on health as well as on aggressive behavior. For example, the Maoa mutant causes trembling and difficulty in righting by pups,[39] the Nos1 mutant causes enlarged stomachs with hypertrophy of the pyloric sphincter,[52] and the adenosine 2a receptor mutant causes high blood pressure and elevates heart rate.[50] When a mutant has an adverse effect on health, it should be determined whether or not these mediate the effects of a gene on aggressive behavior.

For all of the reported chromosomal and genetic effects on aggressive behavior, the allelic differences are present from birth and in all tissues. Without further information, it would not be possible to know when, where, and how the gene acts to affect aggressive behavior. However, the times and places of transcription are known for many genes. This limits the possible sites and times for a gene's expression. Also, pharmacological treatments may identify the timing of a gene's effects on aggressive behavior. For example, pharmacological inhibition of Nos1 in adults has the same effect on aggressive behavior as the Nos1 null mutant,[54] and pharmacological treatment of adults with adenosine analogs have the opposite effect on aggressive behavior as the Ar2a null mutant.[51] Both of these findings are consistent with an adult action of these genes on aggressive behavior. Temporal-specific knockouts can also be used to identify when a gene initially acts with regard to its effect on aggressive behavior.[67] Similarly, tissue- or cell-specific knockouts can be used to identify where a gene initially acts with regard to its effects on aggressive behavior.[67] Regardless, there may be a cascade of metabolic effects following the initial action of a gene. The relationship between this cascade and the effects of the gene on aggressive behavior can be further investigated with mutants or treatments that manipulate putative steps in the cascade.

There has also been some concern that all viable knockout mutants will increase aggression, suggesting that these are nonspecific influences.[68] However, some knockouts, (such as for the Estr and Oxt) decrease aggression in males.[30,46] Also, the Estr mutant[30,31] has opposite effects on male and female aggressive behavior. Furthermore, there are more than 250 viable knockouts; effects of aggression have been reported for only 13 of these.[66]

## NATURE AND NURTURE

The fact that there are genes with effects on aggressive behaviors of mice does not imply that there are no environmental effects on aggressive behavior. Rather, the development of aggressive potential in individuals and of differences between individuals is a function of both genes and environments. Elsewhere, I have reviewed effects of prenatal maternal environment, postnatal maternal environment, peer environment, other experiences, and test conditions on the aggressive behaviors of mice.[12] For example, differences in aggressive behavior of inbred strains of mice were first reported in 1942 by Ginsburg[69] and Scott.[70] They also showed that the rank ordering of the inbred strains could be changed by winning or losing fights. The rank ordering for the aggressive behavior of the C57BL10 strain used in both studies was also dependent on whether the males were transferred from cage to cage by forceps or in a small box.[71]

## ANIMAL MODELS OF HUMAN AGGRESSIVE BEHAVIOR

Genes with effects on aggressive behaviors in mice can be used as animal models for some but not all aspects of human aggression. Many genes of mice and humans are homologous.[7,12,66] Thus, if a gene affects variation in a type of aggression in mice, it is conceivable that it may also have an effect on variation in a type of human aggression. Two examples of this have been reported. There is an effect of variants of the monoamine oxidase A gene on mouse and human aggression[39,72] and of a 5-HT1$_B$ gene on mouse and human aggression.[35,73] Further development of such models depends on identifying the similarities and differences between mice and humans for the biological pathways from gene to behavior and of the relationship between the aggressive behavior of mice and humans. Elsewhere, I have considered these issues in more detail.[7,12,66,74]

## OVERVIEW AND CONCLUSIONS

If the reports described above are accepted as valid, genes with effects on steroid hormones, neurotransmitters, second messengers, growth factors, and development are now known to have an influence on offensive attacks of male mice. Regardless, many of the reports on genes and aggressive behavior are consistent with those from neural, endocrine, or pharmacological research on offense in male mice and other animals.

There has been a long-standing debate as to whether the same genes affect aggression of males and females.[75] The studies reviewed here indicate that some genes affect not only male but also female aggressive behavior. The direction of effect is the same in males and females for the Htr1b and Tlx genes, whereas the direction of effect is opposite in males and females for the Estr and Tgfa genes. However, some genes (Nos1, A2ar) appear to affect male but not female aggression. Other genes may have been tested for only aggressive behavior in males. I suggest that all genes with effects on male offense should also be tested for effects on female aggression. Also, they should be tested for effects on other types of aggression in males and females. Only one gene (CamK2a) has been tested for effects on both male offense and defense. Homozygote mutants for this gene reduce both.

In 1993, very few pieces of the mouse genome had been identified as having an effect on male offense. These were the Ar gene, the Tgfa gene, the t region on chromosome 17, and the male-specific region of the Y chromosome. Since then, it has been reported that an additional 13 genes have an effect on male offense. This is about three to four new genes a year. I expect that as many or more will be added each year to this list from studies with linkage mapping, transgenics, and null mutants. As is discussed above, careful and critical consideration must be given for each new gene as to whether it, in fact, influences an aggressive behavior. Then, if there is truly an effect of this gene, the exact behavioral effects of the gene should be determined. Also, the pathway from gene expression to behavioral effect should be specified for the gene itself and for its interactions with other genes and the environment.

## REFERENCES

1. Huntingford, F. and Turner, A., *Animal Conflict*, Chapman and Hall, New York, 1987.
2. Volavka, J., *Neurobiology of Violence*, American Psychiatric Press, Washington, D.C., 1995.

3. Baron, R. A. and Richardson, D. R., *Human Aggression, 2nd ed.,* Plenum Press, New York, 1994.

4. Moyer, K. E., *The Psychobiology of Aggression*, Harper and Row Publishers, New York, 1976.

5. Brain, P. F., *Hormones, Drugs and Aggression*, Eden Press, Montreal, 1979.

6. Archer, J., *The Behavioural Biology of Aggression,* Cambridge University Press, Cambridge, 1988.

7. Maxson, S. C., Potential genetic models of aggression and violence in males, in *Genetically Defined Animal Models of Neurobehavioral Dysfunction*, Driscoll, P., Ed., Birkhauser, Boston, 1992, 174.

8. Svare, B., Recent advances in the study of female aggressive behaviour in mice, in *House Mouse Aggression*, Brain, P. F., Mainardi, D., and Parmigiani, S., Eds., Harwood Academic Publishers, 1989, 135.

9. Maxson, S. C., The genetics of aggression in vertebrates, in *The Biology of Aggression*, Brain, P. F. and Benton, D., Eds., Sijthoff and Noordhoff, Rockville, MD, 1981, 69.

10. Simon, N. G., The genetics of intermale aggressive behavior in mice: recent research and alternative strategies, *Neurosci. Biobehav. Rev.*, 3, 97, 1979.

11. Michard, C. and Carlier, M., Les conduites d'agression intraspecifique chez la souris domestique: differences individuelles et analyses genetiques, *Biol. Behav.*, 10, 123, 1985.

12. Maxson, S. C., Methodological issues in genetic analysis of an agonistic behavior (offense) in male mice, in *Techniques for the Genetic Analysis of Brain and Behavior,* Goldowitz, D., Wahlsten, D., and Wimer, R. E., Eds., Elsevier, New York, 1992, 349.

13. Maxson, S. C., Ginsburg, B. E., and Trattner, A., Interaction of Y-chromosomal and autosomal gene(s) in the development of aggression in mice, *Behav. Genet.*, 9, 219, 1979.

14. Maxson, S.C., Didier-Erickson, A., and Ogawa, S., The Y chromosome, social signals, and offense in mice, *Behav. Neural Biol.*, 52, 251, 1989.

15. Selmanoff, M. K., Maxson, S. C., and Ginsburg, B. E., Chromosomal determinants of intermale aggressive behavior in inbred mice, *Behav. Genet.*, 6, 53, 1976.

16. Shrenker, P. and Maxson, S. C., The Y chromosomes of DBA/1Bg and DBA/2Bg compared for effects on intermale aggression, *Behav. Genet.*, 12, 429, 1982.

17. Stewart, A. D., Manning, A., and Batty, J., Effects of Y-chromosome variants on male behaviour of the mouse, *Mus musculus, Genet. Res. Cambr.*, 35, 261, 1980.

18. Carlier, M. and Roubertoux, P., Differences between CBA/H and NZB mice on intermale aggression, in *Genetic Approaches to Behaviour*, Medioni, J. and Vaysse, E., Eds., Privat, IEC, Toulouse, France, 1986, 45.

19. Guillot, P.-V., Carlier, M., Maxson, S. C., and Roubertoux, P. L., Intermale aggression tested in two procedures using four inbred strains of mice and their reciprocal congenics: Y chromosomal implications, *Behav. Genet.*, 25, 357, 1995.

20. van Oortmerssen, G. A., Benus, R. F., and Sluyter, F., Studies on wild house mice IV: on the heredity of testosterone and readiness to attack, *Aggressive Behav.*, 18, 143, 1992.

21. Sluyter, F., van Oortmerssen, G. A., and Koolhaas, J. M., Studies on wild house mice VI: differential effects of the Y chromosome on intermale aggression, *Aggressive Behav.*, 29, 379, 1994.

22. Weir, J. A., Allosomal and autosomal control of sex ratio and mating behavior in PHH and PHL mice, *Genetics*, 84, 755, 1976.

23. Maxson, S. C., Searching for candidate genes with effects on an agonistic behavior, offense, in mice, *Behav. Genet.*, 26, 471, 1996.

24. Lenington, S., The t complex: a story of genes, behavior, and populations, *Adv. Study Behav.*, 20, 51, 1991.

25. Lenington, S., Drickamer, L. C., Robinson, A. S., and Erhart, M., Genetic basis for male aggression and survivorship in wild house mice *(Mus domesticus), Aggressive Behav.*, 22, 135, 1996.

26. Brown, R. E., *An Introduction to Neuroendocrinology*, Cambridge University Press, Cambridge, 1994.

27. Ohno, S. and Geller, L. N., Tfm mutation and masculinization versus feminization of the mouse central nervous system, *Cell*, 3, 236, 1974.

28. Olsen, K., Genetic determinants of sexual differentiation, in *Hormones and Behaviour in Higher Vertebrates*, Balthazart, J., Prove, E., and Gilles, R., Eds., Springer-Verlag, Berlin, 1983, 138.

29. Simon, N. G., Lu, S. F., McKenna, S. E., Chen, X., and Clifford, A. C., Sexual dimorphism in regulatory systems for aggression, in *The Development of Sex Differences and Similarities in Behavior*, Haug, M., Whalen, R., Aron, C., and Olsen, K. L., Eds., Kluwer, Dordrecht, Netherlands, 1993, 389.

30. Ogawa, S., Lubahn, D. B., Korach, K. S., and Pfaff, D. W., Behavioral effects of estrogen receptor gene disruption in male mice, *Proc. Natl. Acad. Sci. U.S.A.*, 94, 1476, 1997.

31. Ogawa, S., Taylor, J. A., Lubahn, D. B., Korach, K. S., and Pfaff, D. W., Reversal of sex roles in genetic female mice by disruption of estrogen receptor gene, *Neuroendocrinology*, 64, 467, 1996.

32. Roubertoux, P. L., Carlier, M., Degrelle, H., Haas-Dupertuis, M.-C., Phillips, J., and Moutier, R., Co-segregation of intermale aggression with the pseudoautosomal region of the Y chromosome in mice, *Genetics*, 135, 225, 1994.

33. Weerts, E. M., Miller, L. G., Hood, K. E., and Miczek, K. A., Increased $GABA_A$-dependent chloride uptake in mice selectively bred for low aggressive behavior, *Psychopharmacology*, 108, 196, 1992.

34. Robel, P., Young, J., Corpechot, C., Mayo, W., Perche, F., Haug, M., Simon, H., and Baulieu, E. E., *J. Steroid Biochem. Mol. Biol.*, 53, 355, 1995.

35. Saudou, F., Amara, D. A., Dierch, A., LeMeur, M., Ramboz, S., Segu, L., Buhot, M.-C., and Hen, R., Enhanced aggressive behavior in mice lacking $5-HT1_B$ receptor, *Science*, 265, 1875, 1994.

36. Lucas, J. J. and Hen, R., New players in the 5-HT receptor field: genes and knockouts, *Trends Pharmacol. Sci.*, 16, 246, 1995.

37. Cologer-Clifford, A., Smoluk, S. A., and Simon, N. G., Effects of serotonergic 1A and 1B agonists and antagonists versus estrogenic systems for aggression, *Ann. N. Y. Acad. Sci.,* 793, 3339, 1995.

38. Brunner, D. and Hen, R. Insights into the neurobiology of impulsive behavior from serotonin receptor knockout mice, *Ann. N. Y. Acad. Sci.*, 836, 81, 1997.

39. Cases O., Seif, I., Grimsby, J., Gaspar, P., Chen, K., Pournin, S., Muller, U., Aguet, M., Babinet, C., Shih, J.C., and de Maeyer, E., Aggressive behavior and altered amounts of brain serotonin and norepinepherine in mice lacking MAOA, *Science*, 268, 1763, 1995.

40. Hen, R., Mean genes, *Neuron*, 16, 17, 1996.

41. Palmour, R. M., Ervin, F. R., and Mejia, J. M., Effects of monoamine oxidase inhibition during brain development upon aggressive behavior in mice, XIII World Meeting of the International Society for Research on Aggression, Mahwah, NJ, July 12–17, 1998.

42. Grimsby, J., Toth, M., Chen, K., Kumazawa, T., Klaidman, L., Adams, J. D., Karoum, F., Gal, J., and Shih, J. C., Increased stress response and B-phenylethylamine in MAOB-deficient mice, *Nat. Genet.*, 17, 206, 1997.

43. Yanai, K., Son, L. Z., Endow, M., and Watanabe, T., Targeted disruption of histamine H1 receptors in mice: behavioral and neurochemical characterization, *Life Sci.*, 62, 1607, 1998.

44. Koing, M., Zimmer, A. M., Steiner, H., Holmes, P. V., Crawley, J. N., Brownstein, M. J., and Zimmer, A., Pain response, anxiety, and aggression in mice deficient in pre-proenkephalin, *Nature*, 383, 535, 1996.

45. Roubertoux, P. L., Carlier, M., Mortaud, S., Tordjman, S., Moutier, R., and Degrelle, H., Neurobehavioral correlates of the substitution of the specific region of the Y chromosome: SR-Y, *Behav. Genet.*, 22, 750, 1992.

46. DeVries, A. C., Young, III, W. S., and Nelson, R., Reduced aggressive behaviour in mice with targeted disruption of the oxytocin gene, *J. Neuroendocrinol.*, 9, 363, 1997.

47. Winslow, J. T., Young, L. J., Hearn, E., Gingrich, B., and Wang, Z., Phenotypic expression of an oxytocin peptide null mutant in mice, *Soc. Neurosci. Abs.*, 23, 1072, 1997.

48. DeFelipe, C., Herrero, J, F., O'Brien, J. A., Palmer, J. A., Doyle, C. A., Smith, A. J. H., Laird, J. M. A., Belmonte, C., Cervero, F., and Hunt, S. P., Altered nociception, analgesia, and aggression in mice lacking the receptor for substance P, *Nature*, 392, 394, 1998.

49. Bigi, S., DeAcetis, L., Chiarotti, F., and Alleva, E., Substance P effects on intraspecifc aggressive behaviour in isolated male mice: an ethopharmacological analysis, *Behav. Pharmacol.*, 4, 495, 1993.

50. Ledent, C., Vaugeois, J.-M., Schiffmann, S. N., Pedrazzini, T., El Yacoubi, M., Vanderhaeghen, J.-J., Costentin, J., Heath, J. K., Vassart, G., and Parmentier, M., Aggressiveness, hypoalgesia, and high blood pressure in mice lacking the adenosine A2ra receptor, *Nature*, 388, 674, 1997.

51. Palmour, R. M,, Lipowski, C. J., Simon, C. K., and Ervin, E. R., Adenosine analogs inhibit fighting in isolated mice, *Life Sci.*, 44, 1293, 1989.

52. Nelson, R. J., Demas, G. E., Huang, P. L., Flashman, M. C., Dawson, V. L., Dawson, T. M., and Snyder, S. H., Behavioral abnormalities in male mice lacking neuronal nitric oxide synthase, *Nature*, 378, 383, 1995.

53. Kriegsfeld, L. J., Dawson, T. M., Dawson, V. L., Nelson, R. J., and Snyder, S. H., Aggressive behavior in male mice lacking the gene for neuronal nitric oxide synthase requires testosterone, *Brain Res.*, 769, 66, 1997.

54. Demas, G. E., Eliasson, M. J. L., Dawson, T. M., Dawson, V. L., Kriegsfeld, L. J., Nelson, R. J., and Snyder, S. H., Inhibition of neuronal nitric oxide synthase increases aggressive behavior in mice, *Mol. Med.*, 3, 611, 1997

55. Chen, C., Rainnie, D. G., Greene, R. W., and Tonegawa, S., Abnormal fear response and aggressive behavior in mutant mice deficient for α-calcium-calmodulin kinase II, *Science*, 266, 291, 1994.

56. Hilakivi-Clarke, L. A., Arora, P. K., Sabol, M.-B., Clarke, R., Dickson, R. B., and Lippman, M. E., Alterations in behavior, steroid hormones, and natural killer cell activity in male transgenic TGFα mice, *Brain Res.*, 588, 97, 1992.

57. Hilakivi-Clarke, L. A., Arora, P. K., Clarke, R., Wright, A., Lippman, M. E., and Dickson, R. B., Opposing behavioral alterations in male and female transgenic TGFα mice: association with tumor susceptibility, *Brit. J. Cancer*, 67, 1026, 1993.

58. Hilakivi-Clarke, L. and Goldberg, R., Gonadal hormones and aggression-maintaining effect of alcohol in male transgenic transforming growth factor-α mice, *Alcoholism: Clin. Exp. Res.*, 19, 708, 1995.

59. Hilakivi-Clarke, L. A., Corduban, T.-D., Taira, T., Hitri, A., Deutsch, S., Korpi, E. R., Goldberg, R., and Kellar, K. J., Alterations in brain monoamines and GABA$_A$ receptors in transgenic mice overexpressing TGFα, *Pharmacol. Biochem. Behav.*, 50, 593, 1995.

60. Hilakivi-Clarke, L. A. and Goldberg, R., Effects of tryptophan and serotonin uptake inhibitors on behavior in mice transgenic for transforming growth factor α mice, *Eur. J. Pharmacol.*, 237, 101, 1993.

61. Monaghan, A. P., Bock, D., Gass, P., Schwager, A., Wolfer, D. P., Lipp, H.-P., and Schutz, G., Defective limbic system in mice lacking the tailless gene, *Nature*, 390, 515, 1997.

62. Stork, O., Weiz, H., Cremer, H., and Schachner, M., Increased intermale aggression and neuroendocrine response in mice deficient for the neural cell adhesion molecule (NCAM), *Eur. J. Neurosci.*, 9, 1117, 1997.

63. Lijam, N., Paylor, R., McDonald, M. P., Crawley, J. N., Deng, C.-X., Herrup, K., Stevens, K. E., Maccaferri, G., McBain, C. J., Sussman, D. J., and Wynshaw-Boris, A., Social interaction and sensory gating in mice lacking Dvl1, *Cell*, 90, 895, 1997.

64. Gerlai, R., Gene-targeting studies of mammalian behavior: is it the mutation or the background genotype?, *Trends Neurosci.*, 19, 177, 1996.

65. Crusio, W. E., Gene-targeting studies: new methods, old problems, *Trends Neurosci.*, 19, 186, 1996.

66. Maxson, S. C., Homologous genes, aggression, and animal models, *Dev. Neuropsychol.*, 14, 143, 1998.

67. Hen, R., Targeting aggressive behavior: constitutive, inducible, and tissue-specific knockouts, in *What is Wrong with My Mouse? New Interplay Between Mouse Genetics and Behavior*, Takahashi, J. S., Ed., Society for Neuroscience, Washington, D.C., 1996, 50.

68. Good, M., Targeted deletion of neuronal nitric oxide: a step closer to understanding its functional significance, *Trends Neurosci.*, 19, 83, 1996.

69. Ginsburg, B. E. and Allee, W. C., Some effects of conditioning on social dominance and subordination in inbred strains of mice, *Physiol. Zool.*, 15, 485, 1942.

70. Scott, J. P., Genetic differences in the social behavior of inbred strains of mice, *J. Hered.*, 33, 11, 1942.

71. Ginsburg, B. E., Genetic parameters in behavioral research, in *Behavior-Genetic Analysis*, Hirsch, J., Ed., McGraw-Hill, New York, 1967, 135.

72. Brunner, H. G., MAOA deficiency and abnormal behavior: perspectives on an association, in *Genetics of Criminal and Antisocial Behaviour*, Bock, G. R. and Goode, J. A., Eds., John Wiley, New York, 1996, 155.

73. Goldman, D., Lappalainen, J., Knowler, W. C., Hanson, R. L., Robin, R. W., Urbanek, M., Guenther, D., Moore, E., Bennet, P. H., Long, J., Virkunnen, M., and Linnoila, M., Genetic linkage of alcoholism and antisocial alcoholism, XIII World Meeting of the International Society for Research on Aggression, Mahwah, NJ, July 12–17, 1998.

74. Maxson, S. C., Issues in the search for candidate genes in mice as potential animal models of human aggression, in *Genetics of Criminal and Antisocial Behaviour*, Bock, G. R. and Goode, J. A., Eds., John Wiley, New York, 1996, 21.

75. Hood, K. E. and Cairns, R. B., A developmental-genetic analysis of aggressive behavior in mice. II. Cross-sex inheritance, *Behav. Genet.*, 18, 605, 1988.

# 23 Genes Participating in Reproductive Behaviours

*Sonoko Ogawa, C. J. Krebs, Y. S. Zhu, and Donald W. Pfaff*

## CONTENTS

Genetic influences on reproductive behaviors can work through direct or indirect routes (see Chapter 1, Ref. 1). In the field of reproductive physiology, many of the indirect routes most easily demonstrated derive from the phenomena of sexual determination and differentiation. Here, we illustrate the most prominent phenomena of direct gene/behavior causal relationships, primarily in the field of female sex behavior responses, and also draw on examples of indirect routes of genetic influence, primarily regarding male-typical sex behavior.

## IN THE FEMALE

Candidate genes for those which could directly underlie hormonal facilitation of female reproductive behavior are revealed in the list of messenger RNAs turned on by estrogens. The product of those genes are, in turn, facilitatory to female reproductive behavior (Figure 23.1). Several genes on that list have attracted patterns of data which make it clear that they participate in direct causation of female reproductive behavior in the adult brain.

### ESTROGEN-RELATED GENES

Gene expression and new protein synthesis are required for effects of estrogens on female rodent reproductive behavior (reviewed in Ref. 2). The activation of at least some of these genes must

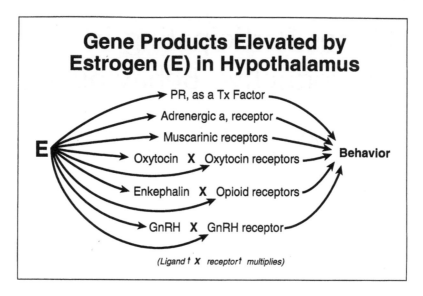

**FIGURE 23.1**  Estrogen administration leads to the upregulation of several transcriptional systems in medial hypothalamic neurons. In turn, the products of those genes foster female reproductive behaviors. Therefore, these genomic actions of estradiol are demonstrated to be on the causal route by which estrogens facilitate behavior. Different gene products have different roles. For example, the progesterone receptor (PR) is itself a transcription factor (Tx Factor). Other gene products facilitate specific neurotransmitter or neuropeptide systems. Note that where estrogen administration would upregulate both a ligand and its cognate receptor, the two hormonal effects could multiply. (From Pfaff, D., The MIT Press, 1999.)

depend upon the binding of estradiol to the classical nuclear estrogen receptor, now designated ER-α, since estrogen receptor blockers can reduce or abolish the facilitatory effects of the hormone.

New insights into the roles that the estrogen receptor, a nuclear protein which acts as a transcription factor, can play in the generation of reproductive behaviors have emerged with the production of the "estrogen receptor knockout" (ERKO) mouse.[3] In this mouse, the function of the classical estrogen receptor, ER-α, was disrupted by the insertion of a neomycin resistance gene. Behavioral assays[4] showed that ERKO females did not show any lordosis behavior in response to stud males. At least two processes were responsible for the absence of lordosis behavior in ERKO females. First, during sexual behavior tests, ERKO females were vigorously attacked by resident stud males immediately after they were introduced into the males' cages, whereas only a very small number of wild-type females were similarly attacked. Since the ERKO females were being treated as intruder males and, indeed, not allowing normal mounts by stud males, lordosis could not occur. Second, during reflex tests, ERKO females showed less immobility and dorsiflexion in response to manual stimulation.[4] This finding was consistent with the observation that in those cases where stud males attempted mounts to ERKO females, the females showed strong rejection behavior consisting of rapid flight and kicking with the hind legs.

To what extent could the behavioral analysis be extended in ovariectomized ERKO mice treated with estrogen, or estrogen plus progesterone?[5] Even though stud males approached and attempted to mount the females regardless of genotype — wild-type, heterozygotes, and homozygous ERKO females — the rejection behaviors of the ERKO females prevented most of these approaches by males from developing into normal mounts and intromissions (Figure 23.2). And, when mounted, ERKO females would not show lordosis. Thus, as in intact ERKO females, female sexual behavior was virtually absent. In contrast, both wild-type and heterozygote female mice, while rejecting male approaches to about the same extent as ERKO females, at least in the first test (with estrogen priming), showed increased levels of lordosis behavior in the second test (with estrogen priming) and the third test (with estrogen and progesterone priming) (Figure 23.2).

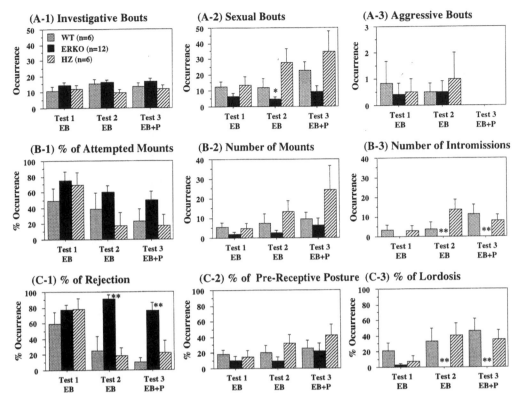

**FIGURE 23.2** Effects of ER-α gene disruption on female sexual behavior in gonadectomized mice. Behavioral interactions were classified as social investigative (A-1), sexual (A-2), or aggressive (A-3). Male mice investigated females of all three genotypes, but they showed only a small number of sexual behaviors toward ERKO females, especially in tests 2 and 3. Moreover, more than 50% of the sexual behavior shown toward ERKO females were attempted mounts (B-1). Males showed very few mounts (B-2) and no intromissions (B-3) toward ERKO females. Most of the time, ERKO females showed rejection of male sexual behavior (C-1), displayed very few pre-receptive postures (C-2), and never showed lordosis (C-3). **$p < 0.05$ vs. WT (wild type) and HZ (heterozygote), $p < 0.05$ vs. HZ. (From Ogawa, S., Eng, V., Taylor, J. A., Lubahn, D. B., Korach, K. S., and Pfaff, D. W., *Endocrinology*, 139, 5074, 1998.)

These behavioral results with ERKO female mice represent one of the few examples where the likely mechanisms for a behavioral phenotype are known. Listed in Figure 1 are several estrogen-stimulated genes whose products foster lordosis. The loss of a functioning estrogen receptor gene would prevent the facilitation of transcription for these messenger RNAs, which, in turn, would diminish several neurochemical products which foster female reproductive behavior. Therefore, the behavior should, indeed, be low or absent in ERKO female mice.

Likewise, the action of another gene, for a different nuclear hormone receptor, on reproductive behavior can be understood by virtue of its interactions with estrogen receptors. Thyroid hormone receptors, liganded by high levels of thyroid hormones, can actually reduce female reproductive behavior.[6,7] Between the two genes which code for thyroid hormone receptors, TR-α and TR-β, the TR-β gene effect appears much more important for thyroid hormone's interference with reproductive behavior.[8] This is because recent data reveal that in female mice with the TR-β gene "knocked out," lordosis behavior is significantly higher than in wild-type females.[9] The mechanisms for the interference by thyroid hormones, working through their nuclear receptors, include competition for DNA-binding sites, exerted against the estrogen receptor, and transcriptional blockage of estrogen-stimulated gene expression.[10,11] Additional mechanisms have not been ruled out.

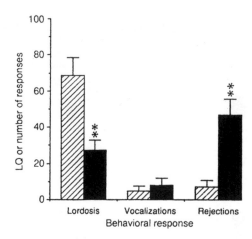

**FIGURE 23.3** In ovariectomized female rats, pretreated with estradiol, microinjection into the hypothalamus of antisense DNA directed against the mRNA for the oxytocin receptor significantly reduced female reproductive behavior (lordosis). LQ = lordosis quotient; ** = significantly different from controls. (From McCarthy, M. M., Kleopoulos, S. P., Mobbs, C. V., and Pfaff, D. W., *Neuroendocrinology*, 59, 432, 1994.)

## Oxytocin Gene

As summarized in Figure 23.1, estradiol turns on the gene for oxytocin in a subset of paraventricular hypothalamic neurons and, in addition, increases messenger RNA levels and oxytocin-binding functions of the oxytocin receptor in the ventromedial hypothalamus. It has been pointed out that estrogenic hormone effects on oxytocinergic systems potentially have a multiplicative character.[12] Not only is estrogen treatment followed by increased oxytocin binding, but also the electrophysiological activation due to oxytocin in hypothalamic neurons is increased.[13] These neurochemical findings do have behavioral importance, since antisense DNA, directed against the messenger RNA for the oxytocin receptor, significantly reduces lordosis behavior (Figure 23.3).[14] The oxytocin knockout mouse is yielding interesting results, since the only deficits in female reproductive behavior detected so far have been under circumstances where the females were treated with progesterone as well as estradiol (Ogawa et al., unpublished data). This feature of the females' behavioral phenotype may be related to the astounding finding[15] that progesterone itself can bind to the oxytocin receptor and affect oxytocin-dependent signal transduction. Behaviors dependent on the genes for oxytocin and its receptor, which bear on female reproductive behavior, are hardly likely to be limited to direct effects on lordosis behavior circuitry. Oxytocin is well known for promoting a variety of affiliative behaviors.[16,17] Recently, we have also found that the oxytocin gene knockout mouse responds differently to stress, and have provisionally concluded that oxytocin, working through its receptor, protects instinctive behaviors related to reproduction, such as parental behaviors, from disruption by mild stress.[18]

## Enkephalin Gene

Another gene which obviously could provide some of the mechanisms by which estrogen, working through estrogen receptor gene products, leads to female reproductive behavior, is that for the opioid peptide, enkephalin (see Figure 23.1). Estrogens reliably turn on the gene for enkephalin in hypothalamic neurons.[19-22] Transcriptional facilitation by the hormone, working through the rat enkephalin promoter, has been demonstrated.[11] Since estrogen also increases the messenger RNA for the delta opioid receptor in the hypothalamus (Vanya Quinones-Jenab et al., unpublished data), the capacity for "multiplicative" actions of the hormone on neural systems for reproductive behavior is once again recognized. The estrogen-induced elevations of enkephalin messenger RNA are closely correlated with the elevation of lordosis behavior.[23] Enkephalin messenger RNA function is important for normal female reproductive behavior, since antisense DNA directed against enkephalin messenger RNA can reduce lordosis performance in estrogen-primed female rats.[24] Permanent absence of the enkephalin gene, as achieved in the enkephalin knockout mouse, shows that the importance of this gene product is not likely to be limited to lordosis itself. Instead, the enkephalin knockout female mouse shows markedly altered responses to mild stress, being much less willing than the wild-type female to venture from safe places toward open or lighted places during open field assays or light/dark transition assays.[25] Thus, one way in which this gene product could promote female reproductive behavior is to allow the female, even under mildly stressful conditions, to engage in proceptive "courtship" behaviors which are a necessary antecedent to sex behavior.

## GnRH Gene

The gene for gonadotropin releasing hormone (GnRH) codes for a protein which includes a decapeptide (also called LHRH), which controls reproduction, as it is responsible for elevating levels of luteinizing hormone and follicle-stimulating hormone released from the pituitary, and for promoting reproductive behavior in both sexes. A mutation discovered in this gene[26,27] led to a failure of GnRH synthesis, and subsequently, a mouse which showed very low levels of gonadotropin release from the pituitary. Reproductive behavior is absent in these mutant mice. In the female, the action of this gene on subsequent physiological events relevant for behavior is manifest in specific portions of the brain in the adult. That is, transplantation of normally expressing GnRH cells into the hypothalamus of the adult hypogonadal mutant female[28] restores female-typical sexual responses.

Modern genetic screening techniques such as differential display PCR (DD-PCR) have allowed additional estrogen-stimulated genes to be identified in medial hypothalamic neurons irrespective of their potential role in sex behavior. Efforts in this area have revealed a variety of transcripts, both induced and repressed by estrogen. The identity of most of these transcripts remains unknown. However, one gene of known identity to emerge from these studies encodes the mitochondrial cytochrome c oxidase subunit I. The role of this gene in sex behavior is unclear; however, Bettini and Maggi[29] have reported that the mitochondrial genome is estrogen-sensitive and therefore plays a subservient role in estrogen responses in the brain.

### PROGESTERONE-RELATED GENES

Under many circumstances, progesterone administered after estrogenic priming enhances the physiological effects of estrogens, including facilitation of reproductive behavior. Occupation of the nuclear progestin receptor is required for hypothalamic neurons participating in facilitation of female reproductive behavior, because the progesterone receptor blocker will interfere with the progestin effect.[30-32] In fact, antisense DNA directed against the messenger RNA for the progestin receptor can reduce lordosis behavior and greatly interfere with progestin-dependent courtship behaviors (Figure 23.4).[33] Subsequent protein synthesis is necessary for the progestin effect, because a synthesis blocker will reduce progesterone's elevation of lordosis behavior.[34] All of this work provided a good background for the demonstration that the gene for the progestin receptor would play an important part in regulation of female reproductive behavior by progesterone. Progesterone receptor knockout (PRKO) female mice do not show a progestin facilitation of female sex behavior.[35] This is not to say that such PRKO females cannot show mating behavior. In fact, the lordosis behavior of PRKO female mice in response to estrogen alone may be slightly greater than that of wild-type females (Ogawa et al., unpublished data). Nevertheless, the abolition of the progesterone facilitation of behavior, in our hands[36] as well as in the initial report of Lydon et al.,[35] demonstrates another direct and powerful effect of a gene on reproductive behavior, consonant with previous neurobiological evidence.

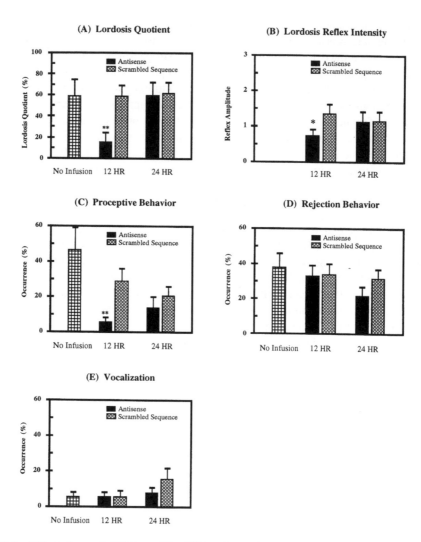

**FIGURE 23.4** Effects (mean ± SEM, $n = 12$) of PR antisense DNA administration on (A) lordosis behavior (number of lordoses/number of mounts × 100), during the mating behavior tests; (B) lordosis reflex intensity; (C) proceptive behavior (number of mounts preceded by proceptive behaviors/number of total mounts × 100); (D) rejection behavior ratio (number of mounts preceded by rejection behavior/number of total mounts × 100); and (E) vocalization ratio (number of mounts preceded by vocalization/number of total mounts × 100), in the mating behavior tests. *$p < 0.05$; **$p < 0.01$ (t-test). No infusion: data from the screening test; 12 HR and 24 HR: 12-h test and 24-h test in which oligonucleotides were infused 12 h or 24 h after estrogen injection, respectively. (From Ogawa, S., Olazabal, U. E., Parhar, I. S., and Pfaff, D. W., *J. Neurosci.*, 14, 1766, 1994.)

Additional genes potentially involved in the progesterone facilitation of female reproductive behavior have been identified using the DD-PCR technique, introduced by Liang and Pardee.[37] This highly sensitive technique reveals transcripts in cells or tissues that are differentially expressed in response to a given stimulus, such as the administration of steroid hormones. It also greatly facilitates the rapid isolation and cloning of the corresponding transcripts as cDNAs. To date, several progesterone-responsive genes have been identified and cloned using this method. Three of the most compelling genes identified thus far include (A) the 73-kDa heat shock cognate protein (hsc73); (B) a membrane-associated progesterone-binding protein (25-Dx);[37a] and (C) a previously unknown gene with a high degree of similarity to members of the secretory carrier membrane protein (SCAMP) family.

## Hsc73

In our characterization of these genes we have discovered that the heat shock cognate protein gene (Hsc73) experiences changes in expression over the course of the estrus cycle in the hypothalamus and other reproductively important tissues. Quantitative *in situ* hybridization analysis of brains from naturally cycling female rats showed that Hsc73 expression throughout many parts of the brain was constant; however, a significant increase of approximately 30% was found in the ventromedial hypothalamus (VMH) and arcuate nucleus of females during the afternoon of proestrus vs. females at diestrus-1. To verify that this increase was steroid hormone-dependent, we compared vehicle treated ovariectomized (ovx) females with ovx females given physiological doses of estrogen and progesterone. Hsc73 expression in the ventrolateral and central aspects of the VMH was responsive to estrogen, while the dorsomedial aspect remained unaffected. In contrast, expression in the arcuate nucleus was not responsive to estrogen alone, but showed a similar increase in Hsc73 message following the subsequent treatment with progesterone. Hsc73 expression in the neocortex did not change, regardless of hormone treatment or stage of the estrus cycle. Hsc73 has been shown to be associated with a variety of cellular trafficking events including (i) the movement of protein aggregates into and out of the nucleus, (ii) the import of proteins into the lysosome for their degradation, and (iii) the uncoating of clathrin-coated vesicles following endocytosis. We are currently investigating the involvement of Hsc73 in these events in the hypothalamus and their potential influence on female sexual behavior.

## 25-Dx

*In situ* hybridization results with 25-Dx, a gene that is repressed by progesterone, revealed its expression in discrete nuclei throughout the hypothalamus. As more information about these and additional progesterone-sensitive genes is gathered in this way, it will help to guide future studies aimed at determining the roles these genes play in facilitating female sexual behavior.

## SCAMP

Characterization of the newly identified SCAMP-like gene by northern blotting indicates that it is more abundant in brain and pituitary than in other peripheral sites including liver, uterus, and heart. Differential slot blotting shows that it is down-regulated by progesterone in the VMH. The finding of an estrogen response element- (ERE)-like sequence near the translational start site suggests that the down-regulation of this gene by progesterone may be similar to the auto-down-regulation mechanism observed for the PR gene.

## IN THE MALE

Because of the aromatization of androgens to yield estrogens, and the importance of this process for the development and adult performance of brain mechanisms for male sex behavior, it would be hypothesized that the ERKO male mouse would have deficiencies in male-typical reproductive behavior. We found[38] that ER gene disruption differentially affected the expression of three discrete components of male sexual behavior: mounts, intromissions, and ejaculations (Figure 23.5). While the numbers of simple mounts and latencies to first mounts were not significantly different among wild-type, heterozygotes, and ERKO males, the frequency of intromissions was reduced in ERKO mice, especially in the first test. Further, the latency to the first intromission was raised. In spite of the fact that some ERKO males could show mounts and intromissions, ejaculations were never observed in ERKO males. In separate experiments,[1] we noticed that ERKO male preferences for female mouse-derived odors were not different from wild-type males. Furthermore, ultrasonic vocalizations emitted by ERKO male mice in response to sexually receptive female mice were not different from that of wild-type mice (Ogawa et al., unpublished observation). Overall, these results

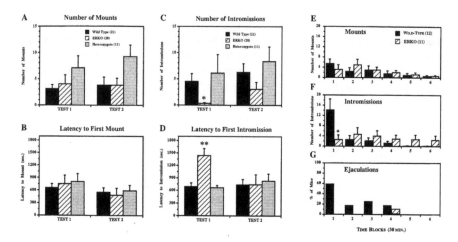

**FIGURE 23.5**   Effects of ER gene disruption on male sexual behavior. During 30-min sexual behavior tests, there were no differences in mean number of mounts (A) and mean latency to the first mounts (B, included only the mice that showed the behavior) between three genotypes. In contrast, there were overall genotype differences in mean number of intromissions (C, $F(2,49) = 3.930$, $p < 0.05$), and ERKO mice showed significantly fewer intromissions compared to HZ (heterozygote), but not to WT (wild-type), mice at $\alpha = 0.05$ (*). Mean latency to the first intromissions (D, included only the mice that showed the behavior) of ERKO mice was significantly longer compared to both WT and HZ mice in Test 1 ($F(2,15) = 11.473$, $p < 0.001$; **significantly different from both WT and HZ at $\alpha = 0.05$), but not in Test 2. Temporal changes of mean number of mounts (E), mean number of intromissions (F), and % of mice ejaculated (G) during 3-h tests were analyzed for six time blocks of 30 min each. ERKO mice showed intromissions continuously during the entire 3-h test period but did not ejaculate, whereas WT (wild-type) mice showed a high number of intromissions (*$p < 0.05$) and ejaculated in the first 30 min. Total numbers of mounts (WT vs. ERKO, mean ± SEM, 13.92 ± 5.60 vs. 14.82 ± 5.49) and intromissions (20.08 ± 5.87 vs. 19.00 ± 8.53) during the 3-h test period were not different between WT and ERKO mice. Vertical bars represent SEM. (From Ogawa, S., Lubahn, D. B., Korach, K. S., and Pfaff, D. W., *Proc. Natl. Acad. Sci. U.S.A.*, 94, 1476, 1997.)

gave the impression that preliminary aspects of male reproductive behaviors, more closely related to sexual motivation, were the least affected in the ERKO males, while the consummatory aspects, especially ejaculations, were greatly reduced following disruption of the classical ER-α gene.

Which of these aspects of the phenotype of the ERKO male mouse might be affected by gonadectomy? Male sexual behaviors were reduced by castration in both wild-type and ERKO mice (Figure 23.6). We then wished to use a hormone replacement treatment which would not depend upon aromatization to estrogens, invoking the loss of the estrogen receptor gene, in order to test the capacity of the ERKO males to show any male sexual behavior following castration. Dihydrotestosterone was the steroid of choice. Dihydrotestosterone restored male sexual behavior, mounts and intromissions, in both wild-type and ERKO males, although the rate of recovery was much slower in ERKO males compared to wild-type animals (Figure 23.6).[39] Even after the fullest recovery we could measure, after 29 days daily injections of dihydrotestosterone, none of the ERKO mice ejaculated. Under this endocrine condition as well, ERKO males could show preliminary aspects of male-typical sexual behavior, perhaps indicative of masculine sexual motivation, but could not complete the full consummatory male-typical response pattern.[39]

Some of the most impressive effects of gene disruption on male-typical behavior are consequences of the disruption of genes important for the development of a full masculine phenotype. Especially when accompanied by abnormalities of the peripheral genitalia, they can lead to an absence of masculine sexual identification, accompanied by the corresponding behavioral changes. The best documented examples derive from genetic damage to the androgen receptor and to an enzyme responsible for reducing testosterone, 5α-reductase type 2.

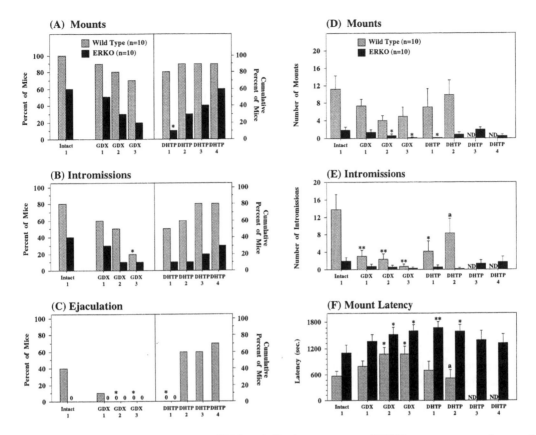

**FIGURE 23.6**  Effects of gonadectomy (GDX) and dihydrotestosterone (DHTP) treatment on male sexual behavior. Gonadectomy reduced sexual behavior and DHTP restored it to the intact levels in both WT (wild-type) and ERKO mice, although ERKO mice showed slower recovery compared to WT mice. The levels of each of three tests after GDX and four tests after DHTP were compared to those of the intact test, using either the binomial test for related samples with small expected frequencies (A–C, two-tailed, *$p < 0.05$ vs. the intact test) or the paired t-test (D–F, two-tailed, **$p < 0.01$, *$p < 0.05$ vs. the intact test). The percent of animals exhibiting the behavior (A–C) after DHTP treatment are shown as a cumulative number because some WT mice were not tested at the third and fourth tests. (From Ogawa, S., Washburn, T. F., Taylor, J. A., Lubahn, D. B., Korach, K. S., and Pfaff, D. W., *Endocrinology*, 139, 5066, 1998.)

## ANDROGEN-RELATED GENES

Defects involving androgen biosynthesis or actions can cause male pseudohermaphroditism (MPH). MPH is characterized by abnormal differentiation of male external genitalia in the presence of testicular tissue.[40,41] There are two major active androgens, testosterone and dihydrotestosterone. Both of them bind to the same androgen receptor, and interactions of the androgen receptor complex with DNA response elements of target genes regulate gene expression in androgen-responsive tissues,[42] which leads to masculinization. The importance of androgens in human male sexual differentiation, male physiology, and male identity is highlighted by studying natural human genetic models — male pseudohermaphrodites with inherited gene defects in androgen receptor, 17β hydroxysteroid dehydrogenase (17β-HSD) and 5α-reductase.

### Androgen Receptor

The androgen receptor (AR) belongs to the nuclear receptor family. Mutations in AR result in androgen insensitive syndrome or testicular feminization, an X-linked form of MPH.[42] This syndrome is classified as complete androgen insensitive syndrome (CAIS) or partial androgen insensitive syndrome (PAIS). Subjects with CAIS, despite a 46XY karyotype with testes and normal to elevated plasma levels of testosterone, have female external genitalia and a female psychosexual orientation.[42] At puberty, breast development occurs, but there is scant to absent pubic and axillary hair. Individuals with PAIS have a broad spectrum of phenotypes, ranging from primarily female, with some virilization such as clitoromegaly, to males with hypospadias or micropenis, and to its mildest form of males with infertility or gynecomastia.

Studies in individuals with CAIS suggest that the androgen receptor plays an important role in development of male identity,[42] even though studies in animals indicate that the androgen effects on male brain and male behavior may be mediated via conversion to estrogen.[43] The exact roles of the sexual steroids and their interactions on the development of male identity are still obscure and remain to be further investigated.

## 17-β-HSD

17β-HSD catalyzes the oxidoreduction of testosterone and androstenedione, androstenediol and dehydroepiandrosterone, and estradiol and estrone.[41] There are at least five isozymes; type 1 to type 5 of 17β-HSD have been cloned.[44] The type 3 isozyme is the primary 17β-HSD catalyzing the biosynthesis of testosterone from androstenedione in the testes.[44,45] Mutation in this isozyme is responsible for MPH due to 17β-HSD deficiency.[45-47] So far, 14 different mutations of the 17β-HSD-3 gene have been detected in 17β-HSD-deficient male pseudohermaphrodites from various regions of the world. These mutations cause impairment of testosterone biosynthesis during the critical period of male sexual differentiation, which results in ambiguous genitalia in affected males.

The affected 46XY individuals with 17β-HSD deficiency have testes and male Wolffian duct-derived structures (epididymis, vasa deferentia, seminal vesicles, and ejaculatory ducts), but their external genitalia are ambiguous and, consequently, many have been raised as girls.[48-50] They usually have a clitoral-like phallus, nonfused labial-scrotum, urogenital sinus, and testes located in the inguinal canal or labial-scrotal folds. The reason why these affected subjects have Wolffian duct-derived structures is not clear although some possibilities have been postulated.[45] This enzyme deficiency has also been reported to cause less severe defects in masculinization of genitalia, such as micropenis.[51]

A gender change from female to male has been reported in various ethnic groups with this condition,[47,50,52,53] including the Arab male pseudohermaphrodites with 17β-HSD-3 deficiency[54] in the Gaza Strip kindred. Many affected subjects with 17β-HSD-3 deficiency were born with severe ambiguity of female genitalia and were raised as girls. During puberty, the affected subjects developed male secondary sexual characteristics with phallic enlargement and abundant body and facial hair, which may result from the normal conversion of androstenedione to testosterone in peripheral tissues by other isozymes.[41] Many of them adopted a male gender role with the physical changes of puberty. These data suggest that some hormonal factor or "biologic force" is strong enough to override the female sex of rearing. The reason why a change in gender identity occurs in this syndrome is still obscure, but androgenic actions during early life or at puberty obviously play an important role since gender change has not been reported in individuals with CAIS.[41,42,50] Masculinization of the brain may also occur *in utero* via conversion of androstenedione to estrogen.[43] Animal studies have demonstrated that androstenedione as well as estrogen can induce male sexual behavior.[43,55]

## 5α-Reductase

In certain androgen responsive tissues, the actions of testosterone are mediated via the conversion to dihydrotestosterone by 5α-reductase.[40,41] 5α-reductase is a NADPH-dependent microsomal enzyme that reduces the double bond at the 4-5 position of a variety of $C_{19}$ and $C_{21}$ steroids including

## TABLE 23.1
### Comparison of Human 5α-Reductase Isozymes

| | Type 1 | Type 2 |
|---|---|---|
| Gene structure | 5 exons, 4 introns | 5 exons, 4 introns |
| Gene, chromosome location | *SRD5A1*, 5p15 | *SRD5A2*, 2p23 |
| Size | 259 amino acids, Mr = 29,462 | 254 amino acids, Mr = 28,398 |
| Tissue distribution | Liver, nongenital skin, prostate | Prostate, epididymis, seminal vesicle, genital skin, uterus, liver, breast, hair follicle, placenta, cerebral cortex, ovary,[a] testis,[a] kidney,[a] adrenal[a] |
| pH optima | Neutral to basic | Acidic or neutral |
| Prostate level | Low | High |
| Activity in 5α-reductase deficiency | Normal | Mutated |
| Finasteride inhibition | Ki ≥ 300 n$M$ | Ki = 3–5 n$M$ |

[a] Controversial.

testosterone. Two isozymes, type 1 and type 2, have been identified and cloned.[56] They have differential biochemical and pharmacological properties (Table 23.1). Mutations in the 5α-reductase-2 gene are responsible for MPH due to 5α-reductase deficiency.[41,47,57,58]

The clinical and biochemical charateristics of 5α-reductase deficiency are best described in a large Dominican kindred.[40,41,59] Most males who are homozygous for 5α-reductase-2 deficiency have normal Wolffian structure, but ambiguous external genitalia with a clitoral-like phallus, severely bifid scrotum, and pseudovaginal perineoscrotal hypospadias at birth. Consequently, many affected males are assigned a female gender at birth and are reared as girls.[59-61]

The onset of puberty in affected males results in an increased muscle mass and deepening of the voice.[41,59] A gender role change to male identity has been reported in most of the affected subjects during or after puberty.[60] The affected subjects have equal height to their unaffected siblings and no gynecomastia in adulthood.[62,63] There is substantial growth of the phallus with rugation and hyperpigmentation of the scrotum. Inguinal testes descending into the scrotum at puberty have been observed in some patients.[62] Libido is preserved and patients are capable of erections.[59,60,62] Although patients are generally oligo- or azoospermic, normal sperm concentrations have been reported in patients without concomitant cryptorchidism[63-66] The function of the sperm appears normal since affected subjects have been reported to father children,[66,67] suggesting that dihydrotestosterone does not play a major role in spermatogenesis and sperm function. These findings suggest that pubertal events, including male sexual function and spermatogenesis, are primarily testosterone mediated.[59,63,65]

The coexistence of both 17β-HSD-3 and 5α-reductase-2 gene defects has been identified in the same individual and in the same kindred.[47,52] The genotypic and phenotypic relationship in these individuals remains to be further explored.

In summary, androgen-related systems as well as estrogen receptor systems are crucial for the development of normal, male typical reproductive behaviors.

## REFERENCES

1. Ogawa, S., Gordan, J., Taylor, J. A., Lubahn, D. B., Korach, K. S., and Pfaff, D. W., Reproductive functions illustrating direct and indirect effects of genes on behavior, *Horm. Behav.*, 30, 487, 1996.
2. Pfaff, D. W., Schwartzgiblin, S., McCarthy, M. M., and Kow, L. M., Cellular and molecular mechanisms of female reproductive behaviors, in *Physiology of Reproduction, 2nd Edition, Vol. 1 and 2*, Knobil, E. and Neill, J. D., Eds., Raven Press, New York, 1994, 107.

3. Lubahn, D. B., Moyer, J. S., Golding, T. S., Couse, J. F., Korach, K. S., and Smithies, O., Alternation of reproductive function but not prenatal sexual development after insertional disruption of the mouse estrogen receptor gene, *Proc. Natl. Acad. Sci. U.S.A.*, 90, 11162, 1993.

4. Ogawa, S., Taylor, J. A., Lubahn, D. B., Korach, K. S., and Pfaff, D. W., Reversal of sex roles in genetic female mice by disruption of estrogen receptor gene, *Neuroendocrinology*, 64, 467, 1996.

5. Ogawa, S., Eng, V., Taylor, J. A., Lubahn, D. B., Korach, K. S., and Pfaff, D. W., Roles of estrogen receptor-α gene expression in reproduction-related behaviors in female mice, *Endocrinology*, 139, 5070, 1998.

6. Dellovade, T. L., Zhu, Y. S., Krey, L., and Pfaff, D. W., Thyroid hormone and estrogen interact to regulate behavior, *Proc. Natl. Acad. Sci. U.S.A.*, 93, 12581, 1996.

7. Morgan, M. A., Dellovade, T., Ogawa, S., and Pfaff, D. W., Female mouse sexual behavior is regulated by thyroid hormones and estrogen, *Soc. Neurosci. Abstr.*, 23, 1853, 1997.

8. Martinelli, G. P., Holstein, G. R., Pasik, P., and Cohen, B., Monoclonal antibodies for ultrastructural visualization of L-baclofen-sensitive GABAB receptor sites, *Neuroscience*, 46, 23, 1992.

9. Dellovade, T., Forrest, D., Vennstrom, B., and Pfaff, D., Two genes coding for thyroid hormone receptor have opposite effect on reproductive behaviors, *Nature*, submitted.

10. Zhu, Y. S., Yen, P. M., Chin, W. W., and Pfaff, D. W., Estrogen and thyroid hormone interaction on regulation of gene expression, *Proc. Natl. Acad. Sci. U.S.A.*, 93, 12587, 1996.

11. Zhu, Y. S., Ling, Q., Cai, L. Q., Imperato-McGinley, J., and Pfaff, D. W., Regulation of preproenkephalin (PPE) gene expression by estrogen and its interaction with thyroid hormone, *Soci. Neurosci.*, 23, 798, 1997.

12. Pfaff, D. W., Multiplicative response to hormones by hypothalamic neurons, in *Recent Progress in Posterior Pituitary Hormones*, Yoshida, S. and Share, L., Eds., Elsevier Science Publishers, B.V. (Biomedical Division), Amsterdam, 1988, 257.

13. Kow, L.-M., Johnson, A. E., Ogawa, S., and Pfaff, D. W., Electrophysiological actions of oxytocin on hypothalamic neurons *in vitro*: neuropharmacological characterization and effects of ovarian steroids, *Neuroendocrinology*, 54, 526, 1991.

14. McCarthy, M. M., Kleopoulos, S. P., Mobbs, C. V., and Pfaff, D. W., Infusion of antisense oligodeoxynucleotides to the oxytocin receptor in the ventromedial hypothalamus reduces Estrogen-Induced sexual receptivity and oxytocin receptor binding in the female rat, *Neuroendocrinology*, 59, 432, 1994.

15. Grazzini, E., Guillon, G., Mouillac, B., and Zingg, H. H., Inhibition of oxytocin receptor function by direct binding of progesterone, *Nature*, 392, 509, 1998.

16. Pedersen, C., Caldwell, J., Jirikowski, G. and Insell, T., Oxytocin in maternal, sexual and social behavior, *Ann. N. Y. Acad. Sci.*, 652, 1, 1992.

17. Carter, C. S., Lederhndler, I. I. and Kirkpatrick, B., The integrative neurobiology of affiliation, *Ann. N. Y. Acad. Sci.*, 807, 13, 1997.

18. Pfaff, D. W., Galindo, K., Larcher, S., Luedke, C., Muglia, L., and Ogawa, S., Role of oxytocin gene in behavioral responses to stress, *Soc. Neurosci. Abstr.*, 24, 1998.

19. Romano, G. J., Mobbs, C. V., Lauber, A., Howells, R. D., and Pfaff, D. W., Differential regulation of proenkephalin gene expression by estrogen in the ventromedial hypothalamus of male and female rats: implications for the molecular basis of a sexually differentiated behavior, *Mol. Brain Res.*, 536, 63, 1990.

20. Romano, G. J., Harlan, R. E., Shivers, B. D., Romano, G. J., Howells, R. D., and Pfaff, D. W., Estrogen increases proenkephalin messenger ribonucleic acid levels in the ventromedial hypothalamus of the rat, *Mol. Endocrinol.*, 2, 1320, 1988.

21. Romano, G. J., Mobbs, C. V., Howells, R. D., and Pfaff, D. W., Estrogen regulation of proenkephalin gene expression in the ventromedial hypothalamus of the rat: temporal qualities and synergism with progesterone, *Mol. Brain Res.*, 5, 51, 1989.

22. Quinones-Jenab, V., Ogawa, S., Jenab, S., and Pfaff, D. W., Estrogen regulation of preproenkephalin messenger RNA in the forebrain of female mice, *J. Chem. Neuroanat.*, 12, 29, 1996.

23. Lauber, A. H., Romano, G. J., Mobbs, C. V., Howells, R. D., and Pfaff, D. W., Estradiol induction of proenkephalin messenger RNA in hypothalamus: dose-response and relation to reproductive behavior in the female rat, *Mol. Brain Res.*, 8, 47, 1990.

24. Nicot, A., Ogawa, S., Berman, Y. E., Carr, K. D., and Pfaff, D. W., Effects of an intrahypothalamic injection of antisense oligonucleotides for preproenkephalin mRNA in female rats: evidence for opioid involvement in lordosis reflex, *Behav. Brain Res.*, 777, 60, 1997.

25. Ragnauth, A., Schuller, A., Chan, J., Ogawa, S., Bodnar, R. J., Pintar, J., and Pfaff, D. W., Female preproenkephalin knockout mice display altered emotional responses, *Soc. Neurosci. Abstr.,* 24, 1998.

26. Charlton, H. M., The physiological actions of LHRH: evidence for the hypogonadal (hpg) mouse, in *Neuroendocrine Molecular Biology,* Fink, G., Harmar, H. A., and McKerns, K. W., Eds., Plenum, New York, 1986, 47.

27. Cattanach, B. M., Iddon, C. A., Charlton, H. M., Chiappa, S. A., and Fink, G., Gonadotropin-releasing hormone deficiency in a mutant mouse with hypogonadism, *Nature,* 269, 338, 1977.

28. Gibson, M. J., Moscovitz, H. C., Kokoris, G. J., and Silverman, A. J., Female sexual behavior in hypogonadal mice with GnRH-containing brain grafts, *Horm. Behav.,* 21, 211, 1987.

29. Bettini, E. and Maggi, A., Estrogen induction of cytochrome c oxidase subunit III in rat hippocampus, *J. Neurochem.,* 58, 1923, 1992.

30. Etgen, A. M. and Barfield, R. J., Antagonism of female sexual behavior with intracerebral implants of antiprogestin RU 38486: correlation with binding to neural progestin receptors, *Endocrinology,* 119, 1610, 1986.

31. Vathy, I. U., Etgen, A. M., and Barfield, R. J., Actions of progestins on estrous behaviour in female rats, *Physiol. Behav.,* 40, 591, 1987.

32. Brown, T. J. and Blaustein, J. D., Inhibition of sexual behavior in female guinea pigs by a progestin receptor antagonist, *Brain Res.,* 301, 343, 1984.

33. Ogawa, S., Olazabal, U. E., Parhar, I. S., and Pfaff, D. W., Effects of intrahypothalamic administration of antisense DNA for progesterone receptor mRNA on reproductive behavior and progesterone receptor immunoreactivity in female rat, *J. Neurosci.,* 14, 1766, 1994.

34. Glaser, J. H. and Barfield, R. J., Blockade of progesterone-activated estrous behavior in rats by intracerebral anisomycin is site specific, *Neuroendocrinology,* 38, 337, 1984.

35. Lydon, J. P., DeMayo, F. J., Funk, C. R., Mani, S. K., Hughes, A. R., Montgomery, C. A., Jr., Shyamala, G., Conneely, O. M., and O'Malley, B. W., Mice lacking progesterone receptor exhibit pleiotropic reproductive abnormalities, *Genes Dev.,* 9, 2266, 1995.

36. Ogawa, S., Lydon, J. P., Conneely, O. M., O'Malley, B. W., and Pfaff, D. W., Effects of progesterone receptor gene disruption on reproductive behavior, in preparation.

37. Liang, P. and Pardee, A. B., Differential display of eukaryotic messenger RNA by means of the polymerase chain reaction, *Science,* 257, 967, 1992.

37a. Krebs, C. J., Jaruis, E. D., and Pfaff, D. W., The 70-kDa *heat shock* cognate protein (Hsc73) gene is enhanced by ovarian hormones in the ventromedial hypothalamus, *Proc. Natl. Acad. U.S.A.,* 96, 1686, 1999.

38. Ogawa, S., Lubahn, D. B., Korach, K. S., and Pfaff, D. W., Behavioral effects of estrogen receptor gene disruption in male mic, *Proc. Natl. Acad. Sci. U.S.A.,* 94, 1476, 1997.

39. Ogawa, S., Taylor, J., Lubahn, D. B., Korach, K. S., and Pfaff, D. W., Modifications of testosterone-dependent behaviors by estrogen receptor-α gene disruption in male mice, *Endocrinology,* 139, 5058, 1998.

40. Imperato-McGinley, J., Male pseudohermaphroditism, in *Reproductive Endocrinology, Surgery and Technology,* Adashi, E. Y., Rock, J. A., and Rosenwaks, Z., Eds., Lippincott-Raven Publishers, Philadelphia, 1996, 936.

41. Zhu, Y.-S., Katz, M. D., and Imperato-McGinley, J., Natural potent androgens: lessons from human genetic model., *Baillieres Clin. Endocrinol. Metab.,* 12, 83, 1998.

42. Quigley, C. A., De Bellis, A., Marschke, K. B., el-Awady, M. K., Wilson, E. M., and French, F. S., Androgen receptor defects: historical, clinical and molecular perspectives, *Endocr. Rev.,* 16, 271, 1995.

43. Goy, R. W. and McEwen, B. S., *Sexual Differentiation of the Brain,* The MIT Press, Cambridge, MA, 1980.

44. Andersson, S. and Moghrabi, N., Physiology and molecular genetics of 17 beta-hydroxysteroid dehydrogenases, *Steroids,* 62, 143, 1997.

45. Geissler, W. M., Davis, D. L., Wu, L., Bradshaw, K. D., Patel, S., Mendonca, B. B., Elliston, K. O., Wilson, J. D., Russell, D. W., and Andersson, S., Male pseudohermaphroditism caused by mutations of testicular 17 beta-hydroxysteroid dehydrogenase 3, *Nat. Genet.,* 7, 34, 1994.

46. Andersson, S., Geissler, W. M., Wu, L., Davis, D. L., Grumbach, M. M., New, M. I., Schwarz, H. P., Blethen, S. L., Mendonca, B. B., Bloise, W., Witchel, S. F., Cutler, G. B., Jr., Griffin, J. E., Wilson, J. D., and Russel, D. W., Molecular genetics and pathophysiology of 17 beta-hydroxysteroid dehydrogenase 3 deficiency, *J. Clin. Endocrinol. Metab.,* 81, 130, 1996.

47. Can, S., Zhu, Y. S., Cai, L. Q., Ling, Q., Katz, M. D., Akgun, S., Shackleton, C. H., and Imperato-McGinley, J., The identification of 5 alpha-reductase-2 and 17 beta-hydroxysteroid dehydrogenase-3 gene defects in male pseudohermaphrodites from a Turkish kindred, *J. Clin. Endocrinol. Metab.*, 83, 560, 1998.

48. Eckstein, B., Cohen, S., Farkas, A., and Rosler, A., The nature of the defect in familial male pseudohermaphroditism in Arabs of Gaza, *J. Clin. Endocrinol. Metab.*, 68, 477, 1989.

49. Saez, J. M., De Peretti, E., Morera, A. M., David, M., and Bertrand, J., Familial male pseudohermaphroditism with gynecomastia due to a testicular 17-ketosteroid reductase defect. I. Studies in vivo, *J. Clin. Endocrinol. Metab.*, 32, 604, 1971.

50. Imperato-McGinley, J., Peterson, R. E., Stoller, R., and Goodwin, W. E., Male pseudohermaphroditism secondary to 17 beta-hydroxysteroid dehydrogenase deficiency: gender role change with puberty, *J. Clin. Endocrinol. Metab.*, 49, 391, 1979.

51. Castro-Magana, M., Angulo, M., and Uy, J., Male hypogonadism with gynecomastia caused by late-onset deficiency of testicular 17-ketosteroid reductase, *N. Engl. J. Med.*, 328, 1297, 1993.

52. Imperato-McGinley, J., Akgun, S., Ertel, N. H., Sayli, B., and Shackleton, C., The coexistence of male pseudohermaphrodites with 17-ketosteroid reductase deficiency and 5 alpha-reductase deficiency within a Turkish kindred, *Clin. Endocrinol. (Oxford)*, 27, 135, 1987.

53. Rosler, A., Belanger, A., and Labrie, F., Mechanisms of androgen production in male pseudohermaphroditism due to 17 beta-hydroxysteroid dehydrogenase deficiency, *J. Clin. Endocrinol. Metab.*, 75, 773, 1992.

54. Rosler, A. and Kohn, G., Male pseudohermaphroditism due to 17 beta-hydroxysteroid dehydrogenase deficiency: studies on the natural history of the defect and effect of androgens on gender role, *J. Steroid Biochem.*, 19, 663, 1983.

55. Gilroy, A. F. and Ward, I. L., Effects of perinatal androstenedione on sexual behavior differentiation in male rats, *Behav. Biol.*, 23, 243, 1978.

56. Russell, D. W. and Wilson, J. D., Steroid 5 alpha-reductase: two genes/two enzymes, *Annu. Rev. Biochem.*, 63, 25, 1994.

57. Andersson, S., Berman, D. M., Jenkins, E. P., and Russell, D. W., Deletion of steroid 5 alpha-reductase 2 gene in male pseudohermaphroditism, *Nature*, 354, 159, 1991.

58. Thigpen, A. E., Davis, D. L., Milatovich, A., Mendonca, B. B., Imperato-McGinley, J., Griffin, J. E., Francke, U., Wilson, J. D., and Russell, D. W., Molecular genetics of steroid 5 alpha-reductase 2 deficiency, *J. Clin. Invest.*, 90, 799, 1992.

59. Imperato-McGinley, J., Guerrero, L., Gautier, T., and Peterson, R., Steroid 5-reductase deficiency in man: an inherited form of male pseudohermaphroditism, *Science*, 186, 1213, 1974.

60. Imperato-McGinley, J., Peterson, R. E., Gautier, T., and Styrla, E., The impact of androgens on the evolution of male gender identity, in *Sexuality: New Perspectives*, DeFries, Z., Friedman, R., and Corn, R., Eds., Greenwood Press, Westport, CT, 1985, 126.

61. Savage, M. O., Preece, M. A., Jeffcoate, S. L., Ransley, P. G., Rumsby, G., Mansfield, M. D., and Williams, D. I., Familial male pseudohermaphroditism due to deficiency of 5 alpha-reductase, *Clin. Endocrinol. (Oxford)*, 12, 397, 1980.

62. Imperato-McGinley, J., Peterson, R. E., Leshin, M., Griffin, J. E., Cooper, G., Draghi, S., Berenyi, M., and Wilson, J. D., Steroid 5 alpha-reductase deficiency in a 65-year-old male pseudohermaphrodite: the natural history, ultrastructure of the testes, and evidence for inherited enzyme heterogeneity, *J. Clin. Endocrinol. Metab.*, 50, 15, 1980.

63. Peterson, R. E., Imperato-McGinley, J., Gautier, T., and Sturla, E., Male pseudohermaphroditism due to steroid 5-alpha-reductase deficiency, *Am. J. Med.*, 62, 170, 1977.

64. Cantu, J. M., Hernandez-Montes, H., Del Castillo, V., Sandoval, R., Armendares, S., and Parra, A., Potential fertility in incomplete male pseudohermaphroditism type 2, *Rev. Invest. Clin.*, 28, 177, 1976.

65. Cai, L. Q., Fratianni, C. M., Gautier, T., and Imperato-McGinley, J., Dihydrotestosterone regulation of semen in male pseudohermaphrodites with 5 alpha-reductase-2 deficiency, *J. Clin. Endocrinol. Metab.*, 79, 409, 1994.

66. Katz, M. D., Kligman, I., Cai, L. Q., Zhu, Y. S., Fratianni, C. M., Zervoudakis, I., Rosenwaks, Z., and Imperato-McGinley, J., Paternity by intrauterine insemination with sperm from a man with 5alpha-reductase-2 deficiency, *N. Engl. J. Med.*, 336, 994, 1997.

67. Ivarsson, S. A., 5-alpha reductase deficient men are fertile [letter], *Eur. J. Pediatr.*, 155, 425, 1996.

# 24 Genetic Influences on Emotionality

*Jonathan Flint*

## CONTENTS

## INTRODUCTION

This chapter examines the genetic influences on emotion first by reviewing the literature on genetic mutations that have been used to dissect the neural systems underlying emotion (a bottom-up approach), and second by the quantitative genetic analyses of emotional reactions to stress (or emotionality) (a top-down approach). As yet, these two approaches have not converged, but it is likely that they will do so in the foreseeable future.

One difficulty facing any researcher in this field is that there is no universally agreed definition of emotion.[1] Furthermore, while quantitative genetic analyses can investigate human subjects, many experimental approaches to the genetics of emotion are only possible in animals that cannot be asked about their emotional state. The two problems of how to define emotion and how animal models of emotion correspond to human emotion are related and best dealt with by considering emotions as reflecting the activity of brain systems. If we accept that the presence of conscious mental processes is not a prerequisite for emotional experience, then it is quite permissible to study emotion in animals[2] and to investigate the biological features common to the different species being studied, such as the neuroanatomy and neurotransmitters involved. A central problem then becomes the elaboration of neural systems mediating emotion.

The inherent circularity of using brain systems to define emotions, yet requiring a model of emotion to investigate the underlying neural system, has been addressed by attempts to subdivide emotion into interdependent, but separable, processes with known relations to brain structures. The dominant paradigms in emotion research from the last century have considered that emotion was a unitary phenomenon that could be typologically subdivided by specific physiological or psychological features. Thus, the James-Lange theory of emotion put forward that the bodily perception of autonomic arousal constitutes the experience of emotion, and Maclean argued that the limbic system was the "visceral brain," the mediator of emotion.[3, 4] However, as with the study of learning and memory,[5-7] there is increasing acceptance that emotion is not a unitary phenomenon.[8-10]

One of the most successful approaches has been to study fear. Like all emotions, fear is a construct, inferred from the concurrence of changes in muscle tension sympathetic and parasympathetic nervous activation, increased respiration, increased reflexes, typical facial expressions, and hormonal changes. These features occur together consistently when subjects are exposed to aversive stimuli and are reasonably enough supposed to indicate a central emotional state (i.e., one that occurs in the head, not the periphery of the organism).

Investigation into the biology (and genetics) of fear proceeds by using specific behaviors to measure the emotion. We subsequently consider two models of fear: fear conditioning and avoidance conditioning. (These are defined and discussed in the material that follows.) The importance of this work is that it has identified putative cellular correlates of some aspects of fear, which in turn suggest molecular mechanisms. Within this context, it is possible to interpret some recent experiments that have begun to identify the genetic basis of fear conditioning.

We then discuss the genetic contribution to individual variation in emotional responses, first in animals, then in humans. Again some of the best data derives from animal models of fear, though these are not the same as those used in fear conditioning. Finally, we discuss how these various results may be reconciled.

## EMOTION AT THE CELLULAR LEVEL: CONDITIONED FEAR

A paradigm that has yielded detailed information relating emotional response to cellular activity and possibly to molecular machinery is conditioned fear. In fear conditioning, the subject (most often a rodent) learns to associate a neutral stimulus (such as a noise or a tone) with a fear-inducing stimulus (an electric shock).[11,12] When the rodent is exposed to the noise or tone (known as the conditioned stimulus) the animal shows a stereotypical pattern of behaviors (the conditioned response): it will freeze, its blood pressure and heart rate will rise, and there will be other physiological changes characteristic of fear (such as the release of "stress hormones").

The neural systems that mediate fear conditioning almost certainly involve the amygdala.[13,14] It has been known for over a century that monkeys subjected to surgical removal of large parts of the temporal lobe become tame, or hypoemotional.[15] Subsequent work defined the amygdala complex as crucial to the development of behavioral changes suggestive of a change in the animal's emotional state[16] and led to the view that parts of the amygdala were involved in assigning emotional significance to events.[17-20] This line of inquiry has progressed to the point where a strong case can be made that the amygdala is essential for the learning and storage of emotional aspects of experience.

Lesions to the amygdala (particularly the central nucleus) before training prevent the acquisition of conditioned fear; stimulation of the amygdala in untrained animals elicits many of the features of conditioned fear.[13,14,21] Work of this sort has been used to delineate a pathway from auditory stimulus via the thalamus to the amygdala and thence to various brain stem regions that evoke the features of fear.[2,8]

However, it is clear that at least two neural processes are involved in fear conditioning. When a shock is given to an animal, not only does it become conditioned to the neutral stimulus (or cue) of the light or tone provided but also to the conditioning chamber and other features that constitute the context of the experiment. Conditioning to the context appears to involve different neural structures from cue conditioning because it is not only dependent on the amygdala, but it is also sensitive to lesions in the hippocampus.[22-24]

The description of neural structures provides the first step toward isolating the cellular mechanisms that might underlie fear. Much interest has focused on glutaminergic transmission and its relation to an electrophysiological phenomenon called long-term potentiation (LTP), first observed in the major synaptic pathways of the hippocampus. LTP is an excellent candidate for learning and memory mechanisms, as it is an activity-dependent form of synaptic strengthening.[25] It is conventionally elicited by giving a 25 to 100-Hz electrical stimulus for about one second to the hippocampal pathway, resulting in a lasting increase in synaptic strength. N-Methyl-D-aspartate (NMDA) receptors are known to be critical for the induction of LTP in the hippocampus where concurrent strong postsynaptic depolarization is required for glutamate to activate NMDA receptors.

Confirming the relation between LTP and learning has not been easy.[26] For instance, changes in hippocampal extracellular field potentials measured while an animal was learning spatial relations have been shown to be due, at least in part, to changes in brain temperature.[27] However a case can be made that a form of LTP is required for emotional learning. LTP is known to exist in the amygdala: high frequency electrical stimulation of inputs to the amygdala have been shown to induce LTP;[28,29] intra-amygdaloid infusions of NMDA receptor antagonists implicate these receptors in the acquisition and performance of conditioned fear;[30-32] and as NMDA receptor antagonists block the induction of conditioned fear, agonists enhance its acquisition.[33]

So, does LTP occur in the amygdala during fear conditioning? LeDoux and colleagues have shown that LTP induction within the amygdala increases auditory evoked responses[34] and have gone on to demonstrate that when an animal is trained to associate an auditory cue with a shock, auditory evoked potentials are altered in the same way.[35] Furthermore, measurements of excitatory postsynaptic currents in coronal slices taken from fear-conditioned rats have been used to show that emotional learning does indeed result in synaptic plasticity.[36] Although it has not been demonstrated that electrically induced LTP is identical to the behaviorally induced synaptic changes, these results substantially strengthen the connection between LTP and emotional learning.

It is quite possible that other mechanisms play a part in fear conditioning. For instance, the central nucleus of the amygdala has a high density of opiate receptors.[37] Local infusion of opiate antagonists into the central nucleus blocks the acquisition of conditioned bradycardia in rabbits, suggesting that the opioid system could be involved in fear conditioning.[38,39] Further complexity is hinted at by the remarkably high concentration of peptides in the amygdala.[40,41]

In summary, careful delineation of the pathways mediating conditioned fear have begun to identify key anatomical components and cellular processes that mediate the emotion. In particular, a form of LTP operating within the amygdala is likely to be part of a neural system that underlies fear.

## EMOTION AT THE CELLULAR LEVEL: AVOIDANCE AND INSTRUMENTAL LEARNING

Learning theorists provide another model of fear based on the view that emotion is a central state elicited by reinforcing stimuli. A reinforcing stimulus is one that alters the probablity that a response (behavior) will occur. It can be either positive (increases the probability of a response) or negative (decreases the probability of a response). Within this conceptual framework, we can measure an animal's fearfulness through its reaction to signals associated with danger.

In avoidance learning, an animal will attempt to escape from an unpleasant emotion (such as fear) either by not doing anything at all (passive avoidance, such as keeping still or else receive an electric shock) or by working to escape the unpleasant sensation (active avoidance, such as running away or receive an electric shock). Since fear conditioning results in avoidance behavior, the distinction between avoidance and fear conditioning is not immediately obvious. However, there is evidence that different neural systems are involved: Anxiolytic drugs (such as the benzodiazepines) do not significantly impair the formation of conditioned fear, but they do weaken avoidance behavior.[42] The similarity between the effect of anxiolytic agents and the behavioral consequences of lesions in the septal area or the hippocampal formation[43] indicates that a septo-hippocampal system mediates avoidance behavior. Further consideration of avoidance behavior led Gray to postulate the existence of a "behavioral inhibition system" (situated in the septo-hippocampal area) that responds to the threat of punishment, the omission of anticipated reward or extreme novelty, by the inhibition of ongoing activity and increased vigilance.[42,43] According to this theory, anxiety reflects activity in the behavioral inhibition system.

One way to conceptualize the two systems is to regard the attachment of emotional signals to a stimulus (i.e., fear) as a process taking place in the amygdala while the behavioral inhibition system is involved with the prediction of aversive events (i.e., anxiety). The hippocampus possibly contributes to anxiety as a cognitive, rather than an affective, process.[13] However, the cellular mechanisms underlying the behavioral inhibition system have yet to be exposed.

Another neural system involved in processing reinforcing stimuli is the orbito-frontal cortex.[9] Stimulation studies in primates show it is involved in correcting responses made to stimuli previously associated with reward or non-reward (i.e., with an emotionally significant stimulus), suggesting that groups of neurons may be involved in evaluating the emotional significance of stimuli.[9] It is likely that the cortical neural system is much more developed in primates than rodents, though lesion studies in the rat show that medial prefrontal cortex is involved in the process whereby a repeated exposure to a conditioned stimulus loses its ability to evoke a response.[44] This work has yet to implicate any specific electrophysiological and neurochemical pathways.

## ENDOCRINE INFLUENCES ON EMOTION

A relationship between hormonal changes and emotion is well-established, but the nature of that relationship is still unclear. For instance, it is not in doubt that variation in reactivity of the hypothalamic pituitary adrenal (HPA) axis is involved in emotional states and that it is partly under genetic control. The older literature on this issue is well summarized elsewhere[42,43,45] and will not be reiterated here. How, though, can the current view of the role of the amygdala incorporate what is known about neuroendocrine changes in emotional states?

As mentioned before, the amygdala has a complex neurochemical anatomy, and it is very likely that some of the compounds found here are involved in the modulation of emotion. In particular, two neuropeptides have been been implicated in the response to stress.[46-48] Corticotropin releasing hormone (CRH), a 41 amino acid hypothalamic peptide, is the prime mediator in stress-induced HPA axis activation.[49,50] Administration of CRH produces many of the behavioral responses associated with fear: It enhances the acoustic startle response[51] and decreases the amount of time spent in the open arms of an elevated plus maze[52] (a validate measure of fearfulness).[53] Furthermore, CRH antagonists reverse these effects.[54] The amygdala is implicated by association: CRH neurons innervate the central nucleus and lesions of the central nucleus blocks potentiation of acoustic startle, mediated by CRH.[55]

There is also evidence that neuropeptide Y is involved in the processing of emotional states in in the amygdala. Neuropeptide Y consists of 36 amino acids and is found at high concentrations within the amydgala.[56] Injections of neuropeptide Y receptor agonists into the amygdala results in anxiolytic behavior;[57] conversely, a decrease in amygdala neuropeptide Y receptors results in anxiogenic behavior.[58] The contrasting effects of the CRH and neuropeptide Y in the amygdala have been intepreted as evidence that they are counter-regulatory signals in the integration of fear.[48] Further work is needed to explore this hypothesis.

## RELATION TO HUMAN EMOTION

How well do these theories about the neural basis of emotion hold up for humans? The equivalent of fear conditioning experiments have been carried out in humans. By associating a loud noise with a neutral stimulus (a monochrome slide) and, after a period of conditioning, measuring skin conductance response to the neutral stimulus in patients with amygdala damage, it has been shown that the amygdala is required for emotional conditioning.[59,60] There is also evidence that the amygdala is involved in processing emotional information. Patients with Urbach-Wiethe disease, a rare hereditary disorder that produces bilateral amygdala damage, have impaired recognition of emotion in facial expressions[61] (though this effect may occur only if lesions occur early in development[62]), auditory recognition of fear and anger,[63] and selective impairment of memory for emotional material.[64] Positron emission tomography of normal subjects viewing photographs of fearful or happy faces also shows that the amygdala is involved in the emotional salience of faces.[65] More recently, functional magnetic resonance imaging has been used to show that the amygdala is involved in emotional memory in humans.[66,67]

In fact, in addition to supporting the animal literature, the available studies on the neural basis of emotional conditioning in humans suggest that emotion may play a greater role in cognition and abstract reasoning than previously thought. Examination of patients with frontal lobe damage who have intact abstract reasoning, memory, and other intellectual skills, but marked impairment in making decisions of a personal or social nature, show a consistent pattern of damage to the ventromedial frontal region,[68] an area implicated in work on primate cortex in emotional processing.[9] Such patients also show the expected defects in emotional processing.

The cerebral basis of emotion almost certainly involves other structures. Nevertheless, the congruence between work on animals and humans in describing the structures involved in processing conditioned fear shows how it is possible to overcome the difficulties of defining emotion and working with animal models. Genetic analysis has built on this foundation.

## THE MOLECULAR BASIS OF EMOTION — EVIDENCE FROM GENE KNOCKOUT EXPERIMENTS

One of the important lessons to emerge from the work on animal models of emotion is that no one brain system can be isolated, even for fear conditioning. Instead, we can see the interaction of several interlocking systems: The amygdala clearly has a role in associating emotional signals with experiences, but which signals become associated with what depends on the activity of other systems. For example, there are different components to the systems mediating contextual and cued fear conditioning. Because of the coordinated integration of neural systems, investigators working on learning and memory have also cast light on emotional systems. One of the most powerful molecular tools applied to studying learning and memory has been gene knockouts in mice.[6,69-71] These experiments also have important lessons for emotion research.

It is now possible to produce mice with specific mutations in any cloned gene and to inactivate that gene in a tissue-specific fashion and at different stages of development.[72,73] Gene ablation is achieved by exploiting homologous recombination in embryonic stem (ES) cells. DNA containing a mutated copy of the gene of interest in tandem with a selectable marker (for instance an antibiotic resistance gene) is introduced into ES cells where in a small number of cases it recombines with and consequently replaces the cell's copy of the gene. ES cells are injected into blastocysts, where they grow into a chimaeric animal, some of whose germ cells may also contain the mutation, thus making it possible to breed from the chimaera to produce animals that are heterozygotes and, by inbreeding, homozygotes, for the mutation.

Inducible gene knockouts again use homologous recombination to replace the gene of interest. But in addition to the mutant gene and selectable marker, the replacement DNA contains a sequence that puts expression of the gene under exogenous control. For instance, gene expression from the construct can then be altered by feeding the animal tetracycline. A slightly different technology is involved in the creation of cell-type restricted knockouts. Here, sequences recognized by a site-specific recombinase are introduced on either side of the gene of interest. Then, mice are crossed with a strain that contains the recombinase, which is itself under exogenous control (generally in a tissue or cell-type specific fashion). In progeny with the gene flanked by recombinase recognition sites, the gene is excised only in those cells in which the recombinase is activated by tissue- or cell-specific factors.

The power of the genetic engineering technology is impressive, but, as with all new technologies, its limitations are still unclear. At least two reservations should be born in mind when considering the results discussed subsequently. First, the effects of any knockout experiment will be present in all tissues and throughout development. Consequently, effects observed in the adult could be the result of developmental abnormalities. The newer tissue-specific knockouts, or those where gene expression is said to be regulated by a tissue-specific promoter, are clearly improvements, but they are not perfect models. Understanding of tissue-specific gene regulation is still rudimentary, and we do not know exactly which sequences are required for controlling gene expression.

Second, the genetic background of a mutant can have an effect on the mutant.[74] Most targeting experiments use the embryonic stem cells derived from substrain 129 mice and then cross chimaeric mice with another inbred line, most commonly C57BL/6. Homozygotes are then generated by producing an F2 intercross, which, on average, will have half the genetic background of C57BL/6 and half from the 129 ES cells. The choice of inbred line into which the mutation is bred can be critical because many inbred lines have specific behavioral phenotypes: DBA show poor hippocampal dependent learning,[75] and C57BL/6 are poor avoidance learners.[76] The data given in Table 24.1 lists the mouse strain used for the ES experiments.

The behavioral mutants that have been constructed with gene knockout technology have been used primarily to answer questions about the neurobiology of memory, particularly to dissect the molecular components of LTP. However, as explained above, LTP is also thought to mediate conditioned fear, so it is to be expected that the transgenic animals created for analysis of memory could cast some light on the genetic determinants of conditioned fear. This has proved to be the case.

A summary of the relevant gene knockout experiments is given in Table 24.1. Although not all animals have been tested with the same protocol so that the data on emotionality is incomplete, enough is known for some general conclusions to be reached. First, it is gratifying to see that the mice with genetic deficits in hippocampal LTP do have abnormal fear conditioning. The prediction that contextual fear conditioning would be affected rather than cued fear is born out. For example, mice that lack one of the isoforms of protein kinase C (PKCγ) have abnormal hippocampal LTP, a mild deficit of contextual fear conditioning but normal tone-dependent (cued) fear conditioning.[77,78] Similarly, deletion of one subtype of metabotropic glutamate receptor results in substantial impairment of LTP, a moderate impairment of contextual fear conditioning, and normal cued fear conditioning.[79,80]

However, as also might be expected, the emerging story is complex. In a demonstration of the power of genetic approaches to dissect behavioral functions, Huang et al.[81] overturned concepts of how hippocampal structure relates to function by specifically ablating mossy fibre LTP. The mossy fibre pathway in the hippocampus had been thought to play an important role in an auto-associative mechanism.[82] Ablation of mossy fibre LTP tested the hypothesis that hippocampal function operates in a serial fashion, passing information from one pathway to the next. The genetic experiment demonstrated that contextual conditioning, behavioral responses to novelty (open field tests), and spatial memory are *un*affected by the elimination of mossy fibre LTP. This suggests that, like the visual system, different regions of the hippocampal system have specialized roles, possibly processing information in parallel.

Most behavioral data are available for mice with mutations in the α calcium calmodulin kinase II gene (αCaMKII). LTP involves activation of NMDA receptors and $Ca^{2+}$ influx, which in turn activates second messenger systems, such as protein kinases. αCaMKII is a serine-threonine protein kinase that can convert to a $Ca^{2+}$ independent form and is therefore a candidate molecular substrate for memory. Deletion of αCaMKII produces deficits in LTP and impairs spatial learning[83,84] but not contextual fear conditioning. Mutation of the gene to a constitutively active form (calcium-independent) confirmed that dissociation of contextual fear conditioning and spatial memory could be achieved. It suggested that different synaptic mechanisms might be responsible for the two forms of learning.[85]

**TABLE 24.1**

**Gene Knockout Experiments Revelant to the Study of Emotion**

| Gene | Method | Strain | Phenotype | Cued Conditioning | Contextual Conditioning | OFA and OFD | Exploratory Behavior | Learning | Resident/ Intruder | Neurological | Active Passive Avoidance | Chromosome | Ref. |
|---|---|---|---|---|---|---|---|---|---|---|---|---|---|
| PKA RI beta and C beta 1 | HR 129 | C57BL/6 | -/- Deficient in mossy fiber LTP, normal Schaeffercollateral LTP | Normal | Normal | Normal | Normal open field rearing, grooming | Normal Morris water maze and Barnes circular maze | | | | 3 | 81, 154 |
| PKA | R(AB) transgene inhibiting PKA | C57BL/6 | | Normal | Deficient contextual conditioning (at 24 h) | | | Impaired spatial (Morris water maze) | | | | | 155 |
| PKC gamma | HR 129 | C57BL/6 | -/- LTP deficient, LTD and PPF normal | Normal | Moderate deficit in contextual conditioning | Normal grooming, feeding, circadian activity, mating | | Morris water maze: Mild spatial learning deficit | | Poor rota rod performance, ataxic gait, hypothemia and loss of righting reflex reduced | | 7 | 77, 78, 156, 157 |
| Fyn, Src, Yes and Abl tyrosine kinases | HR 129 | C57BL/6 | LTP deficient in fyn mice | | | | | Impaired spatial learning in fyn mice | | | | 10 | 158 |
| BDNF | HR 129 | BALB/c | -/- LTP deficient +/- LTP deficient | | | | | | | | | 2 | 159 |
| N-methyl-D-aspartate receptor epsilon1 | HR 129 | C57BL/6 | -/- LTP deficient | | | | | Impaired spatial memory (Morris task) | | | | 16 | 160 |
| N-methyl-D-aspartate receptor R1 | HR 129 | C57BL/6 | -/- Die after birth | | | | | | | | | | 161 |
| N-methyl-D-aspartate receptor R1 | Cre-Lox | C57BL/6 | LTP deficit | | | | | Impaired spatial memory (Morris water maze) | | | | | 72, 162, 163 |
| Metabotropic glutamate receptor 1 | HR 129 | C57BL/6 | -/- LTP reduced, STP and LTD normal | Normal | Moderate deficit in contextual conditioning | Locomotor activity normal (corrected) for gait | | Morris water maze: Mild spatial learning deficit | | Ataxic gait, tremor, loss of righting reflex | | | 79, 80, 164 |
| Glutamate receptor delta | HR 129 | C57BL/6 | -/- LTD impaired | | | | | | | Motor incoordination | | | 165 |

**TABLE 24.1** (continued)
**Gene Knockout Experiments Revelant to the Study of Emotion**

| Gene | Method | Strain | Phenotype | Cued Conditioning | Contextual Conditioning | OFA and OFD | Exploratory Behavior | Learning | Resident/ Intruder | Neurological | Active Passive Avoidance | Chromosome | Ref. |
|---|---|---|---|---|---|---|---|---|---|---|---|---|---|
| Adenylyl cyclase | HR 129 | C57BL/6 | -/- Deficient LTP | | | | | Unimpaired escape latencies, deficit transfer task in Morris task | | | | | 166 |
| Histamine H1 | HR 129 | C57BL/6 | -/- Circadian ambulation ratio disturbed | | | Decreased OFA, increased OFD | Decreased rearing | | | | Normal passive avoidance latency | 6 | 91 |
| Alpha-Ca-Calmodulin Kinase II | HR 129 | | +/- | Reduced | | Reduced OFD | Open field thigmotaxis decreased | | Normal resident attacks intruder, more defensive attacks | | | 18 | 83, 84, 86, 167 |
| | | | -/- | Reduced | | Reduced OFD | Open field thigmotaxis decreased | | Increased resident attacks intruder, reapproach resident after attack | | | | |
| | Transgene: Ca independent kinase activity | | Abnormal LTP at 5–10 Hz | Normal | Normal | | No light dark preference | Deficient contextual memory, normal cued memory (Barnes maze) | | | | | 85, 168 |
| | Forebrain-specific promoter | | Abnormal LTP at 10 Mz | Normal | Normal | | | Impaired spatial memory (Barnes maze) | | | | | 73 |
| Estrogen receptor | HR 129 | C57BL6 | +/- Reduced sexual behavior | | | Non significant increased OFA | Non-significant open field thigmotaxis decreased, rearings increased | | Normal offensive attacks | | | 10 | 169 |
| | | | -/- Reduced sexual behavior | | | Increased OFA, reduced OFD | Open field thigmotaxis decreased, increased rearings | | Reduced resident intruder attacks | | | 12 | |
| Monoamine oxidase A | IFN-b transgene | C3H/HeJ | -/- Mating behavior abnormal | | | | Open field thigmotaxis decreased | | Increased resident intruder attacks | Pup neurological abnormalities | | X | 170 |

| | | | | | | | | | | | | |
|---|---|---|---|---|---|---|---|---|---|---|---|---|
| mu opiate receptor | HR 129 | C57BL/6 | +/– | Reduced nociceptive responses | | Normal | | | | Passive avoidance normal | | 171 |
| | | | –/– | Reduced nociceptive responses | | Normal | | | | Passive avoidance normal | | |
| K+ channel | HR 129 | 129SvEMS | +/– | Normal body weight | | Normal | | Normal rota-rod performance | | Normal active avoidance | | 172 |
| | | | –/– | Reduced body weight | Decreased auditory startle | Normal | | Reduced rota-rod performance, altered skeletal muscle contraction | | Normal active avoidance | | |
| Interleukin 6 overexpression | Transgene | C57BL/6 X SfL | | | | | | | | Reduced active avoidance | | 89 |
| GABA A gamma 2 | HR 129 | C57BL/6 | +/– | Normal behavior | | | | | | | | 173 |
| | | | –/– | Die after birth (18 days max) | | | | | | | | |
| Pre-proen-kaphalin | HR 129 | CD1 | –/– | Reduced nociceptive responses, stress induced analgesia normal | | Reduced OFA | Open field thigmotaxis increased | | Increased residents attack intruders | | 4 | 174 |
| 5-TH1b receptor | HR 129 | C57BL/6 | –/– | | | Normal OFA | | | Increased residents attack intruders | | | 175 |
| Dopamine transporter | HR 129 | C57BL/6 | +/– | | | Non-significant increase in locomotor activity | | | | | | 176 |
| | | | –/– | | | | | Increased spontaneous locomotor activity | | | 13 | |
| Dopamine 1a receptor | HR 129 | C57BL/6 | –/– | Growth retardation (decreased motivation?) | | Normal OFA | Decreased rearing in open field | | | | 13 | 177 |
| | | | | | | Increased OFA | | Cocaine-induced hypoactivity, increased locomotor activity | | | 9 | 178, 179 |
| Dopamine D2 receptor | HR 129 | C57BL/6 | +/– | Normal reflexes, abnormal gait | | Reduced OFA | No rearing in OFA | Hypoactive on rotarod | | | | |
| Neuronal NO synthase | HR 129 | C57BL/6 | –/– | Abnormal male sexual behavior | | Normal OFA | | | Increased male resident attacks intruder | | 5 | 180 |

**TABLE 24.1** (continued)

**Gene Knockout Experiments Revelant to the Study of Emotion**

| Gene | Method | Strain | Phenotype | Cued Conditioning | Contextual Conditioning | OFA and OFD | Exploratory Behavior | Learning | Resident/ Intruder | Neurological | Active Passive Avoidance | Chromosome | Ref. |
|---|---|---|---|---|---|---|---|---|---|---|---|---|---|
| Tyrosine hydroxylase (Dopamine) | HR 129 | C57BL/6 | -/- Decreased spontaneous activity, cessation of eating and drinking | | | | | | | Impaired rota-rod performance | | 7 | 181 |
| Neural-cell adhesion molecule | HR 129 | C57BL/6 | -/- | | | Normal OFA | | Impaired spatial learning (Morris task) | | | | 9 | 182 |
| Nicotinic acetylcholine receptor beta 2 | HR 129 | C57BL/6 | -/- | | | | Normal black & white box | Normal spatial memory (Morris maze) | | | Passive avoidance performance not augmented by nicotine | 11 | 92 |
| cAMP responsive element binding protein | HR 129 | C57BL/6 | -/- LTP impaired | Decreased cue conditioning (at 24 h) | Decreased contextual conditioning at 60 min (normal at 30 min) | | | Impaired spatial memory (Morris water maze) | | | | 1 | 183 |
| Corticotrophin releasing hormone | HR 129 | 129 and CD1 | -/- Basal ACTH secretion reduced | | | Increased OFA | Lower latencies to emerge in black & white box | | | | | | 87 |

Note:    Column 1 (Gene) names the gene that has been inactivated or otherwise altered in the mutant, and the method of genetic manipulation is given in column 2; HR 129 — homologous recombination in ES cells from 129 strain mice; Cre–lox; inactivation using cre–lox recombinase. Column 3 (Strain) indicates the inbred mouse strain into which the mutation has been bred from the 129 donor line. Column 4 (Het) refers to the genotype of the animal that has been tested. –/– refers to the homozygote and +/– to the heterozygote. The next 8 columns describe the phenotype. The first column (Phenotype) gives information about the phenotype in which the investigators were interested (in most cases long term potentiation (LTP)). The following columns describe phenotypes relevant to emotion. OFA: open-field activity; OFD: open-field defecation. The penultimate column gives the chromosome location of the gene.

What makes the αCaMKII gene of particular interest for the study of emotion is that the knockout animals have many behavioral features indicating they are "fearless." When placed in an open field, an exposed arena considered threatening to rodents, mice prefer the relative safety of the periphery, but the knockout mice spent more time in the center of the field. They were also less likely to freeze and defecate during contextual fear conditioning and were much less fearful in tests of aggressive behavior.

Since mice are territorial, one way of measuring aggression is to introduce an "intruder" mouse to a "resident." In this paradigm, knockout mice often re-approached the resident wild-type mouse (in a non-aggressive fashion) even after being attacked, a behavior that rarely occurs in wild-type intruders.[86] The availability of mutants with regulated expression of $Ca^{2+}$ independent αCaMKII has permitted even more detailed investigation of the genetic control of fear related behaviors. Mayford et al.[73] found that mice with high levels of expression of the constitutively active αCaMKII in the lateral amygdala and striatum had severe impairments of cued and contextual fear conditioning that could be reversed by suppressing transgene expression. These results indicate that the genetic effect on fear conditioning is through effects on synaptic plasticity and memory formation, not via alterations of developmental processes. The experiment also lends further weight to the view that the lateral amygdala is the site of synaptic plasticity for fear conditioning, mediated by LTP.

One other mutant is worth discussing at this point: a CRH knockout.[87] As discussed previously, there is evidence that CRH is anxiogenic, and transgenic mice with overexpression of CRH show the expected reduction in exploratory behavior and increased anxiety.[88] It was predicted that deletion of the CRH receptor would reduce emotional responses. The knockout mice tested on a battery of fear-related measures and did indeed show reduced anxiety: They were more likely to explore a novel environment and lower latencies to emerge from a dark box.

Thus, the first feature of importance to emerge from the data collated in Table 24.1 is that animals with knockouts of genes predicted to be involved in emotional processes do show the expected changes.

The second feature of importance is that apparently very different genes can have effects on measures of emotionality. So far, we have discussed genes predicted to have a role in determining fear conditioning from knowledge of the neural systems thought to be involved. It is likely there will be many other genes involved in fear conditioning that operate within other, unknown, neural systems. Indeed the knockout data point in this direction. How, for instance, is the interleukin 6 gene effect mediated?[89] The data suggest that it is involved in active avoidance, another model for emotionality described previously. Transgenic mice expressing the interleukin 6 gene were given an active avoidance learning procedure in which the first arm of a Y maze they entered was deemed the preferred arm and subsequent entries into the "wrong" arm were punished by foot shock. At 12 months, both heterozygous and homozygous mice showed significantly more errors than controls. Are we to conclude that IL6 is a gene involved in anxiety? Similarly it might follow from knockout studies that pre-proenkephalin,[90] histamine H1 receptors[91] and nicotine receptor[92] are candidates for "anxiety" or "fearfulness" genes.

The difficulty here is that the knockouts have not been given a battery of behavioral measures to define fearfulness or emotionality, but rather have been given a test chosen to make a particular point. The gene knockout experiments that investigate LTP have a context. There is a lot of information on how LTP operates and how it relates to memory and fear-related behavior. Thus, specific hypotheses can be tested and elaborated. We cannot so easily assume that because knockout animals show changes in open field behaviors that they show a change in emotionality. As outlined at the beginning of this review, the behavioral phenotype is defined by a series of correlated measures. Without the additional behavioral testing necessary, we cannot conclude that IL6 and other genes have an effect on emotion. We will have to wait for the full significance of many of the knockout experiments to become manifest.

Finally, the data in Table 24.1 indicate that knockouts do *not* have a specific effect on a single behavioral measure. Mutants constructed to investigate LTP have deficits in other behavioral tests, often in unexpected ways. For instance *fyn* kinase mutants have deficits in suckling behavior.[93] Similarly, mice with mutations in genes involved in tissue-specific or developmental processes are found to have behavioral changes. Mice lacking the Dvl1 gene, a homologue of the *Drosophila* segment polarity gene *Dishevelled*, have a complex behavioral deficit: they show striking abnormalities in social behaviors and sensorimotor gating, part of the process by which inhibitory neural pathways enable an organism to attend to a single stimulus by filtering out irrelevant stimuli.[94]

Not only does this imply that genes with behavioral effects are pleiotropic but also that there is considerable redundancy. Indeed, at first sight it is surprising that so many of the knockouts were viable at all. The kinases that have been removed are widely expressed genes, involved in many different cellular and developmental processes. Their absence might be expected to result in far more drastic consequences than low LTP. The implication is that other genes, each of which has a certain degree of redundancy, have compensated for the mutation. In other words, genes influence behavior through their interaction with a network of similar genes, so their effects result from a disturbance in the balance of the system. Each gene may be participating in a number of different systems.

In summary, while genetic manipulation of animals offers a powerful new approach to the analysis of the genetic basis of emotion, gene knockouts have not revealed genes that specifically affect emotion. Nor are they likely to. The emerging picture is of a network of interacting brain systems containing considerable redundancy. The same conclusion has been drawn from the analysis of *Drosophila* learning mutants.[95]

Nevertheless, genetic analysis of wild-type mice does not show that poor performance on measures of fear conditioning means that an animal is poor at learning or memory tasks. Similarly in humans, individual differences in emotions (for example as measured by personality differences) do not correlate highly with other behavioral and cognitive measures. Therefore, it is important to see how the genetic variants responsible for individual differences in emotion relate to the genes identified in the knockout experiments. Advances in gene mapping make it feasible to address this issue and are discussed next.

## QUANTITATIVE GENETIC ANALYSIS OF EMOTIONALITY

A top-down analysis of the genetic influence on emotion starts from the supposition that individuals experience emotions differently (either in intensity or quality) and that these individual differences represent variation in an underlying trait, which may have a genetic basis. Most success to date has been achieved by working with animal models (conditioned fear, active and passive avoidance, and altered behaviors in novel, therefore frightening, environments). These are discussed first. The more contentious work on the genetics of human emotional traits concludes this section.

The results of artificial selection experiments demonstrate that there is a genetic component to fear-related behaviors in rodents.[96-101] The experiments also go some way to validating the concept that the same brain system or systems underlie different but genetically correlated emotional behaviors. Two selection experiments employed measures of fearfulness in a novel environment, the open field, a brightly lit white arena that is potentially frightening to rodents which, not unnaturally, prefer quiet, dark, small enclosed spaces.[96,99] Since strong emotion is known to result in defecation and urination in humans, Calvin Hall used them as a measure of rodent fearfulness.[102] Furthermore, consistent with the view that frightened animals freeze, urinate, and defecate, a negative correlation was found between the animal's level of activity in the open field and its defecation rate, suggesting that the same brain system mediated the fearful behaviors, often called "emotionality" in the rodent literature. Other open field activities have also been used: Frightened animals are considered to spend less time in the center of the arena (thigmotaxis) and to rear less.

Rodent emotionality is a moderately heritable trait. By selectively breeding over many generations, rats with extremes of defecation and mice with extremes of activity in the open field, it has been shown that the heritability of open field activity is about 30% and open field defecation about 10%. The genetic correlation between the two traits is high (about –0.9).[98,99] Rodents selected for high open-field activity have low defecation rates and vice versa.

Artificial selection of rats for a conditioned avoidance task (speed of acquisition of an escape response when a light is paired with a foot shock) provide more evidence that there is a genetic component to a "fearfulness" neural system. The low-avoidance rats freeze in the test paradigm, and the high-avoidance rats have great difficulty learning to inhibit a response,[103,104] indicating that differences in the conditioned avoidance task are due to variation in emotionality rather than differences in learning ability. Consistent with this hypothesis, low-avoidance rats defecate more than high-avoidance rats in the open field[105,106] (although one study has not replicated this[107]) and have lower open-field activity.[100,106,107] Thus the selection experiments not only demonstrate genetic influence on emotional behaviors, but they also show that a series of phenotypically correlated behaviors reflect the action of the same neural system or systems.

Further work on the rat strains supports this view. A large body of data on the action of anxiolytics on animal behavior has been synthesized by Gray[42,43] in support of his theory of a behavioral inhibition system, activity that is thought to represent anxiety. Much of the evidence on which this theory relies derives from the fact that anxiolytic drugs (such as benzodiazepines), produce a behavioral phenotype similar to rats selectively bred for low defecation rates (the Maudsley non-reactive (MNR) rat). For instance, the performance of benzodiazepine-drugged rats improves in active avoidance tasks; the speed of a drugged rat when running to a food reward is less inhibited by a foot shock than is the speed of an undrugged rat. In both cases, the rat's performance approximates to that of the MNR rat (a "non-anxious" rat).

In summary, the genetic basis of one model system of emotion, rodent fearfulness, or emotionality, is well established. The genetic data, in combination with analysis of drug effects, lesion, and stimulation experiments define a neural system that overlaps with the models of conditioned fear discussed previously. Furthermore, the wealth of genetic information and the data on genetic correlations between measures of individual differences of fearfulness make it in principle a very good candidate for gene mapping analysis. Thus, a top-down genetic approach is possible.

The principle of genetic mapping of fearfulness is relatively straightforward. Measures known to have a genetic component are taken from a large number of animals, and the animals are genotyped. A correlation is then sought between genotype and phenotype. In the simplest situation we might compare mean phenotypic scores of homozygotes, and if a significant difference is found, then we have evidence that the locus affects the phenotype.

Initially, mapping behavioral traits like fearfulness appeared to be a very complex undertaking. Indeed, the failure to reproduce putative linkage findings in human behavioral disorders raised the possibility that the genetic basis of psychological traits might be just too complex to yield to molecular approaches.[108] Surprisingly then, the results for mapping behavioral traits in rodents have not born out this pessimism.

Three loci (on chromosomes 1, 12, and 15) have been identified as contributing to almost all the genetic variance of emotionality (defined as open field activity and open field defecation) in a cross derived from BALB/cJ and C57BL/6 strains.[109] In this experiment, a genome search was undertaken for loci-influencing open-field activity. Chromosomal regions exceeding a predetermined threshold were then investigated to see if correlated measures also mapped to these regions. Of six loci examined, significant scores were found toward the telomere of chromosome 1, the centromere of chromosome 12, and the middle of chromosome 15 for open-field defecation and for entry into the open arms of an elevated plus maze (another ethological measure of fear-related behavior). In a separate experiment, open-field activity was mapped in a cross between A/J and C57BL/6J mice, and loci on chromosomes 1, 10, 15, and 19 were found to contribute to about a quarter of the phenotypic variance (hence again almost all the genetic variance).[110] Given the simplicity of the measure (activity in the open field), the use of different crosses, and likely contextual variation in the way the behaviors were measured (e.g., type of apparatus, sequence of testing and other environmental variables), the fact that two loci (on chromosomes 1 and 15) were replicated is striking.

Gershenfeld and Paul[111] have gone on to measure thigmotaxis in the open field and light dark (LD) transitions. Thigmotaxis again mapped to chromosome 1, though more centromeric than the previously identified locus for OFA. In the LD transition test, the animal is placed in a brightly lit compartment from which it may escape into a black box. Using the the number of transitions between light and dark boxes as a measure of exploratory behavior, Gershenfeld and Paul mapped a QTL to chromosome 10.

The intepretation of the LD test is complicated by the fact that each measure was repeated on different days, so that, for instance, they have three scores for LD transitions and an average score over all trials. Moreover, they found low correlations between the LD measures and thigmotaxis (ranging from 0.15 to 0.25). As explained previously, such low correlations make it difficult to argue that their results are relevant to emotionality. Furthermore, LD transitions are yet another measure of activity, so the QTL identified on chromosome 10, which accounts for differences in light dark transitions, vertical rearing, and open field ambulation, is likely to be influencing activity, not emotionality.

By contrast, conditioned fear is independent of activity and, as previously discussed, is correlated with other measures of emotionality. Genetic loci contributing to variation in fear conditioning have now been mapped. The mapping experiment provided an opportunity to test the hypothesis that fear conditioning and open-field behaviors have some neural systems in common: We would expect to find loci that influence both types of behavior presumably containing pleiotropic genes that are involved in determining emotionality. Two F2 intercrosses and one study of recombinant inbred strains identified a locus on chormosome 1 that lies within the region identified by the work on open field behaviors.[112-114] Therefore, it appears very likely that a locus on chromosome 1 determines variability in a neural system mediating fear. Unfortunately, we have no clue what this gene (or genes) may be; no obvious candidates have been mapped to this region of the mouse (or human) genome. Whether homologous loci in the rat genome account for genetic variability in emotionality is not yet known. One study of hyperactivity in the rat identified a locus on chromosome 8 that contributed substantially to the variation in spontaneous activity, rearing and activity in the open field.[115] However, this is not homologous to loci identified in the mouse. A summary of the mapping experiments is given in Table 24.2.

Although it seems surprising that a small number of loci account for the majority of the genetic variance, in fact other complex quantitative traits so far examined consist of a handful of loci that contribute to a large proportion of the genetic variance, while a larger number of loci of small effect account for the remainder. Analysis of traits in a number of animal and plant species has shown that less than ten loci frequently explain more than half the phenotypic variance of a trait (and hence even larger proportions of the genetic variance[116]). The few extant studies of behavior show a similar picture: Two loci have been reported to account for 46% of the genetic variance of alcohol preference in mice[117] and three loci explain nearly 85% of the genetic variance of morphine preference in mice.[118] Moreover, it is possible that loci may be syntenic between species. A study of traits in cereal crops has shown that loci discovered in one species map to homologous chromosomal positions in disparate taxa.[116] It may be possible to use gene mapping and cloning in rodents as a way of isolating homologous human genes. The task of finding genes for susceptibility to emotions in humans may not be as daunting as has appeared to be the case from the attempts to map genes for behavioral disorders.

## GENETIC MAPPING OF EMOTIONS IN HUMANS

A genetic influence on human emotion makes sense only if we consider that the propensity to feel emotion differs between individuals, either in intensity or type of experience. We have a common sense view that some people are more emotional than others, by which we mean they react in

**TABLE 24.2**
**Anxiety-Related Phenotypes Mapped in the Mouse**

| Chr | Position | Phenotype | LOD | Phenotype | LOD | Phenotype | LOD | Phenotype | LOD | Reference | Sytnenic Region |
|---|---|---|---|---|---|---|---|---|---|---|---|
| 1 | 28.7 | Context | 5.14 | Cue | 2.38 | | | | | 114 | 2q33–q34 |
| 1 | 37.2 | Avg. thigmotaxis | 3.97 | | | | | | | 111 | 2q36 |
| 1 | 49.6 | Context | 4.32 | Cue | 2 | | | | | 114 | 2q37 |
| 1 | 67 | Context | 3.8 | Cue | 5.6 | | | | | 113 | 1q31–q32 |
| 1 | 71.9 | Avg. light-dark | 2.52 | | | | | | | 111 | 1q31–q33 |
| 1 | 73.2 | OFA | 7.73 | | | | | | | 111 | 1q31–q34 |
| 1 | 77.3 | Context | 4.76 | | | | | | | 114 | 1q24–q25 |
| 1 | 79.4 | Vertical rearing | 4.47 | | | | | | | 110 | 1q24–q25 |
| 1 | 99.5 | OFA | 13.4 | OFD | 5.6 | EPM | 9 | | | 109 | 1q41–q42 |
| 1 | 100.5 | OFA | 7.08 | | | | | | | 110 | 1q41–q42 |
| 2 | 89.2 | Context | 3 | | | | | | | 113 | 20q11 |
| 3 | 38.2 | Context | 4.1 | | | | | | | 113 | 4q32 |
| 3 | 44.4 | Context | 2.72 | | | | | | | 114 | 1q21 |
| 4 | 31.7 | OFA | 6.2 | | | | | | | 109 | 9q32 |
| 6 | 0 | Avg. thigmotaxis | 2.77 | | | | | | | 111 | 7q22 |
| 6 | 10.2 | Avg. light-dark | 2.91 | | | | | | | 111 | 7q31 |
| 8 | 8 | OFA | 9.5 | | | | | | | 97 | 8p12–p11 |
| 8 | 16 | Context | 2.03 | | | | | | | 114 | 13q14–q21 |
| 9 | 48.6 | Context | 2.7 | Cue | 2.59 | | | | | 114 | 15q24 |
| 10 | 14.2 | Context | 4.7 | Cue | 4.6 | | | | | 113 | 6q25 |
| 10 | 74.1 | OFA | 8.76 | Vertical rearing | 8.51 | Light-dark | 8.94 | Avg. light-dark | 9.46 | 110, 111 | 2q13 |

**TABLE 24.2** (continued)
**Anxiety-Related Phenotypes Mapped in the Mouse**

| Chr | Position | Phenotype | LOD | Phenotype | LOD | Phenotype | LOD | Phenotype | LOD | Reference | Sytnenic Region |
|---|---|---|---|---|---|---|---|---|---|---|---|
| 12 | 46.4 | OFA | 4.3 | OFD | 4.5 | EPM | 2.4 | | | 109 | 14q32 |
| 15 | 31.4 | OFA | 11 | OFD | 3.1 | EPM | 5.4 | | | 109 | 8q24 |
| 15 | 41.5 | OFA | 3.6 | Avg. light-dark | 2.52 | | | | | 110, 111 | 8q24–qter |
| 16 | 47.9 | Context | 4.5 | Cue | 3.7 | | | | | 113 | 21q–q22 |
| 17 | 10.9 | OFA | 3.5 | | | | | | | 109 | 16p13 |
| 18 | 12.2 | OFA | 4 | | | | | | | 109 | 5q31 |
| 19 | 20.6 | Light-dark | 2.51 | Avg. light-dark | 3.95 | OFA | 2.54 | | | 111 | 9p24 |
| 19 | 24 | OFA | 3.15 | Vertical rearing | 3.81 | | | | | 110 | 9p24 |

*Note:* Chr: chromosome. Position: the distance in centimorgans from the end of the mouse chromosome at which the QTL is most likely to be found. Note that the confidence interval for almost all these studies is about 25 centimorgans. The phenotypes are given in the next four columns with the highest LOD (log likelihood ratio) score reported. Only LOD scores over 2 are given. For a complete genome screen, a LOD of 4.3 is regarded as significant. In the phenotypes columns, Context and Cue referred to fear conditioning to the context and cue respectively. Light-dark and Avg. light dark are the number of transitions (and average number of a series of trials) in a light dark box. EPm is the elevated plus maze. See text and the original references for descriptions of phenotypes. The final column gives the homologous human chromosomal region.

predictable ways (for example by crying) in stressful situations. For most purposes, we can equate a propensity to be emotional with temperament or personality.[1] If we accept that some personality traits identified by psychologists measure such a propensity (neuroticism, for instance) and that these traits have a genetic basis, then genetic analysis has the potential to locate genes that govern our propensity to experience different degrees of emotion, or, in short, our level of emotionality.

How good is the evidence that personality traits can be identified and that they have a sound biological basis? Over the last few years, a consensus has emerged about the most appropriate classification of personality types. The identification of personality traits is in principle straight-forward: Take a large number of behaviors or descriptions of personality (such as the 4504 dictionary terms reported[119]) and attempt to find correlations (technically in fact co-variances) between them. This analysis might find that "worrying," "irritability," and "recurrent feelings of guilt" are consistently associated and could be used to define a trait called nervousness. The procedure can be justified by using factor analysis.[120] The assumption that a higher-order description (or factor, such as nervousness in the example given here) can subsume the individual descriptions is supported to the extent that the factor can account for the variance of the items. Unfortunately, it is not straightforward to decide how best to extract factors. Indeed, for some aspects of the procedure, an infinite number of mathematically equivalent solutions may be available, a state of affairs that has not surprisingly given factor analysis a bad name. The appearance of different personality typologies (Cloninger's[121] 7 personality dimensions, Cattell's[122] 16, or Eysenck's[123] 3) does little to instill confidence in the validity of personality trait theories.

Within the last few years, trait psychologists have realized that a more comprehensive model can accommodate the numerous alternative structures.[124-127] The reason for the more optimistic view of personality theories is that similar trait structures have been identified from very different starting points. By including two or more personality assessment procedures in the same study, it has been possible to compare and validate the different approaches. Furthermore, it has been increasingly accepted that higher-order factors represent what is common to lower-order traits and account for the observed correlations between them. So far, robust evidence has been collected for two such super factors: the traits of neuroticism (or negative emotionality) and extraversion, defined, for instance, by different personality schedules such as the Eysenck Personality Questionnaire,[123] Revised NEO Personality Inventory, and Big Five Inventory.[128] The status of other factors is less secure, but there is general agreement that an additional two or three are required to account for the available data on personality traits.[125,127,129,130]

The biological validity of these higher-order factors is attested by the congruence of genetic research. Typically, twin designs have been used to ask how much of the variance of a trait can be accounted for by genetic variation. The heritabilty of a trait is estimated as twice the difference between monozygotic and dizygotic correlations. Results from a variety of twin studies suggest that it lies between 40 and 60% for each personality trait.[131,132]

The importance for emotion research of the emergence of agreement on personality typology is the evidence that personality type is a risk factor for affective disorders. A number of investigators have argued that high scorers on a neuroticism (N) index are at increased risk of both anxiety and depression[131,133] Neuroticism, like other personality traits, is moderately heritable. Five large twin studies put the figure for neuroticism between 40 to 50% and suggest that the genetic effect is primarily additive, although adoption and family studies give lower estimates for additive genetic variance.[131,134-136] The genes that predispose to neuroticism appear largely to overlap with those that predispose to two common affective disorders, depression and generalized anxiety disorder. We know this from investigation of the phenotypic correlations between N, depression and anxiety, and from twin studies. A number of studies have looked at the course of depressive illness with respect to N scores. In prospective studies, N is the only personality trait consistently associated with later depression (studies are not available for the association with anxiety).[137-140] N is also related to the duration and outcome of severe episodes of major depression.[141,142] Measures of neuroticism in subjects when depressed and on recovery have shown that scores are largely independent of the psychiatric state,[137] and follow-up of hospital inpatients with severe depression also suggests relative stability of scores. The relation between anxiety disorders and neuroticism scores has been less thoroughly investigated, but it is known that high neuroticism scores are associated with an increased risk of the common anxiety disorders.[143]

How much of the phenotypic correlation with anxiety and depression represents environmental or genetic correlation with N? Eaves et al.[131] present evidence that individuals with high N have a high chance of suffering from symptoms of anxiety and depression but that environmental experiences will determine which symptoms are manifest. Their data are backed up by other works. The genetic correlation between generalized anxiety disorder (GAD) and major depression is now attested by a number of studies: Results from both the Virginia and Australian NHMRC twin registers indicate that the same genes act to determine generalized anxiety disorder and major depression.[144,145] The magnitude of comorbidity between anxiety and depression is highest for generalized anxiety disorder (GAD),[146] and multivariate genetic analysis of the Virginia sample estimated the genetic correlation between major depression and GAD to be one. Neuroticism is highly correlated with depression. Analysis of the Virginia twin data showed that around 55% of the genetic liability of major depression was shared with neuroticism.[147] Jardine et al.'s multivariate analysis of 3810 twin pairs led to the conclusion that the genetic basis of neuroticism accounted for 92% of the genetic variation of anxiety and 83% of the genetic variation of depression.[145] Furthermore, the high genetic correlation between GAD and depression, the phenotypic comorbidity between anxiety and depression, and the importance of neuroticism as a genetic risk for depression indicate that neuroticism is very likely to be a genetic mediator of the risk of developing GAD.

Molecular genetic analyses have yet to confirm or disprove the findings from quantitative genetic analysis. However, there are already some intriguing results. Lesch and colleagues report an association between a polymorphism in the serotonin transporter (5HTT) gene regulatory region and N scores.[148] An allelic variant at this locus was shown to reduce transcription of the gene in lymphoblastoid cell lines (note that these cells, found in the peripheral blood, may not be typical of how the gene is expressed in the central nervous system), and individuals with either one or two copies of the variant were found to have high N scores. This is odd, since the drugs that inhibit serotonin reuptake (such as fluoxetine (Prozac)) are antidepressants and anxiolytic, whereas the variant that results in poor transporter function (equivalent therefore to ingesting fluoxetine) is associated with a genetic predisposition to depression. The variant was estimated to constitute 7 to 9% of the genetic variance.[148]

More than just genetics can explain a population association of the kind described here. In order to implicate genetic transmission as the cause of the association, a within-pedigree analysis was carried out, which confirmed the finding — with probability of 0.028 — that the null hypothesis of no association could be rejected. An apparent confirmation of the Lesch paper is to be found in a report that the same 5HTT regulatory region variant is found more frequently in patients with depressive disorder than in controls.[149]

Nevertheless, we should be cautious of regarding these results as robust. Two reports dealing with another personality trait appear to make up a replicated finding of an association with a molecular variant and a personality trait. Again a functional polymorphism in a neurotransmitter receptor gene (in this case the dopamine D4 receptor (D4DR)) has been described and the variant observed to occur more frequently in one personality type (extraversion or novelty seeking behavior) than another. In a study of 124 Isreali subjects and in 315 unrelated individuals from the United States, the D4DR variant accounted for 10% of the genetic variance.[150,151] However, a number of studies have failed to find the association between personality traits and dopamine D4 receptor gene polymorphisms.[152,153] Given that the history of psychiatric genetics is littered with unreplicated findings, it is too early yet to be confident that QTLs for personality have been found.

## CONCLUSION

We have dealt in this chapter with the results of gene knockout experiments and the results of quantitative genetic analyses, top-down and bottom-up approaches, respectively, to investigating genetic influence on emotionality. It is perhaps surprising that single mutations should have such specific effects and should have proved so useful in dissecting the molecular pathways of learning and behavior. The expectation that most knockouts would be lethal has not been fulfilled. Instead, they have been at the center of a revolution in our understanding of the relationship between genes, cellular physiological mechanisms, neuronal function, and behavior. Although the knockouts have been designed to answer questions about learning and memory, they have also helped dissect the molecular basis of conditioned fear, a model system for emotion. We can confidently expect that, in the next few years, the use of knockouts will substantially advance our knowledge of the brain systems mediating fear-conditioning and hence give further clues to the genetic influence on emotion.

One of the most tantalizing questions to have emerged from the success of the gene knockout experiments is to what extent the genes found to affect measures of conditioned fear overlap with those that determine variation in fearfulness in the normal animal (rodent or human). The answer to date is they do not appear to be the same at all. Table 24.1 gives the chromosomal locations in the mouse genome of the genes that, when inactivated, lead to changes in the emotionality of the mouse. None map to the end of chromosome 1, the most important locus so far identified,[109-111,113,114] and indeed none map to any of the regions of the genome found in the gene mapping experiments. In a way, this is an exciting finding. It confirms that the molecular systems we know about are only part of the machinery involved in mediating emotionality. No doubt the coming years will provide plenty of surprises.

## REFERENCES

1. Ekman, P. and Davidson, R. J., *The Nature of Emotion,* Oxford University Press, Oxford, 1994.
2. Rogan, M. T. and LeDoux, J. E., Emotion: systems, cells, synaptic plasticity, *Cell,* 85, 469, 1996.
3. McLean, P. D., Psychosomatic disease and the "visceral brain": recent developments bearing on the Papes theory of emotion, *Psychosom. Med.,* 11, 338, 1949.
4. McLean, P. D., Some psychiatric implications of physiological studies on frontotemporal protein of limbic system (visceral brain), *Electroencephalogr. Clin. Neurophysiol.,* 4, 407, 1952.
5. Thompson, R. F. and Kim, J. J., Memory systems in the brain and localization of a memory, *Proc. Natl. Acad. Sci. U.S.A.,* 93, 13, 438, 1996.
6. Bailey, C. H., Bartsch, D., and Kandel, E. R., Toward a molecular definition of long-term memory storage, *Proc. Natl. Acad. Sci. U.S.A.,* 92, 13, 445, 1996.
7. Baddeley, A., The fractionation of working memory, *Proc. Natl. Acad. Sci. U.S.A.,* 93, 13468, 1996.
8. LeDoux, J. E., In search of an emotional system in the brain: leaping from fear to emotion and consciousness, in *The Cognitive Neurosciences,* Gazzaniga, M. S., Ed., The Mit Press, London, 1995, 1049.
9. Rolls, E. T., A theory of emotion and consciousness, and its application to understanding the neural basis of emotion, in *The Cognitive Neurosciences,* Gazzaniga, M. S., Ed., The Mit Press, London, 1995, 1091.
10. Gray, J. A., A model of the limbic system and basal ganglia: applications to anxiety and schizophrenia, in *The Cognitive Neurosciences,* Gazzaniga, M. S., Ed., The Mit Press, London, 1995, 1165.
11. Davis, M., Animal models of anxiety based on classical conditioning: the conditioned emotional response and fear-potentiated startle effect, in *Psychopharmacology of Anxiolytics and Antidepressants,* File, S. E., Ed., Pergamon Press, 1991, 187.
12. LeDoux, J. E., Emotion: clues from the brain, *Annu. Rev. Psychol.,* 46, 209, 1995.

13. LeDoux, J. E., Emotion and the amygdala, in *The Amygdala: Neurobiological Aspects of Emotion, Memory, and Mental Dysfunction,* Aggleton, J. P., Ed., Wiley-Liss Inc., 1992, 339.

14. Davis, M., The role of the amygdala in conditioned fear, in *The Amygdala: Neurobiological Aspects of Emotion, Memory, and Mental Dysfunction,* Aggleton, J. P., Ed., Wiley-Liss Inc., 1992, 255.

15. Brown, S. and Schaefer, A., An investigation into the functions of the temporal and occipital lobes of the monkey's brain, *Philos. Trans. R. Soc. London [Biol],* 179, 303, 1888.

16. Kluver, H. and Bucy, P. C., Preliminary analysis of functions of the temporal lobes in monkeys, *Arch. Neurol. Psychiatr.,* 42, 979, 1939.

17. Weiskrantz, L., Behavioral changes associated with ablation of the amygdaloid complex in monkeys, *J. Comp. Physiol. Psychol.,* 4, 381, 1956.

18. Geschwind, N., The disconnexion syndromes in animals and man. Part 1, *Brain,* 237, 1965.

19. Rolls, E. T., Neural systems involved in emotion in primates, in *Emotion: Theory, Research, and Experience. Vol 3 Biological Foundations of Emotion,* Plutchnik, R. and Kellerman, H., Eds., Academic Press, New York, 1986, 125.

20. LeDoux, J. E., Emotion, in *Handbook of Physiology, Sec1. The Nervous System. Vol 5, Higher Functions of the Brain,* Part 2., Mouncastle, V. B., Ed., American Physiological Society, Bethesda, MD, 1987, 419.

21. Kapp, B. S., Whalen, P. J., Supple, W. F., and Pascoe, J. P., Amygdaloid contributions to conditioned arousal and sensory information processing, in *The Amygdala: Neurobiological Aspects of Emotion, Memory and Mental Dysfunction,* Aggleton, J. P., Ed., Wiley-Liss Inc., New York, 1992, 229.

22. Fanselow, M. S., Neural organization of the defensive behavior system responsible for fear, *Psychonom. Bull. Rev.,* 1, 429, 1994.

23. Phillips, R. G. and LeDoux, J. E., Differential contribution of amygdala and hippocampus to cued and contextual fear conditioning, *Behav. Neurosci.,* 106, 274, 1992.

24. Kim, J. J. and Fanselow, M. S., Modality-specific retrograde amnesia of fear, *Science,* 256, 675, 1992.

25. Bliss, T. V. and Collingridge, G. L., A synaptic model of memory: long-term potentiation in the hippocampus, *Nature,* 361, 31, 1993.

26. Eichenbaum, H., Spatial learning — the LTP memory connection, *Nature,* 378, 131, 1995.

27. Moser, E., Mathiesen, I., and Andersen, P., Association between brain temperature and dentate field potentials in exploring and swimming rats, *Science,* 259, 1324, 1993.

28. Clugnet, M. C. and LeDoux, J. E., Synaptic plasticity in fear conditioning circuits: induction of LTP in the lateral nucleus of the amygdala by stimualtion of the medial geniculate body, *J. Neurosci.,* 10, 2818, 1990.

29. Gean, P.-W., Chang, F.-C., Huang, C.-C., Lin, J.-H., and Way, L.-J., Long-term enhancement of EPSP and NMDA receptor-mediated synaptic transmission on the amygdala, *Brain Res. Bull.,* 31, 7, 1993.

30. Fanselow, M. S. and Kim, J. J., Acquisition of contextual Pavlovian fear conditioning is blocked by application of an NMDA receptor antagonist D, L-2-amino-5-phosphono-valeric acid to the basolateral amygdala, *Behav. Neurosci.,* 102, 233, 1994.

31. Miserendino, M. J. D., Sananes, C. B., Melia, K. R., and Davis, M., Blocking of acquistion but not expression of conditioned fear-potentiated startle by NMDA antagonists in the amygdala, *Nature,* 345, 1990.

32. Kim, M. and Davis, M., Lack of a temporal gradient of retrograde amnesia in rats with amygdala lesions assessed with the fear-potentiated startle paradigm, *Behav. Neurosci.,* 107, 1088, 1993.

33. Rogan, M., Staubli, U., and LeDoux, J. E., Facilitation of AMPA receptors accelerates classical fear conditioning in rats, *Soc. Neurosci. Abstr.,* 20, 1007, 1994.

34. Rogan, M. T. and LeDoux, J. E., LTP is accompanied by commensurate enhancement of auditory-evoked responses in a fear conditioning circuit, *Neuron,* 15, 127, 1995.

35. Rogan, M. T., Staubli, U. V., and LeDoux, J. E., Fear conditioning induces associative long-term potentiation in the amygdala, *Nature,* 390, 604, 1997.

36. McKernan, M. G. and ShinnickGallagher, P., Fear conditioning induces a lasting potentiation of synaptic currents in vitro, *Nature,* 390, 607, 1997.

37. Goodman, R. R., Snyder, S. H., Kuhar, M. J., and Young, W. S., III., Differential delta and mu opiate receptor localizations by light microscopic autoradiography, *Proc. Natl. Acad. Sci. U.S.A.,* 77, 2167, 1980.

38. Gallagher, M., Kapp, B. S., McNall, C. L., and Pascoe, J. P., Opiate effects in the amygdala central nucleus on heart rate conditioning in rabbits, *Pharmacol. Biochem. Behav.*, 14, 497, 1981.

39. Gallagher, M., Kapp, B. S., and Pascoe, J. P., Enkephalin analogue effects in the amygdala central nucleus on conditioned heart rate, *Pharmacol. Biochem. Behav.*, 17, 217, 1982.

40. Gray, T. S., Autonomic neuropeptide connections of the amygdala, in *Neuropeptides and Stress,* Tache, Y., Morley, J. E., and Brown, M. R., Eds., Springer Verlag, New York, 1989, 92.

41. Gray, T. S., The organization and possible function of amygdaloid corticotropin-releasing factor pathways, in *Corticotropin-Releasing Factor: Basic and Clinical Studies of a Neuropeptide,* DeSouza, E. B. and Nemeroff, C. B., Eds., CRC Press LLC, Boca Raton, FL, 1990, 53.

42. Gray, J. A., *The Psychology of Fear and Stress,* Cambridge University Press, Cambridge, 1987.

43. Gray, J. A., *The Neuropsychology of Anxiety: an Enquiry into the Function of the Septo-hippocampal System,* Oxford University Press, Oxford, 1982.

44. Morgan, M. A., Romanski, L. M., and LeDoux, J. E., Extinction of emotional learning:contribution of medial prefrontal cortex, *Neurosci. Lett.,* 163, 109, 1993.

45. Checkley, S., The Neuroendocrinology of Depression and Chronic Stress, *Br. Med. Bull.,* 52, 597, 1996.

46. Koob, G. F., Heinrichs, S. C., Pich, E. M., Menzaghi, F., Baldwin, H., Miczek, K., and Britton, K. T., The role of corticotropin-releasing factor in behavioral responses to stress, *Ciba Found. Symp.,* 172, 277, 1993.

47. Heinrichs, S. C., Menzaghi, F., Pich, E. M., Britton, K. T., and Koob, G. F., The role of crf in behavioral aspects of stress, *Ann. N. Y. Acad. Sci.,* 771, 92, 1995.

48. Heilig, M., Koob, G. F., Ekman, R., and Britton, K. T., Corticotropin-releasing factor and neuropeptide-Y — role in emotional integration, *Trends Neurosci.,* 17, 80, 1994.

49. Vale, W., Spiess, J., Rivier, C., and Rivier, J., Characterization of a 41-residue ovine hypothalamic peptide that stimulates secretion of corticotropin and beta-endorphin, *Science*, 213, 1394, 1981.

50. Owens, M. J. and Nemeroff, C. B., Physiology and pharmacology of corticotropin releasing factor, *Pharmacolog. Rev.,* 43, 425, 1991.

51. Swerdlow, N. R., Geyer, M. A., Vale, W., and Koob, G. F., Corticotropin releasing factor potentiates acoustic startle in rats — blocked by chlordiazepoxide, *Psychopharmacology*, 88, 147, 1986.

52. Baldwin, H. A., Rassnick, S., Rivier, J., Koob, G. F., and Britton, K. T., Crf antagonist reverses the anxiogenic response to ethanol withdrawal in the rat, *Psychopharmacology*, 103, 227, 1991.

53. Pellow, S., Chopin, P., File, S., and Briley, M., Validation of open:closed arms entries in an elevated plus maze as a measure of anxiety in the rat, *J. Neurosci. Methods,* 14, 149, 1985.

54. Heinrichs, S. C., Pich, E. M., Miczek, K. A., Britton, K. T., and Koob, G. F., Corticotropin-releasing factor antagonist reduces emotionality in socially defeated rats via direct neurotropic action, *Brain Res.,* 581, 190, 1992.

55. Liang, K. C., Melia, K. R., Campeau, S., Falls, W. A., Miserendino, M. J. D., and Davis, M., Lesions of the central nucleus of the amygdala, but not the paraventricular nucleus of the hypothalamus, block the excitatory effects of corticotropin-releasing factor on the acoustic startle reflex, *J. Neurosci.,* 12, 2313, 1992.

56. Minth, C. D., Bloom, S. R., Polak, J. M., and Dixon, J. E., Cloning, characterization, and DNA-sequence of a human CDNA-encoding neuropeptide tyrosine, *Proc. Natl. Acad. Sci. U.S.A. — Biol. Sci.*, 81, 4577, 1984.

57. Heilig, M., McLeod, S., Brot, M., Heinrichs, S. C., Menzaghi, F., Koob, G. F., and Britton, K. T., Anxiolytic-like action of neuropeptide-Y — mediation by Y1 receptors in amygdala, and dissociation from food-intake effects, *Neuropsychopharmacology*, 8, 357, 1993.

58. Wahlestedt, C., Pich, E. M., Koob, G. F., Yee, F., and Heilig, M., Modulation of anxiety and neu-ropeptide-Y-Y1 receptors by antisense oligodeoxynucleotides, *Science*, 259, 528, 1993.

59. Bechara, A., Tranel, D., Damasio, H., Adolphs, R., Rockland, C., and Damasio, A. R., Double dissociation of conditioning and declarative knowledge relative to the amygdala and hippocampus in humans, *Science*, 269, 1115, 1995.

60. LaBar, K. S., Phelps, E. A., and LeDoux, J. E., Fear conditioning following unilateral temporal lobectomy in humans: impaired conditional discrimination, *Soc. Neurosci. Absr.,* 20, 360, 1994.

61. Adolphs, R., Tranel, D., Damasio, H., and Damasio, A., Impaired recognition of emotion in facial expressions following bilateral damage to the human amygdala, *Nature*, 372, 669, 1994.

62. Hamann, S. B., Stefanacci, L., and Squire, L. R., Recognizing facial emotion, *Nature*, 379, 497, 1996.

63. Scott, S. K., Young, A. W., Calder, A. J., Hellawell, D. J., Aggleton, J. P., and Johnson, M., Impaired auditory recognition of fear and anger following bilateral amygdala lesions, *Nature*, 385, 254, 1997.

64. Cahill, L., Babinsky, R., Marowitsch, H. J., and McGaugh, J. L., The amygdala and emotional memory, *Science*, 377, 295, 1995.

65. Morris, J. S., Frith, C. D., Perrett, D. I., Rowland, D., Young, A. W., Calder, A. J., and Dolan, R. J., A differential neural response in the human amygdala to fearful and happy facial expressions, *Nature*, 383, 812, 1996.

66. Adolphs, R., Tranel, D., and Damasio, A. R., The human amygdala in social judgement, *Nature*, 393, 470, 1998.

67. Morris, J. S., Ohman, A., and Dolan, R. J., Conscious and unconscious emotional learning in the human amygdala, *Nature*, 393, 467, 1998.

68. Damasio, A. R., *Descartes' Error: Emotion, Reason and the Human Brain,* Grosset/Putman, New York, 1994.

69. Mayford, M., Abel, T., and Kandel, E. R., Transgenic approaches to cognition, *Curr. Opin. Neurobiol.,* 5, 141, 1995.

70. Tonegawa, S., Mammalian learning and memory studied by gene targeting, *Ann. N. Y. Acad. Sci.,* 758, 213, 1995.

71. Abel, T., Martin, K. C., Bartsch, D., and Kandel, E. R., Memory supressor genes: inhibitory constraints on the storage of long-term memory, *Science*, 279, 338, 1998.

72. Tsien, J. Z., Chen, D. F., Gerber, D., Tom, C., Mercer, E. H., Anderson, D. J., Mayford, M., Kandel, E. R., and Tonegawa, S., Subregion- and cell type-restricted gene knockout in mouse brain, *Cell*, 87, 1317, 1996.

73. Mayford, M., Bach, M. E., Huang, Y. Y., Wang, L., Hawkins, R. D., and Kandel, E. R., Control of memory formation through regulated expression of α CaMKII transgene, *Science*, 274, 1678, 1996.

74. Threadgill, D. W., Dlugosz, A. A., Hansen, L. A., Tennbaum, T., Lichti, U., Yee, D., Lamantia, C., Mourton, T., Herrup, K., Harris, R. C., Barnard, J. A., Yuspa, S. H., Coffey, R. J., and Magnuson, T., Targeted disruption of mouse Egf receptor — effect of genetic background on mutant phenotype, *Science*, 269, 230, 1995.

75. Upchurch, M. and Wehner, J., Differences between inbred strains of mice in the Morris water maze performance, *Behav. Genet.,* 18, 55, 1988.

76. Schwegler, H. and Lipp, H.-P., Hereditary covariations of neuronal circuitry and behavior: correlations between the proportions of hippocampal synaptic fields in the regio inferior and two-way avoidance in mice and rats, *Behav. Brain Res.,* 7, 1, 1983.

77. Abeliovich, A., Paylor, R., Chen, C., Kim, J. J., Wehner, J. M., and Tonegawa, S., PKCγ mutant mice exhibit mild deficits in spatial and contextual learning, *Cell*, 75, 1263, 1993.

78. Abeliovich, A., Chen, C., Goda, Y., Silva, A. J., Stevens, C. F., and Tonegawa, S., Modified hippocampal long-term potentiation in PKCγ-mutant mice, *Cell*, 75, 1253, 1993.

79. Aiba, A., Kano, M., Chen, C., Stanton, M. E., Fox, G. D., Herrup, K., Zwingman, T. A., and Tonegawa, S., Deficient cerebellar long-term depression and impaired motor learning in mGluR1 mutant mice, *Cell*, 79, 377, 1994.

80. Aiba, A., Chen, C., Herrup, K., Rosenmund, C., Stevens, C. F., and Tonegawa, S., Reduced hippocampal long-term potentiation and context-specific deficit in associative learning in mGluR1 mutant mice, *Cell*, 79, 365, 1994.

81. Huang, Y. Y., Kandel, E. R., Varshavsky, L., Brandon, E. P., Qi, M., Idzerda, R. L., McKnight, G. S., and Bourtchouladze, R., A genetic test of the effects of mutations in PKA on mossy fiber LTP and its relation to spatial and contextual learning, *Cell*, 83, 1211, 1995.

82. Treves, A. and Rolls, E. T., Computational analysis of the role of the hippocampus in memory, *Hippocampus*, 4, 374, 1994.

83. Silva, A. J., Stevens, C. F., Tonegawa, S., and Wang, Y., Deficient hippocampal long-term potentiation in α-calcium-calmodulin kinase II mutant mice, *Science*, 257, 201, 1992.

84. Silva, A. J., Paylor, R., Wehner, J. M., and Tonegawa, S., Impaired spatial learning in α-calcium-calmodulin kinase II mutant mice, *Science*, 257, 206, 1992.

85. Mayford, M., Wang, J., Kandel, E. R., and O'Dell, T. J., CaMKII regulates the frequency-response function of hippocampal synapses for the production of both LTD and LTP, *Cell*, 81, 891, 1995.

86. Chen, C., Rainnie, D. G., Greene, R. W., and Tonegawa, S., Abnormal fear response and aggressive behavior in mutant mice deficient for $\alpha$-calcium-calmodulin kinase II, *Science*, 266, 291, 1994.

87. Timpl, P., Spanagel, R. S., I., Kresse, A., Reul, J. M. H. M., Stalla, G. K., Blanquet, V., Steckler, T., Holsboer, F., and Wurst, W., Impaired stress response and reduced anxiety in mice lacking a functional corticotropin-releasing hormone receptor 1, *Nat. Genet.*, 19, 162, 1998.

88. Stenzelpoore, M. P., Heinrichs, S. C., Rivest, S., Koob, G. F., and Vale, W. W., Overproduction of corticotropin-releasing factor in transgenic mice — a genetic model of anxiogenic behavior, *J. Neurosci.*, 14, 2579, 1994.

89. Heyser, C. J., Masliah, E., Samini, A., Campbell, I. L., and Gold, L. H., Progressive decline in avoidance learning paralleled by inflammatory neurodegeneration in transgenic mice expressing interleukin 6 in the brain, *Proc. Natl. Acad. Sci. U.S.A.*, 94, 1500, 1997.

90. König, M., Zimmer, A. M., Steiner, H., Holmes, P. V., Crawley, J. N., Brownstein, M. J., and Zimmer, A., Pain responses, anxiety and aggression in mice deficient in pre-proenkephalin, *Nature*, 383, 535, 1996.

91. Inoue, I., Yanai, K., Kitamura, D., Taniuchi, I., Kobayashi, T., Niimura, K., Watanabe, T., and Watanabe, T., Impaired locomotor activity and exploratory behavior in mice lacking histamine $H_1$ receptors, *Proc. Natl. Acad. Sci. U.S.A.*, 93, 13316, 1996.

92. Picciotto, M. R., Zoli, M., Léna, C., Bessis, A., Lallemand, Y., LeNovere, N., Vincent, P., Pich, E. M., Brûlet, P., and Changeux, J.-P., Abnormal avoidance learning in mice lacking functional high-affinity nicotine receptor in the brain, *Nature*, 374, 65, 1995.

93. Yagi, T., Aizawa, S., Tokunaga, T., Shigetani, Y., Takeda, N., and Ikawa, Y., A role for Fyn tyrosine kinase in the suckling behavior of neonatal mice, *Nature*, 366, 742, 1993.

94. Lijam, N., Paylor, R., McDonald, M. P., Crawley, J. N., Deng, C.-X., Herup, K., Stevens, K. E., Maccaferri, G., McBain, C. J., Sussman, D. J., and Wynshaw-Boris, A., Social interaction and sensorimotor gating abnormalities in mice lacking Dvl1, *Cell*, 90, 895, 1997.

95. Davis, R. L. and Dauwalder, B., The *Drosophila dunce* locus: learning and memory genes in the fly, *Trends Genet.*, 7, 224, 1991.

96. Broadhurst, P. L., Application of biometrical genetics to the inheritance of behavior, in *Experiments in Personality*, Eysenck, H. J., Ed., Routledge and Kegan Paul, London, 1960, 3.

97. Broadhurst, P. L., A note on further progress in a psychogenetic selection experiment, *Psychol. Rep.*, 10, 65, 1962.

98. DeFries, J. C., Gervais, M. C., and Thomas, E. A., Response to 30 generations of selection for open field activity in laboratory mice, *Behav. Genet.*, 8, 3, 1978.

99. DeFries, J. C. and Hegman, J. P., Genetic analysis of open-field behavior, in *Contributions to Behavior Genetic Analysis: The Mouse as a Prototype*, Lindzey, G. and Thiessen, D. D., Eds., Appleton-Century-Crofts, New York, 1970, 23.

100. Broadhurst, P. L. and Bignami, G., Correlative effect of psychogenetic selection: a study of the Roman high and low avoidance strains of rats., *Behav. Res. Therapy*, 2, 273, 1965.

101. Bignami, G., Selection for high rates and low rates of avoidance conditioning in the rat, *Behav. Res. Therapeutics*, 2, 273, 1965.

102. Hall, C. S., Emotional behavior in the rat. I Defecation and urination as measures of individual differences in emotionality, *J. Comp. Psychol.*, 22, 345, 1934.

103. Driscoll, P., Woodson, P., Fuem, H., and Battig, K., Selection for two-way avoidance deficit inhibits shock-induced fighting in the rat, *Physiol. Behav.*, 24, 793, 1980.

104. Chamove, A. S. and Sanders, D. C., Emotional correlates of selection for avoidance learning in rats, *Biolog. Psychol.*, 10, 41, 1980.

105. Imada, H., Emotional reactivity and conditionability in four strains of rats, *J. Comp. Physiolog. Psychol.*, 79, 474, 1972.

106. Gentsch, C., Lichtsteiner, M., Driscoll, P., and Feer, H., Differential hormonal and physiological responses to stress in Roman high- and low-avoidance rats, *Physiol. Behav.*, 28, 259, 1982.

107. Durcan, M. J., Wraight, K. B., and Fulker, D. W., The current status of two sublines of the Roman high and low avoidance strains, *Behav. Genet.*, 14, 6, 1984.

108. Pauls, D. L., Behavioral disorders: lessons in linkage, *Nat. Genet.*, 3, 4, 1993.

109. Flint, J., Corley, R., DeFries, J. C., Fulker, D. W., Gray, J. A., Miller, S., and Collins, A. C., A simple genetic basis for a complex psychological trait in laboratory mice, *Science*, 269, 1432, 1995.

110. Gershenfeld, H. K., Neumann, P. E., Mathis, C., Crawley, J. N., Li, X. H., and Paul, S. M., Mapping quantitative trait loci for open-field behavior in mice, *Behav. Genet.*, 27, 201, 1997.

111. Gershenfeld, H. K. and Paul, S. M., Mapping quantitative trait loci for fear-like behaviors in mice, *Genomics*, 46, 1, 1997.

112. Owen, E. H., Christensen, S. C., Paylor, R., and Wehner, J. M., Identification of quantitative trait loci involved in contextual and auditory-cued fear conditioning in BXD recombinant inbred strains, *Behav.Neurosci.*, 111, 1, 1997.

113. Wehner, J. M., Radcliffe, R. A., Rosmann, S. T., Christensen, S. C., Rasmussen, D. L., Fulker, D. W., and Wiles, M., Quantitative trait locus analysis of contextual fear conditioning in mice, *Nat. Genet.*, 17, 331, 1997.

114. Caldarone, B., Saavedra, C., Tartaglia, K., Wehner, J. M., Dudek, B. C., and Flaherty, L., Quantitative trait loci analysis affecting contextual conditioning in mice, *Nat. Genet.*, 17, 335, 1997.

115. Moisan, M. P., Courvoisier, H., Bihoreau, M. T., Gauguier, D., Hendley, E. D., Lathrop, M., James, M. R., and Mormede, P., A major quantitative trait locus influences hyperactivity in the Wkha rat, *Nat. Genet.*, 14, 471, 1996.

116. Paterson, A. H., Lin, Y.-R., Li, Z., Schertz, K. F., Doebly, J. F., Pins, S. R. M., Liu, S.-C., Stansel, J. W., and Irvine, J. E., Convergent domestication of cereal crops by indepedent mutations at corresponding genetic loci, *Science*, 269, 1714, 1995.

117. Melo, J. A., Shendure, J., Pociask, K., and Silver, L. M., Identification of sex-specific quantitative trait loci controlling alcohol preference in C57BL/6 mice, *Nat. Genet.*, 13, 147, 1996.

118. Berrettini, W. H., Ferraro, T. N., Alexander, R. C., Buchberg, A. M., and Vogel, W. H., Quantitative trait loci mapping of three loci controlling morphine preference using inbred mouse strains, *Nat. Genet.*, 7, 54, 1994.

119. Allport, G. W. and Odbert, H. S., Trait-names: a psycholexical study, *Psycholog. Monogr.*, 47, 1936.

120. Eysenck, H. J., The logical basis of factor analysis, *Am. Psychol.*, 8, 105, 1953.

121. Cloninger, C. R., Svrakic, D. M., and Przybeck, T. R., A psychobiological model of temperament and character, *Arch. Gen. Psychiatr.*, 50, 975, 1993.

122. Catell, R. B., Eber, H. W. and Tatsuoka, M. M., *Handbook for the Sixteen Personality Factor Questionnaire (16PF)*, Institute for Personality and Ability Testing, Champaign, IL, 1980.

123. Eysenck, H. J. and Eysenck, S. B. G., *Manual of the Eysenck Personality Questionnaire*, Educational and Industrial Testing Service, San Diego, CA, 1975.

124. John, O. P., The "Big Five" taxonomy: dimensions of personality ih the natural language and in questionnaires, in *Handbook of Personality: Theory and Research*, Pervin, L. A., Ed., Guildford Press, New York, 1990,

125. Watson, D., Clark, L. A., and Harkness, A. R., Structures of personality and their relevance to psychopathology, *J. Abnorm. Psychol.*, 103, 18, 1994.

126. Digman, J. M., Personality structure: emergence of the five factor model, *Annu. Rev. Psychol.*, 41, 417, 1990.

127. Deary, I. J. and Matthews, G., Personality traits are alive and well, *The Psychologist*, 299, 1993.

128. Costa, P. T. and McCrae, R. R., *Revised NEO Personality Inventory (NEO-PI-R) and NEO Five-Factor Inventory (NEO-FFI) professional manual*, Psychological Assessment Resources Inc., Odessa, FL, 1992.

129. Zuckerman, M., Kuhlman, D. M., and Camac, C., What lies beyond E and N? Factor analysis of scales believed to measure basic dimensions of personality, *J. Personal. Soc. Psychol.*, 54, 96, 1988.

130. Cloninger, C. R., Temperament and personality, *Curr. Opin. Neurobiol.*, 4, 266, 1994.

131. Eaves, L. J., Eysenck, H. J., and Martin, N. G., *Genes, Culture and Personality: An Empirical Approach*, Academic Press, Harcourt Brace Jovanovich, London, 1989.

132. Loehlin, J. C., *Genes and Environment in Personality Development*, 2, Sage Publications, London, 1992.

133. Clark, L. A., Watson, D., and Mineka, S., Temperament, Personality, and the Mood and Anxiety Disorders, *J. Abnorm. Psychol.*, 103, 103, 1994.

134. Floderus-Myrhed, B., Pedersen, N., and Rasmuson, I., Assessment of heritability for personality, based on a short form of the Eysenck Personality Inventory, *Behav. Genet.*, 10, 153, 1980.

135. Loehlin, J. C. and Nichols, R. C., *Heredity, Environment and Personality*, University of Texas Press, Austin, TX, 1976.

136. Rose, R. J., Koskenvuo, M., Kaprio, J., Sarna, S., and Langinvainio, H., Shared genes, shared experiences and similarity of personality: data from 14, 288 adult Finnish co-twins, *J. Personal. Soc. Psychol.,* 54, 161, 1988.

137. Clayton, P. J., Ernst, C., and Angst, J., Premorbid personality traits of men who develop unipolar or bipolar disorders, *Euro. Arch. Psychiatr. Clin. Neurosci.,* 23, 340, 1994.

138. Angst, J. and Clayton, P., Premorbid personality of depressive, bipolar and schizophrenic patients with special reference to suicidal issues, *Compr. Psychiatr.,* 27, 511, 1986.

139. Hirschfeld, R. M. A., Klerman, G. L., Lavori, P., Keller, M. B., Griffith, P., and Coryell, W., Premorbid personality assessments of first onset of major depression, *Arch. Gen. Psychiatr.,* 46, 345, 1989.

140. Lewinsohn, P., Steinmetz, J. L., Larson, D. W., and Franklin, J., Depression-related cognitions: antecedent or consequence?, *J. Abnorm. Psychol.,* 90, 213, 1981.

141. Scott, J., Eccleston, D., and Boys, R., Can we predict the persistence of depression?, *Br. J. Psychiatr.,* 161, 633, 1992.

142. Duggan, C. F., Lee, A. S., and Murray, R. M., Does personality predict long-term outcome in depression?, *Br. J. Psychiatr.,* 157, 19, 1990.

143. Andrews, G., Stewart, G., Morris-Yates, A., Holt, P., and Henderson, S., Evidence for a general neurotic syndrome, *Br. J. Psychiatr.,* 157, 6, 1990.

144. Kendler, K. S., Major depression and generalised anxiety. Same genes, (partly) different environments — revisited, *Br. J. Psychiatr.,* 168, 68, 1996.

145. Jardine, R., Martin, N. G., and Henderson, A. S., Genetic covariation between neuroticism and the symptoms of anxiety and depression, *Genet. Epidemiol.,* 1, 89, 1984.

146. Kendler, K. S., Neale, M. C., Kessler, R. C., Heath, A. C., and Eaves, L. J., Major depression and phobias: the genetic and environmental sources of comorbidity, *Psycholog. Med.,* 23, 361, 1993.

147. Kendler, K. S., Neale, M. C., Kessler, R. C., Heath, A. C., and Eaves, L. J., A longitudinal twin study of personality and major depression in women, *Arch. Gen. Psychiatr.,* 50, 853, 1993.

148. Lesch, K.-P., Bengel, D., Heils, A., Sabol, S. Z., Greenberg, B. D., Petri, S., Benjamin, J., Muller, C. R., Hamer, D. H., and Murphy, D. L., Association of anxiety related tratis with a polymorphism in the serotonin transporter gene regulatory region, *Science,* 274, 1527, 1996.

149. Ogilvie, A. D., Battersby, S., Bubb, V. J., Fink, G., Harmar, A. J., Goodwin, G. M., and Smith, C. A. D., A polymorphism in the serotonin transporter gene is associated with susceptibility to major depression, *Lancet,* 347, 731, 1996.

150. Ebstein, R. P., Novick, O., Umansky, R., Priel, B., Osher, Y., Blaine, D., Bennett, E. R., Nemanov, L., Katz, M., and Belmaker, R. H., Dopamine D4 receptor (*D4DR*) exon III polymorphism associated with the human personality trait of novelty seeking, *Nat. Genet.,* 12, 78, 1996.

151. Benjamin, J., Li, L., Patterson, C., Greenberg, B. D., Murphy, D. L., and Hamer, D. H., Population and familial association between the D4 dopamine receptor gene and measures of Novelty Seeking, *Nat. Genet.,* 12, 81, 1996.

152. Jönsson, E. G., Nöthen, M. M., Gustavsson, J. P., Heidt, H., Brené, S., Tylec, A., Propping, P., and Sedvall, G. C., Lack of evidence for allelic association between personality traits and the dopamine $D_4$ receptor gene polymorphisms, *Am. J. Psychiatr.,* 154, 697, 1997.

153. Malhotra, A. K., Virkkunen, M., Rooney, W., Eggert, M., Linnoila, M., and Goldman, D., The association between the dopamine D-4 receptor (D4dr) 16 amino-acid repeat polymorphism and novelty seeking, *Molec. Psychiatr.,* 1, 388, 1996.

154. Brandon, E. P., Zhuo, M., Huang, Y.-Y., Qi, M., Gerhold, K. A., Burton, K. A., Kandel, E. R., McKnight, G. S., and Idezerda, R. L., Hippocampal long-term depression and depotentiation are defective in mice carrying a targeted disruption of the gene encoding the RIβ subunit of cAMP-dependent protein kinase, *Proc. Natl. Acad. Sci. U.S.A.,* 92, 8851, 1995.

155. Abel, T., Nguyen, P. V., Barad, M., Deuel, T. A. S., Kandel, E. R., and Bourtchouladze, R., Genetic demonstration of a role for PKA in the late phase of LTP and in hippocampus-based long-term memory, *Cell,* 88, 615, 1997.

156. Harris, R. A., McQuilkin, S. J., Paylor, R., Abeliovich, A., Tonegawa, S., and Wehner, J. M., Mutant mice lacking the γ isoform of protein kinase C show decresed behavioral actions of ethanol and altered function of γ-aminobutyrate type A receptors, *Proc. Natl. Acad. Sci. U.S.A.,* 92, 3658, 1995.

157. Chen, C., Kano, M., Abelivochi, A., Chen, L., Bao, S., Kim, J. J., Hashimoto, K., Thompson, R. F., and Tonegawa, S., Impaired motor coordination correlates with persistent multiple climbing fiber innervation in PKCγ mutant mice, *Cell*, 83, 1233, 1995.

158. Grant, S. G. N., O'Dell, T. J., Karl, K. A., Stein, P. L., Soriano, P., and Kandel, E. R., Impaired long-term potentiation, spatial learning and hippocampal development in fyn mutant mice, *Science*, 258, 1903, 1992.

159. Korte, M., Carroll, P., Wolf, E., Brem, G., Thoenen, H., and Bonhoeffer, T., Hippocampal long-term potentiation is impaired in mice lacking brain-derived neurotrophic factor, *Proc. Natl. Acad. Sci. U.S.A.*, 92, 8856, 1995.

160. Sakimura, K., Kutsuwade, T., Ito, I., Manabe, T., Takayama, C., Kushiya, E., Tagi, T., Aizawa, S., Inoue, Y., Suglyama, H., and Mishina, M., Reduced hippocampal LTP and spatial learning NMDA receptor ε1 subunit, *Nature*, 373, 151, 1995.

161. Li, Y., Erzurumlu, R. S., Chen, C., Jhaveri, S., and Tonegawa, S., Whisker-related neuronal patterns fail to develop in the trigeminal brainstem nuclei of NMDAR1 knockout mice, *Cell*, 76, 427, 1994.

162. Tsien, J. Z., Huerta, P. T., and Tonegawa, S., The essential role of hippocampal CA1 NMDA receptor-dependent synaptic plasticity in spatial memory, *Cell*, 87, 1327, 1996.

163. McHugh, T. J., Blum, K. I., Tsien, J. Z., Tonegawa, S., and Wilson, M. A., Impaired hippocampal representation of space in CA1-specific NMDAR1 knockout mice, *Cell*, 87, 1339, 1996.

164. Conquet, F., Bashir, Z. I., Davies, C. H., Daniel, H., Ferraguti, F., Bordi, F., Franz-Bacon, K., Reggiani, A., Matarese, V., Condé, F., Colingridge, G. L., and Crépel, F., Motor deficient and impairment of synaptic plasticity in mice lacking mGluR1, *Nature*, 372, 237, 1994.

165. Kashiwabuchi, N., Ikeda, K., Araki, K., Hirano, T., Shibuki, K., Takayama, C., Inoue, Y., Kutsuwada, T., Yagi, T., Kang, Y., Aizawa, S., and Mishina, M., Impairment of motor coordination, purkinje cell synapse formation, and cerebellar long-term depression in GluRδ2 mutant mice, *Cell*, 81, 245, 1995.

166. Wu, Z. L., Thomas, S. A., Villacres, E. C., Xia, Z., Simmons, M. L., Chavkin, C., Palmiter, R. D., and Storm, D. R., Altered behavior and long-term potentiation in type I adenylyl cyclase mutant mice, *Proc. Natl. Acad. Sci. U.S.A.*, 92, 220, 1995.

167. Rotenberg, A., Mayford, M., Hawkins, R. D., Kandel, E. R., and Muller, R. U., Mice expressing activated CaMKII lack low frequency LTP and do not form stable place cells in the CA1 region of the hippocampus, *Cell*, 87, 1351, 1996.

168. Bach, E. M., Hawkins, R. D., Osman, M., Kandel, E. R., and Mayford, M., Impairment of spatial but not contextual memory in CaMKII mutant mice with a selective loss of hippocampal LTP in the range of the θ frequency, *Cell*, 81, 905, 1995.

169. Ogawa, S., Lubahn, D. B., Korach, K., and Rfaff, D. W., Behavioral effects of estrogen receptor gene disruption in male mice, *Proc. Natl. Acad. Sci. U.S.A.*, 94, 1476, 1997.

170. Cases, O., Seif, I., Grimsby, J., Gaspar, P., Chen, K., Pournin, S., Muuler, U., Agyet, M., Babinet, C., Shih, J. C., and De Maeyer, E., Aggressive behavior and altered amounts of brain serotonin and norepinephrine in mice lacking MAOA, *Science*, 268, 1763, 1995.

171. Sora, I., Takahaski, N., Funada, M., Ujike, H., Revay, R. S., Donovan, D. M., Miner, L. L., and Uhl, G. R., Opiate receptor knockout mice define μ receptor roles in endogenous nociceptive responses and morphine-induced analgesia, *Proc. Natl. Acad. Sci. U.S.A.*, 94, 1544, 1997.

172. Ho, C. S., Grange, R. W., and Joho, R. H., Pleiotropic effects of a disrupted $K^+$ channel gene: reduced body weight, impaired motor skill and muscle contraction, but no seizures, *Proc. Natl. Acad. Sci. U.S.A.*, 94, 1533, 1997.

173. Gunther, U., Benson, J., Benke, D., Fritschy, J. M., Reyes, G., Knoflach, F., Crestani, F., Aguzzi, A., Arigoni, M., Lang, Y., Bluethmann, H., Mohler, H., and Luscher, B., Benzodiazepine-insensitive mice generated by targeted disruption of the gamma2 subunit gene of gamma-aminobutyric acid type A receptors, *Proc. Natl. Acad. Sci. U.S.A.*, 92, 7749, 1995.

174. Konig, M., Zimmer, A. M., Steiner, H., Holmes, P. V., Crawley, J. N., Brownstein, M. J., and Zimmer, A., Pain responses, anxiety and aggression in mice deficient in pre-proenkephalin, *Nature*, 383, 535, 1996.

175. Saudou, F., Amara, D. A., Dierich, A., LeMeur, M., Ramboz, S., Segu, L., Buhot, M.-C., and Hen, R., Enhanced aggressive behavior in mice lacking 5-HT$_{1B}$ receptor, *Science*, 265, 1875, 1994.

176. Giros, B., Jaber, M., Jones, S. R., Wightman, R. M., and Caron, M. G., Hyperlocomotion and indifference to cocaine and amphetamine in mice lacking the dopamine transporter, *Nature*, 379, 606, 1996.

177. Drago, J., Geren, C. R., Lachowicz, J. E., Steiner, H., Hollon, T. R., Love, P. E., Ooi, G. T., Grinberg, A., Lee, E. J., Huang, S. P., Bartlett, P. F., Jose, P. A., Sibley, D. R., and Westphal, H., Altered striatal function in a mutant mouse lacking $D_{1A}$ dopamine receptors, *Proc. Natl. Acad. Sci. U.S.A.*, 91, 12564, 1994.

178. Xu, M., Moratalla, R., Gold, L. H., Hiroi, N., Koob, G. F., Graybiel, A. M., and Tonegawa, S., Dopamine D1 receptor mutant mice are deficient in striatal expression of dynorphin and in dopamine-mediated behavioral responses, *Cell*, 79, 729, 1994.

179. Xu, M., Hu, X.-T., Cooper, D. C., Mortalla, R., Graybiel, A. M., White, F. J., and Tonegawa, S., Elimination of cocaine-induced hyperactivity and dopamine-mediated neurophysiological effects in dopamine DI receptor mutant mice, *Cell*, 79, 945, 1994.

180. Nelson, R. J., Demas, G. E., Huang, P. L., Fishman, M. C., Dawson, V. L., Dawson, T. M., and Snyder, S. H., Behavioral abnormalities in male mice lacking neuronal nitric oxide synthase, *Nature*, 378, 383, 1995.

181. Zhou, Q.-Y. and Palmiter, R. D., Dopamine-deficient mice are severely hypoactive, adipsic, and aphagic, *Cell*, 83, 1197, 1995.

182. Cremer, H., Lange, R., Christoph, A., Plomann, M., Vopper, G., Roes, J., Brown, R., Baldwin, S., Kraemer, P., Scheff, S., Barthels, D., Rajewsky, K., and Wille, W., Inactivation of the N-CAM gene in mice results in size reduction of the olfactory bulb and deficits in spatial learning, *Nature*, 367, 455, 1994.

183. Bourtchuladze, R., Frenguelli, B., Blendy, J., Cioffi, D., Schutz, G., and Silva, A. J., Deficient long-term memory in mice with a targeted mutation of the cAMP-responsive element-binding protein, *Cell*, 79, 59, 1994.

# 25 Approaches to Studying the Genetic Regulation of Learning and Memory

*Sheree F. Logue, Elizabeth H. Owen, and Jeanne M. Wehner*

## CONTENTS

## INTRODUCTION

Dissection of genetic regulation and identification of candidate genes involved in learning mechanisms have been facilitated by the development of a number of techniques in the recent past. Although much work has been done in this field with *Drosophila* and other rodent models, this chapter will focus on learning work with the mammalian mouse model. The multiple behavioral tasks that can be used to evaluate learning in the mouse fall into two major categories: simple and complex. Simple learning, such as active or passive avoidance, requires the use of a single piece of information to provide a correct response. Complex learning, such as spatial or contextual learning, requires the integration of multiple pieces of information. Biologically, the distinction between these two types of learning has been described as differences in the degree of hippocampal dependence. Complex learning tasks require the hippocampus, while simple learning can be successfully completed following hippocampal lesions.[1-3] While focusing on both categories of learning, this chapter will illustrate the multiple ways in which learning and its central nervous system mechanisms are being studied. Each paradigm and method of evaluation can yield important information and should be used in conjunction with one another to provide a complete picture.

## BEHAVIORAL PARADIGMS

Both passive and active avoidance are frequently used to evaluate simple learning. In active avoidance, the animal must move to a "safe" area in response to a cue (light or sound stimulus) to avoid an electric shock. Measures of learning can include the latency to escape, the number of successful avoidances, and the number of trials needed to reach a pre-set criteria. In passive avoidance, the animal learns that doing nothing in response to a conditioned stimulus will avoid an adverse consequence. The number of training trials required to reach a pre-set criteria is a measure of the animal's learning. Exact procedures (i.e., one-way vs. two-way avoidance) and the type of response elicited from the animal can vary between experiments. These differences may change the results of any given study and should be considered when reviewing the literature.

Spatial and contextual learning have been examined in a number of paradigms. In the Morris water task,[4] animals must locate an escape platform hidden below the water's surface. To locate the platform from various start locations, multiple cues in the test room are utilized. While latency to find the platform can be measured during training, some strains, such as DBA/2, show a decreased latency to escape without using a spatial strategy.[5] Therefore, spatial learning is measured during a probe trial or a reversal test following training. The platform is removed from the pool for the probe trial or to a different location in the pool for the reversal test. The amount of time spent searching the trained quadrant or the number of times the trained platform site is crossed compared with the other three quadrants is an index of the animal's spatial learning. A spatial strategy would result in more site crossings and more search time in the trained quadrant than in the other three quadrants. In a reversal test where the animal had engaged in a spatial strategy, the latency to find the new platform location would be longer than to find the old location. An alternate version of the task that can evaluate sensory and motor processes needed for its successful completion is an important component of any complex learning task. For example, the visible platform version of the Morris water task uses a well-marked platform whose position is changed throughout training. This manipulation allows assessment of visual acuity and locomotion. The Barnes circular maze[6,7] is similar to the Morris water task; however, the animal is not required to swim. Instead, the mouse navigates across a circular platform that has several holes along its edge. Extra maze cues are used to locate the hole with an escape tunnel beneath it. Measures of learning include latency to enter the tunnel, errors of looking into or moving through an incorrect hole, and the search path of the animal. A reversal trial in which the tunnel is rotated 135 degrees from its training position can also be used to evaluate spatial learning. Contextual fear conditioning[8,9] uses a one-trial Pavlovian learning scheme to train the animal to associate a foot shock with a given test context. The amount of freezing, a natural fear response, to the test context is measured 24 hours later. The alternate version of this task trains the animal to associate an auditory clicker with the foot shock. The amount of freezing to the auditory stimulus in a context altered from the training context is a measure of the animal's ability to associate two events and to display the measured response of freezing.

## BASELINE STRAIN DIFFERENCES

Naturally occurring strain differences can provide invaluable information. The differences can be used to correlate biochemical, molecular, or physiological markers with the behavior of interest. Relationships of this type can identify possible pathways involved in learning mechanisms. For example, the number of successful avoidances on a two-way active avoidance paradigm was related to the number of mossy fiber terminals on hippocampal neurons.[10] Substrains, DBA1/Halle and the DBA2/G, showed differences in active avoidance learning, the presence of spontaneous bursts in the hippocampal CA3 neurons, and stimulation thresholds for evoked EPSPs.[11] Biochemically, learning differences between C57BL/6 mice and DBA/2 mice in the Morris water task are mirrored by the amount of hippocampal particulate protein kinase C. Eleven recombinant inbred strains added power to this correlation.[12] With the new technology of null mutant or transgenic mice, knowledge of baseline differences in performance and learning ability become increasingly important. While the value of genetically altering a single gene is recognized, learning is a polygenic process making the contribution of background genes important to the interpretation of behaviors in mutant mice.[13,14] If the wildtype strain cannot perform the learning task, mutant mice with the same genetic background will not be informative as to deficits caused by the "knockout" gene. Despite the information gained, strain differences are correlative in nature, limiting claims of causal influence. In addition, it must be remembered that while a correlation with biochemical or molecular markers can provide evidence for the involvement of a given system, inbred strains differ in multiple ways that may not be known or controlled for in a learning experiment.

Active avoidance has long been used to evaluate the genetic differences between mouse strains (see review by Wahlsten, 1972[15]). Although strain differences depend on the measures used and environmental components of the task, DBA/2 mice consistently perform better and have more rapid acquisition than other strains.[10,16-18] When training was distributed with an intertrial interval of 24 hours or more, the DBA/2 performance faltered while the C57BL/6 mice performance improved.[16] In addition, C57BL mice from a different source show little to no extinction once the task is acquired.[17] Other strains, such as CBA, AKR, and SM/J, are in the middle in terms of rank order performance on this task. The results for passive avoidance are almost opposite of those for active avoidance, indicating possibly different mechanisms underlying these types of learning. C57BL/6 mice seem to show the best learning with massed training, while DBA/2 animals are the worst.[16] However, individual task conditions can be important for a strain's performance. In a study that used wheel running as the avoidance behavior, there were no strain differences found in general avoidance learning.[19] This conflicting result could be caused by the increased wheel running activity of DBA/2 mice rather than response differences to the conditioned stimulus. For passive avoidance, the foot shock intensity and the age of the animal are important for the animal's performance, but these components' effect on learning is strain-dependent for most tasks.[20] For example, the perception of pain from the foot shock can differ for each strain. Shock perceptions can be measured at various intensities by recording the animal's flinching, running/jumping, and vocalization responses.[21] Equalization of all the strains in terms of response thresholds to foot shock could provide varying results in avoidance behavior.[18]

A strain survey of the Morris water task demonstrated that both the C57BL/6Ibg and DBA/2Ibg animals decreased their escape latencies in the hidden platform version of the task, but only C57BL/6Ibg mice indicated spatial selectivity through a platform reversal test.[5] The other two strains examined, BALB/cByJ and C3H/2Ibg, performed poorly on the visible platform task, indicating a possible visual problem. BALB/cByJ mice are an albino strain, while the C3H/2Ibg animals have been shown to have retinal degeneration (*rd*).[15] A more extensive strain survey was recently completed on 12 inbred strains and 7 $F_1$ hybrids that are commonly used in the study of learning.[14] For the Morris water task, this survey indicates large differences between the strains on a preference score calculated to represent spatial selectivity. Overall, the $F_1$ hybrids showed the best spatial learning, while the albino strains showed the worst. Most of the poor performance in the albino strains can be attributed to visual problems as detected by the visible platform version of the task. In addition, this study illustrated differences that can occur between substrains and under different environmental conditions. The 129/SvJ animals showed no spatial selectivity as measured by a preference score, while the 129/Svev were the best of the inbred strains on this measure. Genetically identical DBA/2 animals from different sources (Jackson Laboratories and the Institute for Behavioral Genetics) performed significantly different in terms of spatial selectivity. Although the various environmental factors were not studied, the importance of early environmental conditions is underscored by this difference.

On contextual learning, strain differences were less pronounced. However, differences between the DBA/2 and C57BL/6 mice in contextual fear conditioning were observed, though their performance on the auditory portion of this task was similar.[22] This contrast remained despite manipulation of several of the test parameters (order of testing, number of training shocks received, and pre-exposure to the context). The robustness of this result may be an indication that contextual fear conditioning is a more widely applicable complex learning behavior. A more extensive strain survey confirmed this difference and illustrated learning differences in contextual fear conditioning between the multiple strains.[14] All of the strains examined, except BuB/BnJ, showed significantly more freezing to the auditory stimulus than to an altered context, indicating that most of the strains examined could hear and learn a simple association to the unconditioned stimulus.

# GENE TARGETING: LEARNING AND MOLECULAR MODELS OF LEARNING

A variety of null mutants, which are lacking a particular gene, and transgenics, which overexpress a particular gene product, have been generated to explore the neural substrates of learning and to evaluate molecular model systems of learning (i.e., long-term potentiation and long-term depression). Some mutants have been tested in several types of learning tasks when the targeted gene is for a general use component that is not known to be brain region specific. The gene manipulation approach is a direct way of assessing the role of different neural substrates in both simple and complex learning. For null mutant mice, issues of gross morphological effects due to the mutation, especially during development, are concerns that must be kept in mind and will be discussed in the subsequent section titled "Issues of Interpretation."

## Assessing Learning with Behavioral Tasks

Learning may involve changes in the morphology of and connectivity between neurons or may rely on interactions between neurons and supportive cells in the brain. Several researchers have targeted genes for components involved in neural plasticity and assessed their impact on learning (see Table 25.1). These studies suggest that cell migration and cell-cell interactions are crucial to neural plasticity. Second, manipulations of neural transmitter systems via pharmacological methods have demonstrated that specific transmitter systems and receptor subtypes are involved in different learning tasks. For example, NMDA glutamate receptors are known to be necessary in learning conditioned fear. The removal of some NMDA or nicotine (ACh) receptors inhibits an animal's spatial learning ability as well as decreases hippocampal LTP (see Table 25.1). Following receptor involvement in the learning pathway, multiple intracellular enzymes and proteins involved in or regulating second messenger pathways are utilized. For example, both the cAMP transduction pathway and multiple protein kinases are implicated as having a role in synaptic plasticity and spatial learning (see Table 25.1).

The synthesis of new proteins is required for long-term memory (see the review by Davis and Squire[47]). The generation of the requisite new proteins would require activation of nuclear systems required for the synthesis of new proteins. To date, gene targeting approaches have looked at several transcription factors and their relationship to learning. For example, the proto-oncogene c-fos as well as the transcriptional factor cAMP-responsive element-binding protein (CREB) show learning deficits (see Table 25.1). Null mutations of the En-2 gene, a member of the engrailed gene family, and the thyroid hormone receptor β, a nuclear hormone receptor that acts as a thyroid hormone dependent transcription factor, also provide insight into possible learning processes.

There are a variety of human disorders that associated with learning impairments and memory dysfunction that are now being modeled using gene targeting techniques. The most prominent of these is the use of transgenic mouse models of Alzheimer's disease (AD). Different researchers have created versions of mice transgenic for the amyloid precursor protein[54-57] using different size isoforms. All the amyloid precursor protein transgenic mice to date have shown some degree of impaired spatial learning, which is also related to the age of the animal in some cases.[56,57]

## Assessing Molecular Models of Learning

There are two primary molecular models of learning: long-term potentiation (LTP) and long-term depression (LTD). Hebb[58] originally postulated that learning and memory involve the long-term strengthening of synapses among neurons that are coincidentally active, and later a form of synaptic

**TABLE 25.1**
**Targeted Genetic Manipulations**

| Protein/Locus | Phenotype | | Mouse | | Human Homolog | | X** |
| --- | --- | --- | --- | --- | --- | --- | --- |
| | Deficits in Learning Tasks | Deficits in Molecular Models | Chrm # | cM | Chrm # | Cytoband | |
| **Structural Neural Plasticity** | | | | | | | |
| urokinase-type plasminogen activator | spatial learning[23] | | 14 | 2.5 | 10 | q24-q24 | T |
| N-CAM (neural-cell adhesion molecules) | spatial learning[24] | | 9 | 28.0 | 11 | q22.2-q22.3 | N |
| gFAP (glial fibrillary acidic protein) | motor learning and eye blink conditioning[25] | cerebellar LTD[25] | 11 | 62.0 | 17 | q21-q21 | N |
| **Receptors** | | | | | | | |
| Glutamate receptor — metabatropic (mGluR1) | context-specific fear conditioning;[26] motor learning and eye blink conditioning[27,28] | hippocampal LTP[26] and cerebellar LTD[27,28] | unknown | | unknown | | N |
| Glutamate receptor — NMDA type 1 tissue specific (CA1 of the hippocampus) | perinatal lethal[29] spatial learning[31] | LTP[32] | 2 | 12.0 | 9 | q34.3-q34.3 | N |
| Glutamate receptor — NMDA type ε1 | spatial learning[33] | LTP[33] | 16 | 3.4 | 16 | p13-p13 | N |
| Glutamate receptor — AMPA GluR2 | increased mortality, but those surviving show reduced exploration and impaired motor coordination[34] | enhanced LTP[34] | 3 | 33.6 | 4 | q32-q33 | N |
| Acetylcholine receptor, nicotinic, β2 subunit | passive avoidance and resistance to the improving effects of nicotine on associative learning[35] | | 3 | 41.8 | 1 | pter-qter | N |
| **Intracellular Systems** | | | | | | | |
| adenylyl cyclase, Type 1 | spatial selectivity in Morris task[36] | hippocampal LTP[36] | 11 | syntenic | 7 | p13-p12 | N |
| protein kinase C, gamma | mild deficit in complex learning[37] and impaired motor coordination[39] | LTP[38] hippocampal LTD is normal[38] | 7 | 2.0 | 19 | q13.4-q13.4 | N |
| calcium-calmodulin-dependent protein kinase II α (α-CaMKII) | homozygotes — impaired in complex and simple learning;[40,42] heterozygotes — impaired in complex but not simple learning | homozygotes — deficit in LTP,[41] heterozygotes — normal LTP but disrupted short-term hippocampal plasticity[43] | 18 | 33.0 | unknown | | N |

**TABLE 25.1** (continued)
**Targeted Genetic Manipulations**

| Protein/Locus | Phenotype | | Mouse | | Human Homolog | | X** |
| --- | --- | --- | --- | --- | --- | --- | --- |
| | Deficits in Learning Tasks | Deficits in Molecular Models | Chrm # | cM | Chrm # | Cytoband | |
| *fyn* (a tyrosine kinase) | Morris task;[45] not radial maze[46] | LTP[45] | 10 | 25.0 | 6 | q21–q21 | N |
| protein kinase A — C$\beta_1$ and RI$\beta$ | neither isoforms has deficits in complex learning[44] | both have deficits in LTP[45] | 3 / 5 | syntenic / 84.0 | 1 / 7 | p36.1–p36.1; pter–p22 | N / N |
| **Genes and Transcription Factors** | | | | | | | |
| *c-fos* (proto-oncogene) | complex learning confounded with sensory deficts; intact simple learning[48] | | 12 | 40.0 | 14 | q24.3–q24.3 | N |
| cAMP-responsive element-binding protein (CREB) | long-term but not short-term memory for complex learning;[21] effects normalized by spaced training[49] | no late phase CA1 LTP21 | 1 | 31.0 | 2 | q32–q34 | N |
| En-2 | motor learning without severe motor deficits[50] | | 5 | 15.0 | 7 | q36–q36 | N |
| thyroid hormone receptor β | no behavioral or neuroanatomical abnormalities[51] | | unknown | | unknown | | N |
| **Other** | | | | | | | |
| nitric oxide synthetase, neuronal | | normal CA1 LTP[52] | 5 | 70.0 | 12 | q24.2–q24.31 | N |
| prion protein | | LTP[53] | 2 | 72.5 | 20 | pter–p12 | N |
| **Disease Models** | | | | | | | |
| amyloid precursor protein | complex learning that is age-related[54,55] | | | | | | T |

** T = transgenic manipulation; N = gene knockout manipulation.

strengthening, LTP, which follows the properties predicted by Hebb, was identified.[59] LTD is a second form of synaptic plasticity that is a model of learning. Originally predicted from theories about cerebellar learning,[60,61] this weakening of synaptic connections in neurons that are coincidenally active has been demonstrated in both the cerebellar[62,63] and hippocampal systems.[64,65] Both LTP and LTD have been well studied using combinations of electrophysiological and pharmacological techniques, and the mechanisms underlying these models are now being studied using the gene manipulation techniques.

Table 25.1 lists the impact of a variety of null mutations in different neural substrates on LTP and LTD. In most cases, the null mutation causes a deficit in the iniation or maintenance of the synaptic plasticity being evaluated, except for the GluR2 null mutant which shows an enhancement of LTP[34]. Another interesting point made by the data needs to be elaborated. In some instances (i.e., the $C\beta_1$ and the $RI\beta$ PKA mutants) the correlation between the impact of the mutation on the learning in the behavioral task and the impact on the molecular model of learning is negative rather than positive, as would be expected if the molecular models truly represent the mechanisms underlying learning. But the lack of correlation between impacts on learning ability and molecular models of learning may actually indicate the complexity of learning at the neural level rather than indicating that the current molecular models of learning are not valid. For example, in the gamma PKC mutants, there is a mild deficit in complex learning and a seeming corresponding deficit in LTP, but normal LTP can be obtained under certain testing parameters. Also, the $\alpha$-CaMKII null mutants in the heterozygote form showed impaired complex learning but normal LTP, although they did have disrupted short-term hippocampal plasticity. Thus, the use of genetically manipulated nervous systems should continue to provide useful infomation for the evaluation and further development of molecular models of learning.

## ISSUES OF INTERPRETATION

Although using null and transgenic mutant mice does provide one of the most direct measures of association between the gene and a behavioral phenotype, there are limitations inherent in the gene targeting technologies that must be considered when the data regarding a particular phenotype are interpreted. The main issues to consider when interpreting the phenotypic data collected from a gene targeted mouse include (1) the relative impact of the specific gene under investigation in relation to the polygenic system that regulates the behavior under investigation, and (2) the impact the gene manipulation has on overall development of the mouse and on other physiological systems (i.e., sensory systems) that may affect the interpretation of the phenotypic data of interest.

Because learning is a complex behavior under polygenic regulation, the deficits in a single-gene null mutation or transgenic mouse can be quantified only when the impact of the gene manipulation is of a magnitude that can be reliably measured in a polygenic system. Two factors that can affect the ability to assess the effect of the gene manipulation are the relative strength of the gene in a polygenic system and the genetic background on which the manipulated gene is being expressed. If the gene under study has only a slight role in the polygenic regulation of a particular phenotypic behavior, then the effect of the gene manipulation on the behavior might be very difficult to ascertain, especially if a mixed genetic background is used because of the overall increase in variability in the phenotype. As for the genetic background, inbred mouse strains vary in their behavioral abilities, and these differences can be used to enhance the possibility of seeing the impact of a gene manipulation on a particular behavior. For instance, a gene manipulation that disrupts learning would be easiest to see on a background with high levels of learning where as a gene manipulation that enhances learning would be easier to detect on a background with low levels of learning. Possible solutions for these problems would be assessment of the gene manipulation on a variety of genetic backgrounds and combinations of single-gene manipulations within the system (i.e., manipulations of receptors and second messenger systems) under study into double knockouts or transgenics to ascertain the role of multiple system components in the behavioral phenotype of interest.

The impact the gene manipulation has on overall development of the mouse and on other physiological systems (i.e., sensory systems) that may affect the interpretation of the phenotypic data of interest can be addressed in a variety of ways. For all gene-targeted mice produced by embryonic stem cell technology, there must be extensive initial characterization of the impact of the gene manipulation on all aspects of development, including central nervous system formation to rule out any gross alterations which could result in changes in the behavioral phenotype (i.e., the gross morphological disorganization of the hippocampus in the *fyn* mutant[45]). More subtle effects such as compensation by other systems (i.e., different subtypes of the same enzyme) must also be evaluated, especially when an expected change in the phenotype has not been seen. At the behavioral level, a battery of different tests should be used to ascertain if there are any deficits in the sensory systems (i.e., the altered pain thresholds in the $\alpha$-CaMKII mice[42]) involved in the learning task under evaluation that could account for the changes seen in the phenotype. For example, deficits in the auditory system would alter the interpretation of data, showing a failure of gene targeted mice to associate an auditory conditioned stimulus with an unconditioned stimulus.

Advances in current gene manipulation technologies such as tissue specific gene manipulation and inducible null mutations that are under specific control by the researcher will help to alleviate some of the problems with interpretation of the phenotypic effect of gene manipulations. The restriction of tissue expression of null mutations is being addressed by using the *Cre-loxP* recombination system of bacteriophage P1.[66,67] In this system, the *Cre* codes for an enzyme that catalyzes the site-specific DNA recombination between two base pair repeats known as *loxP*. To generate a mouse with a tissue specific gene manipulation, two separate strains are first constructed, with one strain containing the *Cre* gene under the control of a tissue-specific promoter and the other strain containing the gene of interest flanked at the 5' and 3' ends with the *loxP* genes. When the two strains are bred, the resulting mice have the two elements of the recombination system. This technique has been succesfully demonstrated with the NMDAR1 gene deletion restricted to the CA1 region of the hippocampus[31,32] as well as with other genes.[68-71] Tissue-specific transgenics are also being developed using tissue specific promotors. Alpha-CaMKII transgenics, which over-express the calcium-independent activated form of the enzyme only in the forebrain, show alterations in LTP, LTD, and spatial representation by place cells.[72,73] A second system being developed for specific control of gene manipulations is the inducible regulation under the control of the tetracycline-resistance gene system.[74] In the tetracycline-resistance gene system, a gene is placed in mice under the control of the *E. coli* tetracycline-resistance operon system. Treating mice carrying this system with tetracycline results in the repression of transcription of the gene controlled by the *E. coli* tetracycline-ressistance operon system. This system has also been used to further demonstrate the specificity of a-CaMKII effects on the LTP, LTD, and spatial representation by place cells.[7]

## SUMMARY

The central nervous systems mechanisms of simple and complex learning tasks are being actively studied using a genetic approach. Learning differences between inbred mouse strains when correlated with brain function suggest which neural substrates are involved in learning. These learning differences and potential neural substrates can be mapped via RI strains and QTL analysis to particular chromosomal regions and eventually specific genes regulating them. In addition to the mapping approach, specific genes thought to be involved in regulating learning processes are being manipulated by null and transgenic mutation techniques in order to assess the role of different genes in learning. Gene mutation studies are providing interesting information on the neural substrates of learning even though these studies require detailed testing of the animals to facilitate accurate interpretation of the results.

## ACKNOWLEDGMENTS

This work was sponsored by MH-48663, MH-53668, a post-doctoral traineeship to S.F.L. (HD-07289), and a pre-doctoral traineeship to E.H.O. (MH-16880).

## REFERENCES

1. Gallagher, M. and Holland, P.C., Preserved configural learning and spatial learning impairment in rats with hippocampal damage, *Hippocampus*, 1, 81, 1992.
2. Logue, S. F., Paylor, R., and Wehner, J.M., Hippocampal lesions cause learning deficits in inbred mice in the Morris water maze and conditioned-fear task, *Behav. Neurosci.*, 111, 104, 1997.
3. Morris, R. G. M., Garrud, P., Rawlins, J. N. P., and O'Keefe, J., Place navigation impaired in rats with hippocampal lesions, *Nature*, 297, 681, 1982.
4. Morris, R. G. M., Spatial localization does not require the presence of local cues, *Learn. Motiv.*, 12, 239, 1981.
5. Upchurch, M. and Wehner, J. M., Differences between inbred strains of mice in Morris water maze performance, *Behav. Genet.*, 18, 55, 1988.
6. Barnes, C. A., Memory deficits associated with senescence: a neurophysiological and behavioral study in the rat, *J. Comp. Physiolog. Psychol.*, 93, 74, 1979.
7. Mayford, M., Bach, M. E., Huang, Y, -Y., Wang, L., Hawkins, R. D., and Kandel, E. R., Control of memory formation through regulated expression of a CaMKII transgene, *Science*, 274, 1678, 1996.
8. Kim, J. and Fanselow, M. S., Modality-specific retrograde amnesia of fear, *Science*, 256, 675, 1992.
9. Phillips, R. G. and LeDoux, J. E., Differential contribution of amygdala and hippocampus to cued and contextual fear conditioning, *Behav. Neurosci.*, 106, 274, 1992.
10. Schwegler, H. and Lipp, H. P., Hereditary covariations of neuronal circuitry and behavior: correlations between the proportions of hippocampal synaptic fields in the regio inferior and two-way avoidance in mice and rats, *Behav. Brain Rese.*, 7, 1, 1983.
11. Yanovsky, Y., Brankack, J., and Haas, H. L., Differences of CA3 bursting in DBA/1 and DBA/2 inbred mouse strains with divergent shuttle box performance, *Neuroscience*, 64, 319, 1995.
12. Wehner, J. M., Sleight, S., and Upchurch, M., Hippocampal protein kinase C activity is reduced in poor spatial learners, *Brain Res.*, 523, 181, 1990.
13. Gerlai, R., Gene-targeting studies of mammalian behavior: is it the mutation or the background genotype?, *TINS*, 19, 177, 1996.
14. Owen, E. H., Logue, S. F., Rasmussen, D. L., and Wehner, J. M., Assessment of learning by the Morris water task and fear conditioning in inbred mouse strains and F1 hybrids: implications of genetic background for single gene mutations and quantitative loci analyses, *Neuroscience*, 80, 1087, 1997.
15. Wahlsten, D., Genetic experiments with animal learning: a critical review, *Behav. Biol.*, 7, 143, 1972.
16. Wimer, R. E., Symington, L., Farmer, H., and Schwartzkroin, P., Differences in memory processes between inbred mouse strains C57BL/6J and DBA/2J, *J. Comp. Physiolog. Psychol.*, 65, 126, 1968.
17. Hamburger, R., Sela, A. and Belmaker, R. H., Differences in learning and extinction in response to vasopressin in six inbred mouse strains, *Psychopharmacology*, 87, 124, 1985.
18. Weinberger, S. B., Koob, G. F., and Martinez, Jr., J. L., Differences in one-way active avoidance learning in mice of three inbred strains, *Behav. Genet.*, 22, 177, 1992.
19. Iso, H. and Shimai, S., Running-wheel avoidance learning in mice (*Mus musculus*) evidence of contingency learning and differences among inbred strains, *J. Comp. Psychol.*, 105, 190, 1991.
20. Sprott, R. L., Passive-avoidance conditioning in inbred mice: effects of shock intensity, age, and genotype, *J. Comp. Physiolog. Psychol.*, 80, 327, 1972.
21. Bourtchuladze, R., Frenguelli, B., Blendy, J., Cioffi, D., Schultz, G., and Silva, A. J., Deficient long-term memory in mice with a targeted mutation of the cAMP-responsive element-binding protein, *Cell*, 79, 59, 1994.
22. Paylor, R., Tracy, R., Wehner, J., and Rudy, J. W., DBA/2 and C57BL/6 mice differ in contextual fear but not auditory fear conditioning, *Behav. Neurosci.*, 108, 1, 1994.

23. Meiri, N., Masos, T., Rosenblum, K., Kiskin, R., and Dudai, Y., Overexpression of urokinase- type plasminogen activator in transgenic mice is correlated with impaired learning, *Neurobiology*, 91, 3196, 1994.

24. Cremer, H., Lange, R., Christoph, A., Plowmann, M., Vopper, G., Roses, J., Brown, R., Baldwin, S., Kraemer, P., Scheff, S., Barthels, D., Rajewsky, K., and Wille, W., Inactivation of the N-CAM gene in mice results in size reduction of the olfactory bulb and deficits in spatial learning, *Nature*, 367,455, 1994.

25. Shibuki, K., Gomi, H., Chen, L., Bao, S., Kim, J.J., Wakatsuki, H., Fujisaki, T., Fujimoto, K., Katoh, A., Ikeda, T., Chen, C., Thompson, R.F., and Itohara, S., Deficient cerebellar long-term depression, impaired eyeblink conditioning, and normal motor coordination in GFAP mutant mice, *Neuron*, 16, 587, 1996.

26. Aiba, A., Chen, C., Herrup, K., Rosenmund, C., Stevens, C. F., and Tonegawa, S., Reduced hippoc-ampal long-term potentiation and context-specific deficit in associative learning in mGluR1 mutant mice, *Cell*, 79, 365, 1994.

27. Aiba, A., Kano, M., Chen, C., Stanton, M. E., Fox, G. D., Herrup, K., Zwingman, T. A., and Tonegawa, S., Deficient cerebellar long-term depression and impaired motor learning in mGluR1 mutant mice, *Cell*, 79, 377, 1994.

28. Conquet, F., Bashir, Z. I., Davies, C. H., Daniel, H., Ferraguti, F., Bordi, F., Franz-Bacon, K., Reggiani, A., Matarese, V., Condé, F., Collingridge, G. L., and Crépel, F., Motor deficit and impairment of synaptic plasticity in mice lacking mGluR1, *Nature*, 372,237, 1994.

29. Forrest, D., Yuzaki, M., Soares, H. D., Ng, L., Luk, D. C., Sheng, M., Stewart, C. L., Morgan, J. L., Connor, J. A., and Curran, T., Targeted disruption of NMDA receptor 1 gene abolishes NMDA response and results in neonatal death, *Neuron*, 13, 325, 1994.

30. Li, Y., Erzurumlu, R., Chen, C., Jhaveri, S., and Tonegawa, S., Whisker-related neuronal patterns fail to develop in the trigeminal brainstem nuclei of NMDAR1 knockout mice, *Cell*, 76, 427, 1994.

31. Tsien, J. Z., Heurta, P. T., and Tonegawa, S., The essential role of hippocampal CA1 NMDA receptor-dependent synaptic plasticity in spatial memory, *Cell*, 87, 1327, 1996.

32. McHugh, T. J., Blum, K. I., Tsien, J. Z., Tonegawa, S., and Wilson, M. A., Impaired hippocampal representation of space in CA1-specific NMDAR1 knockout mice, *Cell*, 87, 1339, 1996.

33. Sakimura, K., Kutuwada, T., Ito, I., Manabe, T., Takayama, C., Kushiya, E., Yagi, T., Aizawa, S., Inoue, Y., Sugiyama, H., and Mishina, M., Reduced LTP and spatial learning in mice lacking NMDA receptor ε1 subunit, *Nature*, 373, 151, 1995.

34. Jia, Z., Agopyan, N., Miu, R., Xiong, Z., Henderson, J., Gerlai, R., Taverna, F. A., Velumian, A., MacDonald, J., Carlen, P., Abramow-Newerly, W., and Roder, J., Enhanced LTP in mice deficient in the AMPA receptor GluR2, *Neuron*, 17, 945, 1996.

35. Picciotto, M. R., Zoll, M., Léna, C., Bessis, A., Lallemanc, Y., LeNovère, N., Vincent, P., Merlo Pich, E., Brulet, P., and Changeux, J-P., Abnormal avoidance learning in mice lacking functional high-affinity nicotine receptor in the brain, *Nature*, 374, 65, 1995.

36. Wu, Z-L., Thomas, S. A., Villacres, E. C., Xia, Z., Simmons, M. L., Chavkin, C., Palmitter, R. D., and Storm, D. R., Altered behavior and long-term potentiation in type I adenylyl cyclase mutant mice, *Proc. Natl. Acad. Sci.*, 92, 220, 1995.

37. Abeliovich, A., Paylor, R., Chen, C., Kim, J. J., Wehner, J. M., and Tonegawa, S., PKCγ mutant mice exhibit mild deficits in spatial and contextual learning, *Cell*, 75, 1263, 1993.

38. Abeliovich, A., Chen, C., Goda, Y., Silva, A. J., Stevens, C. F., and Tonegawa, S., Modified hippoc-ampal long-term potentiation in PKCγ -mutant mice, *Cell*, 75, 1253, 1993.

39. Chen, C., Kano, M., Abeliovich, A., Chen, L., Bao, S., Kim, J. J., Hashimoto, K., Thompson, R. F., and Tonegawa, S., Impaired motor coordination correlates with persistent multiple climbing fiber innervation in PKCγ mutant mice, *Cell*, 83, 1233, 1995.

40. Silva, A. J., Paylor, R., Wehner, J. M., and Tonegawa, S., Impaired spatial learning in α-calcium-calmodulin kinase II mutant mice, *Science*, 257, 206, 1992.

41. Silva, A. J., Wang, Y., Paylor, R., Wehner, J. M., Stevens, C. F., and Tonegawa, S., Alpha calcium/calm-odulin kinase II mutant mice: deficient long-term potentiation and impaired spatial learning, *Cold Spring Harbor Symp. Quant. Biol.*, 57, 527, 1992.

42. Chen, C., Rainnie, D. G., Greene, R. W., and Tonegawa, S., Abnormal fear response and aggressive behavior in mutant mice deficient for α-calcium-calmodulin kinase II, *Science*, 266, 291, 1994.

43. Silva, A. J., Rosahl, T. W., Chapman, P. F., Marowitz, Z., Friedman, E., Frankland, P. W., Cestari, V., Cioffi, D., Südhof, T. C., and Bourtchuladze, R., Impaired learning in mice with abnormal short-lived plasticity, *Curr. Biol.*, 6, 1509, 1996.

44. Huang, Y.-Y., Kandel, E. R., Varshavsky, L., Brandon, E. P., Qi, M., Idzerda, R. L., McKnight, G. S., and Bourtchouladze, R., A genetic test of the effects of mutations in PKA on mossy fiber LTP and its relation to spatial and contextual learning, *Cell*, 83, 1211, 1995.

45. Grant, S. G. N., O'Dell, T. J., Karl, K. A., Stein, P. L., Soriano, P., and Kandel, E. R., Impaired long-term potentiation, spatial learning, and hippocampal development in *fyn* mutant mice, *Science*, 258, 1903, 1992.

46. Miyakawa, T. Yagi, T., Kagiyama, A., and Niki, H., Radial maze performance, open-field and elevated plus-maze behaviors in Fyn-kinase deficient mice: further evidence for increased fearfulness, *Mol. Brain Res.*, 37, 145, 1996.

47. Davis, H. P., and Squire, L. R., Protein synthesis and memory, *Psycholog. Bull.*, 96, 518, 1984.

48. Paylor, R., Johnson, R. S., Papaioannou, V., Spiegelman, B. M., and Wehner, J. M., Behavioral assessment of c-*fos* mutant mice, *Brain Res.*, 651, 275, 1994.

49. Kogan, J. H., Frankland, P. W., Blendy, J. A., Coblentz, J., Marowitz, Z., Schütz, G., and Silva, A. J., Spaced training induces normal long-term memory in CREB mutatant mice, *Curr. Biol.*, 7, 1, 1996.

50. Gerlai, R., Millen, K. J., Herrup, K., Fabien, K., Joyner, A. L., and Roder, J., Impaired motor learning performance in cerebellar En-2 mutant mice, *Behav. Neurosci.*, 110, 126, 1996.

51. Forrest, D., Hanebuth, E., Smeyne, R. J., Everds, N., Stewart, C. L., Wehner, J. M., and Curran, T., Recessive resistance to thyroid hormone in mice lacking thryoid hormone receptor β: evidence for tissue-specific modulation of receptor function, *EMBO J.*, 15, 3006, 1996.

52. O'Dell, T. J., Huang, P. L., Sawson, T. M., Dinerman, J. L., Snyder, S. H., Kandel, E. R., and Fishman, M. C., Endothelial NOS and the blockade of LTP by NOS inhibitors in mice lacking neuronal NOS, *Science*, 265, 542, 1994.

53. Collinge, J., Whittington, M. A., Sidle, K. C., Smith, C. J., Palmer, M. S., Clarke, A. R., and Jefferys, J. G. R., Prion protein is necessary for normal synaptic function, *Nature*, 370, 295, 1994.

54. Yamaguchi, F., Richards, S. J., Beyreuther, K., Salbaum, M., Carlson, G. A., and Dunnett, S. B., Transgenic mice for the amyloid precursor protein 695 isoform have impaired spatial memory, *Neuroreport*, 2, 701, 1991.

55. Muller, U., Cristina, N., Li, Z. W., Wolfer, D. P., Lipp, H. P., Rulicke, T., Brandner, S., Aguzzi, A., and Weissmann, C., Behavioral and anatomical deficits in mice homozygous for a modified beta-amyloid precursor protein gene, *Cell*, 79, 755, 1994.

56. Moran, P. M., Higgins, L. S., Cordell, B., and Moser, P. C., Age-related learning deficits in transgenic mice expressing the 751-amino acid isoform of human beta-amyloid precursor protein, *Proc. Nat Acad. Sci.*, 92, 5341, 1995.

57. Hsiao, K., Chapman, P., Nilsen, S., Eckman, C., Harigaya, Y., Younkin, S., Yang, F., and Cole, G., Correlative memory deficits, Aβ elevation, and amyloid plaques in transgenic mice, *Science*, 274, 99, 1996.

58. Hebb, D. O., *The Organization of Behavior*, John Wiley & Sons Inc., New York, 1949.

59. Bliss, T. V. P. and Lomo, T., Long-lasting potentiation of synaptic transmission in the dentate area of the anaesthesized rabbit following stimulation of the perforant path, *J. Physiol.*, 232, 331, 1973.

60. Marr, D., A theory of cerebellar cortex, *J. Physiol. (London)*, 202, 437, 1969.

61. Albus, J. S., A theory of cerebellar function, *Math. Biosci.*, 10, 25, 1971.

62. Ito, M., Sakurai, M., and Tongroach, P., Climbing fibre induced depression of both mossy fibre responsiveness and glutamate sensitivity of cerebellar Purkinje cells, *J. Physiol. (London)*, 324, 113, 1982.

63. Ekerot, C.-F. and Kano, M., Long-term depression of parallel fibre synapses following stimulation of climbing fibres, *Brain Res.*, 342, 357, 1985.

64. Dudek, S. M., and Bear, M. F., Homosynaptic long-term depression in area CA1 of hippocampus and effects of N-methyl-D-aspartate receptor blockade, *Proc. Natl. Acad. Sci.*, 89, 4363, 1992.

65. Mulkey, R. M. and Malenka, R. C., Mechanisms underlying induction of homosynaptic long-term depression in area CA1 of the hippocampus, *Neuron*, 9, 967, 1992.

66. Abremski, K., Hoess, R., and Sternber, N., Studies on the properties of P1 site-specific recombination: evidence for topologically unlinked products following recombination, *Cell*, 32, 1301, 1983.

67. Sauer, B., Manipulation of transgnes by site-specific recombination: use of Cre recombinase, in *Guide to Techniques in Mouse Development*, Wassarman, P. M. and DePamphilis, M. L., Eds., of *Methods in Enzymology*, Abelson, J. N. and Simon, M. I., Eds., Academic Press, Boston, MA, 225, 890, 1993.

68. Orban, P. C., Chui, D., and Marth, J. D., Tissue- and site-specific DNA recombination in transgenic mice, *Proc. Natl. Acad. Sci.*, 89, 6861, 1992.

69. Lasko, M., Sauer, B., Mosinger Jr., B., Lee, E. J., Manning, R. W., Yu, S-H., Mulder, K. L., and Westphal, H., Targeted oncogene activation by site-specific recombination in transgenic mice, *Proc. Natl. Acad. Sci.*, 89, 6232, 1992.

70. Gu, H., Marth, J. D., Orban, P. C., Mossmann, H., and Rajewsky, K., Deletion of a DNA polymerase β gene segment in T cells using cell type-specific gene targeting, *Science*, 265, 103, 1994.

71. Tsien, J. Z., Chen, D. F., Gerber, D., Tom, C., Mercer, E. H., Anderson, D. J., Mayford, M., Kandel, E. R., and Tonegawa, S., Subregion- and Cell type-restricted gene knockout in mouse brain, *Cell*, 87, 1317, 1996.

72. Mayford, M., Wang, J., Kandel, E. R., and O'Dell, T. J., CaMKII regulates the frequency-response function of hippocampal synapses for the production of both LTD and LTP, *Cell*, 81, 891, 1995.

73. Rotenberg, A., Mayford, M., Hawkins, R. D., Kandel, E. R., and Muller, R. U., Mice expressing activated CaMKII lack low frequency LTP and do not form stable place cells in the CA1 region of the hippocampus, *Cell*, 87, 1351, 1996.

74. Furth, P. A., St. Onge, L., Boger, H., Gruss, P., Gossen, M., Kistner, A., Bujard, H., and Henninghausen, L., Temporal control of gene expression in transgenic mice by a tetracycline-responsive promotor, *Proc. Natl. Acad. Sci.*, 91, 9392, 1994.

# Index

## H

Milton Keynes UK
Ingram Content Group UK Ltd.
UKHW052024071024
449327UK00027B/2411

9 780367 399559